Energy Systems and Sustainability

Energy Systems and Sustainability

POWER FOR A SUSTAINABLE FUTURE

SECOND EDITION

Edited by Bob Everett, Godfrey Boyle, Stephen Peake and Janet Ramage

OXFORD
UNIVERSITY PRESS

Published by Oxford University Press, Great Clarendon Street, Oxford OX2 6DP in association with The Open University, Walton Hall, Milton Keynes MK7 6AA, United Kingdom

Oxford University Press is a department of the University of Oxford. It furthers the University's objective of excellence in research, scholarship, and education by publishing worldwide in

Oxford New York Auckland Cape Town Dar es Salaam Hong Kong Karachi Kuala Lumpur Madrid Melbourne Mexico City Nairobi New Delhi Shanghai Taipei Toronto

With offices in

Argentina Austria Brazil Chile Czech Republic France Greece Guatemala Hungary Italy Japan Poland Portugal Singapore South Korea Switzerland Thailand Turkey Ukraine Vietnam

Oxford is a registered trade mark of Oxford University Press in the UK and in certain other countries.

Published in the United States by Oxford University Press Inc., New York

First published 2003. Second edition 2012.

Edited and designed by The Open University.

Typeset by SR Nova Pvt. Ltd, Bangalore, India.

Printed and bound in the United Kingdom by Halstan Printing Group, Amersham.

This book forms part of Open University teaching materials. Details of Open University modules can be obtained from the Student Registration and Enquiry Service, The Open University, PO Box 197, Milton Keynes MK7 6BJ, United Kingdom (tel. +44 (0)845 300 60 90, email general-enquiries@open.ac.uk).

www.open.ac.uk

British Library Cataloguing in Publication Data available on request

Library of Congress Cataloguing in Publication Data available on request

ISBN 978 0 19 959374 3

2.1

MIX
Paper from responsible sources
FSC® C016278
FSC
www.fsc.org

Preface

Billions of people in developing countries are striving to achieve higher living standards. The rate of energy use of countries such as China and India has grown enormously over the past few decades. If current trends continue, by the end of the twenty-first century the world's population could reach 10 billion.

How can this growth in global energy demand be met cleanly, safely and sustainably?

It is a critical moment for energy policy. On the one hand there are those who say that increasing the production of fossil fuels is essential to continuing economic growth. This conflicts with the need to deal with climate change by reducing greenhouse gas emissions, particularly carbon dioxide. The Intergovernmental Panel on Climate Change have asked for global CO_2 emissions to be halved by 2050. Then there is a further dimension: resource depletion. World oil prices soared to over US$140 a barrel in 2008 and the International Energy Agency has conceded that 'the age of cheap oil' may be over. While supplies of natural gas and coal currently remain plentiful, they are finite resources and cannot be used indefinitely. Nuclear power may offer a partial solution, but it has its own problems of uranium supply, waste disposal and nuclear weapons proliferation.

This book is focused on energy supply, but that is not because energy saving is any less important. It is also focused on the UK, but with comparisons and examples looks at a range of other countries: the USA, France, Denmark, India and China.

Many readers will no doubt want this book to tell them what 'the answer' is. Alas, there is no 'magic silver bullet'. This book attempts to clearly set out some of the energy sustainability problems and explain the issues, the terminology and the basic physics. It provides some technical solutions for a transition to a more sustainable low carbon energy future and will hopefully give a good grounding for reading other books on the subject, particularly the companion volume *Renewable Energy*.

This book and its companion have been written initially for undergraduates studying energy modules at The Open University. They are also aimed at students and staff in other universities and at professionals, policymakers and members of the public interested in sustainable energy futures. We hope that both books will contribute to a better understanding of the sustainability problems of our present energy systems and of potential solutions. We also hope they convey something of the enthusiasm we feel for this complex, fascinating, evolving and increasingly important subject.

Bob Everett
Lecturer in Renewable Technology, The Open University

New to this edition

Drawing on the latest available data, this edition includes energy statistics from India and China to enable comparison with those from the UK, USA, France, and Denmark. US units are now included in the Appendix.

Coverage of contemporary issues, such as 'peak oil', coal-to-liquids, carbon prices, carbon capture and storage (CSS) has been extended; the implications of global increasing energy consumption are also discussed. Examples of newer technologies have been included as part of the update.

Editor biographies

Bob Everett

Bob Everett is Lecturer in Renewable Technology at The Open University's Faculty of Mathematics, Computing and Technology (MCT). He is an electrical engineer by training and took up energy research in 1977 after working in the printing industry developing what is now called desktop publishing. He has worked on a number of housing energy research projects featuring active and passive solar heating and is a keen advocate of combined heat and power generation. Since 1991 he has worked on a number of Open University modules on the environment and sustainable energy.

Godfrey Boyle

Godfrey Boyle is Professor of Renewable Energy in The Open University's MCT Faculty. His main research interests are in solar and wind power, energy systems modelling and energy policy, and he has chaired various Open University modules on renewable and sustainable energy. He is also a visiting professor at The Energy and Resources Institute (TERI) University in New Delhi, India, a Fellow of the Institution of Engineering and Technology (IET) and a Trustee of the UK National Energy Foundation.

Stephen Peake

Stephen Peake is Senior Lecturer in Environmental Technology at The Open University and a Fellow of the Judge Business School, University of Cambridge. He has worked in the field of energy and sustainability for over 20 years, as a Fellow of the Royal Institute of International Affairs in London (including a stint at the Shell International Petroleum Company), as a Fonctionnaire at the International Energy Agency in Paris, and as an official with the United Nations Framework Convention on Climate Change in Bonn, Germany. He is an enthusiastic researcher, teacher and consultant in the field of energy, climate and sustainability.

Janet Ramage

Janet Ramage is a Visiting Lecturer at the Open University. Her academic background is in physics, and after completing her studies in London she became a lecturer at the University of California at Los Angeles (UCLA). Returning to London, she became interested in energy studies in the late 1970s and produced short courses for physics teachers. A sabbatical year at the University of California, San Diego (UCSD) and Harvard led to the publication of her book *Energy, A Guidebook* (Oxford, 1983), which had a second edition in 1997. She has been a contributor to Open University energy module material, including the first edition of this book, since 1992.

Contents

Chapter 1

Introducing energy systems and sustainability

By Stephen Peake, Bob Everett and Godfrey Boyle

1.1 Introduction

This chapter introduces the main themes of the book. It explains what is meant by 'energy systems' and gives an overview of our present global energy system in terms of the primary energy sources it uses. It then elaborates on the concepts of sustainable development and sustainability and continues with an overview of how these might apply specifically to energy systems. In order to explore why sustainable energy matters it then compares three different perspectives on some of the critical factors and forces that may shape energy systems into the future. The chapter ends by briefly outlining three basic options for achieving a transition from our present energy systems to the sustainable energy systems of tomorrow.

1.2 What is an 'energy system'?

The word 'energy', when it first appeared in English in the 16th century, had no scientific meaning at all. Based on a Greek word used by Aristotle, it meant forceful or vigorous *language*. It was not until the early 1800s that the concept of energy in the modern sense was developed by scientists to describe and compare their observations about the behaviour of such diverse phenomena as the transfer of heat, the motion of bodies, the operation of machinery and the flow of electricity. Today, the standard scientific definition is that *energy is the capacity to do work*: that is, to move an object against a resisting force.

In everyday language, the word 'power' is often used as a synonym for electricity or energy – and indeed in this book the three words will occasionally be used in this rather loose way merely as substitutes for each other. But when speaking scientifically, *power is the rate of doing work*, that is, the rate at which energy is converted from one form to another, or transmitted from one place to another. The main unit of measurement of energy is the joule (J) and the main unit of measurement of power is the watt (W), which is defined as a rate of one joule per second. The twin concepts of energy and power will be discussed in more detail in Chapter 2, and the subject of electricity in Chapters 4 and 9.

Whilst the scientific concept of 'energy' dates from the early nineteenth century, concepts such as 'energy policy' and 'energy system' are much more recent. The UK, like many other European countries, has been heavily reliant on coal throughout much of the twentieth century. For most of that time government 'energy policy' meant 'coal policy'. However, after World War 2 there was an era of relatively cheap oil supplies from the Middle East. By the 1960s oil was cheap enough in the UK to displace coal from power station use. Then in 1973 came the Yom Kippur War between Egypt, Syria and Israel. World oil prices rose fourfold in a matter of months (see Figure 1.1). Oil prices rose again in 1979 with revolution and war in Iran.

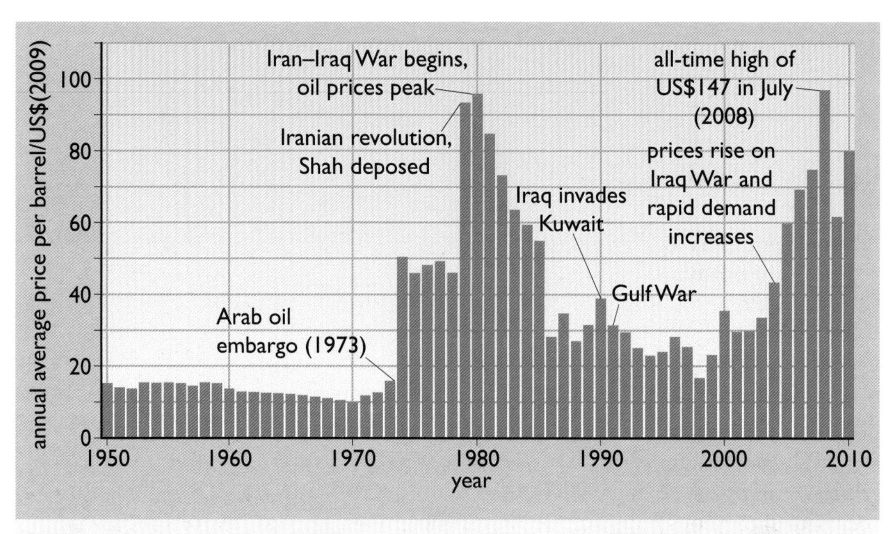

Figure 1.1 Annual average inflation-corrected oil prices (in US$2009) and related global events, 1950–2010 (source: BP, 2010; EIA, 2011)

These price shocks convinced many governments that they should have some kind of 'energy policy'. The Danes, for example, had become highly reliant on imported oil. Their response to the oil shocks involved fuel switching (from oil to coal), energy conservation and the promotion of renewable energy, as will be described in Chapter 3. This thinking required not just a 'fuel policy' but careful analysis of a whole national 'energy system'. The concept of an 'energy system' is described in Box 1.1.

BOX 1.1 Energy systems and boundaries

A system is a set of interconnected entities that are part of an integrated whole. A key element in systems thinking is in choosing where to draw a system boundary for analysis purposes (Figure 1.2). Whatever is outside the boundary is considered part of the system's environment. There may be flows (of materials, pollution or money) in either direction across this boundary between the system and its environment.

A modest **energy system** might, for example, comprise a coal-fired power station fed with coal from a local mine and providing electricity to consumers in a local city. The system boundary might be drawn to include all of these. A *system optimisation* might consider what energy service the electricity actually provides. If it is illumination, it might assess the benefits of using more efficient light bulbs in order to reduce the consumption of coal.

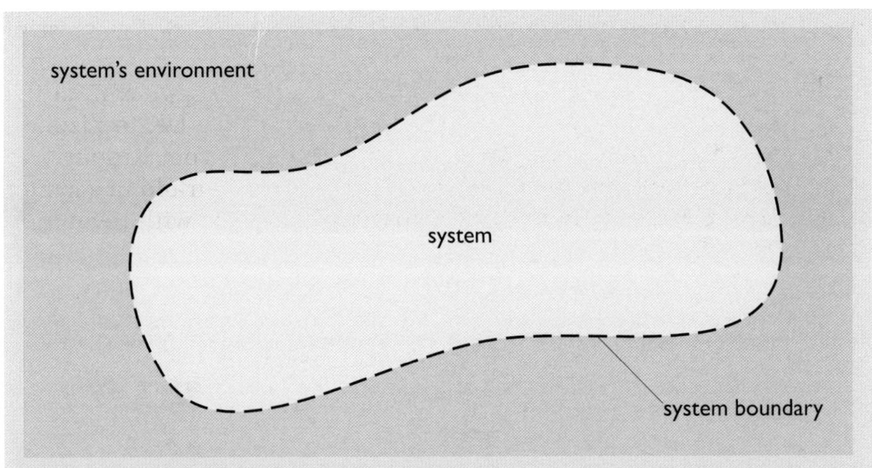

Figure 1.2 A system is distinguished from its surrounding environment by a 'system boundary'.

The **system's environment** is everything that is outside the system boundary. If acid rain pollution from the power station damages trees in a neighbouring country, that is damage done to the 'environment of the system'.

However, normally the word 'environment' is used to mean 'the **natural environment**'; i.e. the Earth's atmosphere, oceans and other parts of the biosphere. This can perhaps be interpreted as everything that is outside the boundary of a very large 'system'; a system that includes all of humanity's energy uses (but humanity itself is part of the biosphere). Where pollution from an energy system damages the natural environment (and that includes the health of people) then an *environmental cost* has been incurred.

This book focuses mainly on the technological aspects of 'energy systems' – the set of energy sources, technologies, transmission, distribution, storage and management systems that we use to produce the energy services that we need (see Figure 1.3).

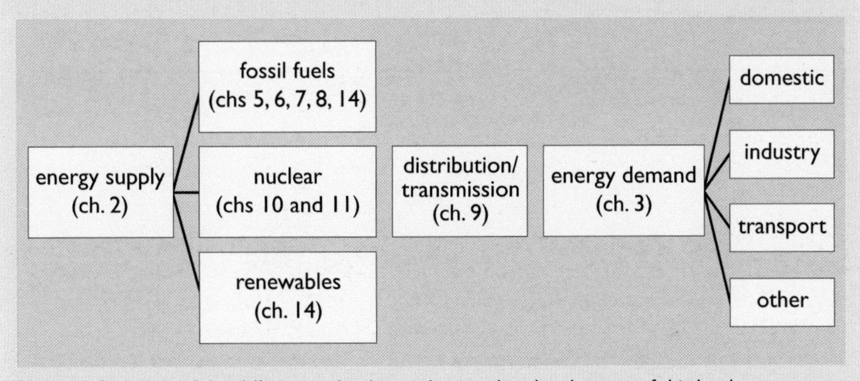

Figure 1.3 A map of the different technologies discussed in the chapters of this book

1.3 The present global energy system

What are the characteristics of today's energy systems? Box 1.2 describes the growth in world primary energy consumption since 1850 which has paralleled a rise in both world population and world GDP.

BOX 1.2 World primary energy consumption

Between 1850 and 2009 world primary energy use rose more than fortyfold (Figure 1.4(a) – see Chapter 2 for an explanation of the term primary energy). Over the same period world population rose sixfold reaching over 6.8 billion in 2010 (Figure 1.4(b)).

The world **gross domestic product (GDP)** increased fiftyfold. This is the total production of the *economic* system, i.e. the monetary value all the goods and services produced (Figure 1.4(c)).

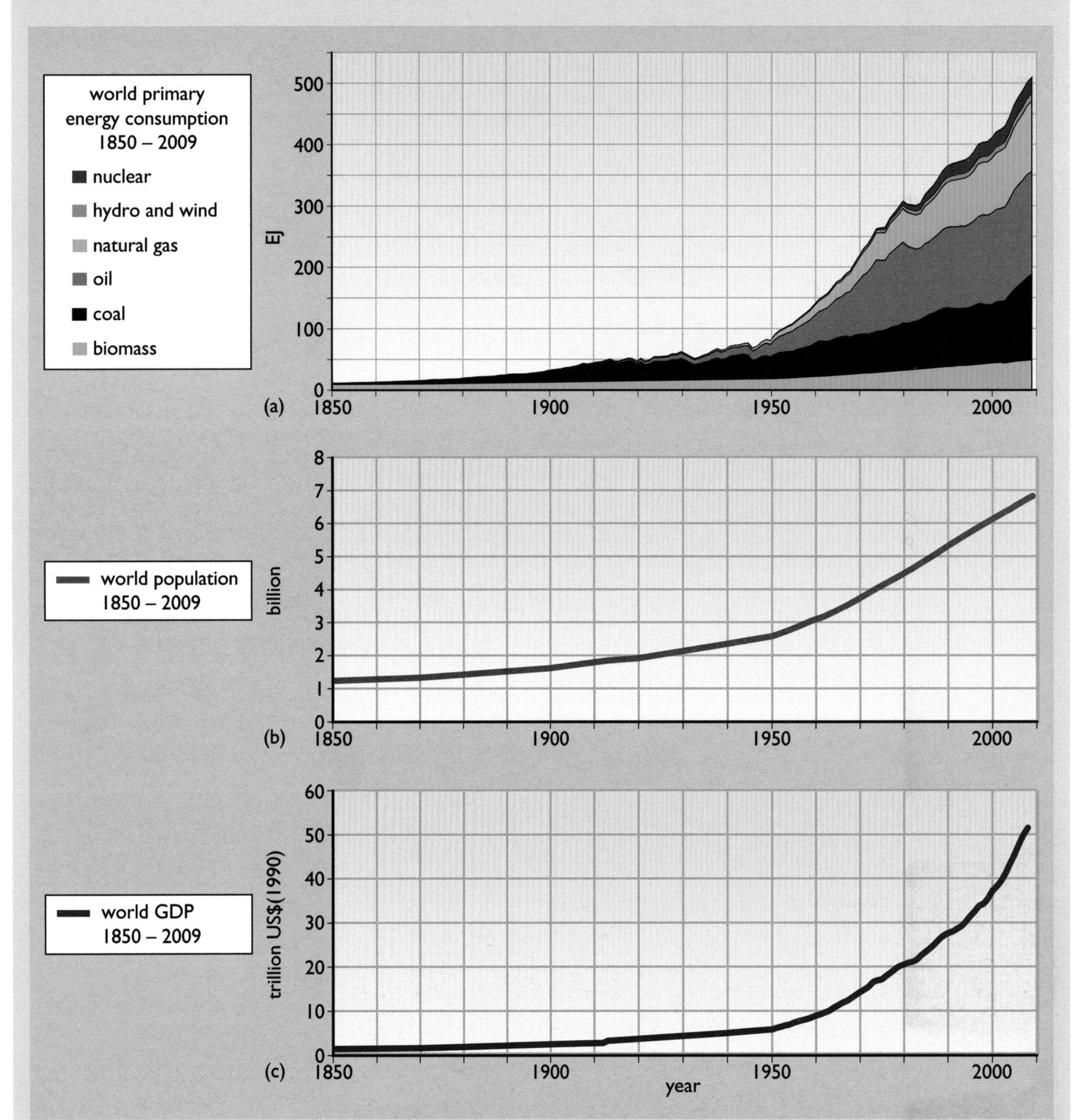

Figure 1.4 Growth in: (a) world primary energy use, 1850–2009; (b) world population, 1850–2009; (c) inflation-corrected world GDP in US$1990, 1850–2009 (sources: (a) Boden et al., 2010; IEA, 2009; IEA, 2010a; LBNL, 2009; Maddison, 2005; Maddison, 2009, (b and c) Maddison, 2009)

For most of the nineteenth century the world's principal fuel was firewood (or other forms of 'traditional biomass'), but coal use was rising fast and by the beginning of the twentieth century it had replaced wood as the dominant energy source. During the 1920s, oil in turn began to challenge coal and by the 1970s had overtaken it as the leading contributor to world supplies. By then, natural gas was also making a very substantial contribution, with nuclear energy and hydro power also supplying smaller yet significant amounts.

As Figure 1.5 shows, total world primary energy consumption in 2009 was an estimated 502 million million million joules, i.e., 502 exajoules, equivalent to the energy content of some 12 000 million tons of oil. (For details of these quantities and units see Chapter 2.)

To put this into some context the supertanker *Sirius Star* (Figure 1.6), a very large crude carrier (VLCC), has a capacity of about 300 000 tonnes, so the *c.*3900 million tonnes of oil used by the world in 2009 (about one-third of total primary energy used) would fill about 13 000 of these.

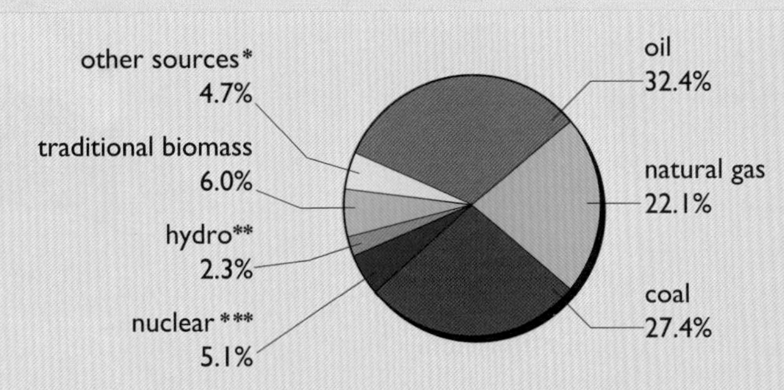

* Includes wind, 'new' biomass, solar and geothermal energy.

** The hydro contribution is the actual electrical output.

*** The nuclear contribution is the notional primary energy that would be needed to produce the actual output at an efficiency of 38%.

Total: about 502 EJ, equivalent to 12 billion tonnes of oil, or an average continuous rate of energy consumption of 15.9 TW.

Figure 1.5 Percentage contributions of various energy sources to world primary energy consumption, 2009. Note that the actual amounts of electricity produced by nuclear and hydro power were almost the same, but owing to a statistical convention in the definition of primary energy, the nuclear contribution is multiplied by a factor of nearly three (see Chapter 2) (principal sources: BP, 2010; IEA, 2009; Maddison, 2005; Maddison, 2009; WWEA, 2009)

Figure 1.6 A US$3m ransom is dropped by parachute onto the oil tanker Sirius Star on 9 January 2009. Somali pirates later released the ship and its crew after keeping them hostage for two months in the Gulf of Aden.

On average, world primary energy consumption *per capita* (i.e. per person) in 2009, including traditional biomass, was about 74 thousand million joules (74 gigajoules), which is equivalent to about 1.75 tonnes of oil per person per year, or about 5.5 litres of oil per day.

But this average conceals major differences between the inhabitants of different regions. As Figure 1.7 illustrates, North Americans annually consume the equivalent of more than 6 tonnes of oil per head, whereas residents of Europe consume about half that amount, and the inhabitants of the rest of the world use only about one-tenth.

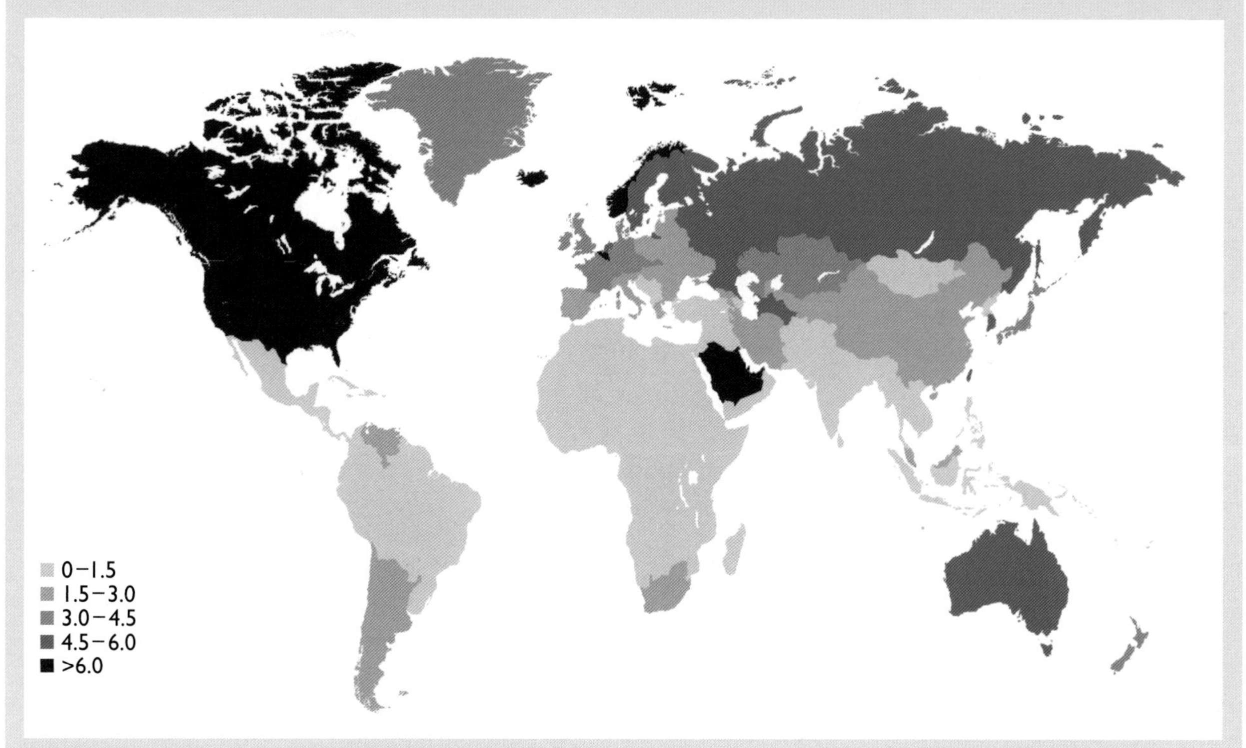

Figure 1.7 Per capita energy consumption in tonnes of oil equivalent per year in 2009 for different regions of the world. Note that these figures include only commercially traded fuels (i.e. they exclude traditional biofuels). The majority of the 1.4 billion or so people living in energy poverty and lacking access to electricity in 2009 live in rural areas in the lighter shaded parts of the map. (source: BP, 2010)

Until the Industrial Revolution at the beginning of the nineteenth century human energy requirements were modest. Most came from natural processes such as the growth of plants. This provided wood for heating and cooking and also food, not just for humans but also for the millions of draft animals used to provide mechanical power. The power of water and the wind was also used to drive simple machinery. The wind was the principal power source for shipping. As described in Box 1.2, there has been an almost continuous growth in world energy use over the past 150 years. There has also been a continuous change in the *proportions* of the different energy sources (Figure 1.8). Coal became the dominant fuel source at the end of the nineteenth century. It was overtaken by oil as the dominant source in the 1960s. The contribution from natural gas and nuclear power has risen in the late twentieth century. It is likely that these transitions from one form of fuel to another will continue in the twenty-first century.

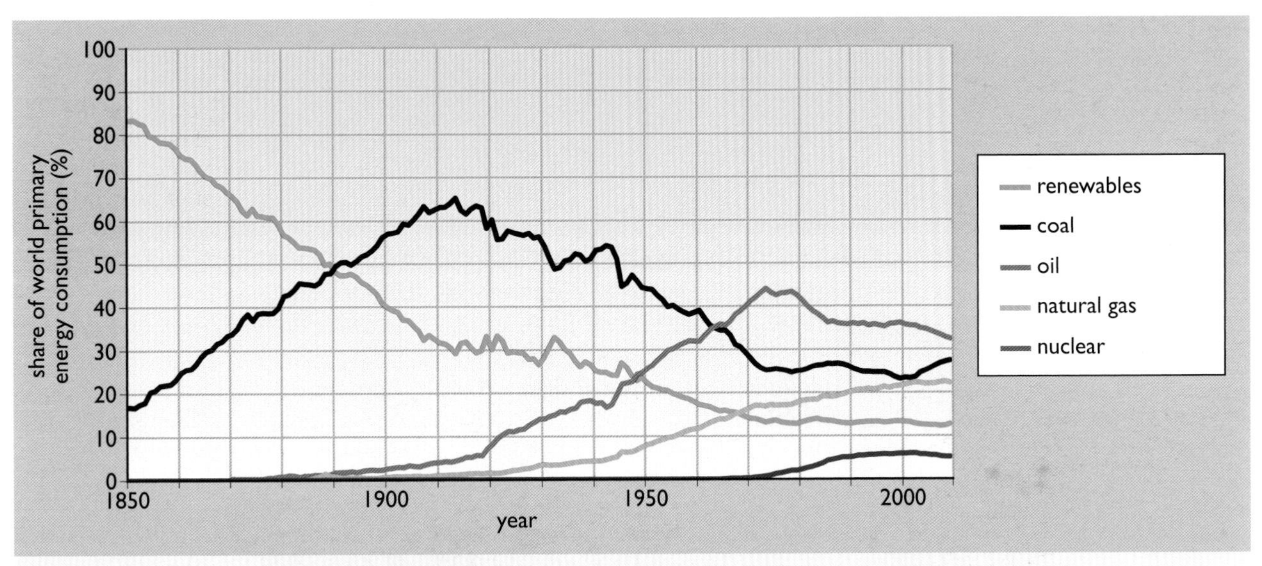

Figure 1.8 Percentage shares of world primary energy consumption 1850–2009. Note that the percentages shown are shares of a *rising* total (sources: BP, 2010; IEA, 2009; Maddison, 2005; Maddison, 2009; Boden et al., 2010)

Fossil fuels

Fossil fuels are composed mainly of carbon and hydrogen, which is why they are called **hydrocarbons**, and they are extremely attractive as energy sources. They are highly concentrated, enabling large amounts of energy to be stored and distributed relatively easily.

During the twentieth century, these unique advantages enabled the development of increasingly sophisticated and effective technologies for transforming fossil fuel energy into other forms of energy and eventually useful energy services. Fossil fuels may be used to provide:

- mechanical power for manufacturing industries
- heat for warm buildings and cooked food
- artificial lighting, to illuminate our buildings and surroundings
- cooling, particularly for the preservation of food
- mechanical power for the transportation of people and goods
- electric power for communication and our digital world.

Today, in the early part of the twenty-first century, fossil-fuel-based systems reign supreme, supplying the great majority of the world's energy.

The fossil fuels we use today originated in the growth and decay of plants and marine organisms that existed on the Earth millions of years ago. Coal was formed when dead trees and other vegetation became submerged under water, did not completely rot, and were subsequently compressed, by geological processes lasting millions of years, into concentrated solid layers below the Earth's surface. Oil and associated natural gas originally consisted of the remains of countless billions of marine organisms that slowly accreted into layers beneath the Earth's oceans and were gradually transformed, through geological forces acting over aeons of time, into the liquid and gaseous reserves we access today by drilling into the Earth's crust.

Nuclear energy

As the name suggests, the release of 'nuclear energy' results from changes in the *nuclei* of atoms. The present nuclear power plants derive their energy from the splitting, or *fission*, of the nuclei of very heavy isotopes such as uranium-235 and plutonium-239. The feature that makes this process attractive is the very high 'energy density' of the fuel. In principle, the complete fission of one kilogram of uranium-235 would release as much energy as the complete combustion of over 3000 tonnes of coal. This is not yet fully achieved in practice, but nevertheless nuclear fuels are much more concentrated sources of the heat needed to produce steam in power stations than fossil fuels. (Later chapters discuss all these aspects in more detail.)

Since the UK opened the world's first grid-connected nuclear power station at Calder Hall in Cumbria in 1956, nuclear electricity generation has expanded to a point where it now accounts for 5% of world primary energy and for approximately 13% of the world's electricity (see Figure 10.1). In some countries, it is the principal source of electricity generation. France, for example, derives nearly 80% of its electricity from nuclear power.

Energy is also released in the *fusion* of two light nuclei. This was the principle of the 'hydrogen bomb', but in that case the energy was released as an explosion - by definition an uncontrolled process. *Controlled fusion* has been achieved, but fifty years of research effort have yet to lead to a system that produces a sustained energy output greater than the energy input needed to run it. So a successful fusion power plant remains elusive, and the present consensus of most experts is that this technology is unlikely to become commercially available for many more decades.

Renewable energy sources

Fossil and nuclear fuels are often termed **non-renewable** energy sources. This is because, although the quantities in which they are available on Earth may be extremely large, they will become increasingly difficult and expensive to extract in the future. Discoveries of new reserves that are cheap to extract are unlikely to keep pace with their consumption. In this sense their supply is finite and therefore non-renewable.

By contrast, hydropower and bioenergy (from biofuels), together with solar, wind and wave power, are examples of **renewable** energy sources – that is, sources that are continuously replenished by natural processes. Renewable energy sources are essentially *flows* of energy, whereas the fossil and nuclear fuels are, in essence, *stocks* of energy.

Worldwide, there has been a rapid rise in the development and deployment of renewable energy sources during the past few decades. This is not only because, unlike fossil or nuclear fuels, there is no danger of their 'running out', but also because their use normally entails only small greenhouse gas emissions and therefore contributes little to global climate change.

The various renewable energy options will be described further in Chapter 14. They include solar power in its various forms, biomass, and hydro, wind, wave, tidal and geothermal energy (Figure 1.9). The 3.7 million EJ per year of solar energy available for use on Earth is over 8000 times the current rate of consumption of fossil and nuclear fuels (around 440 EJ in 2009). Two other, non-solar, renewable energy sources are shown in Figure 1.9: these are the motion of the ocean tides, caused principally by the Moon's gravitational pull (with a small contribution from the Sun's gravity) and geothermal heat from the Earth's interior, which manifests itself in convection in volcanoes and hot springs, and in conduction in rocks.

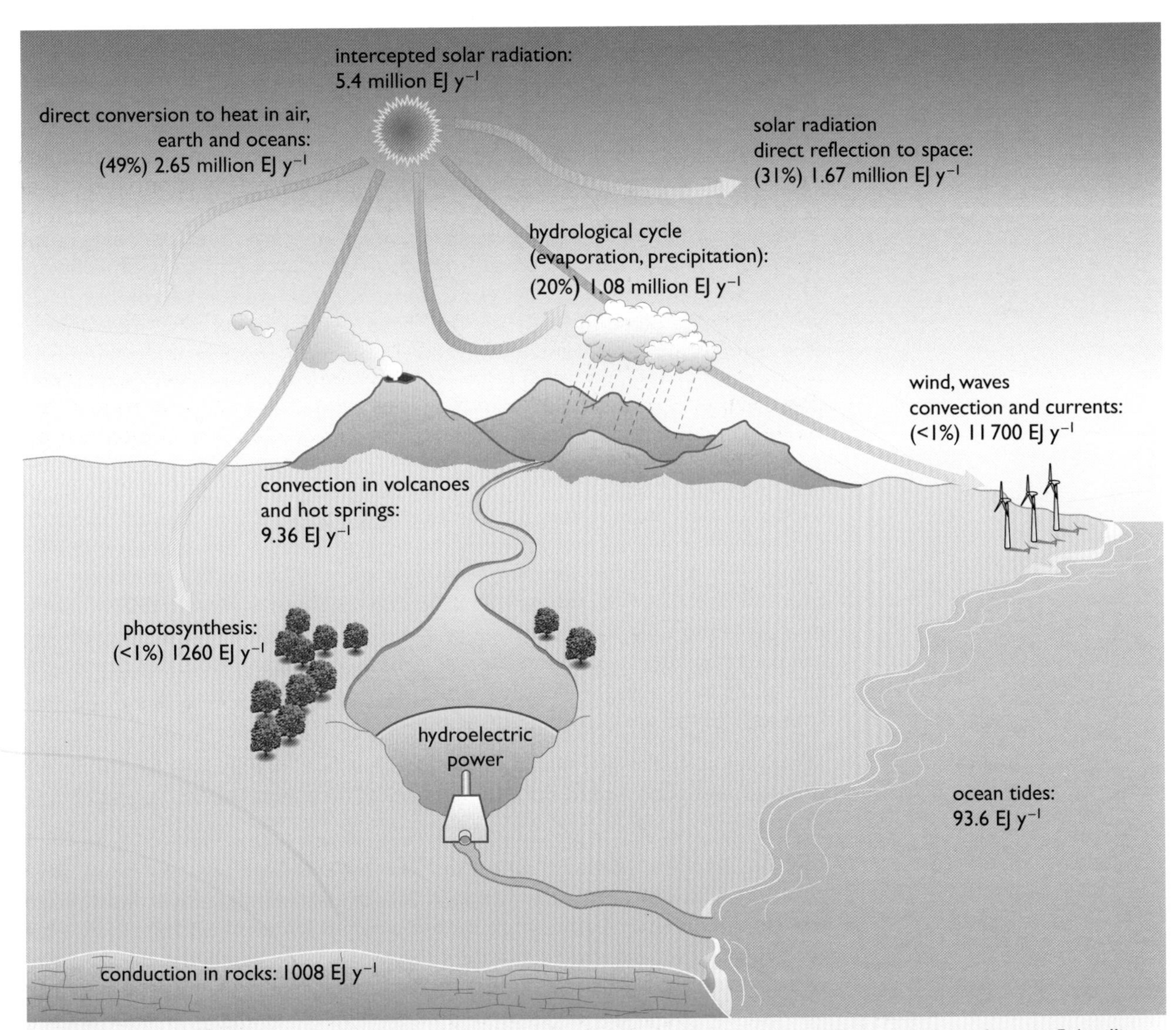

Figure 1.9 The various forms of renewable energy depend primarily on incoming solar radiation, which totals some 5.4 million exajoules (EJ) per year. Of this, approximately 30% is reflected back into space. The remaining 70% is in principle available for use on Earth, as shown, and amounts to approximately 3.7 million EJ.

1.4 What is a sustainable energy system?

The term 'sustainability' entered into common usage relatively recently, following the publication of the report *Our Common Future* by the United Nations' Brundtland Commission in 1987. The Commission defined sustainability, and in particular sustainable development, as 'Development that meets the needs of the present generation without compromising the ability of future generations to meet their needs.' (United Nations, 1987).

In the context of energy, sustainability has come to mean the harnessing of those energy sources that fulfil three requirements:

(1) they are not significantly depleted by continued use;

(2) they do not entail the emission of pollutants or other hazards to human or ecological and climate systems on a significant scale; and

(3) they do not involve the perpetuation of significant social injustices.

These are related to the three dimensions of sustainable development (Box 1.3). Sustainability is, of course, a very broad ideal. Some forms of renewable energy such as wind, solar thermal and solar photovoltaic systems come close to fulfilling the three requirements described above. It is argued that fossil fuel systems could be made more sustainable in the future through technological innovations – for example, by the use of carbon capture and storage (CCS), described in Chapter 14. Nuclear power could be made more sustainable through the development of deep-level storage for nuclear waste (see Chapter 11).

Sustainability is therefore in practice a relative rather than an absolute concept. It is not so much that some energy sources are sustainable and others not; it is more that some energy sources, in certain contexts, are more sustainable than others. Determining the relative sustainability of one energy system vis-à-vis another is usually a complex process, involving detailed consideration of the specific processes and technologies proposed, the context in which they are being used, and the differing values and interests of the various parties involved.

When we speak of 'the future' in the context of a 'sustainable future', what do we mean? The end of the 21st century? The end of the 3rd millennium? Forever?

Ideally, in view of the Brundtland Report's injunction that humanity should not compromise the needs of future generations, we should judge the sustainability of all energy systems on an indefinite time scale – far into the distant future. In practice, however, this might be realistically interpreted as endeavouring to ensure that energy systems become sustainable (or at the very least, much less un-sustainable) over the next century or so.

The long-term ecological and climatic consequences of our present fossil-fuel-dominated energy systems cannot be accurately predicted, but a precautionary approach would suggest that we should be pursuing a strategy of decarbonization and increasing energy efficiency (see Section 1.6).

BOX 1.3 Sustainable development and energy

The concept of sustainable development is frequently unpacked into three dimensions or pillars (IEA, 2001 citing OECD, 2001):

Economic sustainability – encompasses the requirements for strong and durable economic growth, such as preserving financial stability and a low and stable inflationary environment (i.e. low financial inflation – see Chapter 12)

Environmental sustainability – focuses on the stability of biological and physical systems and on preserving access to a healthy natural environment

Social sustainability – emphasizes the importance of well functioning labour markets and high employment, of adaptability to major demographic changes and stability in social and cultural systems, of equity and democratic participation in decision making.

Sustainable development emphasizes the links among these three dimensions, their complementarities, and the need for balancing them when conflicts between them arise. A key question is the extent to which all three dimensions can be pursued within an overall framework of economic, environmental and social policies. Given a choice, policymakers will always support those decisions and investments that offer win-win-win (all three dimensions) or win-win (two out of three dimensions) opportunities. Proponents of government support for the new renewables technologies, for example, point to the triple wins that investment in a new energy sector will deliver. Frequently, however, progress in one of the dimensions of sustainable development implies losses in another. A classic example is reducing energy taxes – this lowers the price of energy, which for example helps poorer people (and is therefore socially 'good'), but may increase overall consumption (and is therefore environmentally 'bad').

Weak and strong sustainability

Those who argue in favour of accepting the trade-offs between objectives in competing dimensions of sustainability are said to be in favour of **weak sustainability.** If the gain in one dimension more than offsets the loss in another, then sustainability can be said to have been enhanced – in a weak sense. Those who argue that there may be circumstances where trade-offs are not acceptable are said to be in favour of **strong sustainability**. Rejecting any policies that allow the Earth's temperature to rise by more than a given amount, say 2 °C, is an example of the strong sustainability approach.

In practice the framework of policies that shapes energy systems and sustainability is underpinned by elements of both weak and strong sustainability. Charging polluters and giving them a choice to change behaviour or technologies is an extremely common 'weak' element. Global caps on emissions of greenhouse gases from groups of countries is an example of a 'strong' element.

Differing weak and strong interpretations of sustainability among developed and developing countries are one reason why there has been persistent global disagreement in international climate change negotiations.

In its purest sense, a globally sustainable energy system would be one that relied on available solar radiation and not on non-renewable resources of fossil fuels or nuclear material. The present system borrows heavily from the distant past – energy from the Sun sequestered in fossil fuels as well as uranium from the geological past; it also borrows from the relatively recent past – we continue to chop down forests faster than we replace them. Global deforestation is not sustainable and in some developing countries it is driven in part by energy poverty itself – the need for cheap, local supplies of fuel for cooking and heating, because rural populations often do not have access to adequate energy supplies.

As Table 1.1 shows, the vast majority of our current primary energy supplies do not meet one or more of the three requirements of a sustainable energy source listed above.

Table 1.1 Mapping primary energy sources against the requirements for a sustainable energy system

Energy source	World primary energy in 2009/%	Significantly depleted by continued use?	Emission of pollutants or other hazards to our human and ecological systems on a significant scale?	Significant social injustices?
oil	32	yes (rapidly)	yes (CO_2 and SO_2 as well as local ecological impacts)	yes
coal	27	yes (less rapidly)	yes (CO_2 and SO_2 as well as local ecological impacts)	yes
natural gas	22	yes (fairly rapidly)	yes (CO_2, leakages of methane (CH_4) as well as local ecological impacts)	yes
traditional biomass	6	yes (rapidly – if not from sustainable sources)	yes (local air pollution and significant ecological impacts if not sustainably produced)	yes
nuclear	5	yes (rapidly – with current technology)	yes (in the case of major accidents and in uranium-mining areas)	yes
new renewables	5	no	few	few
large hydro	2	no	few	some

Given current energy technologies (2010), the category of energy sources that stands out in Table 1.1 as best matching the requirements of a sustainable energy system is new renewables; this includes wind, solar thermal, solar electricity, modern biomass, geothermal, wave and tidal energy technologies. Future technological developments may improve the sustainability of other energy sources. For example the use of carbon capture and storage with coal- or gas-fired power stations should reduce their emitted pollutants (but may make their resource depletion rating *worse*).

Also the three requirements set out in the table do not include any economic considerations. These might include the benefits of a cheap energy supply to those on a low income or the true financial damage costs of pollution, a subject to be discussed in Chapter 13. In this sense, the three criteria represent a 'strong' view of sustainability as applied to energy systems.

1.5 Why sustainable energy matters

Sustainable energy is a topic of great political, social, economic and environmental interest for three key reasons:

(1) There is a very large unmet demand in developing countries for commercially traded forms of energy, and this unmet demand is likely to persist or even increase throughout the course of this century.

(2) The supply of conventional oil and natural gas (and possibly coal) may be insufficient to meet future energy demand; therefore the global mix of energy sources is likely to change in the coming decades.

(3) Our energy options are now severely constrained by the need to reduce the world's emissions of carbon dioxide and other greenhouse gases, as part of our efforts to limit **anthropogenic** climate change (i.e. that produced by human activities).

Each of these reasons represents a particular perspective on the future. Indeed, each reason has behind it a community of champions (often dedicated and enthusiastic) who argue strongly in support of their perspective and advocate how the world should respond. We can label these three communities:

'Growthists', whose focus is on economic or what are sometimes called in this context 'above ground' factors (e.g. what are future energy prices or economic growth rates likely to be or who might want to claim fossil fuel reserves to secure their own national supply?)

'Peakists', whose focus is on geological or 'below ground' factors (e.g. how much coal, oil or gas can we extract?)

'Environmentalists', whose focus ranges from local and regional pollution to global Earth system considerations (e.g. how much CO_2 can the atmosphere, oceans and land vegetation absorb, and at what risk?)

Each of these three perspectives are outlined below. To some extent they represent extremes or simplifications of more subtle views on the future, as discussed in the summary that follows.

Growthist: 'business as usual'

There is a widely held view that there is, and will continue to be, a very large unmet demand for the supply of commercial energy throughout the course of this century. For example, about 1.4 billion people have no access to electricity and over 2.5 billion await clean cooking technologies. In terms of future potential unmet demand, another 2 billion people are forecast by the UN to join the world's population by 2050. By this time average world incomes (i.e. GDP) are expected by mainstream economic forecasters to have risen around fourfold with a consequent increase in rates of material consumption (IEA, 2010a). This 'growthist' view of the future is championed, for example, by the International Energy Agency. Growthists assume a strong relationship between economic growth and energy consumption (though not necessarily energy-related emissions).

Forecasting global energy demand is not trivial. Forecasts require the development of complex mathematical computer models using sets of equations that represent activities in all sectors of the economy, across the range of different fuels (coal, oil, natural gas, nuclear, renewables), and across all regions of the world. The IEA's models calculate energy demand as the product of changes in population, incomes and technological change, using around sixteen thousand mathematical equations. In 2010, the Agency in its annual *World Energy Outlook* (IEA, 2010a) included a 'Current Policies' scenario describing a 'Business As Usual (BAU)' future for the world energy system to 2035 (Figure 1.10). This shows total world energy consumption growing by around 67% by 2035 compared with 2010 and overall fuel shares remaining unchanged.

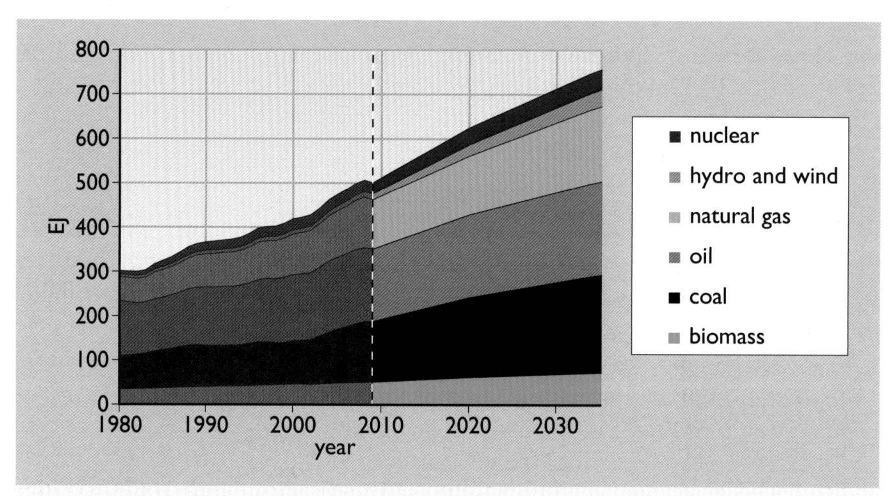

Figure 1.10 A typical 'growthist' or 'Business As Usual' future energy scenario (sources: BP, 2010; IEA, 2010a)

One of the criticisms of the forecasts of mainstream energy–economic models is that they fail to take adequately into account the fact that 'steady exponential growth' is in the long term environmentally unsustainable. Systems can and do exhibit exponential growth from time to time, but cannot do so indefinitely (Box 1.4).

BOX 1.4 Linear growth and steady exponential growth

Linear growth

The graphical representation of any quantity that is increasing by the *same amount* every year (or day, or decade) will of course be a straight line. So this type of growth is referred to as **linear growth**. Figure 1.4, for instance, shows that in the period since about 1960 world population increased more or less linearly from 3 billion to nearly 7 billion in 50 years: an average annual increase of about 80 million people per year.

Steady exponential growth

In analysing a rising (or falling) *curve*, it is useful to start with the concept of a **percentage change**. Per cent (%) means of course 'per hundred', so, for instance, 5% of 3000 is equal to

$$\frac{5}{100} \times 3000 = 150$$

If, for example, in a year notionally called Year 0, world annual oil consumption was 3000 Mt (million tonnes) with consumption rising at a rate of 5% per year, consumption in Year 1 would be:

$$3000 + 150 = 3150 \text{ Mt.}$$

If this 5% **annual growth rate** continues, the next increase will be 5% of the Year 1 consumption:

$$\frac{5}{100} \times 3150 = 157.5$$

which results in a Year 2 consumption of 3307.5 Mt, and so on.

This type of rise, where the value increases by the same *percentage* in each year (i.e. by the same relative amount), is technically **exponential growth** (commonly called **steady exponential growth**). Notice that this is quite different from the linear growth discussed above.

With the same annual growth rate, the *actual* (or *absolute*) annual increase becomes greater each year, so exponential growth will be represented on a graph by an upward curve. Figure 1.11 shows how the above annual oil consumption of 3 Gt (3 gigatonnes, or 3000 Mt) in Year 0 would increase over a period of 60 years, at two different annual growth rates, 5% and 2.5%.

As the name indicates, the **doubling time** for any increasing quantity is the time taken for it to double. It is perhaps less obvious that in exponential growth (because of the unvarying annual growth rate) the doubling time is the same *no matter where you start*. Consider, for instance, the 5% curve in Figure 1.11. The annual consumption in Year 10 is about 5 Gt, and moving forward from this we find that consumption reaches 10 Gt at about Year 25 – a doubling time of slightly less than 15 years. This is confirmed by further doublings at about Year 39 and again at Year 53, the increasing accuracy leading to the conclusion that the doubling time is about 14 years.

It is also worth noting that the doubling time for the 2.5% curve appears to be just under 30 years – twice that of the 5% line. As Table 1.2 shows, there is an inverse relationship between percentage increase and doubling time.

Table 1.2 Annual growth rates and doubling times

Annual growth rate	0.5%	1.0%	2.0%	5.0%	7.0%	10%
Doubling time/years	139	69.7	35.0	14.2	10.2	7.3

The following simple relationship between growth rate and doubling time, although it does not give precise values, can be very useful for rough estimates:

$$\text{doubling time in years} \approx \frac{70}{\text{annual percentage growth rate}}$$

Consequences of steady exponential growth

Continued exponential growth has disturbing consequences if it reflects the consumption of a finite commodity (such as a fossil fuel). Figure 1.12 shows the 5% line of Figure 1.11 redrawn with its time axis expressed not in years but in terms of doubling times. Over the period of the first doubling time, the *total amount of fuel consumed*, the sum of all the individual annual quantities, is represented by the yellow area under the curve.

The amount consumed over the next doubling period is represented by the light red area. This is obviously larger than the yellow area for the previous period, and so on, in successive doubling periods. What is perhaps not so obvious is that the total fuel consumed over the period of the fourth doubling (the black area) is in fact slightly *greater* than the *sum* of the preceding yellow, light red and dark red areas. For another (fifth) doubling to take place at the same growth rate would require more fuel to be available than is represented by *all* of the yellow, light red, dark red and black areas.

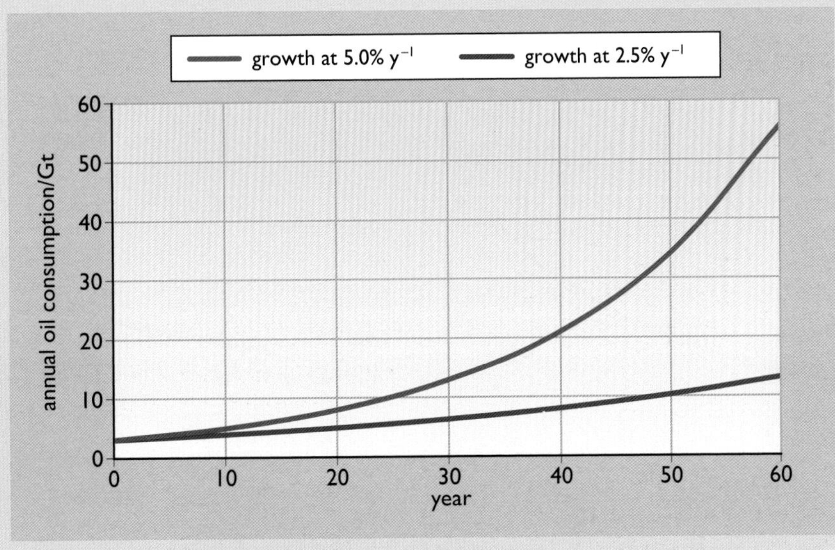

Figure 1.11 Exponential growth at different annual rates

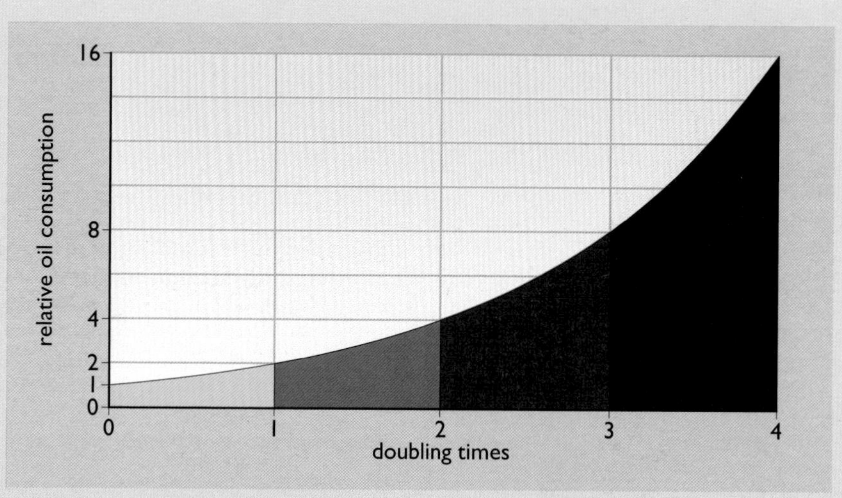

Figure 1.12 Exponentially growing fuel consumption

Put very simply, in this pattern of exponential growth, each successive doubling time consumes as much fuel as the *total used in all the previous doubling times.*

China as an example of exponential growth

In the real world of economics and politics, a long-term annual GDP growth rate of 2% is considered a relatively modest ambition, yet this would double the overall GDP of a nation within 35 years (Table 1.2). As shown in Figure 1.13, between 1981 and 2008 Chinese GDP grew by a factor of nearly 12, an average rate of 9% per year or a doubling time of just under 8 years.

This increase in GDP has been accompanied by a growth in energy demand, particularly for China's indigenous coal. The growth in coal production (Figure 1.14), although obviously not a simple exponential rise, represents overall an average increase of approximately 5.6% per annum, doubling every 12.5 years.

Chinese coal production already accounts for nearly a half of the world's total. A further doubling (in another 12.5 years) would mean that Chinese coal reserves would be being depleted at an alarming rate. It would also require an extraordinary investment in new coal mines – and would result in a dramatic increase in global CO_2 emissions. (For further discussion of China and coal see Chapters 2 and 5.)

It should be obvious from the analysis above that eventually there are likely to be resource limits to growth. This thinking has led some commentators to question the validity of assumptions of steady percentage growth and indeed to question the very idea of long-term *sustainable* growth (see Jackson, 2009).

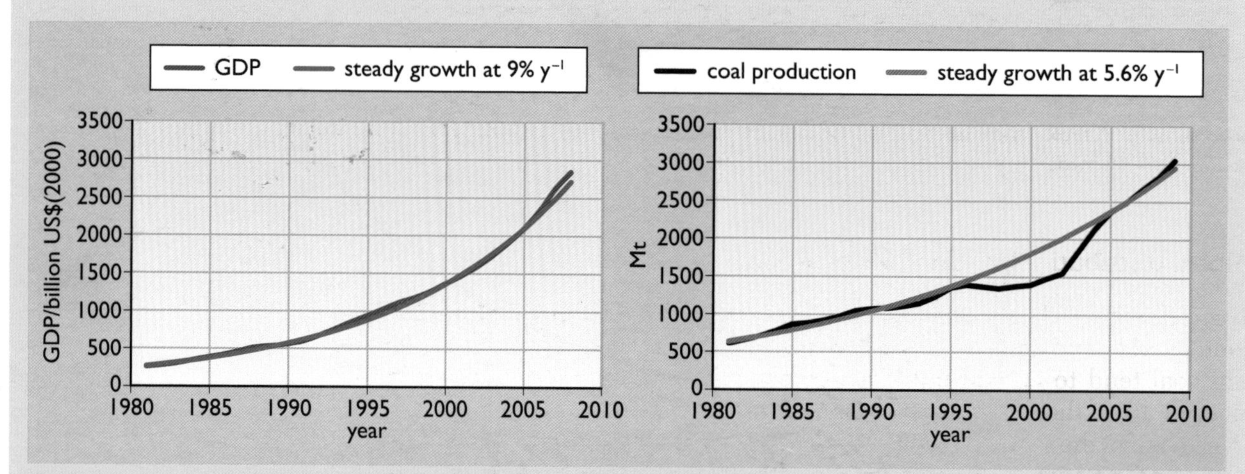

Figure 1.13 Chinese GDP growth 1981–2008 is almost exponential (source: IEA, 2010b)

Figure 1.14 Chinese coal production, 1981–2009 (sources: Aden, 2010; BP, 2010)

Peakist: fossil fuels limit growth

There is much uncertainty over when oil, natural gas and eventually coal production will peak as supplies are depleted. The notion that the world is nearing a peak in oil supplies (in the first instance) is a subject of great current interest and debate. In contrast to the growthists, there is an alternative 'peakist' (after peak oil) view that through resource constraints, higher prices and technical change, future energy demand may not be as high as the growthists predict. Peakists are concerned that although the world has large *resources* of fossil fuel, its *reserves* (i.e. the portion that is economic to extract) are limited. Even more important, the production of many sources is *rate limited* by the availability of financial investment, water or even other fuels needed for extraction. As a result supplies of oil, natural gas and even coal may be insufficient to meet the world's current level energy demand. For a review of studies of 'peak oil' see Sorrell et al., 2009).

Though the world has relatively large stocks of proven coal reserves, some analysts question how long these will last at projected consumption rates. Global reserves of uranium – used in civil nuclear power – are also uncertain but limited. For peakists, we live at a time of increasing concern about peak oil, peak gas and even peak coal and at a turning point in our use of fossil fuels (see Figure 1.15).

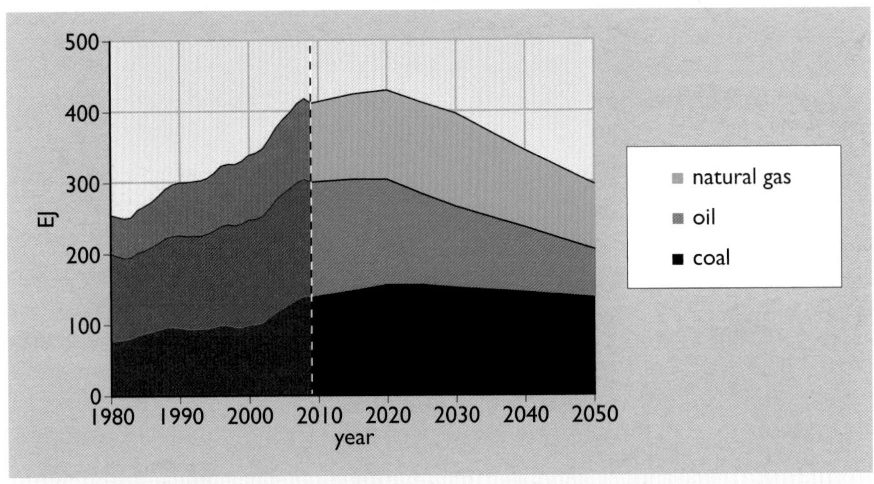

Figure 1.15 A typical 'peakist' scenario of future fossil fuel supplies (sources: BP, 2010; EWG, 2007; Campbell et al., 2009)

Global distribution of fossil fuel reserves

Since the fossil fuels were created in specific circumstances where the geological conditions were favourable, the largest deposits of oil, gas and coal tend to be concentrated in particular regions of the globe (see Figure 1.16), although less appreciable deposits are remarkably widespread. Over a half of the world's conventional oil reserves are located in the Middle East. As for natural gas, over 40% of the world's conventional reserves

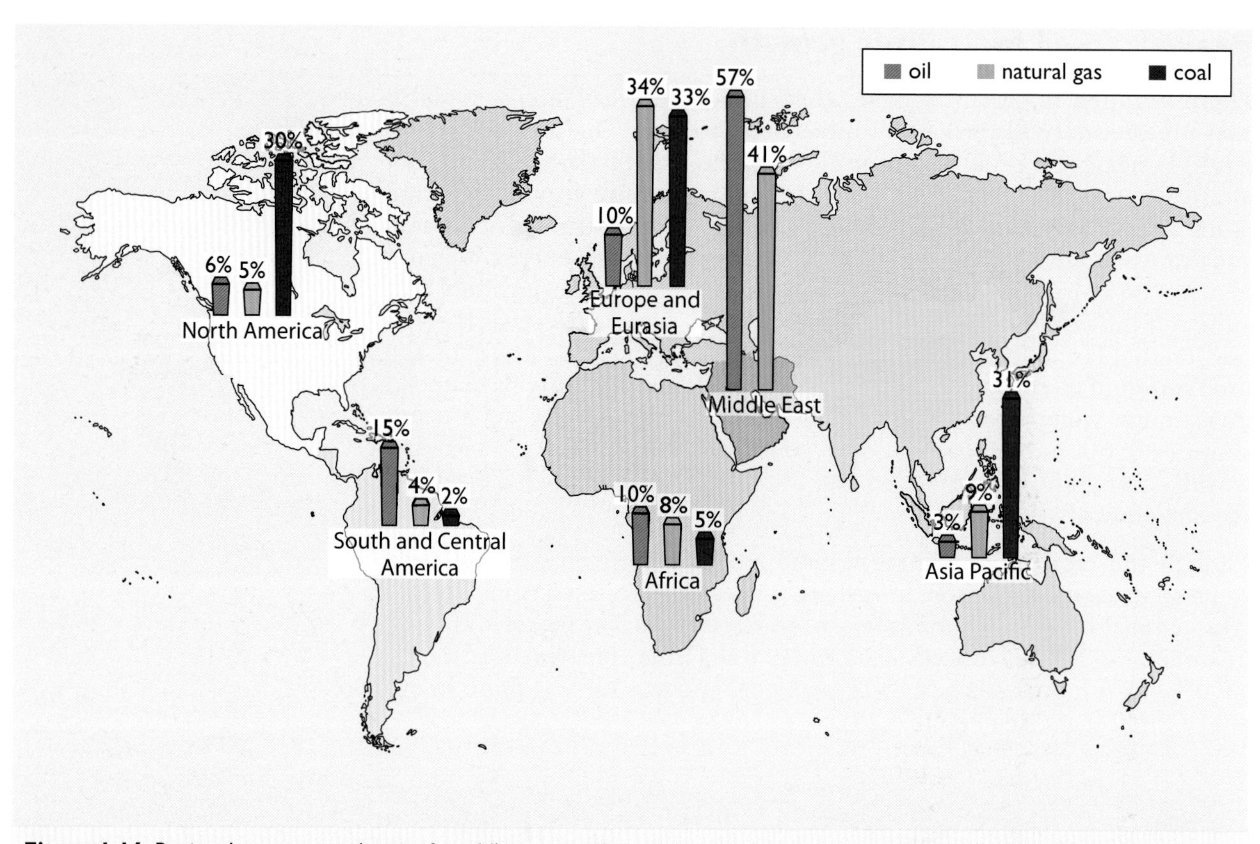

Figure 1.16 Regional percentage shares of world's proven oil, natural gas and coal reserves, 2009 (note that the totals may not add to 100% because of rounding). Totals in 2009: oil – 1333 thousand million barrels (the Middle East holds over half of proven oil reserves); natural gas – 187 trillion cubic metres (the Russian Federation, Iran and Qatar hold over half of proven reserves); coal – 826 billion tonnes in 2009 (source: adapted from BP, 2010)

are in the Middle East, most of it in a single field under the Persian Gulf between Qatar and Iran. A further 30% is in the former Soviet Union (BP, 2010). Although coal deposits are rather more evenly spread throughout the world, seventy per cent of world coal reserves are concentrated in just four countries: Australia, China, Russia and the United States of America. (BP, 2010) (Figure 1.16).

Although human society now consumes fossil fuels at a prodigious rate, the amounts of coal, oil and gas that remain are still very large. One simple way of assessing the size of reserves is called the **reserves/production ratio** (R/P) – the number of years the reserves would last if use continued at the *current* rate (Figure 1.17).

Coal has the largest R/P ratio. Present estimates suggest the world has around 120 years' worth of coal left at current use rates. In 2000 this figure stood at over 200 years. World coal production has in fact gone up since then, but reserves have also been revised *downwards* significantly (see Chapter 5). For oil, current R/P estimates suggest a lifetime of about 45 years at current rates. For natural gas, the R/P ratio is somewhat higher, at around 60 years (BP, 2010).

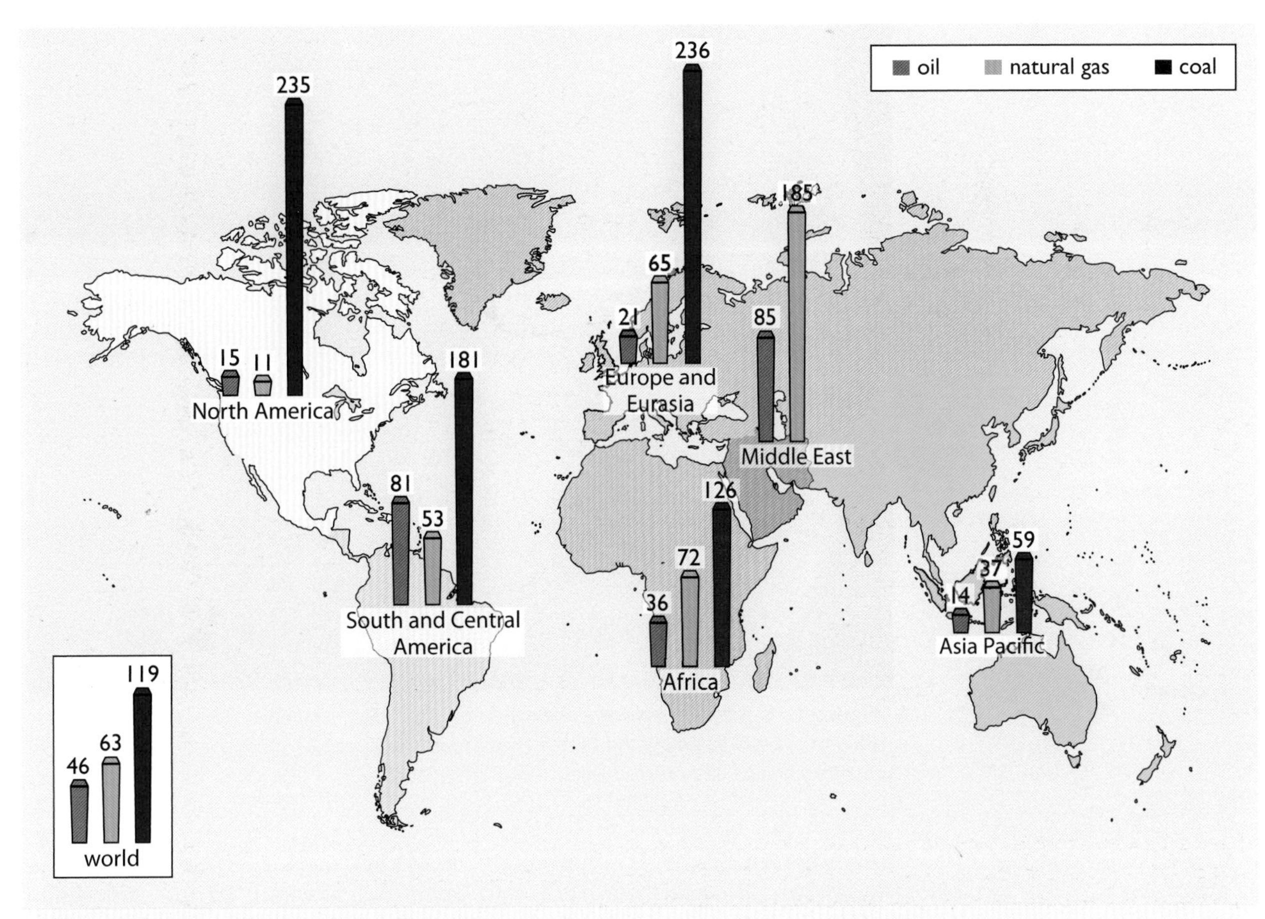

Figure 1.17 Reserves/production (R/P) ratios (in years) for oil, natural gas and coal, 2009, for various regions of the world and the world as a whole; coal reserves in the Middle East are very small indeed (source: adapted from BP, 2010)

Fossil fuel R/P ratios need to be interpreted with great caution, however. They do not take into account the discovery of new proven reserves, or technological developments that enable more fuel to be extracted from the existing resources. The distinction between reserves and resources is described in Chapters 5 and 7. Additionally, they do not take into account future increases in production, which may also overstate future R/P ratios.

Despite these developments, it seems likely that, at least in the case of oil from conventional sources, world production is now near its peak. The IEA's 2010 *World Energy Outlook* suggests that conventional oil production has peaked and will 'plateau' from now on, but that supplies of natural gas liquids and un-conventional oil will continue to increase to 2035.

From now on, although vast *resources* of conventional oil will still remain, the *reserves* are likely to be on a declining curve (see Sorrell et al., 2009). The concentration of scarce fossil fuel resources has already led to major world crises and conflicts, such as the 1973/4 oil crisis and the first and second Gulf Wars of the 1990s/2000s. It has the potential to create similar, or even more severe, problems in the future. Nation states are keen to claim and defend potential new sources of hydrocarbons and other mineral resources (Figure 1.18).

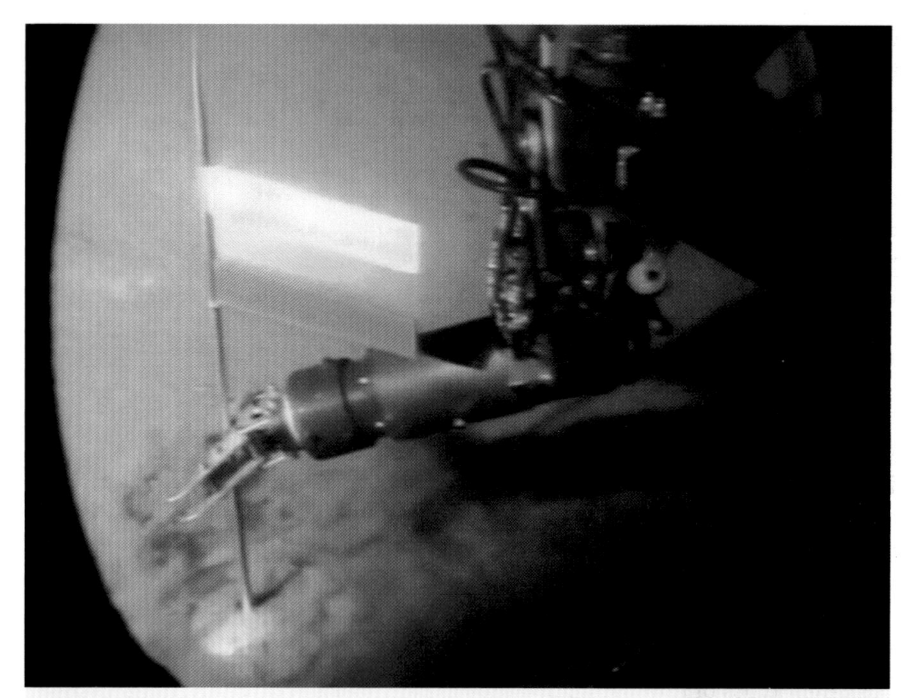

Figure 1.18 Russian explorers plant their country's flag in summer 2007 on the seabed 4200 m (14 000 ft) below the North Pole to further Moscow's claims to the Arctic territory and its natural resources

As shown in Figure 1.1 there were rapid increases in crude oil prices in the 1970s followed by a return to lower prices again in the 1990s. Oil prices began to rise again after 2000, as did others linked to the its price. Substantial rises can cause widespread economic disruption (as they did in 1973 and for many years after). They can lead to widespread protests, as seen in 2007 and 2008 when prices temporarily reached just under US$150 per barrel. Amongst other things, this caused significant increases in the price of basic foods (e.g. grain), which in turn sparked food riots in this period in Eastern Europe, Asia, Africa, and Latin America. The world relies on oil and natural gas to produce most of its food. We use oil in farming and for the distribution of food. Currently, natural gas (the price of which tends to rise and fall in line with that of oil) is used as a hydrogen source to make fertilizer. It is possible that social stability and economic growth are sensitive to energy prices above a certain critical threshold.

Peak uranium?

The nuclear industry has traditionally argued that there is little danger of the world 'running out' of nuclear fuel in the near future; but there is also a peakist perspective on the uranium resource. Although uranium resources have been identified in many countries, roughly two-thirds of the world's recoverable resource and production is in just five countries (see Chapter 11). Current uranium demand is significantly higher than supply and there is a significant chance that future supplies will struggle to follow present levels of demand, let alone be sufficient for a nuclear renaissance.

On the other hand, nuclear proponents argue that advanced nuclear technologies such as the fast breeder reactor (FBR) should enable uranium deposits to be used even more effectively, thus extending the lifetime of reserves. The FBR design relies on converting the non-fissile isotope uranium-238 into fissile plutonium-239 for use as a reactor fuel. Its development has been inhibited by the technical and safety difficulties of fuel reprocessing to extract this plutonium and concerns about the proliferation of nuclear weapons. Given the current low price of uranium, this technology remains economically un-competitive.

Environmentalist: global environment limits growth

The massive use by our society of coal, oil and natural gas has, literally, fuelled enormous increases in material prosperity – at least for the majority in the industrialized countries. But it has also had numerous adverse consequences. Fossil fuel combustion generates large quantities of carbon dioxide (CO_2); nearly 30 billion tonnes in 2009. Different fossil fuels produce different amounts of CO_2 per unit of energy delivered (see Box 1.5).

BOX 1.5 Carbon dioxide emission factors for different fossil fuels

Different types of fossil fuel contain different proportions of carbon, hydrogen and oxygen (and other elements such as sulfur). This is because coal, oil and natural gas have different molecular structures. At room temperature coal is solid, oil is liquid and natural gas is gaseous – this itself gives us a clue as to their different molecular complexities.

Coal contains a complex mixture of different molecules. Although mainly carbon it can also contain complex *aromatic* hydrocarbons which have a hexagonal molecular structure. Oil is a similarly a complex mixture of mainly *aliphatic* hydrocarbons in long chains, while natural gas is a mixture of mainly methane and some ethane which have small molecules. The different fossil fuels have different ratios of carbon to hydrogen (natural gas contains a higher ratio of hydrogen than coal). This ratio governs both their energy density and the amount of carbon dioxide emitted for every unit of heat produced when they are burned. As a fuel, natural gas contains nearly twice as much energy per kilogram as coal, but only emits 60% as much CO_2 when burned (see Table 1.3).

Table 1.3 Carbon dioxide emissions for different fuels

Fuel	Specific carbon dioxide emissions/g CO_2 MJ^{-1}
Coal	93
Fuel oil	78
Natural gas	57

Source: AEA, 2010.

Rising greenhouse gas concentrations are causing the Earth's temperature to increase rapidly – certainly, in comparison with the Earth's natural cycles of climate change recorded over the last few million years. This is highly likely to cause significant changes in the world's climate system (see Box 1.6).

BOX 1.6 The greenhouse effect and global climate change

The **greenhouse effect** in its natural form has existed on the planet for hundreds of millions of years and is essential in maintaining the Earth's surface at a temperature suitable for nearly all life. Without it, we would all freeze.

The Sun's radiant energy, as it falls on the Earth, warms its surface. The Earth in turn re-radiates heat energy back into space in the form of infrared radiation. The temperature of the Earth establishes itself at an equilibrium level at which the incoming energy from the Sun exactly balances the outgoing infrared radiation (Figure 1.19).

If the Earth had no atmosphere, its surface temperature would be approximately minus 18 °C, well below the freezing point of water. However, our atmosphere, whilst largely transparent to incoming solar radiation in the visible part of the spectrum, is partially opaque to outgoing infrared (long wavelength) radiation. It behaves in this way because, in addition to its main constituents, nitrogen and oxygen, it also contains water vapour and other very small quantities of **greenhouse gases**. Put simply, these inhibit the outflow of heat, so keeping the Earth's surface considerably warmer than it would otherwise be. The average surface temperature of the Earth is in fact around 15 °C, some 33 °C warmer than it would be without the greenhouse effect.

The net energy gain due to incoming solar radiation that penetrates the atmosphere as shown in Figure 1.19 is 69 units (100 incoming – 31 reflected). The net loss of infrared energy escaping to space is also 69 units (57 + 12 = 69 units). The incoming and outgoing radiation are therefore balanced, and the Earth is in thermal equilibrium.

The greenhouse gases that contribute most to global warming are water vapour, carbon dioxide and methane.

Water vapour evaporating mainly from the oceans plays a major part in maintaining the natural greenhouse effect, but human activities have very little influence on the vast processes through which water cycles between the oceans and the atmosphere.

Carbon dioxide (CO_2) is primarily generated by natural processes. These include the process of cellular respiration, in which organisms expel carbon dioxide; and the emissions of CO_2 that occur when organisms die and decompose. But since the Industrial

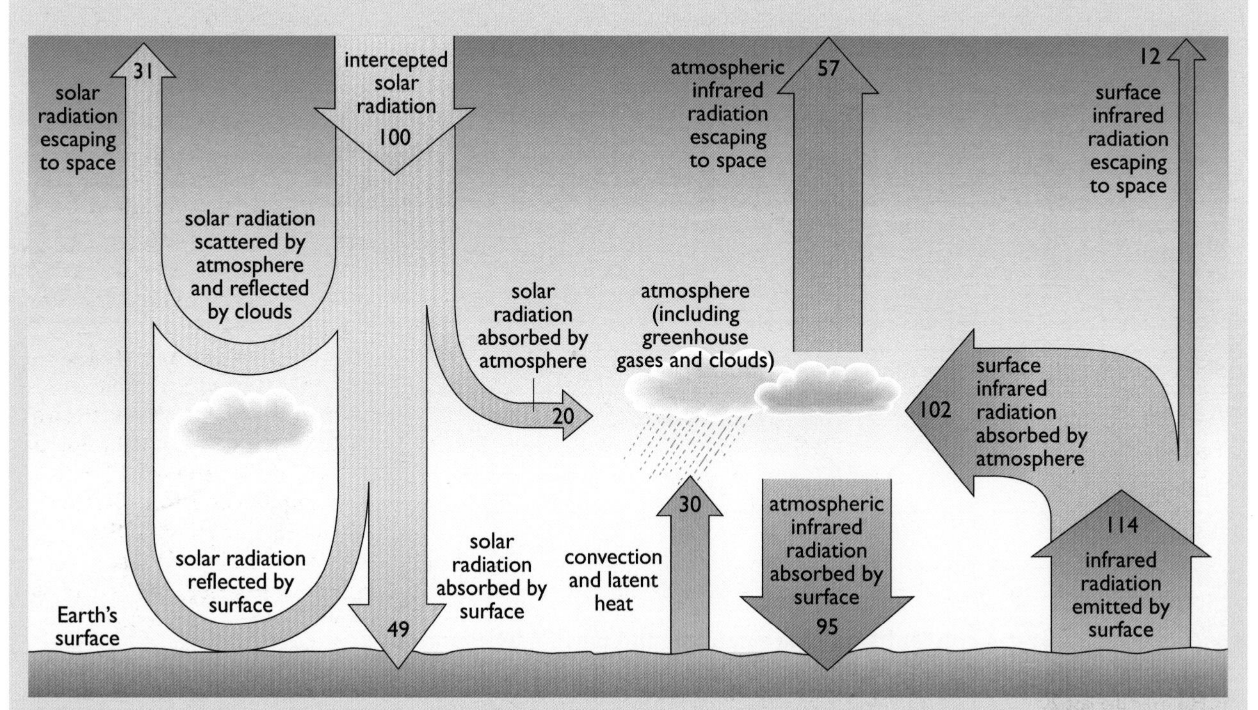

Figure 1.19 The greenhouse effect raises the Earth's temperature by reducing the amount of short-wave solar radiation that is reflected directly to space. Without an atmosphere the average global surface temperature would be lower by about 33 °C

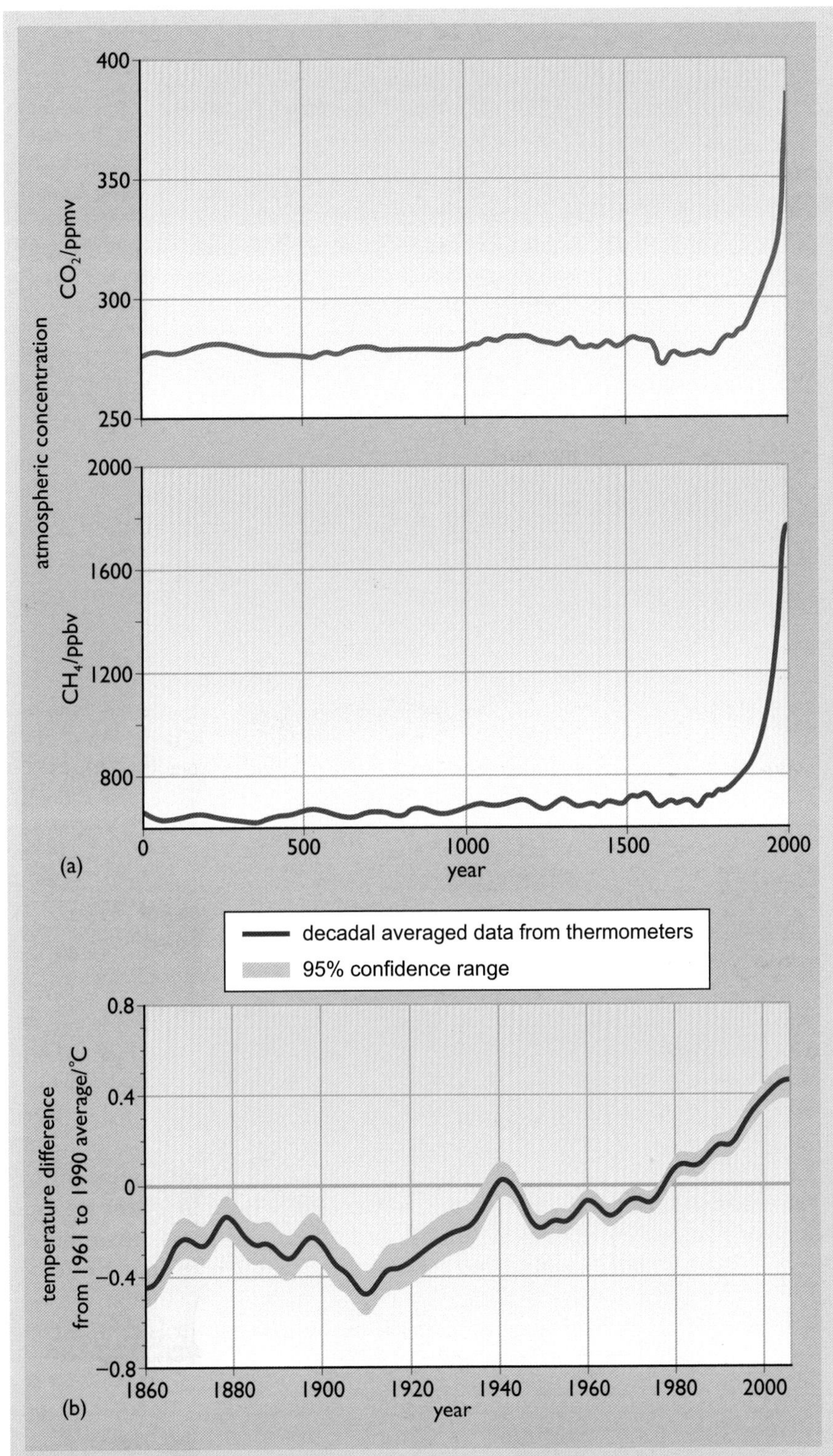

Figure 1.20 (a) Variation of atmospheric concentrations of carbon dioxide (CO_2) and methane (CH_4) over the last 2000 years; (b) estimated global mean temperature variations, 1860–2005 (source: IPCC, 2007)

Revolution, the burning of fossil fuels by humanity has been adding substantial quantities of CO_2 to our atmosphere. The concentration of CO_2 in the atmosphere in pre-industrial times was around 280 parts per million by volume (ppmv) but levels have been steadily rising since then, reaching some 390 ppmv in 2010 (See Figure 1.20(a).

Methane is given off naturally when vegetation decays in the absence of oxygen – for example, under water. However, various human activities, including increasing rice cultivation which causes methane emissions from paddy fields, intensive farming of livestock and leaks of fossil methane from natural gas distribution systems, have caused the levels of methane in the atmosphere to increase sharply. Concentrations have risen from about 750 ppbv (parts per billion by volume) in pre-industrial times to around 1790 ppbv in 2010.

These additional emissions of carbon dioxide and methane (as well as some other gases) are the main causes of the so-called *anthropogenic* greenhouse effect. The increase in carbon dioxide concentration is responsible for just over three times more global warming than the increase in methane concentrations. 'Radiative forcing' is a concept that allows us to sum up the contribution of different quantities of different greenhouse gases to global warming. The comparative radiative forcing of these greenhouse gases, and others, is discussed in Chapter 13.

Unlike the operation of the natural greenhouse effect, which is benign, the anthropogenic greenhouse effect is almost certainly the cause of a global warming trend that will have very serious consequences for the Earth's ecosystems and in turn humanity. Though a small minority dissents, the majority of scientists now believe that the activities of modern humans, acting to enhance the natural processes, has already caused the mean surface temperature of the Earth to rise by about 0.8 °C during the twentieth century (IPCC, 2007) – see Figure 1.20(b). Moreover, if steps are not taken to limit greenhouse gas emissions (i.e. if 'business as usual' policies are pursued), atmospheric CO_2 levels will probably rise by 2100 to around 560 ppmv – around double the pre-industrial levels (or possibly more, depending on the assumptions made). These levels would be likely to lead to rises in the Earth's mean surface temperature of between 1.1 and 6.4 °C by the end of the twenty-first century. These temperature rises would be very likely to result in significant changes to the Earth's climate system. Such changes would probably include more intense rainfall, more tropical cyclones and more long periods of drought, all of which would disrupt agriculture. Moreover, ecosystems will be damaged, with some species unable to adapt quickly enough to such rapid changes in climate.

In addition, at the same time, owing to thermal expansion of the oceans, sea levels are predicted to rise by up to half a metre during the twenty-first century, sufficient to submerge some low-lying areas and islands (Figure 1.21). In the longer term, further sea level rises would result if parts of the Greenland and Antarctic ice sheets were to begin to melt. These problems will be described further in Chapter 13.

Figure 1.21 President Mohammed Nasheed of the Maldives and 13 other government officials hold a meeting underwater to raise global awareness of the plight of low-lying island states in the build up to the Copenhagen Climate Conference in late 2009

Most if not all energy options and technologies have environmental and social impacts of one form or another (e.g. air, water, land pollution, materials resource constraints, social impacts – see Chapter 13). Among these, global climate change has become the main focus of environmentalists' attention as the principal challenge and design constraint on the energy economy of the future. Our energy options are now severely limited by the need to reduce our emissions of carbon dioxide and other greenhouse gases in our efforts to limit global warming and climate change. If we are to meet global objectives to stabilize the Earth's climate, we must dramatically cut our use of fossil fuels or find ways of using them whilst capturing the carbon dioxide released during combustion.

At the international Copenhagen Climate Conference in 2009 a target was agreed to limit global warming to 2 °C above its pre-industrial average. Scientific analysis suggests that if we are to have a 50% chance of stabilizing the global temperature at this level then emissions of CO_2 must peak in the next decade and then decline rapidly to around half of today's levels by 2050 (Allen et al., 2009). The energy use in one such scenario is shown in Figure 1.22. In this, the IEA's 'Blue Map' scenario, although world primary energy use increases between 2010 and 2050, world fossil fuel

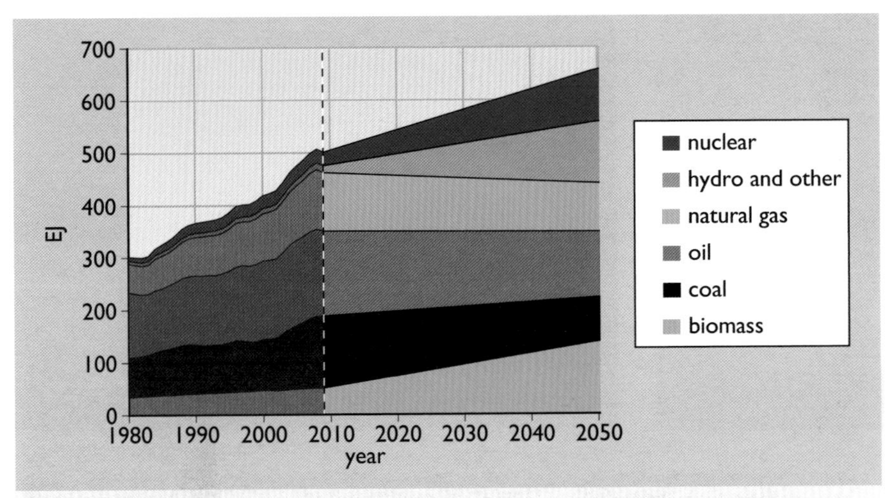

Figure 1.22 An example of a (pro-nuclear) environmentalist perspective: the IEA's 'Blue Map' Energy Scenario for 2010 to 2050 (sources: BP, 2010; IEA, 2010c)

use diminishes. It envisages global emissions of CO_2 declining by 50% by 2050, with the aim of limiting global warming by less than 2 °C – though the IEA notes that deeper cuts will probably be required to achieve this. The environmentalist perspective it illustrates is similar to a peakist one in that fossil fuel use declines. However, in the peakist view it does so from scarcity (and with, no doubt, high prices). In the environmentalist perspective the decline is driven by global policies and in a world where fossil fuel supplies may still be cheaply available. However, it does only describe what *may* happen rather than forecasting what is likely to happen. This scenario is discussed further in Chapter 14.

If fossil fuels were likely to be cheap and in plentiful supply over the course of this century this would present us with a very challenging problem. Perhaps, fortunately, from the point of view of climate change, our fossil fuel resources are finite. Exactly how much we have left is, as we have seen, the subject of much scientific dispute.

All fuels and energy technologies have their negative social and environmental impacts (see Chapter 13) that act to constrain our energy options in a variety of ways on a regional, national and local basis. While globally there is an abundance of renewable energy from the Sun, at a national and local level there are various limits to the amount of renewable energy that can be accessed (for example water and soil nutrient constraints for biomass crops or variations in solar radiation intensity in the case of photovoltaics).

There is no doubt that the extraction, distribution and burning of fossil fuels does significant damage to human health and ecological systems. This may be air pollution from combustion, producing sulfur dioxide, oxides of nitrogen and particulates. It may be due to fires and explosions on oil or gas rigs with resulting oil spills (Figure 1.23) or it may involve mining accidents or conflicts over access to fuel resources. But, perhaps most profoundly, in the form of today's global climate change that is almost certainly the result of increasing atmospheric carbon dioxide concentrations caused by fossil fuel combustion.

(a)

(b)

Figure 1.23 (a) Eleven crew members lost their lives in the explosion on board BP's Deepwater Horizon oil production platform in the Gulf of Mexico in April 2010; (b) a brown pelican caught up in the subsequent oil pollution

Nuclear safety

Nuclear power has grown in importance since its inception just after World War II and now supplies some 5% of world primary energy. A major advantage of nuclear power plants, in contrast with fossil fuelled power stations, is that they do not emit greenhouse gases in their operation. However, greenhouse gas emissions are involved in both the construction of power plants and in uranium mining. Those from mining may increase if lower grades of ore have to be used in the future. The use of nuclear power, as will be described in Chapters 11 and 13, gives rise to particular problems:

- the ever-present possibility of major nuclear accidents which, although rare, can have catastrophic effects.
- difficulties of long-term radioactive waste disposal
- routine emissions of radioactive substances
- dangers from the proliferation of nuclear weapons.

Although the majority of nuclear reactors in most countries have operated without serious safely problems, a number of major accidents, like those at Windscale in the UK in 1957, Three Mile Island in the USA in 1979, Chernobyl in Ukraine in 1986 and Fukushima in Japan in 2011 have created widespread public unease about nuclear technology in general. The Chernobyl accident led to the widespread cancellation of orders for new nuclear power stations in Europe and the USA.

Despite these difficulties, the nuclear industry has been attempting to develop more advanced types of nuclear reactor which, it claims, will be cheaper to build and operate, and inherently safer, than existing designs. These are being promoted as an improved technological option for generating the carbon-free electricity that will be required in the twenty-first century. The moratorium on new nuclear construction following Chernobyl turned into a renewed interest in the first decade of the twenty-first century in many countries, particularly in Asia. It seemed that there might be a 'nuclear renaissance'. In Europe, the construction of two new reactors was started, one in Finland and one in France. The Fukushima accident appears to have swung opinion back against nuclear power, particularly in Germany, Switzerland and Italy in favour of other renewable and sustainable energy alternatives.

Three perspectives: a summary

Each of the perspectives discussed above are to some extent 'extreme' simplifications. They are not mutually exclusive – there are sophisticated arguments about how they constructively or destructively interfere with each other (Figure 1.24). Belief in one, two or all three of these perspectives produces different shades of argument and overlapping sets of policy prescriptions. Some environmentalists are growthists who are pro-nuclear while others are anti-nuclear and pro-renewables. Moreover, not all growthists believe we are far from peak-oil or are sceptical about climate change.

In the absence of a significant technological breakthrough – a technological 'silver bullet' such as cheap nuclear fusion or some as yet undiscovered advanced solar energy technology – average energy prices are likely to increase significantly in real terms. Some may argue that this would be a good thing as it may help limit energy demand; it may also accelerate our search for affordable, cleaner and more sustainable sources of energy. This is particularly important for the two billion or so poorest people in the world that do not have enough energy available to them to lead healthy and fulfilled lives. This is a major barrier to their social and economic development.

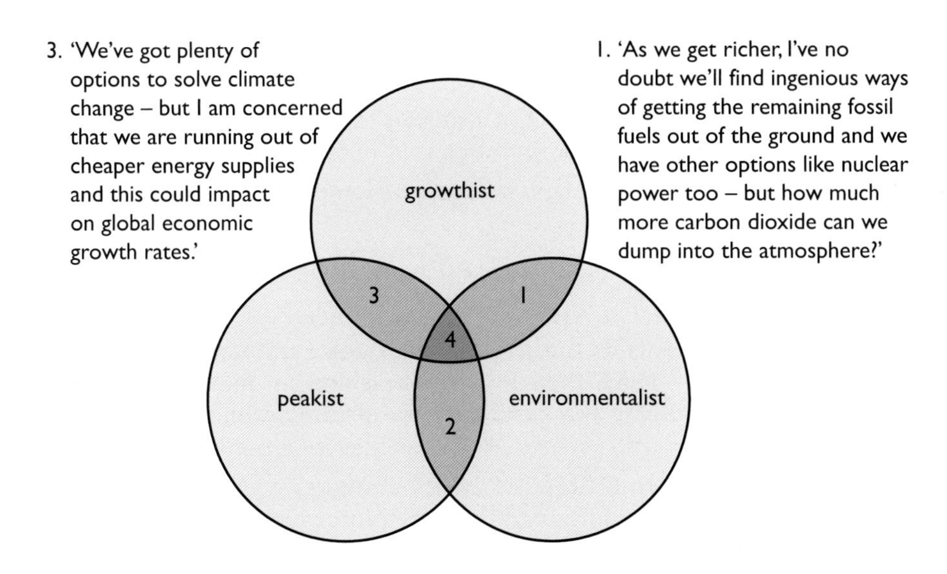

4. 'Economic prosperity is the key to solving the twin problems of maintaining energy supply and combatting climate change.'

2. 'Consumption is at the heart of the matter. We can't go on consuming as we do. The bad news is that we are changing the Earth's climate. The good news at least is that we may be running out of cheap oil and gas, possibly coal.'

Figure 1.24 Four different perspectives on the energy future arising from different degrees of agreement with or belief in the three perspectives

Energy prices are likely to change considerably in the future relative to today. The price differences between fuels are also likely to change. This will cause problems for economic planners attempting to decide on new long-term investments in energy infrastructure such as power stations or energy-intensive factories. Prices may change for reasons of supply and demand. For example, in a 'Growthist' world of high oil demand but limited supply, oil prices are likely to keep increasing. Government-imposed carbon taxes may make high-carbon fuels such as coal more expensive than natural gas or renewable energy sources.

In the longer term it is even possible that severe restrictions on CO_2 emissions from fossil fuel combustion will leave the world awash with unwanted fossil fuels, particularly coal.

Technology and geology together with policy are therefore all sources of considerable uncertainty around the planning of sustainable energy systems. The energy future might be 'business as usual' or we might be entering a period of great turbulence in the world's energy systems.

1.6 How can we achieve the transition to low-carbon energy systems?

There are four main possible approaches to reducing greenhouse gas emissions from the national and world energy systems:

(1) managing human population levels – e.g. speeding up the 'demographic transition' in developing countries so that poverty is reduced, education improved and fertility rates decline

(2) accepting lower growth rates in national/global average per capita GDP

(3) reducing the amount of energy required to produce a unit of GDP in the economy (e.g. through investment in energy efficiency)

(4) reducing the amount of carbon emitted per unit of energy consumed.

The four options can be explored with what has become known as the Kaya Identity (Box 1.7).

Options 1 and 2 are outside the scope of this book, which will concentrate on options 3 and 4. From these two options we can identify three key approaches to improving the sustainability of human energy use in the future:

(a) using energy more efficiently in its conversion, distribution and consumption, thus improving the energy intensity of the GDP

(b) reducing the CO_2 emissions of fossil fuel technologies, i.e. improving their carbon intensity

(c) switching to renewable and/or nuclear energy sources on a much wider scale, thus improving the overall carbon intensity of the energy supply.

BOX 1.7 The Kaya Identity

The Kaya Identity is a simple formula used to explore the options available to reduce CO_2 emissions from energy systems at both the global and the national level. It calculates CO_2 emissions, F, as a product of terms such as the population, P, and income expressed in terms of the gross domestic product, G. (IPCC, 2003).

The Kaya Identity is an expression of the form:

$$CO_2 \text{ emissions} = \text{population} \times \frac{\text{GDP}}{\text{population}} \times \frac{\text{energy}}{\text{GDP}} \times \frac{CO_2 \text{ emissions}}{\text{energy}}$$

This can be written as:

$$F = P \times \frac{G}{P} \times \frac{E}{G} \times \frac{F}{E}$$

where:

F is CO_2 emissions from human sources
P is population
G is Gross Domestic Product (GDP)
E is primary energy consumption

and the ratios are:

G/P is per capita GDP
E/G is energy intensity
F/E is average carbon intensity of energy.

National comparisons of the first two of these ratios will be discussed in Chapters 2 and 3 while the carbon intensity of some different fuels has been described in Box 1.5. Data on national and global energy use, CO_2 emissions, GDP and population can be found in IEA, 2010b.

Approach (a) is known as a 'demand-side' response and can involve both technological and behavioural changes in the consumption part of an energy system. The two other approaches are 'supply-side' responses and are largely technological.

(a) Using energy more efficiently

Many energy technologies have been evolving towards more efficient designs over time. On the supply side, the modern combined cycle gas turbine (CCGT) power plant can generate electricity more efficiently than its coal-fired steam turbine counterpart. The waste heat from power stations (of all sizes) can be used for heating buildings by using combined heat and power generation (CHP). Both of these technologies are described in Chapter 9. On the demand side, the efficiency, or more precisely the *efficacy*, of lighting has improved by a factor of more than 300 since the beginning of the nineteenth century (Chapters 3 and 9). The efficiency of domestic heating has increased from about 25% for a coal fire to over 90% for a modern condensing gas boiler (Chapter 7). The energy demand for domestic heating can also be dramatically reduced through the use of improved insulation (Chapter 14). However, the development of the motor car is perhaps an example of a technology where for many years size, power and speed have taken precedence over energy efficiency. The principles of modern fuel efficient hybrid petrol–electric cars are described in Chapters 9 and 14. This book does not aim to deal with energy efficiency in detail, but more information will be found in Harvey, 2010.

(b) Reducing CO_2 emissions from fossil-fuelled technologies

There are a range of technologies that can be used to reduce these emissions. 'Fuel switching' from coal to natural gas has significantly reduced both the carbon dioxide and sulfur dioxide emissions of UK power stations (Chapter 9). However, given the world's limited supplies of natural gas this cannot be described as being long term a 'sustainable' option. A major future development is that of 'carbon capture and storage' (CCS), technologies which capture the carbon dioxide emitted from power stations and 'sequester' it deep underground, rather than allowing it to escape into the atmosphere. This is of particular interest where coal is used as the fuel. These technologies are described in Chapter 14. Some CCS technologies involve the production of hydrogen that can also be used for electricity generation in fuel cells, which have considerable possibilities for road vehicles. Chapter 14 also describes a possible future 'hydrogen economy' where hydrogen is widely used in power stations, homes, vehicles and even aircraft. Although CCS does significantly reduce the emissions of carbon dioxide and other pollutants, it requires more fossil fuel to produce the same amount of electricity, possibly making future problems of resource depletion even worse.

(c) Switching to renewable and/or nuclear energy sources

The use of renewable and nuclear energy usually involves environmental impacts of some kind, but these are normally lower than those of fossil sources in terms of CO_2 emissions. The environmental and other impacts of various forms of energy will be discussed in Chapter 13. Nuclear energy will be described in Chapters 10 and 11. Renewable energy will be described briefly in Chapter 14.

Changing patterns of UK energy use

Before considering the feasibility, and the plausibility, of radical changes in patterns of energy production and consumption, of the kind that will be needed during the first half of the twenty-first century if we are to progress towards sustainability, it is useful to recall the profound changes that have already occurred in our energy systems during the latter half of the twentieth century.

In the UK just after World War II most homes and other buildings were heated by coal. Most electricity generation was coal-fired, and most rail transport was propelled by coal-burning steam engines. Coal combustion caused major pollution problems, including the notorious London 'smogs'.

Coal miners perished in their dozens, and sometimes hundreds, in mining accidents every year, and many others died slowly of lung diseases caused by inhaling coal dust. Open coal fires in most houses were so inefficient that, despite consuming large quantities of energy, they only heated a few rooms effectively whilst the rest remained cold.

Motor cars were still owned only by a minority and air travel was confined to a small elite. Most people travelled by bus, train, cycle or on foot. Journeys were relatively few, compared with today, and usually over quite short distances.

Since the late 1940s, the UK's energy systems have been transformed. Natural gas, which burns much more cleanly and efficiently, was introduced very rapidly to British homes and buildings from the 1970s, after its discovery beneath the North Sea, and has now replaced coal as the main heating fuel for buildings. Most homes now have gas-fired central heating systems which ensure that the whole house is maintained at a comfortable temperature.

Coal is still used for electricity generation, but flue gas desulfurization and electrostatic precipitators now greatly reduce emissions of sulfur dioxide and particulates. Sulfur is removed from natural gas at source and it is also refined out of motor fuel. In new power stations, coal is increasingly being replaced by natural gas, which can be burned very cleanly and efficiently using combined cycle gas turbines (CCGT). Nuclear power, since its modest beginnings at Calder Hall in 1956, now contributes about one-fifth of UK electricity.

Cars are now owned by the majority, air travel overseas has become a mass market, railways are powered mainly by electricity, and travel overall, measured in passenger-kilometres, has quadrupled since the 1950s (Figure 1.25) After a short 20-year period of being a net oil exporter, thanks to its North Sea reserves, the UK has recently returned to being a net oil importer, as it was before the 1970s.

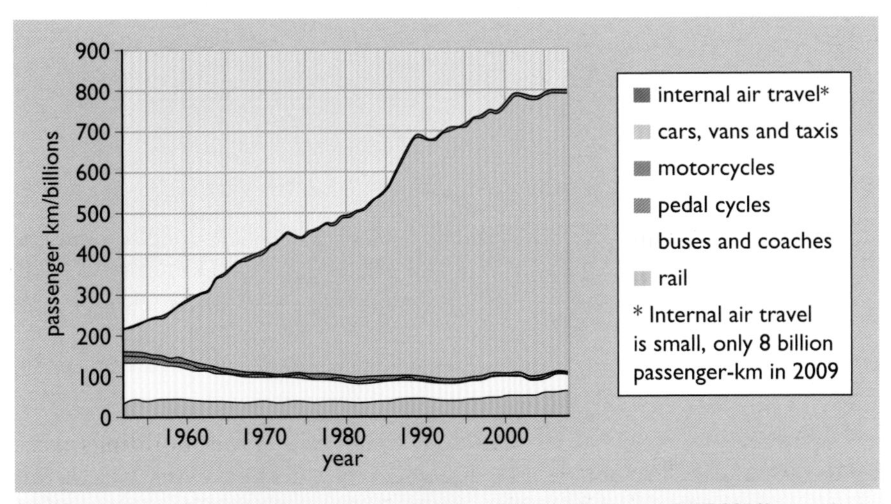

Figure 1.25 UK passenger transport by mode, 1952–2008 (source: DfT, 2010)

The dramatic changes that have occurred in the UK's energy systems during the past 50 years have, broadly, been paralleled in most 'developed' countries over the same period. Examples of changing patterns of energy use in other countries are given in Chapters 2 and 3.

Given the scale and profundity of the changes over the past half-century, it does not seem unrealistic to suggest that equally profound changes could well occur over the next 50 to 100 years, as we attempt to improve the sustainability of our energy systems, nationally and globally.

The UK government is committed to reducing the country's greenhouse gas emissions by 80% from 1990 levels by 2050. Chapter 14 briefly looks at what that might mean for future UK energy use. It also looks in a little more detail at the IEA Blue Map scenario whose projected energy use has been shown in Figure 1.22.

1.7 Summary

This chapter has posed some fundamental questions about energy systems and sustainability and begun to sketch out a framework for developing our understanding of the answers. What is a sustainable energy system?, Why does sustainable energy matter?, How can we achieve the transition to sustainable energy systems? These are profound questions to ask as well as intellectually demanding to answer. The remaining chapters of the book will describe some of the many subtleties and nuances involved in understanding the trade-offs between one source of energy, one technology, or one social, economic or environmental impact and another.

References

Aden, N. (2010) 'Initial Assessment of NBS Energy Data Revisions' [online], Berkeley, CA, Lawrence Berkeley National Labs, http://china.lbl.gov (accessed 10 December 2010).

AEA (2010) *Guidelines to Defra/DECC's GHG Conversion Factors for Company Reporting* (Version 1.2.1 FINAL), produced by AEA for the Department of Energy and Climate Change (DECC) and the Department for Environment, Food and Rural Affairs (Defra). Updated 6 October.

Allen, M., Frame, D., Frieler, K., Hare, W., Huntingford, C., Jones, C., Knutti, R., Lowe, J., Meinshausen, M., Meinshausen, N. and Raper S. (2009) 'The exit strategy' [online], *Nature Reports Climate Change*, http://www.nature.com/climate/2009/0905/full/climate.2009.38.html (accessed 2 February 2011).

Boden, T.A., G. Marland, and R.J. Andres (2010), *Global, Regional, and National Fossil-Fuel CO_2 Emissions* [online], Carbon Dioxide Information Analysis Center, Oak Ridge National Laboratory, U.S. Department of Energy, Oak Ridge, TN., http://cdiac.ornl.gov/trends/emis/tre_glob.html (accessed 11 December 2010).

BP (2010) *BP Statistical Review of World Energy*, London, The British Petroleum Company; available at http://www.bp.com (accessed 10 December 2010).

Campbell, C. (2009) *Newsletter No.100 – April 2009* [online]. The Association for the Study of Peak Oil and Gas, http://aspoireland.org (accessed 10 December 2010).

DfT (2010) 'Transport Statistics Great Britain', London, Department For Transport; available at http://www.dft.gov.uk/pgr/statistics/datatablespublications/tsgb (accessed 14 December 2010).

EIA (2011) 'World Crude Oil Prices' [online], US Energy Information Administration, http://www.eia.doe.gov (accessed 15 February 2011).

EWG (2007) *Coal: Resources and Future Production*, EWG-Paper No. 1/07, Energy Watch Group; available at http://www.energywatchgroup.org (accessed 10 December 2010).

Harvey, L. D. D. (2010) *Energy Efficiency and the Demand for Energy Services: Energy and the New Reality 1*, London, Earthscan.

IEA (2001) *Toward a Sustainable Energy Future*, Paris, International Energy Agency.

IEA (2009) *Key World Energy Statistics*, Paris, International Energy Agency; available at http://www.iea.org (accessed 11 December 2010).

IEA (2010a) *World Energy Outlook 2010*, Paris, International Energy Agency.

IEA (2010b) *CO_2 emissions from fuel combustion*: Annual historical series 1971–2008, Paris, International Energy Agency; available at http://www.iea.org (accessed 11 December 2010).

IEA (2010c) *Energy Technology Perspectives 2010*, Paris, International Energy Agency; available at http://www.iea.org (accessed December 2010).

IPCC (2003) *Climate Change 2003: the Scientific Assessment*, Intergovernmental Panel on Climate Change, Cambridge University Press.

IPCC (2007) *Climate Change 2007: the Scientific Assessment*, Intergovernmental Panel on Climate Change, Cambridge University Press.

Jackson, T. (2009) *Prosperity Without Growth*, London, Earthscan.

LBNL (2009) *China Energy Databook*, Berkeley, CA, Lawrence Berkeley National Laboratory; available only as CD from http://china.lbl.gov/databook.

Maddison, A. (2005) 'World Development and Outlook 1820–2030: Evidence submitted to the Select Committee on Economic Affairs, House of Lords, London, for the inquiry into "Aspects of the Economics of Climate Change"'; available at http://www.ggdc.net (accessed 10 December 2010).

Maddison, A. (2009) 'Statistics on World Population, GDP and Per Capita GDP, 1–2008 AD'; available at http://www.ggdc.net/maddison (accessed 10 December 2010).

OECD (2001) *Analytic Report on Sustainable Development*, ii 'Key Features and Principles', SG/SD(2001)2, Paris, Organisation for Economic Co-operation and Development.

Sorrell, S., Speirs, J., Bentley, R., Brandt, A. and Miller, R. (2009) *Global Oil Depletion: An assessment of the evidence for a near-term peak in global oil production*, Report for United Kingdom Energy Research Centre.

UN (1987) *Our Common Future* (the Brundtland Report), United Nations, World Commission on Environment and Development, Oxford, Oxford University Press.

WWEA(2009) *World Wind Energy Report 2009*, World Wind Energy Association; available at http://www.wwindea.org (accessed 11 December 2010).

Chapter 2

Primary energy

By Janet Ramage

2.1 Introduction

This chapter is probably best treated as a reference source for the remainder of the book. No one could be expected to recall everything in the following pages after a single reading, so the aim might rather be knowing where to find the information when it is needed later.

The chapter falls into two main parts, differing in the nature of their content. Sections 2.2–2.4 establish the basis needed for any serious discussion of the 'world of energy'. This includes *definitions* of the terms used in energy studies, an introduction to the *units* used to specify quantities, and an account of the *conventions* used in the presentation of data, in this book and elsewhere.

With this basis, the aim in Sections 2.5–2.9 is to draw a picture of the world of energy as it is today and how it has changed over recent years. A survey of the situation for the world as a whole is followed by accounts of the energy regimes of six countries, chosen to exemplify not only the major variations worldwide but the significant differences between apparently similar countries.

2.2 Consuming energy

The world population is consuming primary energy at an average rate of about 15.9 terawatts.

This statement raises a number of questions:

- How do we consume energy?
- What is primary energy?
- What are 15.9 terawatts, and how are they related to the world annual consumption of about 12 000 Mtoe quoted in Chapter 1?
- How do we know the world's energy consumption? What are the sources of energy data, and how reliable are they?

This first part of this chapter addresses these essential questions about the nature and basis of our knowledge of energy consumption.

What is energy consumption?

The Law of Conservation of Energy states that energy is conserved. The total quantity stays constant. You cannot create energy or destroy it. If you have ten units of energy at the start, you have ten units of energy – somewhere – at the finish. In this sense, we never consume energy (see Box 2.1).

It is, however, a matter of great practical importance that energy can take many different forms, and what we *can* do – and have done at least since our ancestors first used fire – is devise means of converting one form to another. When we talk of consuming energy this is what we mean: converting the chemical energy stored in fuels (such as wood, coal, oil or gas), the energy stored in atomic nuclei, the gravitational energy of water in a high reservoir, the kinetic energy of moving water or the wind, and the radiated energy of sunlight into heat or electrical energy or light or the kinetic energy of a moving vehicle. All these forms of energy are discussed later, but the first important point is this: *consumption is conversion.*

BOX 2.1 'Conservation of energy'

Why should I switch off lights to conserve energy, when there is a law which states that energy is always conserved?

The question is of course mischievous, deliberately confusing two different meanings of 'conserve'. It is, however, important to appreciate that phrases such as 'conservation of energy' do have these two interpretations, both of which are common in discussions of our uses of energy.

The first interpretation, the Law of Conservation of Energy, is fundamental. It underlies all our reasoning, even when it is not specifically stated.

In the other sense of the term, when we are told to 'conserve energy' by switching off unnecessary lights, we are really being asked to conserve *energy resources*. When we turn off the light we are 'saving energy' by reducing the amount of coal or other fuel burned in a power station.

Fortunately the context usually makes it clear which of the above meanings is intended, and in practice the two senses of the word rarely lead to problems.

What is primary energy?

In brief, if you collect wood for a fire to heat your soup, you are using primary energy directly. If you heat your soup in a microwave oven, you are not.

The word 'primary' means the earliest, or original, and the common feature of all the **primary energy sources** is that they are naturally occurring energy stores or energy carriers. The *fossil fuels* (coal, oil and natural gas) were once plants or small creatures drawing from their surroundings the carbon that makes them fuels, and the *biomass* sources (wood, grasses, seeds, etc.) are those using the same process today. Others, including *wind*, *wave* and of course *solar* power, draw their energy more directly from the solar radiation reaching the Earth, whilst *tidal* power, uniquely, derives its energy mainly from the relative motion of the Moon and the Earth. The remaining two primary resources in current use, *geothermal* energy and *nuclear* energy, both draw on energy originally stored billions of years ago, when matter was being formed.

The term **primary energy** was introduced by energy statisticians to allow comparison of the energy contributions from different primary resources, or estimation of the total primary energy consumption of a country or region – or the world. Suppose a certain small country used ten million tonnes of oil and fifteen million tonnes of coal in one year. Which fuel made the greater contribution to national energy consumption? To answer this question, we need to know the **specific energy** or *energy per tonne* for each fuel. These and similar quantities are introduced in Section 2.3. (The question is answered in Box 2.8.)

'Energy arithmetic'

Any serious discussion of our uses of energy must be quantitative:

> My car uses very little petrol.

In driving a thousand miles, or standing in the garage? Compared with Saudi Arabian oil exports or with a bicycle?

This trivial example illustrates two requirements. In order to compare quantities we must be able to measure them, i.e. we need *units* (in this case litres, gallons or tonnes); and we must know which type of quantity we are discussing (litres per kilometre, litres per year or just litres).

In 1960 the scientific world reached agreement on a single consistent set of units: the Système International d'Unités. The **SI system** uses three main base units: the metre, the kilogram and the second, and the units for many other quantities are derived from these. Some of the derived units, such as *metres per second* for speed, reveal their base units immediately, whilst for others the combination of base units has been replaced by a specific name. The name of the SI unit for energy is the **joule**, abbreviated **J**. (The appendix has more details of this and other SI units.) In everyday terms, one joule is a rather small quantity of energy – roughly the amount needed to raise a medium-sized apple just one metre vertically upwards.

One of the happier consequences of the energy debates of the past few decades has been a growing appreciation of the advantages of using this universal unit for all amounts of energy. Nevertheless, if you open a book or search for data on the Web, you can still find yourself in a rather less tidy world (see Box 2.2). Quantities of energy are quoted in tonnes (or tons) of oil or coal, cubic metres of gas, kilowatt-hours, terawatt-years, therms, quads, calories and Calories, etc., and if we are to follow the real-world debate, we need to come to terms with these. Accordingly, one aim in this chapter is to introduce the art of 'energy arithmetic' – of converting between different ways of specifying quantities of energy.

The need to use extremely large (or extremely small) numbers can also lead to problems. Most of us can visualize a dozen objects, perhaps even a hundred, but few can picture 16 trillion: 16 000 000 000 000 (over two thousand times the population of the world). Such numbers cannot be avoided, but they can be made more manageable by using special names, or more compact ways of writing them. The appendix at the end of this book explains these methods in detail. Table 2.1 is a short summary of the main prefix names used in this book.

Table 2.1 Prefixes

Letter	Prefix	Multiply by[2]	As power of ten
μ[1]	micro	one-millionth	10^{-6}
m	milli	one-thousandth	10^{-3}
k	kilo-	one-thousand	10^{3}
M	mega-	one million	10^{6}
G	giga-	one billion (one thousand million)	10^{9}
T	tera-	one trillion (one million million)	10^{12}
P	peta-	one quadrillion (one billion million)	10^{15}
E	exa-	one quintillion (one billion billion, or one million million million)	10^{18}

[1] Letter *mu* in the Greek alphabet.
[2] Note the general rule that each multiplier is a thousand times the previous one.

BOX 2.2 The use and misuse of m and M

The lower case letter 'm' is probably familiar to most people as the abbreviation for a *metre*, the unit of length – as in a 100 m race. Table 2.1 introduces a second use: as a *prefix* for any unit – as in 50 mW (fifty milliwatts). In later chapters, it will appear in a third role, as the usual *symbol* for mass in an equation, but in this case always in italics – as in $m = 10$ kg, for instance.

The upper case 'M' appears most commonly as the *prefix* mega – as in a 50 MW (fifty megawatt) wind turbine. Occasionally M is used instead of m as the *symbol* for mass.

These are the customary scientific uses of the two forms, and the context usually makes it clear which is meant. Unfortunately, in the real world of energy data, this is by no means always the case. In data from countries that have not adopted the SI system, including the USA, the letter M may have a variety of meanings, and the following are a few examples worth noting.

In US data, '200 Mt of coal' may not mean 200 million tonnes, but a much smaller quantity: 200 *metric tonnes* (as opposed to the customary US unit of 'tons', see Box 2.7). In other cases, the prefix M may stand for one thousand times. More commonly the prefix MM is used for a million times. Readers are therefore warned that care is needed, and cross-checking advisable, in using data from a variety of sources.

Energy and power

The difference between *energy* and *power* is probably the most important distinction in the world of energy studies. Newspapers almost invariably get it wrong, confusing the two terms, and in any case assuming that 'power' means electricity. This section explains the difference between the two quantities and defines some of the units in which each of them is measured.

Watts

A terawatt is one million million watts (Table 2.1) – but what is a watt? The important point is that a watt is a measure of the *rate* at which energy is being transformed or converted from one form to another. Technically a watt is a unit of **power**, of energy per second:

- One **watt** is by definition one joule per second.

Thus a 600 W heater is converting electrical energy into heat at a rate of 600 joules in each second. And we, the population of the world, with our 15.9 TW rate of consumption, are converting 15.9 million million joules of primary energy every second into the forms of energy we want (and a great deal of waste heat).

Kilowatt-hours

The kilowatt-hour (kWh) is a unit of **energy**.

- One **kilowatt-hour** is the amount of energy converted in one hour at a rate of one kilowatt.

The heater in a 3 kW clothes dryer, for instance, used for 40 minutes (two-thirds of an hour), converts 2 kWh of electrical energy into heat energy.

Like any quantity of energy, a kilowatt-hour must of course be equal to a certain number of joules. The reasoning in Box 2.3 shows that one kilowatt-hour is *exactly* the same as 3.6 megajoules.

BOX 2.3 kW and kWh

Note that 1 kW is 1000 watts (Table 2.1), and that there are 3600 seconds in an hour.

Power

1 watt = 1 joule per second
1 kilowatt = 1000 joules per second = 3 600 000 joules per hour

Energy

1 kilowatt-hour = 3 600 000 joules
1 kWh = 3.6 MJ

Multiplying successively by 1000 leads to other useful relationships:

1 MWh = 3.6 GJ
1 GWh = 3.6 TJ
1 TWh = 3.6 PJ

It is important to appreciate that the kilowatt-hour and the watt are *general* units for energy and power respectively. Although many of us meet them first in the context of electricity, they are equally applicable to the energy you use and the power you develop in running up a flight of stairs – or the average rate of energy consumption per person of the world's population (Box 2.4).

BOX 2.4 Per capita consumption

It can be useful to convert very large numbers into more manageable quantities. Instead of world primary energy consumption, the average consumption *per person* might be easier to visualize.

In 2009, the world rate of primary energy consumption was 15.9 TW (see Figure 1.5), which is 15.9 million million watts, and the world population was about 6829 million people. The average per capita rate of consumption was therefore:

$$\frac{15.9 \times 1\,000\,000}{6829} = 2328 \text{ W} = 2.328 \text{ kW}$$

On average, therefore, we are each consuming primary energy at a constant rate of slightly less than two and a half kilowatts.

There are 24 hours in a day, so the average daily consumption per person is about:

$$2.328 \text{ kW} \times 24 \text{ hours} = 55.87 \text{ kWh}$$

This is of course the average daily personal consumption of *all forms* of primary energy. Remembering that 1 kWh is 3.6 MJ, it becomes about 200 MJ, which is the energy content of about five and a half litres of oil (see Section 2.3).

So the average person – man, woman or child throughout the world – uses the energy equivalent of more than a gallon of oil per day. This must, of course, supply *all* our energy needs: food production and a water supply; the provision of housing; heat for cooking and to keep us warm; clothing and manufactured goods; transport of people and freight; communications and entertainment; the medical, educational and other services that we expect.

2.3 Quantities of energy

The publication of national or international energy data was largely a development of the second half of the twentieth century, but records of dealings in *commodities* are as old as trade itself. During the eighteenth and nineteenth centuries, coal became an extremely important commodity for developing countries such as Britain. As it was also the dominant energy source, the data on coal production and consumption came to serve as national energy data for much of the period. When new energy sources such as oil began to appear, it was natural to assess their contributions in terms of the quantity of coal they could replace, and Britain continued to do this into the 1980s, expressing all national energy data in *tonnes of coal equivalent*.

Meanwhile, some of the most accessible international energy statistics were being assembled and published by the major oil companies, and not surprisingly their favoured unit for energy was the *tonne of oil equivalent*. In the UK, where oil has played a major role since the 1970s, the national statistics now mainly use tonnes of oil equivalent.

Units based on oil

When oil is burned, whether in a furnace or an internal combustion engine, its chemical energy is converted into heat energy. One **tonne of**

oil equivalent (toe) is simply the heat energy released in the complete combustion of 1000 kg of oil. This varies between crude oils from different sources, but a commonly used figure for statistical purposes is 41 868 MJ, or 41.868 GJ, per tonne, so:

1 toe = 41.868 GJ

Approximations such as 41.9 GJ or even 42 GJ are often used when this precision is appropriate (see Box 2.5).

BOX 2.5 Significant figures

The calculation in Box 2.4 uses two pieces of information: that in 2009, the world was consuming primary energy at a rate of 15.9 TW, and that world population was about 6829 million. Using these numbers, it concludes that the average daily energy consumption per person was 55.87 kWh.

Assuming that the arithmetic is correct, is this result justified? As the qualifying comments in Box 2.4 suggest, it is not, for the following reason.

World total rate of consumption is given as 15.9 TW. What does this mean? Exactly 15.9 TW? Unlikely, and in that case it should have been expressed as 15.900... with an infinite number of zeros. All that 15.9 TW in fact tells us is that world consumption is believed to lie somewhere in the range between 15.85 TW and 15.95 TW.

Repeating the Box 2.4 calculation using each of these extreme possibilities gives two new values for per capita daily consumption: 55.70 kWh and 56.06 kWh. So the best that can be said on the basis of 15.9 TW is that the actual daily consumption could lie anywhere in this range. It would therefore be more honest to round the result to **55.9 kWh**, rather than pretending that we know it more accurately.

Notes

- The second input to the above calculation, world population, is given as 6829 million, i.e. with four digits. So the uncertainty in this will have a negligible effect compared with the much larger uncertainty in the terawatts.
- However, if the world consumption was known to be 15.90 TW, the daily figure could justifiably expressed as 55.88 kWh.
- It is important to note that the presence, absence or position of a decimal point does not affect the reasoning here. It is entirely a question of the number of digits.
- Exact numbers, such as 3600 seconds in an hour, do not need to be followed by long strings of zeros to indicate that they are precise.

The point of this example is to draw attention to a general rule (particularly important when using a calculator or spreadsheet capable of expressing numbers to as many as fifteen digits). *The number of significant digits in the final result of a calculation should not be greater than the smallest number of significant digits in any of the input data.*

National and international energy data is often expressed in **millions of tonnes of oil equivalent (Mtoe)**. World primary energy consumption of about 502 EJ (Figure 1.4) then becomes just under 12 000 Mtoe (Box 2.6). This of course includes all forms of energy, so the actual world oil consumption in 2009, which was 3882 million tonnes, accounted for nearly a third of the total.

Figure 2.1 Filling barrels at a Pennsylvania oil well in 1870

Another measure of quantity of oil and correspondingly of energy is the barrel. This odd unit, alien in a world of pipelines and super tankers, comes from the size of the barrels used to carry oil from the world's first drilled well in Pennsylvania in the 1860s (Figure 2.1). One barrel is 42 US gallons or 35 Imperial (British) gallons – about 159 litres (Table A7 in the Appendix).

How is a barrel of oil related to a tonne of oil? A barrel is a certain volume, whereas a tonne is of course a mass, and crude oils from different sources have different densities, so more barrels would be needed to hold one tonne of a 'light' crude than for a 'heavier' one. The solution has again been to adopt a world average for statistical purposes: 7.33 barrels to the tonne. The oil industry commonly expresses data in **million barrels daily (Mbd)**. In 2009, for instance, world conventional oil consumption was 78 Mbd, and total world energy consumption was equivalent to about 240 Mbd (Box 2.6).

Finally, we have the everyday units for the fuel used in our vehicles: the litre and the gallon. Petrol (gasoline) has an energy content of about 44 GJ per tonne – slightly higher than that of crude oil. But it has an appreciably lower density, so in terms of *volume*, its energy content is only about 33 MJ per litre, compared with an average of nearly 36 MJ per litre for crude oil.

Units based on coal

One **tonne of coal equivalent (tce)** is the heat energy released in burning one metric tonne of coal, but in this case there is a serious issue. Which coal?

Coal is a much more variable material than crude oil, and worldwide its energy content ranges from less than 15 GJ to over 30 GJ per tonne, a difference that will obviously be reflected in the value of 'one tonne of coal equivalent'.

The widely used *BP Statistical Review of World Energy* (BP, 2010) simplifies the situation by distinguishing between **hard coal**, assigned an energy content of 28 GJ per tonne and **lignite** (a low-quality form of coal; see Chapter 5), with only 14 GJ per tonne. The conversion from tonnes of coal mined in any country to the corresponding total 'coal' energy then depends on the proportions of these two types. (Table 2.2 in Section 2.5 reflects this, with a world average of about 20 GJ per tonne for coal.)

However, national energy statistics often take an even simpler route, assigning just one average value for the energy per tonne of coal. The figure of **28 GJ per tonne** is used in UK statistics, and is the value adopted in this book, unless otherwise specified.

The BTU and related units

Before the general adoption of the joule, the *British thermal unit* (BTU) and its multiples were in common use, in the English-speaking world in particular.

- One **BTU** was originally defined as the heat energy needed to warm one pound of water by one degree Fahrenheit; but to achieve consistency

BOX 2.6 World energy in Mtoe, Mbd and TW

Expressed in terms of the accepted scientific unit for energy, the joule, world primary energy consumption in 2009 was close to 502 EJ (exajoules), or the average *rate* of consumption was **502 EJ per year**.

As mentioned in the main text, this rate of consumption can also be expressed as just under 12 000 Mtoe per year, 15.9 TW and 240 Mbd. The energy arithmetic that relates all these figures is as follows.

Millions of tonnes of oil per year

One tonne of oil equivalent (1 toe) is equal to 41.868 GJ (see main text).
1 Mtoe is therefore 41.868 million GJ
502 EJ is the same as 502 000 million GJ
So in Mtoe this becomes 502 000/41.868 = 11 990 Mtoe
or within the precision of the data, **12.0 thousand Mtoe**.

Millions of barrels of oil daily

There are 365 days in a year, so the daily world primary energy consumption is:

$$\frac{502}{365} = 1.375 \text{ EJ} = 1375 \text{ million GJ}$$

There are 7.33 barrels in one tonne of oil, so the energy per barrel is:

$$\frac{41.868}{7.33} = 5.71 \text{ GJ}$$

and the number of millions of barrels daily is:

$$\frac{1375}{5.71} = 241$$

i.e. in 2009, daily world primary energy consumption was equivalent to **241 Mbd**.

$$1 \text{ Mbd} = \frac{365}{7.33} \approx 50 \text{ Mtoe y}^{-1}$$

Terawatts

The conversion from exajoules a year to terawatts starts with the definition of the watt:

1 watt is 1 joule per second ...
which is 3600 joules per hour ...
or 24 × 3600 = 86 400 joules per day ...
or 365 × 86 400 = 31 536 000 joules per year ...
So a power of 1 watt delivers 31.5 MJ per year.
1 terawatt (TW) is one million million watts ...
which is 31.5 million million MJ per year ...
So a power of 1 TW delivers 31.5 EJ per year.

World consumption of 502 EJ per year can therefore be expressed as an average power of:

$$\frac{502}{31.5} = \mathbf{15.9\ TW}.$$

Figure 2.2 Filling a London coal cellar. Coal was delivered in hundredweight sacks, and the 'coal holes', often with attractive iron covers, can still be identified in many eighteenth- or nineteenth-century streets.

BOX 2.7 Tonnes, tons and short tons

As mentioned above, national or even international energy statistics do not yet appear in one agreed set of units, and whilst the approved SI units for mass are the kilogram and its multiples such as the metric tonne (1000 kg), you will still find other 'tons' in use. These are their relationships:

(1) The **tonne**, or metric tonne, is 1000 kg and is equal to about 2205 lb (pounds).

(2) The **ton**, still used in the pre-metric system of weights and measures of the UK and other countries, is equal to 1016 kg (2240 lb). A **hundredweight (cwt)** is a twentieth of a ton, or just over 50 kg (Figure 2.2).

(3) The **short ton** is still the customary unit for a quantity of coal or wood in the USA and some other countries. One short ton is 907 kg (2000 lb) – about 10% less than a tonne.

BOX 2.8 Two fuels

The comparison

A country used 10 million tonnes of oil and 15 million tonnes of coal last year. Which fuel made the greater contribution to national primary energy consumption?

Data: 1 toe = 42 GJ, 1 tce = 28 GJ

The primary energy contribution from 10 million tonnes of oil is:

$$10 \times 1\,000\,000 \times 42 \text{ GJ} = 420 \times 10^6 \text{ GJ} = \mathbf{420 \text{ PJ}}$$

And the primary energy contribution from 15 million tonnes of coal is:

$$15 \times 1\,000\,000 \times 28 \text{ GJ} = 420 \times 10^6 \text{ GJ} = \mathbf{420 \text{ PJ}}.$$

So the energy contributions from the two fuels were the same.

The switch

To reduce carbon emissions, the country plans to reduce annual coal consumption to 12 million tonnes, compensating by increased oil consumption. If the present total primary energy consumption is to be maintained, what must be the new annual oil consumption, in millions of tonnes?

The contribution from 12 million tonnes of coal will be:

$$12 \times 1\,000\,000 \times 28 \text{ GJ} = 336 \times 10^6 \text{ GJ} = 336 \text{ PJ}.$$

i.e. a reduction of $420 - 336 = 84$ PJ

To maintain the original total energy, oil must therefore now contribute a total of:

$$420 + 84 = \mathbf{504 \text{ PJ}}.$$

which requires 12 million tonnes of oil.

In summary, instead of different quantities of two fuels making equal energy contributions, the country now has equal quantities making significantly different contributions.

with SI units it is now defined as precisely 1055.06 joules. Multiples of the BTU include the **therm** (100 000 BTU) and the quad.

- One **quad** is a quadrillion BTUs (see Table 2.1) and is equal to 1.05506 EJ.

The **therm,** equal to 100 000 BTU, was a widely used unit for gas and national energy statistics in the UK until the 1990s. The BTU still appears occasionally, for instance in the power rating of gas boilers (thousands of BTUs per hour). The BTU is still the basis for energy quantities in the USA. In particular, US data often use the MMBTU (million BTU), or on the larger scale, the quad.

As can be seen from the above definitions, the BTU and the quad (Q) are slightly larger than the kilojoule (kJ) and exajoule (EJ) respectively.

The calorie and related units

In most of Europe, and many other countries, the common unit for heat in the past was the calorie.

- One **calorie** is the heat energy needed to warm one gram of water by one degree Celsius and is equal to 4.19 joules.

For many purposes the **kilocalorie**, written kcal or Calorie (with capital C), has proved more convenient and often remains, together with the kilojoule, as the unit for the energy content of food.

This gives us yet another way of looking at our energy consumption. Nutritionists tell us that the daily *food energy* intake needed to support a normal adult lies in the range 2000–2500 kcal, which is about 8–10 MJ; but Box 2.4 has shown that world average daily *primary energy* consumption per person is about 200 MJ. So it appears that the energy we each use in non-food forms is, *on average*, about twenty times the amount we each need to feed ourselves. This is of course by no means universally the case.

2.4 Interpreting the data

There remains a final question about world primary energy, or indeed any energy data. How do we know? Before venturing further into the sources of energy, we should perhaps discuss the *sources of data*. Where do the figures come from? The first answer is that we find them in official statistics, technical journals and similar publications. However, one shouldn't believe everything one reads in books or on the Web (or anything in newspapers) and care is always needed in interpreting published figures, for reasons which can be characterized under three headings: *definitions*, *conversions* and *conventions*.

Definitions

World data usually start as national statistics, and with some 200 countries it is hardly surprising that the terminology doesn't always match at the seams. Does 'production' include energy used by the producer? Does 'consumption' include energy used for transmission of energy? Unless we know the answers to such questions, how are we to interpret the statement that 82.17% of Britain's coal consumption in 2009 was for electricity generation? In the absence of pages of explanation, it might be better to say, 'About 82% ...', or even, 'roughly four-fifths'.

A further mismatch appears in comparing figures for *production* and *consumption*. One would hope that any difference would be accounted for by changes in stocks, but when production figures necessarily come from producers and consumption figures from consumers this is by no means always the case. Recent world statistics, for instance, include about 19 million tonnes of 'unidentified' crude oil exports. Some of it may be on the high seas – in ships, one hopes – but the figures again illustrate the problem.

Conversions

We have seen several examples of conversion between different energy units, but have not bothered too much about the nature of these relationships. On inspection we find that the term *equivalent* has been used in a number of different ways.

First there are cases where the conversion between units is *exact*. One watt is exactly one joule per second because that is how it is defined; and there are exactly 3600 seconds in an hour, so 1 kWh is therefore exactly 3.6 MJ. Then there are relationships which have been agreed universally. One foot (1 ft), for instance, is now defined as exactly 0.3048 of a metre and one pound (1 lb) as 0.45359237 of a kilogram. (The conversion between joules and British thermal units given above follows from these.)

When we come to quantities such as the heat content of a fuel, matters are not so simple. The heat content of a particular specimen of oil can be measured to great accuracy under laboratory conditions, and with similar care we might measure the solar energy reaching a particular roof in the course of a particular day. But it is hardly practicable to use these methods for the total output during the lifetime of an entire oil well or solar panel. In the real world it becomes essential to use *average* values. The problem is that not everyone uses the same average. If your tonne of oil equivalent and daily solar energy are not the same as mine, our discussion is likely to end in confusion. Once more, the rule is to make sure what the figures mean before using them.

Conventions

Finally, there is the rather different question of the output from power stations. The difficulty is not in measuring it, as most national data include the annual kilowatt-hours produced, and conversion of these to joules is no problem. The question is whether this output should count as 'primary energy'. Shouldn't that be the *input*? Unfortunately there are difficulties with this. Recording the input of coal, oil or gas is relatively straightforward, but measuring the total 'water energy' entering a country's hydroelectric plants in a year, or the total wind energy sweeping across its wind turbin[illegible] is not practicable. Nuclear plants, whose input is the result of a comp[illegible] series of processes, pose a similar problem.

Figure 2.3 shows the essential facts for the world's main types of power station. In most modern **thermal power stations**, where heat from the fuel produces steam or hot gases to drive a turbine, more than half the energy input becomes waste heat (the underlying reasons for this will be discussed in Chapter 6). Hydroelectric plants, the other main contributors, are not subject to this heat loss, so their output is only a little less than the input. This difference has led to a frequently used convention for the main types of power station:

- Assuming that about 38% of the fuel input energy to a modern thermal power station becomes electrical output, a **notional primary energy input** is calculated by dividing the electrical output by 0.38. This convention is commonly used for nuclear plants.
- The primary energy input for hydroelectric plants is taken to be equal to the electrical output.

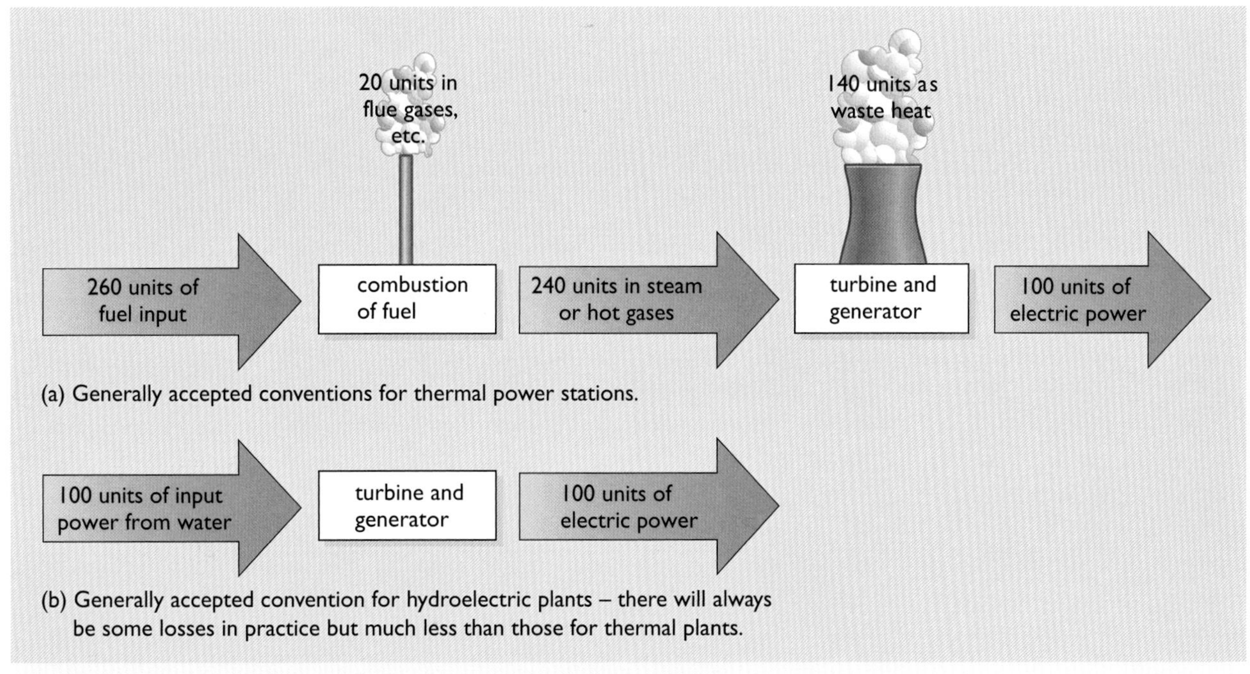

Figure 2.3 Primary energy accounting conventions for electric power

This dual convention has become dominant and is now used by bodies such as the United Nations Department of Economic and Social Affairs and the International Energy Agency (IEA) of the OECD, and in UK energy statistics. However, the *BP Statistical Review of World Energy* (BP, 2010) gives both the actual electrical output in TWh and a corresponding notional input in Mtoe for both nuclear and hydroelectric power. And yet other conventions are still in use in some national or other statistics. Where such sources are used in this book, the conventions are explained in the relevant tables or graphs.

All data in this book have been converted in accordance with the two rules above unless otherwise specified. A consequence of this is that when using the data or looking at diagrams such as Figure 2.4, it is important to bear in mind, for instance, that although the world primary contribution from nuclear power is shown as more than twice that of hydroelectricity, the annual *electrical output* from hydro is in fact slightly greater.

In summary, it is always wise to read the small print when using statistical data, but it should be clear from the above account that it is essential with energy data.

2.5 World primary energy sources

Figure 2.4 reproduces Figure 1.5 of Chapter 1, and Table 2.2 shows the data.

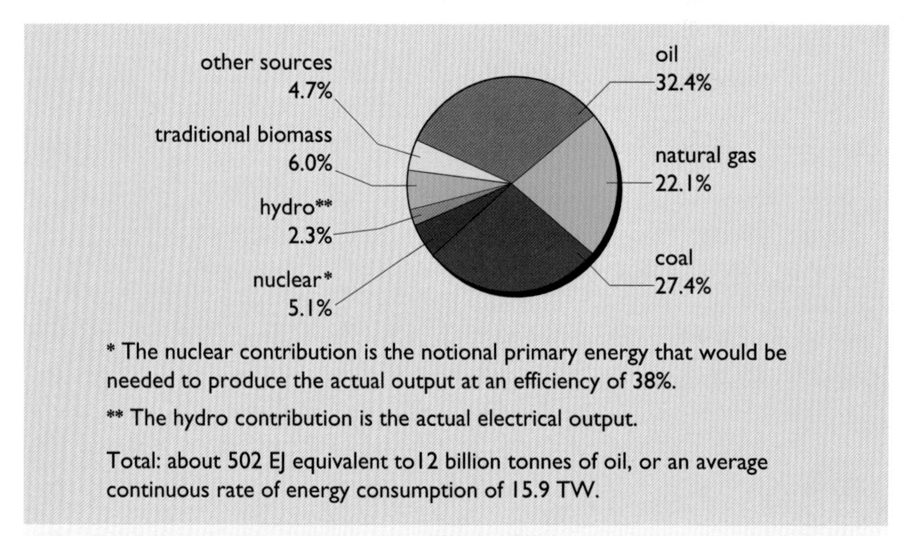

Figure 2.4 World primary energy consumption, 2009 (principal sources: BP, 2010; IEA, 2009; Maddison, 2005; Maddison, 2009; WWEA 2009)

The picture is clear enough. In the year 2009, the three fossil fuels together were supplying over eighty per cent of the world's primary energy. The largest of the other contributions was almost certainly the total energy from biomass (including biofuels), although for the reasons outlined in Box 2.9, its exact magnitude is difficult to establish. Excluding traditional biomass and considering only the 'commercially traded' resources, the dominance of fossil fuels becomes even more striking. They account for nearly ninety per cent of the world's total traded energy.

BOX 2.9 Bioenergy

Bioenergy is the general term for energy derived from biomass: materials such as wood, plant and animal wastes, etc., which – unlike the fossil fuels – were living matter relatively recently. Such materials may be burned directly to produce heat or electric power, but can also be converted into solid, liquid or gaseous **biofuels**.

Estimates of the contribution of biomass to world primary energy are subject to considerable uncertainty. The materials are often 'non-commercial' – they may be gathered in surrounding forests or fields, or arise as waste by-products of other activities, and are often used on site, or bartered for other goods or services. In other words, they are not formally traded, so the economists' methods of keeping track of quantities are not available. These resources are usually called **traditional biomass**, distinguishing them from commercially traded bioenergy, often referred to as **new biomass**, a term that embraces sources such as purpose-grown wood, plants that are processed to produce liquid biofuels, forestry wastes that are sold, etc. (Municipal solid wastes are often included, although this is a subject of dispute.) These 'new biomass' contributions, being commercially traded, are usually treated in energy statistics in the same way as other 'new' renewables such as wind power or solar photovoltaics.

There have been many attempts during recent decades to estimate the energy contribution of traditional biomass, both worldwide and in specific countries. It is generally accepted that its percentage contribution worldwide is gradually falling. Estimates in the 1990s ranged between 10% and 15% of total primary energy, implying an annual contribution of about 50 EJ from traditional biomass, but more recent studies have suggested that biomass in all forms contributes about this total to world primary energy, and that some 60% of this, about 30 EJ, comes from traditional biomass. In Figure 2.5, the biomass data for the past two decades are based on this reasoning. For the earlier years, when traditional biomass dominated, it is assumed that consumption was rising in line with world population.

Sources: IEA, 2008; LBNL, 2008; Maddison, 2005

Table 2.2 World primary energy consumption, 2009

Energy source	Quantity in customary units	Energy /EJ	Percentage contribution
Oil	3882 million tonnes	162.6	32.4%
Natural gas	2940 billion cubic metres	111.1	22.1%
Coal	6675 million tonnes	137.3	27.4%
Fossil fuels percentage			*81.9%*
Nuclear electricity	2698 billion kWh[1]	25.6[2]	5.1%
Large hydro	3272 billion kWh[1]	11.8[2]	2.3%
Traditional biomass[3]		30.0	6.0%
Other sources[4]		23.5	4.7%
Total world annual primary energy consumption		**501.8**[5]	

[1] Actual power station output
[2] See 'Conventions', above
[3] An approximate value see Box 2.9
[4] About 18 EJ from 'new biomass', 1.6 EJ of geothermal heat, 1.5 EJ from wastes, 1.2 EJ of wind power and the remainder from other renewables (mainly solar power).
[5] Total may differ from sum owing to rounding.
Principal sources: BP, 2010; IEA, 2009; Maddison, 2005, 2009; WWEA, 2009.

Figure 2.5 reveals how consumption of the non-renewable and carbon-dioxide-producing fossil fuels has increased since the mid-twentieth century. The dramatic growth in the use of oil between 1965 and 1973 is perhaps the most striking feature, with an annual increase approaching 8% a year. Had this rate been maintained, as seemed likely in the early 1970s,

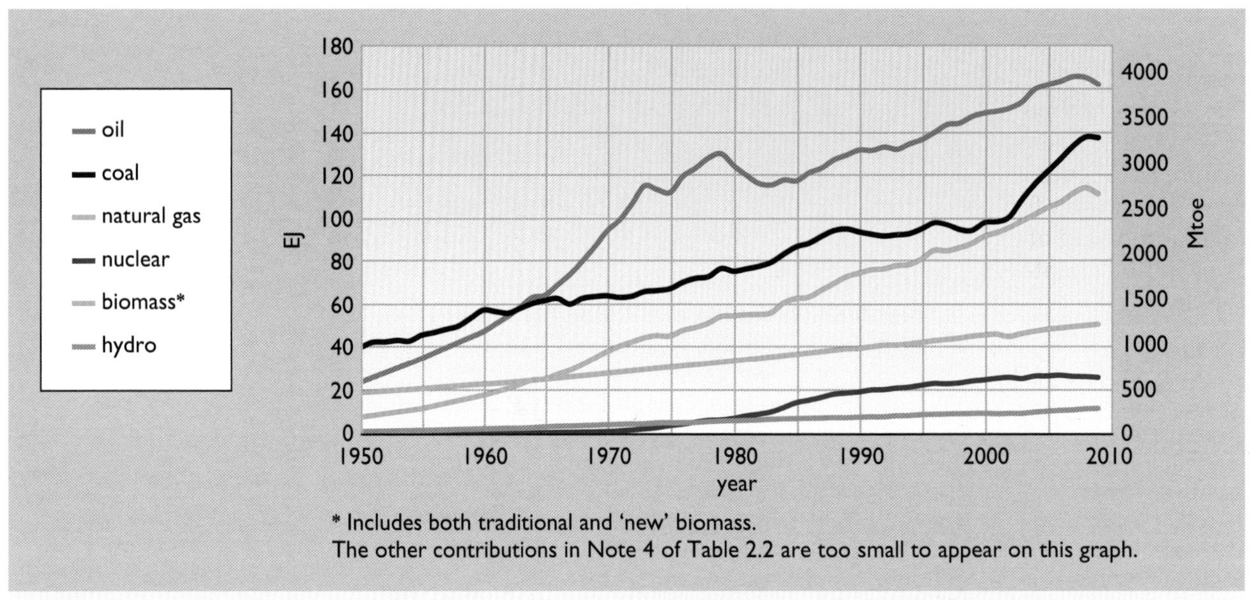

Figure 2.5 World annual primary energy consumption by source, 1950–2009 (principal sources: BP, various years; IEA, various years; Maddison, 2005, 2009; Romer, 1976)

the annual output required by the end of the century would have exceeded 20 *billion* tonnes (about six times the actual output in 2000). It is no surprise that sudden doubts about future supplies led to panic and disarray. The crises of the 1970s, with oil prices doubling in 1973 and rising steeply again at the end of the decade (Figure 2.6), followed by economic recession in the early 1980s, did eventually bring the growth in oil consumption to a halt – but only after some delay, and only temporarily. The mid-eighties saw a return to annual increases, which continued until the economic crises of 2008–2009 again brought fluctuating prices.

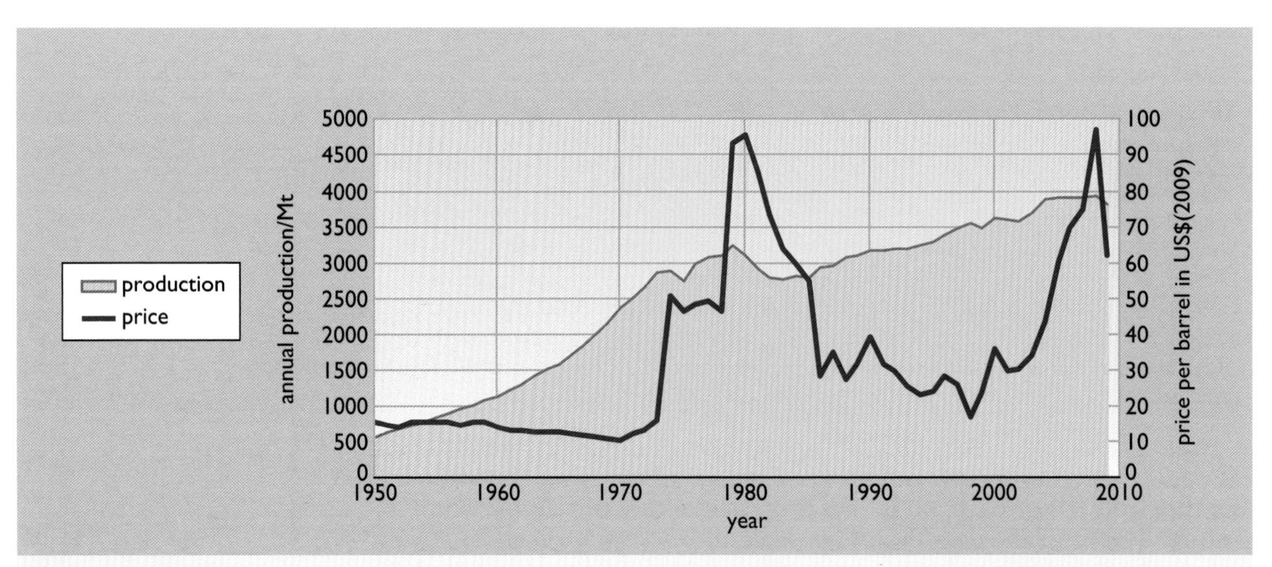

Figure 2.6 World annual oil production and the annual average price of oil, 1950–2009 (sources: pre-1965 Romer, 1976; 1965–2009 BP, 2010)

Coal, the dominant fuel for nearly a century, fell to second place in the 1960s, and with the continuing steady rise in natural gas consumption, many observers in the 1990s expected to see coal in third place within a decade. Events proved them wrong, however, and the sudden change in the fortunes of coal in the first decade of the present century is one of the most striking features of Figure 2.5. One reason for this is obvious. The world reserves/production ratio for coal is several times that for oil or natural gas, and a high proportion of the reserves lies under Asia (Figure 1.16). As discussed later in this chapter, one country, China, has been largely responsible for the reversal in the fortunes of this least desirable fossil fuel.

Natural gas production, with its almost linear growth, stands in marked contrast, and its history is very different. In 1950, its use was almost confined to the USA, and as late as 1970, North America was still responsible for two-thirds of world production. But new sources elsewhere, already being developed in the 1960s, accounted for much of the rise of the next fifty years. Natural gas is relatively widely distributed, with over fifty countries developing the resource at the time of writing (2010); but the USA, with over a quarter of world output, still produces (and consumes) more than any other single country. The Russian Federation and nearby central Asian countries account for about a fifth of the total production, the gas fields of the Middle East supply about a seventh, and the North Sea fields rather less than a tenth.

As Table 2.2 shows, the world's nuclear power plants contribute slightly less than 3000 TWh of electrical output a year, and large-scale hydroelectric plants slightly more than 3000 TWh. Nuclear output, having risen continuously since the first power stations in the 1950s, fell during the first decade of the present century, as the rate of decommissioning exceeded the rate of construction of new plants (see Chapter 10). Annual hydro output is of course subject to variations in rainfall but, over the longer term, output has continued to rise with the construction of new plants. (The commissioning of one new plant in China increased world hydro output by several per cent.) However, hydroelectricity, like nuclear power, has given rise to environmental concerns that have acted, and may in future act, as a brake on further large-scale development.

More detailed discussions of the primary energy sources, the histories of their use and accounts of the associated technologies appear in later chapters.

International comparisons

Dividing 15.9 terawatts equally between the 6800 million inhabitants of the world gives each of us slightly less than two and a half kilowatts of continuous primary power, or the equivalent of about five and a half litres of oil a day (Box 2.4).

It will have been no surprise to learn in Chapter 1 that world energy is not distributed in uniform shares, and the further comparisons in Table 2.3 surely provide food for thought. A seventh of the world's population is currently consuming over two-fifths of the world's primary energy. The average daily energy used by an individual in the wealthiest two dozen countries is more than four times that in the rest of the world. It is a salutary

Table 2.3 International comparisons, 2009

	Percentage of world total			Comparisons	
	Population	**Energy produced**[1]	**Energy consumed**[1]	**Primary energy consumption per capita as a multiple of world average**	**GDP**[2] **per capita as a multiple of world average (2007)**
Wealthiest countries[3]	14%	31%	41%	3.1	3.4
Rest of world	86%	69%	59%	0.7	0.6
Selected regions					
USA and Canada	5%	18%	22%	4.5	4.0
Western Europe	15%	9%	14%	2.3	3.0
Middle East	2%	10%	3%	1.3	0.9
Africa	13%	7%	4%	0.4	0.3
Selected countries[4]					
China	19.7%	18.1%	19.5%	1.1	0.8
USA	4.6%	14.8%	19.3%	4.5	4.1
Russian Federation	2.1%	10.3%	5.6%	2.9	1.3
India	17.5%	3.8%	5.2%	0.3	0.4
Japan	1.9%	0.8%	4.3%	2.5	3.2
Canada	0.5%	3.4%	2.2%	4.9	3.5
France	0.9%	1.1%	2.2%	2.6	3.1
Brazil	2.8%	1.8%	2.0%	0.7	0.9
United Kingdom	0.9%	2.4%	1.9%	2.3	3.3
Australia	0.3%	2.4%	1.0%	3.5	3.5
Poland	0.6%	0.6%	0.81%	1.6	1.6
Greece	0.2%	0.1%	0.27%	1.8	2.7
Bangladesh	2.4%	0.18%	0.22%	0.1	0.2
Switzerland	0.1%	0.11%	0.21%	2.1	3.8
Denmark	0.1%	0.23%	0.16%	2.2	3.5
Kenya	0.6%	0.12%	0.15%	0.3	0.1

[1] Annual primary energy, including bioenergy contributions.

[2] The term *gross domestic product* (GDP) has been described in Box 1.2. The *national* GDP's here are expressed in terms of 'purchasing power parity' (PPP), i.e. they are adjusted to take into account the local purchasing power of the GDP per person in each country. Many goods are normally cheaper in the poorest countries, so the contrasts would be even greater if normal exchange rates were used to convert to US dollars.

[3] USA and Canada, Western Europe, Australia and New Zealand, Japan.

[4] In descending order of total primary energy consumption.

Sources: BP, 2010; IEA, 2009; UN, 2009.

thought that to bring the remaining 150 or so countries even to the present European level of per capita energy provision would require world total primary energy production to rise to over *twice* its current level. It is also worth noting that the percentages in the table include the contributions from traditional biomass, usually an important component of total energy in the less developed countries. It follows that for these countries to reach European levels in the use of 'modern' energy sources would require an even greater increase – and of course a corresponding increase in the environmental consequences of the use of these resources.

The average citizen of the USA earns about twenty times as much as the average Bangladeshi and consumes over forty times as much energy each day. These are extremes, but comparison of per capita *energy consumption* and per capita *income* does show some correlation. However, we should be wary of the easy conclusion that rising living standards necessarily mean the consumption of ever more energy. As later chapters will show, many of our energy systems have improved in efficiency by large factors over the years – supplying the same quantity of heat, light or driving power for a much smaller energy input. Unfortunately, this increased efficiency has usually led to increased levels of demand for the output, rather than a decrease in the demand for energy.

For the world as a whole, total annual primary energy production and consumption differ so slightly that we have implicitly taken them to be equal, but this is by no means the case for individual countries. The difference between energy consumption and indigenous production determines of course whether a country is a net energy importer or exporter, a matter of considerable economic and strategic importance. In the remainder of this chapter we compare the primary energy production and consumption data for a few of the countries in Table 2.3, to see some of the detail behind the differences.

2.6 Primary energy in the UK

The UK pattern of energy consumption shown in Figure 2.7(b) has much in common with the world data in Figure 2.4, reflecting the fact that world consumption is dominated by the industrialized countries. The main

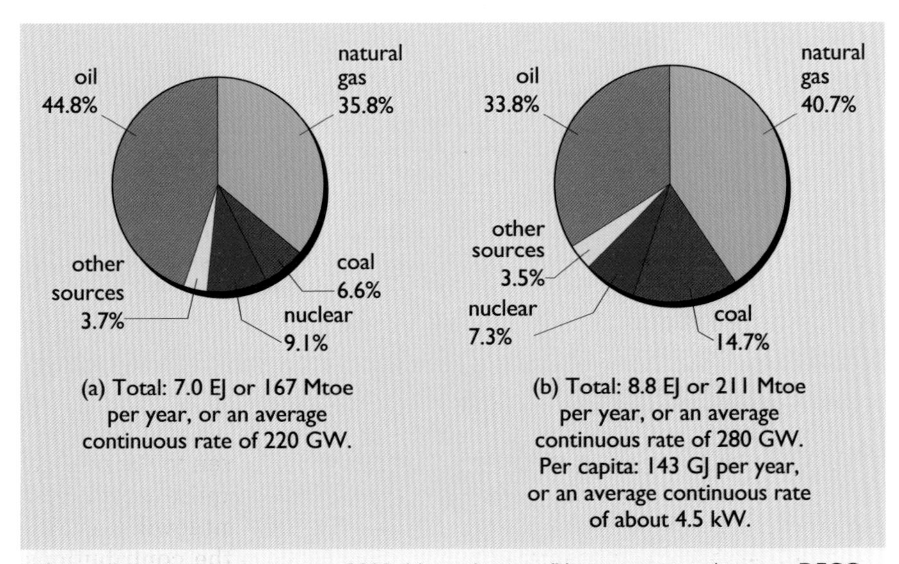

Figure 2.7 UK: primary energy, 2009: (a) production; (b) consumption (sources: DECC, 2010a; DECC, 2010b)

differences in the UK are the absence of two sources that are important on the world scale. Traditional biomass does not normally feature significantly in the national statistics of the more industrialized countries, and the relatively small UK hydroelectricity output (less than from wind energy) is included in 'other sources' in Figure 2.7 along with renewables and wastes, and waste heat from power stations where it is used for heating purposes, rather than in a separate category.

One obvious difference between Figures 2.4 and 2.7 is that the latter needs two diagrams. On the world scale, primary energy consumption in any year differs only slightly from world production, but for a particular country the two may differ widely – and as Figure 2.8 shows, their relationship can vary dramatically over time.

Over the eighty-year period of Figure 2.8, UK energy *consumption* rose fairly steadily, but at a much lower rate than for the world as a whole. In both cases, the Second World War, the energy crises of the 1970s and the economic crisis of the first decade of the present century led to slight reductions. However, the striking feature for the UK is the variation in energy *production*, and its effect on energy imports and exports. The 1930s saw the final years of the long period during which the energy of UK coal

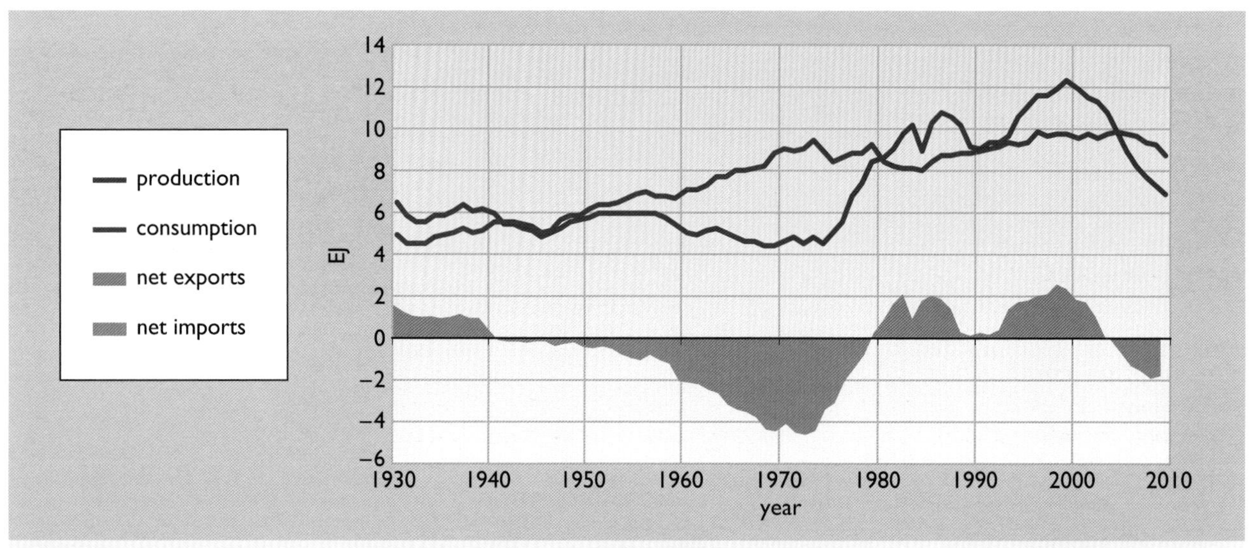

Figure 2.8 UK: annual primary energy production and consumption, and net exports and imports, 1930–2009 (sources: DECC (DUKES), various years; DECC, 2010b and other years)

exports exceeded that of oil imports. A country which had been an energy exporter for well over a century was brought within a few decades to a position where almost half its energy needs were being met by imports. But the final quarter of the twentieth century and the first decade of the twenty-first were to see yet more reversals.

The UK's changing energy scene

To understand the background to these fluctuating fortunes we need to look at the data for individual energy sources (Figure 2.9). Until the 1950s, Britain was not merely a coal-producing country but a coal-*based* country, with almost all primary energy production and over ninety per cent of

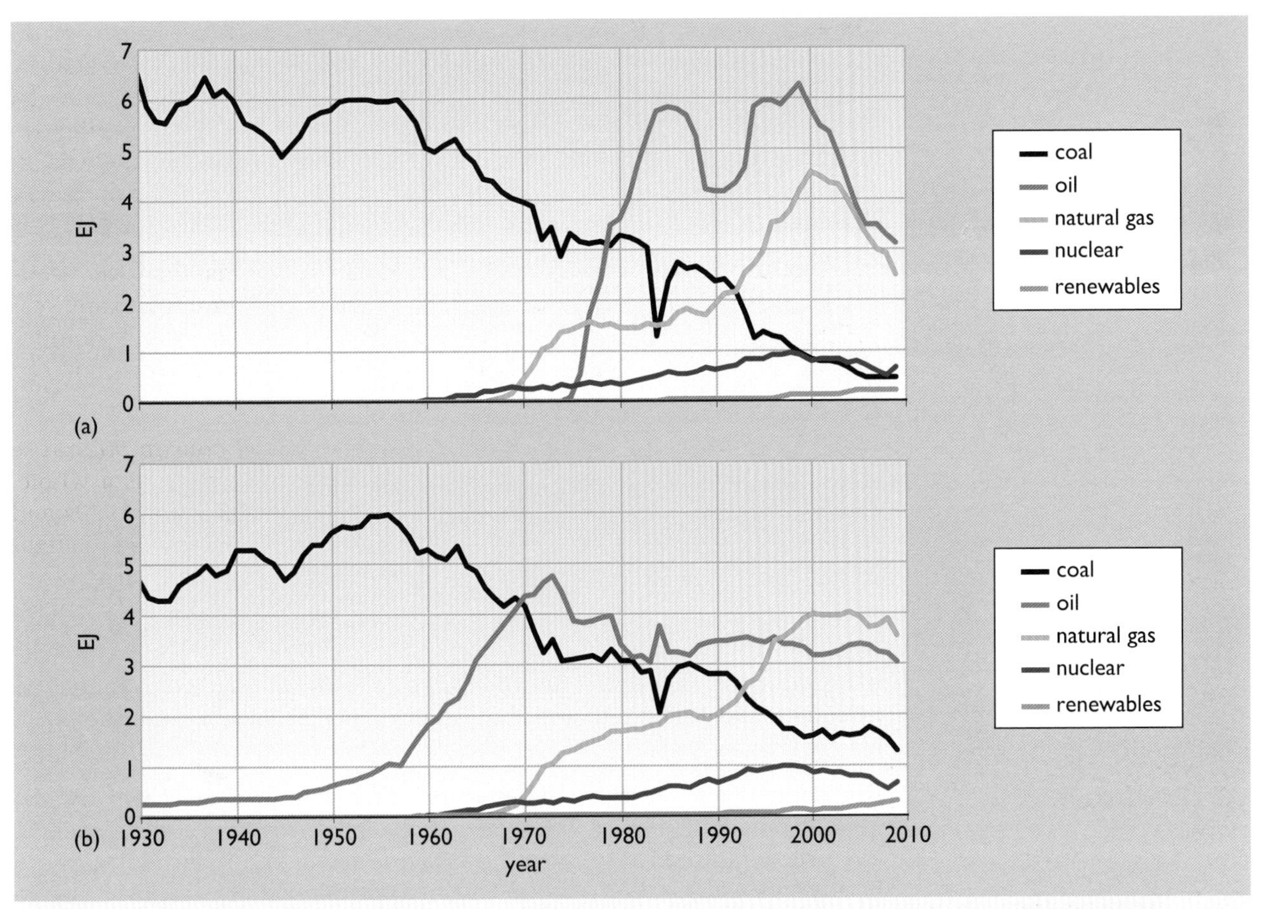

Figure 2.9 UK: annual primary energy by source, 1930–2009: (a) production; (b) consumption (sources: DECC (DUKES) various years; DECC, 2010b and other years)

consumption coming from coal. Coal fuelled the railways, the power stations and industrial machinery and, together with the 'town gas' derived from coal, met almost all the country's heating needs. But this was about to change. Coal production, which had started to rise again after the war years, began its long decline. Meanwhile, as in other industrialized countries, oil consumption in the UK was growing rapidly, with an average annual increase of over 10% a year from 1950 to 1970 (Figure 2.9(b)). These two factors are sufficient to account for the transformation during the 1950s from net energy exporter to net importer, a state that was to continue for more than thirty years.

However, the late 1960s already saw the start of yet another change, with the first contributions from Britain's North Sea gas fields. In the early years, from 1967 to 1972, output more than doubled each year – a remarkable annual average growth rate of over 100%. Consumption rose in step as the change from town gas to natural gas spread across the nation. Coal was of course the main energy source being displaced, but coal production was dropping in parallel with the falling demand. In consequence, the overall effect of Britain's natural gas resource on energy imports was very slight in these early years, and it was only with the discovery of yet another resource that the trend eventually reversed.

The story of the exploration and subsequent development of North Sea oil will be told in later chapters, but its immediate consequences for the UK can be seen in Figures 2.8 and 2.9. The first significant deliveries came in 1976, and within three years oil had outstripped coal in its contribution to primary energy production. (As a proportion of primary *consumption*, oil – imported – had overtaken coal in the late 1960s.) By 1980 Britain was self-sufficient in oil, and despite the continuing fall in coal output, was about to become again a net energy exporter. A brief reversal occurred when oil production fell by a third following the steep drop in world oil price in the early 1980s (Figure 2.6), but as the end of the century approached, output reached its highest level ever, with oil accounting for almost half of the nation's primary energy production. This was not to last, however, as Figure 2.9 shows, and within a decade oil output had halved.

The period of rapidly increasing oil output saw a levelling-off in gas production, and with completion of the national change from town gas, a much slower annual rate of rise in consumption. But then came the 'dash for gas' of the 1990s, as the electricity generating industry started to take advantage of the technical and financial merits of gas turbine plants. This development and its environmental and other consequences will be treated in more detail in later chapters but, as Figure 2.9 shows, the falling UK gas production of the first decade of the present century led to increased imports to meet demand, and the country again became a net importer of energy (Figure 2.8).

Nuclear power is the only energy source other than the fossil fuels to make a significant contribution to UK primary energy. Growth at a rather modest 3% average annual rate over the last quarter of the twentieth century brought its contribution, on the 'notional primary energy input' basis, to almost exactly that of coal in the year 2000. However, as in the world as a whole, UK nuclear output was already falling with the diminishing number of operational plants. (The slight difference in nuclear power production and consumption in Figure 2.9 reflects imports of nuclear electricity from France.)

Renewables in the UK

Biofuels in general had virtually disappeared from Britain's energy supply two centuries ago with the decline of wood as a fuel, but they have reappeared in recent decades, often in very different forms. In the year 2009, renewables and wastes contributed a total of about 290 PJ of primary energy, representing over 3% of the country's total primary energy consumption (Figure 2.7(a)).

As Figure 2.10 shows, about three-quarters of the renewables total comes from organic wastes, the main single contributor being **landfill gas**. For the past century or more, Britain has disposed of up to 90% of its domestic and commercial rubbish in landfills, where the organic component, decaying over a period of years in the absence of air, produces a gas that is relatively rich in methane (the main component of natural gas). Over recent decades this landfill gas has increasingly been collected and used, mainly as fuel for small-scale electric power plants. **Municipal wastes** refers to domestic

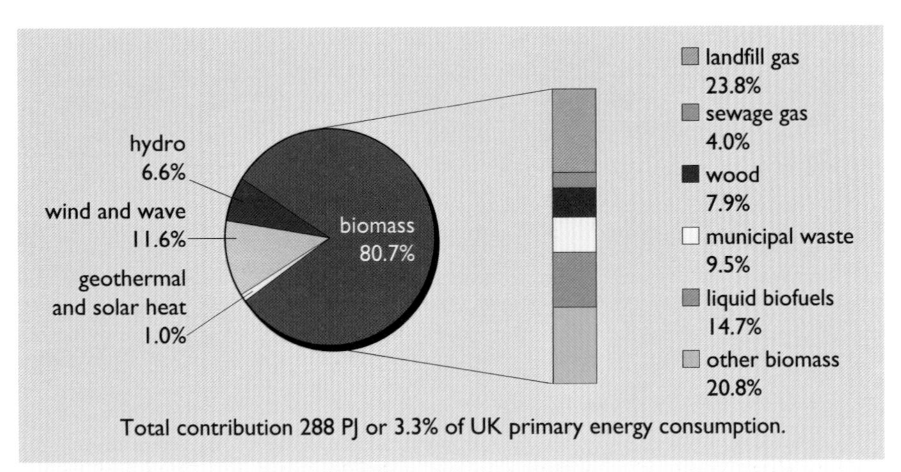

Figure 2.10 UK: energy from renewables and wastes, 2009 (source: DECC, 2010b)

and commercial rubbish when it is burned directly in power plants, whilst **other biomass** includes a variety of animal and vegetable wastes used to produce heat, and often electricity. About two-thirds of the total biomass contribution shown in Figure 2.10 is therefore used for the generation of electricity, the remaining third being the input for liquid biofuels for vehicles and a relatively small fraction used directly for heating.

It is, however, important to note when comparing the different sources in Figure 2.10 that the contributions from hydroelectricity and wind power are their *electrical output*, whilst those for biomass are the *heat content* of the different fuels. As we have seen, the final electrical output from any thermal power plant may be a third or less of the heat input, so the useful output from the two-thirds of the biomass that is used for power generation will be much less than Figure 2.10 suggests. Indeed, contrary to the impression given by the diagram, the joint electrical output from hydro and wind in 2009 was greater than the total electrical output from biomass, and the hydro output alone was greater than the total output of all the landfill gas power plants.

Two contributions in Figure 2.10 are of particular interest. As recently as the year 2000, neither wind energy nor liquid biofuels would have appeared at all in the diagram, as their contributions were far too small to be shown separately. **Wind energy**, contributing less than 1 TWh in the year 2000, saw a rapid growth to nine times this in 2009, achieving an average annual rate of increase of almost 30% per year for nine years. Continuing this rate of increase until the year 2020 could in principle see a wind contribution equal to almost half the total UK electricity demand in 2009.

The role of the liquid biofuels and their use for transport is less clear-cut. The growth in the use of these alternatives to 'fossil fuel' petrol or diesel is clear. In 2003 they accounted for less than one-tenth a per cent of all UK vehicle fuel, a contribution that rose to 1% in 2007 and 3% in 2009. However, these are *consumption* figures, and UK *production* of liquid biofuels fell by the end of 2009 to less than half the 2007 output. Of the one billion litres of biodiesel consumed in 2009, only about a fifth was produced in the UK, the rest having been imported.

Later chapters of this book treat the history of Britain's energy industries in more detail and discuss the social, economic and technical factors that have determined the changes described here.

2.7 Primary energy in Denmark

Comparison of Denmark's primary energy production and consumption in the year 2009 (Figure 2.11) with the UK data in Figure 2.7 reveals a number of similarities and some very important differences. The fossil fuels dominate energy consumption and production in both countries. The total primary energy consumption in Denmark (population 5.54 million) is about one eleventh of the total for the UK (population 61.6 million), so *per capita* consumption is almost the same in the two countries. However, a closer inspection of the two diagrams reveals one significant difference. In Denmark, unlike the UK, total production is greater than total consumption. In other words, Denmark was self-sufficient in energy in 2009 (see Box 2.10), whilst the UK was not.

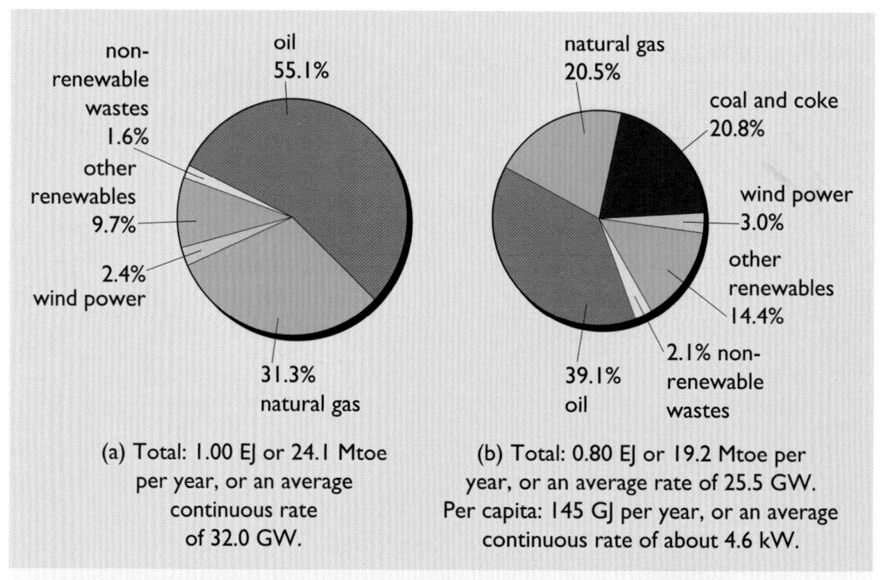

Figure 2.11 Denmark: primary energy, 2009: (a) production; (b) consumption (sources: BP, 2010; DEA, 2009; DStat, 2009; WWEA, 2009)

Comparison of Figure 2.12 with the same fifty-year period of Figure 2.8 reveals two countries sinking into ever greater energy debt throughout the 1960s, but whose fortunes changed with the discoveries of North Sea gas and oil. They differ slightly in that UK coal exports still almost balanced oil imports in 1960, whilst Denmark had almost no indigenous energy resources (see also Figure 2.13). However, the major difference is in the patterns of production and consumption in the two countries *after* 1970.

Denmark's only indigenous primary energy resource in 1960 was a limited amount of lignite, and the country had been almost totally dependent on imported fuel throughout modern times. By 1970, the steep rise in oil consumption (Figure 2.13(b)) was leading to a serious position strategically and economically. The measures taken by successive Danish governments

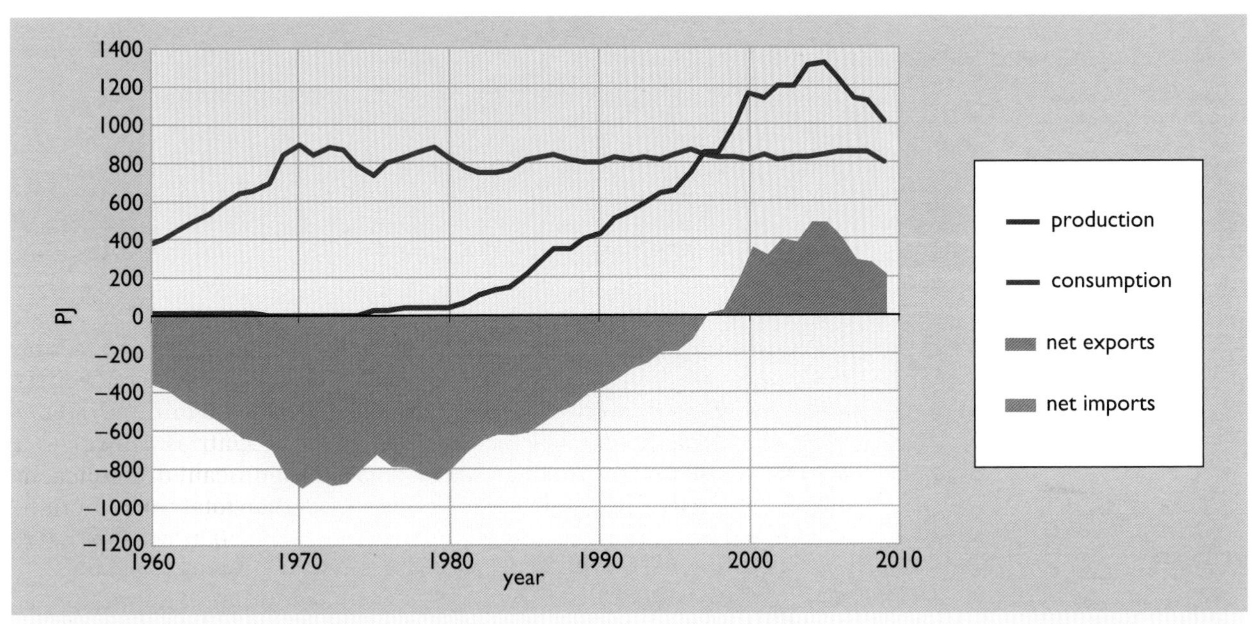

Figure 2.12 Denmark: annual primary energy production and consumption, and net exports and imports, 1960–2009 (sources: BP, 2010; DEA, 2009; DStat, 2009; WWEA, 2009)

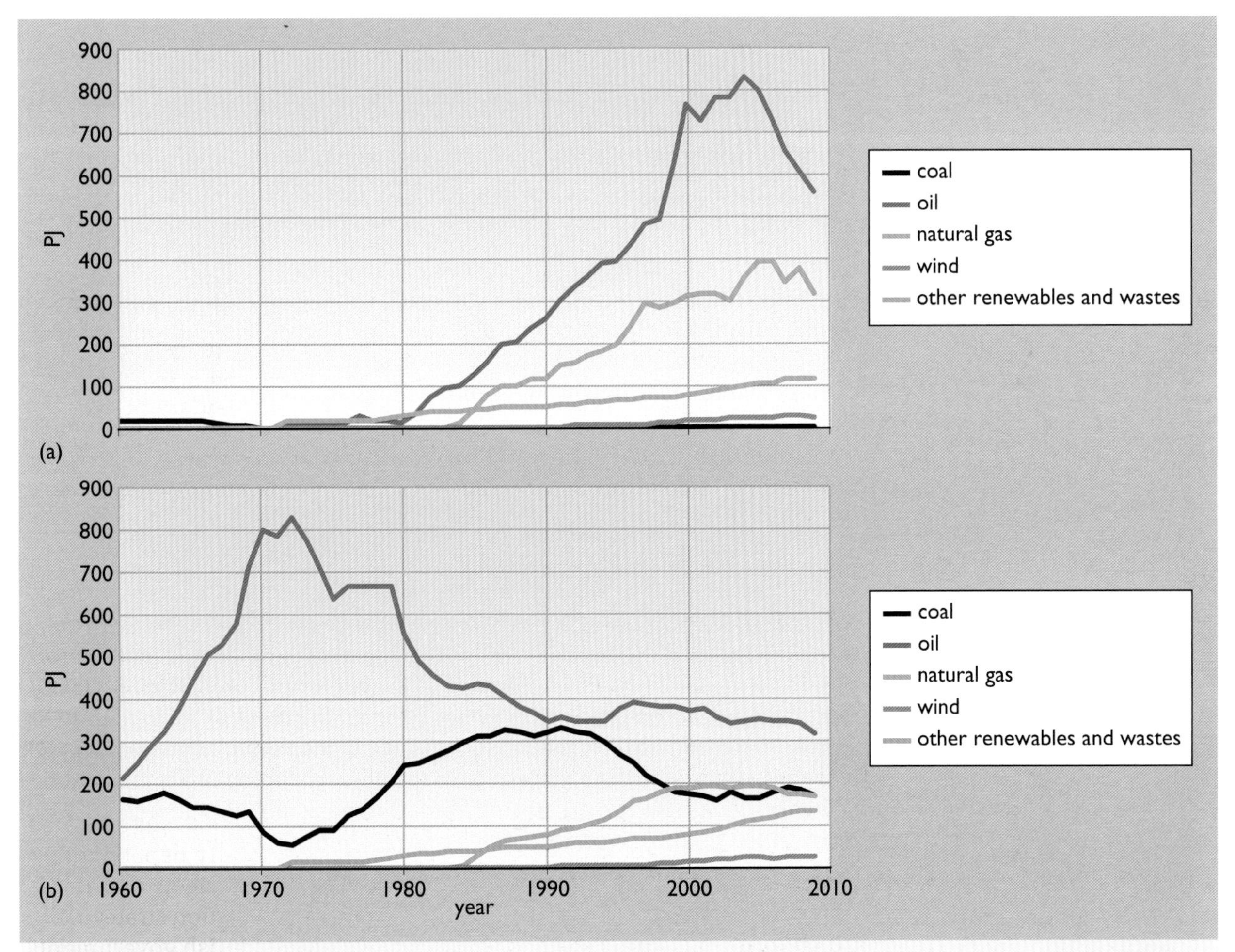

Figure 2.13 Denmark: annual primary energy by source, 1960–2009: (a) production; (b) consumption (sources: BP, 2010; DEA, 2009; DStat, 2009; Eurostat, 1990)

BOX 2.10 Self-sufficiency

A useful measure of the extent to which a country depends on energy imports is its degree of **self-sufficiency**. This is defined as the total primary energy production divided by the total primary energy consumption, expressed as a percentage. A self-sufficiency greater than 100% obviously implies that the country has an energy surplus and is therefore likely to be a net exporter.

This is the case for Denmark, with a self-sufficiency of 130%, but not for the UK, whose self-sufficiency of 80% suggests that 20% of the country's energy consumption is provided by imports. It is, however, important to note that self-sufficiency in an individual energy resource such as oil can be as important as overall self-sufficiency.

to ameliorate this situation will be discussed in Chapter 3, but their main result can be seen in the graphs: a total primary energy demand that hardly changed over four decades, achieved mainly by a remarkable reduction in oil consumption to less than half its peak value. As Figure 2.13 shows, oil was at first partially replaced by increased consumption of imported coal (mainly for electricity generation), but the final decade of the century saw domestic production of oil and gas making an appreciable contribution. In 1997 Denmark became self-sufficient in energy for the first time in modern history, and at the time of writing (2010) still exports a significant fraction of its North Sea oil and gas.

Denmark's development of its natural gas resource came some five years after the oil – a reversal of the UK sequence. Closer study of Figures 2.9 and 2.13 reveals another difference. Since the start of production, Denmark has always exported about half its gas (by pipeline to Germany and Sweden), whilst the UK has supplemented its indigenous supply by importing gas (by pipeline from the Norwegian fields).

However, the main supply-side difference between the two countries in recent decades is the nature of the *non-fossil fuel* contribution to energy production (and consumption). In the UK, as we have seen, most of this comes from nuclear power, with renewables and wastes contributing only about 4% of total primary energy. Denmark, in contrast, has no nuclear contribution, but renewables and wastes supply nearly 14% of total energy production, and almost 20% – a fifth – of total consumption.

Renewables in Denmark

As we have seen, the renewable energy sources play a much greater role in Denmark than in the UK, but in comparing Figures 2.10 and 2.14 several differences need to be borne in mind. A relatively minor point is that the Danish statistics identify two categories: 'biofuels' and 'non-renewable wastes' (combustible waste materials, plastics, etc.), whereas the UK includes all combustible materials in the general 'renewables and wastes' category. It is also worth noting that Denmark imports about 20 PJ of renewables (mainly wood wastes from neighbouring countries), bringing the country's total to about 160 PJ. With a population of only 5.54 million, this means a per capita renewables consumption of almost 30 GJ – nearly six times the corresponding figure for the UK. As shown in Figure 2.14, the

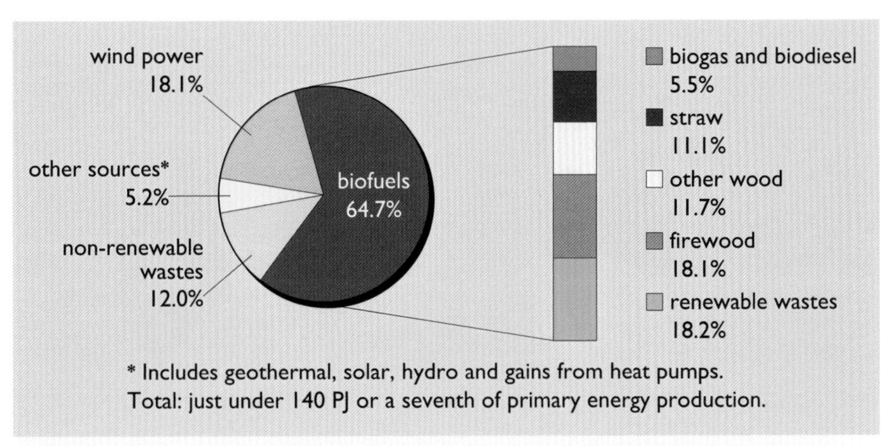

Figure 2.14 Denmark: energy from renewables and wastes, 2008 (source: DEA, 2010)

biofuels that play a major role in Denmark are mainly wastes, rather than purpose-grown crops, and their uses include power generation, heating, and the production of liquid biofuels from straw.

Geothermal energy, heat drawn from below the Earth's surface, contributes a tiny fraction of the renewables total in both Britain and Denmark. The availability of this resource obviously depends on local geology, and neither country expects a major increase in its input. However, the 'other sources' in Figure 2.14 include the energy to be gained from surroundings that are at normal ambient temperature through the use of **heat pumps**, an approach that has attracted increasing interest in several countries in recent years. A heat pump, as the name suggests, 'pumps' heat from a cooler region into a warmer one, against the natural direction of heat flow. The principle will be discussed later, in Chapter 9, but the result is obviously useful, warming buildings in cold weather or, in reverse, cooling them on hot days. Denmark, quite justifiably, includes such gains in the renewables total, and the current contribution of nearly 6 PJ, although a small fraction of the whole, represents an appreciable heat gain. A proportionate annual contribution in the UK would be enough to heat half a million homes.

2.8 Primary energy in the USA

With the USA, we come, not surprisingly, to a very different situation: a country whose average citizen consumes energy at twice the rate of their British or Danish counterpart (Figure 2.15). The USA is the world's major energy importer, and consumes in total a fifth of the world's primary energy production. The pattern of consumption is not, however, very different from the two European countries discussed above, with the dominance of fossil fuels that is common to nearly all the world's industrialized countries, and a percentage contribution from nuclear power similar to that of the UK. (The actual nuclear output is of course much greater than for the UK and at some 850 trillion kWh a year, puts the USA easily in first place worldwide.)

Sixty years ago, the USA was self-sufficient in energy – as it had been throughout most of its recorded energy history. As shown earlier, this was

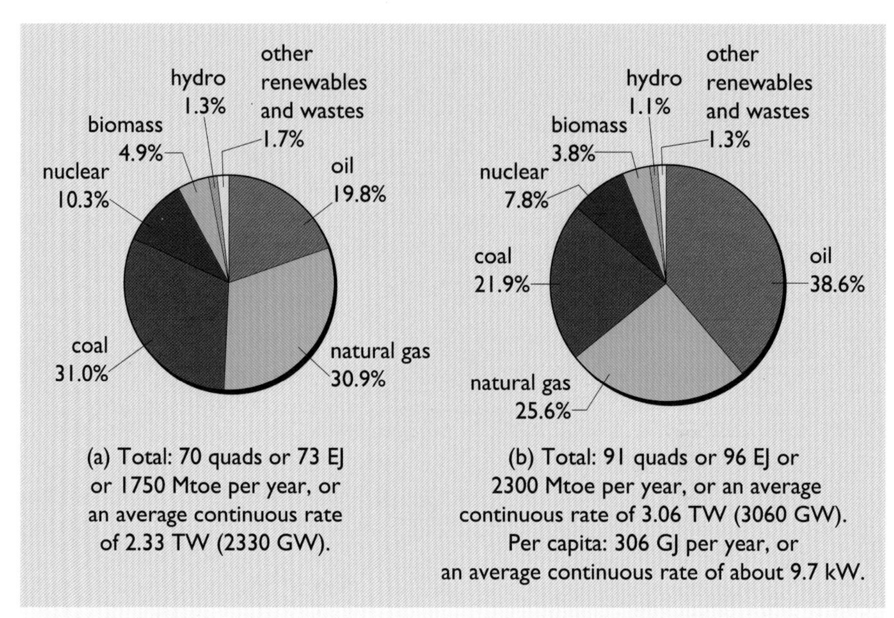

Figure 2.15 USA: primary energy, 2009: (a) production; (b) consumption (sources: BP, 2010; EIA, 2010a; IEA, 2009)

also the case for the UK, and comparison of Figures 2.16 and 2.8 reveals two countries whose thirst for oil in the period from about 1950 to about 1970 led to similarly growing gaps between energy production and consumption. But there was one significant difference. In the early 1970s the UK was starting to develop its indigenous oil resources, whilst in the USA the resources were beginning to be exhausted, as Figure 2.17(a) shows.

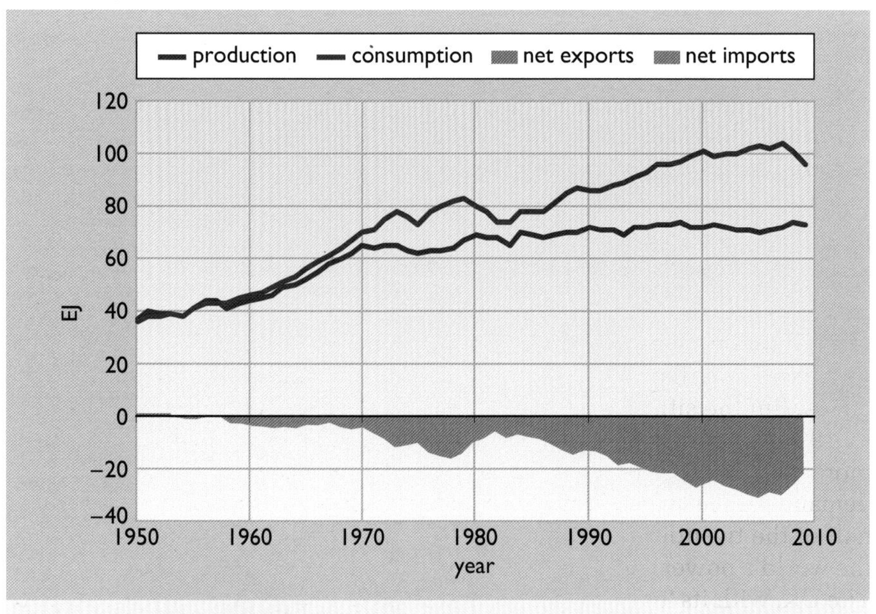

Figure 2.16 USA: annual primary energy production and consumption, and net exports and imports, 1950–2009 (sources: BP, 2010; EIA, 2010a; IEA, 2009)

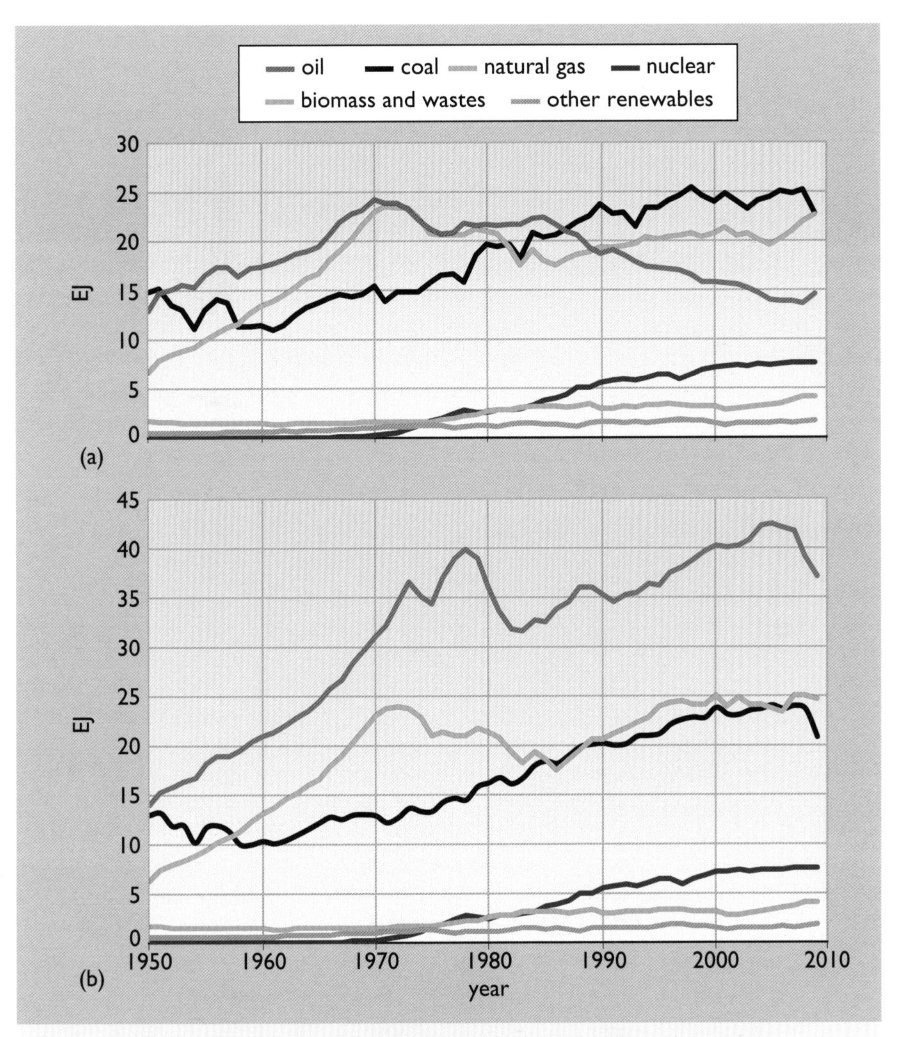

Figure 2.17 USA: annual primary energy by source, 1950–2009: (a) production; (b) consumption (sources: BP, 2010; EIA, 2010a; IEA, 2009)

Nevertheless, at the time of writing (2010), the two countries have one unfortunate feature in common: an overall energy self-sufficiency of *less than 80%*.

However, comparison of Figure 2.17 with Figure 2.9 shows another striking contrast. The UK, once living on coal, has seen a steady decline in production over the past fifty years. Until 1960 the USA seemed to be following a similar path, with first oil and then gas overtaking coal in the 1950s. But the situation then changed. Coal production began to rise, and by 1990 its contribution had overtaken both oil and gas. In this, the USA is more characteristic of the world as a whole. As discussed in later chapters, demand for electric power rose rapidly worldwide throughout the second half of the twentieth century, and coal, still the principal fuel for most of the world's power stations, experienced a corresponding growth in output. The UK, with its 'dash for gas', is therefore an exception, and Denmark, for reasons discussed above and in Chapter 3, is another. France (discussed in Section 2.9) is a third, for yet other reasons.

Another aspect of Figure 2.17 bears closer study. At the start of the 1970s, production from existing oil fields in the USA reached its peak, and output was already falling when the first dramatic increase in world oil prices occurred in 1973. The country's dependence on oil from the Middle East became obvious to everyone, and for a couple of years consumption fell. It soon resumed its rise, however, encouraged in part by the development of the Alaskan oil fields, whose output delayed the fall in national production for about a decade. The further price rise in 1978 and the recession of the early 1980s brought about a more serious fall in consumption, but this again was only temporary, and the resumed steady rise resulted in the greatest shortfall ever in the year 2005. Subsequent economic crises led to reduced energy consumption in almost all industrialized countries, and the USA was no exception, with consumption in 2009 a few per cent lower than the peak of 20 Mbd (million barrels a day) in 2005.

Renewables in the USA

Comparison of Figure 2.18 with Figures 2.10 and 2.14 reveals one common feature and some clear differences. In all three countries, biomass dominates the renewables contribution to national energy production, whilst the very different inputs from hydroelectricity and geothermal energy obviously reflect the natural advantages of the USA. The US wind power contribution is nearly ten times that of Denmark, but in terms of output per square kilometre of land area, the Danish output is about 25 times that of the USA and four times that of the UK.

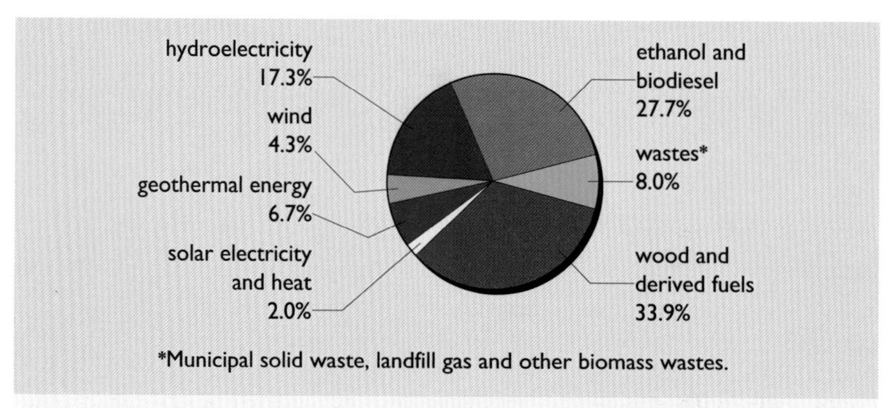

Figure 2.18 USA: energy from renewables and wastes, 2009; total 5.9 EJ or 6.1% of primary energy consumption (source: EIA, 2010b)

A review of Figures 2.7, 2.11 and 2.15 shows that, whilst the total quantities of energy and the degrees of self-sufficiency may be very different for the three countries, their general patterns of consumption are not dissimilar, with each meeting over 80% of its demand for primary energy from the three fossil fuels. And as we have seen, the same is true world as a whole – not surprisingly, as world consumption is dominated by industrialized countries such as these.

However, not every country has chosen – or has been able – to adopt this pattern, and the present survey concludes with brief accounts of three further countries where primary energy consumption includes major contributions from sources that have played only a relatively small role in the countries discussed so far.

2.9 Other countries of interest

France

France differs only slightly from Denmark or the UK in per capita primary energy consumption (Figure 2.19(b)). But similarity with the other two countries ends there, as even a casual glance reveals. One striking difference is the extremely low self-sufficiency – a country consuming twice as much primary energy as it produces.

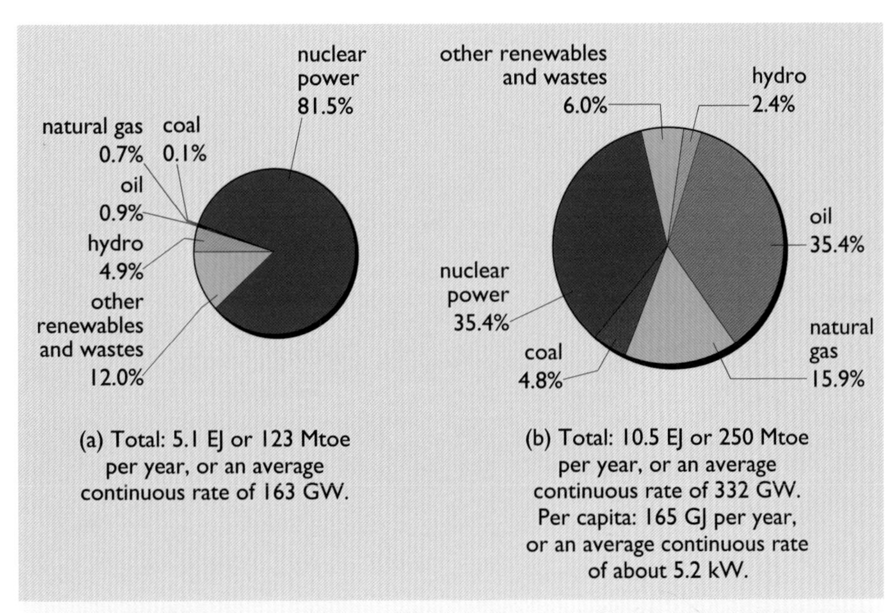

Figure 2.19 France: primary energy, 2008: (a) production; (b) consumption (sources: BP, 2010; DGEC, 2009; Eurostat, 1990)

It is easy to see the reason for this difference. As Figure 2.19(a) reveals, France has virtually no indigenous fossil fuel resources. The more detailed Figure 2.20, with the oil contribution too small even to appear on the production graph, emphasizes the fact that for half a century the growing need for fossil fuels was necessarily met almost entirely by imports. The French government, realizing during the 1970s that, unlike the UK or Denmark, the country was not about to be rescued by the discovery of indigenous oil or gas, made the decision to invest in a national nuclear power industry.

With a rapid rise in construction during the early 1980s, and a continuing increase at a rather slower pace over the following twenty years, by 2005 the country had 59 operational reactors, with a collective power rating of 63 GW. Annual output had reached 450 TWh, accounting for over 80% of France's electricity production, together with exports of some 50 TWh a year to surrounding countries (including up to 12 TWh per year to the UK through undersea cables).

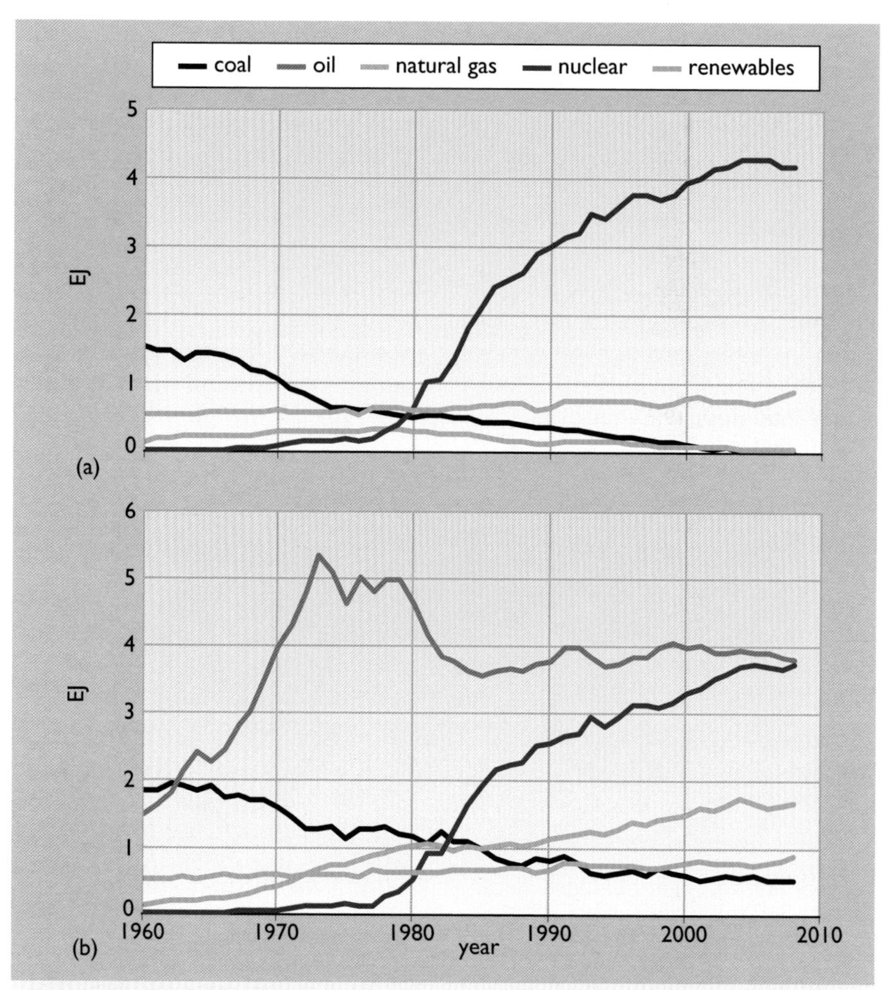

Figure 2.20 France: annual primary energy by source, 1960–2008: (a) production (oil production was too small to show in this graph); (b) consumption (sources: BP, 2010; DGEC, 2009; Eurostat, 1990)

India

India is the world's second most populous country and its fourth largest consumer of energy. As Figure 2.21 shows, the country's overall energy self-sufficiency in 2009 was a reasonable 73%. However, if we omit the biomass contribution (see below), the self-sufficiency in terms of the commercially traded forms of energy falls to 64%. Ten years earlier, this was much higher, at about 80%, and Figure 2.22 shows the main reason for the change: the continuing increase in the consumption of oil with little change in its production. Despite government support, exploration for oil has not resulted in the discovery of any major new fields. Oil production has therefore remained essentially unchanged for twenty-five years, during which time oil consumption has risen by a factor of three.

The increasing use of imported coal revealed by the graphs is perhaps surprising, as India has amongst the world's highest coal reserves. The explanation lies mainly in the quality of the indigenous coal, and the need for the 'hard coal' required by India's growing number of coal-fired power stations and its steel industry (see Chapters 3 and 5).

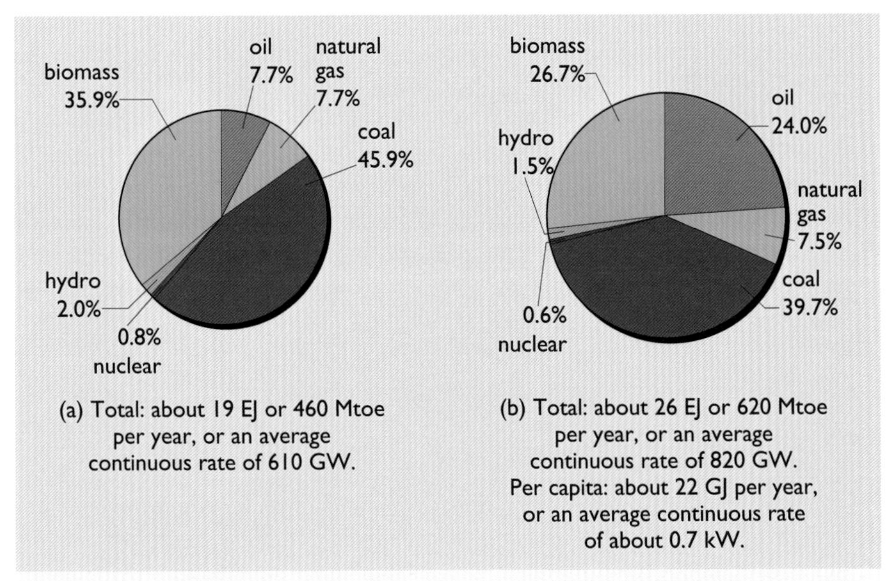

Figure 2.21 India: primary energy, 2009: (a) production; (b) consumption (sources: BP, 2010; Maddison, 2005, 2009; TERI, 2009)

The only other commercially traded energy contributions large enough to appear separately in Figure 2.21 are electricity from hydroelectric and nuclear plants. In the period from 2005 to 2009, India's nuclear plants were producing 16–18 TWh per year, and hydroelectric plants 100–120 TWh. (It is a reflection of the magnitude of India's population that the joint output of these two sources is equivalent to a continuous supply of only 12 watts per person.)

During the 1990s, India's population was increasing at slightly over 2% per year, while the consumption of commercially traded forms of energy was

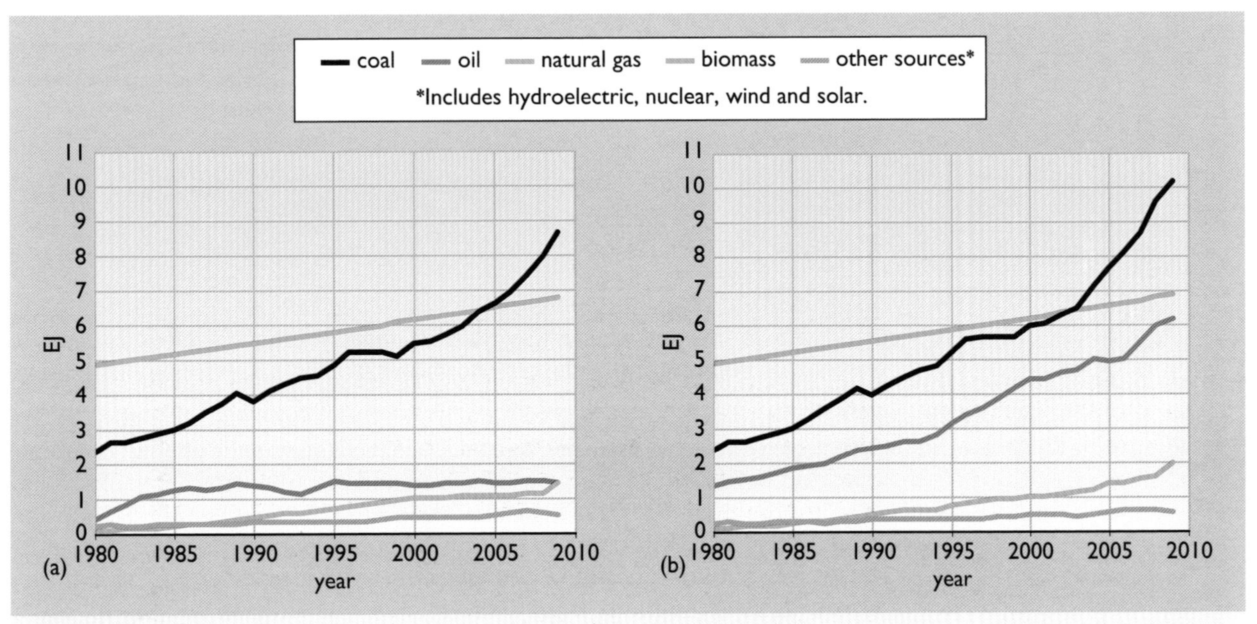

Figure 2.22 India: annual primary energy by source, 1980–2009: (a) production; (b) consumption (sources: BP, 2010; Maddison, 2005, 2009; TERI, 2009)

rising at an average annual rate of over 5%, resulting in an increase in per capita consumption – to be expected in a developing country. The world economic crisis at the turn of the century temporarily reduced growth in energy consumption, in India as elsewhere, but this was only a brief pause. The first decade of the present century has seen the country's commercial energy consumption rising at an average rate of almost 6% per year, while the rate of population growth has slowly been reduced, reaching an annual increase of only 1.8% per year at the time of writing (2010). The implication of these two very different rates of change is an ever more rapidly growing per capita consumption. It is worth noting, however, that if commercial energy consumption continued to rise by 6% per year, even with a stable population it would still take nearly forty years for India to reach the present per capita consumption of a typical Western European country.

Biomass in India

A further obvious difference between India and the four countries studied earlier is the major role played by biomass, or more specifically *traditional biomass*. For many people this resource is at least as important as the fossil fuels – and in rural areas is often the only accessible source of heat and light. (It is estimated that about half the population still have no access to mains electricity.) The traditional resources in India include fuel wood, dung (dried for direct use as fuel or digested to produce 'biogas' for burning), bagasse (sugar cane fibre), rice husks and other agricultural residues. Of these, fuel wood is thought to account for about half the total, with dung and other agricultural residues contributing most of the remainder.

As discussed in Box 2.9, estimates of the present contribution from these often unrecorded energy sources are necessarily subject to considerable uncertainty, and past contributions are even less certain. It is nevertheless of some importance for countries such as India to endeavour to quantify the contribution and its rate of change, if only to assess the consequences of a future shift from traditional biomass to other energy sources. Recent analyses of the energy contributions from this resource, worldwide and for specific countries, have concentrated on establishing relationships between population growth and growth in traditional biomass consumption. The uniform rate of rise in biomass consumption shown in Figure 2.22 is based on these studies.

China

In 2009, China became the world's major consumer of energy, displacing the USA. However, with a population approaching 1.4 billion, China's *per capita* consumption, despite doubling over the previous decade, had reached only a quarter of the US level.

The most obvious feature of Figure 2.23 is the dominance of coal, the main contributor to China's energy production and consumption for many years. Despite the need to import certain types of coal (see Chapter 5), exports were more than sufficient to counterbalance these imports. In the mid-1980s this situation began to change rapidly (Figure 2.24). Rising exports, encouraged throughout the 1990s, reached 93 million tonnes in 2003, resulting in a shortfall in coal for China's growing industries and electric

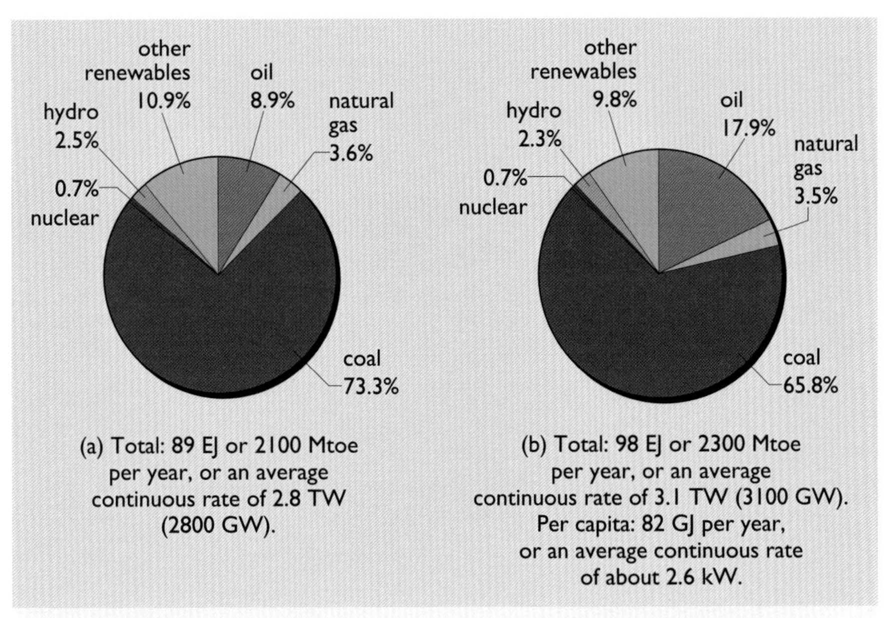

Figure 2.23 China: primary energy, 2009: (a) production; (b) consumption (sources: Aden, 2010; BP, 2010; LBNL, 2008)

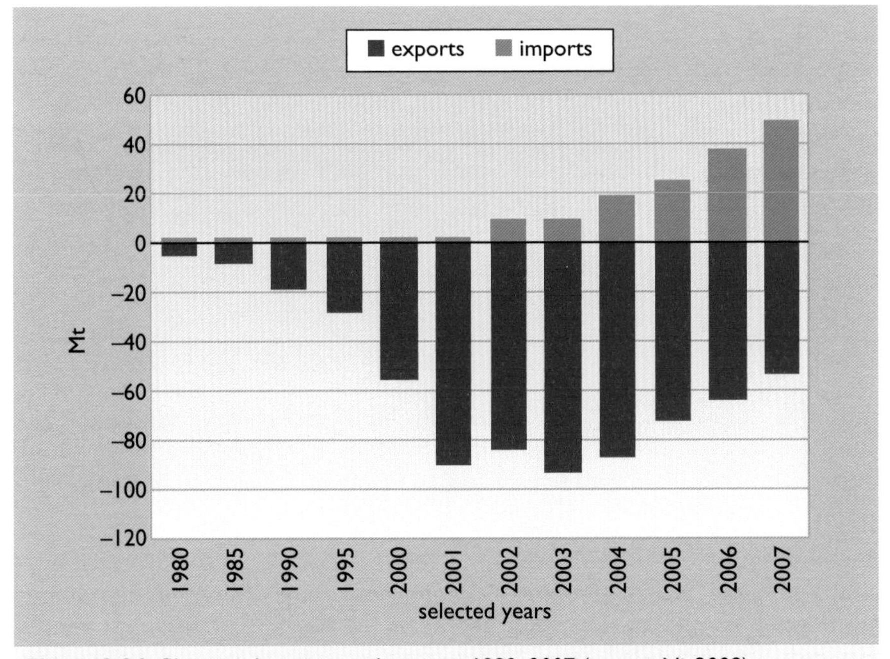

Figure 2.24 China: coal imports and exports, 1980–2007 (source: Ni, 2009)

power infrastructure (see Chapter 3). The government responded to the crisis by imposing coal export controls, and with rapidly rising indigenous production, supply kept pace with or exceeded demand for the next few years (Figure 2.25). This remains the case at the time of writing (2010).

In 2009 oil accounted for less than a fifth of China's total energy consumption – a smaller fraction than for any of the countries studied

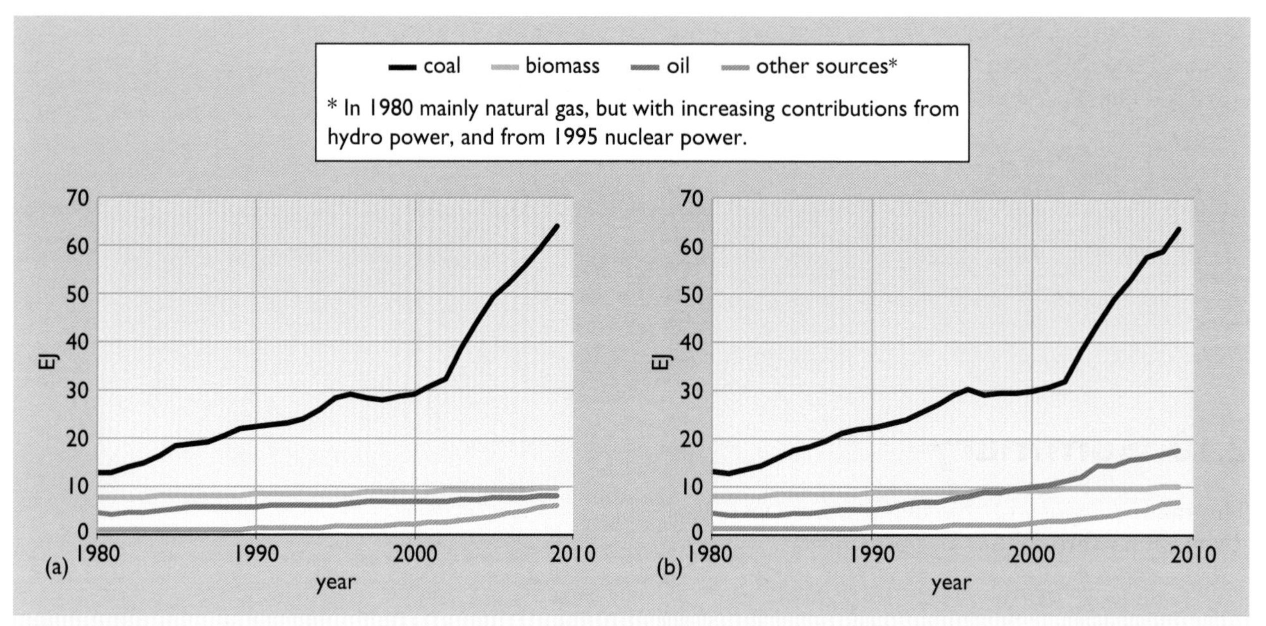

Figure 2.25 China: annual primary energy by source, 2009: (a) production; (b) consumption (sources: Aden, 2010; BP, 2010; LBNL, 2008)

above. Nevertheless, the limited oil resource has led to a gradual increase in imports, and in 1993 the country became a net oil importer. The first nine years of the present century have seen a continuing rise in the imported fraction, from less than a third to more than half (55%) of the country's total oil consumption of 420 million tonnes. (For comparison, the corresponding figures for the USA are 62% of a total of 890 million tonnes.)

As Figure 2.25 shows, in 2009 sources other than coal and oil were jointly contributing less than a fifth to China's total energy consumption. The contribution from traditional biomass had grown only slightly over recent decades, resulting in a diminishing percentage of the total. The very low population growth (1% or less per year) may account for this in part, and other factors are thought to include increased electrification and the introduction of more efficient biomass stoves in rural households (Ni, 2009). Despite recent revisions of the estimated reserves, the output of natural gas has also remained fairly constant for several decades.

Development of hydroelectricity is official policy, and output almost doubled in the five years to 2009; but even the huge 18 GW Three Gorges plant has had a relatively small effect in the context of national energy demand. The 620 TWh total hydro contribution in 2009, a fifth of the world's hydroelectric output, represented only 2.5% of China's annual energy production. In the same year, nuclear power, which entered the field only in the mid-1990s, generated just 70 TWh from its 9 GW of operational capacity. The further ten nuclear plants under construction are expected to double this by the year 2015.

Comparison of Figure 2.25 with the corresponding diagram for the world as a whole (Figure 2.5) reveals the importance of China in the global energy economy. We see one country accounting for virtually the entire growth in world coal consumption in the first decade of the present century. One consequence of this heavy reliance on coal is that China is estimated to be responsible for as much as a fifth of the world's carbon dioxide emissions – although it should perhaps be added that the country is only eightieth in per capita CO_2 emissions. And there remains the issue of China's future oil consumption. A simple calculation reveals that to raise per capita oil consumption to that of countries such as Denmark or France, world oil production would need to rise by about two billion tonnes – a 50% increase. There is little doubt that China's influence on global energy, as in other areas, can only increase in the coming years.

2.10 Summary

The main purpose of this chapter has been to establish a basis for later discussions throughout the book. Not the scientific basis, which comes later, starting in Chapter 4. Here, after introducing the essential terminology and discussing the nature and reliability of the data, we have mainly been concerned to establish a foundation of facts about the quantities of energy used today and the sources of that energy. We have also looked at the past, at how we reached the present situation. And, probably of greatest importance for the future, we have seen the disparities in energy availability between different regions of the world. How the world, and the individual countries discussed above, use the energy is the subject of the next chapter.

References

Aden, N. (2010) 'Initial Assessment of NBS Energy Data Revisions' [online], Berkeley, CA, Lawrence Berkeley National Laboratory, http://china.lbl.gov (accessed 10 September 2010).

Boden, T.A., G. Marland, and R.J. Andres (2010), *Global, Regional, and National Fossil-Fuel CO_2 Emissions* [online], Carbon Dioxide Information Analysis Center, Oak Ridge National Laboratory, U.S. Department of Energy, Oak Ridge, TN, http://cdiac.ornl.gov/trends/emis/tre_glob.html (accessed 11 December 2010).

BP (2010) *BP Statistical Review of World Energy*, London, The British Petroleum Company; available at http://www.bp.com (accessed 26 June 2010).

DEA (2009) *Energy Statistics 2008*, Copenhagen, Danish Energy Agency; available at http://www.ens.dk (accessed 19 August 2010).

DEA (2010) *Monthly Electricity Statistics*, Copenhagen, Danish Energy Agency; available at http://www.ens.dk/en-us (accessed 21 September 2010).

DECC (2010a) *Digest of United Kingdom energy statistics (DUKES)*, Department of Energy and Climate Change; available at http://decc.gov.uk/en (accessed 26 August 2010).

DECC (2010b), *UK Energy in Brief 2010*, Department of Energy and Climate Change; available at http://www.decc.gov.uk (accessed 09 August 2010).

DGEC (2009) *Repères: Chiffres clés de l'énergie*; available at http://www.developpement-durable.gouv.fr (accessed 15 September 2010).

DStat (2009) StatBank Denmark [online], Statistics Denmark, Copenhagen, http://www.statbank.dk (accessed 13 September 2010).

EIA (2010a) *Annual Energy Review 2009*, US Energy Information Administration; available at http://www.eia.doe.gov/emeu/aer/contents.html (accessed 10 September 2010).

EIA (2010b) 'Renewable Energy Consumption and Electricity Preliminary Statistics 2009' [online], Washington DC, US Energy Information Administration, Office of Coal, Nuclear, Renewable, Electric and Alternate Fuels, http://www.eia.doe.gov/cneaf/alternate/page/renew_energy_consump/rea_prereport.html (accessed 30 September 2010).

Eurostat (1990), *Energy 1960–1988*, Luxembourg, Office des publications officielles des Communautés européeannes.

IEA (2008), *World Energy Outlook: 2008 edition*, Paris, International Energy Agency; available at http://www.iea.org (accessed 25 July 2011).

IEA (2009), *Key World Energy Statistics*, Paris, International Energy Agency; available at http://www.iea.org (accessed 21 September 2010).

LBNL (2008) *China Energy Databook*, Berkeley, CA, Lawrence Berkeley National Laboratory; available only as CD from http://china.lbl.gov/databook.

Maddison, A. (2005) 'World Development and Outlook 1820–2030: Evidence submitted to the Select Committee on Economic Affairs, House of Lords, London, for the inquiry into "Aspects of the Economics of Climate Change"'; available at http://www.ggdc.net (accessed 10 December 2010).

Maddison, A. (2009) 'Statistics on World Population, GDP and Per Capita GDP, 1–2008 AD'; available at http://www.ggdc.net/maddison (accessed 29 September 2010).

Ni, C. (2009) 'China Energy Primer' [online], Berkeley, CA, Lawrence Berkeley National Laboratory, http://china.lbl.gov/research/china-energy-databook/china-energy-primer (accessed 1 October 2010).

Romer, R.H. (1976) *Energy: An Introduction to Physics*, San Francisco, W. H. Freeman and Company.

TERI (2009) *TERI Energy Data Directory and Yearbook (TEDDY)*, New Delhi, The Energy and Resources Institute.

UN (2009) *World Population Prospects: The 2008 Revision: Highlights*, New York, Population Division of the Department of Economic and Social Affairs of the United Nations Secretariat; available at http://un.org/esa/population/publications/wpp2008 (accessed 20 August 2010).

WWEA (2009) *World Wind Energy Report 2009*, World Wind Energy Association; available at http://www.wwindea.org (accessed 3 September 2010).

Chapter 3

What do we use energy for?

By Bob Everett and Janet Ramage

3.1 Introduction

The previous chapter has described the flows of large amounts of primary energy, yet what consumers really require are **energy services**, such as warm homes, cooked food, illumination, mobility and manufactured articles. The first part of this chapter takes a tour of all the different ways of using energy and notes how they have changed over the centuries. The second part looks at how these different categories are dealt with in UK national statistics and looks at some key differences in uses between sample countries.

3.2 Primary, delivered and useful energy

Converting the forms of primary energy described in the previous chapter into the energy services listed above may require a long chain of activities. We need to ask some key questions:

How much energy is 'lost' between our primary source and its final use?

Exactly what is the 'energy' in our utility bills?

What is the really essential 'useful' energy?

Lighting is a very good example. If you want to read at night you need reasonable illumination. You could light a candle and produce light directly from fuel (candle wax). Candles are not very bright, of course, and since a candle converts only about one ten-thousandth (0.01%) of its fuel into light, it is not a very efficient way of doing things. You are more likely to turn on an electric light (see Figure 3.1).

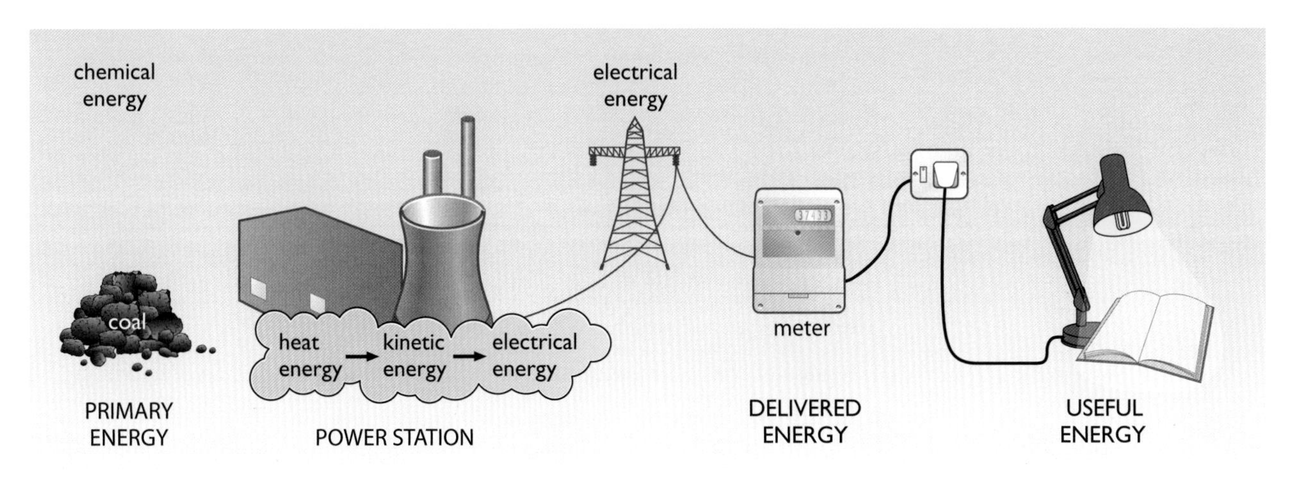

Figure 3.1 Primary energy, delivered energy and useful energy

What happens when you do this is a succession of energy conversion and transfer processes, with energy being lost as waste heat at every stage. At the input stage is the primary energy source, which might be coal, oil, gas or uranium in its naturally occurring state in the ground or equally, wood or solar heat.

Taking coal as the energy source in the example in Figure 3.1, by the time it reaches the power station typically the equivalent of about 2.5% of its available energy has already been used in mining and transporting it. Similar losses apply to oil, which may have been transported thousands of miles by sea and will also have been refined, and to natural gas, which may have been pumped to western Europe through pipelines stretching from distant Siberia.

In the power station the chemical energy of the coal is converted to heat, then into the kinetic energy of the steam turbines and finally into electrical energy. As discussed in Chapter 2, only about 30–40% of the energy of the fuel emerges as electricity. Some of this is likely to be lost as heat in the transformers and wires of the electricity distribution system, before it reaches the consumer's electricity meter. At this point it becomes **delivered energy**. This is what the consumer actually receives and is billed for – about a third of the primary energy extracted at the coal mine.

The wastage does not stop there. The electrical energy has to be converted into light of a suitable combination of wavelengths. This **useful energy** must then be directed onto the pages of the book.

If you were determined to be energy efficient, you could read by the orange light of a sodium street lamp. These can convert nearly 30% of the electrical energy into visible light. But you are more likely to choose the more acceptable white light of a compact fluorescent lamp. This, however, will produce only one watt of visible light for every 10 watts of electric power, the rest being lost as heat. So, overall, for every gigajoule of primary energy that left the coal mine, only about 32 MJ have been converted into light (see Box 3.1); the other 968 MJ have become waste heat. This may sound appallingly wasteful, but is still over 100 times more efficient than using a candle!

BOX 3.1 Conversion efficiency

What is the efficiency of the complete coal-to-light conversion process described in the text?

The **conversion efficiency**, often simply called the **efficiency**, of any energy conversion system is defined as the *useful energy output divided by the total energy input*. In practice it is very common to express this as a percentage of the input:

$$\text{percentage efficiency} = \frac{\text{energy output}}{\text{energy input}} \times 100$$

We'll consider the fate of 1 GJ of primary energy in the form of coal.

Good-quality, hard coal has a typical energy content of 28 GJ per tonne.

The mass of coal containing 1 GJ = $\frac{1}{28}$ tonne = 35.7 kg (i.e. a large sackful),

If 2.5% of this energy is used in mining and transporting the coal, the net energy entering the power station is only 97.5% of 1 GJ:

energy entering power station = 1 × 0.975 = 0.975 GJ

If we take the fuel-to-electricity efficiency of a modern coal-fired power station (commonly referred to as the thermal efficiency) to be 35%:

electrical energy leaving the power station = $0.35 \times 0.975 = 0.341$ GJ

On average, about 7% of this will be lost as heat in transmission in the wires and transformers so the consumer receives only 93%:

delivered electrical energy = $0.93 \times 0.341 = 0.317$ GJ

But a compact fluorescent lamp only turns 10% of this into light:

useful light output = $0.1 \times 0.317 = 0.0317$ GJ, an overall energy efficiency of **3.17%**

This theme of increasing efficiency of energy use is a recurring one. Although it may seem that we are currently living in a 'gas-guzzling' society, the truth is that we would probably be guzzling even more if it hadn't been for many significant improvements in energy efficiency, particularly over the last 150 years.

3.3 The expanding uses of energy

As shown in the previous chapters, the world use of energy has risen enormously over the past two centuries. This has been the result of a growing world population multiplied by increasing energy use per capita (per head of population). For those who live in an industrialized world of cheap, readily available energy, it can be difficult to imagine what life would be like without it. It is worth, therefore, reflecting on exactly what society uses energy for and what changes in these uses have taken place over the broad timescale.

There are many ways of classifying these uses. Modern statisticians like to categorize them by the sectors of the economy; *domestic* (sometimes referred to as *residential*), *industrial, services* (or *commercial*) and *transport*. In some countries agriculture is treated as a separate sector, though in the UK it is statistically bundled with services. We will look at these sectors and the detailed changes in their energy use over the last 40 years will be looked at later in this chapter. Vaclav Smil (Smil, 1994) has made estimates of energy use per capita for different societies in history. He has categorized the uses according to food, household, industry, transport and services, although in practice, as in any other classification system the boundaries may blur into each other. Figure 3.2 shows figures for the per capita energy consumption of the UK in 2000, together with Smil's estimates for a number of past societies:

- Europe in 10 000 BC: a stone-age society of hunter-gatherers living in wooden huts.
- Egypt in 1500 BC: a bronze-age culture with settled agriculture, organized irrigation and enough spare time to build the Pyramids.
- China in 100 BC: the Han Dynasty, another agricultural society, also with organized irrigation, but which had mastered the art of making cast iron.
- Europe in AD 1300: an advanced agricultural society, using wrought iron smelted with charcoal, trading coal by sea and capable of building

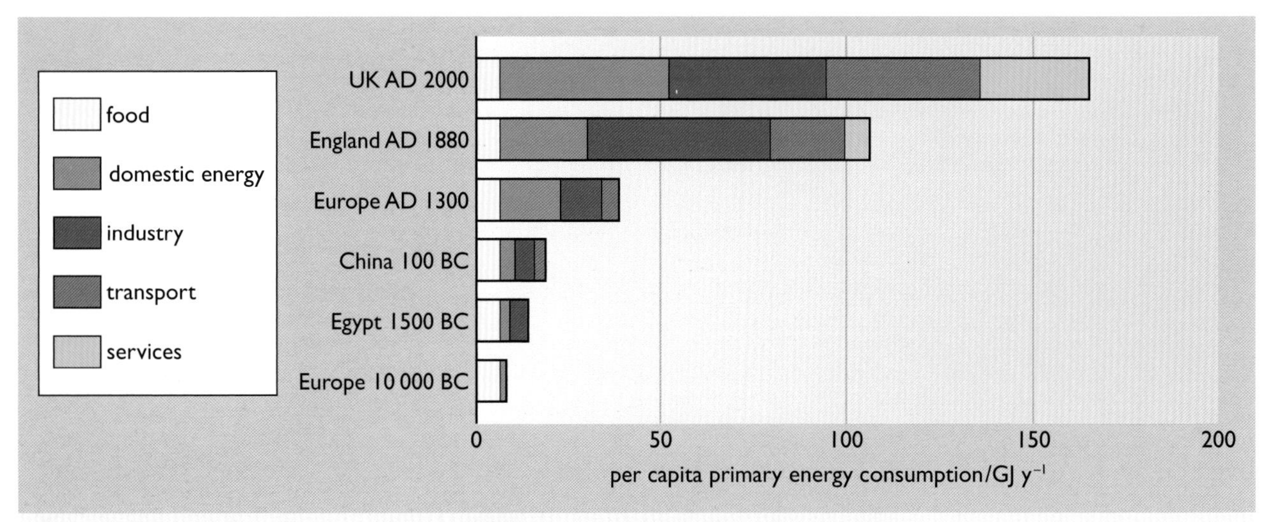

Figure 3.2 Per capita energy consumption for different societies (adapted from Smil, 1994)

large cathedrals. This is the world of Chaucer's *Canterbury Tales*. The UK population was then about 6 million.

- England in AD 1880: Victorian industrial society, fuelled by coal and driven by steam, criss-crossed with new railways, exporting enormous amounts of coal to the world, busy inventing a host of new chemical and industrial processes, and just about to embark on exploiting mains electricity. The UK population was then about 30 million, about half that in 2010.

Each society consumes more energy than the previous one, yet it is not a matter of using more energy for every category of use, rather that new uses arise and expand. Moreover, they are likely to arise in one country, such as agriculture in Egypt or industrial society in Victorian Britain, and then spread across the world. It is worth looking at each category in more detail.

Food

This is the first and ultimately most important category in Figure 3.2. The actual edible food energy or dietary energy needed by a human being is about 10 MJ per day, about 3.6 GJ (or 2400 kcal) per year. Obviously if you are involved in hard physical labour, then you will need more than this. This figure does not normally appear in energy statistics, though the energy to grow, harvest, package, transport and cook food does.

For prehistoric European humans in 10 000 BC, life was a matter of hunting wild animals and gathering naturally growing food plants. The energy used to do this was the effort made by each individual. If a social group could gather enough food energy to be able to continue to survive and reproduce, it prospered; if it couldn't it would die out. In addition to this, the energy of wood was used to cook the food and to heat the relatively primitive homes (wood huts or caves).

The first agricultural societies grew up around 3000 BC in Mesopotamia, planting crops and using domesticated animals such as oxen to carry out the hard work of ploughing and raising water for irrigation. These skills slowly spread to surrounding countries. By 1500 BC, the land of the Nile Valley of ancient Egypt was intensively cultivated, involving ploughing and irrigation. The farming base of society was sufficiently productive to generate a surplus of food for the ruling elite and (amongst others) the army of Pyramid builders.

The Han Dynasty Chinese of 100 BC had a similar intensive agriculture, but had improved the productivity of their ploughing with mass-produced cast-iron ploughshares. The country's extensive canal system allowed the trade of agricultural produce.

Figure 3.3 A traditional-style Greek irrigation windmill at work in Crete

Stepping forward to Europe in AD 1300, we come to another agricultural society, again slightly more advanced, with oxen being replaced by stronger horses and more efficient designs of ploughs. The vast bulk of the population lived on, and worked, the land. They existed in a **subsistence economy**, eating what they grew, and trading what little surplus was left over after the feudal landowners had taken their cut. The basic methods that they used had changed little for over a thousand years, though the quality of tools and ability to harness and use draught animals had improved. By this time, water power had also been harnessed for irrigation and grinding corn and the earliest windmills were starting to become common (Figure 3.3).

By the 1880s English society was radically different. Agriculture was embarking on changes every bit as drastic as those taking place at the same time in industry. The bulk of the English population was living in rapidly growing manufacturing cities, swollen by a continuing drift of manpower from the countryside. The new job opportunities in urban factories, coupled with competition from imported food brought by ship and railway, led to a continuing decline in rural population. Yet all of these people had to be fed. Farming efficiency had been improved in the 18th century by introducing larger and more powerful breeds of horse, and the 19th century saw the introduction of artificial fertilizers and farm mechanization. The idea of modern farms as 'agribusiness' presiding over 'corn deserts' provokes angry debate today, but it is the end product of a set of trends set in motion over a hundred years ago.

Fertilizers

Eighteenth and nineteenth century chemists had discovered that sun and rain were not the only ingredients required to grow crops. Plants also needed phosphorus, potassium and nitrogen in suitable forms.

From the 1840s onwards various deposits of rocks were quarried which could supply the modest amounts of phosphorus and potassium required. Canals and then railways made the transport of these bulk fertilizers practicable, even into rural areas. But supplying large quantities of nitrogen remained a problem. The main supplies in Europe during the late 19th century were natural sodium nitrate (saltpetre) imported by sea from Chile, and ammonium sulfate – a by-product of town gas production from coal (see Chapter 5).

In 1914 the First World War created a nitrate supply crisis for Germany. Its chemical industry came to the rescue with the Haber–Bosch process, manufacturing ammonia (NH_3) from nitrogen in the air and hydrogen from town gas. Initially this was needed for the manufacture of explosives for the war, but afterwards it was seen that the process had a genuine 'swords into ploughshares' application by providing nitrogenous fertilizer. Initially the process was extremely energy intensive. In 1930 an ammonia plant would have consumed the energy equivalent of nine tonnes of oil to make one tonne of ammonia. Since then there have been extensive improvements. A modern plant (2000) now requires only 26 GJ, about 0.6 tonnes of oil equivalent per tonne of ammonia (Smil, 2004). Nevertheless, the energy involved in fertilizer manufacture remains a major input in modern intensive agriculture and about 1.2% of global primary energy use is devoted to it (IFA, 2008).

The immediate rewards of using nitrate fertilizer were increased yields with existing strains of crops. In 1300 typical English yields of wheat were less than one tonne per hectare of land. By 1880 they had been more than doubled. Yet even these yields have been increased by the breeding of special strains of crops capable of making use of very high levels of nitrogen in the soil. UK wheat production reached yields of 4 tonnes per hectare in the 1970s and the figure now stands at almost 8 tonnes per hectare.

This massive use of synthetic nitrogen fertilizer is now worldwide, and it has been estimated that at least one-third of the protein in the current global food supply is derived from the Haber–Bosch process (Smil, 1994). Two of the disadvantages of this high nitrogen input have been the **eutrophication** of watercourses – a form of poisoning resulting from high nitrate levels, and the emission of nitrous oxide – a greenhouse gas. The latter will be discussed in Chapter 13.

Looked at one way, this extra fossil fuel energy input is simply an aid to improving the efficiency of take-up of solar radiation into useful farm produce. A more pessimistic viewpoint would see this as 'a sad hoax, for industrial man no longer eats potatoes made from solar energy, now he eats potatoes partly made of oil' (Odum, 1971). The question of whether or not an agricultural process can produce more energy in the crop than the fossil fuel inputs taken to grow and process it is an important one, particularly for the development of biofuels such as ethanol and bio-diesel.

Farm mechanization

While Victorian industry ran on coal, agriculture ran on horses and human muscle power and this continued into the 20th century. By 1901 Great Britain had 3.5 million horses, of which 1.1 million were employed on farms. About 30% of lowland farm area was devoted to their keep, a third of this just for farm horses. This horse population was matched almost one to one with about a million full-time farm workers plus an army of casual labour, mainly from the cities, at harvest time. Mechanical assistance was limited to a few low-powered steam engines used for threshing and grinding. Although horse-drawn reaping machines had been introduced in the UK in the early years of the 19th century their use was limited by the small size of fields and narrow lanes. In the USA, space was not a problem. By the end of the 19th century fully automatic horse-drawn combine harvesters were in use. The largest needed 40 horses to operate but could harvest a hectare of wheat in 40 minutes.

Farm mechanization in the UK was very slow to develop. By 1920 there were only 10 000 farm tractors in the country. They were outnumbered nearly 100:1 by horses. There was little incentive for mechanization. During the 1920s and 1930s there was high unemployment and labour was cheap. British agriculture was in a very depressed state, competing with cheap imported food.

However, in 1939 all that changed. World War II brought a serious food blockade and food rationing. Meat became a tightly rationed luxury, and land that had been pasture for livestock was ploughed up to grow more vegetables for direct human consumption. The war drained manpower from the land, and productivity could only increase by more mechanization.

After the War, British society had changed. Post-war reconstruction required manpower in factories. Labour prices rose compared with the price of farmland and mass-production techniques had reduced the price of farm machinery. It no longer made economic sense to employ armies of agricultural labourers when a few men and machines could do the same job (Figure 3.4). A modern tractor can plough a field ten times faster than its horse-drawn equivalent. Even so, it is perhaps surprising that it was not until the 1950s that the number of tractors in Britain exceeded the number of farm horses.

Again, looked at one way, the mechanization of farms has meant an injection of fossil fuel energy. On the other hand, removing the horse has freed all their grazing land for further agricultural production. Smil (1994) estimated that feeding America's farm horses required 25% of their cultivated land. One key advantage of the tractor is that you can switch it off. Once the ploughing or harvesting is done it can be put away in a barn for next year. A horse keeps eating, whether or not you use it. Another 'benefit' is that the large workforce is no longer needed, theoretically freeing it for other uses. This is fine if there is alternative employment, but in many cases it has just continued the trend of the depopulation of the countryside and growth of cities.

Figure 3.4 A combine harvester rapidly chews its way through an Essex wheat field

The food shortages of World War II have left a deep mark on UK (and EU) agricultural policies. Extensive subsidies to promote production have led to high levels of energy and fertilizer use. At the time of writing (2010) the UK is self-sufficient in a wide range of basic foodstuffs. The energy price paid for this amounts to a modest 1 GJ per capita per year for the fuel energy to run UK farms and about the same again for manufactured fertilizers.

Domestic energy

Compared with the modern centrally heated, electrically lit society, the past was cold in winter, draughty, smoky, dirty and very dark after sunset. For prehistoric humans, as already mentioned, home was likely to be a cave or a wooden hut. Later societies developed their building skills with bricks and mortar and metal woodworking tools to produce the house as we know it today. Yet central to the home was the *fire* for cooking, heating and lighting.

Heating, washing and cooking

The Roman elite may have had central heating, but most of their expertise disappeared in the Dark Ages. In medieval Britain, the use of fire was a rather unruly and dangerous affair (especially so in a wooden house) and must have been extremely energy inefficient. According to Bowyer (1973), there appears to be no evidence of chimneys in England before the 12th century. Buildings simply had permanently open smoke holes in the roof. The indoor air quality must have left a lot to be desired and cooking must have been a serious fire hazard. The Great Fire of London in 1666, which destroyed over 13 000 houses, started in a pie shop in Pudding Lane. As with other, later, energy-related catastrophes, it led to a series of Acts of Parliament, specifying how chimneys were to be built, at least to be safe, if not necessarily efficient.

The modern fireplace is a product of careful design, originating in the 16th century. The late 19th century fireplace shown in Figure 3.5 is also a masterpiece of cast-iron construction, providing **space heating** (that is heating of the living spaces) as well as an oven for baking and a place to boil kettles and saucepans. **Domestic hot water** for washing had to be produced in this manner – today we expect to have hot water on tap, probably heated by a gas boiler, and are likely to use electric kettles for making tea or coffee.

In rural Britain, wood would have been the normal fuel used in such a fireplace, but by the 17th century, coal was widely used in cities (see Chapter 5). The hearth was the centre of the home, and especially after dark, not least because of the expense of lighting any other room (see below). This fire effectively needed to be kept burning all year round for cooking and water heating. In summer it would make the living room too hot, and in winter it would still be too cold. The joys of warm rooms, and running hot water, although technically feasible, passed most of Britain by until the latter part of the 20th century. This was stoically accepted as part of life. The Reverend Francis Kilvert, a Herefordshire clergyman, describes getting up on Christmas Day, 1871:

Figure 3.5 The coal fire was the heart of the home for heating and cooking – a reconstruction in the Beamish Museum, Co. Durham

> It was an intense frost. I sat down in my bath upon a sheet of thick ice which broke in the middle into large pieces whilst sharp points and jagged edges stuck all round the sides of the tub [...], not particularly comforting to the naked thighs and loins, for the keen ice cut like broken glass. The ice water stung and scorched like fire. I had to collect the floating pieces of ice and pile them on a chair before I could use the sponge and then I had to thaw the sponge in my hands for it was a mass of ice.
>
> (quoted in Plomer, 1977)

The open coal fire, with a thermal efficiency of about 25%, remained the normal mode of heating in UK homes until the 1960s, much to the derision of visitors from the Continent and USA where central heating was considered normal. Central heating only reached the bulk of UK homes after the introduction of North Sea natural gas in the late 1970s.

The introduction of town gas made from coal in the early 19th century was initially for lighting. It soon became apparent that it was an ideal controllable fuel for cooking. Unlike a coal stove, a gas cooker could be turned on and off quickly. Cooking, with all its attendant smells and mess, could be carried out in the kitchen, instead of spilling over into the living room with its coal fire devoted to space heating. This in turn influenced house design, making the kitchen and living/dining rooms separate entities.

The Victorian housewife's life was not easy. Even middle-class families would aspire to have a female domestic servant (or 'skivvy') whose life was even harder. Before the advent of modern detergents and washing machines, cleaning clothes was mostly a matter of hard physical labour, scrubbing them in hot water with soap (see Figure 3.6). Before the electric vacuum cleaner, house cleaning was an endless inefficient chore of chasing after dust and grit with a dustpan and brush.

Figure 3.6 Hard female labour – washday in 1924, before washing machines and modern detergents, meant hard scrubbing with hot water and soap

The First World War in 1914 was a turning point. Millions of men were drafted into the armed forces, creating a severe labour shortage. Women left domestic service to work in the munitions factories, where they enjoyed considerable social equality with men. After the war they did not want to return to the old ways as domestic drudges. There was a 'servant shortage'. The market was ripe for selling 'labour-saving appliances', vacuum cleaners, washing machines, electric irons, etc., a process which has continued to the present day.

Preserving and processing food

Today in the UK, 'food' is something that we buy from a supermarket, usually wrapped in plastic. We are likely to deposit it immediately into our domestic refrigerator and, at mealtimes, transfer it rapidly to the plate via a microwave oven. At the extremes it has become an industrial product, processed and packaged by machines and transported vast distances.

In the past, preserving and processing food largely took place in the home. Before about AD 1000, even basic tasks such as the grinding of corn would have been done by hand in the home with a small hand grindstone. The introduction of water-, wind- and horse-powered mills turned this from a domestic activity to a (local) industrial one. Also, given the fire risks in a medieval home, baking was a task that was safest to leave to a professionally run bakehouse.

Dealing with meat was a particular problem. Animals could either be slaughtered and preserved with salt or kept alive for as long as possible before being served up. In the countryside this didn't pose too much of a problem. In the expanding cities it created chaos. Even up to the mid-nineteenth century herds of cattle were driven on the hoof through the streets to city-centre markets to be sold and then slaughtered. This was a road traffic nightmare, but also meant that a host of food-rendering industries naturally grew up, also in the cities. All these required energy for **process heat**. Animal fats were rendered down to make, amongst other things, tallow for lighting fuel, and bones boiled to make glue. One of the smelliest processes was considered to be the boiling down of blood to make fertilizers.

The arrival of railways and cold storage by the end of the 19th century was a godsend. Cattle markets, the new cold stores and all the smelly processing industries were removed to their current locations in industrial estates well away from the noses of the more well-to-do householders. This freed up city centres for the services sector, in particular banking and shopping.

Although 'machine' refrigeration as we know it today is a late 19th century invention, it did not really take off until the early 20th century because bulk ice was a globally traded commodity. In 1900 the UK imported half a million tonnes of ice from Norway and even some from the USA (Weightman, 2001). The food distribution chain relied on blocks of ice bought along with the food. As refrigeration technology improved, and especially as electricity became cheaper through the 20th century, commercial ice-making plants replaced imported ice. Finally, towards the end of the century, the mass-produced commercial and domestic refrigerator arrived, killing off the ice trade, but contributing to the increase in domestic electricity consumption (see below).

Figure 3.7 A naked gas flame is an extremely inefficient light source

Lighting

The past was very dark because artificial lighting was extremely expensive. This in turn was due to the inefficiency of lighting based on the naked flame. This involved the combustion of a fuel, be it candle wax, lamp oil, or coal gas, so that unburned particles of soot in the flame glowed and gave off a tolerable light (Figure 3.7).

Perceived light intensity is measured in **lumens**. A modern twenty-watt compact fluorescent lamp emits about 1400 lumens, whereas a wax candle only emits about 13 lumens. A lumen is actually a unit of lighting power, but one which uses the human eye as a meter. 'What is the brightness of a light?' may sound a vague question but by careful research it has been pinned down to an 'average' response of a large number of individuals. From this it has been possible to lay down standards of acceptable illumination.

A reasonable modern standard for desk lighting is about 300 lumens per square metre, which would translate into 23 candles per square metre of desk surface! You may like to try reading this book by the light of a single candle to appreciate the problem – in the past you may not have been able to afford more than this. For example a budget study of a 1760s Berkshire family estimated that they would have spent 1% of their annual income to get a mere 28 000 lumen-hours of light (Nordhaus, 1997). That is equivalent to two candles (and probably rather smelly tallow ones) for three hours per day!

Eighteenth and nineteenth century developments of oil and kerosene lamps managed to boost the output of a single lamp to 200 lumens, but this was largely done by increasing the fuel throughput rather than the efficiency.

It did not significantly improve the affordability of artificial light; nor initially did the introduction of lighting by town gas made from coal, which was priced to be competitive with oil and kerosene.

At the end of the 19th century a new technology arrived: the light bulb, the electric incandescent lamp powered by mains electricity. Gas lighting fought back with the Welsbach gas mantle which improved its efficiency by a factor of four (see Chapter 5). In towns and cities these two technologies swept away candles and oil lamps. Gas lights managed to compete with electricity well into the 20th century. Since then there has been a continuous development both in new light sources and in the efficiency of electricity generation used to power electric lamps (topics that will be described further in Chapter 9). The effectiveness of lights or lighting systems is expressed in terms of their **efficacy** (or 'usefulness') in lumens per watt. Table 3.1 charts the progress over the years.

Table 3.1 Efficacies of different lighting sources

Lighting source	Approximate efficacy /lumens watt^{-1}
Early 19th century tallow candle	0.08
Gas mantle circa 1916	0.9
Incandescent electric lamp circa 1990[1]	5
'Low energy' compact fluorescent lamp[1]	25
Low pressure sodium street lamp[1]	60

[1] Per watt of primary energy, taking electricity generation efficiency as 33%.
Sources: Nordhaus, 1997 and manufacturers' data.

It is quite remarkable that the complex process of choosing to burn a litre of kerosene in an engine to drive a generator and so power a fluorescent lamp can produce over a hundred times more useful light than burning the same amount in an oil lamp. The overall result is that the real cost of useful light has fallen dramatically. The 1760s family mentioned above spent 1% of their income to obtain 28 000 lumen-hours of light a year. Since then it is estimated that the real cost of a lumen-hour of artificial light has fallen by a factor of 7 000 (Fouquet, 2008). A modern home is likely to use 10 million lumen-hours per year. Today, nearly 20% of global electricity generation is used for lighting.

Even so, it is worth reflecting that 1.4 billion people in developing countries today do not have electric light (IEA, 2010) and, in this respect, are effectively still technologically in the early 20th century.

Industry

Physical labour

The word 'industry' normally conjures up pictures of smoking chimneys. However, it should not be forgotten that industry also includes a vast range of physical activities, such as digging, sawing, polishing, grinding, which are now carried out by machines.

The scene from the French workshop in Figure 3.8 is dominated by the labourers who are providing the physical work. They are likely to have kept this up for a whole working day, six days a week. Today their function would have been replaced by an 800 watt electric motor (about the rating of a heavy-duty DIY electric drill). Some tasks, such as endlessly sawing tree trunks into planks or polishing flat sheets of glass by hand must have been mind-numbingly dull as well as physically exhausting.

Figure 3.8 Eight men rotating a vertical capstan in a mid-eighteenth century French workshop. The rope is pulling gold wire through a die (source: Diderot and D'Alembert, 1769–72).

The development of textile mills in Lancashire in the 18th century required mechanical energy to drive the spinning machines. Power from water mills competed with horses running in treadmills and the developing steam engines (see Chapter 6). By the 1840s it was clear that steam power was the way ahead and by the 1880s it was the normal source of motive power in UK factories.

But steam engines were heavy and difficult to move. They were ideal for ships, where weight was not a problem, but for motion on land they needed carefully laid rails with no steep gradients. By 1880, the UK was criss-crossed with railways on viaducts and cuttings, but all of this civil engineering work was done by the hand labour of a large army of 'navvies' (or navigators) and thousands of horses. The steam shovel, the bulldozer and the JCB digger are largely 20th century developments.

Neither was the steam engine very suitable for small mechanical tasks. The domestic sewing machine, for example, was traditionally a hand- or treadle-powered device. It had to wait until the mid-twentieth century for the development of small, cheap, electric motors and the availability of cheap electricity to drive it. Now, it seems that almost everything has an electric motor in it. Even the average DIY tool box is now likely to contain an electric drill, an electric saw and an electric screwdriver. All of this has contributed to the rise in electricity demand (see below).

Iron and steel manufacture

The Europeans of 10 000 BC lived in the Stone Age. The Ancient Egyptians were a Bronze Age civilization and we (since about 1000 BC) live in the Iron Age. Extracting metals such as iron from their ores is not easy. It needs a fuel which is both capable of burning to produce a high temperature and of chemically reacting with iron oxide to produce metallic iron. Traditionally this high-quality fuel was charcoal – almost pure carbon made from wood. The charcoal-making process was not very efficient. It took about ten tonnes of wood to make one tonne of charcoal. In the UK the wood supplies held out until the 17th century, when the switch to coal took place. Even then, the coal usually needed to be refined to coke to obtain the high temperatures required.

Most early furnaces could not make pure iron. They produced a lumpy mixture of iron and slag which then had to be physically beaten out with hammers to produce 'wrought iron'. Thus 18th century iron foundries grew up where there were three resources – iron ore, coal, and water power to drive the bellows for the furnaces and the hammers. At that time it took ten tonnes of coke to produce one tonne of iron. Since then there have been continuous improvements in the processes, beginning with the introduction of the blast furnace in the 1830s with steam engines to drive the bellows.

Wrought iron is not suitable for making tough machine components. For that you need steel, which is made from iron by carefully burning off the impurities in high-temperature blast furnaces. In Victorian Britain iron and steel production increased massively, swelling the energy needs of the industrial sector. By 1903 it was using nearly 30 million tonnes of coal a year, over one-sixth of the country's total consumption. By this time it was possible to produce a tonne of iron with under two tonnes of coke. Today, a century on, this figure has dropped to less than one tonne of coke per tonne of iron. However, in terms of output, the UK iron and steel industry has faded to a shadow of its former glory as production has moved to countries such as India and China.

Aluminium smelting

Victorian steel was fine for steam engines, but not much use for 20th century aeroplanes, which required something much lighter. Aluminium as an element was isolated in 1824 but proved extremely difficult to purify chemically. It was found that it could be extracted from aluminium oxide using electrolysis, a process in which electricity is used to drive a chemical reaction. Mass production had to wait for the availability of cheap electricity. In the 1880s smelting one tonne of aluminium metal required more than 50 000 kilowatt-hours of electricity. By the end of the 20th century, this figure had been reduced by more than two-thirds. Even so, the price of aluminium is largely determined by the cost of the electricity used to extract it. The '**embodied energy**' (Box 3.2) of aluminium is amongst the highest of common household materials available today, which is why recycling of aluminium cans is an important energy-saving activity.

BOX 3.2 **What is embodied energy?**

Many manufactured materials require a large amount of energy to produce them. This is called their 'embodied energy'. It can be an important factor in choosing materials or even whole processes. Table 3.2 lists some example values.

Table 3.2 Embodied energy of some materials

Material	Embodied energy /MJ kg^{-1}	Production process
Aluminium	227–342	Metal from aluminium ore
Cement	5–9	From raw materials
Copper	60–125	Metal from copper ore
Plastics	60–120	From crude oil
Glass	18–35	From sand and other materials
Iron	20–25	From iron ore
Bricks	2–5	Baked from clay
Paper	25–50	From standing timber

Source: Smil, 1994.

Processes that need high temperature heat

There are many other industries with a long history that need large amounts of high temperature heat. Brick making, for example, is simple enough, but requires enough energy to drive all the water out of what are essentially just lumps of mud. The other ingredients of traditional building practice are lime mortar and cement. The essential ingredient of lime mortar is quicklime or calcium oxide, made by heating chalk to drive off the carbon dioxide content. Cement is slightly more complex and is made by heating silicate clays. The Romans used cement to build the 43-metre domed concrete roof of the Pantheon in Rome in AD 120, but the secret of its manufacture disappeared and was only rediscovered in the 19th century. Its first major use in the UK was in the construction of the London sewers in the 1860s.

Glass making is another energy-intensive activity that has been practised for over 4000 years. It requires sufficient heat to melt sand. Traditionally this required a high-quality fuel such as charcoal, or later coke, and a lot of hard work with the bellows to fan the flames. Since the whole point of glass is that it should be transparent, it is vital that the soot and grit from the fuel is kept out of the mixture. The ideal modern solution is to use natural gas or electricity to provide the heat.

In addition there are many other chemical processes that need process heat. One of the most important introduced in the early years of the 19th century was the manufacture of soaps and detergents, essential for the booming wool and cotton industries. Soap manufacture needed sodium hydroxide.

This was produced by the Leblanc process from common salt (sodium chloride), limestone, coke and sulfuric acid, using coal for process heat. Ironically, although the end product was supposed to be clean white textiles, this process was notorious for its air pollution, emitting large quantities of hydrochloric acid vapour. It was this problem that led to the formation of the Alkali Inspectorate and some of the first UK legislation on industrial air pollution. Today, sodium hydroxide is made using electricity to separate the sodium and chlorine of common salt. Like aluminium smelting this is an industry that depends for its profits on cheap electricity.

Transport

The modern growth in transport energy use is commented on at the end of this chapter, but it is worth pointing out that this growth has only been possible because we are 'free' to travel. In many past societies (and some modern ones!) this was not an option. In the UK in AD 1300 feudal lords 'owned' the peasants as well as the farmland. They were every bit as tied to the land as the animals. Most people never travelled more than a day's walking distance from their homes in their whole lives. They were also limited by the atrocious state of the roads. The characters in Chaucer's *Canterbury Tales* were the lucky few – a pilgrimage to see Canterbury Cathedral was the closest they would get to a 'holiday'. The energy inputs to the transport sector were simple enough: wind power for sailing ships and copious quantities of hay and oats for horses.

By the 1880s society had changed entirely. People were, theoretically at least, free to travel as they wished. Although road travel had improved during the 18th century, railways swept aside the expensive stage coach competition for long-distance land travel. Steam ships were transporting freight not just on short coastal routes, but also on regular long-distance ocean-going routes. However, in the cities the horse bus and horse tram ruled the road. It was these and competing cheap railway fares that ushered in modern concepts of 'commuting' and 'suburbia'; it was no longer necessary to live next to your workplace. It was also permissible for the urban work force to have 'summer holidays' and they had the spare cash to afford them. The railway and steamship companies were only too happy to provide the transport arrangements to new seaside resorts. The transport energy sources were now still hay and oats for horses, but also enormous amounts of coal: by 1903 UK railways were using 13 million tonnes of coal a year and coastal shipping another 2 million tonnes. Photographs of the main roads at this time show them to be strangely empty by modern standards. Almost everything went by train and continued to do so well into the 20th century.

However, the urban horse did not last into the new century. In cities the electric tram and the petrol-engined bus had almost completely substituted for their horse equivalents by the First World War in 1914. This freed large areas of hayfields around cities, which had provided the transport energy supply, for yet more suburbia and yet more commuting.

In the USA, the motor car took off in 1907 with the famous Model 'T' Ford. Although Adolph Hitler introduced the autobahn and his 'people's car', the Volkswagen, to Germany in the 1930s, the real explosion in car ownership in Europe did not come until after the Second World War. In the UK this was marked by the massive programme of motorway building in the 1960s and 70s. The number of passenger-kilometres travelled by car in Britain rose by a factor of 12 between 1952 and 2008 (DfT, 2009) (see Figure 1.25). Although the growth in energy use for UK land-based transport flattened out in the late 1990s, that for international air transport has risen further, encouraged by cheap fares.

Services

If you live in a subsistence farming economy, as most people did until the 14th century, your lifestyle is very limited and most of the food and goods that you need are supplied locally. There is not much need for trade, distribution or, for that matter, money. By the 19th century, 'England was a nation of shopkeepers' as Napoleon put it. Farm produce had to be sold in cities and manufactured goods made in cities were traded worldwide. Distribution companies, banks and insurance companies became every bit as important as manufacturing industry, employing more and more clerks and increasing volumes of paper. In 1880s Britain, the 'services' or 'commercial' sector was limited but growing and it has continued to grow to the present day. Food and goods are increasingly made by machines, while the sales, distribution and surrounding financial activities are carried out by people in offices and shops.

The services sector includes almost all activities that aren't in the others. It includes all office activities in commercial offices and in public administration. It also includes education (and the writing of books such as this one).

'Services' also include newspapers and mass entertainment. In AD 1300, there were no newspapers (Thomas Caxton only started printing books in the 1470s) and mass entertainment was limited to travelling storytellers. By 1880, the UK had high levels of literacy and a thriving newspaper and book publishing industry, all of which in turn depended on high-volume paper manufacture, itself quite an energy-intensive industry. The development of radio and television in the 20th century has been a major spur to the spread of mains electricity. It is not that the receivers actually require large amounts of energy, it is rather that the alternatives of battery power have always proved expensive and inconvenient.

Telecommunications is another important area in the services sector. By the 1880s the electric telegraph had been developed and criss-crossed Europe. There were even transatlantic cables. The telephone as we know it had just been invented.

Increasingly, society is dependent on various electric and electronic devices for control and regulation. If the traffic lights fail in a modern city, then the result is gridlock and chaos. But these are really only 'automatic traffic cops' substituting energy for physical effort. The early gas-lit version from the 1860s (Figure 3.9(a)) still needed to be rotated manually. While Victorian railways could physically propel 500-tonne trains at 100 kilometres per hour, they could only do so *safely* by the use of signalling, and this only became effective after the introduction of the electric telegraph in the 1850s. Similarly, modern airports can only function through the extensive use of radar and air traffic control. Given that in these examples so much can be achieved with relatively small amounts of energy, it is perhaps amazing that we waste so much very crudely in other applications.

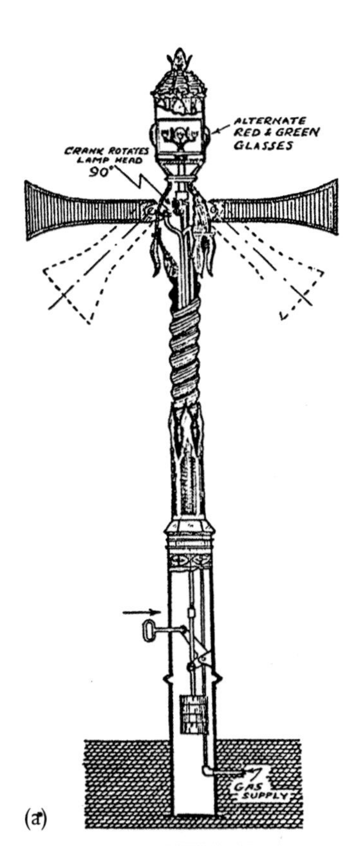

(a)

(b)

Figure 3.9 (a) This early gas-lit traffic light was installed outside the Houses of Parliament in London. The red and green lantern was rotated manually, allowing it to control both traffic and pedestrians; (b) Its modern automatic radio controlled LED equivalent.

3.4 UK energy uses today

At the time of writing (2010):

- Total UK annual primary energy consumption had risen to almost 9000 PJ, equivalent to over 200 million tonnes of oil.
- The population had doubled from its 1880 figure to almost 62 million.
- Per capita annual primary energy use had thus risen to nearly 145 GJ or almost 3.5 tonnes of oil equivalent.

Whilst in the past the UK depended almost entirely on coal, 'diversity of supply' including oil, gas, nuclear power and renewable energy is now a key aim of energy policy.

The previous chapter has described the range of different primary energy sources available, but where is all this energy used? The picture is potentially quite confusing, since there are many transformations that take place within the energy system, notably in the refining of oil and the generation of electricity. Since physics tells us that energy is conserved, the total annual energy consumption of a nation, a factory or a household

BOX 3.3 UK energy balance, 2009 – primary energy production and consumption

Table 3.3 shows a much simplified energy balance sheet for the UK for the year 2009. On the supply side, the two sources are home production and imports. Some can then be subtracted: exports and the non-energy uses of oil and gas (for example as lubricants, or raw material for plastics, etc.) and 'marine bunkers' i.e. the fuel used by ships travelling overseas, which is traditionally not included in national consumption.

Since the aim is to find the amounts of energy used in a particular year, any changes in stocks must be allowed for: subtracting fuel stored and adding fuel used from the stocks. At the bottom line the total for each fuel type is then the UK's actual consumption of each fuel, giving a total national primary energy consumption of 8840 PJ.

Table 3.3 UK primary energy production and consumption, 2009

Category	**Energy source/PJ**					**Total[3]/PJ**
	Coal	**Oil**	**Natural gas**	**Primary electricity[1]**	**Renewables and wastes[2]**	
Production (+)	462	3129	2500	690	206	6988
Imports (+)[4]	1039	3511	1641	24[4]	50	6264
Exports (–)	–25	–3244	–494	–13[4]	0	–3776
Non-energy use (–)	0	–339	–32	0	0	–371
Marine bunkers (–)	0	–110	0	0	0	–110
To stock (+/–)	–178	40	–18	0	0	–155
Total consumption[3]	**1298**	**2988**	**3598**	**700**	**257**	**8840**

[1] Refers to UK nuclear, hydro and wind generators. As explained in the previous chapter, the 'nuclear electricity' is treated as the equivalent primary energy input to produce it at an efficiency of 38%. The wind and hydro contributions are simply their electrical energy output.
[2] Includes other sources of renewable energy that are burned to produce electricity, for example landfill gas and municipal waste. They are treated here as being equivalent to the energy content of the fuel.
[3] Some totals may differ slightly from the sum of the items owing to rounding.
[4] To and from France through a cable under the English Channel.
Source: DECC, 2010a, Table 1.1.

must be equal to its total net annual energy supply. Box 3.3 gives details of UK *primary* energy production and consumption and its imports and exports. Box 3.4 looks at how the country's *delivered* energy is used. There are then the considerable energy losses in conversion to consider.

It can be seen from Boxes 3.3 and 3.4 that there is a discrepancy: only 6021 PJ of delivered energy has been used from a primary energy supply of 8840 PJ. About 5% of the discrepancy is the non-energy use of fuels and that used in international shipping. Most of the large remainder is the energy 'lost' by the energy industries in converting primary energy into the convenient forms of energy that consumers want to use. These **losses in conversion and delivery** consist of, for example, waste heat from power

BOX 3.4 UK energy balance, 2009 – delivered energy consumption

Table 3.4 shows where all this energy was used. This, of course, is now the **delivered energy** as received by the consumer. The categories of consumer are fairly straightforward.

Domestic: this is the energy use of all the buildings that people live in.

Services: the energy use of shops and offices, schools and colleges, museums, etc. Public lighting and energy used in agriculture (except fertilizer production) has been included in this box as well. Note that some countries put the domestic and services sectors together for statistical purposes and in others agriculture is treated as a separate sector.

Transport: this covers both public and private, carrying both people and goods, in road vehicles, trains, and planes.

Industry: this is obvious, but note that it excludes the energy used by the *energy industries* themselves (see below), for example coal mines, power stations and oil refineries.

In Table 3.3 the contributions were those from various *sources*, such as 'coal' and 'oil'. In the final use stage of Table 3.4, the categories of energy are slightly different:

Solid: includes coal and 'manufactured fuels' such as coke and 'smokeless fuels'.

Liquid: now refers to all the oil products (diesel, petrol, heating oil, aviation fuel, etc.).

Natural gas: essentially the same as in the previous table. Here it is being distributed to consumers through the gas supply network.

Electricity: now means all electricity, from power stations and generators of every type (including renewables).

Renewables and heat: includes biofuels and heat from solar panels used directly in each sector (these are sometimes described as 'thermal renewables') and waste heat from power generation where that is put to practical use.

Table 3.4 UK delivered energy consumption, 2009

Sector	Fuel type/PJ					Total[2]/PJ
	Solid	Liquid	Natural gas	Electricity[1]	Renewables and heat	
Domestic	29	126	1206	440	21	1821
Services[3]	4	54	297	335	29	720
Transport	0	2290	0	33	42	2366
Industry	71	226	410	352	54	1114
Total[2]	**105**	**2696**	**1913**	**1160**	**147**	**6021**

[1] Here treated as the energy content of the delivered electricity (i.e. 3.6 PJ for each TWh delivered).
[2] Some totals may differ slightly from the sums of the items owing to rounding.
[3] This includes public lighting and agriculture.
Source: DECC, 2010b, page 10.

stations and the electricity consumed by coal mines and oil refineries. This is discussed in more detail in the next subsection.

The delivered energy consumption figures (Box 3.4) allow examination of final energy use in two ways. Looking down the table by sector, it reveals, for instance, that we use more energy in transporting ourselves and our goods from place to place than we do in manufacturing, or just living at home. Looking across it by fuel, a striking feature is how little anybody wants in the form of solid fuels. This category would have looked completely different back in 1880, but now even industry uses only a tiny 6% of its energy in this form. The demand for liquid fuel for internal combustion engines and the preference for gas central heating in the domestic sector are obvious.

Figure 3.10 shows the data of Tables 3.3 and 3.4 in a graphical form.

The top bar shows the contributions to the primary energy supply. The next three bars show the delivered energy consumption broken down by fuel, by sector and by end use.

The second bar of this chart clearly shows the magnitude of the energy losses in conversion and delivery. The minute amount of solid fuel shown in this bar indicates that the bulk of UK coal consumption goes for electricity generation.

The third bar shows the breakdown into the different sectors. The energy consumption by the services sector is roughly equally divided between commercial offices in the private sector and public administration,

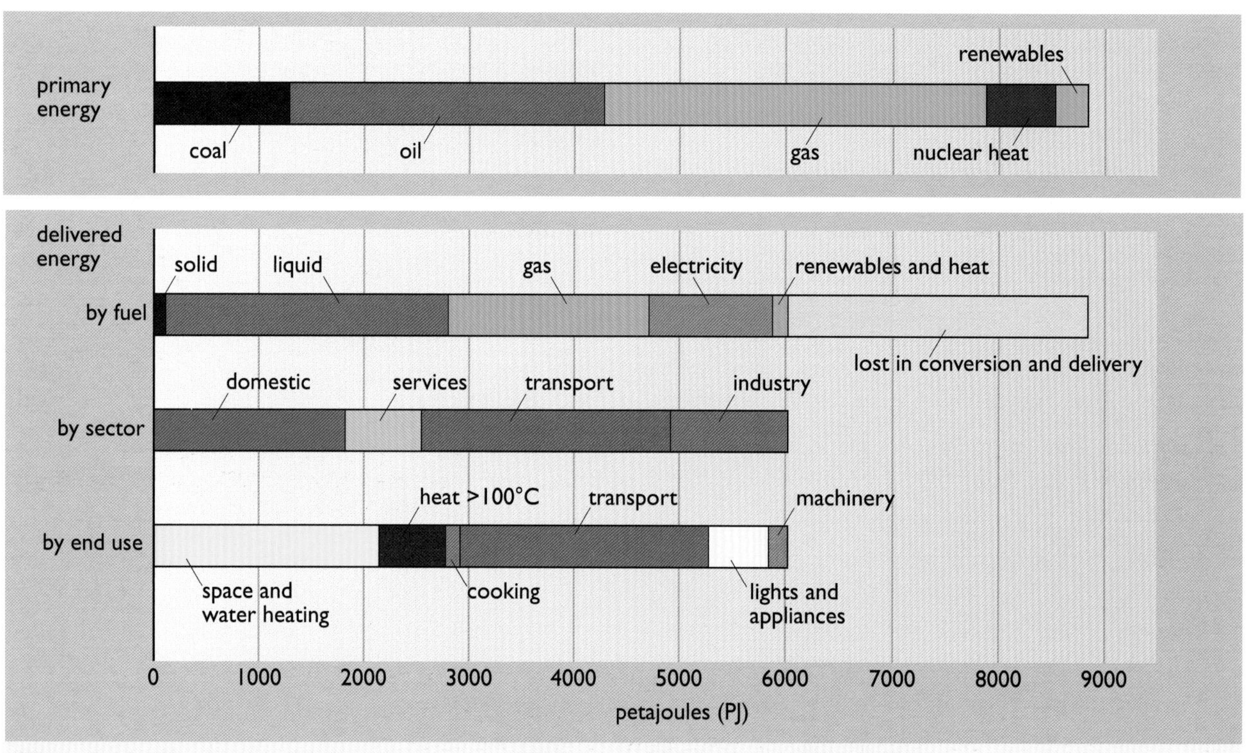

Figure 3.10 UK primary and delivered energy use, 2009 (sources: DECC, 2010a; DECC, 2010b)

government, schools, hospitals, etc. However, it also includes 40 PJ devoted to 'agriculture'. This is the energy required to run the UK's farms, and about 35% of it is electricity. But this is not the whole picture. The energy content of the fertilizers used on farms is probably about the same amount again, but its production is classified under 'industry'.

Transport energy use is significant and will be considered at the end of this subsection.

Modern industry covers a whole range of activities. Their relative energy use is shown in Figure 3.11. Iron, steel and the manufacturing of metal products consumes just over a fifth of the industrial energy use. The chemical industry, which includes the production of plastics, uses a further 16%. The processing of food and drink, which might have been seen as a 'domestic' activity back in the 1300s is now a significant part of 'industry'.

The final bar of Figure 3.10 shows a breakdown by end use. Data like this can only be determined from surveys of sample groups of consumers and the results can be subject to errors. Nevertheless it is useful information. It is, perhaps, striking that about a half of delivered energy is used as heat. For example, in the domestic sector (as with services), the bulk of the delivered energy is used for space heating (see Figure 3.12). It is also remarkable that the amount of energy we use for cooking food is larger than the 100 PJ or so used on farms to grow it.

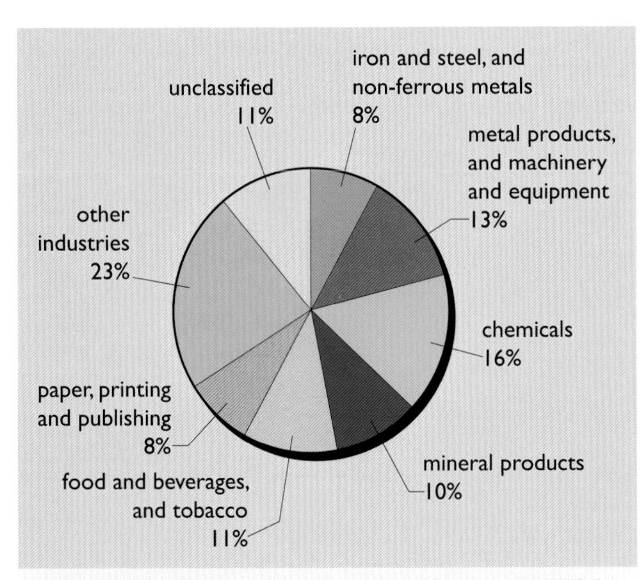

Figure 3.11 UK industrial energy use, 2009 (source: DECC, 2010c, Table 4.3)

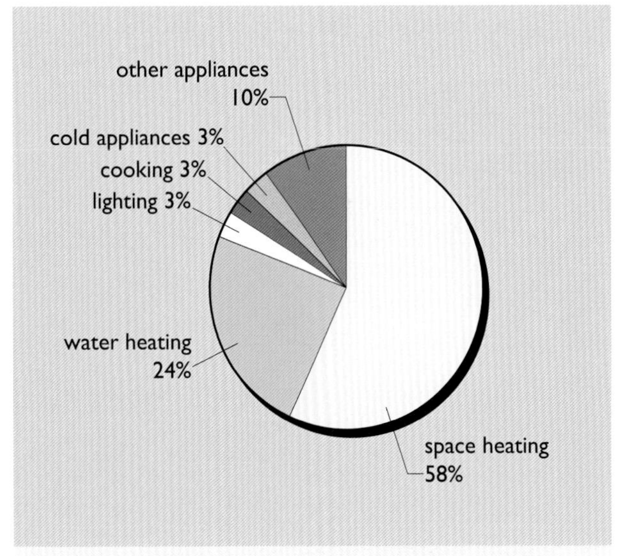

Figure 3.12 UK domestic delivered energy use, 2008 estimate (note that the total is not 100% owing to rounding) (source: DECC, 2010c, Tables 3.6 and 3.10)

UK electricity conversion losses

So far, none of the energy data above explains anything about the reason for the large difference between the total primary energy consumption and the total delivered energy consumption. For that we need more detailed statistics, showing all the conversion processes in the middle. There are many of these, each involving some 'lost' energy. For example, about 10% of the energy content of the petroleum in the 'primary energy' bar of Figure 3.10 is lost in the oil refinery by the time it has been converted to the 'delivered' liquid fuels in the second bar of the chart. Other losses are incurred in the conversion of coal to other solid fuels and pumping gas through pipelines. But the chief culprit is the generation of electricity.

Figure 3.13 gives the details. Notice that it has two horizontal scales, one in petajoules (PJ) for the fuel inputs for comparison with the tables above, and another in terawatt-hours (10^9 kWh) for the electricity generated. The first bar shows a total energy input to all the power stations of about 3290 PJ of fuels used to generate electricity. This can be broken down as:

- a coal input of about 1035 PJ, 80% of the total UK coal consumption shown in Table 3.3
- a small amount of oil, mainly used in generating plants used to meet peak demand
- a large amount of natural gas, 1280 PJ. This represents a 50-fold increase since 1990 – the so called 'dash for gas', a subject returned to in Chapters 7, 9 and 14.
- about 650 PJ of nuclear heat input
- 200 PJ of biomass and waste materials that have been burned to generate electricity
- 65 PJ of electricity directly generated from renewables such as hydro and wind power and pumped storage plants.

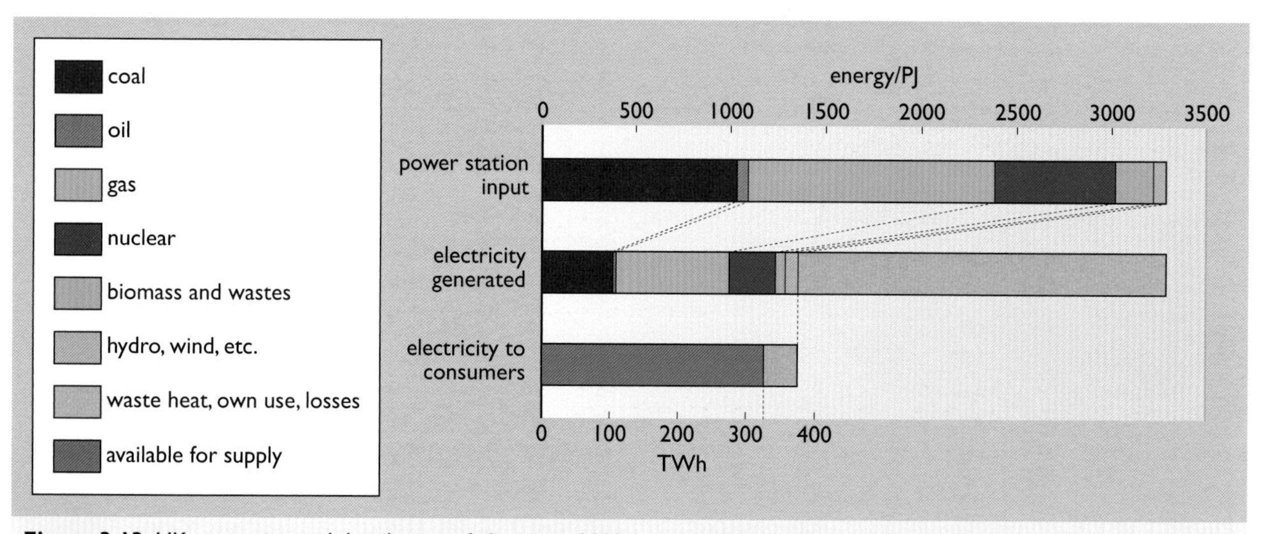

Figure 3.13 UK generation and distribution of electricity, 2009 (sources: DECC, 2010a, Electricity, Tables 5.1.2 and 5.2; DECC, 2010b, pp. 26, 28)

The second bar shows the total generated electrical output: approximately 375 TWh, 1350 PJ or 41% of the input; the remainder has become waste heat either dumped into the sea or into the sky via large cooling towers (Figure 3.14). But that is not all. About 6% of the generated power is used by the power stations themselves, or consumed in pumped storage plants, and therefore doesn't enter the distribution network. Finally, around 7% of the energy literally vanishes into thin air on the way to the consumers, as heat resulting from the electrical resistance of the cables and losses in transformers.

Figure 3.14 The UK electricity industry annually disposes of 2 EJ of waste heat into the sky or the sea

Ultimately, there is a net output of 327 TWh or 1180 PJ. Just over a third of the input energy has become useful output or, to put it another way, the overall 'system efficiency' is only 36%.

A further 3 TWh is imported from France and not actually produced by UK power generators. This gives a total supply of 330 TWh. But Table 3.4 shows consumers receiving 1160 PJ of electricity (322 TWh). Why the discrepancy of 8 TWh? The answer is that it is consumed by the other energy industries, particularly oil refineries. This a different type of loss, the *electricity* required to convert crude oil entered in the primary energy bar of Figure 3.10 into the refined 'liquids' (petrol, diesel, etc.) in the second bar of that chart.

Couldn't the waste heat from power stations be put to use? Indeed it could. In countries such as Denmark, power stations don't just make electricity; their waste heat output is recycled into a massive network of insulated pipes carrying it into about half the buildings in the country. This large-scale distribution of heat is known as **district heating**. When the heat comes from power stations, rather than just large boilers, the process is known as **Combined Heat and Power Generation**, **CHP** or **co-generation**. The beneficial effect of this on the Danish national energy balance is described at the end of this chapter.

Trends in UK energy consumption

As shown in Chapter 2, UK primary energy consumption peaked in about 2005 and has been slowly declining. What about delivered energy? The long-term picture is interesting. As we've seen above, in 2009 the UK used about 6000 PJ of delivered energy. Distributed between the 62 million inhabitants, this works out at about 100 GJ per capita per year. If we look back to the beginning of the 20th century, to the year 1903, the UK used 167 million tonnes of coal and about 2 million tonnes of oil (Jevons, 1915). The major part of this energy, probably about 85%, was delivered energy. The fledgling electricity industry was tiny and the bulk of 'conversion losses' were in the mining industry's own use of coal and in the provision of town gas. Divide this amongst the 42 million inhabitants and the figure for annual per capita delivered energy consumption is 100 GJ, identical to that over 100 years later!

There are key differences between 1903 and 2009. In 1903 the fuel (almost all coal) was delivered to the factory and home and used with the poor efficiencies of the time. The heating efficiency of a domestic coal fire is typically about 25% and we have already seen the poor efficiency figures for gas lighting (Table 3.1). The fuel had to be burned on site and the losses took place right there. In 2009, we were using the convenience fuel of electricity with a high efficiency at the point of use. The useful heat output of an electric fire is 100% of the electrical input. Modern incandescent and fluorescent lamps give far more useful lumens per watt than their gas counterparts. We get far more useful energy service out of the delivered energy. The price we pay is in the enormous energy losses at power stations.

Looked at with a long-term time perspective, UK delivered energy on a per capita basis hasn't changed much in a century, although there have been ups and downs and shifts between the different sectors. Looking at the picture since 1970, total delivered energy consumption rose until 2000, levelled off, and has been declining slowly since 2005.

Figure 3.15 and Table 3.5 show the changes in the sectors – industry has shown the biggest decline and transport the largest growth.

In the domestic sector delivered energy consumption rose by about a quarter between 1970 and 2005 and then slowly declined. This energy rise has reflected increasing standards of heating and comfort in UK homes, spurred on by the availability of cheap North Sea gas. In 1970 only 31% of homes had central heating; by 2007 this figure had risen to 92%, over three-quarters being gas-fuelled. The decline in energy consumption since 2005 may show the influence of energy efficiency schemes, particularly the

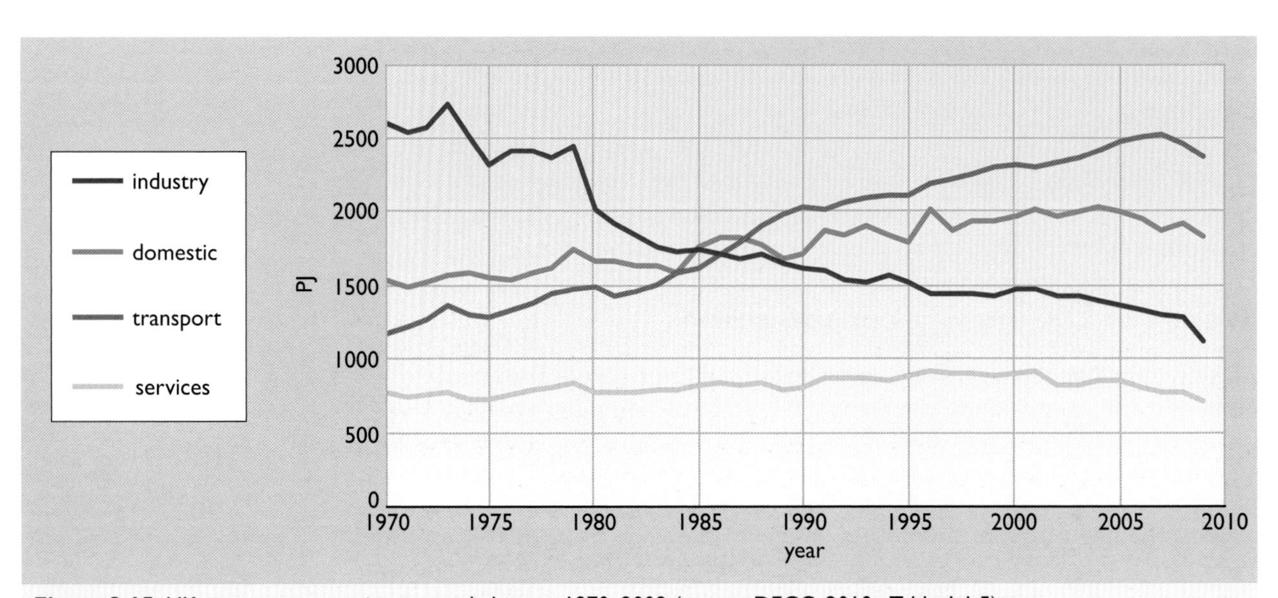

Figure 3.15 UK energy consumption sectoral changes, 1970–2009 (source: DECC, 2010a, Table 1.1.5)

Table 3.5 Trends in UK delivered energy

	Delivered energy/PJ		
Sector	**1970**	**2009**	**Change 1970–2009**
Domestic	1544	1821	+18%
Services	779	720	–8%
Transport	1179	2366	+101%
Industry	2612	1114	–57%
Total	**6114**	**6021**	**–2%**

Source: DECC 2010a, Table 1.1.5.

promotion of loft insulation and high-efficiency gas boilers. Energy used in cooking has fallen steadily since 1970, possibly owing to the increased use of pre-prepared food and 'instant' meals. Total domestic electricity consumption has risen steadily since 1970, feeding a host of new electrical appliances. However, energy efficiency standards have made an impact. The amount of electricity used in 'cold' appliances and that used for lighting may have peaked around 1998. Electricity used for electronics and TVs still seems to be on the way up (DECC, 2010c).

In the services sector delivered energy consumption peaked around 2000 and has since then declined slightly. Almost half of the energy used by this sector is electricity in offices filled with computers and photocopiers; 40% of that is used for lighting. Indeed, in many office buildings, especially those with large areas of glazing, the problem is that of dealing with the surplus of heat. Even in the UK, air conditioning is seen as essential by new office developers, much to the dismay of those interested in energy conservation.

In 1970 industry used over 40% of UK delivered energy. By 2009 this figure was down to under 20%. This reflects the shift away from heavy industry, such as steel making and car manufacturing, to lighter 'high value' industries such as electronics and earning money internationally by providing services such as banking and insurance. This is part of a much longer trend (see below).

The most important change has been the large rise in transport energy use, which has doubled in almost 40 years. What is most noticeable is the enormous increase in travel by car. Figure 3.16 shows the UK's changing transport energy consumption since 1970, including road freight and the energy used in refuelling international aircraft. It can be seen that the bulk of the consumption is now in road and air transport. Water transport and railway energy use are very small by comparison. However, it is worth reflecting that back in 1903 the railways used 13 million tonnes of coal (about 350 PJ). The large decrease in railway energy use reflects both the contraction of the railway network, and the switch from steam to diesel and electric traction that took place in the 1960s (see Chapter 8).

The growth in road transport energy consumption was greatest during the 1980s but has slowed and, at the time of writing (2010), appears to have peaked. High oil prices and campaigns for more fuel-efficient cars have had an effect. Air transport consumption has risen dramatically, encouraged by

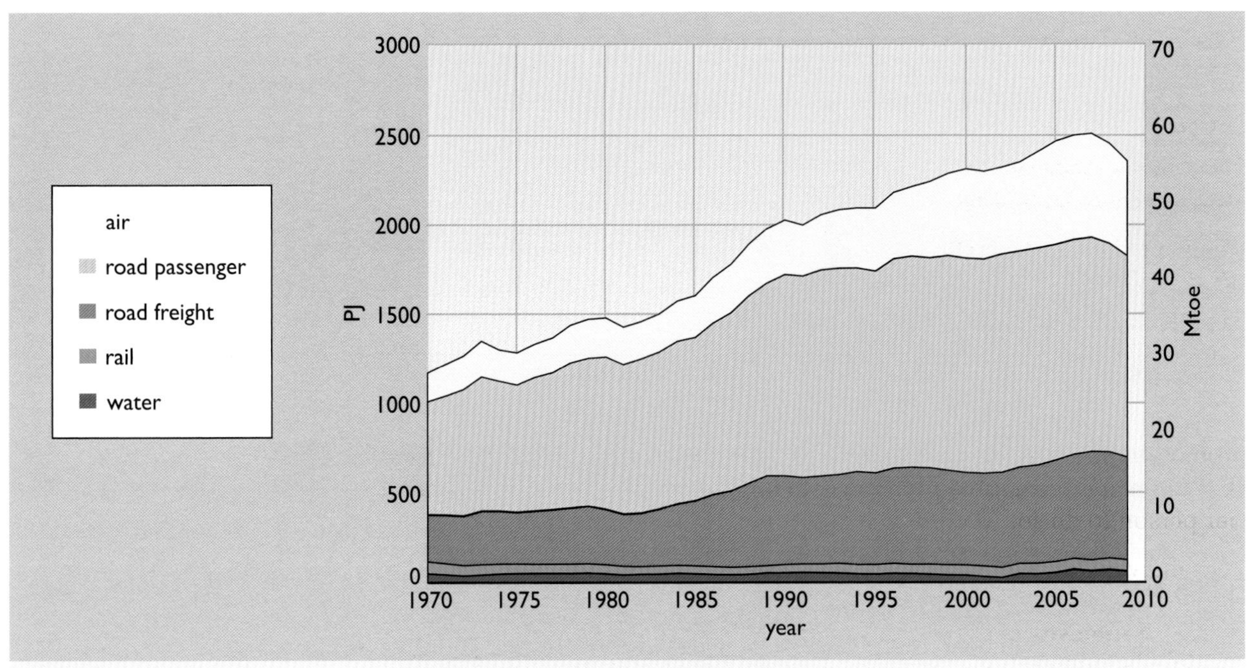

Figure 3.16 UK transport energy consumption, 1970–2009 (source: DECC 2010c)

air fares that are cheap in relation to earnings. In 2008 UK airports handled 235 million passengers (DFT, 2009). However, even UK air transport energy use may have reached a peak in 2006.

International comparisons

Energy and GDP

The Gross Domestic Product (GDP) of a country is the monetary value of the goods and services that the country produces in a year. We would expect that energy consumption and GDP might be related, and indeed it is generally true that the higher the GDP of a country, the higher its energy consumption (see Table 2.3). This has led energy forecasters in the past to assume that growth in GDP must automatically require more energy. Looking at trends over time and making international comparisons requires that GDP is expressed in a common currency (usually the US dollar) and corrected for inflation (see Chapter 12 for details of this).

Table 3.6 shows the relation between energy consumption and GDP on a per capita basis for three of our sample countries.

The GDP figures, originally in the local currency, have been corrected for inflation so that they can be expressed in money of the year 2000. They have then been converted into US dollars using the financial exchange rate. To make the comparisons easier, all the quantities are given per capita, i.e. total divided by population.

Let us look at Denmark and the USA. We see that the goods, services, etc., produced per person in the USA in 2007 were worth US$38 000 and those in Denmark US$32 800, about 15% less. Yet the per capita primary energy

Table 3.6 Energy and GDP for three countries, 2007

	UK	Denmark	USA
Population/millions	61	5.5	302
Annual per capita GDP at exchange rate/US$(2000)	29 050	32 800	38 000
Annual per capita primary energy consumption/GJ	146	151	324
Energy/GDP ratio/MJ US$$^{-1}$	5.0	4.6	8.5

Source: IEA, 2009.

consumption in Denmark was only half of that in the USA. So it appears that Denmark manages to produce goods and services using far less energy per person to do so.

A useful way to compare the 'energy efficiencies' of different countries is to divide the energy consumed by the GDP. The result is the **energy/GDP ratio** or **energy intensity** of the economy, the amount of energy used in producing each dollar's worth of national product in MJ per dollar (MJ US$$^{-1}$). This ratio has been mentioned earlier in Box 1.7. The final row in the table shows this for our three countries. The Danish economy would appear to perform better than those of either the USA or the UK. Why? The following are just a few of many possible reasons:

- The Danes may obtain their energy in forms that involve less wastage than in other countries (combined heat and power rather than conventional fossil-fuelled power stations, for instance).
- They may earn their GDP by less energy-intensive types of activity than others: for example in the UK a pound earned by industry requires three times more expenditure on energy than one earned in the services sector (DECC, 2010c).
- They may use less energy for the same purposes (greater industrial energy efficiency, better insulation of buildings, etc.).
- They may use less energy than others to support economically non-productive activities (e.g. watching TV, going for a drive, reading a book).

The historical picture is also interesting. Figure 3.17 shows the energy intensity of a number of different countries from 1880 onwards, expressed in constant US$ (at 1972 values).

This diagram shows that when countries industrialize, their energy/GDP ratio rises to a peak and then declines. The UK industrialized first and its peak occurred in about 1850. Canada's peak occurred in 1910 and that for the USA occurred in 1920. As other countries such as Japan industrialized, they were able to do so in a more energy-efficient manner by drawing on the past experience of the UK, USA and Canada. Between 1880 and 1990 the UK's energy/GDP ratio fell by a factor of three and has fallen a further 35% since then. Put another way, modern UK citizens manage to produce goods and services of equivalent value using only a fifth of the energy used by counterparts 130 years ago.

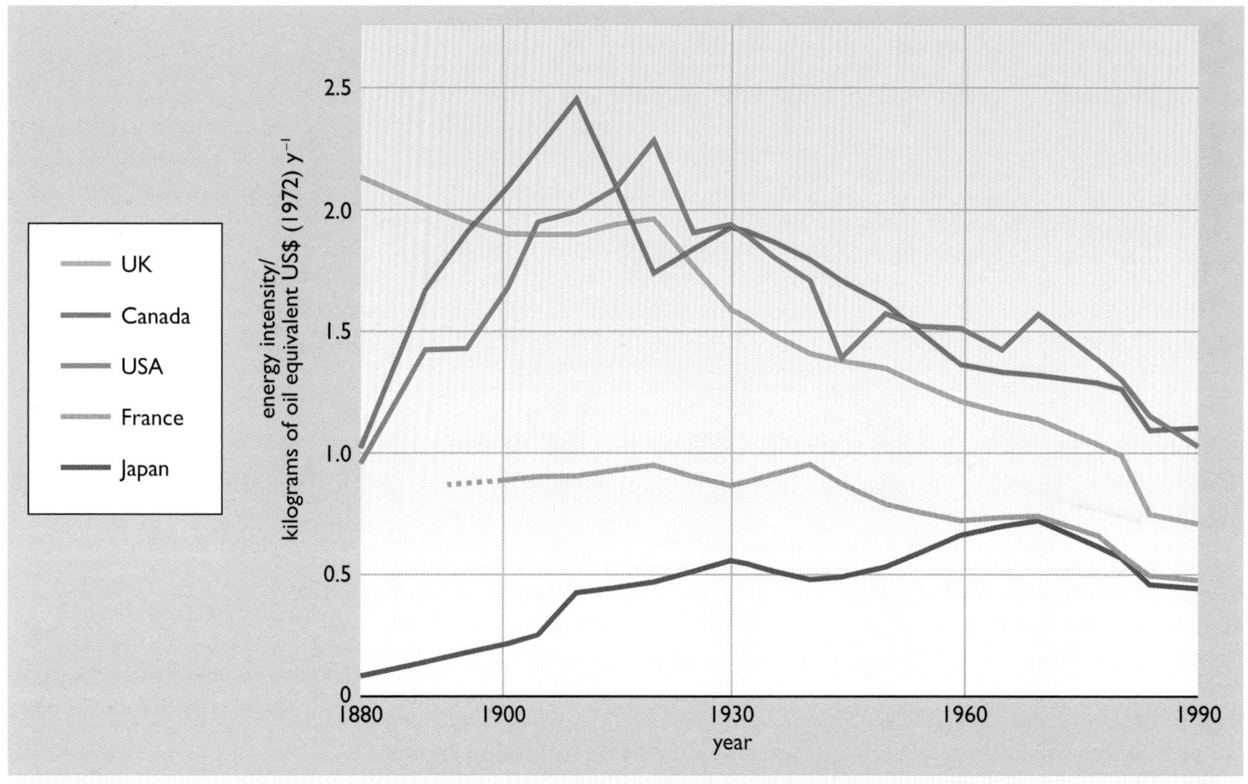

Figure 3.17 The energy intensity of selected economies, 1880–1990 (source: Smil, 1994)

Economic growth does not necessarily happen every year. In 2009, the UK GDP was 5% lower than it had been in 2008 (taking it roughly back to where it had been in 2005). This has no doubt been a factor in the decline in energy use that can be seen in the various charts. However the energy/GDP ratio has still continued to improve.

There is also the issue of the embodied energy in imports. In 1880 the UK was 'the workshop of the world', exporting heavy machinery that required a large amount of energy to produce. That role has now been taken over by Japan, India and particularly China. The current relatively good energy/GDP ratios of the UK and Denmark may be at the expense of a poorer ratio in another country somewhere else in the world.

More and more electricity

It is perhaps extraordinary that although mains electricity was first deployed in the 1880s the world doesn't seem to be able to get enough of it. Figure 3.18 shows how per capita electricity consumption has risen dramatically over the last 50 years, spurred on by falling prices and a host of new electric gadgets and applications.

As might be expected, US per capita electricity consumption is the highest but this is not uniformly so across the country. That in the state of California has been held at around 8 MWh per capita per year since the 1970s by extensive energy efficiency campaigns. The rise in consumption in most western countries has flattened out in recent years as similar efficiency measures have taken effect.

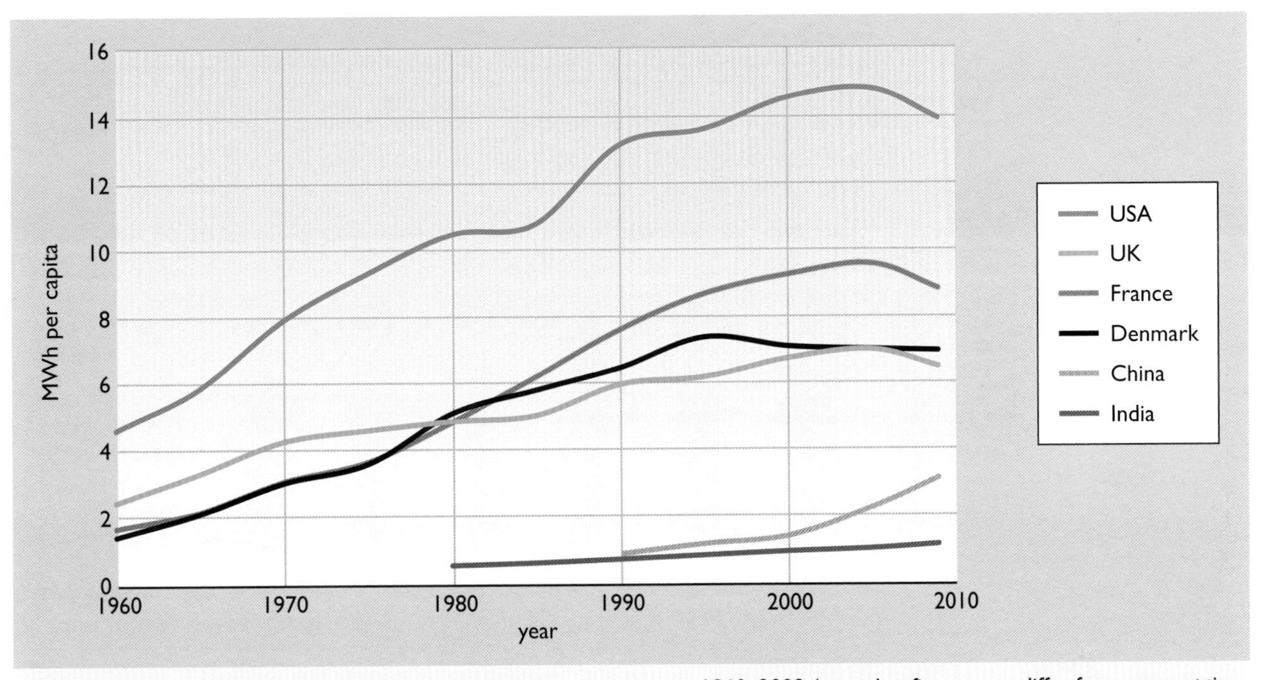

Figure 3.18 Annual per capita electricity generation for six countries, 1960–2009 (note that figures may differ from *consumption* figures given in text owing to losses and exports) (sources: Eurostat, 1990; BP, 2010; Maddison, 2009; national statistical agencies)

France is one country which has been transformed by the availability of electricity from nuclear power stations. Since 1960 per capita consumption has risen by a factor of almost six. The rise for the combined domestic and service sectors of the economy is impressive. In 1960 they used 17.5 GWh; in 2008 they used 289 GWh – over sixteen times as much! This, in part, has reflected the general improvement in national living standards.

The Danish rise in electricity use is particularly remarkable, as since the mid-1970s it has been achieved with hardly any increase in national primary energy consumption (see Chapter 2). A key ingredient of this success has been the increasing use of CHP, making sure that the waste heat from the power stations is put to good use.

Perhaps the most important feature of this chart is the rapid increase in electricity use in China. This has involved the construction of major hydroelectric projects such as the Three Gorges Dam and many large coal-fired power stations. Despite being a large rural country it is estimated that over 99% of the Chinese population have access to electricity (IEA, 2006). Between them China and the USA use about a third of the world's coal production for generating electricity.

In India, electrification has also increased, using coal as the main fuel. However, this has barely kept pace with the growing population, so that the per capita consumption remains low. In 2005 only just over a half of the population had access to electricity.

More and more travel

The story of increasing road transport use, and the consequent increase in demand for oil, described for the UK above, is echoed worldwide. European

Table 3.7 Transport statistics for six countries, to 2008

	UK	Denmark	USA	France	India	China
Transport energy growth, 1980–2006	+69%	+51%	+46%	+38%	+130%	+550%
Per capita transport energy use, 2006/GJ y^{-1}	41	41	102	34	2	4
Cars per 1000 inhabitants, 2008	475	381	780[1]	498	9(2004)	29

[1] includes all 2-axle, 4-wheeled vehicles

Sources: Eurostat, 1990; CEC 2010; Maddison, 2009; Ni, 2009; TERI, 2009; national statistics agencies.

countries have high levels of car ownership, about one car for every two people. In the USA the figure is higher, more like three for every four people. Some sample statistics are given in Table 3.7.

Perhaps the most important numbers are in the right-hand two columns of the table. Both India and China have low levels of transport energy use on a per capita basis, but have seen a phenomenal growth rate. This poses the serious question 'what will happen to world global oil demand and CO_2 emissions if these countries continue to aspire to the standards of *mobility* enjoyed by Europe and the USA?'

Broadly speaking, the patterns of energy use in the UK are not significantly different from those in Denmark, France, or the USA, but there are some differences that are worth pointing out. The patterns in India and China are considerably different.

UK and Denmark

While the UK has had a history of relative fuel security, initially with plenty of coal, and, since the 1970s, an abundance of oil and gas, Denmark has been in a very different position. As pointed out in the previous chapter, in 1972 Denmark was almost totally reliant on imported fuel, 95% of it oil (see Figures 2.12 and 2.13). The world price of this rose by a factor of over 7 between 1972 and 1980 (see Figure 2.6), with serious consequences for the Danish national budget. Drastic measures had to be taken. Initially there was a lot of belt-tightening – at one point the driving of cars on Sundays was banned. Subsequently, a number of more medium- to long-term policy decisions were implemented:

- switching power stations from oil to coal
- expanding the use of combined heat and power generation (CHP)
- phasing out individual oil-fired central heating units, many of which had poor thermal efficiencies, by expanding the district heating networks

- implementing new insulation standards in building regulations
- expanding renewable energy supplies, especially biomass and wind power.

The results have been quite impressive. The rapid growth in national primary energy consumption was halted and consumption has remained almost stable at about 800 PJ per year ever since (see Figure 2.12). By 2007 over 60% of Danish homes used district heating, with the bulk of the heat coming from CHP plants (DEA, 2010). The results of this attention to energy efficiency and renewable energy are shown in Figure 3.19 in a comparison with the UK. Note that these diagrams are equivalent to the second bar of Figure 3.10.

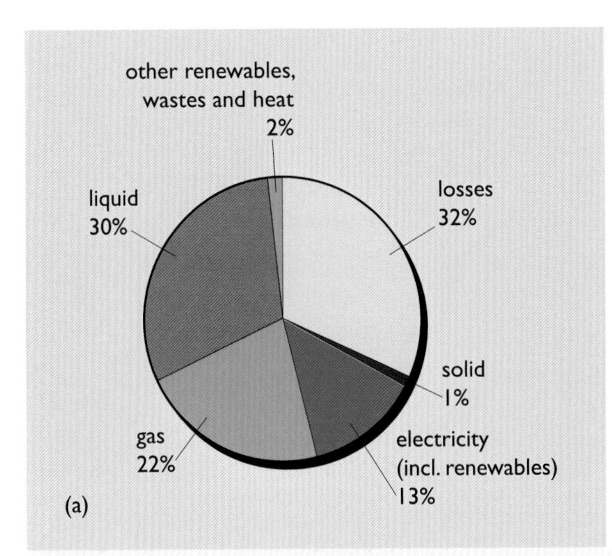

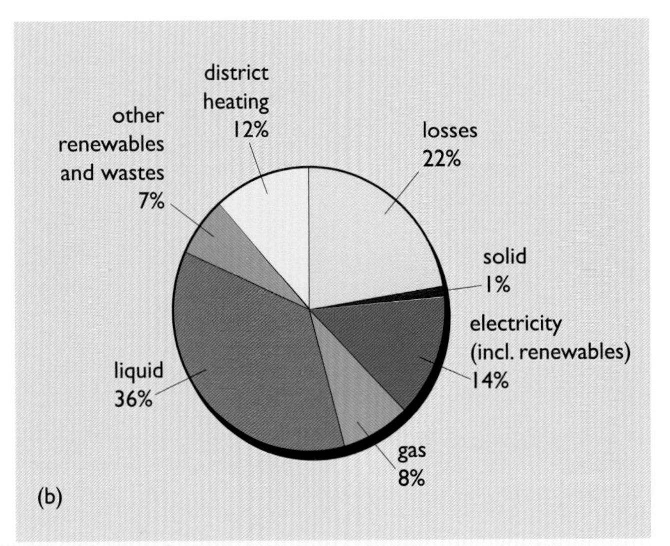

Figure 3.19 Comparison of primary energy breakdowns: (a) UK, 2009 data; (b) Denmark, 2008 data. Note that in both cases 'electricity' includes that from renewable sources (sources: DECC, 2010c; DEA, 2009)

In Denmark district heating accounted for 12% of the delivered energy supply, and the extensive use of CHP kept the national figure for losses down to 22% compared with the UK figure of 32%. Although Denmark is still heavily reliant on oil, mainly for transport, its energy mix is very varied, with an overall 14% contribution to its primary energy from renewable energy.

USA

US per capita energy use is high – and until the 2008 recession continued to grow. Primary energy consumption increased by over 40% between 1970 and 2008 (EIA, 2010). This growth was shared between all sectors of the economy. Energy use in the services sector more than doubled over this period. As in the UK, natural gas is widely used as a heating fuel but, as pointed out above, electricity use has been the growth area. About a half of US electricity comes from coal-fired plants. Coal consumption for electricity generation increased by almost a factor of three between 1970 and 2009. The conversion losses in the US electricity industry are 27 EJ, i.e. nearly three times the total UK primary energy demand.

France

As mentioned earlier, French energy use has also grown significantly. The country's primary energy consumption increased by 75% between 1970 and 2005 but has decreased slightly since then. As in the USA, this growth has been shared across all sectors of the economy (DGEC, 2009). The key factor has, of course, been the increasing use of nuclear power. In 2008 over 75% of France's electricity came from this source. As for delivered fuels, the mix is similar to that of the UK, although there is relatively less use of natural gas and more use of oil, coal, wood and electricity.

India

India is a large country with a population of over a billion people, of whom only about 25% live in cities or large towns (in China it is over 40%). Although average Indian per capita energy and electricity use is very low, it is not uniformly so. The major cities are highly industrialized, with an increasing demand for electricity, cars and transport fuel. However, nearly half of the population, those in rural areas, still lack access to electricity.

Coal is the dominant fuel for electricity generation. The overall system efficiency is low, only about 28% compared with 37% in the UK. This is mainly due to enormous transmission and distribution losses amounting to a quarter of all the electricity generated!

As pointed out in Chapter 2, about a quarter of the country's energy supply consists of biofuel in the form of wood, crop wastes, dung, etc. Since most these materials aren't 'traded' the precise amounts don't enter the normal energy statistics. The figures for the commercially traded fuels are straightforward enough and are shown in Table 3.8.

Table 3.8 India: percentages of commercially traded energy used by different sectors, 2006–7

Sector	Energy used	Breakdown
Residential and commercial	17% (1.5 EJ)	62% petroleum 37% electricity 1% gas (plus a large 'non-traded' biomass contribution)
Transport	19% (1.7 EJ)	98% petroleum 2% electricity
Industry	49% (4.3 EJ)	59% coal 18% petroleum 3% gas 20% electricity
Agriculture	8% (0.7 EJ)	48% petroleum 1% gas 51% electricity (plus a large 'non-traded' biomass contribution)
Others	8% (0.7 EJ)	88% petroleum, 12% electricity

Source: TERI, 2009, Table 1.

Indian industry has much the same mix of iron and steel, chemical and manufacturing components as other countries. Coal, being an indigenous fuel, is widely used. The transport sector uses the same mix of petrol and diesel as in other countries, and the railways consume a modest amount of electricity.

When we come to agriculture, which is treated as a separate statistical sector, the importance of energy use for irrigation is clear. Agriculture uses nearly 20% of all Indian electricity, mainly in over 80 million electric water pumps (2003 figures, TERI, 2009). It is supplied at subsidized prices for this purpose. These are supplemented by a further 70 million diesel water pumps. Diesel fuel is obviously important for the 23 million tractors, but biomass should not be forgotten, providing the 'fuel' for over 100 million animal carts (TERI, 2009).

Although the residential and commercial sector is shown as consuming 17% of the total commercially traded energy, this is only a part of the story. 'Untraded' biomass may make up nearly 80% of domestic fuel use.

In terms of energy use for cooking and lighting, the population may be regarded as three groups of people. The richer city dwellers are likely to cook with liquefied petroleum gas (LPG) or kerosene stoves and have access to electric light. Poorer city dwellers are likely to use LPG or kerosene lamps for lighting and firewood for cooking. At the bottom of the league are the rural poor, who are likely to use firewood or cow dung for cooking and, if they are lucky, may have LPG or kerosene lighting.

The poor efficiencies of fuel-based lighting have already been discussed. The efficiencies of cooking stoves using firewood and dung are also very low, less than 10%, compared with figures of 50% or more for kerosene or LPG stoves.

These basic considerations of energy efficiency have led to national programmes of electrification and the promotion of LPG use.

China

Like India, China is a very large country. In 1980 it had a population of nearly a billion, 80% of whom lived in the countryside in a predominantly rural economy. Between then and 2000 over 200 million people have moved to the cities. By 2006 the population had risen to over 1.3 billion and over 40% were living in cities or towns, particularly the new industrial coastal cities in the south and east.

'Traditional' biomass has always been an important source of energy, and even in 2005 it made up almost a fifth of the primary energy use. Between 1980 and 2006, if biomass is included, Chinese primary energy consumption tripled, but its GDP rose by a factor of over ten (LBNL, 2009). As the country has industrialized it has travelled through the cycle shown in Figure 3.17, with the energy/GDP ratio tripling between 1952 and 1978 and falling by a factor of three since then (Rosen and Houser, 2007).

Chinese government statistics concentrate on the supply side and their categories for sectoral energy use have not been the same as those used in

Table 3.9 China: percentages of commercially traded energy used by different sectors, 2006

Sector	Energy used
Residential and commercial	13% (8.9 EJ)
Transport	7% (5 EJ)
Industry and construction sectors	72% (49 EJ)
Agriculture	4% (2.4 EJ)
Others	4% (2.9 EJ)

Source: LBNL, 2009.

other countries. Interpretation work by the US Lawrence Berkeley National Laboratories has provided some analysis of delivered energy use for 2006 for each sector, as shown in Table 3.9.

As in India, these figures omit large amounts of 'non-traded' biomass (probably about 10 EJ) in the residential and agricultural sectors. Also they are somewhat of a snapshot of a moving target: China's total energy consumption *more than doubled* between 2000 and 2009!

The high proportion of energy consumed by industry is extraordinary. It is comparable to England in the 1880s (see Figure 3.2). Between 1980 and 2006 the sector's energy use rose by more than a factor of four and this accounts for 80% of the total national increase in energy use, and that really means its use of coal. The industrial sector consumes 70% of the country's electricity generation, which is mainly produced from coal.

By 2006 China was producing almost half the world's total cement and flat glass, 35% of the world's steel and 28% of the world's aluminium. Most of this industrial output has been used within the country, creating new factories and infrastructure and particularly homes for the 10 million people a year moving from the countryside to the cities.

In the domestic sector biomass is a key fuel. An estimate of 12.7 EJ for domestic energy use (Zhou et al., 2007) suggests that it provides over a half of the energy needs, particularly for those living in rural areas. The estimated proportion of energy devoted to space heating, 32%, is low compared with the 58% for the UK (see Figure 3.12). City dwellers in northern China are likely to have district heating fed by coal-fired boilers. However, in southern China, buildings traditionally have no provision for space heating, even though it can be quite cold in winter.

In 1980 the transport sector used about 0.8 EJ, less than the UK did in the same year, and nearly half of that was coal for steam locomotives (Ni, 2009). By 2006 the sector used 5 EJ, of which 98% was petroleum.

As in India, the agricultural sector uses relatively large amounts of electricity, 3.5% of national production in 2006. Much of China's farmland is dependent on pumping 'fossil water' from deep underground aquifers.

The question of the moment is how long the expansion of Chinese energy use, particularly its use of coal, can continue.

3.5 Summary

In the few pages of this chapter we have covered over 10 000 years and spanned the economies of a number of different countries.

There are a number of basic trends. As we have risen from the culture of hunter-gatherer through to subsistence farmer and then to modern industrial society, energy sources have been tapped:

- To allow the basic tedious tasks of life to be carried out with less human labour.

  The process of growing food, which even in the 14th century occupied the full-time labour of most of the population, is now carried out intensively in many countries by relatively few people and machines, aided by artificial fertilizers. This has allowed the bulk of the population to concentrate on manufacturing and service activities, and to have time for 'leisure'.

- To allow human activities to be done more easily and comfortably and to carry on into the hours of darkness.

  The efficiency of basic activities and the environment in which they are done, such as space heating and food preparation, have been improved by the development of better technology and fuels. Artificial lighting, which 150 years ago would have been considered extremely expensive, is now extremely cheap.

- To enable new products to be manufactured and distributed.

  The process of industrialization, especially in the UK, has involved the use of a massive amount of fossil fuels. It is important to note that as other countries industrialized they did so in a more energy-efficient manner, leading to falling ratios of energy use per unit of GDP.

- To allow new activities to take place, such as mass travel and communication.

  Over the last hundred years or so, there has been a massive growth in electricity consumption. Electricity appears to be the clean, controllable, end-use fuel of choice, even though its generation may mean the production of considerable amounts of pollution at the power station (out of sight of the consumer) and the disposal of large amounts of waste heat.

  In recent years, however, the main growth area in energy use has been in transport. In Europe as a whole there has been a massive growth in car ownership and use, leading to pollution and congested cities. In the UK the rise in road and air transport energy consumption has recently levelled off.

But a large proportion of the world's population is still living in societies without the benefits of all these energy services.

Overall, looking back in time, we would probably not wish to go back to an earlier era. Indeed there may be an underlying fear that somehow if energy supplies were to run short we would automatically be propelled back to a 14th-century lifestyle. However, when we look back to Victorian society in the 1880s, we can see a culture furiously manufacturing products and inventing exciting new ones. Yet it did so by using large amounts of energy, in what we would now see as an inefficient manner, and creating a pall of urban pollution that we are no longer prepared to tolerate.

It is worth considering whether or not the future inhabitants a century hence might look back to us here at the beginning of the 21st century with similar thoughts.

References

Bowyer, J. (1973) *History of Building*, Crosby, Lockwood & Staples.

BP (2010) *BP Statistical Review of World Energy June 2010*, London, The British Petroleum Company; available at http://www.bp.com (accessed 29 July 2010).

CEC (2010) *EU Energy and transport in figures: statistical pocketbook 2010*, European Commission: Directorate-General for Energy and Transport (DG TREN), Belgium, Office for Official Publications of the European Communities; available at http://ec.europa.eu (accessed 20 July 2010).

DEA (2009) *Energy Statistics 2008*, Copenhagen, Danish Energy Agency; available at http://www.ens.dk (accessed 19 March 2010).

DEA (2010) *Danish Energy Policy 1970–2010*, Copenhagen, Danish Energy Agency; available at http://www.ens.dk (accessed 7 January 2010).

DECC (2010a) *Digest of United Kingdom energy statistics (DUKES)*, Department of Energy and Climate Change; available at http://www.decc.gov.uk/en (accessed 29 July 2010).

DECC (2010b) *UK Energy in Brief 2010*, Department of Energy and Climate Change; available at http://www.decc.gov.uk (accessed 29 July 2010).

DECC (2010c) *Energy Consumption in the United Kingdom*, Department of Energy and Climate Change, *Industrial Table 4.3*; available at http://www.decc.gov.uk (accessed 19 March 2010).

DfT (2009) 'Transport Statistics Great Britain', London, Department for Transport; available at http://www.dft.gov.uk (accessed 29 July 2010).

DGEC (2009) *Repères: Chiffres clés de l'énergie*; available at http://www.developpement-durable.gouv.fr (accessed 31 July 2010).

Diderot, D., and D'Alembert, J. L. (1769–72) *L'Encyclopédie ou Dictionnaire Raisonné des Sciences, des Arts et des Métiers*. Avec approbation et privilege du Roy, Paris.

EIA (2010) *Annual Energy Review 2009*, US Energy Information Administration; available at http://www.eia.doe.gov/emeu/aer/contents.html (accessed 19 March 2010).

Eurostat (1990) *Energy 1960–1988*, Luxembourg, Office des publications officielles des Communautés européennes.

Fouquet, R. (2008) *Heat, Power and Light*, Cheltenham, Edward Elgar.

IEA (2006) *World Energy Outlook*, Paris, International Energy Agency; available at http://www.iea.org (accessed 7 January 2011).

IEA (2009) *Key World Energy Statistics*, Paris, International Energy Agency; available at http://www.iea.org (accessed 19 March 2010).

IEA (2010) *World Energy Outlook 2010*, Paris, International Energy Agency.

IFA (2008) *Food Prices and Fertilizer Markets*, International Fertilizer Association; available at http://www.fertilizer.org (accessed 29th August 2010).

Jevons, H. S. (1915) *The British Coal Trade*, republished in 1969, Newton Abbott, David & Charles (Publishers) Ltd.

LBNL (2009), *China Energy Databook*, Berkeley, CA, Lawrence Berkeley National Laboratory; available only as CD from http://china.lbl.gov/databook.

Maddison, A. (2009) 'Statistics on World Population, GDP and Per Capita GDP, 1–2008 AD'; available at http://www.ggdc.net/maddison (accessed 30th July 2010).

Ni, C. (2009) 'China Energy Primer', Lawrence Berkeley National Laboratory; available at http://china.lbl.gov/research/china-energy-databook/china-energy-primer (accessed 19 March 2010).

Nordhaus, W. D. (1997) 'Do Real-Output and Real-Wage Measures Capture Reality? The History of Lighting Suggests Not' in Bresnahan T. F., *The Economics of New Goods*, London, University of Chicago Press; available at http://cowles.econ.yale.edu (accessed 29 August 2010).

Odum, H. T. (1971) *Environment, Power and Society*, New York, Wiley-Interscience.

Plomer, W. (ed) (1977) *Kilvert's Diary 1870–1879*, Harmondsworth, Middlesex, Penguin Books.

Rosen, D. and Houser, T. (2007) *China Energy: A Guide for the Perplexed*, Peterson Institute for International Economics; available at http://www.iie.com (accessed 19th March 2010).

Smil, V. (1994) *Energy in World History*, Boulder, CO, Westview Press.

Smil, V. (2004) *Enriching the Earth*, Cambridge, MA, MIT Press.

TERI (2009) *TERI Energy Data Directory and Yearbook (TEDDY)*, New Delhi, The Energy and Resources Institute.

Weightman, G. (2001) *The Frozen Water Trade*, London, HarperCollins.

Zhou, N. et al., (2007) *Energy Use in China: Sectoral Trends and Future Outlook*, Berkley, CA, Lawrence Berkeley National Laboratory; available at http://china.lbl.gov (accessed 19 March 2010).

Chapter 4

Forms of energy

by Janet Ramage

4.1 Introduction

The term *energy* has already appeared many times in this book. Chapter 1 discussed *energy systems* and referred to *energy sources* such as fossil fuels, the wind and the Sun. Chapter 2 introduced such terms as *primary energy*, *energy production* and *energy consumption* and the important fact that *energy is conserved*, while Chapter 3 considered our uses of energy in the form of *electricity*, *heat* or *light*.

The purpose of this chapter is to ask some basic questions about the nature of energy. What do all the forms mentioned above have in common, and how do they differ? What makes materials such as coal, oil, gas, wood or peat useful fuels, and why does their combustion inevitably release carbon dioxide? Or in a different context, how does pumping water up a mountain allow us to store electricity, and why may this be more practicable than using batteries? And at the most basic level, how many different '*forms of energy*' are there?

The development of the concept of energy as one entity – one actor who can put on many faces – coincided almost exactly with the nineteenth century, from the first use of energy as a scientific term in 1807 to Einstein's identification of energy with mass in 1905. This same period saw equally major changes in our picture of 'matter'. The idea that all materials consist of *atoms* led to new insight into the nature of heat, and the later discovery that atoms consist of yet smaller particles revolutionized our picture of chemical processes and electricity. These parallel histories of our ideas about energy and matter provide the framework for this chapter.

However, the purpose in tracing the past is to understand the present; so the introduction to each form of energy is followed by a short case study illustrating its use in a present-day energy system. These studies therefore also serve as brief introductions to the principles of systems investigated in more detail in later chapters.

4.2 Kinetic and potential energy

> The grand modern ideas of Potential and Kinetic Energy cannot be too soon presented to the student.
>
> (Peter Guthrie Tait (1831–1901), mathematician and physicist, professor of natural philosophy at the University of Edinburgh)

Figure 4.1 Isaac Newton (1642–1727), born on a farm in Lincolnshire, was a bright child who delighted in 'inventions' and model-making. His later lifetime achievements make an impressive list: the laws of motion and of gravitation, proof that white light consists of many colours, a new type of telescope (which he made, grinding the mirror from an alloy he also invented), and important new concepts in mathematics. He also reformed England's coinage and was a Member of Parliament (twice).

The scientific concept of energy arose from questions about the nature of motion. What is special about a moving object? What keeps it moving? In the seventeenth century, the great period that saw the birth of modern science, people such as Galileo, Huygens and Newton (Figure 4.1) developed the idea that moving objects must possess 'something' that stationary objects lack.

The first real step was Isaac Newton's proposal that the natural behaviour of a moving object is to continue in its current state of motion (**Newton's First Law of Motion**). We shouldn't, it seems, be asking what keeps something moving, but rather what stops it – or more generally, what *changes* its motion. Newton again offered an answer: the motions of objects are changed by *forces* (**Newton's Second Law of Motion**). It is the pull of gravity, a downward force, that causes objects to accelerate towards the Earth.

If these falling objects are gaining 'something' as they speed up, where does this come from? Notice the significance of this question. Once people ask where it comes from, they are beginning to treat it as a real quantity, not just a ghostly entity that might appear or disappear into nothingness. They are starting to think of a *quantity that is conserved.*

This quantity continued to appear under various different names, with the concept gradually becoming clearer, and in 1807 Thomas Young (most famous for demonstrating that light must be a type of wave) proposed the term **energy**.

Early explanations of the energy gained by a falling object suggested that the downward force, the gravitational pull of the Earth, performs *work* on the object, and this work in some way provides the energy associated with the motion. This was a useful concept, but the real revolution in ideas came in the 1850s. It started with the opposite question. An object thrown upwards slows down and eventually stops, losing its 'motion energy'. Where does this energy go? In 1853, William Rankine, observing that the object is gaining the potential to start moving down again, suggested that, 'by the occurrence of such changes, actual energy disappears, and is replaced by Potential or Latent Energy' (Rankine, 1881). Rankine's term was soon adopted, and within a few years **potential energy** came to mean not just a potentiality, but a new form of energy. The energy of a *moving* body, now called **kinetic energy**, was another form, and there were to be still more.

To complete the picture it was necessary to know how to quantify these two forms of energy. How do they depend on the mass, speed or position of an object? The reasoning starts with a return to the idea of the **work done** by a force. By equating the change in potential energy of an object to the work done in lifting it, Rankine related potential energy to mass and height. The fact that total energy must be conserved as the object falls then leads to a formula for kinetic energy in terms of mass and speed (Box 4.1).

Boxes 4.2 and 4.3 are brief introductions to two present-day renewable energy systems in which these two forms, *kinetic energy* and *potential energy*, play major roles.

BOX 4.1 Force and energy

Force, mass and weight

Newton's explanation of the behaviour of falling apples, the Moon and the planets rested on his brilliant separation and clarification of the laws of mechanics and gravitation. Expressed in modern terms, the essential features are that:

- the force (F) needed to accelerate an object is proportional to the mass of the object (m) and the acceleration (a):

  $F = m \times a$

- the gravitational force pulling an object towards the Earth is also proportional to its mass.

Appropriately, the name adopted as the unit for force is the **newton (N)**.

Galileo showed that all objects in free fall, regardless of their masses, have the same downward acceleration. This is due to the gravitational pull of the Earth, and is now called the **acceleration due to gravity** (**g**). In fact it varies slightly from place to place, with an average value at sea level of about 9.81 m s^{-2} (metres per second per second). Informally it is referred to as 'little g', and the approximation g = 10 m s^{-2} is often used when greater precision is not required.

In scientific terms, the **weight** (W) of an object is the downward gravitational force acting on it, so it follows that:

$W = m \times g$

That is, the weight of anything in newtons is about 10 times its mass in kilograms. Or reversing this, we can say that an object whose weight is one newton has a mass of about 0.1 kg – the nice ripe apple in Figure 4.2.

Work and potential energy

Rankine proposed that:

- the *work done* by a force that is moving an object is equal to the force multiplied by the distance moved, i.e. work = force × distance
- the change in potential energy on raising an object is equal to the work done by the force raising it.

To raise anything upwards at a steady speed, you need an upward force that is just equal to its weight (W). So the work done in raising an object against gravity through a height, H, is:

- work = force × distance = $W \times H = m \times g \times H$

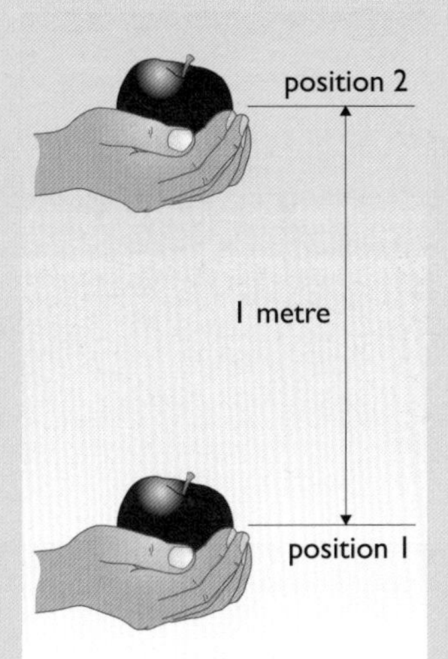

Figure 4.2 One joule is approximately the work required to raise an apple by one metre

This final quantity, often written as just ***mgH***, is therefore the *increase in potential energy*.

It is worth noting that this should strictly be called the **gravitational potential energy**, to distinguish it from other types of potential energy that have subsequently been identified (see Section 4.4).

Kinetic energy

If energy is conserved, an object in free fall, losing potential energy, must gain an equal amount of kinetic energy. If it starts from rest (zero kinetic energy) and falls through a height H, it should therefore acquire *kinetic energy*, E_k, equal to $m \times g \times H$ (as above).

However, to use the idea of kinetic energy in other contexts, we need to know how it is related to the *speed* of the object. The reasoning involves three steps:

- From the definition of acceleration, the final speed of the falling object, (v), is equal to its acceleration multiplied by the time of fall:

  $v = g \times t$

- Speed is distance divided by time, so the average speed during the fall is H divided by t. But we know that if an object starts from rest and has a constant rate of acceleration, its final speed is just twice the average speed, so:

  $v = 2 \times \frac{H}{t}$

- We have seen that the kinetic energy gained by the object must be equal to mgH, but we want it in terms of the speed. From the first equation for v above, t must be equal to $\frac{v}{g}$, and putting this into the second equation leads to the useful result that:

$$g \times h = \frac{1}{2}v^2$$

Finally, therefore, the kinetic energy, $m \times g \times H$, becomes:

$$m \times \frac{1}{2} \times v^2 \text{ (or simply } \frac{1}{2}mv^2\text{).}$$

Summary

The *weight*, (W), of an object is equal to mg – its mass multiplied by the acceleration due to gravity.

The change in *potential energy* of a mass m displaced through a vertical distance H is mgH. If m is in kilograms and H is in metres, then the energy is in joules.

The kinetic energy of a mass, m, moving with speed, v, is $\frac{1}{2}mv^2$.

BOX 4.2 **Using kinetic energy: a wind turbine**

How much power does the wind deliver to a turbine with a diameter of 60 metres when the wind speed is 9 metres per second (about 20 miles per hour)?

The density of air at normal pressure is 1.29 kg m⁻³.

Air has mass, so the wind – air in motion – carries kinetic energy, and a modern wind turbine is designed to transform this into **electrical energy**. In siting a turbine or a wind farm it is obviously important to know how much energy the wind can deliver, and the aim of this case study is to show in principle how this may be calculated.

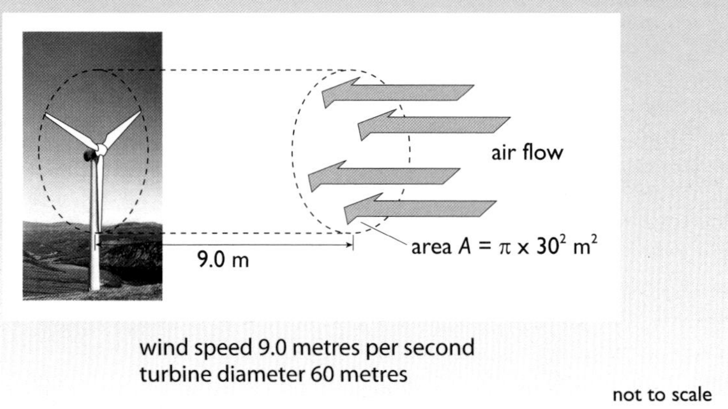

Figure 4.3 The air reaching the wind turbine in one second

First the question needs clarifying. The wind delivers energy continuously, so we need to ask about the *rate* at which energy arrives: not the number of joules, but the number of joules per second. This is technically the *power*, and as shown in Box 2.3 its unit is the *watt* (W), one watt being a rate of one joule per second.

To find the rate at which kinetic energy is delivered by the wind, we need to know the mass of air reaching the turbine per second. The reasoning is as follows.

- If the wind speed (v) is 9 m s^{-1}, all the air in a cylinder with a length of 9 metres will pass through the swept area (A) in each second (Figure 4.3).
- The diameter is 60 metres, so A is the area of a circle with radius 30 metres, which is $\pi \times 30^2$ square metres. So the volume (V) of the 9-metre cylinder is:

$$V = 9 \times \pi \times 30^2 = 25\,447 \text{ cubic metres}$$

- The density of air (the mass per cubic metre) is 1.29 kg m^{-3}, so the mass of air arriving at the turbine per second is:

$$m = 1.29 \times 25\,447 = 32\,827 \text{ kg (over thirty tonnes!)}$$

- The kinetic energy of a mass m moving with speed v is $\frac{1}{2}mv^2$ (Box 4.1).
- So finally, the energy arriving per second, which is the *power delivered by wind*, is:

$$P = \frac{1}{2} \times 32\,827 \times 9^2 = 1\,329\,494 \text{ watts,}$$

or about **1.33 MW**

Note how important the wind speed is in this calculation. Firstly it determines the mass of air arriving, and then the kinetic energy depends on the square of the wind speed (v^2), so the delivered power is proportional to the *wind speed cubed*. It is easy to show that in the present case a relatively small fall in wind speed to 7 m s^{-1} would reduce the input power to less than half the above figure, while an increase to 12 m s^{-1} would raise it to over 3 MW. This strong dependence on wind speed is obviously very important, both in assessing a site and in considering the effects of wind variability.

Of course, not all the power delivered by the wind can be extracted. The air continuing to move downwind of the turbine will carry away some kinetic energy, and in any machine there are energy 'losses' in the form of heat produced by friction. Even under ideal wind conditions the electrical power output of a wind turbine will be less than half the wind power input, and for the 60-metre turbine in the 20 mph wind, the electrical output might perhaps reach 600 kW.

In practice, of course, wind speeds are constantly fluctuating, and any prediction of the expected output from a turbine or a wind farm requires detailed assessment of the wind regime for the particular site.

BOX 4.3 Using potential energy: pumped storage

The upper reservoir of a pumped storage system has a surface area of 8 square kilometres, and lies 370 metres (about 1200 ft) above the lower reservoir. How much energy is stored if overnight pumping raises the water level in the top reservoir by half a metre?

The density of water may be taken to be 1000 kg m^{-3} and the approximation g = 10 m s^{-2} may be used.

The idea of a pumped storage system is to use the surplus overnight output of a power station to pump water into a high reservoir, hold it there until electricity demand rises again and then release it to generate power (Figures 4.4 and 4.5).

The surface area of the reservoir is 8 square kilometres, which is 8 million square metres. So the volume of water needed to raise the level by half a metre is:

$0.5 \times 8 = 4$ million cubic metres, or 4×10^6 m^3

So with a density of 1000 kg per cubic metre, the stored mass is:

volume × density $= 4 \times 10^6 \times 1000 = 4 \times 10^9$ kg (4 billion kg or 4 million tonnes)

The stored energy is the potential energy gained when this mass of water is raised through a height of 370 metres.

Box 4.1 showed that the gain in potential energy when a mass m is raised through a height H is equal to $m \times g \times H$.

Using the approximation g = 10 m s^{-2}, the gain in potential energy in raising each kilogram of water through 370 metres is therefore:

$m \times g \times H = 1 \times 10 \times 370 = 3700$ joules

So the gain in raising the above 4 billion kilograms is:

$4 \times 10^9 \times 3700 = 14.8 \times 10^{12}$ J = **14.8 TJ, or 14.8 million MJ**

Given that one kilowatt-hour (1 kWh) is equal to 3.6 MJ, this stored energy is just over 4 million kWh –

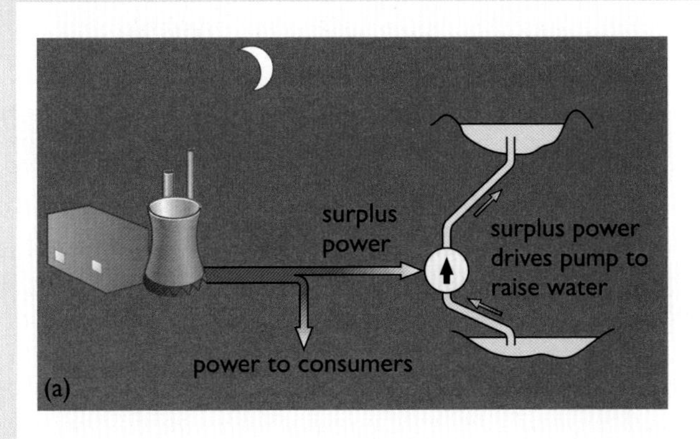

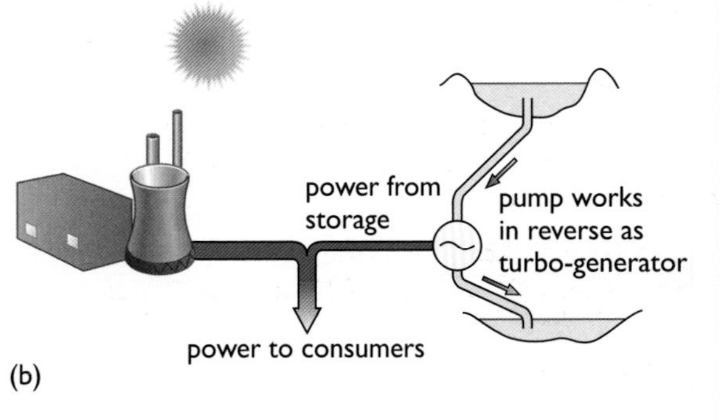

Figure 4.4 Pumped storage system (a) at time of low demand (b) at time of high demand

enough to supply over a third of a million UK households with electricity for a day.

More details of pumped storage appear in *Renewable Energy*.

Figure 4.5 The upper reservoir of the Cruachan pumped storage plant, 1200 ft above Loch Awe in Scotland

4.3 Heat

Figure 4.6 Francis Bacon (1561–1626) led an adventurous life seeking advancement in the courts of Elizabeth I and James I. As Lord Chancellor he was found guilty of taking bribes – despite claiming that they hadn't changed his decisions! Sent to the Tower, he was released when the king paid his fines. He is best known today as one of the first to argue that science should proceed by observing the real world, rather than studying the writings of the ancient Greeks.

Writing nearly 400 years ago, Francis Bacon (Figure 4.6) thought heat to be a 'brisk agitation' of the particles of matter:

> [...] not that heat generates motion or that motion generates heat (though both are true in certain cases), but that heat itself, its essence and quiddity, is motion and nothing else.
>
> (Bacon, 1620)

Beautifully expressed and very much like our present view, but unfortunately Bacon was far ahead of his time. Quite different theories, regarding heat as some kind of fluid, arose during the next 200 years and dominated well into the nineteenth century. The final acceptance of Bacon's view came only during the later years of the century, with the success of the **atomic theory of matter**. The atomic theory held that everything consisted of arrangements of indivisible fundamental particles, called atoms – and that there was only a limited number of types of atom. (The name came from the Greek ἁτομοσ, meaning 'uncuttable'.)

As this picture of matter became established, the **kinetic theory of heat** developed rapidly. A simple gas such as helium came to be seen as consisting of a very large number of identical atoms moving constantly in random directions. When the average kinetic energy of the atoms was found to depend only on the temperature, it was a relatively short step to identifying heat itself as kinetic energy. For more complex materials, such as molecules or metals, the energy associated with the forces between atoms must also be taken into account, but it remains the case that heat energy is not some new *fundamental* form. The 'heat content' of a simple gas is no more than the kinetic energy of its atoms, and an increase in temperature is an increase in the average speed of these atoms: *hotter means faster*.

BOX 4.4 **Using heat: storage heaters**

A small off-peak electric hot water cylinder (Figure 4.7) holds 140 litres of water. Assuming that the water is initially at 10 °C and is heated overnight to 60 °C, how many such cylinders would be needed to store the surplus power station output of Box 4.3 in the form of heat?

The idea of storing surplus energy in the form of heat is probably very old. (Did hunter-gatherers use rocks heated by the sun or on a fire to keep themselves warm in their caves?) The use of cheaper 'off-peak' electricity overnight to provide space heating or hot water for the following day was introduced in the UK in the 1950s. More recently, growing interest in solar heating has generated many ideas for storing heat in the masonry of buildings.

The above question is not of course suggesting a practicable scheme, but a useful comparison of the energy storage capacities associated with two different forms of energy.

First, some data on water and a definition:

- The density of water is 1000 kg m^{-3} (kilograms per cubic metre). There are 1000 litres in one cubic metre, so the mass of water stored in one 140 litre cylinder is 140 kg.
- The heat energy needed to raise the temperature of 1 kg of any substance by 1 °C is called its **specific heat capacity**, C. For water, the value of C is 4200 J kg^{-1} K^{-1} (joules per kilogram per kelvin).

The accepted scientific unit for temperature is the kelvin (K), but an increase of 1K is exactly the same as an increase of 1 °C, so this more familiar unit can be used in the calculations here. (The kelvin scale of temperature will be introduced in Chapter 6.)

It follows from the definition of specific heat capacity, C, that the heat energy needed to raise the temperature of M kg of a substance from a temperature T_1 to a temperature T_2 is equal to:

$$C \times M \times (T_2 - T_1)$$

So the heat energy needed to raise the temperature of 140 kg of water from 10 °C to 60 °C is:

4200 × 140 × 50 = 29 400 000 J, which is **29.4 MJ**.

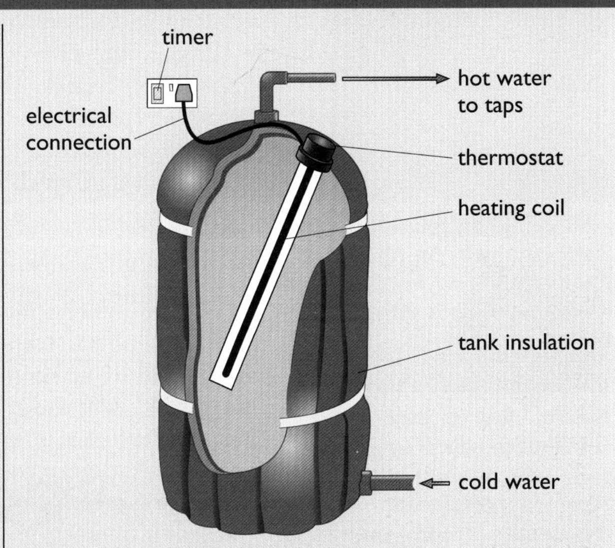

Figure 4.7 Off-peak hot water cylinder

One kilowatt-hour is 3.6 MJ, so in household terms the cylinder is storing just over 8 kWh.

Returning to the question. The reservoir of Box 4.3 stored 14.8 million MJ of energy. So the number of these cylinders needed to store this would be:

14.8 million divided by 29.4, which is about **half a million**!

However, it is interesting to compare the volumes of water needed in the two cases. Half a million 140 litre cylinders hold a total of 70 thousand cubic metres of water, which is obviously very much less than the 4 million cubic metres to be raised by pumping.

So heating a cubic metre of water by 50 °C stores a great deal more energy than lifting the same cubic metre 370 m up a mountain. If the stored energy is so much greater, and electrical heating is technically simpler, why is this method not used to store the power station output, instead of pumped storage? One answer comes when we ask how the electrical energy is to be retrieved, and lies with the Second Law of Thermodynamics, to be discussed in Chapter 6.

4.4 Electrical energy

James Joule (Figure 4.8) was a man with a passion for science. Working alone in his private laboratory, he carried out a series of beautifully precise experiments in the 1840s. In the best known of these he used a falling weight to drive a paddle wheel whose rotation heated a fluid, and established

Figure 4.8 James Joule (1818–1889), almost entirely self-educated in science, made little contact with the scientific establishment until 1847, when he took his results to a meeting at Oxford. The chairman told this unknown person to be brief, and tried to prevent any discussion, but (by Joule's later account) a young man rose and 'created a lively interest'. This was William Thomson, Lord Kelvin (Figure 6.9), whose own account tells of 'a very unassuming young man, who betrayed no consciousness that he had a great idea to unfold. I was tremendously struck with the paper ...'

that the heat produced was proportional to the work done by gravity in pulling down the weight. Another version was even more ingenious: the falling weight drove a small, electric generator that effectively ran a simple immersion heater. The relationship between *work* and *heat* was the same in both experiments, and the same again in others. By the end of the 1840s Joule was convinced that the heat output was not merely related to the work input, but that work was being converted into heat.

The second of Joule's experiments is extremely interesting because instead of direct conversion from work to heat, there is an intermediate stage where electrical energy must be playing a role.

Electrical effects were already familiar in the mid-nineteenth century. It was long known that there are two types of electricity: rubbing a cat's fur with an amber rod leaves the rod with one type and the cat with the other. Both rod and cat have acquired electric charge, but of opposite types – which came to be called positive and negative. Delicate experiments showed that objects with the same type of charge repelled each other whilst those with opposite charges attracted. The similarity with the gravitational force acting between two masses was recognized, and led to the idea of **electrical potential energy** arising from forces between two or more electric charges.

Electrical circuits, involving moving rather than static electric charges, were also investigated. Alessandro Volta invented the battery (the 'voltaic cell') in 1799, and about twenty years later, Georg Ohm began to create order from rather diffuse ideas of what we now call **current** and **potential difference** (see 'Electrical circuits', below). An **electric current**, he said, is a flow like the flow of heat, and potential difference is like the temperature difference that causes the flow. The analogy is good, but further clarification had to wait for the next advance in theories of matter: the demise of the indivisible atom.

Electrons

The death knell of the 'uncuttable' atom came with the discovery of the **electron**. Between about 1870 and 1890, experiments with a forerunner of the cathode ray TV tube revealed that a red-hot wire emitted particles with very interesting properties.

- They are very light, with about one two-thousandth of the mass of the lightest atom.
- They are all completely identical: the same mass and the same electric charge.
- Their electric charge is of the type called negative.

The discovery of these very light, electrically charged particles which must come from *inside* the atoms dramatically changed the picture of matter. It revealed that atoms must be composite, rather than single entities. Some other component (or components) must account for most of the mass of the atom, and because a complete atom is electrically neutral, this must have a *positive* electric charge. One early idea was the 'plum pudding' model: an atom consisting mainly of a single positively charged mass, with the light, negatively charged electrons distributed throughout it. Further evidence led ultimately to a different arrangement (see Section 4.6), but regardless of the detailed picture, it was soon recognized that electrons could offer an explanation of the flow of electric current in a metal.

The suggestion is that each atom in a metal can relatively easily release an electron (or in some cases more than one) and that these **free electrons** are the carriers of electric currents. The metal atoms, now with surplus positive charge, will have become *positive ions*. (An **ion** is the electrically charged particle that results when an atom or molecule has gained or lost one or more of its normal quota of electrons.) The picture of any piece of metal is therefore of a lattice of positive ions through which the free electrons are constantly moving, in random directions and at extremely high speeds (Figure 4.9). When a length of wire is connected to an electrical supply, the relatively slow drift of this entire fast-moving swarm of electrons along the wire is the electric current whose effects we observe.

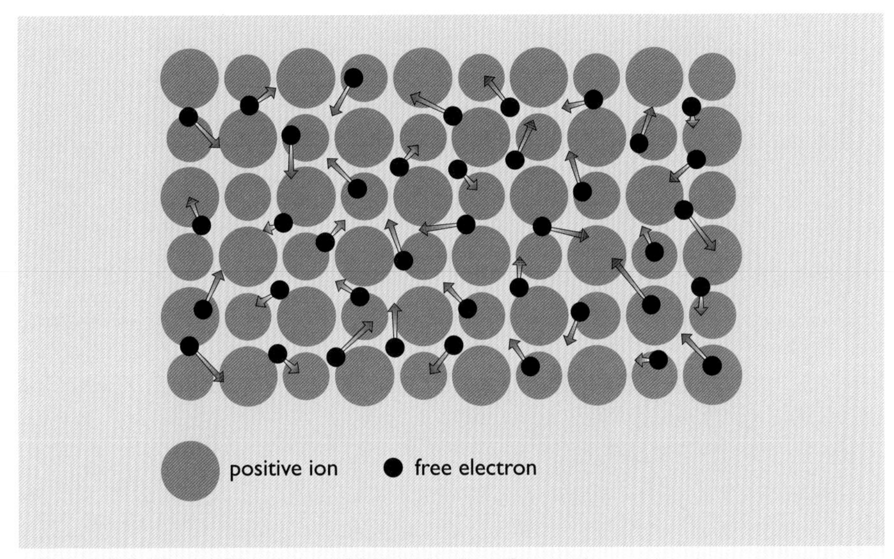

Figure 4.9 The positive ions and free electrons of a metal

Ohm famously determined that the *current in a wire is proportional to the potential difference* across it (**Ohm's Law**), and Joule almost equally famously discovered that the rate at which heat is developed by a flowing current is proportional to the *current squared* (the phenomenon known as Joule heating). Together, these results lead to useful expressions for electrical energy and power.

Electrical circuits

Current

Electric current could in principle be measured by counting the number of electrons per second passing any point, but even a modest current would involve numbers in the trillions. The normal *unit* for current is the more practical ampere, or **amp** (**A**), with its sub-units milliamp (**mA**) and microamp (**μA**). The usual *symbol* for electric current is I, as in the equations below.

BOX 4.5 A power supply

It is important to appreciate that an electrical *supply* does not supply electrons; it recycles them.

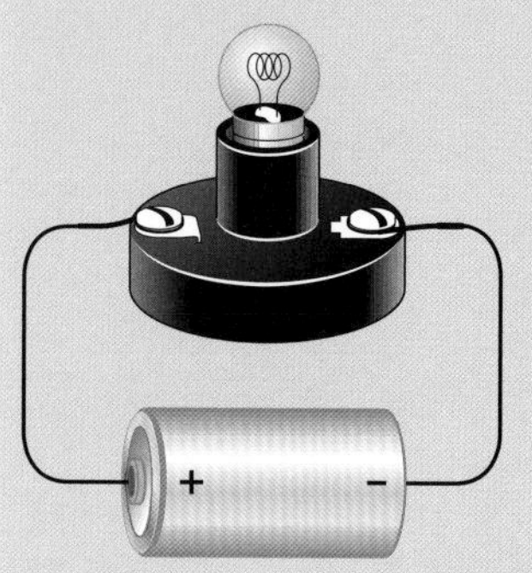

Figure 4.10 A torch bulb circuit

Consider a simple circuit such as a battery and torch bulb (Figure 4.10). When the battery is connected, the entire swarm of fast-moving free electrons drifts along the wire filament of the bulb, constantly colliding with the ions of the metal. In these collisions the ions gain energy, which we observe as heating of the metal. The electrons, collectively losing energy as they pass along the wire, eventually return to the battery, which replaces the lost energy. As long as the circuit is complete, the battery supplies energy continuously, and this is the real function of any electrical supply. It is, correctly, a **power supply.**

Voltage

The quantity commonly referred to as the **voltage** of a supply is the difference in *electrical potential* that it maintains between its terminals: a measure of the energy supplied to electric charges as they pass through. The *unit*, the **volt (V)**, is defined so that if a current of one amp flows for one second from a one-volt supply, the energy supplied is exactly one joule. Rather confusingly, the usual *symbol* for voltage in an equation is also V.

The potential difference between the two ends of a wire such as the bulb filament in Figure 4.11 is commonly referred to as *the voltage across the wire*. In the circuit shown this is equal to the supply voltage. If two or more such components are connected in a simple loop with a supply, the *sum* of the individual component voltages must be equal to the supply voltage.

Resistance

If two quantities are proportional to each other, the result of dividing one by the other is always the same. Ohm's Law therefore allows us to define

BOX 4.6 The torch bulb

If the bulb is labelled 1.5 V, 0.3 A, what are its resistance and its power when connected to the 1.5 V battery?

Figure 4.11 shows the torch bulb circuit with the conventional symbols for the battery and a resistor (the bulb filament).

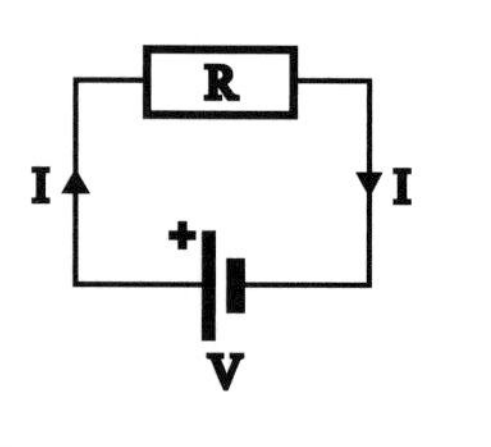

Figure 4.11 The bulb circuit with conventional symbols

The **resistance** of the bulb is the voltage divided by the current:

$$R = \frac{V}{I} = \frac{1.5}{0.3} = \mathbf{5.0\ \Omega}$$

Note that this is the resistance of the bulb filament when it is very hot, which could be several times its resistance at room temperature.

The power is the voltage multiplied by the current:

$$P = V \times I = 1.5 \times 0.3 = \mathbf{0.45\ W}$$

Simple calculations like this can be important in considering circuits in which the power supply is an array of photovoltaic cells (see Box 4.8).

the resistance (R) of a component as the voltage (V) across it divided by the current (I) flowing through it:

$$R = \frac{V}{I}$$

The *unit* for resistance is the **ohm**, abbreviated as **Ω** (Greek letter omega).

The resistance of a particular component, defined in this way, may not in practice have a constant value. Increasing the current in a wire, for instance, may cause its temperature to rise, and this usually increases its resistance. But even if this occurs, it remains useful to define resistance as voltage divided by current, and indeed, many scientists, if asked to 'state Ohm's Law', will offer a well-known re-arrangement of the above definition:

$$V = I \times R$$

Electric power

As we have already seen in many cases, power is the rate at which energy is transformed, and its unit is the watt (W). So Joule's rule for the heating effect can be reformulated: the *power dissipation* in a component is proportional to the square of the current (I^2). This rule only holds, however, if Ohm's Law also holds. The more fundamental general rule, combining Joule's rule and the above definition of a volt, is that the *power in watts is equal to the voltage in volts times the current in amps*:

$$P = V \times I$$

This relationship holds equally for the electric power provided by a supply or the power being dissipated as heat in a component such as a lamp filament.

BOX 4.7 Storing electrical energy: a car battery

A heavy-duty 12-volt car battery has a storage capacity of 100 ampere-hours. How much energy does it store when fully charged?

Figure 4.12 A car battery

The most commonly used systems for storing electrical energy are rechargeable batteries. Strictly, the electrical energy is converted into chemical energy as a battery is charged, with the reverse process taking place when it is used to deliver electric power. Nevertheless, it seems appropriate to consider this method of storage in the context of electrical rather than chemical energy.

Batteries, or sets of batteries, offer a wide range of energy storage capacities and usually have the advantage of being portable. This case study considers the potential of the common lead–acid car battery in two very different contexts.

The storage capacity of a car battery is usually expressed in ampere-hours. A battery with a capacity of 100 ampere-hours could in principle supply a current of 2 amps for 50 hours, half an amp for 200 hours, and so on.

Take the first case. If the current is 2 A and the voltage is 12 V, the power output (see main text) is:

$$P = V \times I = 12 \times 2 = 24 \text{ W}$$

One watt is one joule per *second*, so the energy supplied in 50 hours is:

$$24 \times 50 \times 60 \times 60 = 4\ 320\ 000 \text{ J} = \mathbf{4.32\ MJ}$$

This therefore is the energy stored by the fully charged 100 ampere-hour 12 V battery.

It is, of course, also equal to:

100 ampere-hours × 12 volts = 1200 watt-hours or **1.2 kWh**

Comparison of the above 4.32 MJ with the 1500 MJ stored by the contents of a full 45-litre (~10-gallon) petrol tank highlights the major problem for electric cars. The achievement of a reasonable range of travel requires the development of rechargeable batteries with much greater storage capacity per kilogram (or per cubic metre of occupied space).

It is also worth noting the impracticability of present-day battery storage for really large-scale systems. Storing the overnight power-station output of Box 4.3 (about 4 million kWh) would require the equivalent of about *three and a half million* car batteries.

Figure 4.13 Michael Faraday (1791–1867), the son of a London blacksmith, became a bookbinder's apprentice at the age of 13. Reading books he bound, he became fascinated by science. He was delighted when a customer gave him tickets for four lectures by Sir Humphrey Davy, and made careful notes afterwards. Determined to become a scientist, he wrote to the President of the Royal Society, but received no reply. Eventually he appealed to Davy, sending his written-up notes as support. Davy was impressed, and took him as his laboratory assistant at the Royal Institution. Eventually succeeding Davy as Director, he was a popular lecturer, founding the Friday evening discourses, which continue today.

4.5 Electromagnetic radiation

In the early nineteenth century, long before the discovery of the electron, experiments began to show that there was an extremely close relationship between *electricity* and *magnetism* – two effects that had previously been regarded as quite independent. In 1820 Hans Oersted showed that a compass needle was deflected when brought near an electric current. In the same year André Ampère demonstrated that there is a *magnetic* force between two parallel wires carrying electric currents. In 1821 Michael Faraday (Figure 4.13) showed that a wire carrying a current would spontaneously rotate around a magnet – the first electric motor. In 1832 he found that moving a bar magnet rapidly through a coil of wire caused a current to flow briefly in the coil – the first generator (see Chapter 9).

In 1864 James Clerk Maxwell (Figure 4.14) published his theory of electromagnetism, drawing together all these results as a set of mathematical equations. Its consequences were revolutionary.

Maxwell's laws of electromagnetism are arguably as great an achievement as Newton's laws of motion. (Ludwig Boltzmann, another great scientist, was prompted to quote Goethe, '*Was it God who wrote these lines* ...'.) Maxwell adopted the idea of **fields of force**, spreading throughout space. An **electric field** surrounds any electric charge, and another charge experiences a force if it comes into this field. A **magnetic field** surrounds a magnet or an electric current, and another magnet or current experiences a force if it comes into this field. Faraday had shown that a voltage, which implies an *electric field*, can be produced by a *changing magnetic field* and Maxwell introduced the converse effect, that a *magnetic field* is produced by a changing electric field. Expressing all these results in mathematical form, he showed that they led to a remarkable prediction. There should be **electromagnetic waves** that consist of nothing but oscillating electric and magnetic fields, and these would be able to travel through totally empty space. Crucially, the theory also predicted the speed of the waves – and it was equal to the speed of light.

Figure 4.14 James Clerk Maxwell (1831–1879) was at first taught at home by his mother, but sent away to school when she died. It was not a success. He was eight, shy and wore strange clothes, and the boys called him 'dafty'. Then, to general amazement, he began to win prizes, and at 14 his first mathematical paper was published. He continued to win prizes (and write poetry) and in 1860 became professor at Kings College London, where he developed his main ideas on electromagnetism. He resigned in 1865, partly through ill-health, and his great book on the subject was written mainly in Scotland. In 1871 he was persuaded back to Cambridge, where he devoted his remaining few years to establishing the Cavendish Laboratory.

The nature of light had been debated for centuries, most famously between Newton, who thought it to be a stream of particles, and others who supported a wave theory. By the nineteenth century the evidence in favour of waves was incontrovertible, but the nature of these waves remained unclear. Maxwell's theory now identified them as electromagnetic waves. The question then arose of the possibility of other types of electromagnetic wave. The radiation lying just outside the visible spectrum (ultraviolet and infrared at shorter and longer wavelengths respectively) was presumably **electromagnetic radiation**, and heat radiation (now known to lie mainly at still longer wavelengths) could be included. Then in 1887 Heinrich Hertz demonstrated the first man-made 'signal' carried by electromagnetic waves – the precursor of radio, TV and microwave transmissions. The full range of today's known electromagnetic spectrum, from the gamma rays of radioactivity to the longest radio waves (Figure 4.15), would appear only gradually during the next century.

The sources of *all* electromagnetic waves are oscillating electric charges. In the case of light these are the electrons in atoms; but any oscillating charge or alternating current necessarily generates electromagnetic radiation. As Maxwell's theory predicted, the waves carry *energy*. Indeed, one might regard electromagnetic radiation, travelling through space in the absence of any material substance, as 'pure energy'.

There was still to be another twist. Max Planck in 1900 (Figure 4.16) and Albert Einstein in 1905 showed that the energy carried by electromagnetic radiation is *quantized*. It is transmitted or received not as a continuous flow, but in discrete well-defined energy units or *quanta*, called **photons**. The energy of each photon depends only on the rate at which the wave oscillates. A photon of red light carries less energy than one of blue light, which has a shorter wavelength and so oscillates at a higher frequency, and a photon of ultraviolet is more energetic still – matters of some importance for photovoltaic systems. (Perhaps the great Newton was right after all, and light really is a stream of particles? Unfortunately not, and we must live with the fact that light behaves sometimes as a wave and sometimes as discrete packets of energy. It *literally* depends on how you look at it.)

One final, rather theoretical point may be worth noting. Electromagnetic radiation and the 'electrical energy' discussed in Section 4.4 are both included in Maxwell's theory, so they could perhaps be described as different aspects of a single form of energy. However, it remains convenient for many practical purposes to treat them as separate forms, as done here.

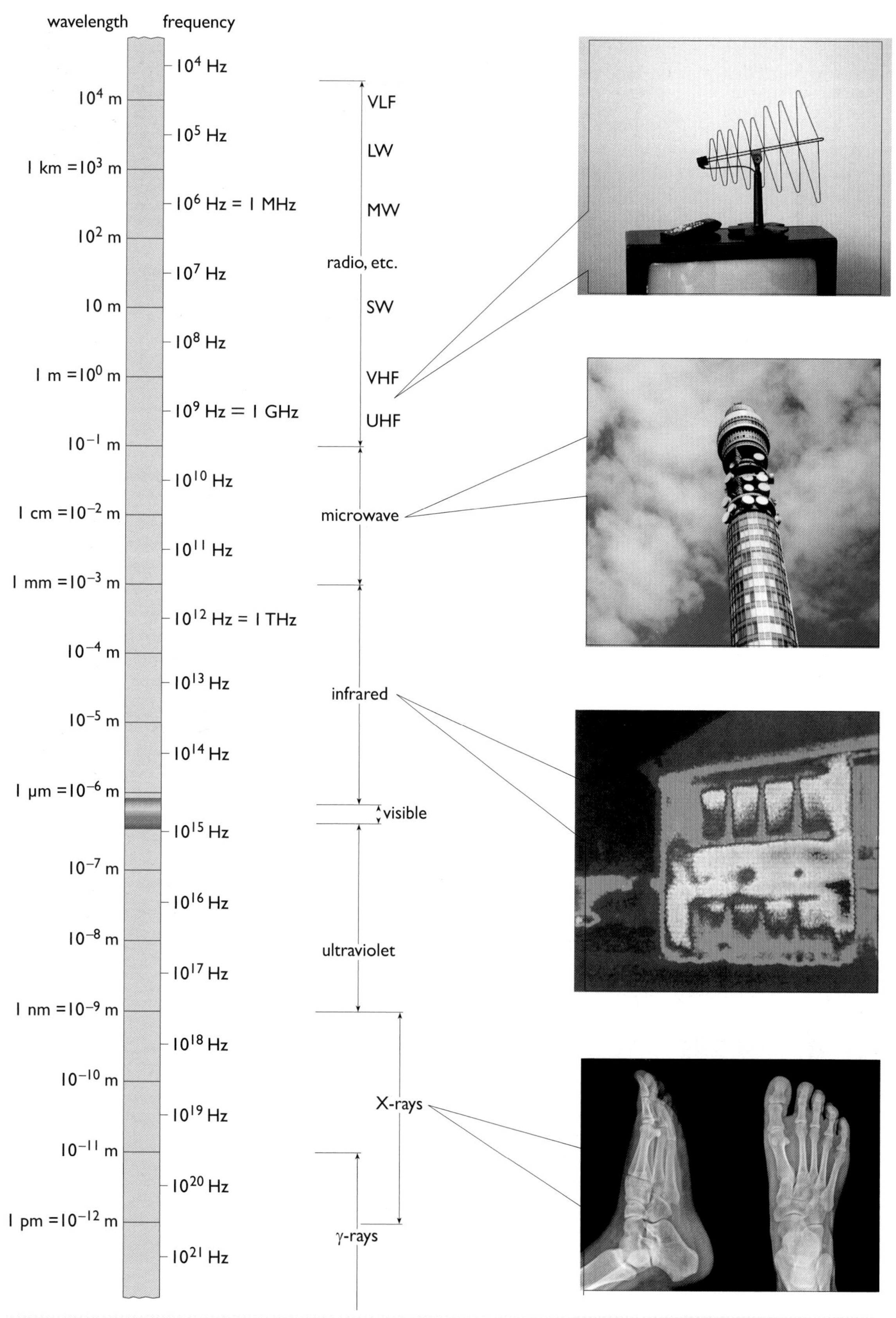

Figure 4.15 The electromagnetic spectrum; 1 Hz = 1 Hertz = 1 cycle per second

BOX 4.8 Using electromagnetic radiation: solar PV cells

The power supply for a small device (Figure 4.17) is an array of thirty-two PV cells, connected as shown in Figure 4.18. In bright sunlight, each cell can maintain a voltage of 0.5 V while producing a current of 60 mA (milliamps). How much power is being supplied to the device?

A **photovoltaic cell**, as its name suggests, uses the energy of light, or other radiation near the visible region of the spectrum, to generate electric power. Figure 4.18 shows the thirty-two PV cells specified in the question, arranged in four *parallel* rows each with eight cells in *series*.

The cells are all identical, so an easy way to find the total power is to calculate the power from a *single* cell and multiply by thirty-two. Each cell maintains a voltage of 0.5 V and supplies a current of 60 mA. As shown in Section 4.4, the electric power P, in watts (W), delivered by any supply is equal to its voltage V, in volts, multiplied by the current I, in amps, that it is supplying.

One milliamp is one thousandth of an amp, so 60 mA is 0.060 A, and the power supplied by one cell is:

$$P = V \times I = 0.50 \times 0.060 = 0.030 \text{ W}$$

Figure 4.17 A small battery charger incorporating the photovoltaic array shown in Figure 4.18

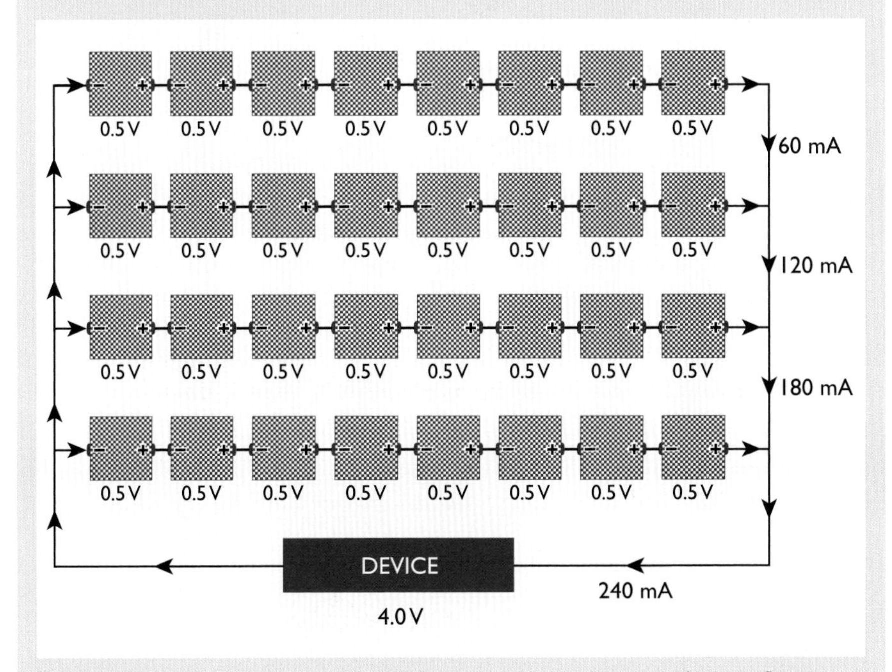

Figure 4.18 Electrical connections of the PV array, showing the directions of current flow

Figure 4.16 Max Planck (1858–1947) was a gifted student of both music and physics. He chose the latter as his career, and won the Nobel Prize in 1918 for the concept of energy quanta. He was president of the leading German scientific society, but crossed swords with the Nazi governments of the 1930s, and eventually resigned in protest at the treatment of Jewish academics. His eldest son was killed at Verdun in 1916, and the second was executed following the 1944 plot against Hitler. In 1945, escaping on foot from the final battles of the war, Planck, aged 87, and his wife were robbed and beaten. Found by Americans, they were taken to safety in Göttingen, where he resumed teaching.

The total power from the thirty-two cells is therefore:

$$32 \times 0.030 = 0.96 \text{ W, or just under one watt}$$

However, it may be worth looking at the system from the point of view of the *device*. Many devices require their power to be delivered at a particular *voltage*, and the arrangement of the cells will affect this, as follows.

In the arrangement shown, each parallel row has eight cells, and each cell behaves like a small battery with a voltage of 0.5 volts. So if they are connected correctly, positive-to-negative in series along the row, the total voltage across the row will be:

$$8 \times 0.5 \text{ volts, or } 4.0 \text{ V}$$

This is of course the same for each row, and is therefore also the voltage across the device. Note that, differently arranged, the same 32 cells might still be able to supply the same power, but would not necessarily produce the correct voltage for the device.

The total current being supplied to the device can also be calculated from the diagram. The 'boost' from each cell comes from its 0.5 V step up in voltage, but the same 60 mA current must flow through all eight cells in any row (there being nowhere else for it to go!). So the current flowing into and out of the top row of cells is 60 mA. On its way to the device, this receives an increase of another 60 mA from the second row, and so on, so that the total current flowing into the device is:

$$4 \times 60 \text{ mA} = 240 \text{ mA, or } 0.24 \text{ A}$$

The complete system, from the point of view of the device, is thus a *4.0 V power supply providing 0.24 A*, and the power delivered is:

$$P = V \times I = 4.0 \times 0.24 = \mathbf{0.96\ W}$$

As we should expect, the result is the same as that calculated above from the data for individual cells.

4.6 Chemical energy

The chemical elements

During the nineteenth century chemistry had become a very well-established science. An important step was the recognition that the chemical **elements** were those substances that consist of just one type of atom, and by the mid-century more than fifty of them had been identified. Careful measurement of the relative masses when these elements combined to form **compounds** led to the construction of a list placing all the known elements in order of their 'atomic weight' relative to hydrogen, the lightest element.

An interesting feature of this list was that different elements that were known to have similar chemical properties often appeared at regular intervals. To display this property, the idea arose of arranging the elements, again in order of their atomic weights, but in horizontal rows, starting a new row whenever the properties started to repeat. The result was a table in which elements with similar chemical properties appeared vertically above each other, an arrangement that came to be called a **periodic table of the elements**.The most famous of these tables, and the model for many later versions, was proposed in 1869 by Dmitri Mendeleev, who cleverly left gaps where an element with the required properties had not yet been identified.

By the year 1900 most of these gaps had been filled; but some elements remained persistently absent, and it was only in the 1940s that the last few of the 92 **naturally occurring** elements were identified and given names. (The last to be found, atomic number 61, was named promethium, after the titan who stole fire from the gods to give it to mankind.) In the same period, developments in nuclear physics were leading to the creation of new 'man-made' elements beyond uranium (see Chapter 10). The periodic table in Figure 4.19 shows all the elements that have been identified and given names at the time of writing, with six more whose status is not yet fully established. Table 4.1 lists the elements in alphabetical order.

The assignment of an **atomic number** to each element, identifying its position in the complete sequence, remained controversial well into the twentieth century (Scerri, 2007), and was resolved only with the development of a more detailed picture of the structures of atoms (see 'The nuclear atom', below).

Using chemical energy

Once it was established that a *chemical element* is a substance that consists of just one type of *atom*, it followed that a **chemical compound,** formed by the combination of two or more elements, must have a basic unit, a **molecule**, that is a cluster of the corresponding atoms.

With this picture, a **chemical reaction** between two substances can be visualized as the reassembly of their constituent atoms into different molecules, a process that almost always involves either an input of energy or the release of energy. Reactions that require an energy input are called **endothermic**; but those of interest in the present context are the **exothermic** processes, and in particular, those in which the chemical compounds we call **fuels** react with oxygen from the air to release energy in the form of heat. We know that energy is conserved, so this heat must come from changes in the energies of the molecules present before and after the combustion. But the exact nature of the 'chemical energy' stored in these molecules could only be established as more became known about the structure of atoms, a topic to be resumed in the next subsection.

One feature of all common fuels: coal, oil, gas, wood, etc., is that they are (or include) compounds involving the elements carbon and hydrogen. Box 4.9 looks in more detail at the processes involved in their combustion.

1 H 1.008																	2 He 4.003
3 Li 6.939	4 Be 9.012											5 B 10.81	6 C 12.01	7 N 14.01	8 O 16.00	9 F 19.00	10 Ne 20.18
11 Na 22.99	12 Mg 24.31											13 Al 26.98	14 Si 28.09	15 P 30.97	16 S 32.06	17 Cl 35.45	18 Ar 39.95
19 K 39.10	20 Ca 40.08	21 Sc 44.96	22 Ti 47.90	23 V 50.94	24 Cr 52.00	25 Mn 54.94	26 Fe 55.85	27 Co 58.93	28 Ni 58.71	29 Cu 63.54	30 Zn 65.37	31 Ga 69.72	32 Ge 72.59	33 As 74.92	34 Se 78.96	35 Br 79.91	36 Kr 83.80
37 Rb 85.47	38 Sr 87.62	39 Y 88.91	40 Zr 91.22	41 Nb 92.91	42 Mo 95.94	43 Tc (98)	44 Ru 101.1	45 Rh 102.9	46 Pd 106.4	47 Ag 107.8	48 Cd 112.4	49 In 114.8	50 Sn 118.7	51 Sb 121.8	52 Te 127.6	53 I 126.9	54 Xe 131.3
55 Cs 132.9	56 Ba 137.3	see below	72 Hf 178.5	73 Ta 181.0	74 W 183.5	75 Re 186.2	76 Os 190.2	77 Ir 192.2	78 Pt 195.1	79 Au 197.0	80 Hg 200.6	81 Tl 204.4	82 Pb 207.2	83 Bi 209.0	84 Po (209)	85 At (210)	86 Rn (222)
87 Fr (223)	88 Ra (226)	see below	104 Rf (267)	105 Db (268)	106 Sg (271)	107 Bh (272)	108 Hs (270)	109 Mt (276)	110 Ds (281)	111 Rg (280)	112 Cn (285)	113 Uut (284)	114 Uuq (289)	115 Uup (288)	116 Uuh (293)	117 Uus (294)	118 Uuo (294)

Lanthanoids	57 La 138.9	58 Ce 140.1	59 Pr 140.9	60 Nd 144.2	61 Pm (145)	62 Sm 150.4	63 Eu 152.0	64 Gd 157.3	65 Tb 158.9	66 Dy 162.5	67 Ho 164.9	68 Er 167.3	69 Tm 168.9	70 Yb 173.0	71 Lu 175.0
Actinoids	89 Ac (227)	90 Th 232.0	91 Pa 231.0	92 U 238.0	93 Np (237)	94 Pu (244)	95 Am (243)	96 Cm (247)	97 Bk (247)	98 Cf (251)	99 Es (254)	100 Fm (257)	101 Md (258)	102 No (255)	103 Lr (256)

Figure 4.19 The Periodic Table of the Elements. The number above the symbol for each element is its atomic number, reflecting its position in the table, and the number below is the relative atomic mass. An alternative 'long' version of the table places the lanthanoids and actinoids (previously called lanthanides and actinides) in the main sequence. A mass in brackets indicates that the available sample is too small for comparison by chemical methods (in some cases just a few atoms). This is often the case for the transuranic 'man-made' elements, which fall into two classes: those whose existence has been recognized and the name approved by IUPAC (The International Union of Pure and Applied Chemistry), and those for which this is not yet the case. The latter are now given temporary names in sequence, from ununtritium (113) to ununoctium (118) at the time of writing. (A useful website showing the properties and present status of all the elements is WebElements™ (Winter, 2010).)

Table 4.1 The named elements with their symbols and atomic numbers

Element	Symbol	Atomic number	Element	Symbol	Atomic number	Element	Symbol	Atomic number
actinium	Ac	89	gold	Au	79	promethium	Pm	61
aluminium	Al	13	hafnium	Hf	72	protactinium	Pa	91
americium	Am	95	hassium	Hs	108	radium	Ra	88
antimony	Sb	51	helium	He	2	radon	Rn	86
argon	Ar	18	holmium	Ho	67	rhenium	Re	75
arsenic	As	33	hydrogen	H	1	rhodium	Rh	45
astatine	At	85	indium	In	49	roentgenium	Rg	111
barium	Ba	56	iodine	I	53	rubidium	Rb	37
berkelium	Bk	97	iridium	Ir	77	ruthenium	Ru	44
beryllium	Be	4	iron	Fe	26	rutherfordium	Rf	104
bismuth	Bi	83	krypton	Kr	36	samarium	Sm	62
bohrium	Bh	107	lanthanum	La	57	scandium	Sc	21
boron	B	5	lawrencium	Lr	103	seaborgium	Sg	106
bromine	Br	35	lead	Pb	82	selenium	Se	34
cadmium	Cd	48	lithium	Li	3	silicon	Si	14
caesium	Cs	55	lutetium	Lu	71	silver	Ag	47
calcium	Ca	20	magnesium	Mg	12	sodium	Na	11
californium	Cf	98	manganese	Mn	25	strontium	Sr	38
carbon	C	6	mendelevium	Md	101	sulfur	S	16
cerium	Ce	58	mercury	Hg	80	tantalum	Ta	73
chlorine	Cl	17	molybdenum	Mo	42	technetium	Tc	43
chromium	Cr	24	meitnerium	Mt	109	tellurium	Te	52
cobalt	Co	27	neodymium	Nd	60	terbium	Tb	65
copernicium	Cn	112	neon	Ne	10	thallium	Tl	81
copper	Cu	29	neptunium	Np	93	thorium	Th	90
curium	Cm	96	nickel	Ni	28	thulium	Tm	69
dubnium	Db	105	niobium	Nb	41	tin	Sn	50
darmstadtium	Ds	110	nitrogen	N	7	titanium	Ti	22
dysprosium	Dy	66	nobelium	No	102	tungsten	W	74
einsteinium	Es	99	osmium	Os	76	uranium	U	92
erbium	Er	68	oxygen	O	8	vanadium	V	23
europium	Eu	63	palladium	Pd	46	xenon	Xe	54
fermium	Fm	100	phosphorus	P	15	ytterbium	Yb	70
fluorine	F	9	platinum	Pt	78	yttrium	Y	39
francium	Fr	87	plutonium	Pu	94	zinc	Zn	30
gadolinium	Gd	64	polonium	Po	84	zirconium	Zr	40
gallium	Ga	31	potassium	K	19			
germanium	Ge	32	praseodymium	Pr	59			

BOX 4.9 Combustion of a fuel

The simplest hydrocarbon fuel is methane, the main component of natural gas, and this simple chemical compound can be used as an example of the combustion of a fuel. A single molecule of methane is one carbon atom with four hydrogen atoms attached. This could be represented in various different ways as shown in Figure 4.20.

As can be seen, all three representations describe the same compound. The 'graphic' is indeed graphic, and can be very useful in visualizing complex molecules, but is hardly for everyday use with pen and paper (or mouse). The second is the easiest for such purposes, but says nothing about the spatial arrangement of the constituents, so the additional information in the third is often used, particularly again for more complex molecules.

However, the molecules involved in the combustion of methane are all fairly simple (Figure 4.21(a)), so only the graphical representations and chemical formulae are needed to describe their interaction during combustion.

Note that oxygen is shown as a molecule consisting of two atoms. This is its normal form in the atmosphere, which means that the oxygen atoms in any combustion equation must always enter as pairs. (Before continuing, you may want to identify the other two molecules shown.)

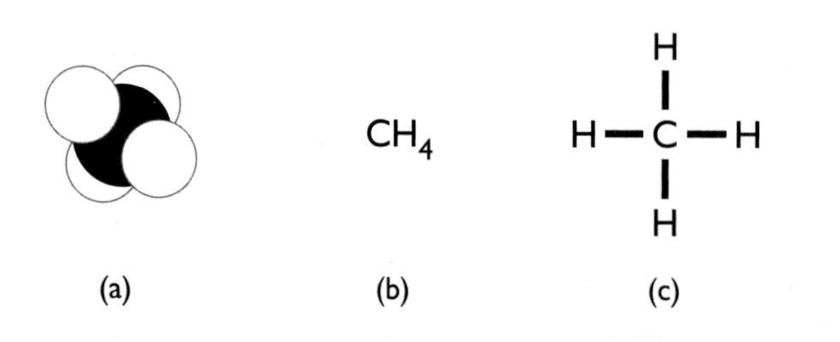

Figure 4.20 Three different representations of a molecule, in this case methane: (a) graphic; (b) the standard chemical formula; (c) structural formula

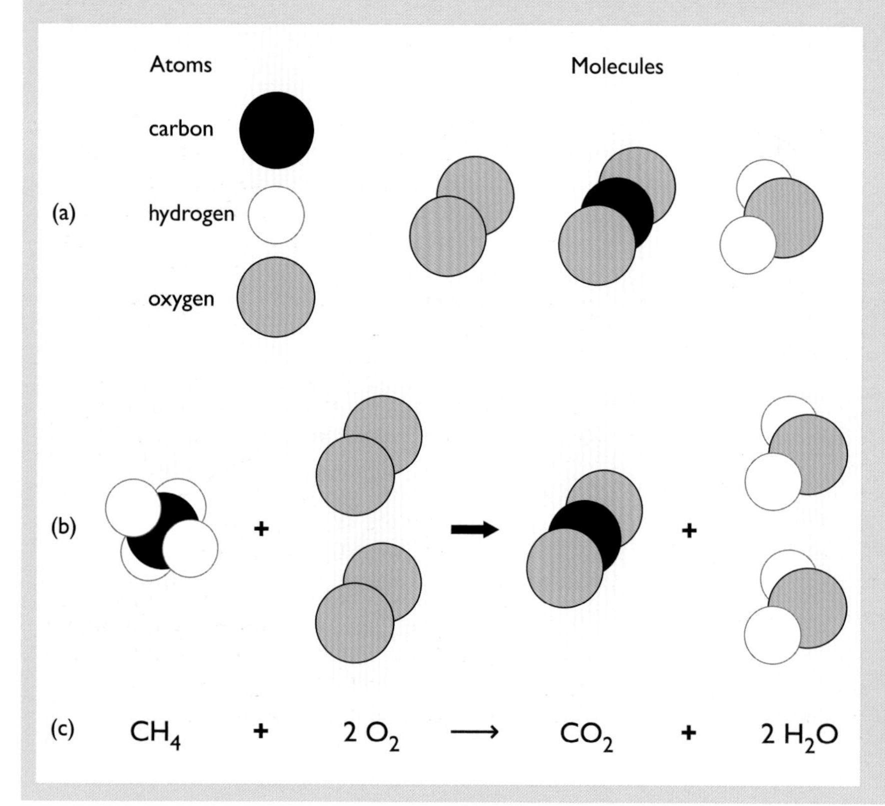

Figure 4.21 The combustion of methane: (a) the chemical components of the process; (b) graphical representation; (c) chemical equation

Atoms do not of course vanish, so any complete representation of a chemical process must have the same number of each type of atom before and after the event.

Finally, then, the combustion of methane may be represented as a graphic and a conventional chemical equation as in Figure 4.21(b) and (c). The number of oxygen molecules needed is determined as follows.

- The single carbon atom needs two oxygen atoms for conversion to carbon dioxide.
- The four hydrogen atoms need two oxygen atoms to become two water molecules.
- So a total of four oxygen atoms are needed – the two molecules shown.

Given that the relative atomic masses of hydrogen, carbon and oxygen are approximately 1, 12 and 16 respectively, it can be seen that burning 16 grams of methane releases 44 grams of CO_2 – or nearly three tonnes of CO_2 for each tonne of natural gas burnt. However, as will be described in Chapter 13, one tonne of free methane in the atmosphere is about 25 times more potent as a greenhouse gas than one tonne of CO_2, so the unintended release of natural gas during extraction can be as serious a concern as its actual use.

BOX 4.10 Using chemical energy: fuels as energy stores

Table 4.2 shows the energy content of five fuels. How much of each of these would be needed to supply energy equal to that stored by the pumped storage system of Box 4.3?

Any fuel can be regarded as an energy store. To be precise, the 'stored energy', the energy released in burning the fuel, is the difference between the initial total chemical energy of the fuel and oxygen and the final total chemical energy of the combustion products. However, it is commonly referred to as the **energy content**, **heat of combustion** or **heat value** of the fuel itself.

Heat values

All the fuels listed in Table 4.2 contain hydrogen, and this means that one combustion product is water (as in Box 4.9). In the heat of a fire or furnace this water appears as hot steam, which can carry off a significant fraction of the energy released in the combustion. The useful energy, the heat value of the fuel, will be increased if this steam is condensed.

Any fuel therefore has two 'heat values', differing by the energy gained by condensing the steam. Unfortunately each is commonly called by a variety of names:

- Condensing the steam gives the **high heat value (HHV)**, also called the **higher calorific value (HCV)** or **gross calorific value (GCV)**.
- Not condensing the steam leads to the **low heat value (LHV)**, also commonly called the **lower calorific value (LCV)** or **net calorific value (NCV)**.

The figure normally quoted for the 'energy content' or specific energy of a fuel is usually the LHV for coal and oil, but the HHV may be used for others (see Table 4.2).

Table 4.2 Fuels as energy stores

Fuel	Energy content /MJ kg^{-1}
Coal (LHV)	28
Oil (LHV)	42
Methane (HHV)	55
Hydrogen (HHV)	140
Wood (HHV)	15

Quantities needed to supply 15 TJ

The stored energy in the case study in Box 4.3 is about 15 TJ, which is 15 million MJ. So the required quantity of each fuel can be calculated using the data in Table 4.2. The mass of coal, for instance, is 15 million MJ divided by 28 MJ per kilogram, which is about 540 000 kilograms. One metric tonne is 1000 kg, so this can conveniently be expressed as about 540 tonnes of coal.

Table 4.3 shows the approximate masses for each of the five fuels, together with data from the storage systems discussed earlier in this chapter. The mass of one cubic metre of water is one tonne, and to complete the picture the mass of a lead–acid car battery is taken as 10 kg. Uranium (discussed in Chapter 10) is also included for comparison.

Although it is unlikely that all the items in Table 4.3 will in practice be competing in any one situation, the hierarchy does illustrate rather vividly the different 'densities' of the forms of energy: gravitational, thermal, electrical, chemical and nuclear.

Table 4.3 Masses to store 15 TJ

Storage medium	Tonnes
Hydrogen	110
Methane	270
Oil	360
Coal	540
Wood	1000
Lead–acid batteries	35 000
Water heated by 50 °C	70 000
Water pumped up 370 metres	4 000 000
Natural uranium	25 kg!

The comparison is of course unfair for a number of reasons. If electric power is the aim, the batteries and the pumped storage plant can store and supply it with relatively little loss, whilst all the others involve appreciable conversion losses. Then we should certainly distinguish between 'storage systems' where the energy can be recycled, or is naturally recycled on a relatively short timescale, and those such as the fossil fuels or uranium where this is not the case. These issues will be discussed in more detail in later chapters.

The nuclear atom

Figure 4.22 Ernest Rutherford (1871–1937), born and educated in New Zealand, joined the Cavendish Laboratory in 1895 and after periods in Montreal and Manchester returned as Director. A large, exuberant man, he attracted lively groups of students and co-workers. The radical change from a picture of matter consisting of many different 'uncuttable' atoms to one in which all atoms are built from just a few fundamental particles was largely his, and he is regarded as the founding father of nuclear physics.

By the year 1900, most scientists were agreed that atoms must be composite, and unravelling the mysteries of their structure became the central scientific problem of the early twentieth century.

The first breakthrough came in 1911, when Ernest Rutherford (Figure 4.22) demonstrated that nearly all the mass of any atom is concentrated into a tiny central **nucleus** with only a ten-thousandth of the diameter of the complete atom. It was almost immediately recognized that the rest of the atom must consist of the light electrons that had been discovered twenty years earlier (Section 4.4). These were known to have *negative* electric charge, but a complete atom is electrically neutral, so the nucleus must have a corresponding *positive* charge, and the resulting electrical force would hold the fast-moving electrons in motion around the nucleus, leading to a stable atom (Figure 4.23).

On this model, the electric charge of a nucleus must be an *exact multiple* of a basic positive charge in order to balance the total negative charge of all the electrons. This led rather naturally to the idea that all nuclei might consist of different numbers of identical heavy, electrically positive particles. These particles, given the name **protons**, should account for most of the mass of any atom. So perhaps the nuclei of the successive atoms in the periodic table (Figure 4.19) simply have increasing numbers of protons – one for hydrogen, two for helium, etc. – each nucleus then being surrounded by a corresponding number of the much lighter electrons. With this idea, the atomic number of an element, its position in the periodic table, is simply the number of protons in its nucleus.

Does this model explain the *masses* of the atoms? Unfortunately not. The second element, helium, which should have two protons, has four times the mass of hydrogen, and carbon, the sixth element, has twelve times the hydrogen mass, and so on.

In 1920 Rutherford offered a partial solution. He proposed a second nuclear particle, the **neutron**, with almost exactly the same mass as a proton but no electric charge. The hydrogen nucleus might then consist of just one proton, with no neutrons, the nucleus of helium would have 2 protons and 2 neutrons, carbon would have 6 of each, and so on. The sequence of atomic numbers would remain the same, but the *relative atomic mass* of any element would now be determined by the total number of **nucleons** (protons plus neutrons) in its nucleus, a quantity called the **mass number** of the nucleus (or sometimes the atomic mass number or nucleon number).

However, even this new interpretation could not explain all the data in the periodic table. How could copper, for instance, have a nucleus consisting of sixty-three and a half nucleons? There must be some other factor.

Isotopes

Neutrons do in fact help to explain seemingly anomalous relative atomic masses such as 63.5 for copper. Natural copper, we now know, consists of two main **isotopes**: atoms whose nuclei have the same number of protons but different numbers of neutrons. Both have 29 protons – essential if the atom is to be copper – but one has 34 neutrons and the other 36. Roughly seven of every ten copper atoms are the lighter **nuclide** $^{63}_{29}Cu$ (with an atomic mass number of 63) and the remainder are the heavier one $^{65}_{29}Cu$ (with an atomic mass number of 65). These proportions of the two isotopes, copper-63 and copper-65, lead to the 'average' relative atomic mass of 63.5.

Isotopes are not uncommon, with more than three-quarters of the stable elements existing as two or more. Even carbon includes a tiny fraction (about 1%) of the isotope carbon-13 (Figure 4.24).

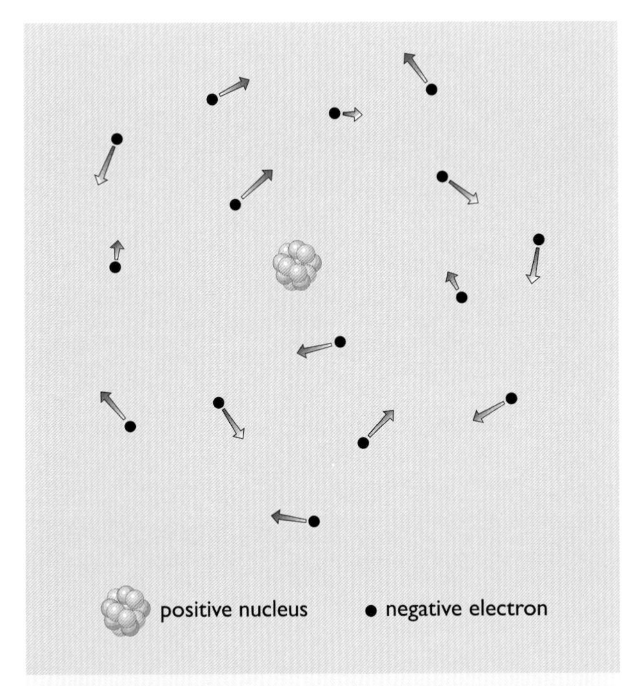

Figure 4.23 The nuclear atom showing the central nucleus and surrounding electrons

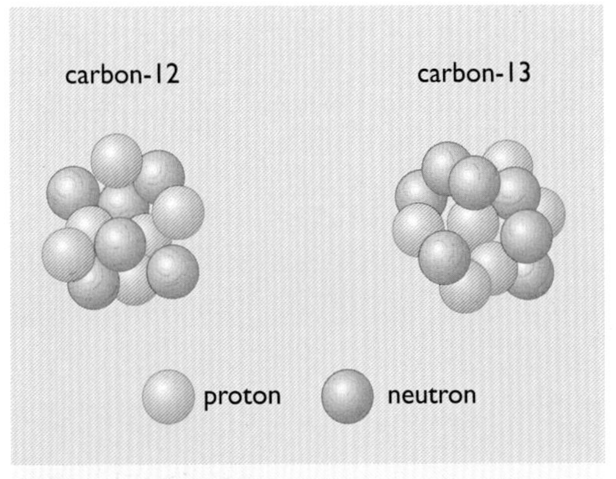

Figure 4.24 The nuclei of two isotopes of carbon

The energy

In a complete atom, the number of protons must of course be matched by an equal number of surrounding electrons, and these almost entirely determine the chemical characteristics of the element. In a chemical process such as the combustion of a fuel, the nuclei of the atoms remain essentially unchanged. The energy that is released comes from changes in the total *kinetic energy* and *electrical potential energy* of the electrons when the atoms combine into different molecules. 'Chemical energy', like 'heat energy', does not therefore represent a *fundamentally* new form of energy. However, in a world where the major part of energy demand is met by the combustion of fuels, it is undoubtedly an important form.

The consequences of this rather simple picture of matter do seem remarkable. A look at the periodic table reveals that the mere change from an atom with ten electrons to one with eleven is responsible for the radically different properties of neon (an inert gas) and sodium (a chemically very active metal). The 'uncuttable' atom may have been abandoned, but its purpose was in a sense achieved, reducing the whole of chemistry and almost all the physical properties of materials to the behaviour of just three basic particles: the proton, the neutron and the electron.

4.7 Beyond chemistry

The nuclear force

A look at the world around us reveals that most atomic nuclei are extremely stable. Radioactive materials are the exceptions (discussed in Chapter 10), but in everyday life we don't normally find elements constantly changing into different elements. The alchemists hoped to turn lead into gold, but our experience is that lead stays as lead and gold remains gold. Iron Age tools may rust, but that is a *chemical* reaction, a linking of iron and oxygen by a rearrangement of the electrons of their atoms. The iron *nuclei* are the same as when the iron was extracted, thousands of years ago.

So an important question remains. What holds a nucleus together? Particles with the same type of electric charge repel each other, so a cluster consisting only of positive protons and neutral neutrons should instantly fly apart. Another force is needed to bind the nucleus. (Gravity attracts all objects towards each other, but the gravitational force between nucleons is far too small.) The required force needs to be very strong at the tiny distances within a nucleus and to attract every nucleon (proton or neutron) to every other nucleon. In 1935 the Japanese theoretical physicist Hideki Yukawa (1907–1981) developed a theory of a completely new force that would achieve this. The details have changed over the years, but this **strong nuclear force** remains, completing a trio of *fundamental forces*: the gravitational, electromagnetic and strong nuclear forces.

Mass and energy

A quite different problem disturbing scientists, even before 1900, was that the two great theories, Newton's laws of motion and Maxwell's laws of electromagnetism, led to conflicting predictions in some extreme cases.

Albert Einstein (Figure 4.25) was troubled by this for some time (since he was 16, he once mentioned), and in 1905 he proposed a solution. This replaced Newton's laws by a new theory, now known as the Special Theory of Relativity. For ordinary behaviour of everyday objects, this gave the same results as the well-tested laws of Newton. But it was consistent with Maxwell's theory, and at a fundamental level it radically changed the way we think about space and time. Its stranger predictions were confirmed in many ways throughout the twentieth century.

Figure 4.25 Albert Einstein (1879–1955) was born in Ulm but studied in Zürich and later was employed in the Swiss patent office. In 1905 he published three revolutionary papers. One explained the quivering movement of small particles (Brownian motion) as the result of bombardment by individual molecules – the first 'direct' evidence that these existed; the second explained the photoelectric effect as direct evidence of Planck's quanta of light energy; and the third was on the Special Theory of Relativity. Later, in Berlin, his General Theory of Relativity unified the laws of motion and gravitation. Dismissed by the Nazis, he spent the rest of his life in Princeton, continuing his search for one theory that would embrace all areas of physics

One consequence of the theory is that the mass of an object is no longer regarded as a fixed quantity. An object has a certain **rest mass** when it is stationary, but when moving it behaves in all respects as though it has a greater mass. This increase in mass is the equivalent of kinetic energy in the new theory. Indeed it can be said that the kinetic energy *is* the difference between the mass of the moving object and its rest mass. *Mass and energy are the same.* (In practice, the separate names have proved too useful to abandon, and the quantity properly called *mass-energy* continues to be given the name and unit that seem most appropriate.)

The customary unit for mass is the *kilogram* and the unit for energy is the *joule*; so if mass and energy are the same, these quantities must be convertible into each other. They are indeed, and the conversion has become rather famous: $E = mc^2$.

The quantity c is the **speed of light** in empty space, approximately 300 million metres per second (3×10^8 m s^{-2}), and this is why we don't normally notice an increase in mass when the speed of an object changes. For example, the kinetic energy (E_k) of a ten-tonne truck travelling at 90 mph is about 80 MJ, but expressed as mass ($m = \frac{E}{c^2}$) it is only about a millionth of a microgram. In such cases we can safely continue to use Newton's laws and treat mass as a fixed quantity – only where objects approach the speed of light is the mass change significant. But the mass-energy equivalence is not confined to kinetic energy. When a fuel is burned in air in a container through which the resulting heat energy can escape, the total mass-energy inside the container decreases. Again, we don't normally notice this because the mass change is very small indeed. Releasing the heat produced in burning a tonne of coal, for instance, reduces the total mass by just under a third of a milligram!

There are however cases where the energy is large enough and the original mass small enough for the change to be detected, and these are explored further in Chapter 10, 'Nuclear energy'.

4.8 Summary

The essentials

This chapter has introduced six forms of energy: *potential energy*, *kinetic energy*, *heat*, *electrical energy*, the energy carried by *electromagnetic waves*, and *chemical energy*. (A seventh form, *nuclear energy*, is the subject of later chapters.)

In many ways, the essential points to take from the chapter lie in the case studies illustrating present-day uses of each form of energy. In practice, of course, no real system involves just one form of energy, so we might usefully reverse the process, starting with a particular system and asking which forms of energy play a role in it.

If, for instance, the wind turbine discussed in Box 4.2 is supplying power for the lights in a remote dwelling, the forms of energy involved will include not only the original kinetic energy of the wind and the delivered electrical energy, but heat and ultimately electromagnetic radiation – or if the power is used to warm up a pie in a microwave oven, we might reverse the order of the last two items.

Analysing existing or proposed energy systems in this way can throw light on new options, or the real sources of inefficiencies, and many examples of this process will be found in later chapters.

In conclusion

This chapter has been a journey through time and an exploration of ideas. In setting out to see how the concepts of the different forms of *energy* have developed, it became evident that these are inextricably entwined with the developing picture of the fundamental particles that make up all *matter*, and the *forces* that act between these particles.

But science does not stand still. New experiments are carried out and new theories emerge. During the past half-century two major aims have been to integrate all the different forces into one theory and to delve further into the nature of the fundamental particles. In this continuing search for the 'theory of everything', even the distinction between forces and particles begins to disappear – but these developments are beyond the scope of this book.

References

Bacon, Francis (1620) 'First Vintage Concerning the Form of Heat' in *Novum Organum*, Aphorisms (Book Two), XX; available at http://www.constitution.org/bacon/nov_org.htm (accessed May 2010).

Rankine, W. (1881) *Miscellaneous Scientific Papers,* (ed. W. J. Millar), London, Charles Griffin and company, p. 203; available at http://www.archive.org/stream/miscellaneousscі00rank#/page/203 (accessed July 2010).

Scerri, E. R. (2007) *The Periodic Table: Its Story and Its Significance*, New York, Oxford University Press, pp. 278–80.

Further reading

Winter, M. (2010) 'WebElements: the periodic table on the web', University of Sheffield and WebElements Ltd, UK, http://www.webelements.com (accessed 15 July 2010).

Chapter 5

Coal

By Janet Ramage and Bob Everett

5.1 Introduction

The next four chapters will discuss probably the oldest of all technologies: the burning of fossil fuels to provide heat. Already familiar to our pre-historic ancestors, the combustion of these fuels still accounts for over eighty per cent of our present-day use of primary energy. The term combustion, however, has a much wider meaning now than in the past. An internal combustion engine or gas turbine may seem very remote from a log fire, but the fuels used in these modern systems have much in common chemically with wood, and their combustion involves essentially the same basic processes, a chemical reaction between oxygen, normally from the surrounding air, and the constituent elements of the fuel – mainly carbon and hydrogen. The reaction leads to the release of energy in the form of heat which is carried away initially by the combustion products. (As will be described in Chapter 10, the 'burning' of nuclear fuel is an entirely different process.)

Fossil fuels were originally living matter: plants or animals that were alive hundreds of million years ago. With the passage of time, their remains have undergone chemical changes leading to the solid, liquid or gaseous fuels that we extract today. Oil and natural gas have been called the *noble* fuels, by analogy with the noble metals (silver, gold, etc.). They are amongst the most concentrated natural stores of energy. Oil, in particular, is relatively easy to move from place to place and both are very convenient to use. It is easy to see why today we choose gas rather than coal for household heating and hot water. Coal, by contrast, might be called the *ignoble* fuel. Compared to natural gas it produces up to twice the amount of carbon dioxide for the same useful heat, and its sulfur content can contribute to smogs and acid rain.

As has been described in the first two chapters, world coal production increased by a factor of four between 1920 and 2009, yet its contribution as a fraction of total energy use has fallen from 60% to just over a quarter.

This chapter starts by briefly looking at the history of the rise of coal as a substitute for wood and then considers the nature of coal, its different forms and its combustion process. It then turns to the uses of coal. Today, most coal is simply burned for electricity generation, but the manufacture of coke from coal for iron and steel production is still an important use. Although industrial chemical feedstocks are now mainly derived from oil this has not always been so, and the chapter briefly describes the nineteenth century production of town gas which made full use of the considerable potential of coal. Next, the chapter looks at the processes of mining coal and considers both deep mining and surface mining. Addressing concerns about fossil fuel depletion, the chapter then looks at global reserves of coal, and the necessary distinction between the terms *reserves* and *resources*. It briefly describes the status of the coal industry in several key countries and then turns to the efficient combustion of coal, describing modern power station systems, the pollutants produced and ways of dealing with them. Finally, it briefly considers the future for coal. Given its high relative CO_2 emissions, this might seem bleak. However, gasification techniques, and the production of oil from coal (see Chapter 7) and carbon capture and storage (see Chapter 14) may open up new possibilities.

5.2 From wood to coal

The early years

For early humans, wood was the naturally available fuel. As today, fuel was needed for cooking food, for heat and light, and for industrial production. (The need for fuel for transport, another present-day requirement, arose much later, in the nineteenth century.) The smelting of metal ores in particular required the use of **charcoal**. This is a secondary fuel produced by a process called **pyrolysis**, the thermal decomposition of organic material at high temperatures in the absence of oxygen. Charcoal was (and still is in many countries) traditionally produced in the forests where the wood was cut. A 'kiln', consisting of stacked wood covered with a layer of earth, would be lit and allowed to smoulder for a few days with a restricted supply of air (Figure 5.1). The interior would reach a temperature of 300–500 °C. The moisture and volatile matter of the wood would be driven off (as a large plume of smoke), leaving a residue of dry charcoal. This is almost pure carbon with twice the energy density of wood. It burns cleanly without smoke and remains the barbecue fuel of choice to this day. However, to produce one tonne of charcoal requires between four and ten tonnes of wood.

Figure 5.1 A traditional- style charcoal kiln in operation in the UK in Cornwall (medieval kilns encased the smouldering wood in a mound of earth)

Given the limited supplies of wood, it is not surprising that alternatives have been sought. There is evidence of coal burning in Britain as long ago as the Bronze Age, and in China and the Middle East over two thousand years ago.

In medieval cities such as London the inhabitants depended on simple open fires for their domestic heating and cooking, and three fuels were available. There was wood, imported initially from the surrounding forests and later, by sea, from more distant sources. There was charcoal, commonly known simply as 'coal'. From about the year 1200, there was coal itself, extracted in County Durham in the north-east of England from surface beds or shallow mines. Carried 500 km by sea to London, it was known as 'sea-coal' to distinguish it from charcoal. It was delivered directly to wharves on the Thames, close to the potential users. The fact that this was considered economic is a testament to the atrocious state of roads at the time. It was far from being a perfect substitute for wood, since Durham coal contains about 1% sulfur. Although wood fires produced smoke, coal fires in addition produced the acidic gas sulfur dioxide (SO_2). They were thus ill suited to wooden buildings whose only 'chimney' was a hole in the roof, and wood remained the preferred household fuel.

The development of chimneys as we know them in the UK had to wait for the re-establishment of brick making in the fifteenth century (it had almost died out with the end of the Roman Empire, although it survived elsewhere in Europe).

Figure 5.2 Tudor chimneys at Hampton Court Palace near London

The new brick chimneys (such as those shown in Figure 5.2) were resistant to the hot, acidic smoke from coal fires (which was now carried high above the buildings). In a short time, the wealthy were building brick houses with many fireplaces, and coal was their favoured fuel. As the city of London grew, so did its coal consumption: 24 000 tons in 1585 and 216 000 tons in 1650. By 1680 there were 1400 ships simply carrying coal from Newcastle to London (Brimblecombe, 1987).

All this had a terrible effect on urban air quality. In the mid-seventeenth century the environmental campaigner John Evelyn wrote that London was cloaked in:

> Such a cloud of sea-coal, as if there be a resemblence of hell upon earth, it is in this volcano in a foggy day: this pestilent smoak, which corrodes the very yron [iron], and spoils all the moveables, leaving a soot on all things that it lights: and so fatally seizing on the lungs of the inhabitants, that cough and consumption spare no man.
>
> (quoted in Brimblecombe, 1987)

Evelyn petitioned King Charles II unsuccessfully for what today would be called 'zoning': to have industry moved out of the city and further down the Thames. The tide of coal, soot and pollution (Box 5.1) was yet to get worse as the Industrial Revolution arrived, leading to the noxious 'London fogs' of the nineteenth and twentieth centuries.

BOX 5.1 Urban air pollutants in coal-fuelled cities

Soot is the most visible part of smoke, consisting of particles of unburned carbon and tars, but also contaminated with other chemicals. Large particles of soot may be obvious as 'smuts' on clean clothing.

Smoke and particulate matter (PM). Smoke may also contain very fine 'particulates', such as the unburned greasy soot from a candle flame (or today from a car engine). Modern statistics concentrate on **PM_{10}s**, particulates smaller than 10 micrometres (μm) in diameter, which can be breathed in.

Sulfur dioxide is a major combustion product of fuels such as coal (and oil) which may contain up to 5% sulfur. In the human body this attacks the lining of the lungs. It also attacks fabrics and building materials, particularly limestone. If emitted in large quantities it can give rise to an increase in the acidity of rainfall (acid rain), which can damage vegetation and fish stocks in freshwater lakes over a very wide area. This will be discussed further in Chapter 13.

The Industrial Revolution

'The Industrial Revolution' is a term coined by Friedrich Engels in 1844. For our present purposes, it is worth identifying four important starting points:

- 1698: Thomas Savery's first steam engine
- 1709: Abraham Darby's use of coke for smelting iron

- 1733: John Kay's invention of the flying shuttle
- 1790s: the development of gas lighting by Lord Dundonald and William Murdoch

The significance of Savery's steam engine was that it used heat from burning fuel to produce a continuous (or at least repetitive) *mechanical* driving force. However, as will be described in the next chapter, half a century of technical development was needed before steam engines were to play a major role.

Kay's flying shuttle was an early element in the development of an industrialized textile industry. New spinning and weaving technologies were developed in the second half of the eighteenth century. Although the great textile mills were initially powered by water they eventually adopted coal power in the form of steam engines as they became more efficient and reliable.

The use of coke instead of charcoal for smelting iron was of great importance for the development of the coal industry. In the late 1600s, a renewed shortage of wood for charcoal was bringing Britain's iron industry to the verge of closure. Pure carbon in the form of charcoal was vital for the smelting of iron ore (essentially iron in chemical combination with oxygen). Firstly, a high-quality fuel is necessary to get the high temperature required (over 1500 °C) to produce liquid iron. Secondly, the carbon has to be physically in contact with the ore. Under the right circumstances this is reduced ('de-oxidized') to metallic iron, the oxygen combining with carbon to produce carbon dioxide. Over the years, various attempts had been made to use coal instead of charcoal, but its variable composition, inert mineral content and particularly its sulfur content led to poor-quality iron.

However, if coal was subjected to a pyrolysis process similar to that of charcoal production, the heat could drive off the sulfurous impurities and leave a residue of carbon (although still contaminated with a lot of ash). By about 1680 this new, cleaner fuel produced by the partial combustion of coal had acquired a name: coke.

Abraham Darby was a maker of brass cooking pots in Bristol. Developing an interest in the potential of coke for iron smelting, he moved to Coalbrookdale, a small Shropshire village close to one of the largest coal fields of the time. Within a few decades this was transformed into a major industrial centre (Figure 5.3). He used coke to produce cast iron, not just for cooking pots but also for large cylinders for the new steam engines. In 1780, his grandson, also called Abraham, completed the world's first cast-iron arched bridge, spanning 30 metres. By the end of the eighteenth century, Britain's iron industry was flourishing, and the country's coal production had reached an estimated ten million tons a year.

Right at the end of that century, working quite independently of each other, Lord Dundonald in Scotland and William Murdoch, a Scot installing steam engines in Cornwall, started to develop gas lighting. As pointed out in Chapter 3, at this time lighting by candles was extremely expensive. They showed that by heating coal in a closed vessel (such as a kettle) a gas could be produced that burned with a bright luminous flame. In the early years of the nineteenth century Murdoch started commercially lighting factories in Birmingham and Manchester, while a German entrepreneur

Figure 5.3 A painting of the iron furnaces in Coalbrookdale in 1801

who anglicized his name to Frederick Winsor gave a demonstration of gas street lighting in London.

The nineteenth century

At the start of the nineteenth century, steam engines fuelled by coal were already supplying power in mills, mines and factories, coke was in general use for iron smelting, and domestic coal fires and kitchen stoves were common. But Admiral Nelson's fleet was still wind powered, transport and agriculture were still horse powered, and houses were still lit by oil lamps or candles. A hundred years later, the start of the twentieth century saw a very different world – in those countries wealthy enough to adopt the new technologies. Power from coal had become dominant in nearly all areas of life. (Agriculture, still mainly dependent on the horse, was an exception.) Railways and steamships were providing transport, coal gas was lighting buildings and streets, and the by-products of coal gas production were providing agricultural fertilizer and chemical feedstocks for a new chemical industry. Electricity, the 'new' form of energy that was beginning to play a role in the industrialized countries, would also depend largely on coal.

The story of the nineteenth century developments that led to this changed world is mainly told elsewhere in this book. Chapter 4 has followed the evolution of the scientific ideas, Chapter 6 will describe the development of the steam engine and the steam turbine and Chapter 9 that of the electrical generator. The production of town gas from coal, an example of a technology-based industry that grew to great importance, only to be superseded, is described in Section 5.4.

One significant consequence of the changes described above was the continued growth of coal output throughout the century. Britain's annual production, rising at 3.5% a year, had grown from ten million tons in 1800 to over 80 million tons in 1861 (Figure 5.4). The population had also grown by about 50%. This troubled the economist William Stanley Jevons who,

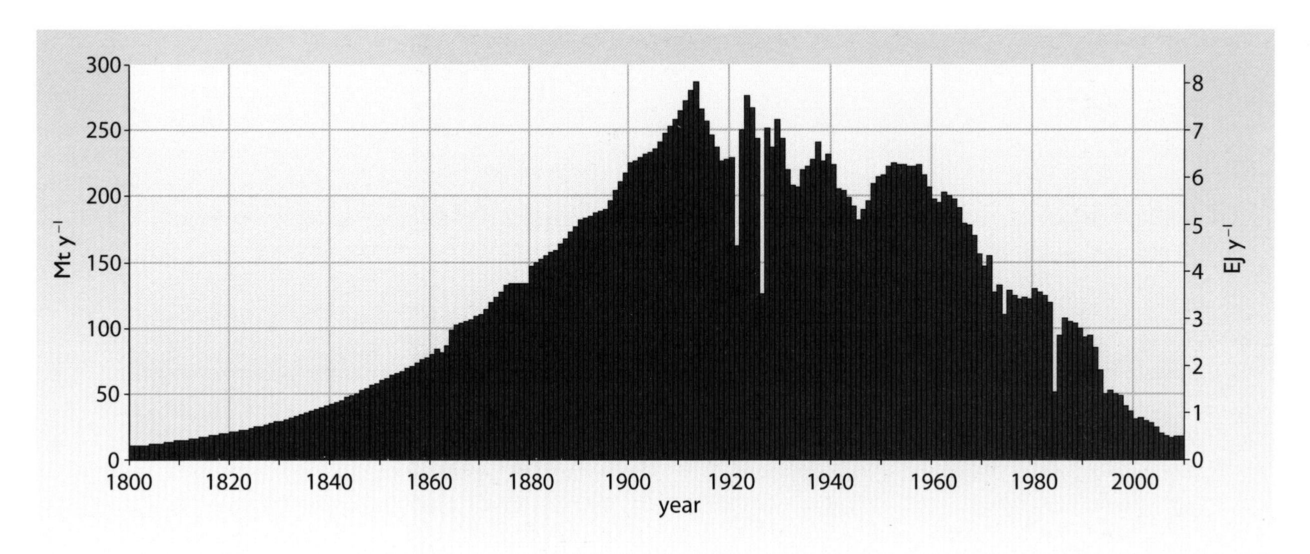

Figure 5.4 UK annual coal production, 1800–2009 (source: DECC, 2009)

writing in the 1860s, wondered how long such exponential growth could continue. His 'growthist' scenario for a future UK in 1961 saw it with a coal consumption of 2.6 billion tons per year, by which time the known UK coal reserves would have been exhausted (Jevons, 1865). Perhaps fortunately, the growth in UK coal production came to an abrupt halt with the First World War. Peak production (nearly 300 million tons) was in 1913 and a third of that went for export.

It is an indicator of British dominance during the Industrial Revolution that coal production in the United States reached one million tons for the first time only in 1840. However, growing demand and an increasing shortage of wood for charcoal led to a remarkable annual growth rate of more than 9% a year, bringing US production to about the same level as the UK by the end of the century. The only other country with an output approaching these was Germany, producing a little over 100 million tons in 1900. World production was then an estimated 800 million tons and rising at about 5% a year. In 1905, it would exceed a billion tons.

5.3 The nature of coal

Coal may be the least desirable fuel, but there is a great deal of it and there are many different types, with different potential uses.

Types of coal

The formation of **coal seams** begins with the preservation of waterlogged plant remains to produce **peat** and then slow compression as the peat is buried. As it rots, much of the hydrogen and oxygen content of the plant material is released as water, carbon dioxide and methane, leaving a high concentration of carbon. About 10 metres of peat will compress down to form about one metre of coal. Seams throughout the world that have been mined at different times have ranged in thickness from as little as 300 mm (one foot) to more than 30 m (100 feet), and can lie more than a thousand metres below the surface or so close that they almost penetrate it. These features are of considerable economic importance. Coal, like all minerals, is essentially a free resource, waiting to be used. Its cost arises in the processes of extraction, treatment and transport to the user, and the cost of these three stages depends on the form and location of the coal seams.

The nature of the coal itself is also important. The earliest geological strata in which coal has been identified are nearly four hundred million years old (Devonian), and coal is present in all strata down to less than a hundred million years ago (Tertiary). The largest quantities are thought to lie in the Tertiary rocks, but this 'young' coal is less valuable than the deposits of earlier periods. As shown below, the different types of coal can be characterized by the extent to which the original plant material has been physically and chemically changed. Obviously, the age of the coal is a significant factor in this, but other geological processes, such as subjection to heat and/or pressure are important. Depending on all these factors, the resulting 'coal' can range from substances in which the original plant material can still be identified to those which are almost pure carbon. Not surprisingly, the heat value – the heat energy released in combustion (as described in Box 4.10) – varies considerably, from little more than ten gigajoules per tonne to well over thirty.

In discussing the resource, it is customary to divide this range of types into two main categories. The terminology varies from country to country, but the most common name for the 'upper' (and older) end of the range is **hard coal** – although Australia, currently the world's main exporter, uses the term **black coal**, and in some national or international data the more specific **anthracite** and **bituminous coal** appear. The 'lower' (or younger) coals are generally referred to as **brown coal** or **lignite**, with *sub-bituminous* coal sometimes included. A further sub-division of the hard coals, obviously of great practical importance, is based on their potential use: *steam coal* (or thermal coal) mainly for power stations; *coking coal*, mainly for the iron and steel industries; and *anthracite*, for direct use as a natural 'smokeless fuel'. The uses of these different types of coal are described in Section 5.4.

Composition of coal

Many of the unattractive features listed in the introduction to this chapter are due to the extremely complex nature of coal. As in all fossil fuels, the constituents that are important for combustion are carbon and hydrogen, but coal, unlike gas or oil, cannot be analysed at the molecular level into relatively straightforward hydrocarbon compounds. Rings of six carbon atoms play an important role in the structure, forming layered arrangements that incorporate not only hydrogen, but significant amounts of oxygen and nitrogen. The structure also includes differing quantities of sulfur and traces of other environmentally undesirable elements, and coal always contains some inert mineral material with no fuel value at all, destined to remain as ash. Finally, all coal incorporates some moisture within its structure.

A full chemical analysis of a sample of coal, called an **ultimate analysis**, lists the proportions by mass of the main elements that are present, usually carbon, hydrogen, oxygen, nitrogen and sulfur. These analyses are often in terms of dry, ash-free samples, excluding any moisture and inert matter.

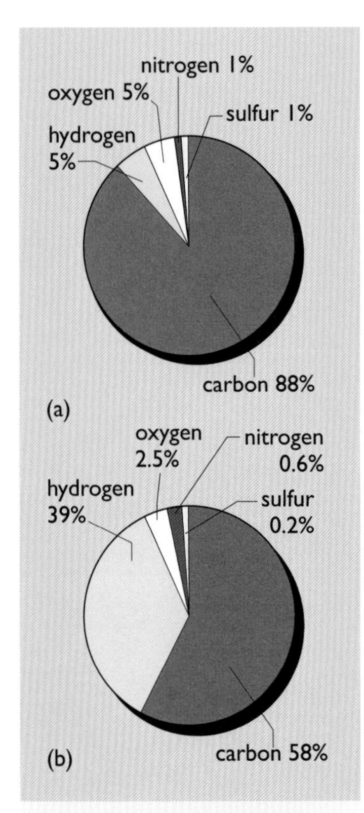

Figure 5.5 Ultimate analysis of a sample of dry, ash-free, medium-sulfur, good-quality bituminous coal: (a) percentages by mass; (b) percentages by number of atoms

Figure 5.5(a) is an example of such an analysis, for a type of good-quality bituminous steam coal that might have been used in a nineteenth century factory or a modern power station. However, at the most basic level, it is the relative proportions by atoms of the various elements that matter, and as we have seen, different atoms can have very different masses. (Section 4.6 has described how, for instance, a carbon atom has twelve times the mass of a hydrogen atom.) Figure 5.5(b), taking this into account, shows the percentages of the different atoms and gives a more informative picture of their significance in the combustion of the coal.

The ultimate analysis is undoubtedly of interest to the coal scientist, but for the technologist a different analysis is needed; one that reflects the value of the coal to its users. Heating is the essential feature common to all uses of coal, leading either to full combustion or part combustion or to the production of other fuels, and it is the sequence from heating to combustion that provides the basis for the most common characterization of different coal types.

The combustion process

The amounts of inert material and moisture are obviously significant in determining the heat values of different types of coal, but there are other features that play important roles. They are best explained by considering the series of processes leading to the complete combustion of a coal sample.

- In the early stages of combustion, as the coal is heated, any moisture in the structure is also heated and evaporates. The moisture content can vary from one or two per cent to as much as a tenth of the total mass in hard coals, and twice that in brown coals. The heating and evaporation both use some of the energy of the coal, but as Box 5.2 shows, this 'loss' is only a small fraction of the total energy released in combustion, even for coal with relatively high moisture content.
- As the temperature continues to rise, the **volatile matter (VM)** in the coal, a range of hydrocarbons or 'bitumens' that give this type of coal its name, boil off and break down. They contain most of the hydrogen and oxygen in the coal and some of its carbon. They break down to produce a range of gases such as hydrogen, carbon monoxide (CO) and methane (CH_4). These are fuels, releasing heat as they burn, and as much as half the heat energy from the coal may appear in this form. Anyone who has watched a coal or wood fire will have seen the spurts of intense flame from these little jets of gas.
- The combustible part of the material remaining after the volatile matter has gone is the remainder of the carbon, known as the **fixed carbon (FC)**. This is effectively charcoal, or coke, which can burn at a high temperature in oxygen from the air, producing carbon dioxide:

  $$C + O_2 \rightarrow CO_2$$

  Depending on the type of coal, the fixed carbon can account for virtually all the heat output or no more than half.
- Finally, with all the fuel burnt, any inert material remains as **ash**. This is likely to consist mainly of a mixture of silicates and oxides of iron and

aluminium. A high ash content is obviously undesirable, and the best coals have less than ten per cent ash. However, coals with up to 15% are fairly common, and in some countries as much as 40% is tolerated when the priority is to use local rather than imported coal.

This analysis carries a number of lessons for the designer of furnaces burning solid fuel, a subject to which we return shortly. It is also the basis of the customary specification of different types of coal.

BOX 5.2 **Removing the moisture**

Energy is needed to heat the moisture to its boiling point and then to evaporate it. For simplicity, we assume that this 'moisture' has essentially the thermal properties of water, and that it is initially at 20 °C. The energy required to heat one kilogram to the boiling point (100 °C at atmospheric pressure) can then be calculated using the data and method of Box 4.4.

The specific heat capacity of water is 4200 J kg^{-1} K^{-1} and it has to be heated through a temperature difference of (100 °C – 20 °C) = 80 K

The energy needed = 1 × 4200 × 80 = 336 000 J kg^{-1} = 0.336 MJ per kilogram.

The energy needed to *evaporate* 1 kg of any liquid is its **specific latent heat of vaporization**, and its value for water in the current situation is 2.258 MJ per kilogram.

The total energy required per kilogram of 'moisture' is therefore 2.594 MJ, or about 2.6 MJ.

Suppose that the coal has 10% moisture. One tonne (1000 kg) of coal will contain 100 kg of moisture, and the energy needed to heat and evaporate this will be **260 MJ**.

Assuming that the coal has a 'typical' heat value of 28 MJ kg^{-1} (as given in Table 4.2) the total energy released will be **28 000 MJ**.

Comparison of the above two figures shows that, even for this coal with this high moisture content, the energy 'lost' is less than 1% of the total. (It should also be borne in mind that coal is often transported thousands of kilometres from the mine to the consumer and that this 'useless' moisture implies an extra transport cost.)

Proximate analysis

Compounds consisting mainly of carbon, hydrogen and oxygen are characteristic of living materials, and in this sense coal is closer to its origins than the hydrocarbons of oil and gas. Indeed, the different types or **ranks** of coal, from brown coal and lignite at one extreme to anthracite at the other, can be regarded as members of a sequence which starts with wood and peat. Figure 5.6 shows some members of this series, analysed in terms of the four constituents appearing in the account of combustion above. This **proximate analysis** reveals that the percentage of fixed carbon is a factor in determining the rank. As can be seen, the heat of combustion (given at the foot of each column) also tends to increase with the rank of the coal.

Properties other than those shown in Figure 5.6 are relevant in the selection of coal for a particular purpose. For example, anthracite is a naturally

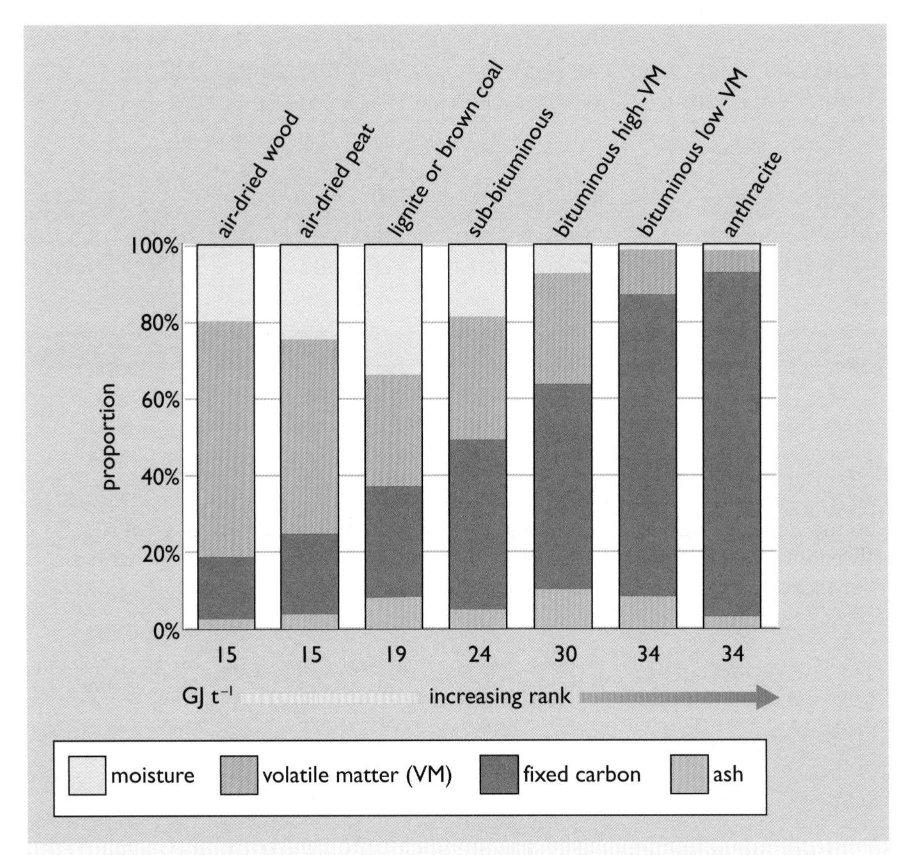

Figure 5.6 Proximate analysis of coal and other solid fuels. Note: the relative proportions of fixed carbon and volatile matter shown here are generally indicative of the different ranks; the ash and moisture content can vary widely between fuels from different sources and the heat values will vary accordingly

smokeless fuel. Low-VM bituminous coal is a steam coal whose ultimate analysis appeared in Figure 5.5. The high energy density of both of these made them the fuels of choice for nineteenth century naval ships and railway locomotives. A bituminous coal might also be a coking coal (see below), but the proximate analysis alone does not indicate whether or not it has the right qualities.

Combustion products

In the full combustion of coal, all the constituent elements shown in Figure 5.6, and some of the others present as minor impurities, undergo chemical change in the heat of a coal fire or furnace. Carbon and hydrogen are the main constituents, and their reactions with oxygen lead to the main combustion products: carbon dioxide and water (initially in the form of steam). The quantities of these products can easily be calculated if we know the percentages of carbon and hydrogen in the coal and appropriate relative atomic masses. As Box 5.3 shows, the combustion of one tonne of coal with the composition shown in Figure 5.5 releases about three tonnes of carbon dioxide into the atmosphere. In terms of the electrical output of a power station using this coal, this represents about one kilogram of CO_2 per kilowatt-hour. (The corresponding figure for a gas combined cycle gas turbine plant is about one half of a kilogram per kilowatt-hour of electrical output.)

BOX 5.3 Carbon dioxide released in coal combustion

How much carbon dioxide is released in the combustion of one tonne of coal, and what are the consequences of this for the emission of CO_2 from a coal-fired power station?

Assuming that all of the carbon in the coal reacts with oxygen to form carbon dioxide:

$$\underset{12}{C} + \underset{2 \times 16}{O_2} \rightarrow \underset{44}{CO_2}$$

The relative atomic masses of carbon and oxygen are 12 and 16 respectively, so the numbers under the chemical equation are the relative masses of the three items, and it can be seen that 12 kg of carbon produce 44 kg of carbon dioxide. Each kilogram of carbon in the coal therefore releases 44/12 = **3.67 kg of CO_2**.

Assuming the coal to be bituminous low-VM coal, the sixth bar of Figure 5.6 shows that 10% of this type is made up of moisture and ash, so each tonne (1000 kg) becomes only 900 kg when dry and ash-free. Figure 5.5(a) shows that 88% of this is carbon, or **792 kg**.

So the mass of CO_2 released in the combustion of one tonne of this type of coal is:

$$792 \times 3.67 = 2904 \text{ kg} = \mathbf{2.90 \text{ tonnes}}$$

The heat value for this coal is given as 34 GJ per tonne (Figure 5.6). Suppose now that it is used in a power station. Chapters 2 and 3 showed that the electrical energy finally reaching the consumer is about one third of the fuel energy input, so for each tonne of coal used, the final delivered electrical energy is about:

$$\frac{34}{3} = 11.33 \text{ GJ} = 11\,330 \text{ MJ}$$

One kilowatt-hour is 3.6 MJ, so this is equal to **3150 kWh.**

Burning one tonne of coal therefore releases 2.90 tonnes of carbon dioxide and produces 3150 kWh of delivered electricity: *just under one kilogram of CO_2 per kilowatt-hour.*

A similar calculation can be done showing that the combustion of the 5% of hydrogen in one tonne of the bituminous coal as shown in Figure 5.5 will result in the production of about 400 kg of steam. If the coal is burned in a power station this will be part of the 'flue losses'. Chapter 4 has pointed out the difference between the higher and lower heat values of a fuel. If somehow this steam could be condensed, its latent heat would yield an extra 1 GJ per tonne of coal, about 3% of the heat value of the coal. In practice this is almost never done. Since coal contains so little hydrogen, the distinction between its higher and lower calorific values is mostly assumed to be unimportant, though for other fuels such as natural gas the difference can be quite significant.

As well as CO_2 and steam, there are other combustion products. Nitrogen can combine with oxygen to produce oxides of nitrogen. These compounds (N_2O, NO, NO_2, etc.), known generically as NO_x, are produced mainly by nitrogen from the coal, but nitrogen from the air may also contribute. Any sulfur present in the coal will readily form sulfur dioxide (SO_2). All coals contain some sulfur, and it can account for as much as 5% of the total mass. Some coals may contain chlorine, particularly in the form of common salt, sodium chloride (NaCl). This can produce acidic hydrogen chloride gas (HCl) as a combustion product.

The quantities of sulfur, chlorine and ash in coal can be reduced by washing it. However, this requires large supplies of clean water, essentially trading a problem of air pollution for one of water contamination.

Coal may also contain various toxic heavy metals such as arsenic, cadmium, lead and mercury. These may end up in the ash, but significant amounts of mercury can be released as a vapour in the flue gases.

The problems of dealing with unwanted pollutants in the flue gases of power stations are discussed further in Sections 5.6 and 5.7.

5.4 Uses of coal

In 2008, about two-thirds of world coal production was burned for electricity generation (IEA, 2010). In the UK the figure was over 80%. This perhaps seems a rather crude use of the material, given its possibilities. In the nineteenth century the development of its use for town gas and coke production explored the full range of its potential.

The nineteenth century gasworks

As described above, experiments with commercial gas lighting started around 1800. Heating coal in the absence of air could boil off the volatile matter, producing a useful illuminating gas, leaving a residue of equally useful coke. However, this process was not straightforward. As pointed out in Chapter 3, a bright yellow flame was the product of hot unburned particles of carbon within a flame. A gas that burned *brightly* meant somewhat perversely that it should *not* quite burn *completely*, but equally it should not actually produce a smoky flame. Even worse, the gas contained all the sulfur impurities of the coal, present as the toxic gases hydrogen sulfide, which smells of rotten eggs, and nitrogen in the form of ammonia, which smells of onions. Although the first market was street lighting (replacing equally smelly cheap fish oil) manufacturing a product with a guaranteed illuminating power that could be piped into the homes of the rich required a careful choice of coal and thorough gas cleaning. Even though this gas was initially expensive, its potential as a controllable fuel for cooking was soon appreciated, increasing demand.

By the end of the nineteenth century virtually every town and city across the industrialized world had a gasworks, producing town gas and coke. The gas was made by heating coal at a high temperature in a closed retort, breaking down much of the tarry bituminous content of the coal and driving off its volatile matter. Figure 5.7 shows one of a set of retorts originally used to light the city of Athens in Greece, but of a standard pattern used in many countries.

Bituminous coal was loaded by hand (using specially long shovels) into long horizontal cast iron retorts which were sealed with doors (missing in this museum example). A furnace fuelled by coke (with the square door in Figure 5.7) supplied the necessary heat. The impure gas rose up through the vertical iron tubes into the first of a set of clean-up processes. Firstly the gas was bubbled through water. Here the ammonia produced from the nitrogen content of the coal dissolved in the water, and the oils and tars from the volatile matter of the coal condensed out as a liquid layer on top. Next the hydrogen sulfide was removed by passing the gas over iron oxide, which absorbed the sulfur. The gas might then go on to a final 'scrubbing' (washing) process to condense out any remaining volatile liquids, such as benzene. The final relatively clean gas consisted mainly of a mixture of carbon monoxide, methane and hydrogen. However, the components that gave the gas its desirable bright illuminating flame were the hydrocarbons ethylene and acetylene.

When all the gas had been driven off from a batch of coal, the residue of coke would be removed to cool. Some of it could then be used to fuel the furnace and the rest, after cooling down, could be sold as heating fuel elsewhere.

The gas could then be conveniently stored at low pressure in telescopic gas holders (such as that shown in Figure 5.8). The common name for these is gasometers, a term attributed to William Murdoch (they aren't a meter at all). The gas was then piped at low pressure to the surrounding consumers.

Figure 5.7 A small retort for making town gas; an example from a museum in Athens, Greece

Figure 5.8 A typical telescoping gas holder, Greenwich No.1, built in 1886 and still used for storing natural gas in London

Everything reeked of sulfur. Once a batch of the clean-up agent iron oxide had absorbed as much sulfur as possible, it would then be removed and left exposed to the air (usually within the gasworks premises). This oxidized the iron sulfide back to iron oxide (which could be reused) leaving a residue of pure sulfur. The gas still contained traces of toxic hydrogen sulfide and other sulfur compounds, the smell of which was useful in detecting leaks. This was just as well because carbon monoxide, a key ingredient of the gas, is also toxic, but odourless. Modern natural gas, which is also odourless, has a 'sulfur' smell in the form of ethyl mercaptan deliberately added for leak detection.

Although gas and coke made up 90% of the output, the remaining 10% had saleable end products as well (Figure 5.9). The ammonia was treated with sulfuric acid (produced from the recovered sulfur) and sold as ammonium sulfate for fertilizer. This was a major source of nitrogen for agriculture (the Haber–Bosch process for making ammonia from nitrogen in the air did not appear until the early twentieth century). The coal tars could be distilled into a range of 'oils'. Unlike their counterparts produced from petroleum, which consist mainly of chain molecules of carbon and hydrogen, these were **aromatic** compounds consisting of cyclic rings, of which benzene is the simplest example. These became the feedstocks for a new chemical industry. The residue of thick tars and pitch was sold for road-making and weatherproofing roofs.

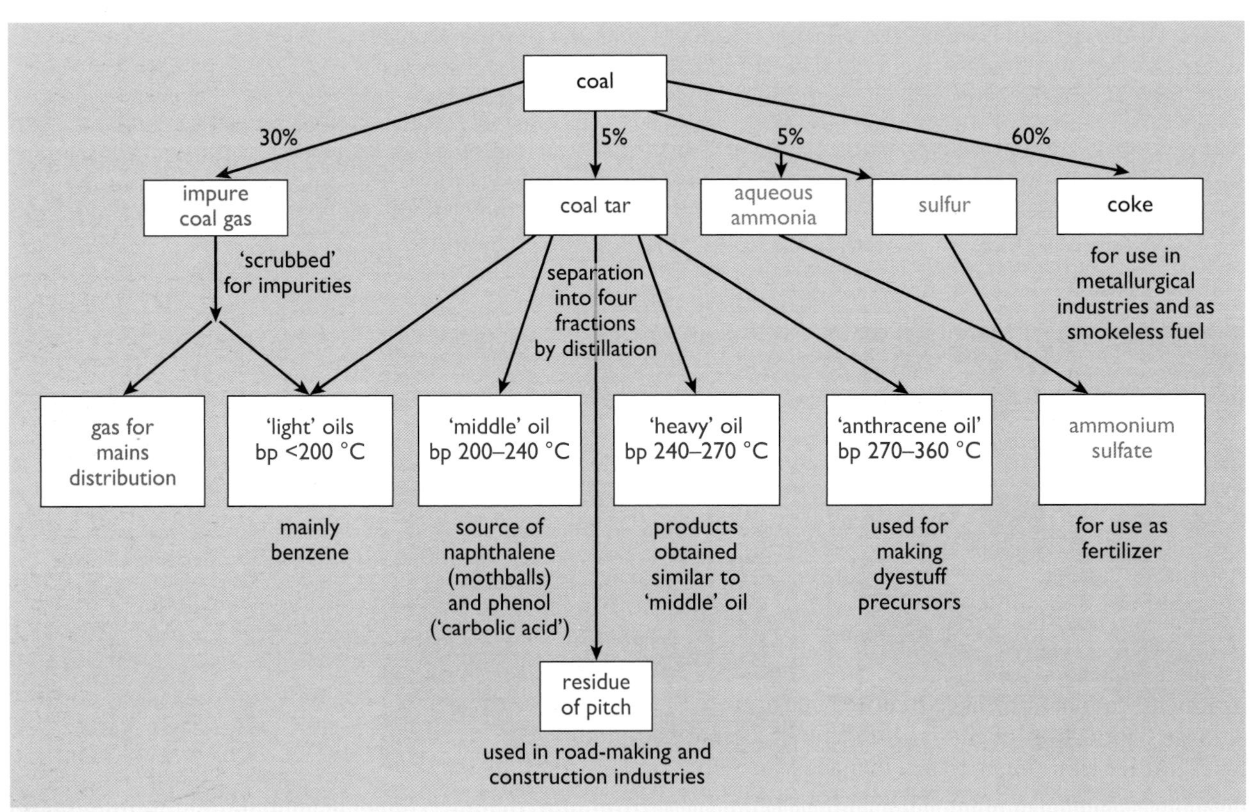

Figure 5.9 The products of coal distillation as practised for the production of town gas and coke (bp = boiling point)

In the 1880s, the gas lighting industry faced a crisis: the invention of viable electric lighting (described in Chapter 9). Just when the electricity entrepreneurs thought that they could sweep up the market in street lighting, gas lighting fought back. The Austrian chemist Carl Auer von Welsbach (Figure 5.10) found that thorium and cerium oxides would glow brilliantly when heated in a gas flame. In 1886 he developed the **gas mantle**, a small fabric sock coated with these chemicals to be fitted over a gas burner. Although the fabric burned away, the residue retained sufficient strength to form a bright glowing mesh. It was extremely fragile, but it improved the lighting efficacy by a factor of about four over the normal gas lamps of the day. By 1895 gas mantles were selling in Britain at the rate of 300 000 per year. This invention allowed domestic gas lighting to last well into the twentieth (and even the twenty-first) century (Figure 5.11), a serious blow to the electricity industry which was forced to diversify into other applications. The gas mantle is still widely used today and appreciated by the 1.4 billion people worldwide who do not have electric light and are still dependent on lighting using lamps running on kerosene or liquefied petroleum gas. Its invention also removed the need for gas to have an 'illuminating' flame based on particles of unburned carbon. The formulation of town gas was changed to ensure complete combustion.

Figure 5.10 The Austrian chemist Carl Auer von Welsbach (1858–1929), who specialized in studying the 'rare earths' using spectroscopy, the fine details of the light given off when chemicals are heated in a flame. In addition to inventing the gas mantle, he also invented the first metal filament electric light bulb and supplied Marie Curie, the early researcher into radioactivity, with radium

The distillation of bituminous coal was not the only way to make a useful gas. **Water gas** could be made by spraying water onto red-hot coke. This produced a relatively clean mixture of carbon monoxide and hydrogen.

$$C + H_2O \rightarrow CO + H_2$$

The basic coal distillation process was changed to ensure that the bituminous component of the coal was broken down at high temperature, producing more gas, but less in the way of tars. This allowed a gas production process that made full use of its coal, reducing gas prices. By the 1920s completely combustible town gas was being metered and sold on its heat content alone, as natural gas is today.

Today, town gas has been almost totally replaced by natural gas. This has the advantage of a higher energy density (allowing the same distribution pipework to carry more energy). It also does not contain toxic hydrogen sulfide or carbon monoxide. However, the possibilities of producing gas from coal remain, in the form of synthetic natural gas and also underground coal gasification, both described in Chapter 7.

Figure 5.11 A gas mantle street light in London (using natural gas) photographed in 2002

Modern coke production

Coke is still vital to the production of iron and steel. In 2007 globally almost 800 Mt of coking coal was used by that industry, approximately half of it in China. In 2009 in the UK nearly 12% of coal use was in coke ovens and blast furnaces. With present-day technology each tonne of steel produced in blast furnaces requires an input of 600–800 kg of coking coal. A large proportion of the world's steel is produced in this way, with the rest coming from electric arc furnaces, in which scrap iron and steel are melted together. As the electricity for these is often from coal-fired power plants, coal is again an important requirement.

Coke for modern steel-making needs to be as free of sulfur and ash as possible. It is produced by heating finely divided coal in the absence of air. This process is similar to that described for the production of town gas above, but it is optimized for large-scale metallurgical coke production. Coal particles, no more than 3 mm in size, are heated in coke ovens to over 1200 °C over a period of up to 36 hours. During this time the more volatile constituents are driven off as coke oven gas, and the remainder, almost pure carbon, becomes liquefied. At the end of this process the coke is discharged ('pushed') from the oven (Figure 5.12). This liquid then fuses ('resolidifies') to form a hard porous coke. Not all types of coal have the right properties for this. Low-sulfur, low-ash **coking coals** may sell for twice the price of ordinary steam coal. The coke oven gas is used to provide the energy for heating the process and if the coke oven is situated inside a steelworks it may be used for heating other processes. It can also be used to provide chemical feedstocks as in the town gas process above.

Figure 5.12 A 'push'– a steel industry term for coke being discharged from a coke oven

Smokeless fuels

'Smogs' (a toxic mixture of fog and acidic smoke) have become features of many large industrial cities, particularly those dependent on coal. The 'London fog' probably reached its worst in the 1890s. However, it took a serious smog in 1952, as a result of which at least 4000 died, to get the finger of blame pointed at smoky domestic coal fires. The resulting pressure for effective legislation led to the 1956 Clean Air Act, which promoted the use of 'smokeless' fuels such as anthracite and coke.

As late as 1970 coal and coke made up nearly a half of UK domestic heating fuel consumption (DECC, 2010a) and the manufacture of coke for domestic heating use was a major industry. Today less than 2% of UK domestic heating is supplied by coal and coke, the demand having been swept away by the widespread availability of that clean-burning low-sulfur fuel natural gas.

Coal for electricity

While the manufacture of coke and smokeless fuels requires specific types of coal, electricity can be generated by burning virtually any type. In a competitive market with other fuels what matters is the overall cost to the consumer of the final electricity. There has thus been a rise in the production of 'lower' sub-bituminous coals and those with higher ash content. If the power stations are sited close to the users, this coal (with its ash and moisture content) may have to be moved hundreds, if not thousands, of kilometres, by rail or sea. An alternative, practised in many countries, is to site the electricity generation in the mining area and transmit the electricity over high-voltage cables: 'coal by wire' as it has been called. This saves on transport costs (including the rail and road infrastructure), and slightly reduces the ash disposal problem, but requires long-distance electricity transmission lines instead, a topic that will be revisited in Chapter 9.

5.5 The coal resource

Mining the coal

This book is essentially concerned with the nature of our present fuels, the ways in which we use them, and the consequences of these uses. Details of the geological processes which led to their existence and of the technologies involved in locating and extracting them are therefore generally beyond its scope. However, the extraction of coal – or any other mineral resource – is an important determinant of its cost and also has environmental, health and social implications. The following is therefore a brief outline of the two main types of coalmine, usually referred to as **deep mines** (worked by tunnelling and which may go down 1000 m) and **surface mines** (essentially large, open pits in the earth). The depth of the coal below the surface is not necessarily the distinguishing criterion, and some present-day 'surface' mines are extracting coal at greater depths than 'deep' mines in the past or in other parts of the world.

Deep mines

The more comprehensive term, **underground mining**, used in the USA, is a better description in this case, as the category includes not only the deep mines reached by vertical shafts, but others reached by a long sloping tunnel or drift. Getting the coal is of course less simple than on the surface, and various methods of extraction have been used over the centuries since the first underground mines. The **longwall** method, already in use in Britain as long ago as the seventeenth century, has proved the most suitable for mechanization. Two long parallel tunnels or *roadways* are driven and these define the *parcel* of coal to be worked. A tunnel connecting these two roadways then exposes the coalface to be worked. In the past, before mechanization, the coal was extracted by hand, cutting a deep slot at the base of the face, along its entire length, and then bringing down the coal with the aid of picks, drills and sometimes explosives. Working narrow seams required miners to work in cramped conditions (see Figure 5.13). They would then hand-load the coal for transport back to the mineshaft.

The first cutting machines had appeared in the mid-nineteenth century, and a hundred years later, Britain's first cutter loader was installed; but manual operations, and in particular manual loading, remained common well into the second half of the twentieth century.

The fully mechanized, present-day underground mine is very different. An **armoured face conveyor** perhaps 250 metres long stretches along the length of the face, and the cutting machine works its way along on this. The cutting machine, or **power loader**, has a large rotating drum armed with steel picks (Figure 5.14). The cut coal falls on to the conveyor and is carried away along the face. In the most recent installations the conveyor is articulated, and advances automatically, together with the roof supports, as the power loader moves along. By the end of each traverse the face is ready for the next run in the opposite direction. Obviously such machines require unbroken, horizontal seams, preferably of constant thickness and quality.

Figure 5.13 A miner lying in a slot just over 30 cm high, hacking out coal with his pick in the Lilley Drift Mine in 1953. Even at that relatively recent date a fifth of Britain's coal was still dug by hand.

Figure 5.14 A modern longwall mining machine shears coal from a thick coalface; it drops on to a moving conveyor belt

The main adverse environmental effects of deep mining have been land subsidence and waste tips. Tips can leak dangerous chemicals into the ground, and also present other dangers. In 1966, an unstable waste tip from the Merthyr Vale mine in Wales released an avalanche of sludge over the village of Aberfan, engulfing not only houses but also the primary school, killing 114 children. Legislation and technical advances have brought improvements in more modern mines, and the closure of many of Britain's older mines appears in some cases to have been followed by successful land rehabilitation.

Accidents and lung diseases have always led to high rates of death or disability amongst miners. In mid-nineteenth century Britain, one in every five underground workers was likely to be killed in an accident before completing a full working life. By 1912, when Britain had about a million miners, the likelihood had fallen to about one in twenty over a working life, although 1913 saw Britain's worst mining disaster, with 439 miners killed in an explosion at the Senghenydd mine in Wales. Towards the end of the twentieth century, the death rate in the USA, with some 160 000 underground workers, was still more than one in fifty over a working lifetime.

It has also been well known for many years that miners, or retired miners, suffer higher than normal rates of respiratory diseases compared with the population at large. Establishing the exact relationship between these rates and exposure to the atmosphere underground is obviously more difficult than counting the deaths and injuries in accidents. Nevertheless, it is perhaps surprising that only in 1950 was pneumoconiosis formally recognized in the USA as a work-related disease – and that another nineteen years passed before the enactment of legislation on dust exposure.

Surface mines

The term *opencast*, commonly used in the UK, better describes the essential characteristic of 'surface' mining. The coal seam is accessed by removing the earth and rock above it (known as the 'overburden'), a process that can involve some of the world's largest machines (Figure 5.15). The coal is then broken up, possibly with the aid of explosives, and removed by mechanical shovels. Various forms of excavation are adopted, depending on the nature, depth and position of the seam. On relatively flat ground, the opened areas can be in the form of long rectangles, with the topsoil and overburden placed to one side for later replacement. Or in some cases a single large pit is opened and the overburden is then moved around within it to gain access to different parts of the coal seam. In mountainous areas, ledges contouring the slopes may be opened. The ultimate depth of excavation can exceed 100 metres.

Figure 5.15 The Bagger 288 bucketwheel excavator is used to remove overburden in German opencast coal mines. It weighs 13 500 tonnes and can remove 240 000 cubic metres of overburden per day. Here it is seen in a rare move between two mine sites.

In 2009 about 60% of the UK's coal production came from opencast mines (in 1950 it was only 6%). In 2008 the figure for the USA was about 70%, but there are significant variations between different areas of the country. The quantity of coal extracted per worker-day in surface mining can be many times that in deep mines – a factor that is reflected in the generally lower cost of surface-mined coal (and a higher safety level amongst workers).

The most obvious deleterious effect of surface mining is on the local landscape. In principle the overburden and topsoil can be replaced and vegetation re-established, but this increases the cost of the coal, and has not always been carried out in the past. Most major coal-producing countries have legislation requiring rehabilitation, or laws prohibiting surface mining on land where rehabilitation would not be possible, but these have not always proved effective – particularly when they conflict with economic or strategic considerations. At the time of writing (2010) coal production using 'mountaintop removal' in the Appalachian states of the USA, West Virginia and Kentucky, is highly controversial, particularly where the spoil is used to fill in the river valleys between the mountains.

Resources, reserves and production

Coal is a widespread resource, mined in every continent except Antarctica (see Figure 5.16).

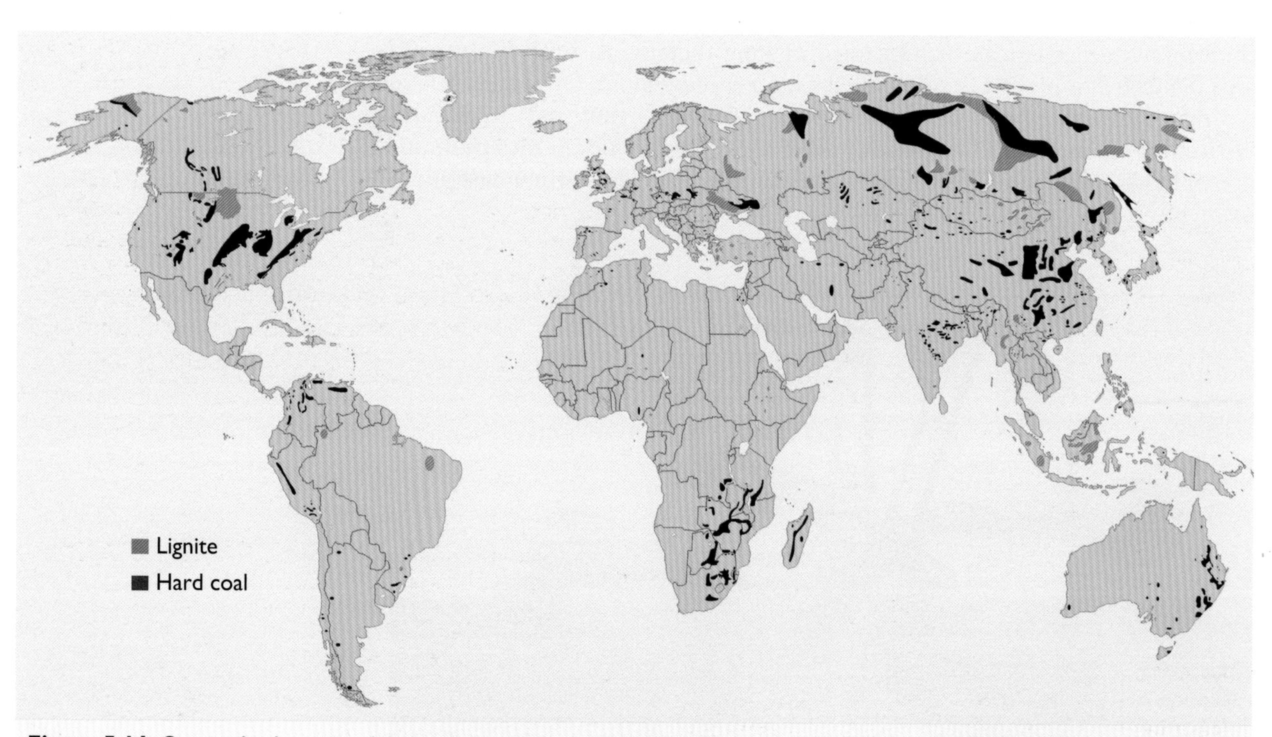

Figure 5.16 Geographic location of the most important coal deposits/basins of the world (source: BGR, 2009)

The global resource of coal is very large and it is often variously said that there is enough coal for 100, 200 or even 1000 years. However, it is necessary to distinguish between the two terms *resources* and *reserves*.

Resources are those quantities of coal (or any other mineral) that are likely to actually exist and could possibly be extracted at some time in the future. **Reserves** are those that are known with considerable certainty to exist and would be economic to extract with today's technology and in today's market. The distinction between the two is often explored with the aid of a McKelvey diagram (see Box 5.4).

BOX 5.4 Resources and reserves

The amounts of a particular mineral such as coal, oil, gas or uranium in the Earth's crust can be described in terms of their *resources* and their *reserves*. It is important to understand the difference.

The category of 'resources' is the larger of the two. It contains concentrations of useful materials in such a form and amount that profitable extraction is either currently feasible or potentially feasible, given reasonably foreseeable changes in techniques and/or price. It is the realm of 'reasonable possibilities'.

The term 'reserves' refers to a subset of resources that are reasonably known to exist and can be extracted profitably and practically under *existing* conditions.

Two important parameters are used to define these categories more precisely. First there is the degree of certainty that a deposit actually exists. This is based on information about geological, physical and chemical characteristics – the grade or quality, tonnage, location and depth of the material. The other parameter is profitability, which depends on the price received, balanced against the extraction costs. This may vary with changes in technology, shifts in world prices and competition from other sources.

These conditions may be tempered by 'above ground' factors such as the necessary ownership, permissions and access to exploit the resource. They may be limited by environmental factors, such as locations within environmentally sensitive areas. The profitability may also be restricted by long transportation distances to reach potential markets.

These two parameters can be used as the axes of a simple box diagram, sometimes called a McKelvey diagram (Figure 5.17); the certainty of existence decreases from left to right, and the relative profitability of extraction decreases downwards.

The top-left segment of Figure 5.17(a) represents reserves. It is bounded at its base by the current limit of commercial profitability. At its right, the reserves box is limited by the degree of certainty about the existence of the material. In order to be classified as reserves the materials must have been identified with sufficient certainty *and* would have to be profitable to extract with current technology, under present economic, technical, legal and political conditions.

Resources that have been identified in an area, but are uneconomic (or legally or politically unavailable) at present, are termed **conditional resources**; that is, they may become reserves if circumstances change. Even limited geological knowledge of an area may make it possible to estimate what resources might be

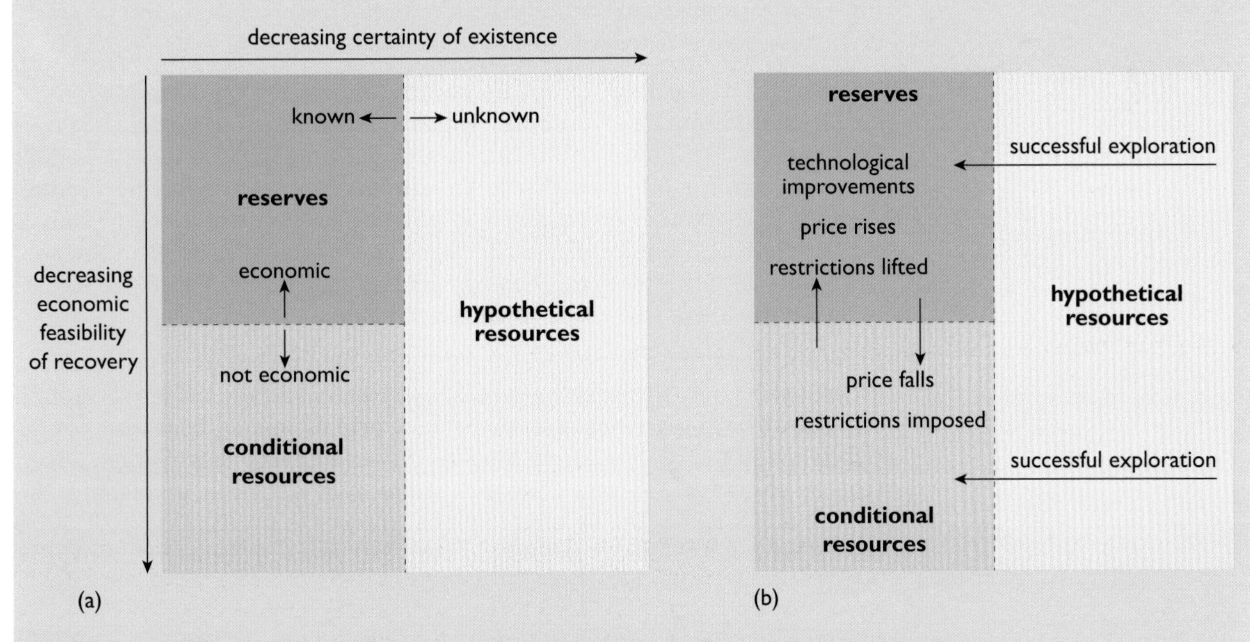

Figure 5.17 The McKelvey diagram: (a) criteria for separating reserves from resources. The field of resources covers the whole box, while reserves cover only the top left-hand corner; (b) how the estimates of reserves can be altered by various factors

present, by comparing the area with similar ones that have already been mined. Such estimates are termed **hypothetical resources**, meaning that they are as yet undiscovered, but they might eventually turn out to include both reserves and conditional resources. These are represented in the right-hand part of the diagram.

The precise terminology of reserve and resource descriptions varies between coal, oil, gas and uranium, but the principles remain the same.

What general factors could enable conditional or hypothetical resources to be reclassified as reserves and vice versa? The possibilities are summarized in Figure 5.17(b).

Successful exploration and discovery could transfer some hypothetical resources from the right-hand area into the left-hand area. If they are deemed economic to extract they will become 'reserves'; if not they will move into the lower-left 'conditional resources' category.

Increases in market price could transfer conditional resources in the lower-left area up into the reserves box. Conversely a fall in price may result in some reserves being reclassified as resources. Similarly new environmental restrictions may make some reserves 'unobtainable' and lead them to be reclassified as resources.

Technological improvements may lower the costs of extraction and thus move some conditional resources into reserves. However, improvements in one country, for example in surface-mined coal in Australia, may make large quantities of material available cheaply on the world market. In a global competitive market, this may mean that marginally economic reserves elsewhere, such as deep-mined coal in the UK and Germany, become 'uneconomic' and have to be reclassified as conditional resources.

Although national tables of coal reserves can be produced, in many cases the figures should be treated with care. Past estimates of reserves may have assumed that the deep mining of narrow coal seams as shown in Figure 5.5 would continue. In practice the increased use of automated machines and surface mining has concentrated production on thick, horizontal and easily accessible seams. This has also been spurred on by competition from cheap oil and gas. Other narrow, inaccessible or otherwise difficult coal seams have increasingly been rendered uneconomic to work and should properly be reclassified as resources rather than actual reserves.

The ease of assessment is not helped by the range of different types of coal with different heat contents. Approximately a half of the world's coal reserves are low-grade sub-bituminous coal and lignite. For statistical purposes 'coal' is often simply broken down into 'hard coal' and 'lignite'. A tonne of 'hard coal' is assumed to have a heat value of 28 GJ t^{-1} and is equivalent to 1.5 tonnes of lignite. Some sets of statistics refer to the coal 'as mined', others refer to their energy content expressed in terms of 'tonnes of coal equivalent' (tce – i.e. 28 GJ/tonne coal).

In recent years there have been major downward revisions of coal reserve estimates in countries such as the USA, Germany and the UK (EWG, 2007). Despite upward revisions elsewhere the world estimated coal reserves (as published in the *BP Statistical Review of World Energy*) have fallen. Given that, as shown in Chapter 2, world production has been increasing, the overall effect is that the estimated global reserve to production (R/P) ratio has fallen from 233 years in 2000 to only 119 years in 2009 (BP, 2001; BP, 2010).

Figure 5.18 shows the regional distribution of coal reserves and current production for the year 2009. It is significant that over a half of the total production takes place in Asia, particularly in China, India and Indonesia.

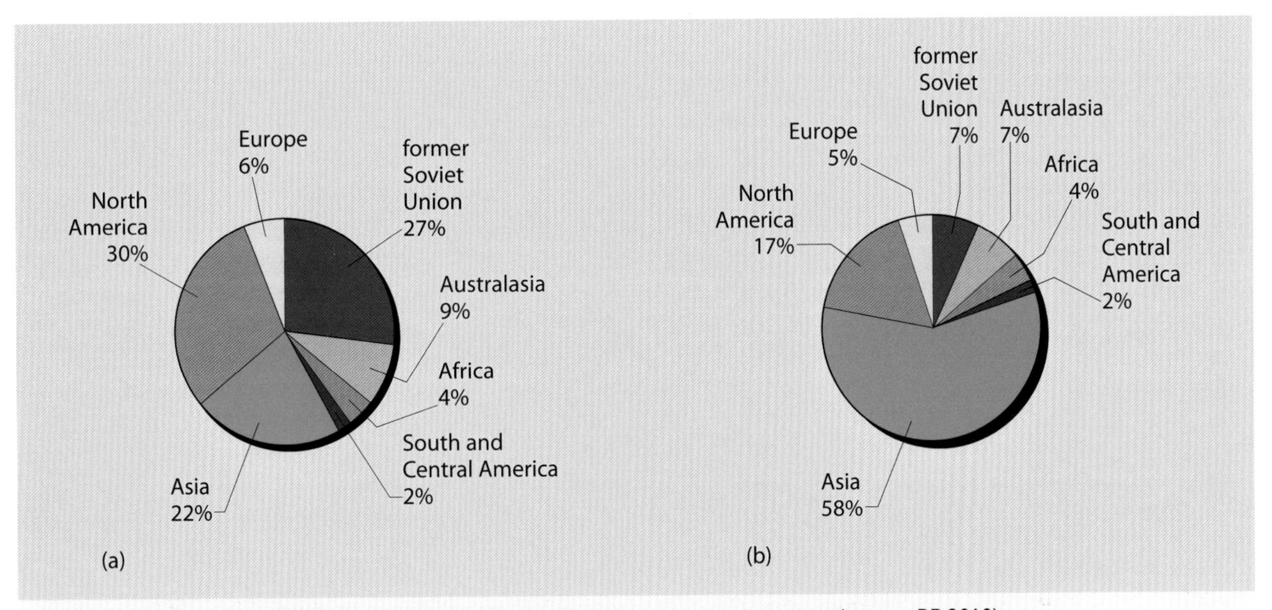

Figure 5.18 World coal, 2009: (a) reserves by region; (b) total production by region (source: BP, 2010)

Table 5.1 shows statistics for the countries with the largest resource, together with others of interest.

Although there are many countries in the world with significant coal reserves, the top five countries in the table account for over 80% of the world's reserves and nearly 80% of world production. Most coal is used

Table 5.1 Coal reserves, annual production and R/P ratios for selected countries, 2009

Country	Reserves /1000 Mt	Percentage of world reserves	Annual production /Mt[1]	Percentage of world production[2]	R/P ratio /years
USA	238	28.9%	973	15.8%	245
Russia	157	19.0%	298	4.1%	>500
China	115	13.9%	3050	45.6%	38
Australia	76	9.2%	409	6.7%	186
India	59	7.1%	558	6.2%	105
Ukraine	34	4.1%	74	1.1%	460
Kazakhstan	31	3.8%	102	1.5%	308
S. Africa	30	3.7%	250	4.1%	122
Poland	8	0.9%	135	1.7%	56
Germany	7	0.8%	184	1.3%	37
Indonesia	4	0.5%	253	4.6%	17
UK	0.2	0.02%	18	0.3%	9
World	826	100%	6941	100%	119

[1] Actual tonnage produced, regardless of type.
[2] Expressed in energy terms, taking into account different grades of coal.
Source: BP, 2010.

within the producing country. The world's top exporters of coal are Australia, Indonesia and Russia. In 2007 just over 600 Mt of coal was traded by sea.

Coal in the USA

Over 90% of US coal is used for electricity generation. The country has about a quarter of the world's reserves and despite downward reserve revisions still has an R/P ratio of over 200 years. Its coal lies mainly in two broad areas, east and west of the Mississippi. Eastern coal makes up about 40% of total production and includes high-quality coking coal from Pennsylvania. More than half of the eastern coal comes from deep mines. Although the mining costs are relatively high, they are offset by a high energy content per tonne and the proximity to industrial customers. In contrast, about 90% of the western coal is surface mined, in some cases from very large beds. The state of Wyoming alone produces over 40% of all US coal (EIA, 2010). Although western coal has a low sulfur content it is classified as 'sub-bituminous' and has a low heat content. The cost of long-distance rail transport to the electricity generating plants around the country is thus a significant factor in the final electricity price.

Coal in China

The rapid expansion of Chinese coal production over the past decade, as described in previous chapters, and its current dominance of world production, obviously pose a serious global problem in terms of CO_2 emissions. Also, given that Table 5.1 shows a reserve estimate of 115 billion tonnes and a reserve/production ratio of only 38 years, it is not clear how long this expansion can continue. In 2003 the Chinese government produced an estimate of 189 billion tonnes of 'economic' reserves, with a resource estimate of about 1000 billion tonnes, but as pointed out in Chapter 2, China is starting to import considerable amounts of coal.

Coal production also poses considerable logistical problems within the country. Most of the coal mines are located in the north and west while the coal demand is in the industrialized cities of the south and east, 1500 km away. Although long-distance transport can be done by rail, many mines do not have rail access and coal has to be transported to the railhead by truck (Aden et al., 2009).

More than 90% of Chinese mines are deep mines. The industry has a very poor safety record with over 5 000 mining deaths per year in the 1990s, approximately ten times worse, per tonne of coal mined, than UK standards. The government has sought to improve this by closing a large number of smaller mines.

In northern China, there are also many areas where uncontrolled underground coal fires have been burning (see Box 5.5).

Acid rain from coal combustion is a serious problem in many Chinese cities. Chinese coal has a sulfur content of up to 5% but this increases with depth. This suggests that as the more economic lower sulfur coal closer to the surface is used up, the average sulfur content of the coal produced may increase.

BOX 5.5 Underground coal fires

Coal seams close to the surface can ignite spontaneously or if struck by lightning. If supplied with sufficient air they can then burn for decades. In China in particular hundreds of burning areas are known in which 10 to 20 Mt of coal burn annually emitting methane, SO_2 and carbon dioxide. Extinguishing these fires is difficult but can be achieved by covering the area with a thick layer of loam or clay to cut off the air supply (BGR, 2009).

Coal in Russia

Russia is a major coal producer and exporter. However, its production in 2009 has fallen to about three-quarters of that in the late 1980s, a result of the availability of natural gas as an alternative fuel. Although the country has large coal reserves and a high reserve/production ratio, as shown in Figure 5.16, much of these lie in remote and inaccessible parts of Siberia, with potentially high transport costs.

Coal in India

Although India has the world's fourth largest coal reserves, the coal is mostly of poor quality with a high ash content and an average heat value of only 19 GJ t^{-1} (IEA, 2007). The main coal mines are in the east of the country, creating problems of transportation to the major cities of Delhi and Mumbai to the west. India imports higher-quality coking coal from Australia.

Coal in Australia

Australia has less than 10% of the world's coal reserves, but much of these are conveniently located in the states of Queensland and New South Wales with good rail access to ports on the coast. About three-quarters of the coal is produced from surface mines. Coal is Australia's largest export (approximately three-quarters of the country's coal production was exported in 2009). Forty per cent of the exports went to Japan and a further eight per cent (mainly coking coal) to Europe. The remainder went to other countries in the Middle East and Asia including coking coal to India and China.

The UK's shrinking coal industry

The history of British coal production offers an interesting example of the problems of forecasting and of the need to interpret resource data with care. As mentioned earlier, in the 1860s, William Stanley Jevons wrote of his worries about the sustainability of the exponential growth in UK coal production. In 1871 a Royal Commission estimated British coal reserves at almost 150 billion tonnes, but this included seams down to one foot (30 cm) thick to a depth of 4000 feet (1200 m). At the time, this would have given an R/P ratio of over 1000 years.

Figure 5.4 showed that UK coal production peaked in 1913 at nearly 300 million tonnes, with 3000 deep mines producing coal. In 1915, Jevons' son predicted that in 2001 Britain would consume 350 million tonnes of coal and export a further 400 million tonnes (Jevons, 1915). In fact, by 2001, as shown in Figure 2.9, oil, natural gas and nuclear power had supplanted coal as a fuel. Including coking coal, the UK only used about 60 million tonnes (DECC, 2009).

Given a global market in coal, the availability of thick, easily mined coal seams in other countries and cheap sea freight steadily eroded the UK's coal exports during the 1920s and 30s. By the 1970s it was becoming cheaper to import coal. This has left much of the UK deep mining considered uneconomic. In 2009 UK coal production was only 17 million tonnes, of which about 60% was from surface mines. At the end of that year there were only 13 deep mines left in operation (DECC, 2010b). Far from *exporting* coal, the UK *imported* a further 38 million tonnes. Steam coal was imported from Russia, the USA and Colombia and coking coal from the USA and Australia.

What of the UK's coal reserves? They are quoted in the 2010 *BP Statistical Review of World Energy* (BP, 2010) as being 155 Mt, about *one thousandth* of the figure given by the 1871 Royal Commission. Where has all this coal gone? It hasn't exactly disappeared. In the terms of the McKelvey diagram of Figure 5.17, it has simply been reclassified out of 'reserves' and back into 'conditional resources'.

5.6 Fires, furnaces and boilers

The design of furnaces is an important one in the world of energy. As you have seen, all but a small fraction of our primary energy comes from fuels – fossil fuel or biofuel – and the destination of three-quarters of these is a fire, furnace or boiler. (The remainder is used in the internal combustion engine, to be described in Chapter 8.) The efficiency with which a boiler extracts energy from the fuel is a critical factor in determining the fuel consumption of systems, from a domestic heating boiler to a 1000 MW power station. Better design means a lower fuel requirement, lower costs and lower emissions of pollutants.

This is also true at the very small scale. A better design of cooking stove can mean a lower wood requirement and less destruction of valuable vegetation. This example can illustrate the problem facing anyone trying to design a really efficient solid-fuel boiler. Exactly how much wood is needed to bring one litre of water to the boil on a fire? Common experience will give a rough answer, but Box 5.6 shows a calculation based on known data. The stove shown might have a fuel-to-useful heat efficiency of about 10%. This should be compared with figures of well over 80% for a modern domestic gas boiler or that of a modern coal-fired power station.

Power station boilers provide a useful study for several reasons:

- They can be amongst the most efficient fuel-burning systems we have.
- Their waste products create some of the world's major pollution problems.
- Interesting technological solutions to some of the problems exist or are being developed, but are still to be universally adopted.

BOX 5.6 Boiling a litre of water

Assuming an idealized efficiency of 100%, how many cubic centimetres of wood are needed to bring one litre (one kilogram) of water to the boil?

Data

Specific heat capacity of water	= 4200 J kg^{-1} K^{-1}
Heat value of wood (Figure 5.6)	= 15 MJ kg^{-1}
Density of wood	= 600 kg m^{-3}
1 cubic centimetre (1 cm^3)	= 10^{-6} m^3

Calculation

Heat energy needed to heat 1 litre of water from 20 °C to 100 °C	= 80 × 4200 J
	= 336 kJ
Heat energy released in burning 1 cm^3 of wood	= 15 × 600 × 10^{-6} MJ
	= 9.0 kJ
Volume of wood required	$= \frac{336}{9.0}$ cm^3
	= **37 cm^3**

This suggests that one thin stick about a foot (30 cm) long should be adequate. Any reader who has tried boiling water on a simple open fire will no doubt find this surprising, but the well-designed stove of Figure 5.19 would perhaps only need ten such sticks per litre of water.

Figure 5.19 Boiling water on a clay stove

Power-station boilers

The starting-point for the furnace designer must be the sequence of events during combustion described in Section 5.4. To extract the maximum energy from the solid fuel, both the fixed carbon and the volatile matter must be burned completely. As one part is solid material and the other a stream of gases, this is not simple. Modern plants operate with a continuous feed of fuel, so the solids and gases must burn completely at about the same rate. The purpose of the plant is to produce steam, so another requirement is for the best possible *heat transfer* from the burning fuel into the circulating water and steam within the boiler, which may be at a temperature of over 500 °C. Finally, the system should minimize the production of undesirable by-products and include a method for dealing with the unavoidable wastes: ash and flue gases.

Although coal is the main subject of this chapter, the following brief accounts include references to other fuels. A power station furnace may need to be very versatile and accept not only a wide range of types of coal, but also other fuels such as co-fired biomass or waste petroleum coke from oil refineries. The basic principles also apply to the generation of electricity from fuels such as municipal solid wastes (MSW).

Grate boilers

None of the boilers to be described here is fed with large lumps of coal. The 'stoker', a person armed with a shovel, disappeared long ago. Modern boilers require that the coal is crushed before burning. In a **grate boiler** (Figure 5.20(a)), the fuel is in pieces a few millimetres across, which are fed in from a hopper or on a conveyor belt, and move across the grate in an upward flow of air. With this arrangement, the fixed carbon burns on the grate and the volatile matter in the space above. Radiant heat from both reaches an array of tubes through which the water circulates, while the hot gases from the combustion pass between another set of tubes. Note that the few tubes shown in the diagram represent the large, much more complex array used in the actual system. Boilers of this type are still used for coal, but mainly for biofuels such as wood chips, processed domestic wastes and other materials that are not suitable for pulverized fuel boilers. Increasingly, however, they are being replaced by the cleaner, more efficient fluidized bed boilers (see below) when systems are upgraded.

Pulverized fuel boilers

Where coal is the sole fuel these are also known as **pulverized coal combustion (PC** or **PCC)** plants. The plant shown schematically in Figure 5.20(b) is the most common boiler type by far in present-day coal-fired power stations. Plants like this have been in use for half a century or more, and when well run can transfer over 90% of the energy content of the coal to the circulating water (or steam).

The essential feature of this boiler is that it uses **pulverized fuel (PF)**, i.e. it is finely ground up before burning. This does not have to be solely coal; it can also be co-fired with wood or other suitable fuels.

The fuel enters the furnace in the form of particles less than about 100 microns (0.1 mm) in size – a dust so fine that it floats. This dust is swept in a controlled flow of air to the burner jets. With tiny particles, the fixed carbon burns out completely in a very short time, so the volatile matter and fixed carbon burn together in roughly the same part of the furnace, increasing the efficiency of heat transfer.

A serious disadvantage of PF boilers is that the ash is a fine dust, and without measures to prevent it, this **fly ash** will be carried into the atmosphere with the flue gases (a topic which will be discussed further at the end of this section).

Fluidized bed boilers

The third type of plant in Figure 5.20 uses a rather different principle. **Fluidized bed combustion (FBC)**, also called **atmospheric fluidized bed combustion (AFBC)**, is regarded as offering solutions to some of the pollution problems of coal combustion, and in addition, the possibility of burning other fuels cleanly. FBC plants have been in use for some time in various industrial processes; the first use in electrical power plants being in the 1980s. By the end of the twentieth century a few thousand were in operation, burning not only coal but other fuels including a variety of wastes.

The essential feature of an FBC plant is a thick layer of inert material – sand or gravel – with particle size in the range 0.3–2.0 mm. This lies on a base plate that has many small apertures through which jets of air are blown. At a certain air speed the whole mass of material expands to a depth of a

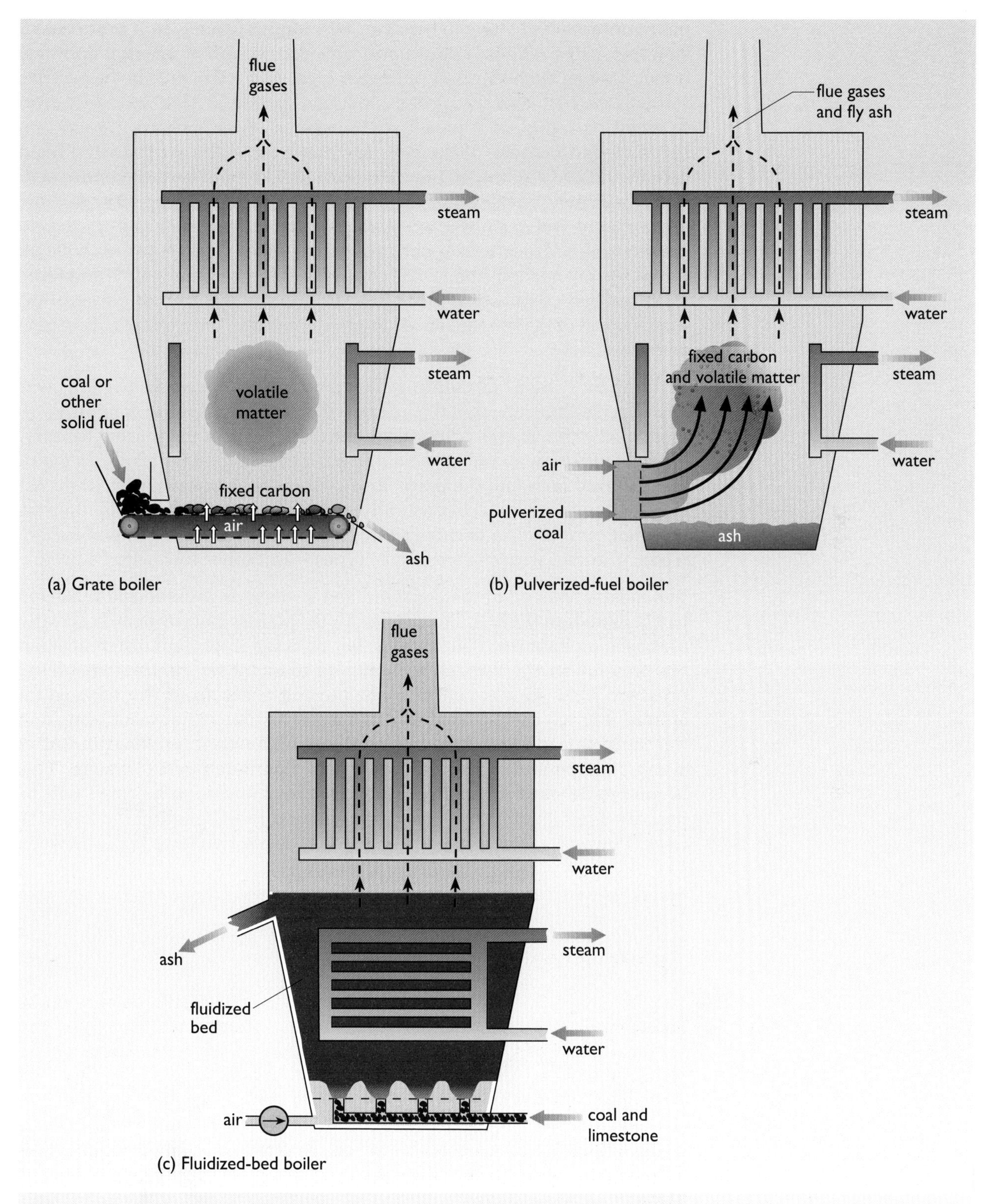

Figure 5.20 Three types of solid fuel power station boiler: (a) grate boiler; (b) pulverized-fuel boiler; (c) fluidized-bed boiler

metre or more and starts to behave like a liquid. Objects in it will float or sink, just as in a liquid. The significant feature for combustion and heat transfer is that as the air flow is further increased it forms bubbles that rise through the bed, and the particles bounce around as if they were indeed in a boiling liquid. In a power-station system, fuel particles are fed into this bed, and because of the constant motion and the air flow, the fixed carbon and the volatile matter both burn quickly and heat the entire bed. The tubes carrying the water to be heated are buried in the bed and/or the containment walls, and the excellent thermal contact with the constantly moving material means that good heat transfer does not require the very high temperatures needed in an 'open' furnace. This process, called **bubbling fluidized bed combustion (BFBC)**, is used in the majority of present FBC plants, most of them in the small-to-medium range, with electrical outputs below 100 MW.

A different system, **circulating fluidized bed combustion (CFBC)**, was developed a little later, but the number of plants grew relatively rapidly – to more than a thousand worldwide in the early years of the present century. With this system, an increased air flow drives some of the particles out of the fluidized bed, into the space above, where they behave more like a hot gas. A circulating system constantly returns them to the bed, maintaining the high temperature and increasing the time available for combustion, which allows a wider range of coal types – and other fuels – to be used.

The most advanced of the fluidized bed systems is **pressurized fluidized bed combustion (PFBC)**. The present PFBC plants operating commercially are based on bubbling bed systems, but with the important difference that the pressure in the furnace is increased to about ten times atmospheric pressure. This means that the hot burned gases from the furnace (at around 900 °C) can be used to drive a gas turbine. The waste heat from this is then used to assist in raising steam for a steam turbine, running at a lower temperature. This is known as a 'combined cycle' system. This is commonly used in gas-fired power stations and such a design will be described in Chapter 9. A good example of a PFBC plant is the 250 MW Osaki Power Station in Japan (Figure 5.21). It has two large beds, between

Figure 5.21 The 250 MW Osaki power station in Japan uses pressurized fluidized bed combustion to drive both gas and steam turbines

2 m and 4 m high when fluidized, and a furnace pressure of about 1 MPa or 10 atmospheres. (Hitachi, 2001). It has an overall fuel to electricity efficiency of 42%.

Pressurized systems using circulating beds are also under development, but have not reached full commercial production at the time of writing. (See also DTI, 2000.)

Flue gases

Section 5.3 described a range of combustion products that are the inevitable result of burning coal. A modern 660 MW power station might consume over 250 tonnes of coal an hour. Below are the quantities of combustion products that would be released into the atmosphere in one hour in the absence of any pollution reduction system. They are approximately:

- 2000 tonnes of nitrogen. This is three-quarters of the air drawn in for combustion and has passed largely unchanged through the whole system, except that heating it accounts for about half the energy loss in the boiler. Apart from its potential oxidation to NO_x it is harmless.
- About 650 tonnes of carbon dioxide produced in combustion. This is of major concern for climate change.
- 150 tonnes or more of steam. This is the other combustion product, and some also comes from moisture originally in the coal. Not condensing this steam accounts for the other half of the lost energy, but the flue gases need to stay hot if they are to rise out of the chimney.
- A tonne or more of oxides of nitrogen (NO_x). Some of this results from the combustion of nitrogen compounds in the coal. The rest is produced by the high-temperature oxidation of the nitrogen in the air. These contribute to acid rain and are damaging to health, but at present the worldwide contribution from power stations is considerably less than that from internal combustion engines.
- From one to twenty tonnes of sulfur dioxide (SO_2). If the coal contained 1% sulfur then the figure would be about five tonnes of SO_2 per hour. Sulfur dioxide is a major contributor to acid rain.
- Up to 20 tonnes of particulate matter in the form of fly ash. Its main visible feature is that it is very dirty, but the tiny particles can damage the lungs, and may also contain poisonous impurities.
- About 30 grams of mercury vapour. The mercury is a trace element in the coal and can contribute to the poisoning of fish in rivers.

Below are some basic solutions for the most important of the emission problems.

Reducing SO_2 emissions

Perhaps the simplest solution to this is to burn low-sulfur coal. However, this is in limited supply and consequently more expensive. Or, as pointed out earlier, some of the sulfur content can be removed by washing the coal before use. This has cost implications and requires a good supply of water. In practice power stations tend to burn cheap, medium-sulfur coal and then remove the sulfur dioxide after combustion.

The main method used to reduce SO_2 emissions is to allow it to react with calcium carbonate (limestone or dolomite) to produce solid calcium sulfite and carbon dioxide:

$$CaCO_3 + SO_2 \rightarrow CaSO_3 + CO_2$$
$$\text{Calcium carbonate} + \text{sulfur dioxide} \rightarrow \text{calcium sulfite} + \text{carbon dioxide}$$

The calcium sulfite is often further oxidized to calcium sulfate ($CaSO_4$) or gypsum. In fluidized bed boilers the limestone can be mixed with the coal and the gypsum will remain in the ash. Alternatively, the flue gases can be 'scrubbed' by passing them through a limestone spray. This is called **flue gas desulfurization (FGD)**. A 'wet scrubber' such as that shown in Figure 5.22 uses a spray of limestone slurry and a sludge of calcium sulfate emerges from the bottom. This process also traps much of the mercury vapour in the flue gases. Alternatively, a 'dry scrubber' process can be used. In both cases a concentrated stream of calcium sulfite is produced; this can be oxidized to useful gypsum and sold for use in plasterboard manufacture. Otherwise it is sent to landfill. Typically 90–95% of the SO_2 can be removed.

All this requires a steady input of limestone, which has to be specially mined. For every tonne of a coal containing 1% sulfur burned, about 30 kg of limestone will be required and 40 kg of gypsum will be produced.

If the power station is close to the sea, an alternative method is to scrub the flue gases with seawater, which contains a certain amount of dissolved calcium carbonate. The resulting calcium sulfite is carried away in the sea.

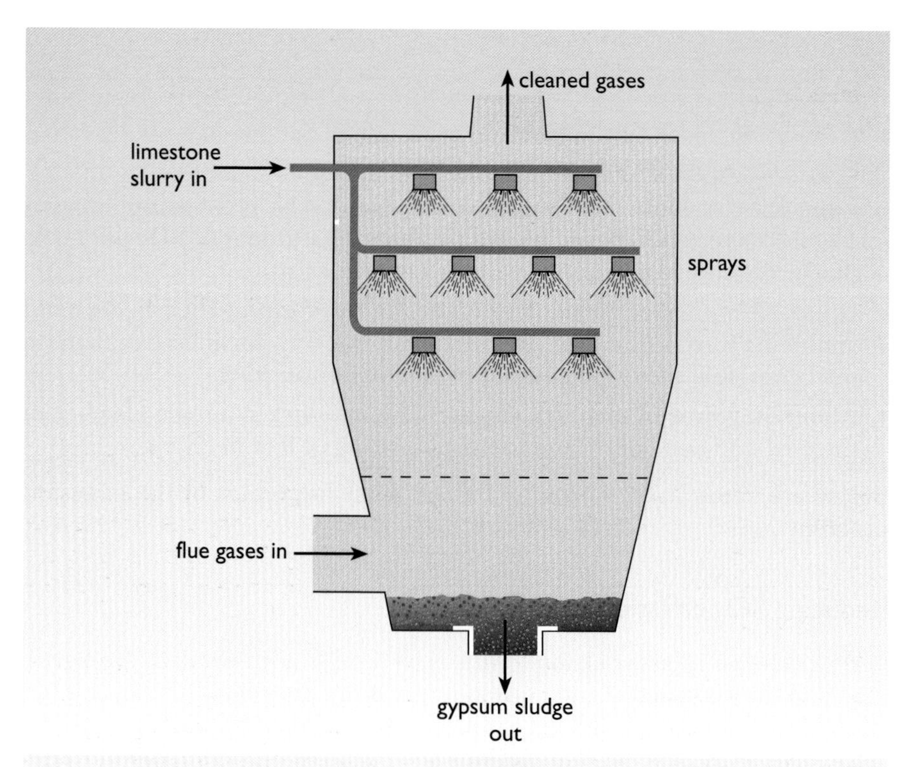

Figure 5.22 Flue gas desulfurization: a wet scrubber

Reducing NO_x emissions

The amount of NO_x that can be produced by oxidation of the nitrogen in the combustion air is highly dependent on both the combustion temperature and the amount of time that the flue gases are held at a high temperature. It is thus desirable to keep the fluidized bed or flue gas temperatures below 1000 °C. There is obviously a trade-off between a high combustion temperature and a good heat transfer to the steam (which may be at over 500 °C). The rate of NO_x production can also be reduced by making sure that combustion is rapid and that the flue gases are cooled quickly. This is one advantage of the pulverized fuel boiler. NO_x production may be more of a problem with the pressurized fluidized bed combustion (PFBC) technology where the aim is to produce a stream of gas hot enough to drive a gas turbine.

There may still be other NO_x produced from the combustion of the nitrogen compounds in the coal itself. A careful control of the combustion air supply can help ensure that this is produced mainly as the oxide NO, rather than the more acidic NO_2. However, too little air can mean that not all of the fuel is burned, resulting in poor fuel efficiency and carbon monoxide being emitted in the flue gas. This requires careful control and monitoring of the fuel flow and air supply.

There are more advanced techniques for removing this remaining NO_x. In **selective catalytic reduction (SCR)** ammonia is injected into the flue gas stream which is then passed over a catalyst made of tungsten or vanadium. The nitrogen oxides are then reduced back to nitrogen, for example:

$$\begin{array}{ccccccc} 4NH_3 & + & 6NO & \rightarrow & 5N_2 & + & 6H_2O \\ \text{ammonia} & + & \text{nitric oxide} & \rightarrow & \text{nitrogen} & + & \text{water} \end{array}$$

This technique is not just used on boilers; it is also being introduced in the USA on the road for diesel trucks.

Gas reburn involves the injection of a small amount of natural gas into the boiler flue gas stream after the main coal combustion. This initially burns by using the oxygen from the nitrogen oxides, converting them to nitrogen. Further air is then added to allow the complete combustion of the natural gas. This process has been fitted to the large Longannet coal-fired power station in Scotland.

Reducing particulate matter

These can be reduced by passing the flue gases through a **cyclone** where they are forced to spin. The centrifugal force separates out the heavier particles which are first flung outwards from the spinning gas stream and then fall to the bottom. An alternative is to use **bag filters**. Finer particles can be separated out using an electric field in an **electrostatic precipitator (ESP)**. A typical aim is to collect 99% of the particulates.

All the above processes of course involve both capital and operating costs. They also use electricity and reduce the overall efficiency of the plant, effectively increasing the fuel cost per unit of output. The issues raised in balancing these costs against the 'costs' of the deleterious effects will be discussed in Chapter 13.

Disposing of the ash

Depending on the ash content of the coal, the power station may also produce 25 tonnes or more of ash per hour, which will need to be disposed of. In many countries large accumulations of unwanted coal ash are a matter of concern because of their potentially poisonous heavy metal content which could leach into water supplies. In the UK and other European countries **pulverized fuel ash (PFA)** is commonly used for construction purposes, such as being mixed with cement and turned into lightweight concrete blocks (commonly called 'breeze blocks'). This effectively 'locks up' much of the heavy metal content.

5.7 A future for coal?

Coal has many disadvantages as a fuel compared to oil and natural gas; it has poor CO_2 emissions and its solid nature makes it difficult to transport and distribute. Its sulfur content also creates pollution problems. It might, therefore, be thought that the future prospects for coal are poor. However, concerns about the limited future supplies of oil and natural gas in the future may renew interest in coal's possibilities.

Nevertheless, four key questions have to be answered:

Can this inconvenient solid fuel be converted into liquid and gaseous fuels for which there is a substantial demand?

Can the process of converting the energy in coal into the desired energy services of warm homes, cooked food, illumination and mobility be carried out with maximum efficiency?

Can the CO_2 produced by the burning of coal be removed and prevented from entering the atmosphere?

Is there enough coal?

These questions will be addressed in turn in later chapters.

The nineteenth-century gasification of coal has been dealt with in this chapter, but more recent coal gasification technology will be described in Chapter 7. This plays a key role in the manufacture of substitutes for oil and natural gas.

The efficient generation of electricity is a key step in turning the energy in coal into useful energy services. The next chapter will discuss the importance of achieving high combustion temperatures for this. The Osaki power plant described above is one possibility for an efficient coal-based combined cycle generation plant. This technology is commonly used, with natural gas as a fuel, as described in Chapter 9. The Integrated Gasification Combined Cycle (IGCC) power plant generates electricity by combining coal gasification with combined cycle technology. It is one of a number of so-called 'clean coal' technologies. Carbon capture and storage is another. The 'carbon capture' process may take the form of scrubbing the CO_2 from the flue gases of relatively conventional power stations or it may require the gasification of the coal before combustion. Once separated the CO_2 could be stored, for example, in saline aquifers under the North Sea. All of these technologies will be described in Chapter 14.

In the longer term, there remains the question of how long coal reserves will actually last, particularly given the current expansion in use in China and India. The problem is nowhere near as serious as for oil (to be discussed in Chapter 7). However, the possibility of a peak in world coal production due to lack of reserves has been suggested, possibly within 20 years (EWG, 2007).

There are limits to current surface and underground mining technologies. Other coal resources may require the development of alternative systems such as underground coal gasification, which is also described in Chapter 7.

5.8 Summary

Like the previous chapter, this one has been a journey through time. Both chapters have traced the development of new concepts, but here they have been technological ideas. The 'history of coal' is the history of our uses of coal, and we have seen how the inventions of the Industrial Revolution led to an almost total reliance on this energy source in those countries that were able to adopt the new machine-dependent lifestyle.

In the nineteenth century much coal was gasified and a range of chemicals extracted as feedstocks for the chemical industries. These are now sourced from oil and the bulk of world coal production is crudely burned for electricity generation.

Although high-quality coal is still in demand for metallurgical coke production, the electricity industry demands very large quantities of cheap coal, which may often be of low quality. In many countries the deep mines of the nineteenth century have given way to large surface mines.

Although the coal resource is widely spread across the world, the problems of transporting it by rail or sea limit what can be considered as 'reserves'.

The study of power station furnaces shows that it is possible to burn coal cleanly and deal with its sulfur content.

The future of coal will probably depend on the development of 'clean coal' technologies, which will be described in later chapters.

References

Aden, N., Fridley D. and Zheng, N. (2009) 'China's Coal: Demand, Constraints and Externalities' [online], LBNL-2334E, Lawrence Berkeley National Laboratory, http://china.lbl.gov/publications (accessed 21 January 2011).

BGR (2009) *Energy Resources 2009*, Bundesanstalt für Geowissenschaften und Rohstoffe; available at http://www.bgr.bund.de (accessed 18 January 2011).

BP (2001) *BP Statistical Review of World Energy 2000*, British Petroleum.

BP (2010) *BP Statistical Review of World Energy, 2010*, London, The British Petroleum Company; available at http://www.bp.com (accessed 18 January 2011).

Brimblecombe, P. (1987) *The Big Smoke: A History of Air Pollution in London Since Medieval Times*, London, Methuen & Co. Ltd.

DECC (2009) *Historical Coal Data: Coal Production, 1853 to 2008*, (spreadsheet), Department of Energy and Climate Change; available at http://www.decc.gov.uk (accessed 10 January 2011).

DECC (2010a) *Energy Consumption in the United Kingdom*, statistical tables, Department of Energy and Climate Change, available at http://www.decc.gov.uk (accessed 10 January 2011).

DECC (2010b) *Digest of United Kingdom energy statistics (DUKES)*, Department of Energy and Climate Change; available at http://www.decc.gov.uk (accessed 23 January 2011).

DTI (2000) *Fluidised Bed Combustion Systems for Power Generation and other Industrial Applications*, Department of Trade and Industry; available at http://www.berr.gov.uk/files/file19290.pdf (accessed 20 January 2011).

EIA (2010) 'Coal Production and Number of Mines by State and Mine Type' [online], available at http://www.eia.doe.gov (accessed 22 January 2011).

EWG (2007) '*Coal Resources and Future Production*, EWG-Paper No. 1/07, Energy Watch Group; available at http://www.energywatchgroup.org (accessed 18 January 2010).

Hitachi (2001) 'A Large-Capacity Pressurized-Fluidized-Bed-Combustion Boiler Combined-Cycle Power Plant', *Hitachi Review*, vol. 50, no. 3, pp. 105–9; available at http://www.hitachi.com (accessed 19 January 2011).

IEA (2007) *World Energy Outlook 2007*, Paris, International Energy Agency; available at http://www.iea.org (accessed 22 January 2011).

IEA (2010) *World Energy Outlook 2010*, Paris, International Energy Agency.

Jevons, H. S. (1915) *The British Coal Trade*, republished in 1969, Newton Abbot, David & Charles (Publishers) Ltd.

Jevons, W. S. (1865) *The Coal Question*, London, Macmillan.

Further reading

CEC (1995) *ExternE: Externalities of Energy, iii: Coal and Lignite*, EUR 16522EN, Luxembourg, European Commission DG XII.

EIA (2006) *Coal Production in the United States: An Historical Overview*; available at http://www.eia.doe.gov (accessed 23 January 2011).

Chapter 6

Heat to motive power

By Janet Ramage

6.1 Introduction

Fossil fuels and biomass jointly account for about 90% of the world's primary energy, and the first stage in our use of most of these resources is *combustion*, because this is how their stored chemical energy is released. Whether the means is a furnace or fire, an internal combustion engine or a gas turbine, the initial product is energy in the form of *heat*. (This is also the case for nuclear reactors, which supply 5% of the world's primary energy, although the source of the heat is very different.)

So it is true to say that we use virtually all the world's primary energy to produce heat. True, but misleading, because what we ultimately want in many cases is not heat at all. We want *electricity* for lighting, television, computers, kitchen appliances and its many other uses; and we want means of *transport*, to travel rapidly and in comfort. An essential step in meeting almost all these needs is the conversion of heat into **motive power**: the driving power of machines.

Any system designed to obtain continuous motive power from heat is called a **heat engine**, a term that embraces not only the power station steam turbines that are the main subject of this chapter but the internal combustion engine and the gas turbine.

The chapter starts with the early steam engines, the first **prime movers** in the modern sense (Box 6.1). These have, of course, long been superseded – in many cases by very different machines – but their relatively simple modes of action provide a useful basis for the discussion of the general principles governing the performance of *all* heat engines. Continuing into the great period of steam engines in the 19th century, we see the development of these principles into the subject of thermodynamics, and with it an increasing understanding of why there is an unavoidable 'heat loss' from any heat engine.

BOX 6.1 Prime movers

The term 'prime mover' has a long history. In medieval astronomy the *primum mobile* was the outermost crystal sphere that carried the stars on their daily rotation around the Earth. Gradually the term came to mean any natural source of energy such as moving water or the wind. Then a final shift led to the present meaning: a machine that converts the energy of any natural source into motive power.

The final two main sections of the chapter look in more detail at the turbines of today's power stations. More than a third of the world's commercially traded primary energy is used to generate electricity. Assuming an average power station efficiency of 35% (as in Box 3.1) this means that roughly 100 EJ y^{-1} of this primary energy becomes *waste heat*. (The other major user of fossil fuel, the internal combustion engine, is even worse, as Chapter 8 will reveal.) So, until we can replace heat engines by some other means for the extraction of motive power from fuels, improving the efficiencies of these machines is a matter of considerable importance.

6.2 Steam engines

The early years

The idea of using steam to drive a machine predates the Industrial Revolution by at least 1700 years. Figure 6.1 shows the *aeolipyle* devised by Hero, an engineer working in Alexandria about 2200 years ago. It is the earliest known example of a **reaction turbine**. Steam shooting out of a jet in one direction produces a reaction force in the opposite direction, causing the sphere to spin. It does not seem to have been developed into an operating machine, and remained just one of Hero's many 'executive toys'.

Some 1400 years later, renewed interest in science and technology and the enthusiasm for the classical world that characterized Renaissance Italy led Leonardo da Vinci and others to propose many different prime movers using the power of steam. Amongst those inspired by Hero's devices was Giovanni Branca, who designed, in the 1620s, what was probably the first steam-operated **impulse turbine** (Figure 6.2), using the force or impulse exerted by the jet of steam striking the vanes or blades of a wheel. Branca's proposal was that the rotating wheel would drive a machine for crushing ore, but with the technology of the time, it is very unlikely that the force from the steam jet would even have rotated the wheel, and it is not surprising that little more was heard of steam turbines for another 500 years.

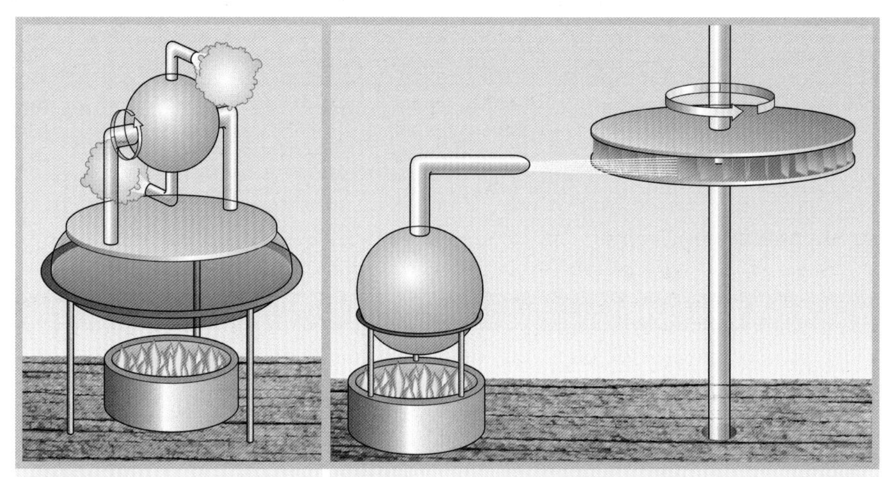

Figure 6.1 Hero's aeolipyle – the first steam-operated reaction turbine?

Figure 6.2 Giovanni Branca's 'prime mover' – the first steam-operated impulse turbine?

When the first practical steam-driven machine did appear, it operated on entirely new principles. Thomas Savery's 1698 machine (Figure 6.3) is worth a close look, because rather than directing the steam as a jet, it operates by *condensing* it, and this has two important results:

- the energy stored in the steam (Box 6.2) is released
- the volume occupied is dramatically reduced.

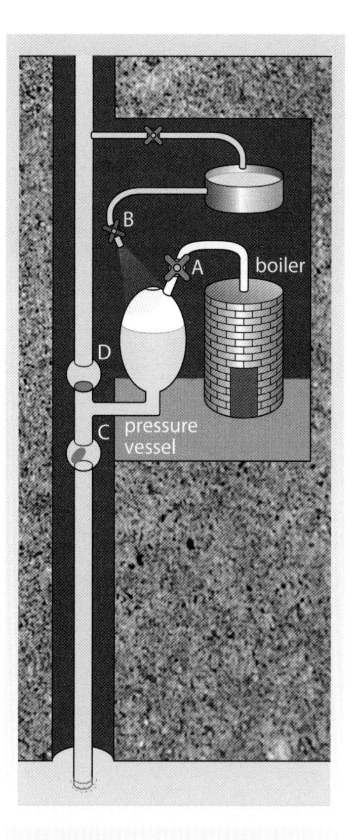

Figure 6.3 Savery's 1698 steam engine, 'The Miner's Friend'. Steam under pressure was used to raise water up the vertical pipe. Valves A and B are manually operated. Valves C and D are one-way flap-valves that only allow water to flow upwards.

BOX 6.2 Steam as an energy store

In order to produce steam under normal conditions, it is often necessary first to heat water from room temperature, say 20 °C, to its normal boiling point of 100 °C .

Box 4.4, in suggesting that heating water is one possible way to store energy, stated that the *specific heat* of water, i.e. the heat energy required to raise the temperature of 1 kg by 1 °C, is 4200 J kg^{-1} K^{-1}.

So the energy input needed to heat 1 kg of water from 20 °C to 100 °C will be:

$$80 \times 4200 = 336\,000 \text{ J} = \mathbf{0.336 \text{ MJ kg}^{-1}}$$

It is common experience that if heat is continually supplied to *water* at 100 °C, it vaporizes and becomes steam. Box 5.2 stated that the *latent heat of vaporization*, the heat needed to convert 1 kg of water at 100 °C into steam at that temperature, is **2.258 MJ kg^{-1}**.

So the total heat input needed to convert 1 kg of water at 20 °C into steam at 100 °C is:

$$0.336 + 2.258 = \mathbf{2.594 \text{ MJ kg}^{-1}}$$

The important point to note here is that the conversion to steam accounts for nearly 90% of the total. This, of course, means that in the reverse process – extracting the stored energy – the main contribution will come from *condensing the steam*.

'The Miner's Friend' was a **steam pump**, developed to meet the problem of flooding in mines. Its operation was simple. With tap A open, steam from the coal-fired boiler fills the pressure vessel. Tap A is now closed and B opened, spraying cold water on to the pressure vessel to condense the steam. The resulting drop in pressure draws water up the pipe from the mine below through the flap-valve C and into the vessel. Tap B is then closed and A opened, and the renewed steam pressure drives the water out of the pressure vessel and up through the flap-valve D to the surface. The pressure vessel is now full of steam and ready to repeat the process.

In overall energy terms, the input for the pump is the *chemical energy* stored in the coal and the output is the *gravitational potential energy* gained by the raised water (see Chapter 4). The data on Savery's engine were not very encouraging. The output power was about that of a fit person running up a flight of stairs; and for each joule of mechanical output, it needed about 300 joules of coal energy input: an overall efficiency of one-third of one per cent! There was some truth in the saying that you needed an iron mine to build one of these early machines and a coal mine to run it.

Savery's concept was, in any case, ahead of the technology of the day, and the vessel and pipe-work frequently burst; but he seems to have been

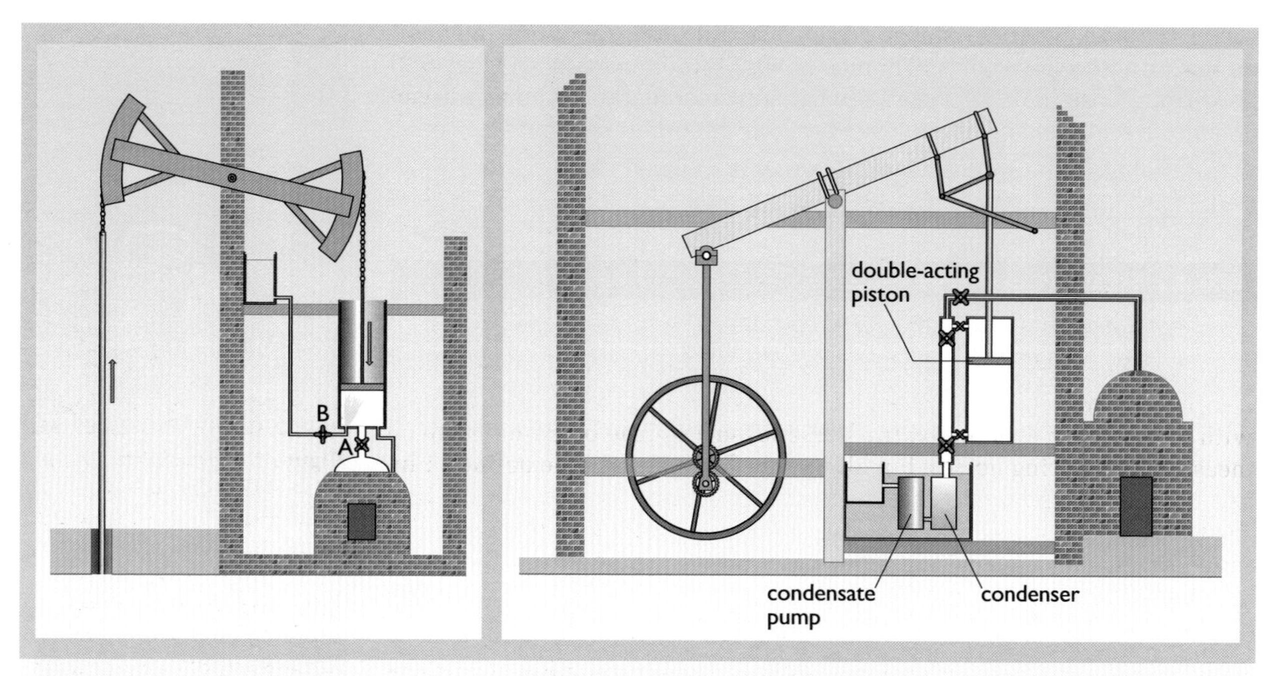

Figure 6.4 Principle of Newcomen's 'Atmospheric engine', 1712

Figure 6.5 The main features of a Boulton and Watt steam engine, circa 1790; it featured a separate condenser

a successful entrepreneur and sold a number of machines. It is unclear whether any operated successfully in coal mines, but there was a growing fashion for water gardens with fountains. Every large estate had to have one, and a Savery pump could be employed to lift water into a high tank to provide the necessary 'head'.

In 1712 a Dartmouth ironmonger, Thomas Newcomen, produced a really successful pump (Figure 6.4). A rocking beam was operated by a piston in a vertical cylinder. Steam was drawn up at low pressure through valve A from the boiler at only slightly above atmospheric pressure, greatly reducing the risk of burst pipes. The steam was then condensed by spraying water into the cylinder via valve B. This resulted in a near-vacuum under the piston, which was then pushed down by *atmospheric pressure*. The opening and closing of the valves was linked to the rocking beam giving continuous automatic operation. The load pulling down on the other end of the beam is sufficient to raise the piston to the top of the cylinder.

The efficiencies of Newcomen engines were probably still less than 1%, but they were far more reliable than Savery's, and over the next 60 years hundreds were built. There were some improvements during that period, but the next real revolution came only with James Watt.

James Watt

From his experiments with model steam engines, Watt realized that the Newcomen engine was wasting a great deal of heat by cooling the whole cylinder and piston each time the steam was condensed, requiring energy from the next steam input to heat it again. He proposed a **separate condenser**, which could be kept cool all the time, while the cylinder, well insulated, could remain hot. He also introduced a condenser **pump**, driven by the main engine, to extract the water and as much air as possible at the end of each cycle. And he realized that the steam supply could be closed off before the piston reached the end of the cylinder, allowing the existing steam to continue to expand, giving up more of its energy as useful work. The steam turbines of today's power stations may be very different from Watt's 1790 machine (Figure 6.5), but it is still possible to trace in them the descendants of these three major advances.

Figure 6.6 James Watt (1736–1819) was born in Greenock, in Scotland. Contrary to popular opinion, there is no evidence that he was inspired by steam from his mother's tea-kettle. An instrument maker at Glasgow University, he was given a model Newcomen engine to repair, and began to wonder why the little machine used so much steam. His results (see main text) led in 1769 to his first patents. When these eventually expired in 1800, he retired to his house in the country, devoting his time to 'mechanical pursuits and inventions'.

In order to specify the output of his steam engines, Watt (Figure 6.6) also defined one of the earliest units for *power*, cleverly choosing one with meaning for everybody: the **horsepower (hp)**. He based its value on measurements using horses; but always the businessman, he defined 1 hp as *1.5 times* the measured 'horse average' so that his customers would find their new 6 hp machines comparing favourably with six real horses. The modern unit for power is the watt (of course), and 1 hp is equal to 745.7 W, or roughly three-quarters of a kilowatt.

Is there anything that Watt did *not* invent? One instance is rather surprising. The efficiencies of his machines were still less than 2%, and he knew that using high-pressure steam would improve this; but he remained cautious about the dangers. Other engineers, realising that metals technology was improving, wanted to move in that direction, but could do nothing (legally) without Watt's consent. His 1769 patents, later held by the firm he established with businessman Matthew Boulton, covered any steam engine with a separate condenser – effectively all steam engines! And the patents would expire only in the year 1800.

Nevertheless people were already experimenting. Pumps were needed for the mines in Cornwall extracting tin, copper and other metals, but there was no local coal to fuel them, and importing it from elsewhere in the country was expensive. So any gain in pump efficiency would be welcome. There were other factors too. A high-pressure system could be more compact, and might even be *portable*.

Richard Trevithick, a Cornishman and a brilliant and adventurous engineer (see Section 6.4), was looking at many options – and attracting threats of lawsuits from Boulton and Watt. Nevertheless, he continued his developments, and on Christmas Eve 1801, a startling result emerged into the twilight (Figure 6.7). The age of steam locomotion had begun.

Figure 6.7 Trevithick's steam carriage

6.3 The principles of heat engines

Chapter 4 described how the world's present *electric power* industry has its roots in fundamental scientific discoveries. No one was trying to build generators in 1800 because until Michael Faraday's experiments in the 1830s there was no reason to believe that rotating a coil of wire inside a magnet would produce anything.

As shown above, the case of *steam power* was completely different. Hero was already trying to produce motion from steam at a time when 'the elements' meant fire, air, earth and water. Scientific knowledge had, of course, advanced greatly by the 18th century, but Savery's engine came from *practical experience* rather than theories of heat, and later developments owed more to advances in metals technology than revolutionary scientific ideas. The situation in the 1840s illustrates the point. Hundreds of steam engines worldwide were supplying power in mines, factories and even on railways, while James Joule was still trying to convince the scientific world that heat was a form of energy.

Carnot's law

Figure 6.8 Sadi Carnot (1796–1832) was born in Paris. His father was a military engineer, a man of principle whose life in those turbulent times alternated between high political position and prison. Graduating from the Ecole Polytechnique in 1814, Sadi also joined the Corps of Engineers. On a visit to Magdeburg in 1821 he discussed with his exiled father a new steam engine imported by Guericke. Back in Paris, he immersed himself in the subject and within a year or so came close to the view of heat and mechanical work that Joule was to develop 25 years later. But his famous book appeared only in a small private edition, and he died at the age of 36 in a cholera epidemic.

A theory of heat engines *had* been developed, twenty years earlier, by Sadi Carnot, a young captain in the French army (Figure 6.8). In 1824 he published a little book with the title *Réflexions sur la Puissance Motrice du Feu et sur les Machines Propres à Développer cette Puissance.* It would become one of the most famous scientific texts, probably best known today in its many English translations, usually titled *Reflections on the Motive Power of Fire* (Carnot, 1992). But although it was well received by Carnot's French scientific colleagues, the book was essentially ignored in other countries and initially had little practical effect. Its importance gradually came to be recognized, but was only fully appreciated in the English-speaking world when William Thomson (Lord Kelvin) read it in the late 1840s. Kelvin (Figure 6.9) evidently had a talent for recognizing genius. He was introduced in Chapter 4 as the person who drew attention to James Joule's ideas, and he seems to have been equally impressed by Carnot's book:

> Never before in the history of natural science has such a great book been written!
>
> Lord Kelvin quoted in Strandh, 1989

What was the great significance of Carnot's work? He didn't construct any machines or carry out experiments, and he still held the view that heat was a sort of fluid, called **caloric**. But as the title of his book suggests, he *thought* about heat engines. He was aware that even the best steam engines of the time were wasting a great deal of their heat input. Efficiencies were improving, but very slowly, and it seemed that there must be something that made it impossible even to approach the perfect heat engine – in modern terms, a machine that would take in heat energy and convert all of it into mechanical energy. So he proposed a simple law: *It is impossible to have a perfect heat engine.*

The most remarkable part of Carnot's study was the result that followed from his law. He was able to show by careful mathematical and logical reasoning that this simple statement led to a formula for the *maximum possible efficiency* of any specified heat engine. Carnot's reasoning was very detailed and precise, and expressed in terms of the caloric theory of heat, so the following account is simplified, abbreviated and 'modernized'.

The Carnot engine

Carnot took as his basis for discussion an extremely idealized version of a steam engine (Figure 6.10). The **working fluid** of his heat engine goes through a continuous cyclic process, taking in heat from the 'boiler' at an *input temperature*, T_1, producing a mechanical output (W), and rejecting waste heat to a 'condenser' at a lower *exhaust temperature*, T_2. Everything is ideal: no friction, perfect insulation, no sudden changes of any sort.

Figure 6.9 William Thomson, Lord Kelvin (1824–1907) grew up in Glasgow. In 1841 he moved to Cambridge, and also visited Paris to work with the great French mathematical physicists. Back in Glasgow as professor, he contributed to many areas of physics, including the first successful transmission of telegraph signals by transatlantic cable.

Suppose the quantity of heat taken in from the boiler by each kilogram of working fluid is Q_1 and the quantity of 'waste heat' rejected to the condenser is Q_2. Because energy is conserved, the mechanical output, the work done, must be equal to the difference:

$$W = Q_1 - Q_2$$

The efficiency of the engine, the output divided by the input, is then:

$$\text{efficiency} = \frac{W}{Q_1} = \frac{Q_1 - Q_2}{Q_1} = 1 - \frac{Q_2}{Q_1}$$

There are other details in the full specification of the idealized **Carnot engine**, but for the purposes of this discussion only the results need to be considered, the most important of which was Carnot's proof that, for his engine:

$$\frac{Q_2}{Q_1} = \frac{T_2}{T_1}$$

i.e. the waste heat divided by the input heat is exactly equal to the condenser temperature divided by the boiler temperature.

This leads at once to a new simple formula for the efficiency of the Carnot engine:

$$\text{efficiency} = 1 - \frac{T_2}{T_1}$$

or an alternative that is often useful in practice:

$$\textbf{efficiency} = \frac{T_1 - T_2}{T_1}$$

(This of course needs to be multiplied by 100 to give the percentage efficiency.)

Carnot then proved the important point that any real heat engine operating between the same two temperatures must be less efficient than his idealized one. In other words, the formula above allows us to calculate the *maximum possible efficiency* for any type of heat engine if boiler and condenser temperatures are known. Box 6.3 discusses the important issue, recognized by Carnot, of the type of *temperature scale* that must be used in his formula.

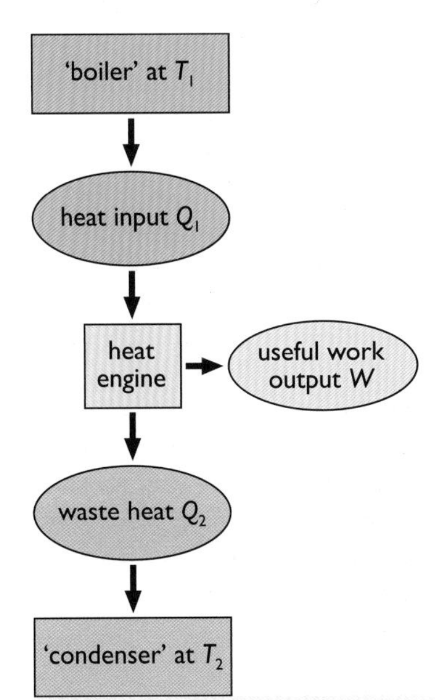

Figure 6.10 Carnot's heat engine

BOX 6.3 Temperature scales

On close inspection, there is something odd about Carnot's formula. Firstly, the calculated efficiency seems to depend on the choice of **temperature scale**. Suppose the input and condenser temperatures are 100 °C and 40 °C respectively. The calculated Carnot efficiency is then 0.60, or 60%. However, if the *same* two temperatures are expressed in degrees Fahrenheit (212 °F and 104 °F), the calculated efficiency becomes about 51%. They obviously can't both be right – and as shown below, both are in fact wrong.

The resolution of these problems is that the formula does not allow any arbitrarily chosen scale of temperature. It applies only to **absolute temperature scales**, i.e. those on which zero degrees is the **absolute zero** of temperature. The concept of absolute zero is discussed later, in Box 6.4, but for the present a brief account of the most common absolute scale will allow some calculations.

The **Kelvin scale** of temperature is defined so that the *size* of each degree is exactly the same as in the familiar Celsius scale. (For instance, the number of degrees between melting ice and boiling water is 100 on both scales.) But the *zero* on the Kelvin scale (**0 K**) is far down at absolute zero, which is 273.15 degrees below 0 °C. So 'ice temperature' on the Kelvin scale is 273.15 K and water normally boils at 373.15 K. In general, the conversion is simple:

temperature in kelvins (K) = temperature in °C plus 273.15

(The approximate value, 273, is often used.)

Notice the useful point that because the size of each degree is the same in both scales, temperature *differences* are always exactly the same in degrees Kelvin or Celsius.

Using Carnot's formula

Carnot's formula can be applied to an early steam engine. Assume that the steam enters the cylinder at 100 °C, and is cooled in a condenser to the temperature of warm water, say 40 °C. The corresponding Kelvin temperatures are therefore approximately $T_1 = 373$ K and $T_2 = 313$ K, and the percentage efficiency will be:

$$\text{percentage efficiency} = \frac{T_1 - T_2}{T_1} \times 100 = \frac{60}{373} \times 100 = 16.1\%$$

This obviously goes some way towards explaining the very low efficiencies of early heat engines, and Carnot's result remains an important guide in the continuing search for improved efficiency. The very high input temperature (T_1) of a combined-cycle gas turbine, for instance, results in greater efficiency than a steam turbine can achieve (see Section 6.7). There have even been proposals for heat engines in space, exhausting to a temperature T_2 that is only a few degrees above absolute zero. This could lead to a very high efficiency indeed!

It should be noted, however, that although Carnot's result is useful in setting an absolute limit to the efficiency of any heat engine, the actual cycle of processes followed by the working fluid of a particular type is likely to be very different from the idealized Carnot cycle. This is certainly the case for the power station steam turbines to be described below and the gas turbines and internal combustion engines discussed in Chapter 8. (More detailed accounts of the working fluid cycles of all these heat engines may be found in Sorensen, 2004.).

The laws of thermodynamics

The first use of the word *thermodynamic* (from *thermo*: concerning heat, and *dynamic*: concerning force or power) was in an 1849 paper by William Thomson (again!). Before long, the term **thermodynamics** came to mean the whole field of study, and within 50 years its three laws, essentially the laws that govern all energy technologies, had been established. What do these laws say?

- The ***First Law of Thermodynamics*** is essentially the *law of conservation of energy*, already discussed (and in many cases assumed) in earlier chapters. It states that you cannot create energy or destroy it. In any change from one form of energy to other forms, the total quantity of energy remains constant. This rules out perpetual motion machines, and can be summarized as 'you can't get something for nothing in the world of energy'.
- The ***Second Law of Thermodynamics*** imposes more severe constraints on any heat engine. It says that when *heat* is the input you can't break even. There must inevitably be some wasted heat energy, so the efficiency of any heat engine is always less than 100%. The Second Law and its consequences are the main subjects of this chapter.
- The ***Third Law of Thermodynamics*** states that there is a lowest possible temperature, an *absolute zero* below which nothing can be cooled. Its origins are briefly discussed in Box 6.4.

BOX 6.4 Absolute zero

The picture of heat as the energy of particles in motion offers an obvious interpretation of the idea of an **absolute zero of temperature**: it is the temperature at which all motion stops. Every particle is frozen in position, and there is no more energy to be extracted. (The modern picture differs slightly, because quantum mechanics does not allow the energy of a system of particles to become exactly zero, but the concept of an unattainable lowest temperature remains.)

The Third Law of Thermodynamics, the law that says that the absolute zero of temperature is unattainable, was formulated in 1906 by Walter Nernst, and is also known as **Nernst's Heat Theorem**. Nernst was Director of the Institute for Experimental Physics in Berlin, a person of great importance at the University, and a grave figure, treated with respect by all. It is said that in his lectures he always carefully referred to the Third Law as 'My Heat Theorem'.

Nernst showed that as the temperature of any object is reduced towards zero, the drop in temperature resulting from the extraction of a given quantity of heat becomes ever smaller and smaller. Effectively, the specific heat capacity becomes greater and greater, with the consequence that although substances can be cooled to temperatures nearer and nearer to absolute zero, nothing will ever reach it.

Atoms in motion

Many people find the Second Law of Thermodynamics rather strange on a first encounter, so it may be worth looking at it from a different viewpoint. Its essential feature is the distinction between *heat* and all the other forms of energy, and the limit this puts on the conversion from heat to these other forms. Common experience shows that there is no similar constraint in the reverse direction. It is only too easy to turn other forms of energy into heat. You bring the car to a skidding halt, converting its kinetic energy into heat, mainly in the brakes and tyres; but the Second Law says that you can't convert all this heat into power to set the vehicle in motion again. Similarly, the electrical energy supplied to a kettle is all converted into heat; but converting all the heat back into electric power is not possible. (This is why the hot water of Box 4.4 is a very poor temporary store for surplus electricity.)

What is it that distinguishes heat from other forms of energy? Chapter 4 showed that the *temperature* of anything is a measure of the kinetic energy of its constantly moving atoms and molecules, and that *heat* is the energy added or removed to change the average speeds that characterize this thermal motion. With this in mind, we can see how heat differs from all other forms of energy: the characteristic quality of thermal motion is that it is *disorganized*. The molecules of the air, for instance, are constantly moving, with a variety of speeds and in all directions, bouncing off each other and all surfaces in the room, changing both speed and direction all the time. And a similar picture of random motion holds for the vibrating atoms or molecules of a solid material.

If this is heat, then the principle that all energy tends to become heat is no more than a law that we all know: *things tend to become disorganized.* (Expressed in rather more formal terms, this becomes an alternative version of the Second Law.)

Consider, for instance, a stationary vehicle as an assembly of atoms moving at random. In a similar vehicle moving north-east at 50 km per hour all the atoms have a *directed* motion superimposed on their random movement, and this gives the kinetic energy of the moving vehicle. When it stops, this extra energy must go somewhere. It becomes heat – further *disordered* motion of the atoms of the brakes and tyres. It is now a warmer, stationary vehicle. To reverse the process and recoup the heat energy requires part of the random motion to become ordered again, something which is very unlikely to happen spontaneously. It is easy to produce chaos. Order presents more of a problem.

A more rigorous approach to thermodynamics introduces the useful concept of **entropy**. In terms of our picture of heat as random motion, the entropy of a system may be regarded as a measure of its degree of disorder. This leads to yet another version of the Second Law: *The entropy of a closed system tends to increase with time.* In other words, if a system is totally isolated from its surroundings (no exchange of heat, no work done on or by the system) then any change within it results in an increase in its entropy. This has led people to refer to entropy as 'the arrow of time': the direction of increasing entropy is the direction in which time is flowing. We all recognize when a film of a breaking wave is being played backwards because we see disorder turning into order. Entropy is decreasing – and things don't happen that way.

Heat flow

A one-way process that we meet every day is the flow of heat from a hot object to a cooler one. You pour hot coffee into a cold cup and the result is cooler coffee and a warmer cup. There is a flow of heat energy from the hotter coffee to the colder cup. You would be very surprised if the reverse were to occur, with the coffee becoming even hotter and the cup ice-cold. Yet, provided the heat energy gained by the coffee is the same as that lost by the cup, energy would still be conserved in this bizarre case. Once again, we must look to the Second Law of Thermodynamics to tell us the direction in which things happen.

The key point is that *hotter means faster*. The atoms of a hot object have a greater average speed than those of a cold object. To take another domestic example, suppose you were to observe that a metal fork lying on the table was spontaneously becoming red-hot at one end while ice crystals appeared at the other end. This would mean that the faster atoms had spontaneously assembled at one end and the slower ones at the other – an extremely improbable event. A simple lottery draw will serve as an example of such improbability. Five red and five black balls are shaken in a closed box and then allowed to run out one at a time. What are the odds that the first five balls drawn will all be red? About 30 000 to one against is the answer; but for ten red balls from a total of twenty red and black, this becomes 700 billion to one against. A typical fork consists of a very large number of atoms (about 10^{24}), so the probability of the rearrangement described above is vanishingly small. It is clear that the Second Law has something to do with all this as the issue is once again order versus chaos. The atoms in the fork don't *spontaneously* adopt the more ordered arrangement: *heat will not of itself flow from a colder to a hotter body*. The natural way is the reverse, the direction of increasing chaos.

Heat *can* be transferred from a colder to a hotter body. Every domestic refrigerator does it, pumping out heat from the cold interior into the warmer surroundings. But it needs an external energy supply to do it. The refrigeration unit is an example of a **heat pump**, a category of machines to be discussed in Chapter 9.

6.4 The age of steam

The age of steam had indeed arrived with the 19th century, but it didn't end with it. Steam-driven trains and ships remained dominant until the mid-20th century, and today, in the early twenty-first, 'steam engines' still drive four-fifths of the world's electricity generators. But these power station machines are not the cylinder-and-piston engines discussed above. As will be described in Chapter 8, the large diesel engine has replaced many applications of reciprocating steam engines. The world's railways have switched from steam to diesel power with significant overall energy savings. The following accounts of the great age of steam and the development of steam-driven vehicles are therefore rather brief, concentrating on the factors that remain relevant to today's steam plants.

Improving the efficiency

As discussed above, the innovations of the 18th century did improve performance, but the real increases in efficiency appeared only when the expiry of Watt's patent in 1800 freed people to experiment with new ideas.

Trevithick and other Cornish engineers were at the forefront, investigating the improvements brought by high pressures, horizontal cylinders, more compact beam-less engines, compound engines, new boiler designs, new linkages for rotary motion, and so on. (The next subsection, *Mobile Power*, discusses some of these.) Efficiencies increased to over 10%, and by the 1840s, steam pumps could achieve over 30 times the pumping power of Savery's machine for the same coal consumption.

Two developments in particular are worth noting for their continuing importance today. As mentioned earlier, it was already recognized before 1800 that **high pressure** steam improves the efficiency of a steam engine. More precisely, the pressure of the steam at input should be as high as possible and the exhaust pressure as low as possible. Why is this the case? A relatively simple answer emerges if we bring together Carnot's rule and the ideas in Box 6.5. At very high pressures, the steam must be at a high temperature in order to be steam at all. So the use of high pressure necessarily meant an increase in input temperature and therefore, by Carnot's rule, greater efficiency.

A second steam engine innovation with lasting importance was the **compound engine**. The idea, proposed by Jonathan Hornblower as early as 1781, was to have two or more cylinders, with the exhaust steam from the first cylinder, instead of being condensed, becoming input steam to the second, and so on. In terms of Carnot's rule there is no gain in this, as the maximum overall efficiency is still determined by the overall initial and final temperatures. There is, however, a gain in practical terms, because each cylinder is exposed to a smaller *range* of temperatures during its operating cycle, and this reduces the heat wasted in re-heating it. For a compound engine to work, each cylinder must be bigger than the previous one, because the volume of steam flowing through during each cycle increases as its pressure falls. There are two main arrangements of the cylinders: side by side or in line, called respectively **cross-compound** and **tandem-compound**. Hornblower's idea initially came to nothing, because his proposed engine used a condenser and air pump, which attracted the attention of Boulton and Watt and their patent lawyers. However, as shown later in Figure 6.21, the concept came to be fully realized in the steam turbines of today's power plants.

During the 19th century, steam engines were finding many new uses: in the manufacturing industries, in transport (see below), and in pumping the water supplies needed by the growing populations of towns and cities. In 1849, Britain supplied three monster pumping engines to drain the Dutch polders, while a growing demand from new mining regions such as South America and Australia was met by machines from Britain and other European countries – and increasingly from the USA, where the famous Corliss steam engine of 1876 was the largest of its time (Strandh, 1989). Its mechanical output power was 1400 hp – just over 1 MW; but it was over 13 m high and weighed 600 tonnes, giving it a power-to-weight ratio of about 1.7 W per kg. (For comparison, a typical modern car engine might achieve 600 W per kg.)

The advances were not without penalties, however. Steam pressures were rising above five atmospheres, and some of James Watt's caution was proving to be justified. In Britain in the 1850s, boiler explosions were

BOX 6.5 Fluids under pressure

Pressure

It is not at all obvious to most of us that we live with a force equal to the weight of a one-kilogram object pushing on every square centimetre of our bodies.

The pressure acting on any surface in contact with a fluid is the result of constant bombardment by fast-moving atoms or molecules. Each particle briefly exerts a force as it bounces off the surface, and the result of all these random events is a net force that is *always perpendicular to the surface – no matter which way the surface is facing*. The pressure is defined as the net force per unit area: the number of newtons per square metre. As with other units such as the newton and the joule, the SI unit for pressure has been given a name, the **pascal** (**Pa**). One pascal is exactly the same as one newton per square metre.

Normal atmospheric pressure at the surface of the Earth, often referred to as 'a pressure of one atmosphere', is approximately 100 000 Pa. This is one-tenth of a megapascal (MPa), leading to the following useful approximate relationship:

1 MPa ≈ 10 atmospheres

(In terms of the older unit for pressure, the **pound per square inch** (lb/sq in or **psi**), one atmosphere is usually taken to be 14.7 psi. A third unit, often used for power station steam pressures, is the **bar**. One bar is exactly 100 000 Pa – effectively about one atmosphere.)

In the following accounts of steam engines pressures are expressed in MPa, but with occasional informal phrases such as 'a pressure of 20 atmospheres', to give a feeling for the extreme pressures involved.

Pressure, temperature and density

The view of pressure as the result of molecular collisions leads to several useful results. The pressure, the net force on a square metre of surface, will depend on:

- the *number* of particles striking per second
- the *mass* of each particle
- the average *speed* of the particles.

In other words, the pressure depends on the *density* of the fluid (the mass of a particle times the number of particles per cubic metre) and the *temperature* (related to their speed – see Section 4.3). In practice, the quantities that we are normally able to control directly are the pressure and temperature, so it makes more sense to reverse this statement and say that the *density of a fluid (liquid or gas) depends on its pressure and temperature*.

In the case of liquids, with their closely packed molecules, the density is not greatly affected by changes in pressure or temperature. Common experience tells us that water is almost incompressible, and that its density does not change significantly with temperature. (We don't normally try to squash more water into the kettle, or leave space to allow for its thermal expansion.) Even under extreme conditions of temperature and pressure, the changes in the densities of liquids are rarely more than a few per cent – but the situation is very different for gases.

Hot steam

Dry steam is the invisible gas at the spout of the kettle. At 100 °C and normal atmospheric pressure, its density is just over half a kilogram per cubic metre. If heated further, like most gases steam will expand. At 200 °C, for instance, and normal atmospheric pressure, its density will have fallen by about a fifth.

Increasing the *pressure* has more startling effects. Most people are familiar with the fact that water boils at a lower temperature at the top of a mountain where atmospheric pressure is lower. This is one aspect of the more general rule that *the boiling point of any liquid depends on the pressure* (the principle of the domestic pressure cooker, and as we shall see, the power-station boiler). Under a pressure of 2 MPa (20 atmospheres), for instance, water remains liquid at 100 °C and boils only when its temperature reaches 212 °C.

Steam heated above the boiling point is referred to as **superheated**. At normal atmospheric pressure, steam at 150 °C is therefore superheated steam. But above 2 MPa pressure, 150 °C is well below the boiling point of 212 °C, and the fluid will still be water, becoming steam only at 212 °C, and superheated steam at temperatures above that.

Ultimately, at a **critical pressure** of about 22 MPa and the corresponding boiling point of 374 °C, called the **critical temperature**, the fluid is said to be at its **critical point**. Above this, if either the pressure or the temperature is increased further, the *distinction between water and steam disappears*. There is no boiling point, just a **supercritical** fluid whose density depends on its temperature and pressure. The 'steam' produced by the boilers of many power stations is in this supercritical state, with a density that can reach 50 kg m^{-3}.

The steam conditions of the plant described in Section 6.7, 25 MPa and 560 °C, are *supercritical*. Modern plants using temperatures above 565 °C are described as **ultra-supercritical**.

resulting in some 500 deaths a year (von Tunzelmann, 1978). This seems to have been the worst period, and by the end of the century, although pressures in some machines exceeded 1 MPa (ten atmospheres), the annual death rate from all types of steam explosion was 'only' a hundred or so (Figure 6.11).

Figure 6.11 The path taken by the boiler after an explosion, circa 1890

Figure 6.12 Richard Trevithick (1771–1833) was the son of a Cornish mine manager. Over six feet tall and very strong, he was as adventurous in his life as in his engineering, and both were sequences of successes and catastrophes. After his 'locomotives' had come to grief (see the main text), he returned to stationary machines. Still failing to make money in Britain, he went to Peru as an engineer in the silver mines. There he prospered – until a revolution forced him to leave the country, abandoning all his property. Penniless in Colombia, he met Robert Stephenson, who gave him his fare back to England, where he continued to generate ideas but no money. He died a pauper in Dartford.

Mobile power

In the early days of steam engines, obtaining more useful work from each tonne of coal was a prime objective, and the emphasis throughout the 18th century was consequently on improvements in efficiency. In the years after 1800, however, other criteria became increasingly significant.

The world's first car (Figure 6.7), which was built by Richard Trevithick (Figure 6.12), had a rather brief life. Accounts of its second outing differ, but they do agree that he parked it in a shed – some say because it had broken down – while he and his friends visited a local inn (presumably either to celebrate or to drown their sorrows). Unfortunately he had failed to extinguish its fire and the engine boiled dry, destroying both machine and shed.

Nevertheless Trevithick and others remained persuaded of the possibilities of steam as power for a *vehicle*. It was obvious to everyone that in this case the size – both the dimensions and the mass – of the engine would be important. A further consideration was that the complete machine, including the furnace and boiler and all the mechanical linkages and steam pipes, must be able to withstand the accelerations and decelerations (including the jolts and turns) of a moving vehicle.

The ingenious mechanical devices in Trevithick's vehicle are beyond the scope of this account, but his steam engine itself is important because it embodied new features that are still seen in modern machines. The steam was no longer produced by lighting a fire under the boiler. Instead, the furnace together with the flue pipes that carried away the hot gases were *inside* the horizontal cylindrical boiler. This idea, of maximizing the heat exchange by passing the hot gases in tubes through the water (or later, flowing the water in tubes through the hot gases) has remained in the **tubular boilers** of present-day steam plants. The single cylinder was horizontal – an arrangement that has continued in most steam locomotives – and there was no condenser, another obvious weight-saving measure.

Trevithick's second vehicle, built in 1803–4, was very similar to the first. The engine is said to have delivered 3 hp (2.25 kW) when propelling the vehicle at just under 8 kph (5 mph), and its coal-to-motive power efficiency under these conditions appears to have been about 4%. He crashed this machine in London, and seems then to have abandoned the idea of a steam-powered road vehicle. He did however achieve the first commercial use of a **steam railway**. Confounding the critics who 'knew' that metal driving wheels on a metal track would be sure to spin, his 1804 locomotive successfully replaced the horses hauling wagons from an ironworks in Wales. Successfully but briefly, because the seven-tonne locomotive proved too heavy and broke the rails. A second similar venture came to the same end. Then in 1808 came 'the first passenger-carrying train'. Running on a little circular track about 10 metres in diameter, near the present Euston Square in London, it was more an entertainment than a transport system, but it was a success with the public – until it derailed and crashed. Not surprisingly, Trevithick was unable to find financial supporters for further experiments with steam vehicles, and he returned to stationary engines.

Figure 6.13 George Stephenson (1781–1848) was born in the pit village of Wylam in Northumberland, where his father was a colliery fireman. He started work in the colliery at the age of ten, but learned to read and write in evening classes and in 1812 was appointed enginewright at a salary of £100 a year. His interests were not limited to trains, and in a conflict over the invention of the miners' safety lamp, Sir Humphrey Davy called him 'a thief and not a clever thief'. Undeterred, he continued his innovations, and by 1814 his 'locomotive' Blücher, the first to use flanged wheels, was operating between the colliery at Killingworth and the port some five miles away. Following the success of the Liverpool–Manchester line (see main text), he and his son Robert, while still developing new engines, began also to play major roles in the construction of railways throughout the world.

Trains

Trevithick may not have continued with steam traction, but George Stephenson (Figure 6.13) and others did, and were more successful. The second decade of the 19th century saw several steam engines hauling trucks on rails, including *Puffing Billy* (Figure 6.14) and Stephenson's *Blücher*. The development of successful engines was obviously important, but Stephenson took a wider interest in the growth of a British railway system. In 1822 he persuaded the directors of the proposed Stockton and Darlington railway to use steam rather than horse power, and at the opening of the line three years later his engine *Active* (later re-named *Locomotion*) drew the world's first regular passenger-carrying steam train. His conviction that a 'locomotive' should be used on the new Liverpool to Manchester route led to the famous Rainhill trials of 1829. He had persuaded the directors of the company to offer a prize (£500) for the engine that best met their performance specifications, and at the end of the trials, there was just one survivor, *The Rocket*, built by George Stephenson and his son Robert. They received a contract for eight locomotives for the new line.

The Rocket incorporated developments of a number of Trevithick's innovations, but its power transmission was much simpler, with one cylinder on each side and the pistons connected directly to the driving wheels – essentially the principle of almost all later steam locomotives (Figure 6.15). Unfortunately their efficiencies also showed relatively little change. Writing in the late 1920s, an American railway engineer (Muhlfeld,

Figure 6.14 *Puffing Billy*

Figure 6.15 The *Flying Scotsman* leaving London on her first non-stop run to Edinburgh, 1 May 1928

1929) deplored the reduced attention to efficiency and the increasing requirements placed on the steam – providing power for automated labour-saving systems and heat for a variety of purposes including keeping the passengers warm, or cool – thus reducing the already poor engine efficiency to an overall system efficiency of perhaps 5%. (The multiple energy uses and resulting efficiencies of some present-day automobiles come to mind.)

Ships

The early history of marine steam engines closely mirrors that of rail locomotives. The first commercially successful steamboat was the *Clermont*, built in 1807 by Robert Fulton and his partner Robert Livingston to carry passengers on the Hudson River in New York State. Her Boulton and Watt engines came from England (on a sailing ship, of course). In 1815, towards the end of his life, Fulton constructed the first stream-driven warship, an armed vessel of 40 tons with side paddle wheels.

The following decades saw increasing numbers of steam-assisted ocean-going sailing ships, but credit for the first fully steam-driven Atlantic crossing probably goes to the *Curaçao*, a 400-ton Dutch paddle steamer which made the passage to the West Indies in 1827, a voyage that took one month. In 1838, the steam ship *Sirius*, 700 tons, offered the first transatlantic passenger service, but was followed into New York within hours by Isambard Kingdom Brunel's 1400-ton *Great Western*, which had made the crossing in 15 days. In general, however, a fast sailing ship with a good wind would easily outpace any steam vessel, and virtually all ocean-going steam vessels of the mid-19th century were still fully rigged for sailing.

The second half of the century saw many advances in marine engine technology. Ocean liners, cargo vessels and the ships of the world's navies were growing ever larger, and from its introduction in 1854, the double-compound (and later triple- and even quadruple-compound) reciprocating steam engine gradually became the standard. By the 1870s, the efficiencies of the best engines had risen to 15%.

Another innovation was the screw propeller. The first operational systems were developed independently by John Ericsson, a Swede living in London, for a naval vessel in the USA, and the Englishman Francis Pettit Smith for a steam ship aptly called *Archimedes*. The efficiency advantages of the propeller over paddle wheels were immediately obvious, and led Brunel to change the plan for his next ship, *Great Britain*. Completed in 1844, she was the largest ship of her time and one of the first to be made of iron rather than wood.

The combination of double- or triple-expansion reciprocating steam engines and screw propellers was to have a very long life. By the early 20th century, with very efficient tubular boilers producing steam at pressures above ten atmospheres, and effective condensers to maintain low exhaust pressure, efficiencies had reached 20%, amongst the best ever achieved by reciprocating steam engines. Cargo ships with these propulsion systems continued to be built, and many remained in service into the 1950s and beyond. But the first of their successors had already appeared more than half a century earlier – a very small ship driven by *steam turbines*.

6.5 Steam turbines

One class of prime mover that had virtually reached its present-day form by the end of the 19th century was the turbine. But this was the ***water turbine***. The name, from the Latin *turbo*, 'something that spins', was coined in about 1830 to describe these new machines. With a series of curved blades enclosed in an outer casing, they soon began to replace slow-moving water-wheels. Their high speeds of rotation made them the ideal machines to drive the new *electrical generators* (see Chapter 9), and the Niagara Falls generators, commissioned in 1894, with outputs of 5 MW, are an indication of the scale that water power had achieved by the end of the century. (For more on water turbines, see Boyle, forthcoming.)

Reciprocating steam engines were also providing power for generators at this time, but many people in the industry were aware that these massive, relatively slow machines were far from ideal for the task. It had also become obvious that they could not compete with the new internal combustion engines as mobile power sources for road vehicles (see Chapter 8), so it is not surprising that experts were predicting the disappearance of the entire steam industry within a few decades.

Steam, speed and rpm

With the example of water turbines before them, why did the steam engineers fail to follow suit? As discussed at the start of this chapter, the idea of steam turbines was very old indeed; but in the first detailed analysis, in the early 1800s, James Watt had concluded that a simple turbine driven by steam jets was *technically impossible*.

Box 6.6 shows the reasoning, and Watt identified two problems. Firstly, to achieve any reasonable efficiency, the boiler would need to produce steam at a pressure of at least 20 atmospheres. And secondly, with the metals technology of the time, a turbine rotating at 30 000 rpm would tear itself to pieces.

However, by the late 1800s high pressures were already being used in conventional steam engines. And in Sweden, the engineer Gustav de Laval, with the aid of many ingenious design innovations, was showing that turbines *could* run at extremely high speeds of rotation, up to 40 000 rpm. His impulse turbines, patented in 1887, proved almost twice as efficient as the average steam engine of the time, and had a degree of success for some years. But they were fated never to dominate the market, because another engineer was already designing a very different turbine.

Parsons' turbo-generator

Charles Parsons completed his first small **turbo-generator** in 1884, and it continued to operate, generating electric power, until 1900. It involved ideas adopted from the increasingly sophisticated water turbines of the time, but totally new in the context of steam engines. Instead of nozzles producing jets, a ring of fixed **guide vanes** on the inside of the stationary outer casing directs the incoming high-pressure steam onto a matching ring of **moving blades** on the rotor (Figures 6.17 and 6.18).

The resulting smooth flow of the steam reduces energy losses, leading to better efficiency than a jet; but it is Parsons' second innovation that solved the problem of very high speeds. He realized that if the single set of blades was replaced by a number of these rings of fixed and moving blades mounted side by side along the axis, each feeding steam to the next, the pressure could drop in *relatively small steps* along the complete turbine. The steam speeds could then be much lower, with the important consequence that the optimum speeds of rotation could also be lower (Parsons, 1911).

BOX 6.6 Steam jets

To produce a jet of steam through a nozzle there must obviously be a greater pressure on the input side of the nozzle than at the output. The process is effectively converting the *thermal energy* of hot, high-pressure steam into *kinetic energy* of the directed jet. So we might expect that the Second Law of Thermodynamics will impose a limit on the efficiency of the process. This is indeed the case, and the lower curve in Figure 6.16 shows how the calculated *jet efficiency* depends on the input pressure, when the output pressure is 0.1 MPa (one atmosphere). It is obvious that for reasonable efficiency very high input pressures are needed.

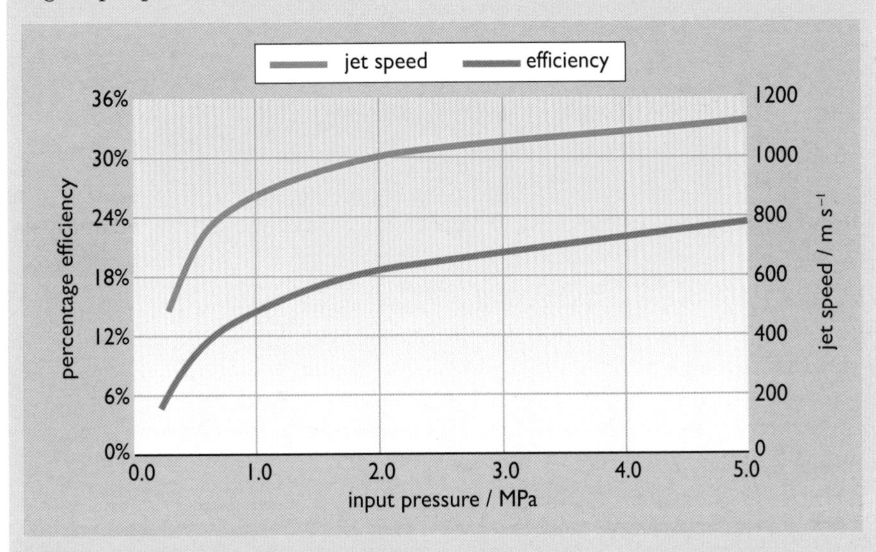

Figure 6.16 Pressure, energy conversion efficiency and jet speed

As the upper line on the graph shows, very high pressures also mean, not surprisingly, very high jet speeds, and a second problem arises when considering how much of the jet energy can be transferred to the rotating turbine. The following analysis, essentially Watt's reasoning, shows that if the speed of the turbine blade is just right, a jet could in principle give up all its energy – but with an unfortunate practical consequence.

Suppose the jet is travelling at 1000 m s^{-1}. If the blade of the rotating turbine is moving at exactly half this speed, i.e. 500 m s^{-1}, the jet will be approaching the blade at a *relative speed* of 500 m s^{-1}. Assuming a 'perfect bounce' the steam will then leave travelling at 500 m s^{-1} *backwards relative to the blade*. But the blade is moving forwards at exactly this speed, which means that the steam has been brought to rest, transferring all its energy to the turbine!

This seems ideal – until we consider an actual turbine. Suppose it has a diameter of 0.3 m (about one foot). In one complete rotation, a blade tip moves a distance of about 1 metre ($\pi \times 0.3$). So for the above blade speed of 500 metres per second, the turbine will need to spin at about 500 rotations per second – which is 30 000 rpm.

The output of Parsons' first small machine was only 10 hp (7.5 kW): much less than the largest reciprocating engines of the time. But the turbine, about half a metre long and 15 cm in diameter, was tiny compared with conventional steam engines. The working section consisted of 14 rings of fixed blades alternating with the 14 rings of moving blades on the rotor.

Figure 6.17 Parsons' first turbo-generator. The upper part of the casing has been removed to show the small rotor with its 14 rings of blades. The piece of shaft shown is about 30 cm long.

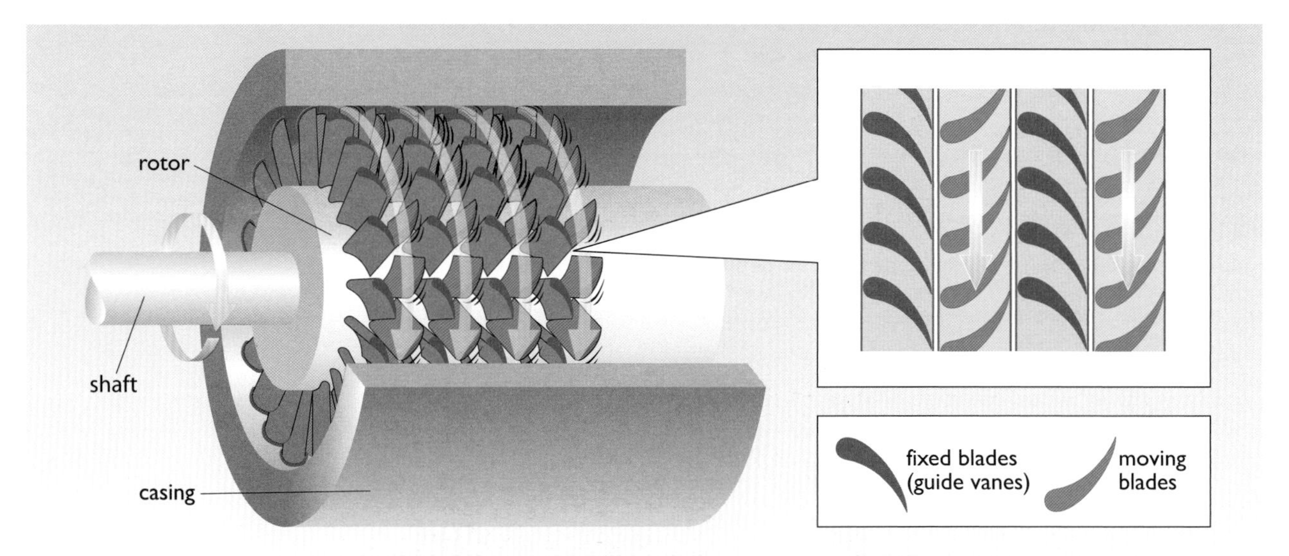

Figure 6.18 Diagramatic representation of Parsons' thoughts on possible vane and blade shapes, 1906

The steam entered at a pressure of 0.66 MPa and was exhausted at normal atmospheric pressure (0.1 MPa), so the average fall in pressure through each of the 14 stages was only two-fifths of an atmosphere (0.04 MPa), and the tiny turbine could rotate at a reasonable 18 000 rpm.

How efficient was this new form of steam engine? One answer comes from the basic principles discussed in Section 6.3. Assume that the input temperature is the boiling point of water under the pressure of 0.66 MPa, which is 163 °C (436 K), and that the departing steam exhausts into the atmosphere at 100 °C (373 K). Carnot's formula then tells us that the maximum possible efficiency is about 14.5%. On average, each of the 14 little rings of blades would be extracting about 1% of the input energy as useful work. However, no actual heat engine achieves its nominal Carnot efficiency, and it seems unlikely that this first turbine even achieved an efficiency of 5% – considerably less than the best reciprocating steam engines of the time.

Nevertheless, Parsons continued to develop the new machine, and within five years several hundred were in operation. Although wider recognition in other countries took longer, it is an indication both of the acceptance of steam turbines for electricity generation and their rapidly increasing size that in 1898 Parsons & Co. were invited to supply a 1 MW turbine for a power station in Elberfield, Germany.

The idea of forming steam jets through nozzles was not entirely abandoned, and turbines were developed that had a ring of fixed nozzles creating jets that gave up their energy through impulse forces as described in Box 6.6. Both the *reaction* and *impulse* principles (Figures 6.1 and 6.2) are retained in the steam turbines of present-day power stations.

Marine engines

The high rotational speeds meant that the steam turbine never became a serious option as a propulsive system for road vehicles; nor did it displace the reciprocating steam engine for locomotives. It did, however, find one other major application, as a marine engine. Parsons foresaw this, and realized that its adoption for naval ships would speed its acceptance in the world of merchant shipping. He also knew that the British Admiralty had a history of resisting all innovation. (They had notoriously insisted on wooden hulls when others were already using iron, and had initially rejected steam engines.) Parsons therefore decided on a demonstration, and constructed a remarkable vessel. *Turbinia* (Figure 6.19) was 30 m (100 ft)

Figure 6.19 Parsons' 1897 *Turbinia*, equipped with a 2300 hp steam turbine engine and capable of 33 knots

long by less than 3 m beam, and less than 1 m draught. She was driven by three turbines, in triple cross-compound arrangement, delivering a total power of 2300 hp (1.7 MW) to her screw propellers. Taken in secrecy to Cowes, she roared out into the 1897 Jubilee Review of the Fleet at Spithead with Parsons at the controls and travelling, it is said, at the unheard-of speed of 33 knots (38 mph).

The Admiralty did adopt turbines, and 1901 saw the first turbine-powered merchant ship, the *King Edward*, carrying passengers on the Clyde. Others followed, and in 1907 came one of the most famous of many Cunard 'liners', the *Mauretania*. She was the largest ship of her time, and her 68 000 hp quadruple-screw direct-coupled turbines gave her a maximum speed of 29 knots. From 1910 until as late as 1932 she held the 'Blue Riband' for the fastest Atlantic crossing. (For a remarkable photograph of *Mauretania* with *Turbinia* alongside, see Parsons, 1911.)

However, as marine engines driving propellers, steam turbines suffered two major disadvantages. One was the mismatch between their rate of revolution, typically thousands of rpm, and the hundred or so rpm of a large screw propeller. A gearing system helped to solve this problem, but many ships retained direct drive and a consequently lower efficiency. Another problem was that a turbine cannot be put into reverse. A commonly adopted solution was to have a separate 'astern' turbine. (You will appreciate that ships do not have brakes, so a reliable astern drive is essential.)

Oil-fired steam turbines were adopted by the British navy for their fast, light destroyers as early as 1908. The cost per unit of energy was greater than for coal, but the savings in other respects were significant. Fuel could be loaded more quickly and stored more compactly, and the tenfold reduction in engine-room manpower meant not only a saving in space, but appreciably lower running costs. Although some 60% of the world's ocean-going tonnage was still coal-fired in the late 1920s, the situation was changing, and oil-fired, high-pressure steam turbines provided the power for the great ocean liners of the 1930s, including the *Queen Mary* (1934) and *Queen Elizabeth* (1938). Many steam turbine ships remained in use into the late 20th century. (The Royal Yacht *Britannia*, launched in 1953, was powered by two oil-fired steam turbines delivering 12 000 hp, and decommissioned only in 1997.) But these were a vanishing, if distinguished, minority. As early as 1912 a ship with a strange appearance had been launched in Denmark. This was the *Selandia*, the first diesel-powered ocean-going vessel, proudly announcing her new-style engines by having no funnels. By 1927, there were more diesel than steam vessels under construction worldwide.

6.6 Power station turbine systems

The operating principles of a modern power station are essentially those of Parsons' original turbo-generator. Steam entering the turbine is directed by curved guide vanes on to the curved faces of the first set of blades, where it is deflected and passes to the next set of guide vanes to repeat the process. As it is deflected, the steam exerts a reaction force on each blade, delivering energy to maintain the rotation of the turbine, which is in turn continuously delivering energy to the generator.

So how does the present-day plant, more than a century later, differ from its progenitor? This section looks briefly at the turbines and associated parts of the plant, and Section 6.7 then considers a specific coal-fired plant, showing how Carnot's theoretical concepts translate in practice into flows of materials and energy. (The *generators* driven by these systems are the subject of Chapter 9.)

The turbines

The present-day turbines are undoubtedly very much larger than Parsons' original, and with increasing size, it becomes necessary to allow for expansion of the steam as the pressure falls along the turbine. In each individual turbine, therefore, the lengths of the blades increase in succession from the steam inlet point to the exhaust point. (This explains the conventional trapezium symbol ( ) for a turbine.)

As Figure 6.20(b) shows, the resulting blades can look very different from the little stubs in Parson's first turbine. The rate of rotation of a modern turbine tends to lie in the range 1500–3500 rpm, a fraction of the 18 000 rpm of Parsons' machine. But size matters. The tip of a metre-long blade mounted on a metre-diameter rotor spinning at 3000 rpm is moving at 470 m s^{-1} (about 1000 mph). This is about 1.4 times the speed of sound, which means that the outer parts of these blades must pass through a 'sound barrier' as the turbine is brought up to speed.

Turbine systems

As well as causing the rotor to rotate, the steam striking the blades of any turbine also exerts an unwanted force *along the axis*. To balance this axial force, the larger turbines are often mounted in a **double-flow** arrangement, with identical pairs facing in opposite directions ( ) on the same axis (Figure 6.20(a)). They must of course *rotate* in the same direction!

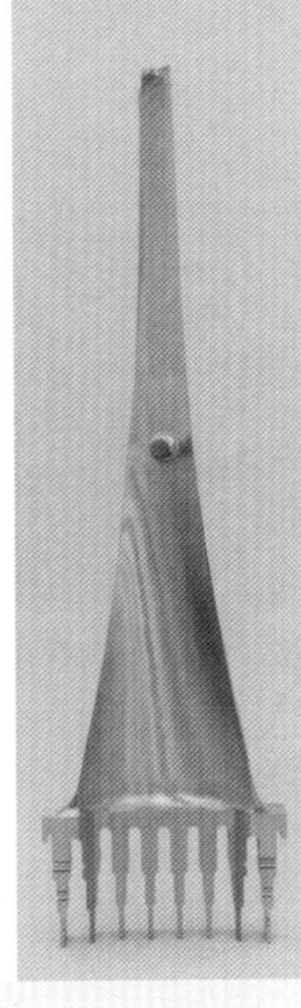

Figure 6.20 (a) One set of turbine blades for a 1000 MW tandem-compound steam turbine; (b) 1.1 metre long last stage blade of a turbine set

A modern system will almost certainly use a *compound* arrangement, often triple-compound, with **high-pressure (HP)**, **intermediate-pressure (IP)** and **low-pressure (LP)** turbines, and for the reason mentioned above, the sizes of the turbines increase through this sequence.

The tandem-compound arrangement (Figures 6.20(a) and 6.21(a)) is common, with all the turbines on a single shaft and driving a single generator. Cross-compound systems (Figure 6.21(b)) are also used, particularly in plants with large power output, as this arrangement permits two smaller generators on the separate shafts instead of a single very large one. (The IP and LP turbines in these diagrams are all double-flow, as described above.)

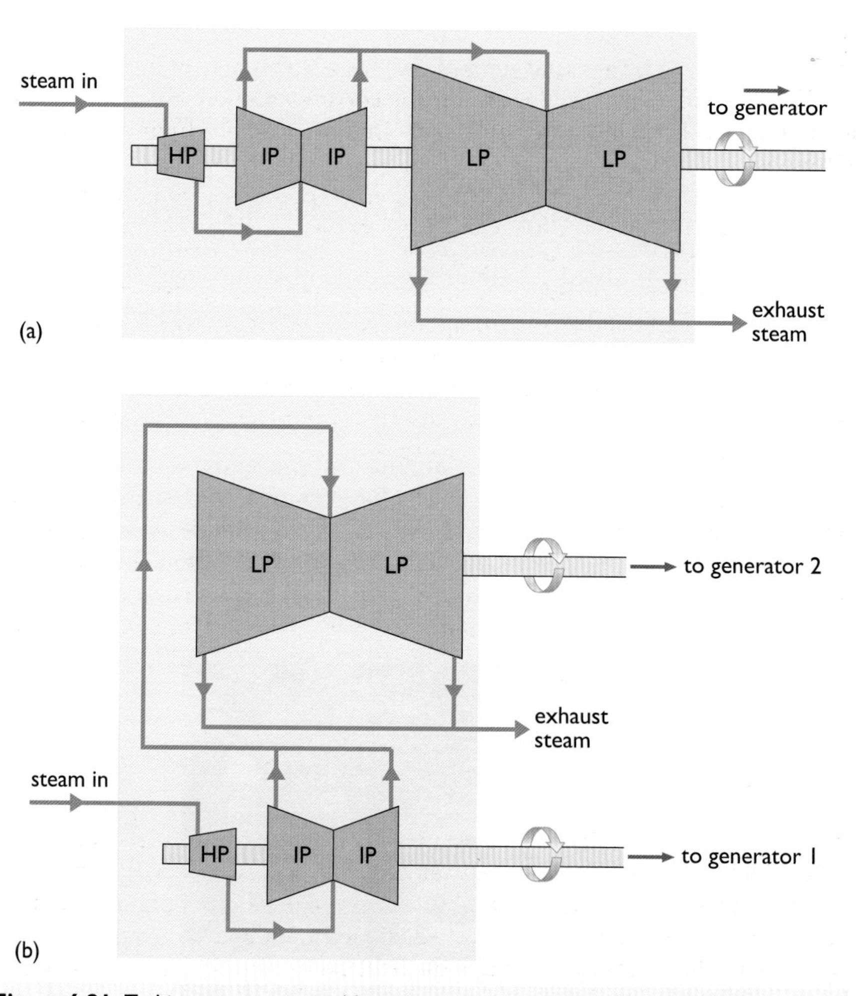

Figure 6.21 Turbine arrangements: (a) tandem-compound; (b) cross-compound

Steam pressures

The steam pressures and temperatures used today are, of course, a great deal higher than those of Parsons' first machine. The steam condition varies between different types of power plant, depending on the size of plant and the nature of the heat source. For a typical large coal-fired plant, the pressure might be as high as 25 MPa, with temperatures in the range 500–550 °C, i.e. *supercritical* steam conditions (Box 6.5). Modern *ultra-supercritical* power station designs are looking to higher temperatures of 600 °C or even 700 °C.

Current nuclear reactors, constrained to work under less extreme conditions (see Chapter 10), commonly supply steam at 4–6 MPa and temperatures in the region of 300–400 °C. The boiling point of water at 5 MPa is about 240 °C, so this is *superheated* steam. At the speeds involved in the turbine, any tiny water drops will damage the blades, so it is important to minimize condensation. Ideally, the steam should exhaust to the condenser as *dry* steam.

The boiler

The design of the boiler that produces the steam is an important factor in the efficiency of the plant. Three centuries of development since the earliest steam engines have led to today's coal-fired boilers, producing thousands of tonnes of steam per hour, with a fuel-to-steam efficiency as high as 90% (see Chapter 5). A modern water-tube boiler is a complex structure, designed to maximize the heat exchange between the burning fuel or hot gases and the water. It will normally have several separate sets of heat exchangers, designed to increase the overall efficiency. These will include one or more **superheaters**, to raise the final temperature of the steam leaving the boiler, and **reheaters**, in which the steam re-visits the boiler between successive turbines in a compound system. The hot flue gases may pass through an **air preheater** to heat the incoming air, or an **economizer** to heat the incoming water, again reducing the fuel requirement.

There must, of course, be control systems in place to monitor and govern the rate of steam supply. The purpose of the turbine is to drive an electrical generator, and this brings its own special requirements. The rate of rotation must remain constant within a very narrow permitted range, and the entire system must be able to respond effectively to the variations in demand – sometimes very rapid variations – that are a feature of the life of any power station.

The condenser

Unlike Parsons' original turbine, which exhausted the steam to the surroundings, the modern power-station plant (following Watt) includes a separate condenser. Carnot's rule shows that the lower the final temperature the greater the efficiency, so a well-designed condenser is essential. The aim, as in the boiler, is to maximize the heat exchange – in this case between the steam leaving the final turbine and a flow of **cooling water**. A common arrangement is for the cooling water to flow through many small tubes surrounded by the exhaust steam. The steam condenses on the surfaces of the water tubes, and the resulting **condensate** (water) is pumped away from the condenser to return eventually to the boiler. The pump (again following Watt) is important in maintaining as low an exit pressure as possible. This pump, and another to circulate the cooling water, will use one per cent or so of the electrical output of the power station, a loss that is usually justified by the resulting improvement in efficiency.

Where a plant is near the sea or a large river, the cooling water can be drawn from this and returned, usually 10–20 °C warmer, to its source. Otherwise a **cooling tower** (Figure 6.22) is needed. In these large, familiar structures the warmed cooling water is sprayed down through a rising current of air. Some of the water evaporates, and the loss of latent heat needed for this process cools the remainder. (An effect familiar to anyone who has stepped out of a hot shower into cool dry air.) The plume seen above the cooling tower is produced when the water vapour carried up by the air stream cools and condenses in the colder air outside. The evaporated water, a few per cent of the total, must of course be constantly replaced as the cooled water returns to the condenser.

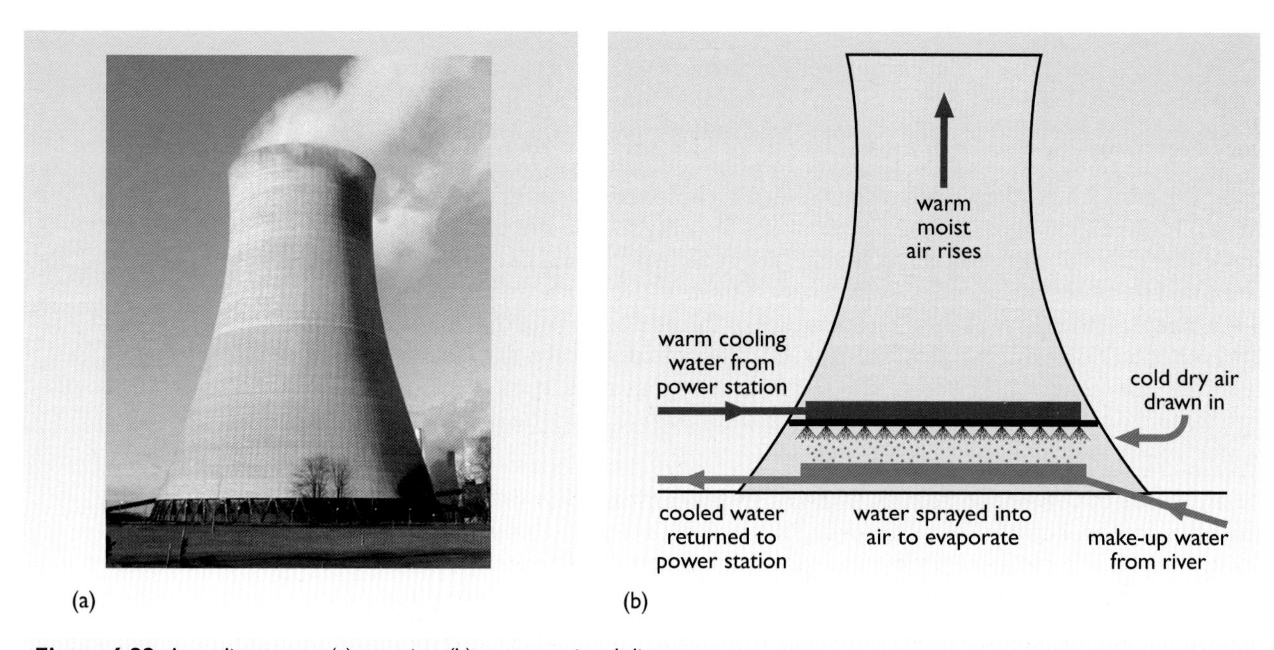

Figure 6.22 A cooling tower: (a) exterior; (b) cross-sectional diagram

In arid areas, where air may be the only available coolant, the necessary good thermal exchange can be achieved by a **fan coil** system. In this very large scale version of a car radiator, the low pressure steam leaving the final turbine is condensed as it passes through a massive assembly of finned pipes cooled by a forced draught of air that carries away the heat.

Materials

The physical properties of materials have played an important and recurring role throughout the history of steam engines. Savery's burst pipes and Trevithick's series of catastrophic failures had their immediate effects on subsequent developments; but ultimately it has been Watt's more cautious approach to high pressures and high speeds that has generally been adopted, leading to today's extensive development and testing of materials before their use under ever more severe conditions.

In recent decades the search for higher efficiencies has gathered impetus, with concerns about fuel resources (and fuel cost) reinforced by the need to reduce carbon emissions. Coal-fired plants under construction at the time of writing (2010) are expected to achieve efficiencies above 45%, but this performance comes at the cost of steam temperatures above 600 °C at pressures approaching 300 atmospheres. The resulting combination of mechanical stresses and exposure to attack by the supercritical fluid can have seriously deleterious mechanical and corrosive effects on the steel of the boiler tubes, turbines and other components.

Advances towards greater efficiency have therefore necessarily been accompanied by the development of new materials, and in particular steels for specific purposes within the plant. Steel is basically iron with a small percentage of carbon; but alloys of iron with other metallic elements (chromium, nickel and other neighbours of iron in the periodic table) can be designed to meet different requirements, such as the toughness and corrosion resistance needed by turbine blades, or the resistance to very high temperatures at the inlets of HP and IP turbines. Each of these improvements, of course, carries a financial cost, which for the plant operators must be balanced against the saving in fuel costs – unless other incentives are offered.

6.7 Flows in a 660 MW turbine system

Figure 6.23 shows schematically the turbine system of a single power plant. In practice, of course, the plant will be much more complex than this, incorporating the features described above and others, designed to maximize both the efficiency and the working life of the component parts. However, this simplified example gives an indication of the flows of *materials* and *energy* that characterize a plant of this size, delivering an output power of 660 MW to the generator.

As Figure 6.23(a) shows, the working fluid (water/steam) flows in a *closed* loop at a rate of about 1800 tonnes per hour, or half a tonne per second. To follow the successive processes, we might start at the point where it leaves the condenser. At this stage it is warm water, maintained at a low pressure (10 kPa) by the pump that draws it along. At this pressure – only about a tenth of an atmosphere – the temperature of 45 °C is just a little below the boiling point of the water.

The pump

The first stage in converting this warm, low-pressure water into very hot, high-pressure steam is the pump. In this system, the water pressure is increased from a tenth of an atmosphere to about 250 atmospheres, and each tonne of water passing through requires just over 25 MJ of energy from the pump for this. With a flow rate of 1800 tonnes per hour, and assuming a pump efficiency of about 90%, the required electrical input to the pump becomes about 50 GJ per hour (14 MW). This is one of several electrical inputs that have to come from the final generator output and must be taken into account in assessing the overall performance of the system.

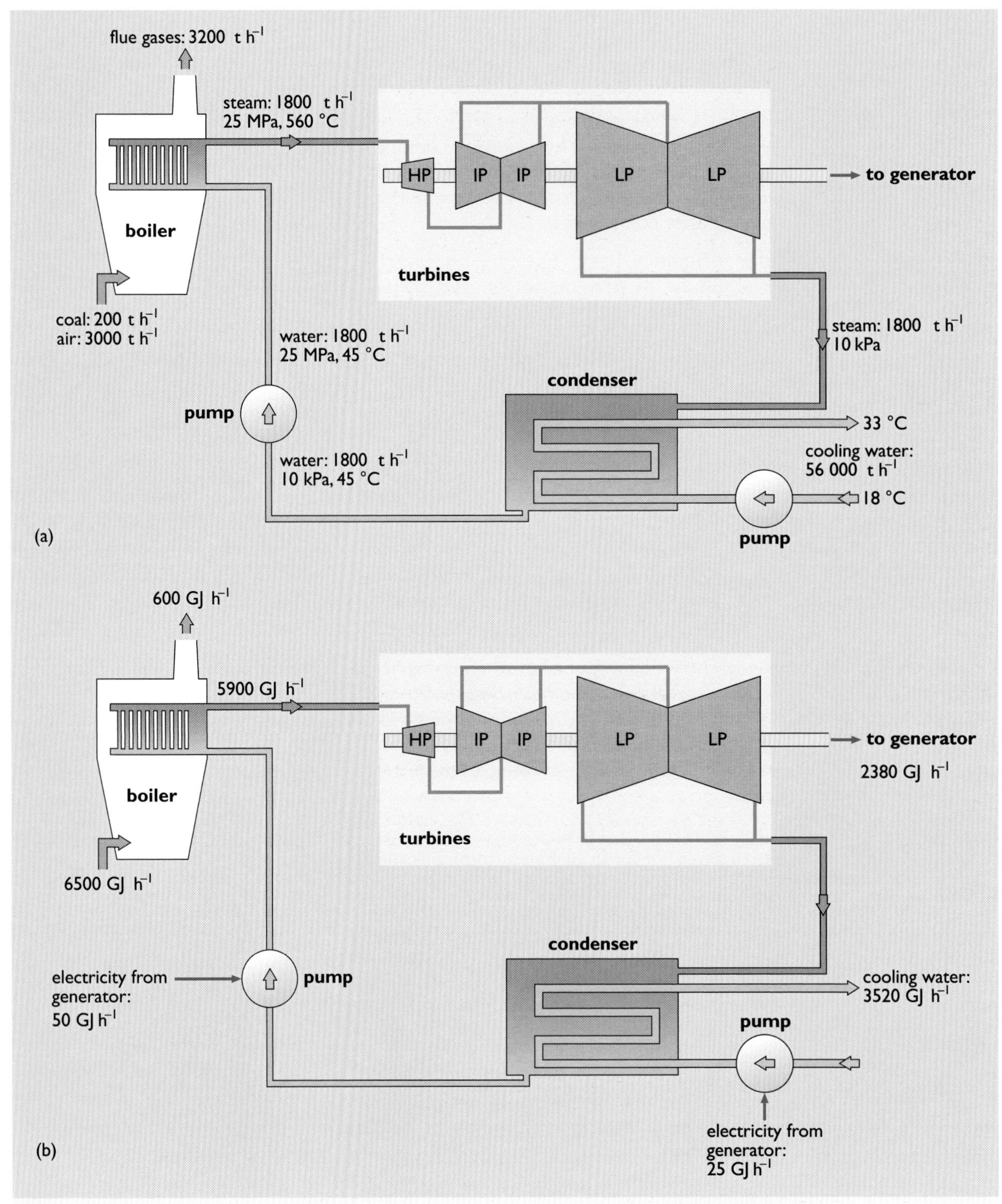

Figure 6.23 Flows in a power station steam turbine system: (a) materials; (b) energy; 1 MPa ≈ 10 atmospheres

The boiler

To raise the temperature of the high-pressure water from 45 °C to the required 560 °C needs a heat input of about 3.3 GJ for each tonne of water. The boiler must therefore supply heat at a total rate of $1800 \times 3.3 = 5900$ GJ per hour. If the fuel-to-heat efficiency of the coal-fired boiler is 90%, an input of about 200 tonnes of coal an hour will be needed – and to allow full combustion, at least 3000 tonnes an hour of air! In a pulverized-fuel boiler, most of this entire mass will exit as flue gases (see 'Flue Gases' in Chapter 5). As mentioned in Section 6.6, in practice some of the heat carried by the flue gases may be used to preheat the incoming air or the feedwater, and the boiler may also include reheat systems. If these features serve their intended purpose of increasing the efficiency, they should, of course, lead to a lower fuel requirement than that shown in the diagram.

The turbines

The turbine system in Figure 6.23 is the tandem-compound arrangement of Figure 6.21, but can be treated as a single entity in terms of the input and output flows. The steam enters at 560 °C, which means that it is a supercritical fluid (Box 6.5). At the pressure of 25 MPa, its density is about 70 kg m^{-3} (over a hundred times that of ordinary steam). As it passes through the turbines, the steam pressure falls as it gives up energy to the blades, delivering in this case a total of 2380 GJ per hour to be transmitted to the generator.

By the time it exhausts to the condenser, at a pressure of only 10 kPa (one-tenth of normal atmospheric pressure), the volume occupied by each tonne of steam has increased by a factor of about a thousand. If the mass flow rate is to remain the same throughout the system, it follows that the volume available in the final stages of the low-pressure turbines must be a thousand times that in the first stages of the high-pressure turbine. This is achieved in part by the two-way double-flow system, but it is still necessary for the turbine diameters to increase by a factor of ten or so between the input and exhaust ends of the complete system.

The condenser

The fluid entering the condenser is *steam* at the low pressure of about 10 kPa, and the fluid leaving it is *water* at approximately the same pressure. As its name suggests, the function of the condenser is to condense the steam, extracting its *latent heat* and maintaining as low a temperature as possible. The heat released must, of course, be carried away by the cooling water, but this water itself must remain appreciably cooler than the condensate, which is at about 45 °C. These factors determine the necessary flow of cooling water, as follows.

In this system, heat must be removed at a rate of 3520 GJ h^{-1} to condense the 1800 tonnes per hour of low-pressure steam. Suppose that the available cooling water enters at 18 °C, and its temperature should not rise above 33 °C. The heat energy needed to raise the temperature of 1 tonne of water by 1 °C is 4.2 MJ (Box 6.2), so the 15 °C permitted rise allows each tonne of cooling water to carry away 63 MJ. Extracting the 3.52 million megajoules an hour therefore needs about 56 000 tonnes an hour of cooling water. (This could be visualized as a flow at about 5 m s^{-1} through a pipe 2 m in diameter.)

The pumps circulating this large volume of cooling water must supply enough power to pump it through the condenser pipes and raise it through any necessary height differences. The total power demand will also depend on whether the water is returned to a river or the sea, or cooled in cooling towers. Again the power for the pumps must come from the generator output. The rather arbitrary figure of 25 GJ h^{-1} used here is equivalent to a power of just under 7 MW, or about 1% of the gross output.

Efficiency

What is the heat-to-work efficiency of this system?

Using the formula from Section 6.3, it is easy to calculate the efficiency of a Carnot engine working between the input and output temperatures of 560 °C (833 K) and 45 °C (318 K), respectively:

$$\text{percentage efficiency} = \frac{T_1 - T_2}{T_1} \times 100 = \frac{515}{833} \times 100 = 61.8\%$$

But no real machine would be expected to achieve this, even if there were no heat or frictional losses. The data in Figure 6.23 reveal just how far the actual steam cycle of this turbine system is from the idealized Carnot cycle, as follows.

Considering just the turbines, the diagram shows a heat supply from the boiler of 5900 GJ h^{-1}, and a work output to the generator of 2380 GJ h^{-1}. The heat-to-work efficiency of the turbines is therefore:

$$\text{percentage efficiency} = \frac{2380}{5900} \times 100 = 40.3\%$$

This is a more realistic figure, but so far only the turbines have been considered. There is also the 600 GJ h^{-1} heat loss from the boiler and the 75 GJ h^{-1} needed by the two pumps mentioned above. These three items together would reduce the overall efficiency to little more than 35%. But this is now unrealistically low for a modern plant of this size – a result of the over-simplification of our model. A plant with steam reheat and the other efficiency-improving methods mentioned earlier could increase the overall turbine efficiency by as much as five percentage points, bringing it back to about 40%.

As shown in Chapter 9, generator losses will reduce this by a few percentage points, leading to an overall fuel-to-electricity efficiency of perhaps a little less than 40% for the system described here. The average figures of 33–35% for power station efficiencies quoted throughout this book are lower than this, in part because not all turbines are designed to operate at the temperatures and pressures used here, but also for the obvious reason that power stations – like every other energy conversion system in the world – are not run constantly at optimum efficiency.

6.8 Summary

The essentials

This chapter can be summarized in two ways:

- It is an account of heat engines, and in particular the steam turbines of modern power stations.
- It is an introduction to the laws of thermodynamics and their consequences.

Different readers will no doubt have their preferred interpretations, but whether it is regarded as a description of the way we currently use a large proportion of the world's primary energy or an explanation of the science that governs all energy systems, it remains central to where we are and where we might go in the world of energy.

Like any other scientific theory, thermodynamics could be superseded in the future; but for the present, designers of machines are well advised to take note of its laws. Looking at the consequences, we might well return to the question at the start of this chapter, and ask why we continue to use valuable primary energy in this wasteful way.

It is important to note that we don't always do this. Processes in which heat does not play an essential part are not subject to the same constraints. A hydroelectric plant can transform the kinetic energy of a running river into electrical energy with very little loss. (It might be thought that the same should apply to wind turbines but in this case there are aerodynamic constraints.)

It is also worth noting that the heat-generating renewable energy sources, whilst they may have great environmental advantages, are still subject to the laws of thermodynamics, and the Carnot limits to efficiency apply equally to a wood-fired power plant or a bio-fuelled vehicle. There is however one exception, where mankind has for centuries achieved large-scale direct conversion from chemical to mechanical energy: an *explosion*. But our concern is with the *continuous* and *controlled* use of energy.

The future

Is there a future for *steam*? This chapter has shown that the coal-fired steam engine was the enabling technology throughout the Industrial Revolution, and remains an essential element in the generation of electricity. However, coming chapters will show that the other fossil fuels have followed very different routes, with very different technologies. When *motive power* is the requirement, the direct use of oil or gas – in the internal combustion engine and the gas turbine – dispenses with the need for steam as an intermediary. And for *heating*, these 'convenience fuels' frequently occupy the place once held by coal.

Does this mean that the futures for steam power and coal are destined to remain inextricably linked? Not necessarily. Coal doesn't *have* to produce steam. As shown in the next chapter, it can be converted into the more desirable fluid fuels that use the other technologies mentioned above.

If this route were to be adopted, the coal-fired power station as we know it might eventually disappear from the scene – although there will need to be compelling reasons to replace well-established existing technology by these more complex processes.

It is even more obvious that steam does not need coal. A sixth of the world's present electricity comes from nuclear power stations, and a smaller fraction from oil-fired plants, all of them using steam turbines. Steam also has a role to play in the generation of electricity from renewable resources such as geothermal energy, wood or other biofuels, and even solar energy. And there is the combined-cycle gas turbine (CCGT), the current system of choice for power stations in countries where natural gas is available. In this *compound* arrangement of a gas turbine followed by a steam turbine, the very high input temperature results in an overall efficiency of over 50% (see Chapter 8). It seems unlikely, therefore, that the steam turbine will disappear in the near future.

In the much longer term, the key question is whether we shall continue to use heat engines at all. Shall we eventually avoid the need to convert heat into motive power? The present situation, described in the opening paragraph of this chapter, is very far from this; but there are two possible routes for change. We could find ways to convert the chemical energy of fuels directly into electricity (as in the fuel cells to be discussed in Chapter 14). Or we could replace the fuels by primary sources that provide motive power directly: hydro power, wind, wave and tidal power – all, as it happens, renewable resources.

References

Carnot, N. L. S. (1992) *Reflections on the Motive Power of Fire*, Dover Books on Physics.

Muhlfeld, J. E. (1929) 'Locomotive', *Encyclopaedia Britannica*, 14th edn, London.

Parsons, C. A. (1911) *The Steam Turbine* [online], Cambridge, Cambridge University Press, http://www.history.rochester.edu/steam/parsons (accessed 10 January 2011).

Sorensen, B. (2004) *Renewable Energy,* 3rd edn, Amsterdam, Boston, Elsevier Academic Press.

Strandh, S. (1989) *The History of the Machine*, London, Bracken Books.

von Tunzelmann, G. N. (1978) *Steam Power and British Industrialization to 1860*, Oxford, Oxford University Press.

Further reading

Boyle, G. (ed) (forthcoming) *Renewable Energy: Power for a Sustainable Future*, 3rd edn, Oxford, Oxford University Press/Milton Keynes, The Open University.

The University of Rochester, NY has facsimiles of the books and writings of Hero, Savery (worth reading for his sales talk), Carnot, Parsons and others on their website at: http://www.history.rochester.edu/steam/

Chapter 7

Oil and gas

By Bob Everett and David Crabbe

7.1 Introduction

It might seem that oil and natural gas are completely different fuels and ought each to have a separate chapter. However, they are usually found in association with each other and their production and processing have a lot in common. This chapter is thus wide ranging and covers many topics:

- the original formation of oil and gas
- the discovery of oil and gas fields and the extraction of their contents
- oil refining and matching the demand for 'light' fuels (petrol and diesel) to the 'heavy' crude oil supply
- current oil reserves and production around the world
- the changing pattern of oil use in the UK since 1950, particularly since the price rises of the 1970s
- the impact of North Sea gas on the UK energy economy
- current gas reserves and production around the world and the problems of transporting natural gas
- the rise (since the 1970s) and fall (since 2000) of UK North Sea oil and gas production
- a short reflection on why oil and gas are considered such special fuels
- the chemistry of making oil (and substitute natural gas) from coal, building on what has already been described in Chapter 5
- unconventional sources of oil and gas, such as heavy oil, shale oil, tar sands and shale gas
- 'Peak Oil' and the problem of the finite nature of the world's oil (and gas) resources.

7.2 The origins and geology of petroleum

The word 'petroleum' means 'rock-oil'. In its commercial sense it usually refers to crude oil although a whole range of potential products from natural gas right through to solid deposits, such as bitumens and waxes, have similar origins.

It is now generally accepted that petroleum deposits arise from the decomposition of aquatic, mainly marine, animals and plants successively buried under layers of mud and silt several hundred million years ago. They thus differ from coal, which is predominantly formed from the remains of land-based plants.

The essential pre-conditions for petroleum formation include an abundance and diversity of plant and animal life, and a seabed environment that discourages immediate breakdown of dead organic material by bacterial action or oxidation. The burial of such hydrocarbon-containing plant and animal material under sea-borne mud and silt 'preserves' the material for the next stage of the petroleum-forming process. Over time, thick layers of marine silt are laid down and compressed over this 'source' material, and as the thickness and depth of these deposits increase so do the pressure and temperature. A chemical maturation process then takes place, with the peak oil formation taking place at about 100 °C. Higher temperatures favour the formation of gas.

But the actual *formation* of petroleum is only part of the picture; it must also be effectively contained. The current location of petroleum deposits may not actually be in the source rock where they were originally formed. Given the high temperatures and pressures underground they are likely to migrate into water-bearing aquifers in permeable rock – what the oil industry calls **reservoir rock**. Any natural tendency of the petroleum and natural gas to seep upwards must also have been prevented by an impermeable seal known as **caprock**, and any lateral migration must have been prevented, or greatly reduced, by the natural occurrence of **geological traps** within the reservoir beds themselves. There are different types of geological trap, but the most important in terms of oil and gas accumulation is the **structural trap**.

These can be further divided into three types, the **anticline**, **fault trap** and **salt dome** (Figure 7.1). The anticline, which is the most common, accounts for around 80% of the world's oil and gas resources. These are formed by folds in the earth's geological strata and when occurring in large structures can hold appreciable quantities of petroleum in place. The world's largest oil field, Ghawar in Saudi Arabia, is over 200 km long and 15 km wide.

A fault trap, by comparison, is formed when reservoir rock is brought into contact with impermeable strata by movement along a geological fault within the earth's crust. A salt dome, as the name suggests, is a dome-shaped formation of rock salt that has been forced upwards through overlying strata until it lies under caprock.

So, to summarize, there are four essential prerequisites for accumulations of petroleum to occur:

(1) There must have been a suitable source of marine deposits contained within the source rocks.

(2) These rocks must have allowed for the lateral movement of petroleum deposits via a permeable pathway to layers of porous and permeable reservoir rock.

(3) The reservoir must be sealed at the top by an impermeable layer or caprock.

(4) A trap must be present to prevent the lateral migration of the petroleum away from the reservoir.

And so, in the most general geological terms, where do modern-day petroleum deposits occur? The answer is: in sedimentary basins – areas within the Earth's crust where layers of rock and marine sediment are known to have accumulated over time. The global distribution of such areas is actually quite widespread. With the geology of land-based sedimentary

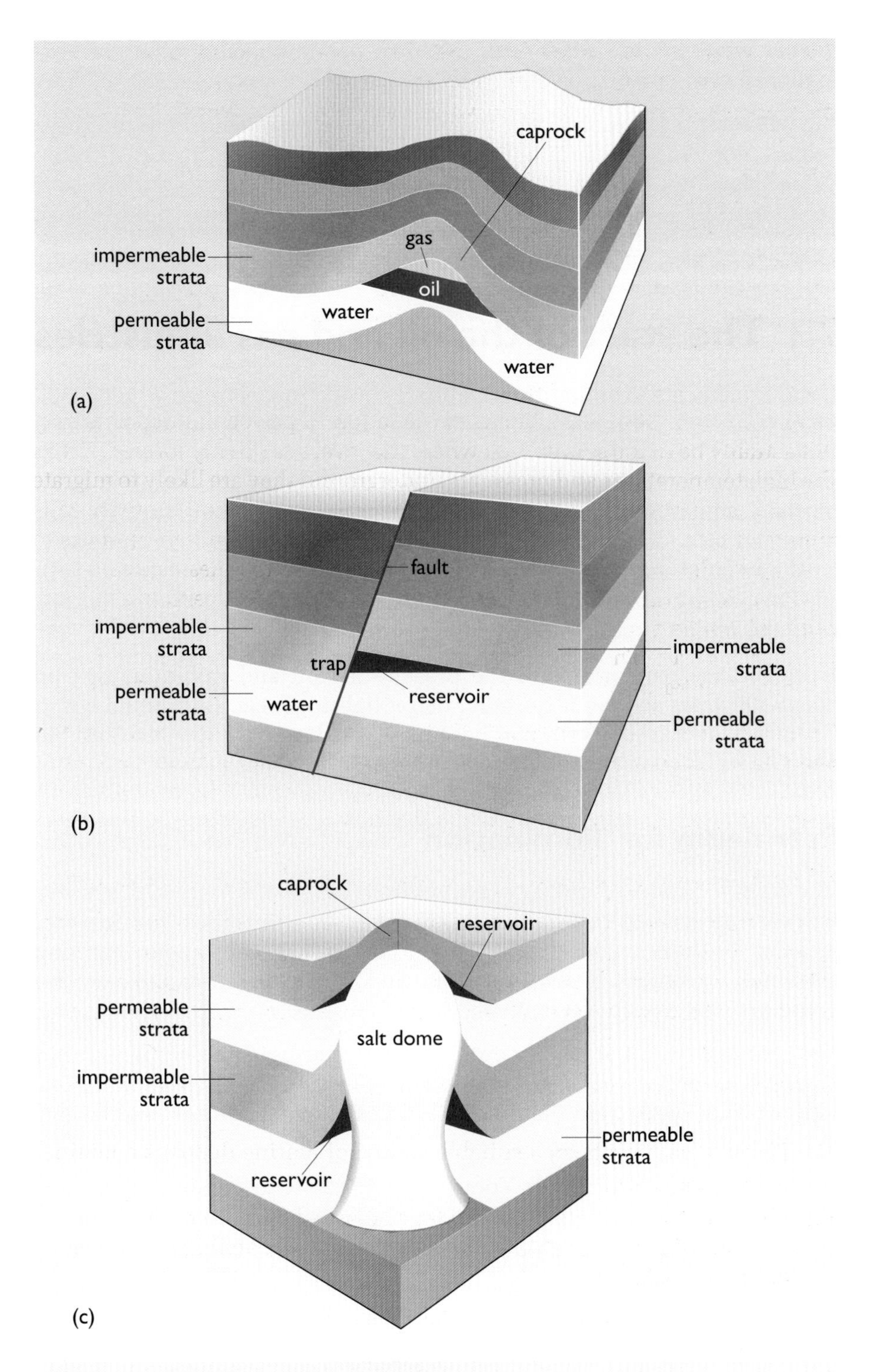

Figure 7.1 Traps: (a) anticline; (b) fault; (c) salt dome

basins, such as the Middle East, increasingly well mapped and explored, much of current oil exploration is now focused on underwater sedimentary formations that extend outward from a continental shelf. Exploration in the Gulf of Mexico led the way in this regard some 60 years ago, to be followed by drilling in the North Sea in the early 1970s, and more recently deeper water off the coast of Brazil. The Arctic appears to be the next exploration area.

But wherever petroleum may be found, we do well to remember its primitive origins, stretching back into deep geological time, with organic material being deposited and transformed deep beneath the Earth's surface over an immensely long period. This contrasts with the extreme rapidity with which it is currently being extracted.

7.3 The start of the oil and gas industries

Local small-scale exploitation of naturally occurring 'petroleum' goes back many centuries. Bitumen has been found in Babylonian monuments. Marco Polo, on his travels through Asia in the late 13th century, commented that people travelled great distances to collect oil from the oil springs at Baku on the Caspian Sea (now in Azerbaijan) as it was good to burn. He also remarked on a local temple in which a sacred flame burned continuously, fired by a local seepage of what we call natural gas. Even early travellers in the Caribbean commented on the Trinidad Pitch Lake – Sir Walter Raleigh used the pitch to caulk the hulls of his ships.

However, the true origins of the present-day oil and gas industries lie about 200 years ago in the quest for better lighting fuels. As pointed out in Chapter 3, at that time artificial lighting was extremely expensive and the cheaper fuels, such as tallow candles, were unpleasant to use.

Petroleum for illumination

The quest to produce lighting using gas from coal in the early 19th century has been described in Chapter 5. This was fine for cities, but in rural areas or for those without access to a piped supply, substandard lighting remained a problem. It was a lucrative opportunity awaiting anyone who could provide a good alternative.

Methanol or 'wood alcohol' could be produced by distilling wood in a similar manner to coal gas production. In Germany in the 1830s, Carl Von Reichenbach researched the production of various 'paraffins' and tars by distilling beechwood.

However, it was a Scot, James Young (Figure 7.2), who was able to produce a superior illuminating oil of consistent quality using a process that we would recognize today as industrial oil distillation and refining. See Box 7.1 for an account.

Figure 7.2 James 'Paraffin' Young

It was during the period when Young's company was at the height of its commercial powers that naturally occurring crude oil was first discovered through drilling at Titusville, Pennsylvania in 1859 by Edwin Drake (see Figure 7.3). This date is generally taken as the commencement of the modern petroleum extraction industry. As with Young's work, the principal outcome of this activity was the isolation of an illuminating oil. Within a few years drilling for oil had spread from Pennsylvania to many other parts of the USA as well as its 'old world' home in Baku. The petroleum industry was born.

BOX 7.1 James Young's oil refinery

James Young is held by many to be the founder of the world's oil refining industry. He was born in Glasgow in 1811 and studied chemistry in evening classes at what is now the University of Strathclyde. In the 1840s, while working for a Scottish chemicals manufacturer he was sent a sample of oil from a spring in a Derbyshire coal mine (Kerr, 1994). The thin, treacle-like oil could be distilled to a clear liquid. The mine owners were not interested in exploiting it themselves because of the limited supply (about 300 gallons per day), but allowed Young to set up and operate a small processing plant in 1848. The main product was lubricating oil for the Manchester cotton mills and a solid paraffin wax could also be separated from the oil.

The Derbyshire venture rapidly became uneconomic because of a reduced flow of oil but Young had the idea that the oil had been naturally distilled out of coal by volcanic activity long ago. He therefore began investigating the distillation of different coals, not for their illuminating gas properties (as was being done by the now well-established 'illuminating gas' industry), but for its potential oil yield. Prime candidates were the so-called cannel (candle) coals which had a bright flame when they burned. Boghead 'Parrot' coal (so called because of the noise made when it burned), obtained from a Scottish mine near Bathgate, proved to give the most oil: over 500 litres per tonne of coal.

In 1850 Young effectively patented oil refining. He took out patents in both Britain and the USA for a low-temperature process to obtain both oil and paraffin wax from bituminous coal. From then on he devoted himself entirely to oil production, becoming one of the first in the world to produce oil and refine it on a commercial scale. In 1851, Young opened Britain's first commercial mineral oil refinery at Bathgate. It distilled local coal in retorts similar to those shown for the manufacture of town gas in Chapter 5 (see Figure 5.7). It also produced gas, which was used to light the factory and the surrounding town, as well as a residue of coke.

Initially the works produced around 4000 litres per week of lubricants and solvents for the paint and rubber industries. He marketed one fraction from his distillery as 'Paraffine Oil', a burning oil for lighting, and developed a lamp to burn it. Before long one quarter of the lamp oil used in London came from the Bathgate Works. In addition the works produced high-quality wax that was sold to candle manufacturers. This was a considerable improvement on candle wax made from animal fat.

Although Young started by refining oil from bituminous coal, when supplies of this ran out he turned his attention to local deposits of oil shale for use as a raw material. These continued to be exploited in central Scotland until the 1960s. Perhaps the most prominent legacy of his early oil industry is the enormous Grangemouth oil refinery on the Firth of Forth to the north of Bathgate.

Figure 7.3 Colonel Edwin Drake (foreground right, wearing a top hat) talking to an engineer in front of his oil well at Titusville, Pennsylvania. They struck oil at a depth of only 22 metres on 27 August 1859

But apart from 'kerosene' for lamps and stoves the only other petroleum product for which there was any demand was lubricating oil for machinery. The remainder of the petroleum, what we would now refer to as the lightest and heaviest fractions, was discarded, including 'gasoline', which was initially deemed too dangerously volatile to use. Within 40 years, however, the oilmen had discovered that it was worthwhile to carefully separate out their hard-won petroleum into a dozen or more fractions, each with their own special uses.

Petroleum for transport

At this time there was no market for oil in transport systems. Even in the early 1880s the roads were still the province of the horse, and the railways the province of the coal-fired steam locomotive. Steam-powered road vehicles were finding a role in local, heavy-duty work, but they were cumbersome. Attempts had been made to introduce them for long-distance road transport but these were unsuccessful given the speed of the railways.

However, the entire situation was about to change. The story of the development of the 'motor car' is told in Chapter 8. 'Gasoline', which was being sold as a cleaning fluid in chemists shops, now had a potentially large-scale application. But it was more than land-based transport that was to be revolutionized in this way. The Wright brothers' flight in 1903 was powered by gasoline, and as will be shown in Chapter 8, oil products have remained the sole fuel for aircraft for over 100 years.

The potential of 'heavy oil' as an alternative to coal was also being appreciated, particularly by the navies of the world. The terms 'heavy' and 'light' in this context refer to the viscosity of the oil and really mean 'thick' and 'thin'. In the extreme, heavy oil may be nearly solid and has to be heated up in order to be pumped. The ability to pump it allowed the automation of the stoking of ships' boilers and made refuelling them much easier. The British navy adopted oil-fired steam turbines for their lightest and fastest warships as early as 1908. Oil-fired steam turbines became the dominant form of propulsion, particularly for large ships. The development of the diesel engine in the 1890s allowed 'heavy oil' to be used with high efficiency. The decade following World War I saw oil gradually replacing coal in both naval and merchant vessels. By 1930, about half of the world's shipping was powered by oil. Oil-fired steam turbines remained dominant, particularly for large ships, into the 1950s, but the second half of the 20th century increasingly saw their replacement by large diesel engines or even gas turbines fuelled by light fuel oil. During the 20th century oil became a strategic resource over which international conflicts would break out. In little over a century the oil industry grew from the tiniest beginnings to arguably the world's most important industry, controlled by a handful of extraordinarily powerful companies.

International conflicts in the 1970s had a profound influence on global oil price and subsequent oil use (see Box 7.2). Despite these, crude oil production grew to about 80 million barrels a day in 2009, or over 30% of the world's total primary energy consumption. In the UK consumption in 2009 was 1.6 million barrels a day, around one-third of the country's primary energy needs.

BOX 7.2 The oil price shocks of 1973 and 1979

The USA was the world's largest oil producer from 1900 right through to the 1970s. In 1935 it produced about 60% of the world's oil and was a major exporter. The world oil price was effectively controlled by a very few large (and mostly US) oil companies (see Drollas and Greenman, 1989). In 1960 some other producers – Iran, Iraq, Kuwait, Saudi Arabia and Venezuela – set up the **Organisation of Petroleum Exporting Countries (OPEC)** to 'secure fair and stable prices for petroleum producers'. They were later joined by a number of other, mainly Middle Eastern, countries.

As can be seen from Figure 1.1, in 1970 the world oil price was low, in fact in real (i.e. inflation-adjusted) terms the lowest it had been since the Depression year of 1931. World annual primary consumption had quadrupled between 1950 and 1970 (see Figure 2.5). OPEC's concerns reached a head in 1973. US oil production had peaked and the USA was now having to import oil. Was oil underpriced? Were the members of OPEC getting a fair price?

In October 1973 a long-simmering war between Israel and its neighbours Egypt and Syria broke out, the Yom Kippur war. This was of global importance since Israel had the backing of the USA while Egypt had the backing of the Soviet Union. The Arab members of OPEC decided to use 'the oil weapon' in support of Egypt and Syria. They announced price rises and production cuts, which would increase every month. A peace deal between the warring countries was eventually brokered in January 1974. Some of the production cuts were lifted but oil prices did not fall. It had become obvious that the world could afford to pay more for its oil. The average price for 1974 was three times what it had been in 1973.

Then in 1979 came the Iranian Revolution. The country's oil output (about 8% of the global total) dropped dramatically. Matters got worse in the following year, 1980, when the neighbouring country of Iraq attacked Iran, destroying much of the oil infrastructure. The world oil price peaked in January 1981 at nearly ten times in real terms what it had been in the summer of 1970.

Although subsequent extra supplies of oil from the North Sea, Russia and Alaska reduced the world oil price, the price rises of 1973 and 1979 have had a profound effect on oil consumption across the globe, as can be seen in the consumption charts in Chapter 2 and in the more detailed descriptions in this chapter.

The natural gas industry

Natural gas has a rather different story to that of oil. Although **associated gas** is found with petroleum deposits, natural gas can also be found on its own. This **non-associated gas** is believed to have its origins in freshwater rather than marine plant material. Associated gas was once thought of as a nuisance, either being flared off or being re-injected into the wells to flush out further oil. Indeed, in parts of the world where there is either no immediate market or means of transportation, flaring at the oil well still occurs, sometimes in enormous quantities. In general every effort is now made to bring natural gas to market.

'Conventional' gas generally flows freely from the reservoirs. There is also a considerable resource of 'unconventional' gas, which is more difficult to extract. This is in the form of tight and shale gas, coal bed methane and methane hydrates. These are described later in Section 7.11.

Like coal and oil, natural gas may not be completely pure. Often associated gas is 'wet' and wells produce a mixture of natural gas, condensate and **natural gas liquids (NGL)**. **Condensate** is a light oil that may contain hydrocarbons such as pentane (C_5H_{12}) and is often regarded as part of 'conventional oil'. The natural gas liquids are lighter hydrocarbons such as propane (C_3H_8) and butane (C_4H_{10}). These have sufficiently high boiling points that they can be liquefied at room temperature under pressure and are then sold as **liquefied petroleum gas (LPG)**.

In contrast to oil markets, the market for natural gas evolved only slowly, initially in the USA in the 1920s, where it was sold both as a heating fuel substitute for manufactured town gas, and also as a power station fuel. However, as late as 1970 two-thirds of the world's production and consumption of natural gas was within the USA.

In post-war Europe discoveries such as the huge Groningen non-associated natural gas field in the Netherlands in 1959 and BP's 'West Sole' field in the southern sector of the North Sea in 1965 have transformed the energy economies of most of the countries of north-west Europe. The old coal-derived 'town gas' was replaced with a cheaper alternative. These supplies have been reinforced by Russian gas piped all the way from Siberia and liquefied natural gas shipped from North Africa and the Persian Gulf. In 2009 natural gas accounted for over 20% of the global primary energy consumption and about 40% of the UK's primary energy consumption.

7.4 Finding and producing petroleum

Oil prospecting

Historically, the presence of oil in any particular location was detected by identifying surface seepages (even underwater seepages noticed by fishermen have been very fruitful). In many cases early prospectors found that drilling close to these sources would result in their locating commercially attractive petroleum deposits. The earliest US and Canadian finds came about this way, though like much else in exploration the early indications did not always bear fruit and great patience proved necessary. Some US fields were up and running very quickly, but despite many years of effort the first significant oil discovery was not made in Canada until 1902.

Drilling for oil is very expensive, so today a range of geophysical techniques are employed to decide whether an area is likely to contain suitable oil-bearing rocks. Only if these analyses give promising results will the drilling phase begin. The initial analysis of a likely area may begin with a **gravimetric survey**. Oil-bearing, sedimentary rocks are generally far less dense than other types and effectively reduce the Earth's gravitational pull within the locality. A substantial thickness of sediments can thus be located by measuring the Earth's gravitational field and searching for a low reading. Gravimetric surveys are far from infallible, but they are cheap and speedy to conduct.

In a similar way a **geomagnetic survey** of a region may be carried out. Non-sedimentary base rocks frequently contain large amounts of iron-rich minerals and are consequently magnetic. As such they can easily be distinguished from sedimentary rocks. Finally, there is the **seismic survey**. Gravity and magnetic surveys may accurately detect the presence of sedimentary formations but they cannot detect the presence (or absence) of oil-bearing traps within the rock strata. The seismic method is based on recording the time taken for sound waves to travel from a source at the Earth's surface down into rocks below, reflect off a rock boundary, and travel back to surface detectors. Figure 7.4 shows an underwater seismic

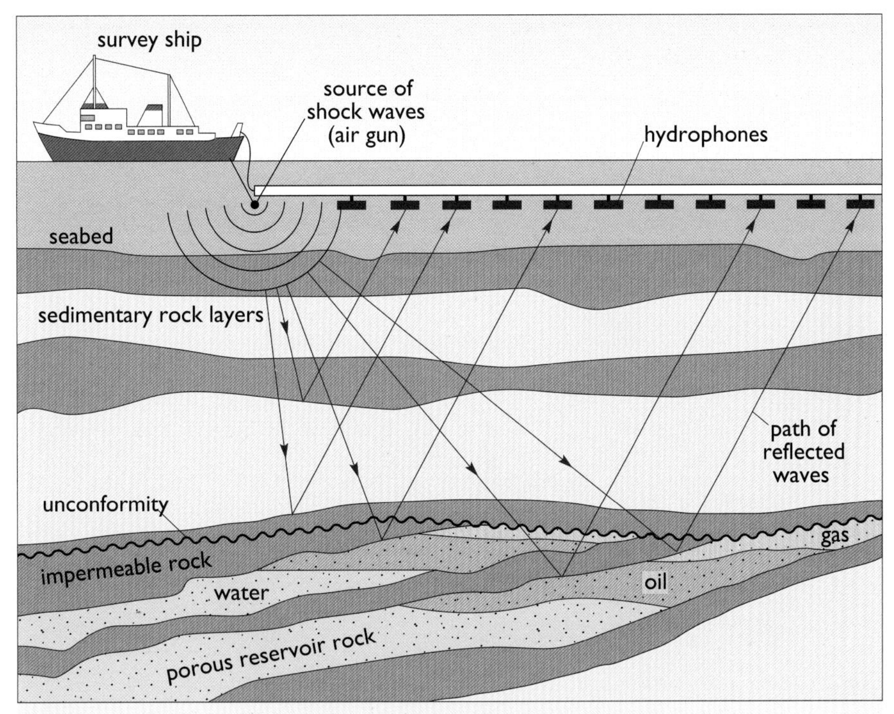

Figure 7.4 Marine seismic survey – pulses of sound penetrate the rock and are reflected back towards the hydrophones

survey being made from a ship, but similar surveys can be carried out on land.

The source of these sound waves was originally a small charge of dynamite, but other techniques, such as compressed air guns, are used in marine environments. A succession of passes over a promising area can allow a three-dimensional map of the underground geology to be drawn up, identifying the location of possible traps.

In the final analysis, however, the presence of oil can only be confirmed by physical exploration itself – drilling. This has always been expensive and has become even more so in recent times with the need to explore less hospitable areas such as the Arctic, the North Sea, the Gulf of Mexico and even deeper waters off the coast of Brazil.

If an initial drilled well confirms the existence of an oil-bearing structure deep below the Earth's surface – and the depth can be anything up to 6500 metres, depending on the geological age of the rock strata – the next stage is to attempt to determine the volume of oil contained, the extent of the field, and its likely productivity. To achieve this it is necessary to drill further appraisal wells. A number of these may be required in order to obtain sufficient information. Should full development then be considered appropriate, a series of production wells will be sunk. In the case of the North Sea, where oil fields have been largely exploited from a few fixed-well platforms, carefully targeted deviation drilling of multiple wells is carried out in order to maximize the physical (and financial) return from the discovery (see Figure 7.5).

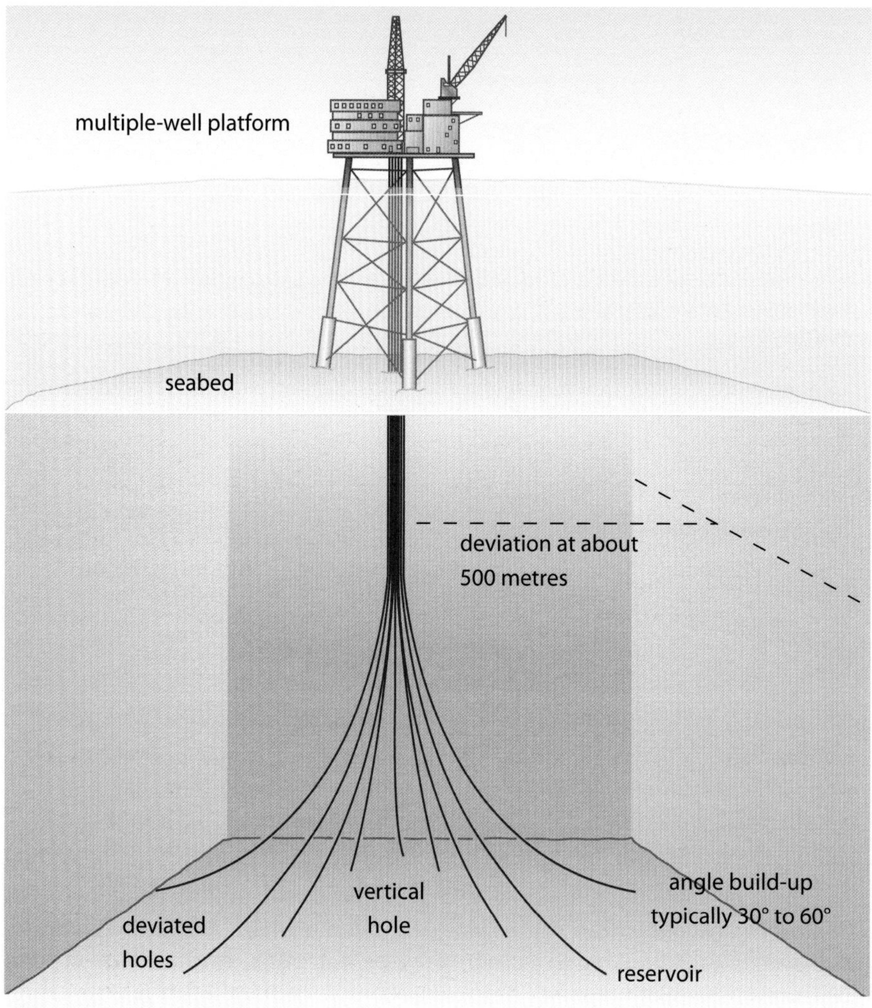

Figure 7.5 Deviation drilling

The development of offshore oil fields in deeper and deeper water has required the development of a range of platforms as shown in Figure 7.6. Conventional fixed platforms fabricated from welded pipe and pinned to the sea floor are commonly used in water depths up to 500 m. At the other extreme are totally floating drilling and production systems. These are only attached to the sea floor by the flexible drill pipe and have to maintain their surface position precisely using satellite positioning and steerable propellers.

Given the large expenditure of capital involved in exploration and drilling, it is hardly surprising that the oil exploration companies seek the highest possible levels of initial production in order to boost their income stream. The problem is always one of obtaining enough money to drill 'the next well' and search for 'the next oil field'.

The amount of oil or gas reserves controlled by a particular company has a considerable effect on its share price and its ability to raise investor capital. Reserve figures are thus carefully worded in terms of the probability of production (see Box 7.3).

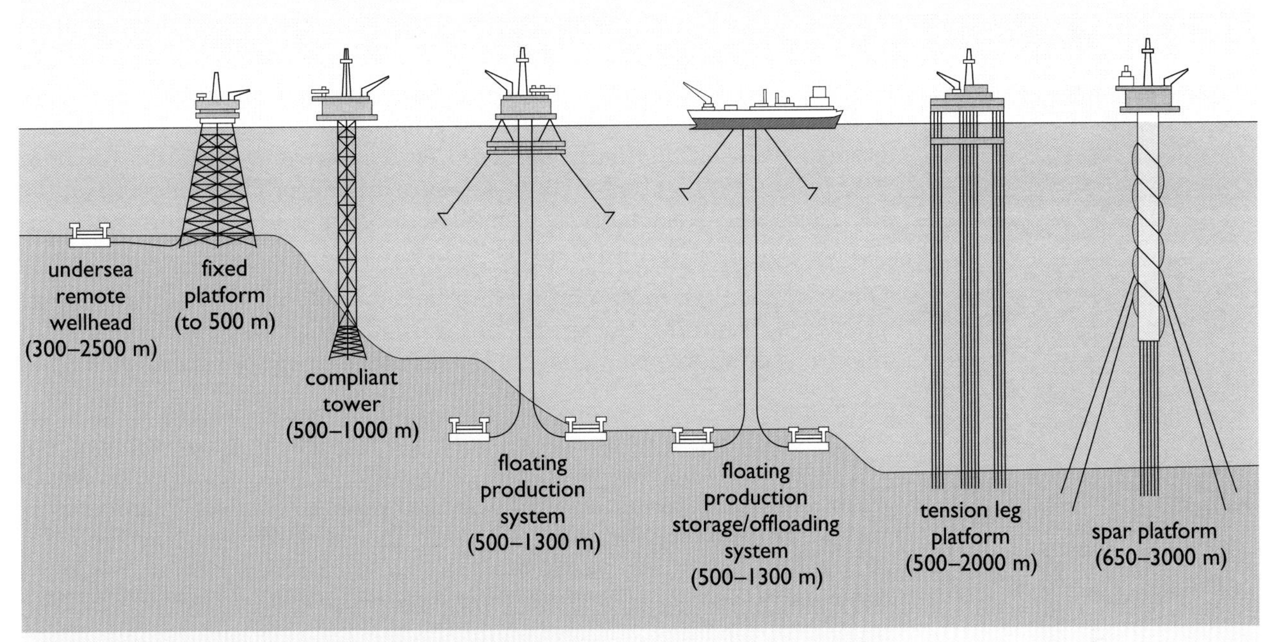

Figure 7.6 Different types of offshore development rigs

BOX 7.3 The three Ps – oil and gas reserves terminology

The McKelvey diagram was introduced in Chapter 5. Terminology in the oil and gas industry further splits reserves into three levels of confidence, illustrated in Figure 7.7:

- **proven reserves (1P or P90):** those with a high probability (>90%) of being economically recovered
- **proven and probable reserves (2P or P50):** the total reserves, including the proven (IP) reserves, with a medium overall probability (>50%) of being economically recovered
- **proven, probable and possible reserves (3P or P10):** the total reserves, including the 2P reserves, with only a low overall probability (>10%) of being economically recovered.

The 'undiscovered' resources at the right are often called 'yet-to-find'.

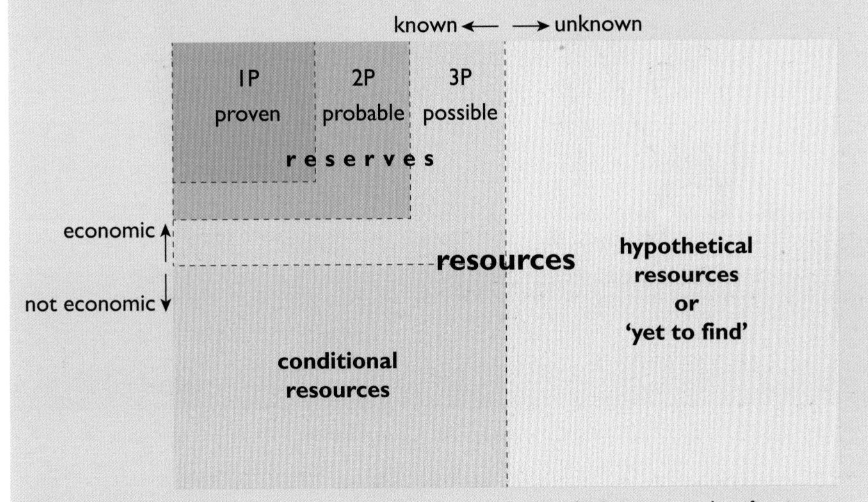

Figure 7.7 A McKelvey diagram of the three Ps of oil and gas reserve classifications (source: adapted from Sorrell et al., 2009)

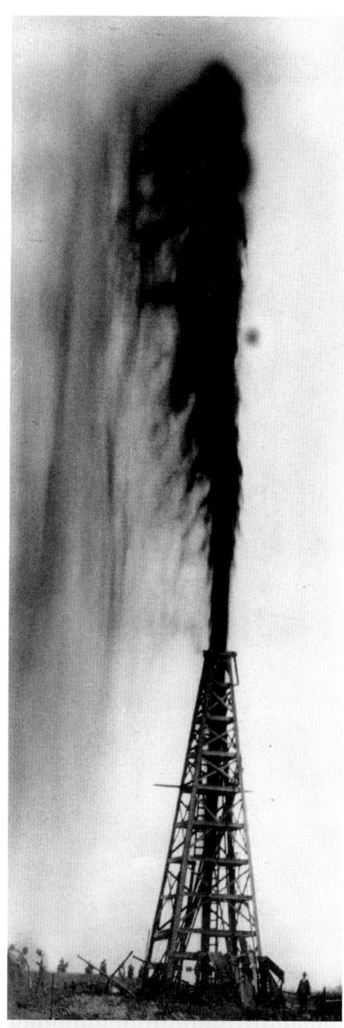

Figure 7.8 Oil spurting from the Spindletop 'gusher' under the reservoir pressure in Texas in 1901. This pressure is used to pump the oil in 'primary recovery'

Production

When oil or gas are first struck in a well, they can be under considerable pressure, leading in the past to spectacular 'gusher' oil strikes (Figure 7.8). All that is then necessary is that the well is 'capped', which may be difficult. Modern wells are normally equipped with a blow-out preventer (BOP) to contain any sudden rush of high-pressure oil or gas. The explosion on the BP Deepwater Horizon oil rig described in Chapter 1 (see Figure 1.23) was in part due to the failure of a BOP to operate.

Once capped then reservoir pressure can be used to drive the distribution pipework, pumping the oil to storage tanks that may be tens or even hundreds of kilometres distant. This is termed **primary recovery**.

At some point in time, possibly after several years, the pressure in the reservoir will fall and various measures will need to be taken to extract the oil, such as raising it with pumps or injecting it with (locally unwanted) natural gas or water to raise the reservoir pressure. This is referred to as **secondary recovery**. Primary and secondary recovery together typically succeed in extracting 30–50% of the oil in place within a reservoir. Secondary recovery may be a long, slow process over many years. The traditional 'nodding donkey' pump (Figure 7.9) operates a pump at the bottom of a well, lifting the oil maybe only a few bucketfuls at a time. Obtaining the necessary water for reservoir injection can pose problems and create conflicts with agriculture. In Saudi Arabia seawater is pumped hundreds of kilometres inland for such purposes.

The term **tertiary recovery** or **enhanced oil recovery (EOR)** refers to more extreme techniques such as injecting steam, carbon dioxide or even nitrogen into a reservoir. It typically allows a further 5–15% of a reservoir's oil to be extracted. It is potentially energy intensive and since the remaining oil is likely to be thick and viscous the rate of oil extraction

Figure 7.9 A traditional 'nodding donkey' pumpjack operates an oil-lift pump at the bottom of the well via a long cable

may be slow. If steam is used then, as with secondary recovery, a good supply of water is necessary.

This use of carbon dioxide (CO_2) for injection can be thought of as a stepping stone on the way to carbon capture and storage (CCS), a topic that will be described in Chapter 14. It can be a by-product of refining or chemical processing, as in the Great Plains Synfuels plant described in Section 7.10. Although the primary intention of its use is simply to flush oil (or natural gas) from a well, it is likely that the bulk of the CO_2 will remain below ground for many and possibly hundreds of years.

Where used with a 'normal' reservoir, EOR is perhaps still within the realms of 'conventional' oil. Techniques for extracting 'unconventional' oil and gas will be described later in the chapter.

7.5 Oil refining and products

Introduction

Since its earliest commercial application, the real technical genius of the oil industry has been to maximize the yield of useful products from a given amount of crude petroleum by oil refining. Crude oil is essentially a complex mixture of hydrocarbons with different chemical compositions. Unlike the aromatic hydrocarbons found in coal (and described in Chapter 5) those in oil and natural gas are mainly chain molecules, known chemically as alkanes or paraffins. Their basic properties are described in Box 7.4.

BOX 7.4 The alkanes or paraffins

Alkanes are the main components of natural gas and petroleum. Methane, CH_4, is the simplest possible alkane and is the main constituent of natural gas. Others consist of chain molecules, though these are not necessarily all straight chains. For many alkanes there may be different *isomers* with the same chemical formulae but with different arrangements of branching chains and with slightly different properties. Generally the larger the molecule, the higher the boiling point and the lower the energy density. Table 7.1 shows the basic properties of some of the more commonly encountered alkanes. They have a general chemical formula C_nH_{2n+2}.

Table 7.1 Basic properties of selected alkanes

Alkane	Formula	Boiling point	Higher heating value/MJ kg^{-1}
Methane	CH_4	−162 °C	55.5
Ethane	C_2H_6	−89 °C	51.9
Propane	C_3H_8	−42 °C	50.4
Butane	C_4H_{10}	−0.5 °C	49.5
Octane	C_8H_{18}	+125 °C	47.1

The different boiling points of the alkanes in crude oil allow it to be separated by distillation into various oil fractions, a process also known as **fractionation**.

In the early days, refining stills were simply large tanks in which crude oil was heated so that the volatile components boiled and vaporized. This hot vapour would rise, cool and condense, and could thus be collected. This process was repeated at different temperatures to separate different fractions. Following development of the internal combustion engine the need for improved product separation led to the use of simple 'fractionating columns' (Figure 7.10).

Hot crude oil would be fed into the bottom of a column. As it rose it cooled and different liquids would condense and run off at different levels (and temperatures). This allowed different boiling point 'cuts' or 'fractions' to be separated out in a single process.

There are several ways of classifying different useful fractions obtained from crude oil; one of the simplest is a division into three: 'light', 'middle' and 'heavy' distillate. Each of these can then be sent for further refining. It is important to remember, however, that familiar oil products are not single chemical entities but mixtures carefully blended for a particular

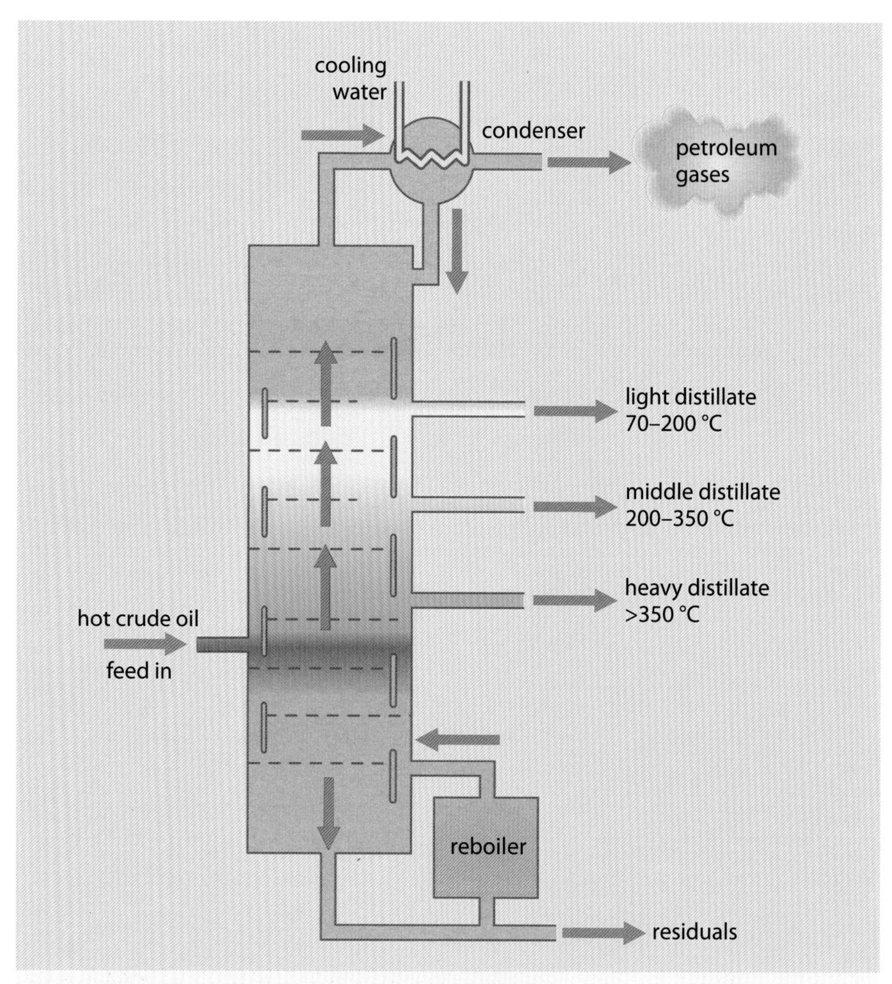

Figure 7.10 Oil distillation in a fractionating column

end purpose, and this is reflected in the broad ranges and overlap of their respective boiling points.

Light distillate is arguably the most important of the petroleum fractions. It has products with boiling points in the range 70–200 °C, such as petrol (or motor spirit or gasoline), the light chemical feedstock called naphtha (pronounced 'naftha'), kerosene, jet fuel and paraffin.

The **middle distillate** range is defined by the products with boiling points in the temperature range 200–350 °C. The two most recognizable oil products falling within this range are diesel fuel, colloquially referred to as 'DERV' (Diesel Engined Road Vehicle), and an overlapping fraction known as 'gas oil', which is mainly used as burning oil for domestic and light commercial heating. Gas oil, as its name suggests, was originally used to provide hydrocarbon enrichment in the manufacture of town gas.

The least volatile of the petroleum fractions is referred to as **heavy distillate**, those constituents of crude oil with a boiling point above 350 °C. Heavy distillate is either solid or semi-solid normal temperatures and may require heating to encourage free flow. Diesel fuel for large ships' engines – marine diesel – falls on the boundary of middle and heavy distillate. In volume terms the most significant product falling within this range has been 'fuel oil' used both for power stations and the largest industrial boilers.

There are, of course, two further components to consider. At the top of the column there are gases too volatile to condense (e.g. propane, butane). At the bottom are the 'residuals' consisting of the very heavy tars remaining after the 'fractions' have been distilled off. These are too dense to rise up the column. They are far from useless, however, and indeed some of the most vital of all oil products are obtained from the very bottom of the barrel, usually by steam or vacuum distillation. They include paraffinic waxes and various grades of lubricating oils. Nothing is wasted, and what then remains is used as either heavy fuel oil, bitumen for road-making, or solid 'petroleum coke', useful either for simple steam raising or even as a source of graphite.

Getting more of what you want

The above description of oil refining is very much a simplified version of what happens in reality. Crude oils themselves are very different mixtures, depending on which part of the world, or indeed which particular oil field, they originate from. North Sea and North African crudes, for example, tend to be 'light', i.e. containing more of the lighter fractions, and 'sweet' i.e. the sulfur content is low, whereas Venezuelan oil and some Middle East crudes are much heavier and may be 'sour', containing greater amounts of sulfur and other unwanted chemical impurities.

Before the discovery of North Sea oil, UK refineries were geared up for dealing with heavier feedstocks, whereas over the past 40 years much of their throughput has been of a lighter variety. Now, with the North Sea oil production in decline, the pendulum has swung back to heavier grades of oil. At the same time product demand has changed markedly, a subject we will be returning to in the next section.

So how do the oil companies take into account both changing crude inputs and changing product demand in the refining process?

A fundamental problem is that the greatest demand is for light products, such as petrol, while the crude oil supply tends to consist more of lower, heavier constituents. So refining techniques known as **cracking** were devised to break down the larger (heavy) hydrocarbon molecules into smaller (lighter) ones. Where this is used for 'heavy oils' and 'oil sands' it is sometimes called **upgrading**.

Figure 7.11 A catalytic cracker – it may be 40 metres high or larger

Much of this is done by a process invented in Russia in the 1890s, **thermal cracking**, applying heat to the heavy distillate and residuals. Later techniques employed catalysts such as aluminium silicates in **catalytic cracking** to reduce the temperatures required (Figure 7.11). (A catalyst is a substance that assists a chemical reaction, but remains unchanged itself.)

The overall hydrogen content of the products for which there is a demand is higher than that of crude oil. Thus some process of 'adding hydrogen' is necessary. As described in Chapter 5, hydrogen can be made by spraying water onto red-hot coke (it can be petroleum coke in this case). By heating heavy oil to a high temperature under pressure in the presence of hydrogen and using a suitable catalyst, new lighter fractions of oil can be produced. This is known as **hydrocracking**.

Also, increasingly the demand is for very low sulfur fuels, while a supply of 'sour' crude may contain 5% sulfur. **Hydrotreating**, a similar process to hydrocracking, converts the sulfur content to the gas hydrogen sulfide. This can be removed and treated, ending up as elemental sulfur, which can be used in fertilizer manufacture. Oil refining and natural gas processing are the world's largest commercial sources of sulfur.

The amount of hydrogen added to oil in a refinery can be significant and gives rise to **refinery gain**, with more barrels of product emerging from a refinery than the number of barrels of crude oil that entered it.

A refinery may produce a range of possible hydrocarbons from basic refining and cracking. The end products are likely to be achieved by **blending** mixtures of more than one fraction or subfraction in order to optimize the characteristics. For example, as will be described in Chapter 8, petrol must have the best possible octane rating, while diesel fuel must have the correct compression ignition characteristics.

Table 7.2 gives a listing and brief description of the types of product available from the various fractions obtained by the distillation of crude oil. This should convey the extraordinary range of products that can be obtained from this one naturally occurring fossil fuel. In practice, given the limitations of fuel storage and distribution one grade of fuel may serve two purposes. In the UK, diesel (DERV) and gas oil are virtually identical and sometimes just referred to as 'gas diesel oil'.

Table 7.2 Products obtained from distillation of crude oil

Product	Description	Uses
Methane, ethane	Gases	Petrochemical feedstock components of natural gas
Propane, butane	Gases – can be liquefied under mild pressure	Heavier gases liquefied for use as bottled liquefied petroleum gas (LPG); petrochemical feedstock
Aviation spirit	Light distillate blend with well-defined spark-ignition characteristics	Piston-engine aircraft
Motor spirit (petrol, gasoline)	Light distillate blend with well-defined spark-ignition characteristics	Motor vehicles (petrol engines)
Naphtha	Light distillate mixture with bp[1] 70–200 °C	Petrochemical feedstock
Industrial spirit	Light distillate mixture with bp 100–200 °C	Industrial solvent
White spirit	Light distillate mixture with bp 150–200 °C	Paint thinner, dry cleaning
Aviation jet fuel	A kerosene with bp within the range 150–250 °C	Jet engines
Burning oil (paraffin, kerosene)	An oil at the upper end of the middle distillate range, bp up to 250–300 °C	Lamps, heating stoves
Diesel fuel – DERV	Middle distillate blend suitable for compression ignition engines	motor vehicles (diesel engines)
Gas oil	Middle distillate within bp range 200–350 °C	Domestic, commercial and industrial boilers for heating/process heat; feedstock for 'cracking' at refineries
Marine diesel oil	Middle/heavy distillate mixture for use in large engines	Ship engines
Fuel oil	Heavy distillate and residual oils with bp in excess of 350 °C, classed typically as 'light', 'medium', and 'heavy'	Fuel for power stations; feedstock for 'cracking' at refineries
Lubricating oils	Oils obtained by vacuum distillation of petroleum residues	Used to reduce friction and wear on moving surfaces; lubricants are classed according to their viscosity
Waxes	Hydrocarbons of high molecular weight obtained from primary lubricating oils by crystallization	Waterproofing, polishes, insulators, candles
Bitumen	Solid/semi-solid residue remaining after crude oil distillation	Waterproofing, road-making
Petroleum coke	Solid residue left behind after crude oil distillation	Steam raising, source of graphite

[1] bp = boiling point at normal atmospheric pressure.

7.6 Oil today

Reserves and production

Since Drake's gusher in Pennsylvania first produced oil some 150 years ago, the question has always been 'where next?' The remainder of the 'lower 48' states, Canada and Mexico were well prospected in succeeding years; Alaska was to wait until the more advanced searches were undertaken a century later. The early industry also found plentiful supplies in Baku, today in Azerbaijan. The Middle East first figured in oil production in 1911, a few years after the discovery of oil in Iran. This in turn led to exploration and discoveries in both Iraq and Saudi Arabia, where production for export got under way on a fairly modest scale in the 1930s. By then, the USA was out on its own in terms of production and consumption, having sufficient excess to support a large export industry. Table 7.3 shows historical figures for oil production. It highlights the declining proportion of North American production and the rising importance of the Middle East.

Table 7.3 World oil production 1900–2000[1]

Year	World production /Mt	Percentage from:		
		USA, Canada and Mexico	Middle East	Former Soviet Union
1910	44	65%	–	22%
1930	201	66%	3%	15%
1950	534	57%	17%	8%
1970	2357	28%	29%	15%
1990	3172	21%	27%	18%
2009	3821	17%	30%	17%

[1] Figures include crude oil, shale oil, oil from tar sands, 'heavy' oil and natural gas liquids.
Sources: Shell, 1966; BP, 2010.

World oil production figures are normally given in either million barrels per day or tonnes per year. Given that the term 'oil' covers such a range of possible products, the conversion between these different values and with energy content is imprecise. Box 7.5 describes the problems.

BOX 7.5 Oil types and units – rather slippery to pin down

The terms 'oil' and 'petroleum' cover a wide range of products.

Oil, as shown in the statistics of Tables 7.3 and 7.4, consists of crude oil, natural gas liquids (NGLs), condensates and unconventional oil, but doesn't include biofuels or refinery gains.

Crude oil made up about 84% of the oil produced in 2009; it is a mixture of hydrocarbons that exist in a liquid phase at atmospheric pressure and room temperature.

NGLs are light hydrocarbons that are produced along with natural gas, such as ethane, propane, butane and condensates. They make up about 14% of current 'oil' production. The ethane is often left in the natural gas, mixed with the methane.

Condensates are light hydrocarbons, such as pentane, which are liquid at room temperature but highly volatile.

Liquefied petroleum gases (LPG) are those such as propane and butane which can be liquefied under mild pressure at room temperature. They are commonly sold for cooking, lighting and transport uses.

Unconventional oil includes extra-heavy oil, natural bitumen (oil sands) and shale oil. Together these made up about 3% of 2009 'oil' production. The term also includes oil from gas-to-liquids (GTL) and coal-to-liquids (CTL) plants though the (small) contribution from these isn't included in the statistics in Table 7.3.

Biofuels are liquid fuels derived from biomass, including ethanol and biodiesel. In 2009 world production was equivalent to just over 1% of the other forms of oil.

As explained in Chapter 2, oil is normally measured in barrels, a unit of volume. Other statistics quote it in tonnes, a unit of mass, and yet others in tonnes of oil equivalent (toe) which is a unit of energy. It would be convenient if there was just one set of conversion factors between these three. Alas the wide variety of oil types and products make this impossible.

The term *density* describes the *mass per unit volume* of a substance. Its inverse, the *volume per unit mass* is known as the **specific volume**. That for crude oil can range between 6 and 8 barrels per tonne, yet light natural gas liquids may only have a specific volume of 11 barrels per tonne. Most statistics quote an 'average' conversion factor of 7.33 barrels per tonne for crude oil, but the average conversion factor for the world mix of different types of oils as quoted for 2009 in Table 7.3 turns out to be closer to 7.4.

The relationship between mass and energy content is just as potentially confusing. The *energy* content of an 'average' tonne of crude oil is *1 tonne of oil equivalent*. As pointed out in Chapter 2, this is equal to 41.868 GJ (often rounded to 42 GJ though many sources use 41.88 GJ). However, this is only an average. One tonne of 'light' oil is likely to contain more than this. This effect isn't sufficient to offset the density variation described above, so typically one barrel of 'light' oil still contains less energy than one barrel of 'heavy' oil. A litre of 'light' petrol contains less energy than a litre of slightly heavier road diesel fuel (DERV).

For the purposes of this chapter:

1 tonne of oil (mass) contains 7.3 barrels (volume) and has an energy content of 1 tonne of oil equivalent or 42 GJ.

An oil production rate of:

$$
\begin{aligned}
\text{1 million barrels per day (bpd)} &= \text{0.365 billion barrels per year} \\
&\approx \text{50 million tonnes per year} \\
&\approx \text{2.1 EJ per year} \\
&\approx \text{2 quads per year.}
\end{aligned}
$$

But when in doubt, look carefully at the energy density values for the appropriate petroleum product. Some of these will be described in more detail in Section 7.9.

The distribution of conventional oil reserves in key regions around the world has already been described briefly in Chapter 1 (Figure 1.16). Figure 7.12 shows these again together with the distribution of production.

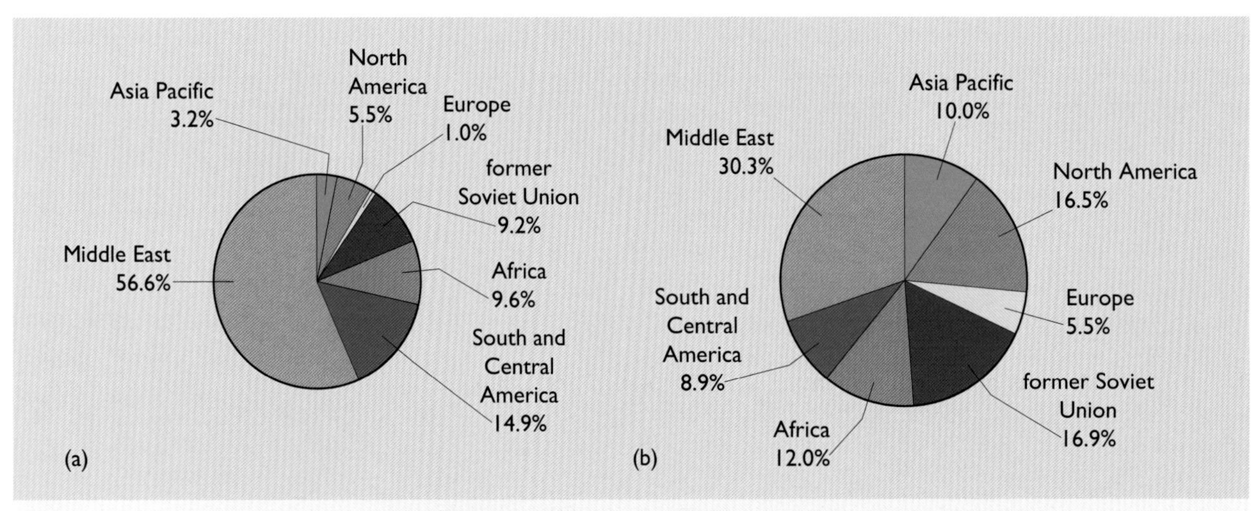

Figure 7.12 Distribution of the world's oil, 2009: (a) conventional reserves; (b) production (source: BP, 2010)

Table 7.4 shows some key oil statistics for the countries with the largest proven reserves (i.e. 1P in the terms of Box 7.3), plus some other countries of interest.

Although production takes place widely across the world, over 30% of the production is in the Middle East and that area accounts for 35% of world oil exports. Over half of the world's oil reserves are in the Middle East and over three-quarters are in OPEC member states.

Table 7.4 Oil reserves, production and reserves/production (R/P) ratios for selected countries, 2009

Country	Proven reserves /1000 Mt	Percentage of world reserves	Annual production /Mt	Percentage of world production	R/P ratio /years
Saudi Arabia	36.3	20.0%	460	12.0%	75
Venezuela	24.8	13.7%	125	3.3%	200
Iran	18.9	10.4%	202	5.3%	89
Iraq	15.5	8.5%	122	3.2%	127
Kuwait	14.0	7.7%	121	3.2%	115
United Arab Emirates	13.0	7.2%	121	3.2%	108
Russia	10.2	5.6%	494	12.9%	21
USA	3.4	1.9%	325	8.5%	11
China	2.0	1.1%	189	4.9%	11
India	0.8	0.4%	35	0.9%	21
UK	0.4	0.2%	68	1.8%	6
World[1]	181.7	100.0%	3821	100.0%	46

[1] BP also lists a further 23.3 Mt of 'unconventional' Canadian oil sand reserves, not included in the world total.
Source: BP, 2010.

Many countries appear to have quite healthy reserves to production (R/P) ratios of 100 years of more. Indeed the overall world R/P ratio of 46 years might suggest that oil supplies should be secure for many years. Section 7.12 describes why this might be an optimistic view.

Worldwide in 2007 there were estimated to be about 70 000 oil fields in production (IEA, 2008). However, the 20 largest giant and supergiant fields produced over a quarter of world production. One field alone, Ghawar in Saudi Arabia, discovered in 1948, produced 7% and the Cantarell field in Mexico produced 2%.

Despite its own large oil production, in 2009 the USA was the world's largest oil importer, followed by China, Germany, Japan and India.

The changing pattern of oil use

The oil industry has come a long way in its 150-year history, not only in terms of its size and importance but also in the way in which the focus of demand for its different products has changed. Initially it was for illumination and lubrication and now it has expanded into transport, bulk heating and chemical feedstocks. However, the oil price rises of 1973 and 1979 (and more recent ones) have had a profound effect. It is instructive to look at the changing pattern in delivered petroleum fuels in the UK from 1950 to 2009 (see Figure 7.13).

Between 1950 and 1973 there was a sixfold increase in overall demand, followed by a slow decline. It is useful to consider statistical snapshots for the years 1950, 1970 and 2009. Table 7.5 gives some key statistics for these years. Many of the historical comments that follow will apply equally in other European countries.

In 1950, oil was cheap, US$1.70 a barrel (US$15 in 2009 equivalent prices), but the UK had a serious import/export balance of payments problem.

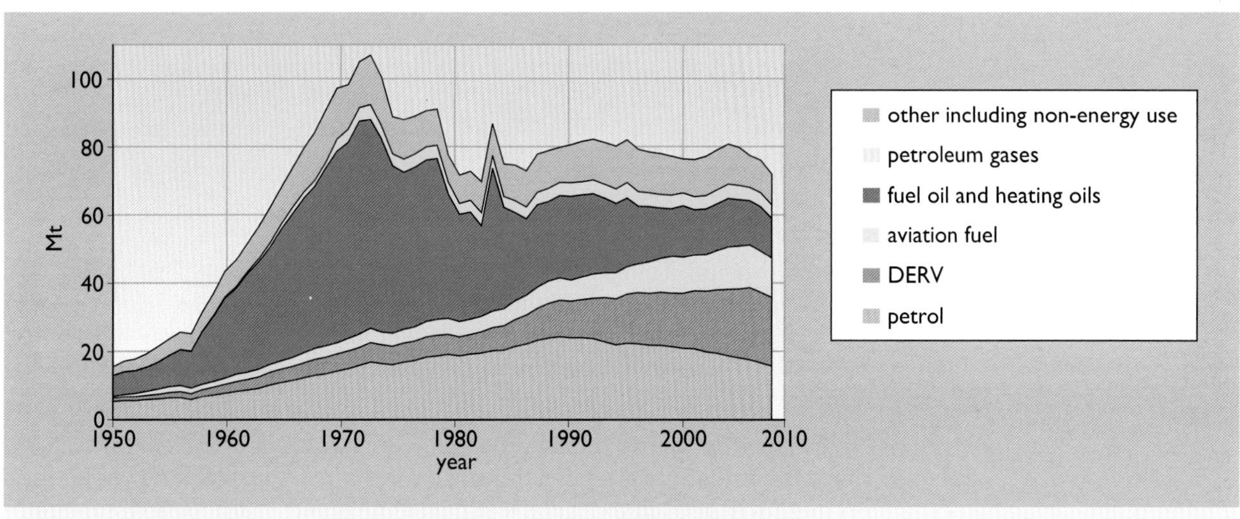

Figure 7.13 UK petroleum use 1950–2009 (sources: DECC, 2009; DECC 2010a)

Table 7.5 Key UK petroleum demand statistics

Year	1950	1970	2009
Oil price (US$ 2009)	15.2	9.9	61.7
UK population/million	50	56	61
Car ownership cars/1000 people	102	214	476[1]
Total internal travel/billion passenger-km	218[2]	403	797
Petrol consumption/Mt	5.3	14.2	15.8
DERV consumption/Mt	1.1	5.0	20.1
Aviation fuel consumption (including overseas flights)/Mt	0.5	3.3	11.5
Fuel use for electricity generation/Mt	Small	12.6	1.6
Naphtha for gasworks/Mt	–	4.6	–
Other heating and fuel oil consumption/Mt	6.1	39.0	10.1
Total petroleum consumption for energy purposes/Mt	14.6	87.1	64.7
Total petroleum consumption for all purposes/Mt	15.5	97.2	72.0

[1]2007.

[2]1952.

Sources: BP, 2010, DECC, 2009, DECC, 2010a, DECC, 2010b.

Wartime petrol rationing only ended in that year. Car ownership was low and air travel expensive. The amount of travel in passenger-km was far lower than today (see Figure 1.25, which goes back to 1952) and the travel that did take place was mainly by train, bus and coach. The UK was still predominantly a coal-based country and coal supplied most chemical feedstocks. Coal also powered the railways, which transported both people and freight.

By 1970, the picture had changed completely. The economy was booming and oil was cheap, US$1.80 a barrel (still under US$10 in 2009 terms). In real terms it was cheaper than it had been in 1950. It was both available and affordable. Car ownership was rising and car travel increasing. Jet airliners were making air travel increasingly affordable. Diesel and electricity had displaced coal from the railways. Naphtha was being used instead of coal for making town gas (see Section 7.10). Oil was also being widely sold for heating purposes, replacing coal.

Existing oil refineries, such as Fawley near Southampton (Figure 7.14), had been enlarged considerably to cope with the increased demand. Oil was even being used in large-scale power stations to generate electricity. A 2 GW power station to burn low grade fuel oil was added to the Fawley refinery in 1971.

After 1970 the picture changed again. By 1980, in real terms, oil was ten times the price it had been a decade earlier. It was no longer a cheap fuel. Its use for electricity generation declined (apart from an extraordinary burst in 1984 during a coal miners' strike) and its use for heating purposes

(a)

(b)

Figure 7.14 Fawley oil refinery near Southampton: (a) aerial view showing Fawley power station in the distance; (b) oil refinery by night

was challenged in the UK and much of north-western Europe by the new cheap fuel, North Sea gas. Despite the oil price collapse after 1986, oil has steadily lost out to gas in the heating market. Although electricity generators turned back to coal (particularly new cheap sources), UK industry turned to gas. By 2009, UK sales of fuel oil were actually lower than they had been in 1950. One small but important area where oil has held its own for heating since 1970 is in the domestic sector, in those homes beyond the reach of the gas grid. In 2009 it still supplied about 7% of the domestic sector's total delivered energy, in the form of gas oil and liquefied petroleum gas.

The overall result is that oil has become predominantly a transport fuel. Over three-quarters of the UK's delivered petroleum fuels are used for this purpose. Many of the petrochemical feedstocks that are not considered as 'energy uses' are still likely to end up as lubricants for the motor or aviation industries.

The rise of the different types of transport fuels can clearly be seen in Figure 7.13. The increase in the use of DERV in part shows the rise of freight transport by road. This makes up about a third of UK road energy use (see Figure 3.16). Also, since 1990, there has been a fall in petrol consumption as new diesel cars have been successfully marketed. Rising oil prices after 2005 have noticeably reduced the overall demand for transport fuel both on the ground and in the air.

Another key change from 1970 has been the tightening of environmental legislation concerning emissions of sulfur dioxide right across Europe (and in many other countries). All European motor fuel now has to be 'low-sulfur'. In 1970 it was still acceptable to burn high sulfur heavy oils in power stations and factories. Today it is not. Removing sulfur has become an essential part of oil refining.

The decline in UK oil demand since 1970 has meant the closure of a number of refineries. Demand has also changed away from the 'heavy' end to the 'light' end of the spectrum of products (Table 7.2). This could be met by investing in catalytic crackers to reprocess the heavier oils into lighter ones. In fact much of the mismatch between the supply of, and demand for, the different fuels is met by international trade. The UK exports fuel oil and petrol and imports DERV and aviation fuel.

What of the future? The UK government is committed to an 80% cut in greenhouse gas emissions from 1990 levels by 2050. This may mean reducing UK petroleum demand back to something like its 1950 level by then. The decline in demand since 2005 is perhaps fortunate since it has partly matched the UK's declining oil production.

What might be done? One possibility is the increased use of transport biofuels. In 2003 a European Union Directive asked member countries to ensure that 5.75% of transport fossil fuels (i.e. petrol and DERV) would be replaced with biofuels by 2010. In the UK this has involved the blending of biodiesel with DERV and the inclusion of up to 5% ethanol in petrol. In 2009 the UK used about 1 million tonnes of liquid biofuels, mainly biodiesel, and much of it imported, making up 2.5% of the country's transport fuel use.

Tackling future transport emissions is a potentially difficult task. Some technical possibilities are:

- more efficient road vehicles and aircraft
- more use of biofuels and a possible future 'hydrogen economy'
- increased use of electric vehicles and a future fully electrified railway system, particularly for freight.

Then there is the deeper question: Is all this travel actually necessary?

Finally, there is still a small heating use of oil. This could possibly be replaced by biofuels, such as wood, or even electric heat pumps.

Some of these topics will be covered in Chapters 9 and 14.

7.7 Natural gas

An energy revolution in the UK

A look at the recent energy history of the UK shows the profound impact of the introduction of natural gas from the North Sea. This story has been matched in many other European countries.

If we look back to the year 1960, 'gas' in the UK meant town gas, mainly produced from coal in relatively small or city-sized gasworks. Many of these were little changed from the 19th century version described in Chapter 5. A process to produce town gas from cheap imported light oil was under development (see Section 7.10).

UK gas consumption was then only about 300 PJ per year (Figure 7.15). In the domestic and services sector it was widely used for cooking, but there was little use for space heating, where coal and smokeless fuels were still the dominant fuels. It was estimated that even in 1967 only 1 million UK households had gas-fired central heating. The iron and steel industry manufactured gas for its own use from coal, but apart from that there was relatively little use in the industrial sector.

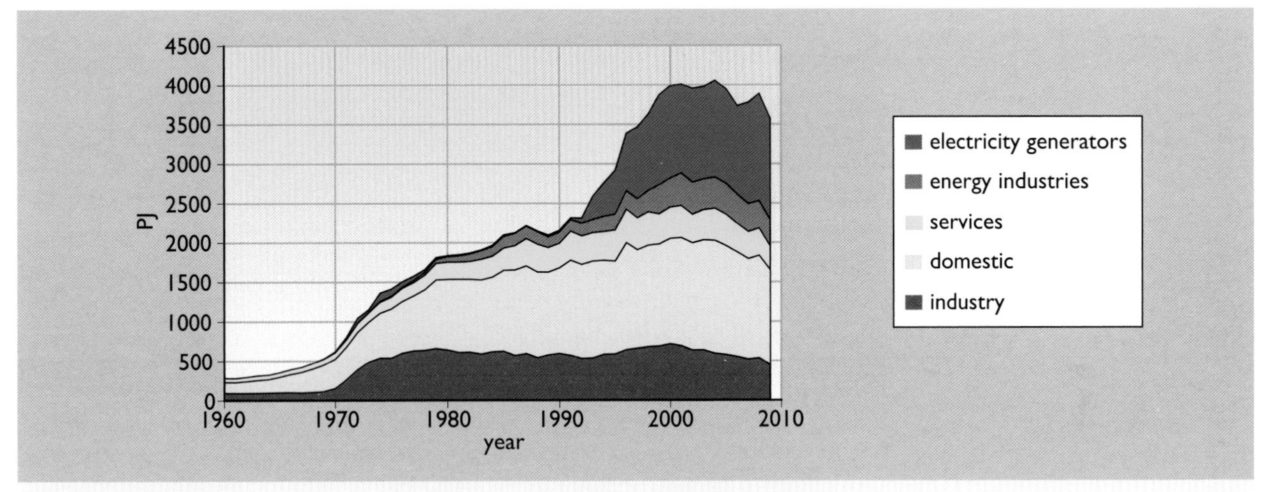

Figure 7.15 UK gas consumption, 1960–2009 (sources: DECC 2009; DECC, 2010a)

As mentioned earlier a giant non-associated gas field was discovered in 1959 in the Netherlands followed by offshore gas fields in the North Sea. These pre-dated North Sea oil discoveries by about a decade. Natural gas is about 90% methane and has about twice the calorific value of town gas. Its large-scale availability thus posed something of a problem. The choice had to be made between converting this fuel into a compatible town-gas type mixture or fitting new burners to every gas-using appliance in Great Britain to enable the use of methane. The latter course was expensive, labour-intensive and intrusive to customers' homes, but the higher calorific value of the gas meant that the existing infrastructure could carry twice as much energy as before.

And so began the North Sea gas conversion process. Over a period of ten years from 1967 about 40 million gas appliances were fitted with new burners in the homes of 13 million customers, including almost 2000 different designs of gas cookers. It cost British Gas £563 million – about £5 billion in 2009 prices. North Sea gas was sold at a lower price than town gas, and was priced to be competitive with coal for heating (see Chapter 12 for more on the relative costs).

Local gasworks were closed down and giant pipelines to distribute gas from the North Sea terminals were laid across the countryside (Figure 7.16).

Although the old local gasworks has gone, some processing of natural gas is still necessary (see Box 7.6).

Figure 7.16 A high-pressure gas pipeline being laid; such pipes can carry the equivalent of several gigawatts of power

BOX 7.6 **Sweet and sour gas**

Natural gas can't be distributed straight from the well. As pointed out earlier it may contain useful natural gas liquids that are condensed out. There may also be some ethane (C_2H_6) present. This is commonly extracted and used as a feedstock for plastics, but often a certain amount is left in the natural gas mixture.

Another useful contaminant is helium, which is widely used in medical equipment and for low temperature applications, such as superconducting magnets. The natural gas industry is the world's largest supplier.

Other contaminants are not so benign. Like town gas made from coal, **sour gas** contains sulfur in the form of hydrogen sulfide (H_2S). **Acid gas** contains CO_2. These contaminants can be removed by washing the gas with organic amine chemicals, such as monoethanolamine (MEA), which react with them. This is known as **amine scrubbing**. The amines are pumped away and then regenerated by heating, releasing the H_2S for conversion into sulfur and CO_2. The sulfur can be sold for fertilizer use and the CO_2 used for reinjection into the gas well to flush out yet more gas. Their removal also reduces the amount of corrosion to steel pipework. Gas that doesn't have these contaminants is known as **sweet gas**.

A new primary fuel had suddenly entered into the UK energy market. By 1980 gas consumption had increased sixfold. It was rapidly adopted by industry and more slowly by the domestic and services sectors as a heating fuel, almost completely replacing coal-based fuels. As pointed out in Chapter 3, by 2007 92% of UK homes had central heating and three-quarters of them used gas as the fuel.

In addition, after the oil price rises of 1973 and 1979, natural gas started to be used by the energy industries themselves, replacing coal and oil.

Then after 1990 came 'the dash for gas' in electricity generation. Prior to this an EU directive had been in force prohibiting the use of such a 'high value' fuel for such a relatively crude use. However, the availability of large quantities of gas in the North Sea and efficient generation technology in the form of Combined Cycle Gas Turbines (CCGTs), to be described in Chapter 9, changed the picture. By 2000 nearly 40% of the country's electricity was being generated from gas, mostly in brand new CCGT stations.

One consequence of using gas as a heating fuel is that the UK demand peaks strongly in the winter. The gas wells and the pipeline systems, including the international connections, thus have to cope with a highly variable flow. A large amount of **gas storage** is obviously highly desirable as well. This can be achieved by re-injecting gas into a depleted field. The UK's main storage is the offshore Rough field, off the north-east coast of England, which can hold up to 2.8 billion cubic metres of gas. This may sound a lot but is only about 3% of the UK's annual demand.

Another way of coping with peaks of demand, such as during spells of bad weather, is the use of **interruptible contracts** for industry and other large users. They are offered gas at a reduced price, but on condition that the supply can be turned off, usually with several days notice. Often these users keep their own back-up supplies of alternative fuels such as heating oil.

Since 2004 UK gas consumption has declined slightly. Declining production in the North Sea has contributed to higher prices and there has been a continuing use of coal for electricity generation. Energy efficiency campaigns for loft insulation, double glazing and domestic condensing gas boilers (Box 7.7) have also had an effect.

BOX 7.7 Domestic condensing gas boilers

Natural gas consists of 90% methane, CH_4. Its combustion (described in Box 4.9) produces a large amount of water vapour:

$$CH_4 + 2O_2 \rightarrow CO_2 + 2H_2O$$

A traditional gas boiler consists of a gas burner heating the water circulated in a central heating system. The hot flue gases are normally just vented and the hot water vapour from the combustion of the gas is lost.

One of the attractions of natural gas is that it is sufficiently clean-burning that this water vapour can be condensed, recovering its latent heat of vaporization. In a condensing boiler (Figure 7.17) this is done by making the heat exchanger sufficiently large so that the return water from the heating system, which may be at about 40 °C, can cool the flue gas to below 50 °C. This can increase the amount of heat extracted from the gas by 11% (the difference between the lower and higher heating values (LHV and HHV – see Box 4.10) and increase the overall boiler efficiency to over 90% (as calculated using the HHV).

Condensing boilers are also available for use with heating oil (gas oil) but the efficiency improvement is not quite so marked since the difference between the LHV and HHV is only about 6%.

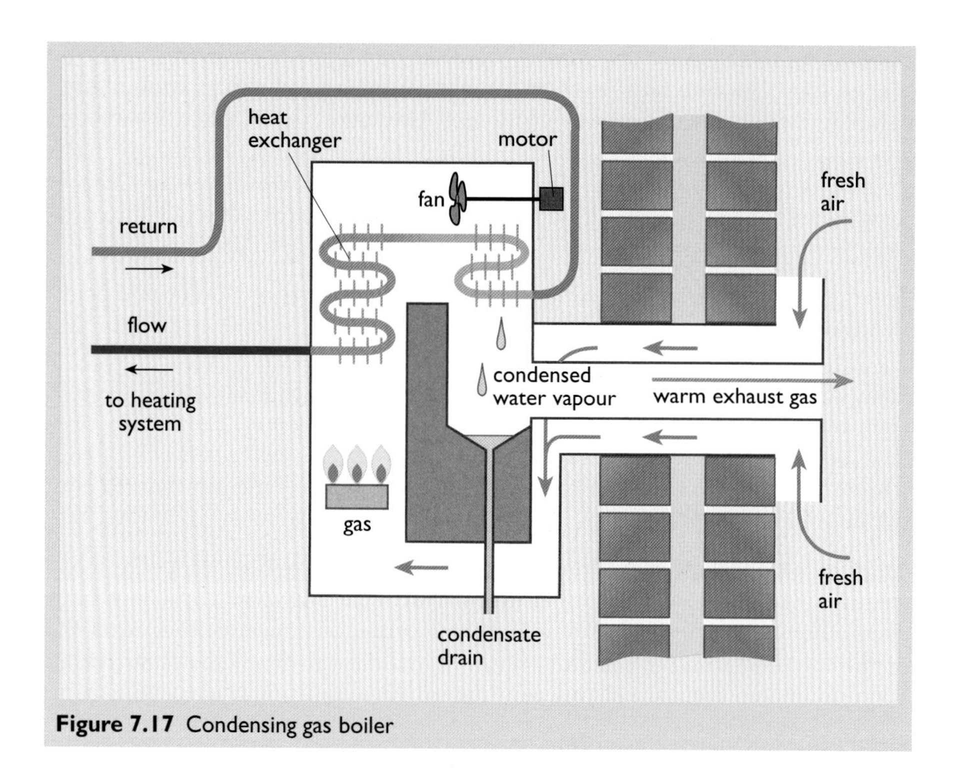

Figure 7.17 Condensing gas boiler

Global gas reserves and production

As pointed out in Section 7.3, gas as a fuel has been a latecomer to the world stage. As late as 1970 two-thirds of world production and consumption was in the USA. Since then, many other countries have developed fields and, most importantly, found ways to market the gas. Natural gas statistics are usually quoted in terms of cubic metres, which may seem slightly odd when the units should be ones of energy. Domestic gas meters also actually measure volume, even though utility bills are always expressed in terms of energy content. Box 7.8 describes the different units and their conversions.

Figure 7.18 shows the distribution of conventional natural gas reserves and production around the world.

Table 7.6 shows key statistics for the countries with the largest reserves, together with some others of interest.

Forty per cent of the world's reserves are in the Middle East. The bulk of it is in a single formation under the Persian Gulf between Qatar and Iran. The Qatar part is known as the North Dome, while the Iranian part is called South Pars. This field has only been developed since the 1970s. Because of the lack of pipelines to distribute the gas, Qatar has become a major exporter of liquefied natural gas (LNG), described in Box 7.9, as well as investing in gas-to-liquids plants to manufacture liquid fuels (see Section 7.10).

BOX 7.8 Gas units

In its earliest days 'illuminating (town) gas' was sold by volume but with a specified illuminating power. As described in Chapter 5 the invention of the gas mantle removed the need for gas to have a glowing flame. All that was necessary was that it produced heat. From the 1920s onwards town gas was measured by volume (measured at atmospheric pressure) but sold by 'the amount of heat contained'. The calorific value of samples of gas was regularly checked. This still carries on and a careful study of the fine print of a UK gas bill will reveal a statement of the calorific value, which changes minutely between successive bills.

The final distribution mixture of natural gas varies from country to country. That in the UK has a higher calorific value of about 39 MJ m^{-3}. The gas tariff, however, is expressed in pence per kilowatt hour (kWh). This makes it easier to understand and compare with electricity bills in the same units.

However, the UK has only recently switched from a unit based on the British Thermal Unit (BTU). Before 1999 the *therm* was the statutory UK unit for heat. Gas bills and even national energy statistics were expressed in these.

1 therm = 100 000 BTU = 29.3 kWh = 105.5 MJ ≈ 2.7 m^3 natural gas

In the USA the energy unit MMBTU is widely used, as is the cubic foot for gas volumes:

1 m^3 = 35.3 ft^3

1 MMBTU = 1 000 000 BTU = 293 kWh = 1.055 GJ ≈ 27 m^3 or 1000 ft^3 natural gas

At the larger scale of national and international statistics:

1000 m^3 natural gas ≈ 39 GJ = 10.8 MWh = 0.93 tonnes of oil equivalent (toe)

or in very round numbers:

1000 m^3 natural gas ≈ 40 GJ ≈ 10 MWh ≈ 1 toe.

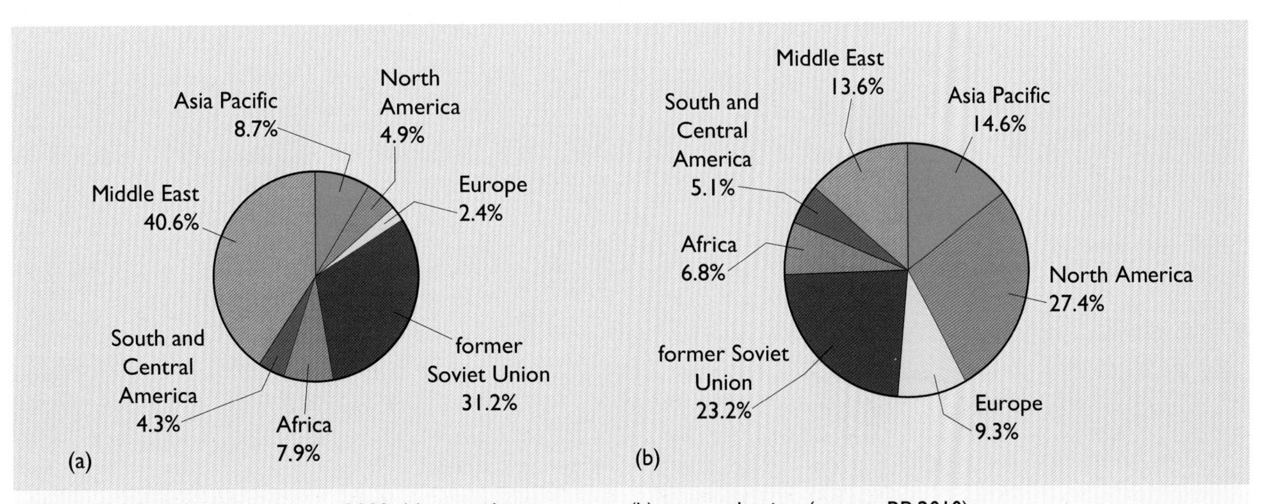

Figure 7.18 Global distribution, 2009: (a) natural gas reserves; (b) gas production (source: BP, 2010)

Table 7.6 Natural gas reserves, production and R/P ratios for selected countries, 2009

Country	Proven reserves /trillion m^3	Percentage of world reserves	Annual production /billion m^3	Percentage of world production	R/P ratio /years
Russia	44.4	23.7%	528	17.6%	84
Iran	29.6	15.8%	131	4.4%	226
Qatar	25.4	13.5%	89	3.0%	284
USA	6.9	3.7%	593	20.1%	12
China	2.5	1.3%	85	2.8%	29
Norway	2.1	1.1%	104	3.5%	20
India	1.1	0.6%	39	1.3%	28
UK	0.3	0.2%	60	2.0%	5
World	187.5	100%	2987	100%	63

Source: BP, 2010.

BOX 7.9 Liquefied natural gas (LNG)

Natural gas has a high energy density in terms of its mass (i.e. MJ per kilogramme). However, there are only two ways of getting a good energy density per unit volume, compressing it and liquefying it. Compressed natural gas (CNG) is often used in road vehicles at pressures of up to 300 atmospheres. Natural gas can be liquefied by cooling it to –162 °C. Although this is an extremely low temperature it can then be stored at atmospheric pressure. LNG has only about half the energy density of crude oil, but it can be transported in insulated tanker ships (Figure 7.19(a)) and stored in large insulated tanks, such as those at Grain in Kent (Figure 7.19(b)). Although this requires considerable capital investment in refrigeration plant and storage, it does allow cheap long-distance transport by sea.

Many impurities must be removed from the gas before cooling, not least those that would actually freeze solid at the extreme temperature. The refrigeration process of course requires energy, equivalent to about 10% of that of the final LNG.

At the receiving end further energy is required to convert the liquid back into a gas before it can be distributed. At Grain this energy comes in the form of waste heat from an adjacent gas-fired power station.

This extra energy use means that LNG has a higher overall CO_2 emission factor than natural gas produced closer to the consumer, but not sufficiently high to make it unattractive.

(a)

(b)

Figure 7.19 (a) The LNG tanker *Arctic Princess* docks at the Grain LNG terminal in Kent; (b) insulated LNG tanks at the Grain terminal

Russia has almost a quarter of the world's gas reserves. The Urengoy field in western Siberia is the world's second largest gas field and it started production in 1978. Distributing this gas has required the construction of enormous long-distance pipelines. The gas is pumped through these by gas turbines actually installed within the pipe, burning air as 'fuel' within a methane 'atmosphere'. The resulting CO_2 only slightly dilutes the gas. In 2009 Russia produced nearly 18% of the world's natural gas and it is exported by pipeline to as far afield as Greece and western Europe. The size of the European gas grid (Figure 7.20) is quite extraordinary, particularly since most of it has only been built in the past 30 years.

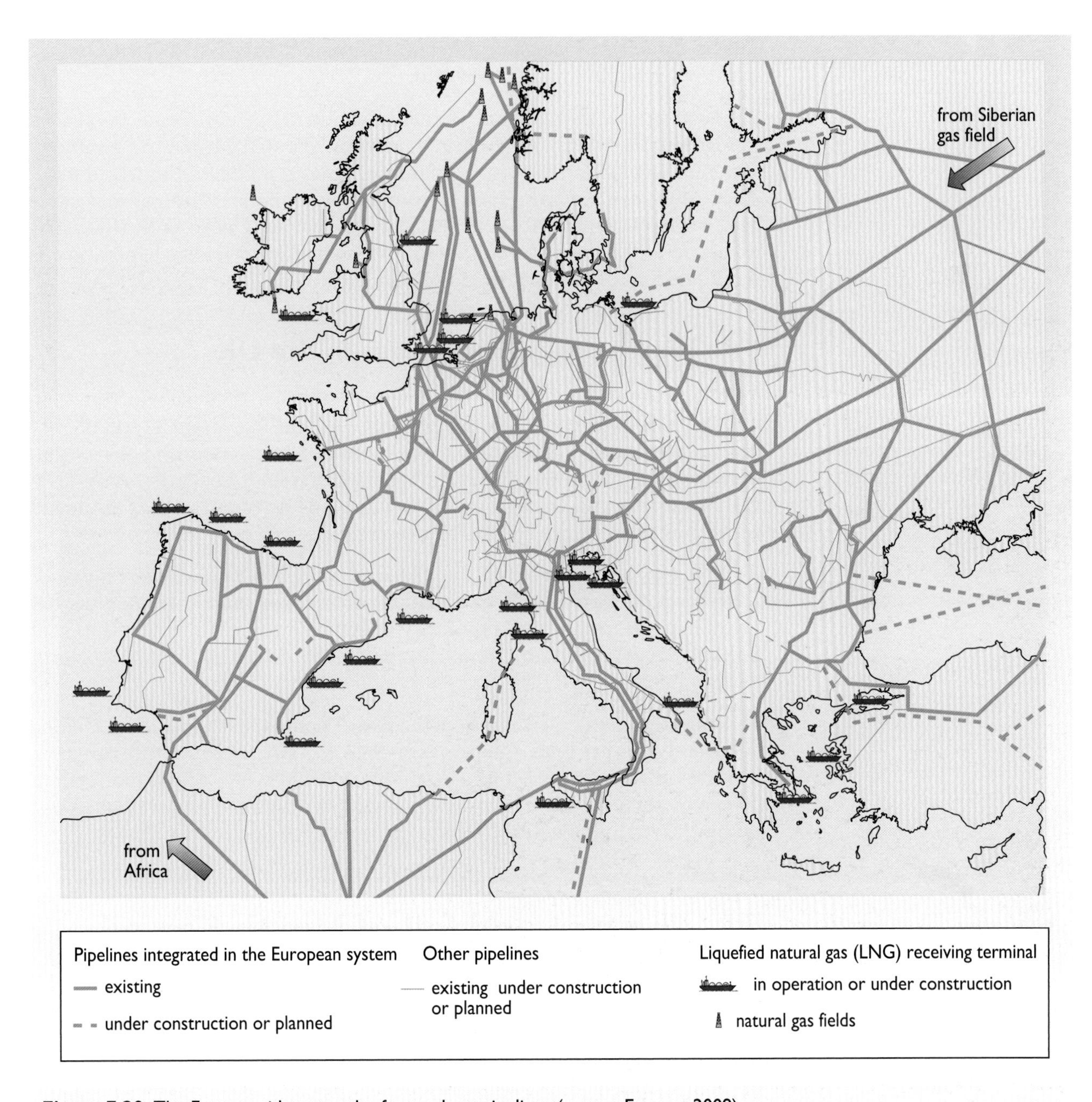

Figure 7.20 The Europe-wide network of natural gas pipelines (source: Eurogas, 2009)

In the north-east, gas flows from Siberia, in the north-west it flows from North Sea fields and in the south from fields in Algeria. In addition there are numerous LNG tanker terminals around the coasts, trading in gas from North Africa and from the Persian Gulf.

The fact that pipelines have to cross national borders creates interesting problems. The bulk of Russian gas exports to Europe have to pass through Ukraine or the neighbouring state of Belarus. There have been regular disputes between Russia, Ukraine and Belarus over the price of gas supplies to them with threats to 'turn the tap off' if payment isn't made. This is difficult since contracted gas supplies have to continue to those countries on the far side who are prepared to pay their bills. A notable new feature of the gas map is thus a new proposed pipeline under the Baltic Sea from Russia directly to Germany. When completed this will be the longest undersea gas pipeline in the world.

US reserves now make up only a small fraction of the world total. As shown in Figure 2.17, US gas production peaked in about 1971, but has remained at a high level since then, now making up about a fifth of the world total. About 40% of US production is from unconventional tight gas and shale gas (Section 7.11) (EIA, 2010a). Natural gas is widely used in the industry, domestic and services sectors and although US electricity generation is dominated by coal, gas provides about 20%.

The future prospects for gas in the UK

The UK's gas production has been in decline since 2000 (see Section 7.8), and the country has increasingly become a net gas importer since 2005. For the moment there is still plenty of gas in the adjacent Dutch and Norwegian sectors of the North Sea fields, but there has been a heavy investment in LNG terminals and storage. In 2009 LNG made up about a third of the net gas imports, and these seem likely to increase.

As with oil, the UK's commitments to reducing CO_2 emissions should see a further reduction in total gas demand as the programmes to increase the use of insulation and high-efficiency heating continue. However, these are unlikely to proceed fast enough to match the declining output of North Sea gas fields.

This puts the country, like most of the rest of Europe, in the position of being at the end of a very long pipeline from Siberia, with other supplies arriving by LNG tanker from the Middle East. This may not be very comfortable from an energy security point of view. Indeed, it is possible that by 2050 its use in the UK could have dramatically declined (see Section 14.7).

7.8 The rise and fall of North Sea oil and gas

North Sea oil has had a profound effect on the UK's energy situation since the 1970s. Other countries such as Norway, Denmark and the Netherlands have also benefited enormously from their shares of the North Sea fields.

But oil and gas fields don't last forever and it is worth reflecting on how the UK's reserves and production have changed over the past few decades.

Figure 7.21 shows a typical production pattern over the life of an oil field.

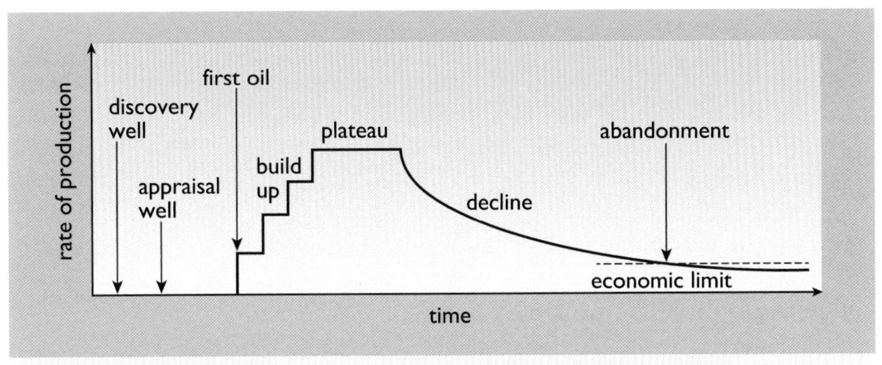

Figure 7.21 Stylized production cycle of an oil field (source: Höök, 2009)

The timescale depends on the size of the field and its location. Obviously developing oil fields in the North Sea is a lot more time-consuming than doing so on land. The first major oil discoveries in the North Sea were made late in 1969, but large-scale production did not start until about 1975. This was fortunate timing, since it cushioned the UK from the worst of the 1970s oil shocks.

There was then a build-up of production, a 'plateau' of maximum production running from about 1983 through to 1999 (see Figure 7.22 (a), which shows the same data as Figure 2.9(a)). The large dip in production around 1990 was a consequence of the Piper Alpha gas platform disaster in 1988, when 167 people died. Although this took place on a gas platform, new safety standards, including equipment modifications, were introduced for all North Sea oil *and* gas platforms.

Although it is possible that new oil discoveries might be made, UK production appears to have entered its 'decline phase' since 1999. It has roughly halved in ten years. This ten-year 'halving time' has similar properties to the 'doubling time' described in Box 1.4, and suggests a decline rate of 7% per year.

The picture for the UK's gas production is similar. Gas discoveries were made around 1965, large-scale production began in 1967 leading to an initial plateau of production between 1975 and 1990, rising to a peak in 2000 and followed by a rather more rapid decline than for oil (Figure 7.22 (b)).

The story can be filled out by looking at the reported reserves. The UK government publication *Energy in Brief* has for many years published steadily updated charts of 'remaining oil and gas reserves' and 'cumulative production' (Figures 7.22 (c) and (d)).

The reserve charts may appear slightly unusual, but the 'top line' shows the estimate in a particular year of the sum of the cumulative production to date and the remaining reserves, i.e. the total oil and gas believed to be originally in place in the ground. New discoveries give a rising trend (and just occasionally, there are downward revisions). A flattening out of the line indicates that there are no new discoveries being made. Note that these

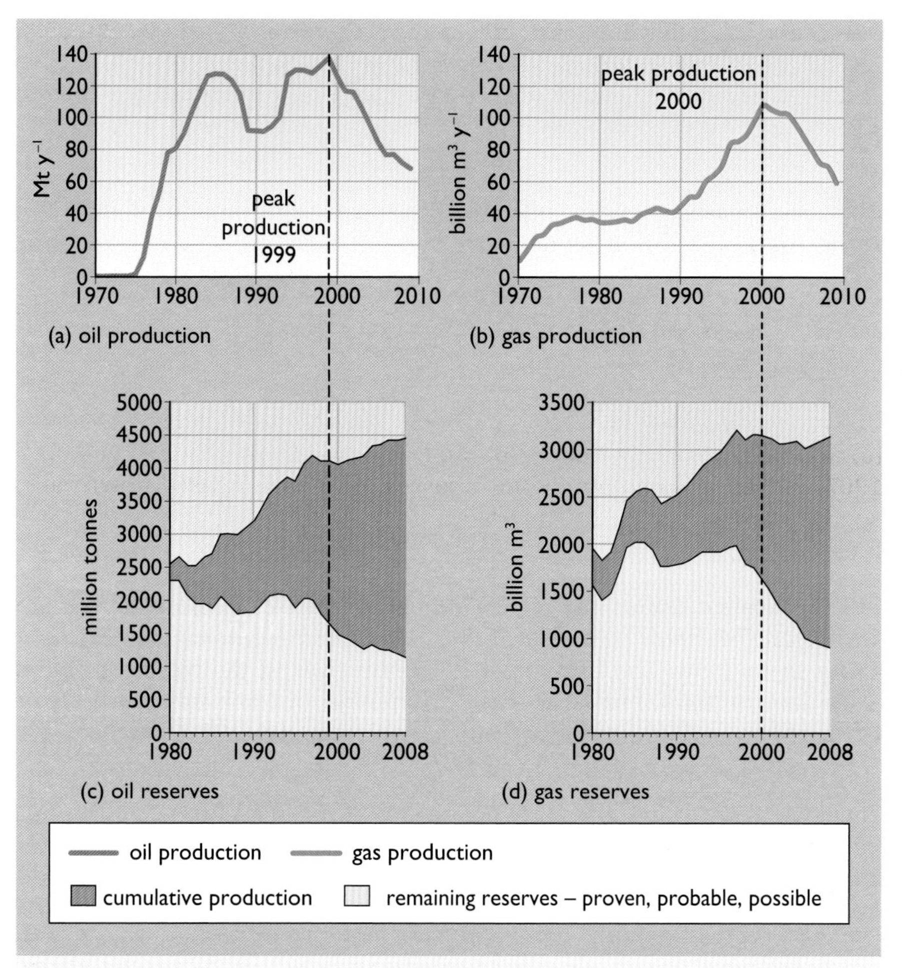

Figure 7.22 (a) UK oil production 1970–2009; (b) UK gas production 1970–2009; (c) and (d) UK remaining oil and gas reserves 1980–2008, showing years of peak production (sources: (a, b) BP, 2010; (c, d) DECC, 2010c)

reserve figures are for proven, probable and possible (i.e. 3P) reserves and thus perhaps optimistic.

The central line in the two reserves charts shows the remaining reserves, as production removes the oil and gas. If no future discoveries or reserve revisions are made then the 'top line' will just remain flat. Just how much of the 'remaining reserves' can actually be extracted remains to be seen.

The reserve charts shown only start in 1980, by which time an enormous amount of oil and gas exploration and field evaluation had taken place. By 1980 it was estimated that UK 3P total discoveries stood at around 2500 million tonnes (see Figure 7.22 (c) at the left). Large-scale production was only just starting (Figure 7.22 (a)) and by that time only about 200 million tonnes had been produced. The remaining reserves were thus 2300 million tonnes. As the years have passed, new oil discoveries have been made, but the production rate has also been increased. Between 1985 and 1997, new discoveries were being made at the same rate as oil was being extracted. By 1997 the figure for total discoveries stood at 4200 million tonnes but the remaining reserves had stayed at about 2000 million tonnes for over a decade. But after 1998 the rate of discovery slowed down. In 1999 production entered a decline phase. By 2009 the picture had become one of falling production and falling reserves. However, the falling demand, shown in Figure 7.13, has meant that the UK hasn't so far had to become a major net oil importer.

It may seem strange that the decline should start so early. In 1999 there were still 1700 million tonnes of remaining reserves. Couldn't it just be pumped out faster? The answer is 'not cheaply'. By then the first wells to be drilled were becoming seriously depleted and the reservoir pressures were falling, requiring secondary and even tertiary recovery, which is expensive and slow.

The picture for gas reserves and production is very similar (see Figure 7.22 (b) and (d)). There have been virtually no additions to the reserves since 1997, indeed at some points the reserves have been revised downwards slightly. UK gas production peaked in 2000 and has been declining since then.

This, of course, poses the immediate local question: where will the UK get its energy from in 10 or 20 years time?

On the wider world stage, the North Sea is an example of the cycle of discovery and production of both oil and gas fields:

- a rapid phase of exploration and discovery
- a rapid phase of production (to extract the maximum financial value)
- a declining phase of discovery
- falling reserves and declining production.

It is also worth reflecting on the use of the term 'reserves to production ratio' (R/P). In 1999, the year of peak production, the R/P ratio for UK oil stood at five years. This sounds as if production might be able to continue at the 1999 level and then stop completely. In 2009 the R/P ratio was much the same, six years. This might be taken to mean that the total oil reserves were in the same state as they were in 1999, or even that new oil must have been discovered. In fact the reserves and production have fallen in tandem so that the ratio between them remains approximately the same.

The topic of the state of global world oil and gas reserves and production will be revisited later in Section 7.12.

7.9 Why are oil and gas so special?

It is perhaps worth pausing at this point in the chapter to reflect on what exactly is so special about oil and gas as sources of primary energy. They:

- have a high energy density
- are clean to burn (compared to coal)
- are convenient and easy to use
- are easy to distribute/store/carry about
- are readily available and (for the moment) cheap.

High energy density – clean to burn

Chapter 4 reflected on the different ways of storing energy. Box 4.10 described the use of fuels for this purpose. A litre of diesel fuel contains about 10 kWh of chemical energy. It is sufficient to propel a car a distance of 15 km or more. In terms of energy stored per unit mass, the liquid transport fuels – diesel, aviation jet fuel and petrol – are nearly twice as good as coal and have lower CO_2 emissions (see Box 7.10). Natural gas has an even higher energy density and lower CO_2 emissions, but being a gas is only used for transport applications as compressed natural gas (CNG).

BOX 7.10 Properties of fuels

Table 7.7 gives the energy densities and specific CO_2 emissions of different fuels. Although all fuels have higher and lower heating values, in practice the higher heat values are only quoted for fuels where condensing of the flue gases is likely to take place, such as in condensing boilers. Somewhat confusingly, the efficiency of a condensing boiler running on light heating oil will be calculated using the HHV. The efficiency of a diesel engine running on DERV (i.e. the same fuel) is normally calculated using the LHV.

Table 7.7 Heating values and specific CO_2 emissions of some fuels

Fuel	Lower heating value/MJ kg^{-1}	CO_2 emissions LHV/g CO_2 MJ^{-1}	Higher heating value/MJ kg^{-1}	CO_2 emissions HHV/g CO_2 MJ^{-1}
Coal (electricity generation)	23.8	93	–	–
Fuel oil	40.8	78	–	–
DERV/Light heating oil	42.9 (9.9 kWh $litre^{-1}$)	74 (2.6 kg $litre^{-1}$)	45.6	69
Aviation jet fuel (kerosene)	43.9	72	–	–
Petrol	44.7 (9.1 kWh $litre^{-1}$)	70 (2.3 kg $litre^{-1}$)	–	–
LPG	46.0	64	49.2	59
Natural gas/CNG	47.8	57	52.8 (39.4 MJ m^{-3})	51

Source: DECC, 2010a, AEA, 2010.

Coal and fuel oil are likely to contain sulfur, typically 1%, but it can be 2% or more. In the other fuels the sulfur is likely to have been refined out, either because of emissions regulations or to prevent corrosion in the fuel distribution system.

Just how 'clean-burning' these fuels are in terms of emissions of nitrogen oxides and carbon monoxide depends on how they are burned. The topic of these emissions from car engines will be dealt with in Chapter 8.

Convenient and easy to use

Compared to lighting a coal fire, or even worse, starting up a steam engine, oil and gas are extraordinarily easy to use (and to turn off).

Ease of distribution, storage and portability

Liquid fuels can be pumped (though really heavy fuel oil may need to be heated first). Oil pipelines run for thousands of kilometres across the globe including under the sea. Oil can be transported long distances cheaply by tanker (a sea journey of 5000 km may cost only about US$1 per barrel).

A flammable gas may not at first sight seem the easiest of fuels to distribute or store, but the technology of distributing piped gas in towns is now 200 years old. Gas pipelines now run from Siberia to Europe. Where these aren't available, natural gas can be liquefied, even though the temperature required (–160 °C) is extraordinarily low.

Readily available and (for the moment) cheap

Crude oil and natural gas are globally traded commodities, as are standard petrol and diesel fuels. They have been readily available for many years. The UK has been particularly fortunate in having North Sea oil and gas during the years after the 1970s oil price rises. They are also, seen from a historical perspective, cheap (a topic which will be discussed further in Chapter 12).

Given all these advantages, it is easy to see why there is a certain unwillingness to obtain the essential 'energy services' in radically different ways. The following sections look at techniques of manufacturing 'synthetic' fuels and producing 'unconventional oil and gas', many of which have been used in times of shortage in the past.

7.10 Conversion technologies

While oil and natural gas are currently widely available, they are not uniformly available across the globe. At points in history countries have found themselves short of one key fuel, but with a plentiful supply of another. For example, during the petrol shortages of World War II town gas made from coal was available in the UK, but storing it in a suitable form for motor cars to run on was a problem (Figure 7.23).

Figure 7.23 Filling a car's gas bag with town gas during World War II. It gave a range of about 20 miles

At various times industrial chemists have been asked if it is possible to make:

- town gas from oil
- petroleum, synthetic natural gas or hydrogen from coal
- petroleum from natural gas.

Essentially the answer to all of these questions is 'yes – but at a price'.

Coal-to-liquids (CTL) in particular has been carried out on a large scale in Germany during the 1930s and 1940s and in South Africa from the 1950s to the present day. The manufacture of synthetic natural gas (SNG) from coal is carried out in the USA, and in Malaysia a gas-to-liquids (GTL) plant turns a surplus of natural gas into petroleum.

The next section contains rather a lot of chemistry, but in many cases the processes are similar. A key central step in many of them is the production of synthesis gas or syngas, a mixture of carbon monoxide and hydrogen that can be chemically manipulated to give a wide range of products.

Town gas from oil

Chapter 5 described the basic pyrolysis of coal to produce a fuel whose main ingredients were carbon monoxide and hydrogen. This process has now been swept away by the availability of cheap natural gas. However, this was not achieved in a single step. In the 1960s a process was developed to produce town gas from naphtha. This is one of the light distillates produced from the refining of oil and consists of molecules containing between five and twelve carbon atoms.

By treating these with high-pressure steam at a high temperature (around 700 °C), a process called **steam reforming**, the carbon–carbon bonds could be sheared completely, producing a gaseous mixture containing carbon monoxide and hydrogen (and a little methane). For example, starting with pentane:

$$\begin{array}{lclcl} C_5H_{12} + 5H_2O & \rightarrow & 5CO & + & 11H_2 \\ \text{pentane} + \text{steam} & \rightarrow & \text{carbon monoxide} & + & \text{hydrogen} \end{array}$$

This is one route for the manufacture of the *synthesis gas* mentioned above. It is similar to *water gas* (also carbon monoxide and hydrogen) which was mentioned in Chapter 5, being manufactured there by spraying water onto red-hot coke.

The next step in oil-based town gas production involved converting some of this mixture into methane and CO_2, a methanation process that will be described a little later. The end product was a sulfur-free substitute for town gas consisting of a mixture of hydrogen (up to 50%), methane, CO_2 and relatively small amounts of carbon monoxide and nitrogen.

In the UK, this process was used in the 1960s. It died out as natural gas was introduced into England, Wales and Scotland in the 1970s but production in Northern Ireland continued until the 1990s. Elsewhere in the world (such as in Hong Kong) it continues to this day.

Synthetic natural gas from coal

Today in most countries the gas distribution networks have been almost totally converted for dealing with natural gas rather than town gas. Thus if coal were once again to be the starting point in gas manufacture then the end point would have to be either a synthetic natural gas, almost pure methane or, given concerns over CO_2 emissions, a gas with a high hydrogen content.

Given the current widespread availability of cheap natural gas there is relatively little interest in the manufacture of methane from coal. However, a US pilot plant in North Dakota (Figure 7.24) has been operating since 1984 producing about 0.25% of the USA's natural gas from lignite from an adjacent surface mine. As with the 19th century gasworks, this late-20th century one has a range of products: chemical feedstocks; ammonium sulfate fertilizer; rare gases, such as krypton and xenon as by-products from its oxygen-making plant; CO_2, which is separated and piped to oil fields in Canada for enhanced oil recovery (US DoE, 2006).

Figure 7.24 The Great Plains Synfuels plant in North Dakota manufactures synthetic natural gas from coal

It is worth considering the processes involved since they are similar to those needed for producing large quantities of hydrogen from coal, or indeed oil from coal.

The first step in the production of either methane or hydrogen is the partial combustion of coal in a limited supply of air or oxygen. The main product is carbon monoxide:

$$\begin{array}{cccc} 2C & + O_2 & \rightarrow & 2CO \\ \text{carbon} & + \text{oxygen} & \rightarrow & \text{carbon monoxide} \end{array}$$

This is an exothermic reaction, i.e. it creates heat (see Box 7.11).

BOX 7.11 Types of chemical reaction

Chemical reactions are of two types:

exothermic reactions are accompanied by the *release of* heat energy

endothermic reactions require an *input* of heat energy in order to proceed.

The complete combustion of carbon, for example, is highly exothermic (this is what makes it a good fuel):

$$C + O_2 \rightarrow CO_2$$

carbon + oxygen → carbon dioxide

Even this reaction requires an input of heat to initiate the process.

In contrast, endothermic reactions usually need a high temperature in order to supply enough heat for the reaction to occur. They can also be *reversible*, i.e. the reaction can proceed in either direction, depending on the conditions. The chemical equation is then written with a two-way arrow. In most cases the forward reaction occurs at high temperatures while the reverse reaction occurs at low temperatures.

In practice this partial combustion also drives off the volatile matter from the coal in the same manner as happened in the 19th century gasworks described in Chapter 5. The resulting mixture is called **producer gas**. In addition to the carbon monoxide it contains useful methane and some hydrogen, but also a large number of undesired components: oils; tars; hydrogen sulfide; ammonia.

Since producer gas is easily made from coal, wood or charcoal, it was widely used to fuel petrol road vehicles during World War II (see Figure 7.25). However, if air is used as the source of oxygen, then the gas will be diluted with unwanted nitrogen.

Figure 7.25 A wartime petrol bus in the UK towing a producer gas trailer fuelled by anthracite. The gas was of poor quality because it was diluted with nitrogen in the combustion air

Modern gasification plants, such as the US one above, thus use pure oxygen for combustion, even though this means going to the lengths of liquefying the air (cooling it to almost –200 °C) to separate the oxygen from the nitrogen.

If a supply of steam is added to the oxygen then the water gas reaction can also be carried out:

$$C + H_2O \leftrightarrow CO + H_2$$

carbon + water ↔ carbon monoxide + hydrogen

This is an endothermic reaction, i.e. it *absorbs* energy and has the effect of cooling the coal down. It needs a high temperature of around 1000 °C to proceed satisfactorily. Since this is a potentially reversible reaction, it is written here with a 'two-way arrow'.

These processes are often carried out on an industrial scale in a **Lurgi gasifier**, developed in Germany in the 1930s (Figure 7.26).

This is a high-temperature, high-pressure gasifier suitable for varying grades of coal, which was originally developed for the conversion of brown coal, or lignite. In the Lurgi gasifier, coal is fed automatically into the top of the

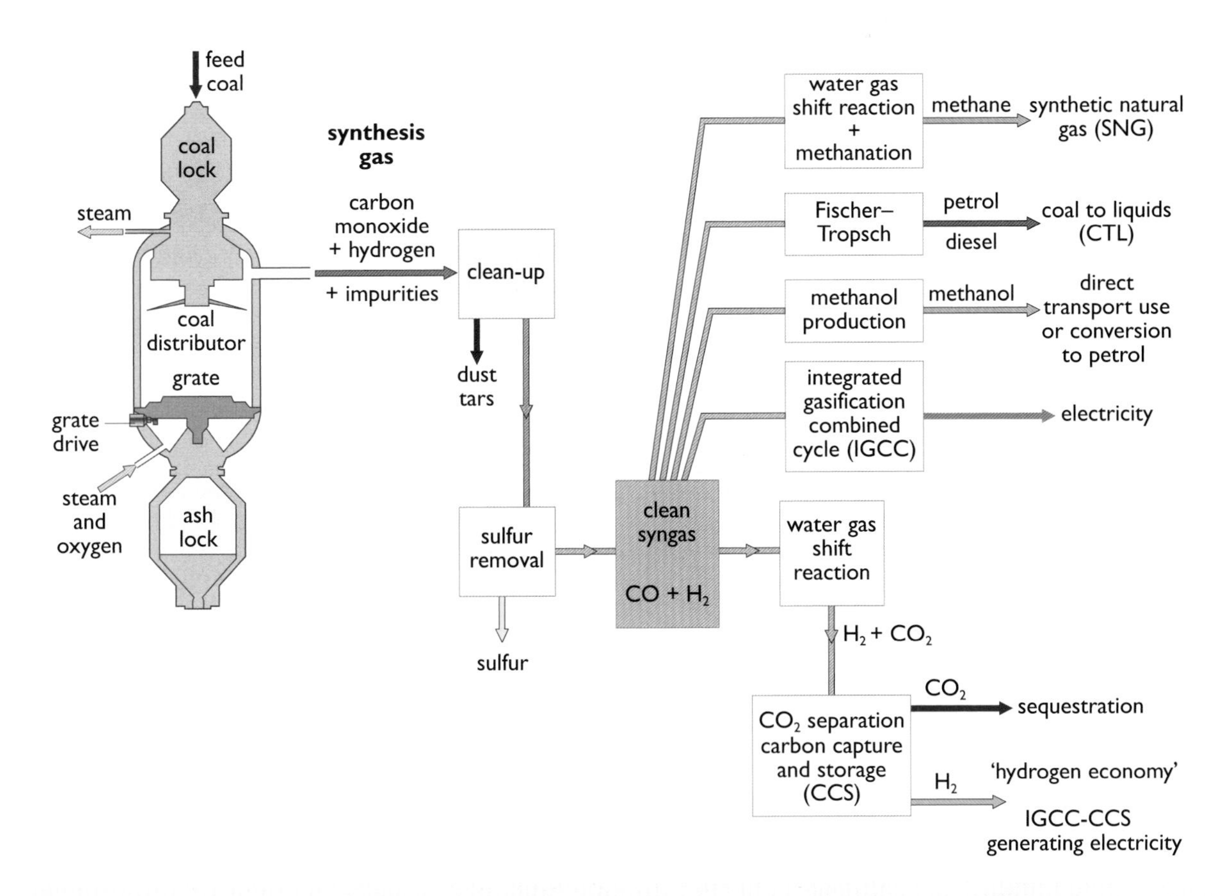

Figure 7.26 The Lurgi gasifier and the uses of synthesis gas (source: adapted from Crawford et al., 1997)

pressure vessel (through an airlock) and distributed evenly through it by a coal distributor. A mixture of steam and oxygen is introduced through a rotating grate at the bottom of the gasifier, where ash is also removed through another airlock. The conditions in the gasifier will depend on the type of coal used. In the North Dakota plant the gasifier operates at about 1300 °C and 30 atmospheres pressure.

The resulting gas is then cooled and the unwanted ammonia, tars and oils are removed. What remains is useful synthesis gas or 'syngas' (carbon monoxide and hydrogen) plus some methane and CO_2 and unwanted sulfur compounds, such as hydrogen sulfide. These sulfur compounds are removed by amine scrubbing or a similar process as described earlier in Box 7.6.

The clean synthesis gas can be used to make a whole range of possible end products as set out in Figure 7.26 some of which will be discussed further in Chapter 14.

In the route to make methane or hydrogen, the next stage is the so-called **water gas shift reaction,** in which some of the carbon monoxide is reacted with steam to produce CO_2 and hydrogen:

$$\underset{\text{carbon monoxide}}{CO} + \underset{\text{water}}{H_2O} \leftrightarrow \underset{\text{carbon dioxide}}{CO_2} + \underset{\text{hydrogen}}{H_2}$$

This is a mildly exothermic reaction but the forward process needs a temperature of around 400 °C and the assistance of a metal catalyst.

If hydrogen was the desired end product, then this forward reaction could be encouraged to make sure that as much of the carbon monoxide as possible was converted to CO_2. If methane is the desired end product then some of the carbon monoxide needs to be retained.

The CO_2 can be separated out using amine scrubbing (in practice the sulfur removal process described earlier may be combined with this at this point).

In the final stage, **methanation**, carbon monoxide is reacted with hydrogen in the presence of a nickel catalyst to produce methane and steam:

$$\begin{array}{ccccccc} CO & + & 3H_2 & \rightarrow & CH_4 & + & H_2O \\ \text{carbon monoxide} & + & \text{hydrogen} & \rightarrow & \text{methane} & + & \text{steam} \end{array}$$

The reaction is exothermic and proceeds spontaneously.

The overall effect of these three steps can be summed up in a single equation. Taking into consideration the atomic weights of the constituents, in a similar manner to Box 5.3, the overall process can be summed up thus:

$$\begin{array}{ccccccc} 2C & + & 2H_2O & \rightarrow & CH_4 & + & CO_2 \\ \text{carbon} & + & \text{steam} & \rightarrow & \text{methane} & + & \text{carbon dioxide} \\ \text{24 tonnes} & + & \text{36 tonnes} & \rightarrow & \text{16 tonnes} & + & \text{44 tonnes} \end{array}$$

Half of the original carbon in the coal ends up as methane and the rest becomes CO_2. A more concentrated fuel has been produced. Methane has a calorific value of 55.5 MJ kg^{-1} but the calorific value of pure carbon is only 32.8 MJ kg^{-1}. Taking into account the different molecular weights of carbon and methane, it would appear that the methane contains almost 90% of the energy in the original carbon. In practice the process has a limited efficiency and more coal would have to be burned (and CO_2 produced) to provide the heat to sustain the various reactions.

It is obvious that coal is not the only major input to the North Dakota plant; large amounts of water are also required. Also, every tonne of carbon input into the plant produces nearly two tonnes of CO_2. In this case much of this is piped to Canada to be used for enhanced oil recovery.

Coal to liquids (CTL)

The production of oil from coal has been practised for over 150 years, so it is not surprising that a range of different approaches have been tried. All three techniques described below were in use in Germany in the 1930s and 1940s. Coal to liquid plants produced over a half of that country's oil during World War II.

The hydrogen content of coal is typically 5% (see Figure 5.5); that of crude oil is around 10% and for transport fuels it is between 12% and 14%. Hence it is either necessary to 'remove carbon' from or 'add hydrogen' to the original source coal in order to convert it to useful liquid hydrocarbon material.

Removing carbon

Figure 7.27 Filling a car with 'Coalene' petrol, produced from coal, in 1935. Its production was encouraged to reduce the UK's reliance on imported oil

The coal distillation process described in Chapter 5 is an example of this. It produced town gas containing hydrogen, carbon monoxide and methane and a range of hydrocarbon oils, while leaving behind a residue of coke containing the 'subtracted carbon'. James Young's paraffin distillation process, described in Section 7.3, has a similar result. This type of process is known as **low-temperature carbonization** and was further developed in the 1920s and 1930s. It was used in several countries, including Germany, the UK, USA and Australia. In the UK it produced a good-quality aviation fuel for the RAF as well as a motor spirit (petrol) which was marketed as 'Coalene' (Figure 7.27). The remaining coke, which was not of a type suitable for metallurgical use, was marketed for domestic heating as 'Coalite'. In order to be economic, the process needed both a reasonable oil price and a demand for coke. In the UK, the availability of cheap North Sea oil and the decline of solid fuel heating rendered this uneconomic in the 1960s.

This technique is also used for the upgrading of heavy oils and bitumen from oil sands, see Section 7.11, where it is known as 'coking'. The remaining petroleum coke may be used for process heating or simply discarded as waste.

Adding hydrogen – direct liquefaction

The earliest process for **direct coal liquefaction** was patented by Bergius in Germany in 1913 and commercialized in the early 1920s. It involved mixing finely ground coal with coal tar oil, heating it to 300 °C or more under pressure with an iron catalyst and feeding in hydrogen, also produced from coal (Bergius, 1932). The result was a mixture of hydrocarbon liquids, essentially a synthetic crude oil (complete with many impurities). The heavier liquids were recycled for further treatment while the lighter ones could be refined into different fractions in the same way as conventional crude oil. This process is similar to the 'hydrocracking' of oil described earlier.

Large-scale production was carried out in Germany in the 1930s and 1940s. In the UK a commercial direct liquefaction plant came into operation at Billingham converting coal into 150 000 tonnes of petrol per year. Cheap post-war oil from other sources rendered this uneconomic in the 1960s. However, the high oil prices of the late 1970s revived interest within Germany in the 1980s. A plant using the Kohleoel (coal–oil) process was built at Bottrop. This was operational from 1981 to 1987 processing 200 tonnes per day of coal (DTI, 1999). More recently a large plant has been constructed in the Inner Mongolia province of China by the state-owned Shenhua Corporation.

The overall efficiency (i.e. the ratio of total energy output to energy input) of this kind of direct conversion process has been put at between 60% and 73% (Höök and Aleklett, 2010).

Adding hydrogen – the Sasol process

An alternative to direct liquefaction was first developed in Germany in the 1930s. Rather than adding hydrogen directly to coal, this first breaks the coal

Figure 7.28 A Sasol plant in South Africa

down into synthesis gas and then builds up heavier hydrocarbon molecules from it using the Fischer–Tropsch (FT) process. The process was further developed after World War II in South Africa, which has considerable coal reserves (see Chapter 5). Their first plant, Sasol 1, went into production in the mid-1950s with a capacity of 6000 barrels per day of petrol or diesel. The much larger Sasol 2 and 3 plants, completed in 1980 and 1982 respectively, were designed to produce 50 000 barrels per day (Figure 7.28).

1 barrel per day ≈ 50 tonnes per year.

The **Sasol process** starts with the gasification of coal in a Lurgi gasifier to produce a stream of synthesis gas, carbon monoxide and hydrogen, as described earlier. The second stage is the remarkable Fischer–Tropsch synthesis route by which a whole range of liquid hydrocarbon products may be obtained. This involves passing the carbon monoxide/hydrogen mixture over a catalyst, typically iron-based, at pressures of 20–30 atmospheres and temperatures between 200 °C and 350 °C (DTI, 1999). Under these conditions hydrocarbon chain molecules can be produced. At lower reaction temperatures paraffins and waxes can be made, while at higher temperatures lighter products suitable for the manufacture of petrol are produced.

Because the reaction produces a whole range of chemicals, they have to undergo subsequent refining to separate out particular fractions.

The overall process is not particularly efficient, with Sasol 2 and 3 having estimated practical efficiencies of around 55%. However, because various pollutants such as sulfur compounds and ammonia are carefully removed from the synthesis gas stream, the final petroleum product can be of very good quality.

Gas to liquids (GTL)

Liquid fuels can be easily transported by tanker. Natural gas requires pipelines or expensive liquefaction plants to manufacture LNG. There are many places in the world that have natural gas available, but which don't have the capacity to export it. Gas to liquids conversion has become increasingly attractive as oil prices have risen.

The section on SNG production above has described the methanation reaction. However, it can be persuaded to work in reverse. Methane can be converted to synthesis gas by *steam reforming*:

$$CH_4 + H_2O \rightarrow CO + 3H_2$$
$$\text{methane} + \text{steam} \rightarrow \text{carbon monoxide} + \text{hydrogen}$$

This is similar to the reaction for producing synthesis gas from pentane described in the subsection above on making town gas from oil.

One convenient transportable liquid chemical that can then be produced from this is methanol (CH_3OH). This is a valuable chemical feedstock and can be used directly as a vehicle fuel (although it is highly toxic) and it can also be further processed into petrol. This process has been used in New Zealand, where gas from the large offshore Maui field was piped ashore and made into petrol at Mobil's Methanol-To-Gasoline (MTG) plant. From the mid-1980s until 1997 (when the natural gas supplies started to run out) it produced some 15 000 barrels daily, which represented over 10% of the country's oil needs.

An alternative approach is to use the synthesis gas to feed a Fischer–Tropsch reaction, as in the Sasol CTL plants. This is done in Shell's Middle Distillate Synthesis (SMDS) process in use in Bintulu in Malaysia, capable of 14 700 barrels a day. Shell is also building a very large plant in Qatar for producing 140 000 barrels a day of liquids. Since, as can be seen from the chemical equation above, every molecule of methane requires a molecule of water to reform it, the availability of fresh water is very important (something in rather short supply in Qatar). Fortunately the overall process of gas-to-liquids is one of 'removing hydrogen' rather than adding it, as in CTL, so the water should in theory become available at the tail end of the process. A key element of the Shell plant is thus a water recovery system.

Is CTL the answer to peak oil?

The prospect of an impending 'supply crunch' in world oil supplies has renewed interest in CTL (and GTL) schemes. Their large-scale adoption poses several key questions.

Is it likely to be economic?

CTL and GTL plants are expensive, but not excessively so. There are considerable economies of scale for both, and the minimum economic size for a CTL plant is estimated to be about 30 000 bpd (EIA, 2006). A 50 000 bpd plant might cost US$3 billion (£2 billion).

One barrel of quality product (say petrol or diesel) represents about 160 litres. A production of 1 barrel per day is thus equivalent to an energy flow rate of about 66 kW continuously. The overall energy flow from a 50 000 bpd plant would be about 8 million litres of petrol a day, or over 3 GW. This is the kind of energy flow that might come (in electricity) from a large coal-fired power station and the investment, US$3 billion, is on the same scale.

One half of the capital cost is reckoned to be in the coal handling. GTL plants are slightly cheaper than CTL ones because the input fuel arrives in a more convenient gaseous form.

Studies of CTL costs suggest a break-even price of US$48–75 a barrel (Höök and Aleklett, 2010). A glance at Figure 1.1 will show that this means it would have been uneconomic from 1986 to about 2004, but of considerable interest since then. The costs of liquid fuel from the new Chinese direct conversion plant have been put much lower, at about US$24 a barrel.

While cheap coal is widely available, natural gas is more difficult to transport and GTL plants are really only economic for those who have a surplus, such as Qatar. Even then, the development of the world trade in liquefied natural gas may make direct gas sales more profitable.

What are the CO_2 emission consequences?

A litre of petrol or diesel emits CO_2 when it is eventually burned. The difficulty with CTL and GTL projects is that even more CO_2 is released in the manufacture of the fuel. The term **well-to-wheels** is used to describe the full pathway of production and use of a vehicle fuel. Fuel from CTL is likely to have at least twice the CO_2 emissions per litre of conventional fossil-derived fuel. The figure for GTL is about 50% extra. However, the indirect CTL process (using synthesis gas) does offer the possibility of separating some of the CO_2 making it available for sequestration.

Is there enough water and coal?

The indirect CTL process is one of 'adding hydrogen' and it requires a large amount of good quality water to supply it. Other water (which doesn't have to be so clean) is also used in cooling towers to cool the gases as they pass through the chemical processes and to condense steam. The water consumption of a 50 000 bpd plant is estimated to be 50 000 tonnes per day, i.e. similar to that required by a large power station. Furthermore any process water that is discharged from the plant must be free from pollutants.

World crude oil production in 2009 was about 80 million barrels per day. What would be needed if, say, 10% of that, i.e. 8 million bpd, had to come from CTL plants, and perhaps within 20 or 30 years? Setting aside the existing CTL production (about 200 000 bpd in 2009), it would require the construction of 160 plants each producing 50 000 bpd and the expenditure of nearly US$500 billion (about £300 billion). Assuming that one tonne of coal is required for every two barrels of oil produced (NPC, 2007), then the total coal requirement is 4 million tonnes per day or nearly 1.5 billion tonnes per year. This would represent an increase of about 20% on the 7 billion tonnes of coal that were mined in 2009 (see Table 5.1).

To summarize, it would seem that the large-scale adoption of CTL could have serious environmental effects on both global CO_2 emissions and water resources, and could place a considerable strain on global coal supplies. On the more positive side there are possibilities of using carbon capture and storage in indirect conversion plants.

7.11 Unconventional oil and gas

Section 7.4 has described the principles of primary, secondary and tertiary recovery from conventional oil reservoirs. Crude oil recovered by all other methods and from other sources is referred to as 'unconventional' or 'non-conventional'. Three such sources are considered here – heavy and extra-heavy oil, shale oil, and oil or tar sands, which have many features in common.

There is yet one more 'unconventional' form of oil, and that takes the form of biofuels, such as bioethanol and biodiesel. They will be discussed briefly in Chapters 8 and 14.

As for gas, 'conventional gas' is methane that can be extracted from the reservoirs shown in Figure 7.1. It may be 'associated gas', accompanying oil deposits, or separate 'non-associated' gas. Initially it flows under the reservoir pressure, but then, as with oil, secondary or tertiary recovery using injected water or CO_2 may be necessary. Four types of unconventional methane gas are considered in this section: tight gas and shale gas, coal bed methane and seabed methane hydrates. In addition underground coal gasification could provide a source of synthesis gas, carbon monoxide and hydrogen.

Heavy oil and extra-heavy oil

This can be considered to be a buried petroleum deposit which is so thick and viscous that it will not flow to the Earth's surface under natural reservoir pressure. Instead, as in tertiary recovery of lighter crude oil, it is necessary to inject steam into the production wells to heat the oil and to force it to the surface, after which it can be further processed and refined. The world's principal source of this heavy oil is the Orinoco basin of Venezuela, where there is an estimated resource of 500 billion barrels or 70 billion tonnes (USGS, 2009), i.e. larger than the oil resource of Saudi Arabia. The country has a production capacity of about 600 000 barrels per day from this source.

Much of this oil also has a high sulfur content which has to be refined out. Attempts to market the unrefined oil in the 1980s and 1990s as 'Orimulsion', an emulsion with water, as a fuel for power stations, were largely unsuccessful because of the high sulfur emissions.

Shale oil

Section 7.2 has described how organic material in sedimentary source rocks is heated and eventually matures into oil or natural gas. However, there are many deposits of hydrocarbons embedded in rocks in an 'immature' state, as a waxy solid known as **kerogen**. Shale rocks originally consisted of layers of mud, which have been compressed into flat, hard sheets. Where kerogen remains trapped between these sheets the rocks are known as **oil shales**. The overall energy content of the rock is relatively low, between 5 and 15 GJ t^{-1} (EASAC, 2007) and can be thought of as being similar to a poor quality lignite with a very high ash content.

Section 7.3 described the early use of oil shale in Scotland and how it was mined and 'retorted' in a similar manner to coal for town gas production.

Figure 7.29 Shale bings between Broxburn and Winchburgh, West Lothian, Scotland

This continued at a consumption rate of 2–3 million tonnes of shale per year from 1880 through to 1940 and at a lower level until 1962. Yields of oil per tonne of shale vary; those worked in Scotland averaged between 90 and 180 litres per tonne of shale. Even in its best years the industry only processed 3 million tonnes of shale to produce the equivalent of 5000 barrels per day of oil (about 0.3% of the UK's current oil consumption).

The industry has left a legacy of waste. Up to its final closure it processed an estimated 180 million tonnes of rock. The retorting process had the unfortunate consequence of expanding the volume of the rock, hence 'refilling the mine' was not an option. The result has been that the landscape west of Edinburgh is littered with iron-red spoil heaps covered in sparse vegetation, known locally as 'bings' (Figure 7.29).

Larger beds of oil shale have been worked in Estonia since the 1920s using both surface and underground mining. This reached a peak of about 30 million tonnes of shale per year in the early 1980s. By 2004 it had fallen to only 14 million tonnes and 80% of that was burned as power station fuel, the remainder being processed for oil. There are concerns about the CO_2 emissions of this shale, since the source rock itself contains CO_2 bound up in the form of carbonates (limestone, for example, is mainly calcium carbonate). Thus burning the shale gives an emission factor of 106 g CO_2 MJ^{-1}, i.e. 14% higher than that for coal. The total world production of oil from shale in 2005 was about 700 000 tonnes (BGR, 2009), about half of which came from Estonia, other major producers being China and Brazil.

There are other oil shale deposits in various parts of the world, but the largest by far is the Green River formation in the USA, where the borders of the states of Utah, Wyoming and Colorado come together. This has been the subject of underground pyrolysis experiments (described later in Box 7.13), but for the moment, conventional oil is considered more attractive. In Australia a pilot scheme, the Stuart Project, was started in 2000, but suspended in 2004 over concerns about its economic viability and CO_2 emissions.

Extracting oil from such sources is not easy and raises the question of the overall energy balance of the whole process, the energy return over energy invested (EROEI). This is discussed in Box 7.12.

BOX 7.12 Energy return over energy invested (EROEI)

This topic, which is also sometimes called EROI – energy return on investment – considers the overall energy benefits of a whole fuel production *system*. Obviously the processing of 'difficult' resources such as oil shales or tar sands requires energy. In its simplest form EROEI addresses the question 'in obtaining a particular useful fuel, what is the ratio of the net energy produced to energy used in the extraction process?'.

$$\text{EROEI} = \frac{\text{Net energy extracted}}{\text{Energy used}} = \frac{\text{Energy output of fuel} - \text{Energy used}}{\text{Energy used}}$$

Source: US DoE, 2007

For coal mining in the USA it has been suggested that the ratio is about 80 (Cleveland, 2005), i.e. for every 1 GJ of energy expended in mining, 80 GJ of energy can be produced in the form of coal. However, as pointed out in Chapter 5, most of the world's coal production is burned to produce electricity. Further energy has to be used in transporting the coal to the power station. This may require a journey of thousands of kilometres by train and/or ship. The example in Box 3.1 assumed that 2.5% (or 1/40th) of the energy of the coal had already been used in mining and transporting it when it arrived at the power station. Thus 80 GJ of 'useful' coal has required 2 GJ of energy expenditure. The *EROEI at the power station gate* is thus only about 40.

What about oil? In the 1930s early oil wells produced light crude oil that required a minimum of pumping and refining. The EROEI for this has been put at over 100. However, as the 'easy oil' has been used up, this figure has declined to about 20. The figures for unconventional sources are lower still, as shown in Table 7.8.

However, even this is only part of the story. As pointed out earlier, the demand is for high quality light (and low sulfur) fuels, such as petrol and DERV. Refining and upgrading the oil, and removing the sulfur, adds a further energy expenditure, lowering the EROEI yet again. A figure of 6 to 10 has been estimated for US petrol from conventional sources (Cleveland, 2005).

Table 7.8 EROEI of some unconventional oil sources

Resource and process	EROEI
US oil shale (surface)	>10
US oil shale (in-situ, non-electric heat)	6.9
US oil shale (in-situ, electric heat)	2.5
Alberta oil sands (surface)	7.2
Alberta oil sands (in-situ) (similar to heavy oil)	5.0

Source: US DoE (2007).

Oil or tar sands

In contrast to oil shale, where the kerogen is not a fully matured petroleum, deposits of so-called oil sands (or tar sands) consist of loose-grained rock bound together by heavy matured bitumen. The recovery of oil from tar sands is less difficult than from oil shales and it has thus received more commercial interest.

The world's deposits are dominated by the Athabasca area of northern Alberta in Canada. About 27 billion tonnes are considered as reserves (BGR, 2009). Of this about 16 billion tonnes may be suitable for surface mining. In 2007 production was about 75 million tonnes, of which just over half was done using mining, the rest using in-situ techniques.

Mining takes place with enormous machinery including some of the largest trucks in the world (Figure 7.30). The extracted tar/sand mixture is put into a stone crusher and hot water added. This creates a slurry that can be piped away for further treatment. This involves adding more hot water and injecting air, allowing a bitumen froth to float on top, while the heavier sand sinks to the bottom. The remaining sand/water mixture is pumped away to tailing ponds to separate out and the water can be reused. This process recovers over 90% of the bitumen (Englehart and Todirescu, 2005).

Figure 7.30 A massive digger at Athabasca. Each truck can hold up to 400 tonnes of oil sand and the tyres are nearly 4 metres in diameter

Deeper deposits can be extracted by drilling boreholes and injecting steam, allowing liquid bitumen to be extracted but with a lower recovery factor. Most of the bitumen is then sent to a special refinery where it is 'upgraded' or 'cracked' into lighter fractions as described in Section 7.5.

Water use is a key environmental issue even where water recycling is used. The net water demand for surface mining is from 2.2 m^3 to 4.4 m^3 for every cubic metre (or roughly one tonne) of bitumen. For in-situ operations it is about 0.2 m^3 to 0.3 m^3 per cubic metre of bitumen.

At present most of the energy for extracting the tar sands is provided by natural gas, which is locally in short supply. There is thus interest in various forms of underground pyrolysis where some of the oil could be burned in-situ to provide the necessary heat. This has much in common with ideas for dealing with oil shale and inaccessible coal seams, and is discussed in Box 7.13.

BOX 7.13 **Underground pyrolysis**

Mining and processing oil shale, tar sands and even coal, is a difficult and expensive process. What if these sources could be raised to a sufficiently high temperature underground to break them down and extract a stream of hot light oil and gas?

Since 1997, the Shell oil company has been carrying out small-scale experiments to this end in Colorado (the Mahogany Project). Their full scale scheme (so far unimplemented) proposes heating oil shale underground using electric heaters to between 550 °C and 750 °C for a period of *several years* in order to break down the kerogen, effectively maturing it into liquid oil and gas that could then be pumped out. The choice of a high-value fuel such as electricity may seem odd, but it allows the temperature to be tightly controlled. The problem of stopping the products leaking away sideways would be solved by setting up a *freeze wall* around the zone. This would involve drilling vertical holes all around the zone, inserting refrigeration pipes into these and then cooling the ground around them to −40 °C producing an impermeable barrier of ice (Shell, 2006). This project is obviously targeted at using nuclear electricity to produce what is essentially a transport fuel, requiring an estimated 1.2 GW of continuous heating to produce 100 000 bpd (5 Mt y^{-1} of oil (IEA, 2008)). Others have suggested using steam from concentrating solar collectors instead.

Similar thinking is involved in the experimental Toe to Heel Air Injection (THAI) process for processing tar sands. However, in this case, heat is provided by burning part of the tar sands underground. In this concept a well is drilled downwards and curved to run horizontally under the section of tar sands to be processed. This forms the 'heel'. At the far end, the 'toe', another vertical well is drilled downwards and is used to inject air into the sand. A portion of the tar sand is ignited and the heat is used both to melt the bitumen and to break it down, forcing the products towards the production well. Slowly, over many months, it is suggested that the burning zone could propagate right along the 'foot' until the whole section is heated and the oils and gas driven off.

Finally **underground coal gasification** has been tried on an industrial scale in the former Soviet Union. Two boreholes are drilled down to a coal seam. Oxygen is pumped down one borehole and the coal ignited. The partial combustion products (including carbon monoxide, methane and sulfur compounds) can be drawn out of the other borehole. If the coal can be raised to a sufficiently high temperature, then steam can be added to the oxygen to produce 'water gas', carbon monoxide and hydrogen. The technique could have possibilities for tapping currently inaccessible coal seams, such as those under the North Sea.

All of these techniques pose questions about their overall CO_2 emissions, in part because of the large heat losses into the surrounding ground and also from the breakdown of any carbonates in the source rock itself. There is also the possibility of contaminating drinking water aquifers with methane or poisonous benzene and sulfur compounds. This is one reason for the tightly controlled conditions in the Shell project. The latter two ideas also require that once the air or oxygen supply is turned off, the combustion should stop. As pointed out in Chapter 5, uncontrolled underground fires in coal seams can be extremely difficult to extinguish.

Are unconventional sources the answer to peak oil?

There are undoubtedly enormous reserves of unconventional oil in the world. However, there are considerable differences in the possible rates of production between these and conventional sources. Perhaps this is best illustrated by comparing the conventional oil gushing freely from the well in Figure 7.8 under the reservoir pressure with the digger shown laboriously shovelling tar sand into a truck in Figure 7.30 (after which the bitumen has to be washed out of the sand and then sent to an 'upgrader' to be thermally cracked). The IEA has suggested (IEA, 2010a) that by 2035:

- Canadian oil sands production might be expanded from 1.3 million bpd (65 Mt y^{-1}) in 2008 to 4.6 million bpd (230 Mt y^{-1})
- Venezuelan extra-heavy oil production might reach 2.3 million bpd (115 Mt y^{-1})
- World oil shale production might reach 0.5 million bpd (25 Mt y^{-1}).

Taken together, these only represent under 11% of current world oil production.

Tight gas and shale gas

Just as sand and shale deposits may contain oil, they can also contain natural gas tightly bound in the rock, i.e. it has not 'migrated' away to become freely flowing conventional gas in a conventional reservoir in relatively porous rock as shown in Figure 7.1. Gas trapped in sandstone rock is described as **tight gas** while that in shale (essentially thin layers of compressed mud) is known as **shale gas**. Getting gas to flow from 'tight' rock into a well is difficult. Since these are sedimentary rocks, the layers of gas-bearing sandstone and shale are likely to be horizontal. Recent developments in drilling technology building on the directional drilling shown in Figure 7.5 have allowed long *horizontal* wells to be drilled, essentially covering a considerable length of a seam (up to 3 km). The rock can then be fractured or 'fracked' by pumping in a high pressure 'fracking fluid'. This is mostly water. In shale rock a propping agent or 'proppant' such as sand is included in the fluid to hold the fractures open. However, various chemicals are also included to act as lubricants. The long drill bore effectively creates a new 'reservoir' into which methane from the surrounding rock can diffuse to be extracted. Despite its length, a single borehole can only tap the gas within a relatively small radius, so in order to get a significant gas flow large numbers of wells must be drilled. Typically about six wells might be drilled from a single 'wellpad'. The life expectancy of the wells may also be much shorter than for conventional gas wells (Stevens, 2010).

Tight gas has been extensively developed in the USA over the past 20 years and in 2009 made up 28% of US consumption (EIA, 2010a). There are also large shale gas reserves particularly in the Marcellus shales in New York state and the Barnett shales in Texas. Shale gas production has increased from virtually nothing in 1990 to about 14% of US consumption in 2009 and EIA projections suggest the figure could be 45% by 2035. A recent EIA assessment of world shale gas resources (EIA, 2011) suggested that the technically recoverable *resource* for the USA could be nearly four times as large as the country's proven gas *reserves* (as shown in Table 7.6).

In China the potential resource may be over ten times their current proven gas reserves. In Europe there may be considerable potential in France and Poland.

What of the UK? While the EIA study above suggests a resource of over 500 billion m^3, a more cautious UK estimate gives a figure of 150 billion m^3 (DECC, 2010d). However, this only represents about 6 EJ or under two years' current UK gas consumption. If sufficient viable reserves could be found, then it has been estimated that producing shale gas at a level of 10% of the UK's 2008 gas consumption, i.e. about 400 PJ y^{-1}, would require 2500–3000 horizontal wells spread over an area of 140–400 km^2, about the size of the Isle of Wight (Tyndall Centre, 2011).

As with any energy technology there are pollution issues, particularly over possible methane leakages from the many thousands of wells and contamination of drinking water aquifers by leaking fracking agents. The precise nature of these agents is currently a matter of commercial secrecy for the drilling companies. At the time of writing (May 2011), the US Environmental Protection Agency is carrying out a study of these potential problems.

Coal bed methane (CBM)

This is methane trapped in underground coal seams and which can be tapped using similar drilling techniques to shale gas. The methane is bonded to the surface of the coal, and seams may contain significantly more gas than conventional rocks. The seams are also likely to be saturated with water. Pumping out the water reduces the pressure level and allows the release of the methane. As with shale gas, there has been considerable development in the USA and in 2009 it made up 8% of US gas supplies though it is not projected by the EIA to grow beyond this. A small number of CBM projects are ongoing in the UK.

Methane hydrates

The temperature of the world's oceans decreases with depth. At depths exceeding 300–500 m, the sea floor is at about 1–2 °C and the pressure is several hundred times higher than that of the atmosphere. Under these conditions, methane can combine with water to form solid, ice-like crystalline compounds, methane hydrates (also known as methane clathrates). They can also occur in deep lake sediments and in very cold onshore permafrost zones in Arctic Canada and Russia. Current estimates are that these deposits globally may contain 1000–5000 trillion m^3 of methane, i.e. far greater than the remaining proven reserves of conventional gas. There has been interest in Japan and China in developing methods for using this methane. On the negative side, given the strong greenhouse gas potential of methane, its uncontrolled release from frozen hydrates as a result of a warming world poses a serious global hazard.

7.12 The 'spectre of peak oil'

Figure 2.6 has illustrated that world oil production after slow growth through the 1980s and 1990s may have reached some kind of plateau. The oil price has risen spectacularly since 1998 and at the time of writing (August 2011) stands at over US$100 a barrel, following political turmoil in Egypt and Libya. Two questions have to be asked:

- Has oil production actually reached its all time peak?
- Is the world actually running out of oil?

Hubbert's shocking scenario

Back in 1956 oil was cheap and oil demand was soaring, particularly in the USA. Yet a much-respected oil industry geophysicist, M. King Hubbert (Figure 7.31), chose to present a rather gloomy paper to the American Petroleum Institute in Texas (Hubbert, 1956). In it he reflected on the problems of steady exponential growth in fossil fuel use (as described in Box 1.4). He also made the extraordinary prediction that US crude oil production would peak in the early 1970s (Figure 7.32 (a)). At the time US oil production was steadily rising, with plenty of spare production

Figure 7.31 The controversial US oil geologist M. King Hubbert

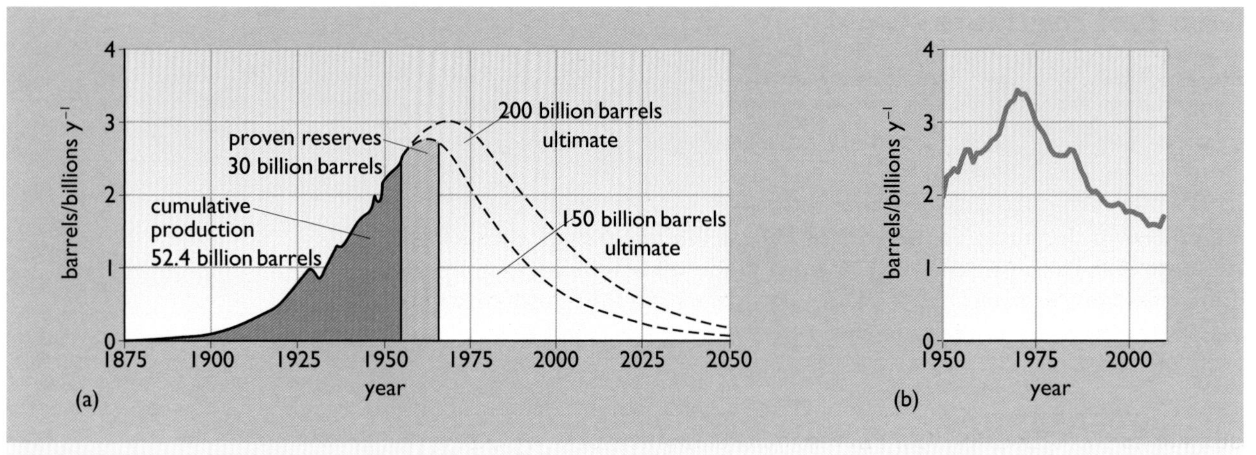

Figure 7.32 (a) Hubbert's original 1956 prediction for US oil production; (b) what actually happened (sources: (a) redrawn from Hubbert, 1956; (b) EIA, 2010b)

capacity – the future seemed assured. According to Hubbert, few had noticed that the *rate of consumption* was greater than the *rate at which new reserves were being discovered.*

In the event, Hubbert was proved correct. US oil production from all sources (including condensate and natural gas liquids) peaked in 1970 (Figure 7.32 (b)). Even allowing for new supplies from Alaska and offshore in the Gulf of Mexico, production has declined dramatically since then. This has left the USA, once a major oil exporter, in the difficult position of becoming a major oil importer.

How did Hubbert arrive at his remarkable prediction? Basically he examined in detail the figures for cumulative discovery and production from the very beginnings of the oil industry in 1859, right up to 1955. He noted that in the early years, US oil production was growing exponentially with a doubling time of 8.7 years but that after 1930 the growth rate had slowed down. He pointed out that there was only a finite amount of oil in the ground, the **ultimate recoverable resource (URR)**, as it is called today. He gave examples of the pattern of depletion of oil fields shown earlier in Figure 7.21 and suggested that overall US production might follow a symmetrical bell-shaped curve as shown in Figure 7.32(a), eventually returning close to zero around 2050. The peak production would come when one half of the total original endowment of crude oil had been used up. This peaking process has come to be known as **Hubbert's peak**.

His choice of curve was to a certain extent one of mathematical convenience given that in those pre-computer days calculations had to be done graphically on log-linear graph paper.

His chart shows both cumulative production up to 1956 and then a possible scenario as the reserves are used up, starting with the 'proven' or 1P reserves. He also had a good idea of the magnitude of the 'proven, probable and possible' (3P) reserves. But any estimate of the 'ultimate' oil in the ground must also include oil that hasn't been discovered yet, the **yet-to-find** oil. In the McKelvey diagram shown in Figure 7.7 in Box 7.3, this is on the right-hand side, undiscovered *resources*. However, they do also need to be commercially viable.

Hubbert's chart shows two suggestions for the 'ultimate' oil, one of 150 billion barrels (Gb) and the other for 200 Gb. The lower figure gives a peak in about 1962, the larger a peak in about 1968, only six years later. This illustrated one of the problems of exponential growth: a large addition to the URR would only make a few years' difference in the date of peak production.

More recently similar calculations have been carried out with the benefit of another 40 years of actual production and exploration data. These give estimates for US 'ultimate' conventional crude oil of 210–220 Gb (Campbell, 1997; Deffeyes, 2001). The cumulative production to the year of peak production, 1970, was about 90 Gb, i.e. just under half of this estimated 'ultimate' figure.

In 2000, the US Geological Survey published a much more optimistic estimate for the URR of 362 Gb. This estimate contains much larger amounts of 'reserve growth' (including tertiary recovery) and of yet-to-find oil. It is unlikely that all this 'extra' oil could ever raise US production back to its heights of 1970 production. It is more a question of whether or not the final part of Hubbert's curve, a long, further decline to almost zero in 2050, will come about quite so quickly.

First the USA ... now the world

If the calculations for the USA above are considered acceptable, then it should also be possible to carry them out for the world as a whole.

In his 1956 paper Hubbert made an estimate of 1250 Gb for the world's crude oil URR. His scenario suggested that production would increase for a while and then peak in around 2000 at a production level of about 35 million bpd. In fact world oil production continued to grow and showed little sign of slowing (see Figure 2.6). In 1969 he produced two further estimates: one for a URR of 1350 Gb with a peak at 65 million barrels per day in 1990 and another for a URR of 2100 Gb peaking at 100 million bpd in 2000.

In practice, the effects of the oil price rises of 1973 and 1979 intervened in the tidy mathematical world of smooth curves. Global oil production remained at about 60 million bpd for much of the late 1970s and early 1980s before climbing slowly to its current plateau of about 80 million bpd after 2005.

1 million bpd $\approx$ 50 Mt y^{-1}
$\approx$ 2.1 EJ y^{-1}

'Modelling the world' is not an easy task. Oil companies (and individual oil-rich states) are notoriously secretive about the true state of their oil reserves, yet at the same time want to know about those of all the rival oil companies. Out of this has come a large independent oil field database, originally owned by Petroconsultants, SA, and now by Information Handling Services (IHS). In 1998 two experienced oil analysts wrote an influential *Scientific American* article: 'The end of cheap oil', (Campbell and Laherrère, 1998) based on their work with the Petroconsultants database. This revisited Hubbert's ideas and pointed out that global oil discovery had peaked in the 1960s. They estimated that the world had used about 800 Gb of oil by then and had about another 1000 Gb to go, i.e. it was about halfway through its URR of 1800 Gb. They also pointed to the dubious nature of some of the stated reserves. Section 7.8 has described the well-documented decline of the UK oil reserves, yet many other countries have been quoting the same reserve figures for years, even in the face of continued production. Also between 1987 and 1990 six OPEC countries added a total of 300 Gb of oil to their stated reserves without claiming any discoveries of new fields. The quoting of oil reserves must thus in part be seen as a 'political' statement, rather than a purely scientific one.

More recent estimates of the remaining conventional oil range from 1100 to about 1330 Gb (IEA, 2008) for 1P or 2P *reserves*. The world figure quoted in Table 7.4 is equivalent to 1330 Gb. However, there might be another 1000 Gb of 'possible' oil in the 3P reserves or in the oil *resource*. An extra 1000 Gb would only last the world another 33 years at the current rate of consumption.

But really the picture should be seen more as one of 'oil at a price'. There is potential to coax more oil out of existing reservoirs using enhanced oil recovery (EOR) or to extract from tar sands and heavy oil. It is just that this will be more expensive. Figure 7.33 gives an indication of the size of the potential oil production for conventional oil and other petroleum technologies. The seemingly large potential for synthetic gas-to-liquids (GTL) fuels and coal-to-liquids (CTL) fuels does not take into account the fact that both gas and coal will be in demand for other purposes in the future, and that their prices may rise.

Although this chart shows 'costs' of only US$10 to US$40 a barrel for production, this does not necessarily mean that the oil will necessarily be sold for this. There is a distinction to be made between a 'cost' and a 'price', which is something determined in a competitive marketplace. Oil is a finite resource and it can only ever be sold once.

However, as Campbell and Laherrère pointed out in their 1998 article, the real problem is not how much oil is left, but the *rate at which oil production can be sustained*. Global production might peak in the same way as US

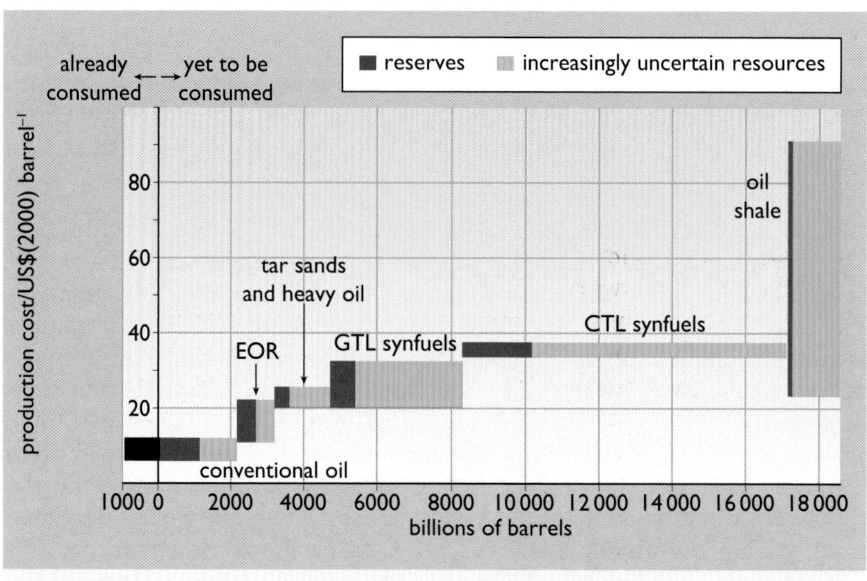

Figure 7.33 Potential for liquid hydrocarbon production (source: redrawn from Sorrell et al., 2009)

production did in 1970 and UK production in 1999 even though there was obviously still plenty of oil left in the ground.

The subject of peak oil is one that many official organizations have been unwilling to discuss. This is despite dire warnings of the effect of a supply crisis and high oil prices on the global economy (for example Hirsch et al., 2005). The 'Hirsch Report' also pointed out that the world would need 20 years' notice of a future peak in order to invest in alternative fuels, more efficient vehicles, etc. But few have perhaps wanted to emulate US President Carter, who in 1977 solemnly announced that the world was running out of oil only to find that ten years later oil was in abundance and prices had collapsed (Carter, 1977).

For many years the International Energy Agency (IEA) has produced 'growthist' scenarios of future global energy and oil use (such as that shown in Figure 1.10). It has used the economists' argument that increased demand will produce higher prices, which in turn will bring more oil to market, as indeed happened in the 1980s.

However, in 2010 in the face of rising oil prices with no obvious matching increase in production, it was forced to concede that there is a problem. The IEA's *World Energy Outlook, 2010* contains a projection of world oil production to 2035 for their 'new policies' scenario (Figure 7.34). This might be described as 'slightly growthist' suggesting a 10% increase in oil production between 2009 and 2035. At the top it shows a small but significant contribution from natural gas liquids (a by-product of gas production) and unconventional oil, both of which are shown as growing only slowly.

However, the chart does clearly show the estimated decline in production from currently producing fields (dark blue). The quantities are staggering. The estimated production from existing fields will have *halved* between 2009 and 2025, a space of only 16 years. Somehow this lost production, 35 million bpd or 1750 million tonnes per year, has to be made up from *new fields* and enhanced recovery from existing ones. This is equivalent to finding *four times* the current output of Saudi Arabia.

The chart has a grey area labelled 'fields yet to be developed' which might be brought into production. It also contains a red area of 'fields yet to be

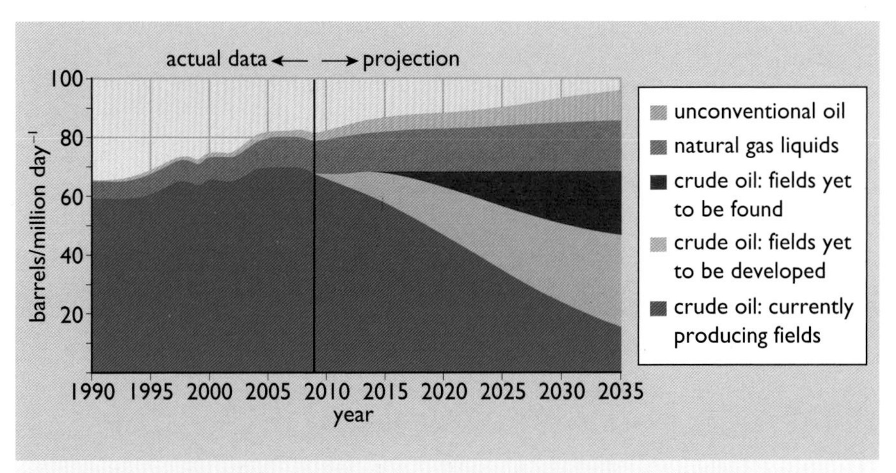

Figure 7.34 A 2010 IEA 'new policies' projection of future world oil supply (source: redrawn from IEA, 2010a)

found'. Where might these be? There are very few prospective oil-bearing areas of the world that have not been appraised. Although Siberia and Central Asia might be potential targets for exploration, to date most of the new discoveries there have been natural gas. The South China Sea, offshore West Africa, and even the Falkland Islands all remain possibilities. Almost literally, the ends of the earth have been explored for crude oil and it may be reasonable to assume that most of it has now been found. The melting Arctic ice cap has led to a rush for exploration there (as described in Chapter 1) and new discoveries have been made in very deep water off the coast of Brazil. Both of these are likely to be expensive to develop. It is likely that world production will become increasingly reliant on the existing reserves in OPEC countries. The IEA suggests that by 2035 over a half of world oil production will come from OPEC.

All that is required to turn this IEA chart from a mildly 'growthist' one to a 'peakist' one is to reduce the size of this 'yet-to-be found' area. A recent review of the literature on peak oil (Sorrell et al., 2009) concluded that there is likely to be a peak in production somewhere between 2009 and 2031. It also pointed out that large resources of conventional oil may be available, but these are unlikely to be accessed quickly and may make little difference to the timing of the peak.

Peak supply or peak demand?

The IEA's past vision of continuous growth in oil production has obviously changed. It now seems that this is neither sustainable from a supply point of view, nor from one of cutting global CO_2 emissions.

'The age of cheap oil is over' was a phrase used at the launch of their 2010 *World Energy Outlook* report (IEA, 2010b). This report offered choices. On the one hand, if nations made cuts in world oil demand in order to limit the global CO_2 concentration to 450 ppm, then the oil price might stay steady at about US$90 a barrel. On the other hand, the world could pursue a 'current policies' (i.e. business as usual) scenario (illustrated in Figure 1.10). This would mean that global oil consumption would continue to rise, and so would the projected oil price, up to US$135 a barrel in 2035. To quote the report:

> If governments act vigorously now to encourage more efficient use of oil and the development of alternatives, then demand for oil might begin to ease quite soon and we might see a fairly early peak in oil production. That peak would not be caused by any resource constraint. But if governments do nothing or little more than at present, then demand will continue to increase, the economic burden of oil use will grow, vulnerability to supply disruptions will increase and the global environment will suffer serious damage. The peak in oil production will come then not as an invited guest, but as the spectre at the feast.
>
> IEA, 2010a

What is the future for natural gas?

The situation is only slightly better for natural gas. As can be seen from Table 7.6, the world reserve/production ratio is 63 years. There are large amounts of conventional gas around the world, but often they are in places beyond the reach of the current pipeline network. The offshore oil and gas fields in Nigeria are a good example. These fields await the construction of liquefied natural gas plants to bring the gas to market.

Obviously any major increase in gas-to-liquids production as a result of an oil shortage will lead to an even faster eventual depletion of gas reserves. The topic of 'peak gas' has not been studied with the same urgency as 'peak oil' and it is not clear how much supplies of shale gas and other 'unconventional gas' will help. One suggestion is that global natural gas production might peak in around 2030 at about 15% above its current (2009) level (Laherrère, 2001).

Hubbert's solution

What was Hubbert's answer to finite fossil fuels? In 1956 he thought it was nuclear power, particularly using thorium (see Chapters 10 and 11). In later life he changed his mind and turned to renewable energy.

7.13 Summary

Oil and gas is a wide-ranging subject. Although these two energy sources might at first sight seem to be totally different, they share many features in common. As described in Section 7.2, they are both formed by the decay of layers of dead marine animals and plants over millions of years in sedimentary rock.

Unlike coal, 'rock-oil' was little used before 1800. Section 7.3 described the beginnings of the shale oil industry in Scotland in the early 19th century. More conventional oil production followed in the USA and Baku, now in Azerbaijan, later in that century. Oil demand for transport grew enormously during the 20th century. By the 1970s the world was so dependent on oil that the price rises of 1973 and 1979 were a considerable economic shock. In contrast, natural gas has been rather a latecomer to global energy supplies.

Section 7.4 described the processes of finding and producing petroleum and gas including primary, secondary and tertiary recovery. It also looked at the terminology of reserves and resources introduced in Chapter 5 describing 'the three Ps' of oil and gas reserves: proven, probable and possible.

Section 7.5 looked at the basic process of oil refining using fractionation to separate different distillates. It went on to the basics of cracking or upgrading the heavier fractions of oil into lighter ones. It also gave a brief description of the wide range of different petroleum products and their uses.

Section 7.6 described the current global distribution of oil reserves and production. Over half of the world's reserves are in the Middle East. It then looked at the changing pattern of oil use in the UK since 1950, including the very marked effects of the oil price rises of the 1970s. Oil has increasingly become just a transport fuel, but also, with the demise of coal on railways, the dominant transport fuel.

Section 7.7 dealt with natural gas, starting with the profound effects that the introduction of North Sea gas had on UK energy use. Initially it served as a replacement for town gas, but eventually it became the dominant heating fuel. After 1990 it also became a major fuel for electricity generation (a topic to be discussed in Chapter 9). The section then looked at global gas reserves and production. The largest reserves are in Russia and in a single enormous gas field under the Persian Gulf. Transporting these gas supplies to western Europe has required the construction of an enormous gas distribution grid and the use of liquefied natural gas.

Section 7.8 looked at the rise and fall of the UK's North Sea oil and gas production between 1965 and 2009. It is an example of an overall cycle of discovery, production and slow decline. A key point is that the decline phase started while there were still large reserves left in the ground.

Section 7.9 briefly reflected on what is so special about oil and gas. They are clean to burn (compared to coal), have a high energy density, are convenient to distribute, store and carry about, and are readily available and, for the moment, cheap. Their long-term future availability and price is perhaps a matter for concern.

The final sections of the chapter have looked at the problems of possible future scarcity. What will we do when the oil and gas runs out? This is a question that various nations have already had to face in the past and industrial chemists have been happy to provide solutions, though usually not particularly cheap ones.

Section 7.10 described the chemistry of making synthetic natural gas and oil from coal. While coal to liquids (CTL) processes are being pursued by countries with a shortage of oil, gas to liquids (GTL) processes are being pursued by those with a surplus of natural gas. Both of these technologies are likely to result in increased CO_2 emissions. However, the chemical processes involved are similar to those used in carbon capture and storage (CCS) systems to be described in Chapter 14.

Section 7.11 looked at a variety of unconventional sources of oil and gas: heavy and extra-heavy oil; shale oil; oil or tar sands; tight gas and shale gas; coalbed methane and methane hydrates. These could make up some of the world's future energy supply, but their rate of production might be limited.

Section 7.12 described the politically difficult issue of peak oil. The problem is not how much oil is left in the ground, but that of maintaining current production levels given declining production from existing oil fields. Two key questions were posed:

- Has oil production actually reached its all-time peak? – The answer is: if it hasn't already it is extremely likely to do so by 2030.
- Is the world actually running out of oil? – It would seem that 'the age of cheap oil is over', but there seems to be plenty of more expensive oil from a range of sources.

According to the IEA, if world oil demand can be cut as part of a policy to lower CO_2 emissions, then future oil prices might stay at current levels. Otherwise they are likely to rise, 'vulnerability to supply disruptions will increase and the global environment will suffer serious damage'.

References

AEA (2010) *Guidelines to Defra/ DECC's GHG Conversion Factors for Company Reporting* (Version 1.2.1 FINAL), produced by AEA for the Department of Energy and Climate Change (DECC) and the Department for Environment, Food and Rural Affairs (Defra). Updated 6 October.

Bergius, F. (1932) 'Chemical reactions under high pressure', *Nobel Lecture*, 21 May 1932; available at http://nobelprize.org (accessed 26 January 2011).

BGR (2009) *Energy Resources 2009*, Bundesanstalt für Geowissenschaften und Rohstoffe; available at http://www.bgr.bund.de (accessed 19 April 2011).

BP (2010) *BP Statistical Review of World Energy*, London, The British Petroleum Company; available at http://www.bp.com (accessed 29 July 2010).

Campbell, C. J. (1997) *The Coming Oil Crisis*, Brentwood, Multiscience Publishing and Petroconsultants S.A.

Campbell, C. J. and Laherrère, J. H. (1998) 'The end of cheap oil', *Scientific American*, vol. 3, pp. 59–65.

Carter, J. (1977) 'Address to the Nation on Energy', Miller Center of Public Affairs; available at http://millercenter.org/scripps/archive/speeches/detail/3398 (accessed 28 February 2011).

Cleveland, C. J. (2005) 'Net energy from the extraction of oil and gas in the United States', *Energy*, vol. 30, pp. 769–82.

Crawford, H. B., Eisler, W. and Strong L. (1997) *Energy Technology Handbook*, McGraw-Hill.

DECC (2009) *Historic Inland Deliveries of Petroleum Products 1870 – 2008*, Department of Energy and Climate Change; available at http://www.decc.gov.uk (accessed 31 January 2011).

DECC (2010a) *Digest of United Kingdom energy statistics (DUKES)*, Department of Energy and Climate Change; available at http://www.decc.gov.uk (accessed 31 January 2011).

DECC (2010b) *Energy Consumption in the UK*, statistical tables, Department of Energy and Climate Change; available at http://www.decc.gov.uk (accessed 31 January 2011).

DECC (2010c) *Energy in Brief*, Department of Energy and Climate Change; available at http://www.decc.gov.uk (accessed 31 January 2011).

DECC (2010d) *The Unconventional Hydrocarbon Resources of Britain's Onshore Basins – Shale Gas*, Department of Energy and Climate Change; available at https://www.og.decc.gov.uk/upstream/licensing/shalegas.pdf (accessed 12 February 2011).

Deffeyes, K. J. (2001) *Hubbert's Peak – The Impending World Oil Shortage*, Princeton, Princeton University Press.

DTI (1999) *Coal Liquefaction: Technology Status Report 010*, Department of Trade and Industry; available at http://www.berr.gov.uk/files/file18326.pdf (accessed 24 January 2011).

Drollas, L. and Greenman, J. (1989) *The Devil's Gold*, London, Duckworth.

EASAC (2007) *A Study on the EU oil shale industry – viewed in the light of Estonian experience*, European Academies Science Advisory Council; available at http://www.easac.eu (accessed 11 February 2011).

EIA (2006) 'Nonconventional Liquid Fuels' [online], US Energy Information Administration, http://www.eia.doe.gov/oiaf/aeo/otheranalysis/aeo_2006analysispapers/nlf.html (accessed 27 January 2011).

EIA (2010a) 'Annual Energy Outlook 2011 Early Release Overview', US Energy Information Administration; available at http://www.eia.doe.gov (accessed 12 February 2011).

EIA (2010b) 'Annual Energy Review 2009', US Energy Information Administration; available at http://www.eia.doe.gov/emeu/aer/contents.html (accessed 22 February 2011).

EIA (2011) 'World Shale Gas Resources: An Initial Assessment of 14 Regions Outside the United States', US Energy Information Administration; available at http://www.eia.doe.gov (accessed 8 May 2011).

Englehart, R. and Todirescu, M. (2005) *An Introduction to Development in Alberta's Oil Sands*, Alberta, University of Alberta; available at http://www.beg.utexas.edu/energyecon/thinkcorner/Alberta%20Oil%20Sands.pdf (accessed 12 February 2011).

Eurogas (2009) *Eurogas Annual Report 2008–2009*; available at http://www.eurogas.org (accessed 18 April 2011).

Farrell, A. E. and Brandt A. R. (2006) 'Risks of the oil transition', *Environmental Research Letters*, vol. 1, no. 1.

Hirsch, R. L., Bezdek, R. and Wendling, R. (2005) *Peaking of World Oil Production: Impacts, Mitigation, & Risk Management*, National Energy Technology Laboratory; available at http://www.netl.doe.gov (accessed 28 February 2011).

Höök, M. (2009) 'Depletion and decline curve analysis in crude oil production', Licentiate thesis, Uppsala University; available at http://www.tsl.uu.se/uhdsg/Personal/Mikael/Licentiat_Thesis.pdf (accessed 24 August 2011).

Höök, M. and Aleklett, K. (2010) 'A review on coal to liquid fuels and its coal consumption', *International Journal of Energy Research*, vol. 34, no. 10, pp. 848–64; available at http://dx.doi.org/10.1002/er.1596 (accessed 27 January 2011).

Hubbert, M. K. (1956) 'Nuclear energy and the fossil fuels', *Proceedings of the Spring Meeting of the American Petroleum Institute, 1956*, San Antonio, Texas, pp. 7–25; available at http://www.hubbertpeak.com/hubbert/1956/1956.pdf (accessed 28 February 2011).

IEA (2008) *World Energy Outlook 2008*, Paris, International Energy Agency; available at http://www.iea.org (accessed 26 February 2011).

IEA (2010a) *World Energy Outlook 2010*, Paris, International Energy Agency.

IEA (2010b) *World Energy Outlook 2010: presentation to the press, London, 9th November 2010*; available at http://www.iea.org (accessed 26 February 2011).

Kerr, D. (1994) *Shale Oil: Scotland. The World's Pioneering Oil Industry*, Edinburgh.

Laherrère, J. H. (2001) 'Forecasting future production for past discovery', *OPEC Seminar*, 28 September.

NPC (2007) *Coal to Liquids and Gas: Topic Paper #18*, US National Petroleum Council; available at http://www.npc.org/Study_Topic_Papers/18-TTG-Coals-to-Liquids.pdf (accessed 27 January 2011).

Shell (1966) *Petroleum Handbook*, London, Shell International Petroleum Co. Ltd.

Shell (2006) *Plan of Operations: Oil Shale Test Project*, Shell Frontier Oil and Gas Inc.; available at http://www.blm.gov/pgdata/etc/medialib/blm/co/field_offices/white_river_field/oil_shale.Par.79837.File.dat/OSTPlanofOperations.pdf (accessed 11 March 2011).

Sorrell, S., Speirs, J., Bentley, R., Brandt, A. and Miller, R. (2009) *Global Oil Depletion: An assessment of the evidence for a near-term peak in global oil production*, Report for United Kingdom Energy Research Centre; available at http://www.ukerc.ac.uk (accessed 25 February 2011).

Stevens, P. (2010) *The Shale Gas Revolution: Hype and Reality*, Chatham House; available at http://www.chathamhouse.org.uk (accessed 8 May 2011).

Tyndall Centre (2011) 'Shale Gas: a provisional assessment of climate change and environmental impacts', *Tyndall Centre Technical Reports*, Norwich, Tyndall Centre for Climate Change Research; available at http://www.tyndall.ac.uk (accessed 12 February 2011).

US DoE (2006) *Practical Experience Gained During the First Twenty Years of Operation of the Great Plains Gasification Plant and Implications for Future Projects*, US Department of Energy Office of Fossil Energy; available at http://www.netl.doe.gov (accessed 25 January 2011).

US DoE (2007) *Fact Sheet: Energy Efficiency of Strategic Unconventional Resources* [online], US Department of Energy Office of Petroleum Reserves, http://www.fossil.energy.gov/programs/reserves/npr/Energy_Efficiency_Fact_Sheet.pdf (accessed 11 February 2011).

USGS (2009) *An Estimate of Recoverable Heavy Oil Resources of the Orinoco Oil Belt, Venezuela*, US Geological Survey; available at http://pubs.usgs.gov/fs/2009/3028/pdf/FS09–3028.pdf (accessed 10 February 2011).

Chapter 8

Oil and gas engines

By Bob Everett

8.1 Introduction

Chapter 6 showed how the reciprocating steam engine developed from its rudimentary beginnings in the late 17th century to a reliable machine capable of powering factories, trains, ships and electric power stations by the end of the 19th century. This required continued effort in improvements in design and the deployment of new materials and progressively higher quality and precision in the manufacturing of components. This brought rewards in successively higher operating efficiencies. During the 1880s the steam turbine was invented (see Chapter 6), giving a much improved performance compared with the reciprocating steam engine. The steam turbine remains to this day a key component of practically all fossil-fuelled electric power stations.

But steam technology is that of the **external combustion engine**. The fuel is burnt in a furnace to heat water in a boiler producing steam which drives the engine. Today, about a quarter of the primary energy of the UK is consumed in **internal combustion engines**, where the fuel is actually burned inside the engine. Development of these had to wait for the widespread availability of suitable fuels: first coal gas for stationary engines and then petrol, kerosene and diesel fuel for transport applications.

This chapter first looks at three internal combustion engine designs which have been extensively developed over the 20th century: the petrol or spark ignition engine, the diesel or compression ignition engine, and the gas turbine or turbojet engine. Finally, it examines the Stirling engine, an external combustion alternative to the steam engine. Each of these engines is a product of a particular time. Although the petrol engine was invented in the 1880s it did not really start making an impact on transportation until the first years of the 20th century. The diesel engine was invented in the 1890s, yet was not seriously used for road vehicles until the 1930s. The gas turbine in its jet engine form was a product of the late 1930s and World War II. Its development since then has depended on the availability of new materials capable of withstanding high temperatures. The Stirling engine was used as an alternative to the steam engine right through the 19th century. Although it is still largely ignored, it could have an interesting future using high temperature steels and innovative modern designs.

8.2 The petrol or spark ignition engine

Although it was obvious by the early 19th century that the steam engine did indeed work, it wasn't clear whether or not the need to use fire to heat

Figure 8.1 Nikolaus August Otto (1832–1891) the German engineer who devised a commercially successful four-stroke internal combustion engine

water wasn't just an unnecessary complication. Why couldn't fuel be used directly inside a cylinder to drive an 'internal combustion engine'? The newly available fuel town gas (see Chapter 5) was ideal for such a device and various inventors produced machines. In 1862 a Frenchman, Alphonse Beau de Rochas, suggested that for a successful design:

(a) the mixture of fuel and air should be compressed as much as possible before ignition,

(b) the maximum expansion of the gases should be achieved after ignition.

The German engineer Nikolaus August Otto (Figure 8.1) produced an unusual engine to achieve these goals. In it gas and air were compressed in a cylinder and ignited, firing a heavy piston vertically upwards as far as it would go. On the downward stroke the piston engaged a rack and pinion, turning a flywheel as it descended. This engine was extraordinarily noisy but was much more efficient than competing designs. It was awarded a gold medal at the 1867 World Exhibition in Paris. By 1876 Otto had come up with a revolutionary new (and quieter) design, the first commercial four-stroke engine. This was immediately nicknamed the 'Silent Otto'. Although this engine ran on town gas as a fuel, and had a curious ignition device involving a small pilot flame on the outside of the engine, the principles are the same as in a modern car engine (see Box 8.1). In the first ten years more than 30 000 of these were produced by Otto's company.

BOX 8.1 Four-stroke and two-stroke engines

Modern petrol engines take two main forms, the four-stroke using the Otto cycle and the two-stroke cycle. Both of these use a piston which is driven up and down inside a cylinder and connected to the drive section by a rotating crankshaft.

At the top of a **four-stroke engine** there is a cylinder head containing a number of valves controlling the flow of gas in and out. The four 'strokes' are: induction, compression, power and exhaust, illustrated in Figure 8.2.

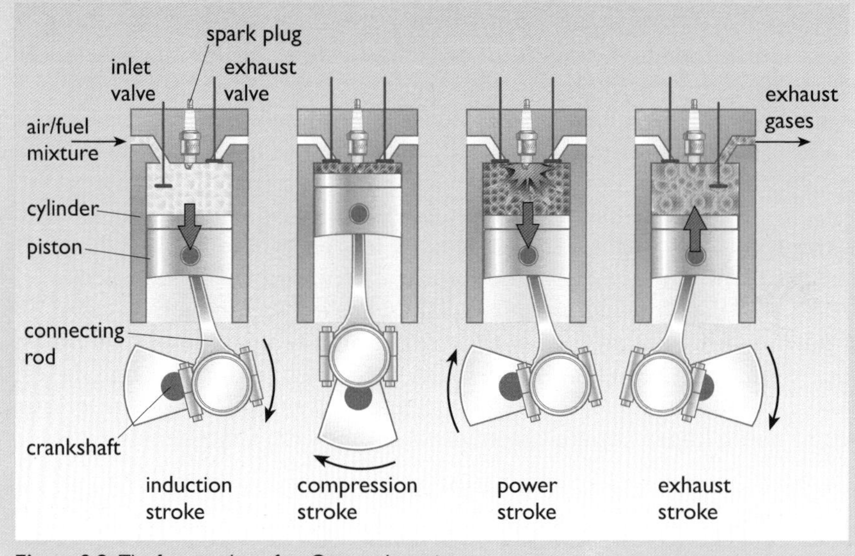

Figure 8.2 The four strokes of an Otto cycle engine

On the induction stroke a small amount of fuel and air is drawn into a cylinder through the open inlet valve, which then closes. On the next stroke this air–fuel mixture is then compressed into typically one-tenth of its original volume, creating a highly inflammable mixture which is then ignited using an electric spark on a sparking plug. The gases then burn very rapidly reaching a high temperature (750 °C or more) and expand, pushing down the piston on the power stroke. Finally, on the exhaust stroke, the burnt gases are pushed out into the exhaust system through the open exhaust valve. The whole cycle then repeats.

The reduction in volume during the second stroke is a rather critical factor called the **compression ratio**. If the volume of the cylinder is 300 cc when the piston is at the bottom of its stroke and the mixture is compressed down to only 30 cc when the piston is right at the top, then the compression ratio is 300:30 or 10:1. This is a typical figure for a modern car engine.

In a **two-stroke engine**, the engine has a sealed crankshaft casing (or crankcase) which allows the bottom side of the piston to function as a pump. The air–fuel mixture travels through the crankcase and up into the cylinder where it is burned. The cycle proceeds as follows, as shown in Figure 8.3, The burning fuel–air mixture, ignited by a spark, pushes the piston down, compressing a new batch of fuel and air in the crankcase (a). When the piston has almost reached the bottom of its travel (b), it uncovers an exhaust port in the side of the cylinder. This allows the burned exhaust gases to escape into the exhaust system. As the piston descends right to the bottom of the stroke (c), it opens a transfer port. This allows the compressed air–fuel mixture to travel from the crankcase into the working cylinder above. Then the rising piston closes off the ports and compresses the mixture. When the piston reaches the top of its stroke the mixture is ignited and the cycle repeats.

Most designs of small two-stroke petrol engines need to have oil mixed with the petrol in order to lubricate the bearings of the crankshaft. A small amount does indeed lubricate the bearings, but the rest is burned, leaving an unpleasant trail of white smoke from the exhaust pipe. Tightening emission regulations have meant that this style of engine has largely been abandoned in favour of four-stroke designs; however, its diesel equivalent is still widely used.

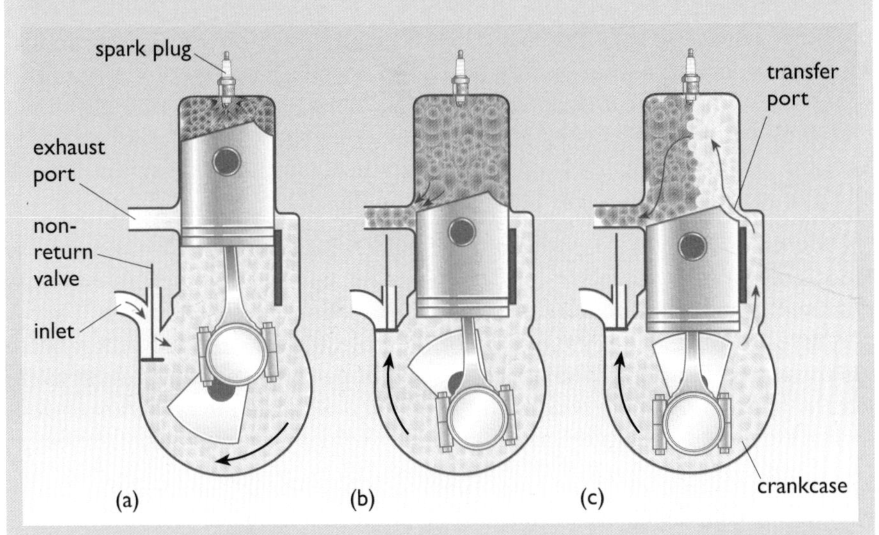

Figure 8.3 The two-stroke engine (source: adapted from Rogers and Mayhew, 1980, Figure 17.3)

Others felt that the four-stroke engine could be improved on. It was suggested that it was inefficient for the same cylinder to function as a pump on one revolution and as a working cylinder on the next. In 1878, the Scotsman Dugeld Clerk built a two-stroke engine, in which the fuel was fed in by a separate pump cylinder. In later designs the two functions were combined, so that the top side of the piston carried out the working cycle, while the bottom side acted as the pump. Although it is difficult to get this two-stroke engine (see Box 8.1) to perform as well as the four-stroke engine (both in terms of efficiency and exhaust emissions) this alternative design remained popular right through the 20th century, especially for motorbikes, and also for large ships' diesel engines.

Figure 8.4 Gottlieb Daimler (1834–1900) received his technical education in Stuttgart and then went to England to study the development of steam cars. He returned to work for Nikolaus Otto. In 1886 he tried his lightweight high-speed petrol engine first on a bicycle, and then a four-wheeled 'car'. In 1890 he founded the Daimler motor company.

The birth of the car engine

As demand for Silent Otto engines grew, Otto's factory took on two new engineers, Gottlieb Daimler (Figure 8.4) and Wilhelm Maybach. In 1882 the pair set up their own workshop outside Stuttgart where they carried out the development of a small high-speed engine to run on volatile petroleum fuel rather than town gas. At that time petrol was usually bought at chemists' shops for use as a cleaning fluid.

For this engine Maybach developed a key component, the **carburettor**. This is a device which vaporizes the petrol and delivers a precise mixture of petrol and air to the engine over a wide range of engine loads.

Daimler's engine also used a high speed of rotation, 900 rpm, rather than the Otto engine's normal 200 rpm. This is important because it is not *the absolute size of the engine* that determines its output power, so much as *the rate of throughput of fuel and air*. The output power depends on both the amount of fuel burned in the engine in a given time and its efficiency. A small engine could potentially have a large power output if it could be persuaded to burn fuel efficiently at a high rotational speed. By 1889 Daimler had produced a two-cylinder design which he started manufacturing in quantity.

There were rivals. In Germany in 1886, Karl Benz patented a three-wheeler car fitted with a petrol engine and by 1893 was producing a more practical four-wheeler with a 3 hp engine and a top speed of nearly 18 km per hour (kph). The French firm of de Dion Bouton, which had been making steam cars, also became interested in small petrol engines. Their 3.5 hp 'Petite Voiture' was the world's best-selling car in 1900 and for a while France was the world's leading car producer.

Meanwhile, in 1890, Daimler and Maybach had set up the Daimler car company. Daimler preferred to concentrate on engine manufacture for larger commercial vehicles. In 1904 a 24 hp, 34-seater, double-decker motor bus was displayed at the Crystal Palace Motor Show in the UK. By 1913 there were 2000 motor buses in service in London. The pattern was repeated in cities across the world. A new generation of petrol buses and electric trams swept away the horse buses and trams that had ruled city streets in the 1880s.

In 1899 the Daimler company received a significant order from Émile Jellinek, a wealthy Czech diplomat. He wanted a light car with a powerful engine for touring and racing. He would buy 36 cars if the first was delivered by October 1900 and if the car was named after his daughter, Mercedes.

The German Daimler factory used the name Mercedes for all its passenger cars from then on.

This new car was a great success and won its first race at an average speed of 56 kph. It was not cheap, costing £2000 (equivalent to about £160 000 at today's prices). Two years later came the 60 hp model, capable of over 100 kph with a touring body and over 125 kph when stripped for racing. There was the slight problem of starting the 9 litre engine with the crank-handle at the front, which required strong muscles. These were cars for the rich, and concerns about fuel efficiency or the price of petrol were largely irrelevant.

Remember 1 horsepower (hp) = 746 watts

These new cars were far faster than existing speed limits and it was necessary to convince police and local authorities to allow them to be driven at high speeds on the public highway, a battle that continues to the present day. See Box 8.2.

BOX 8.2 Power and speed

The 1903 Mercedes set the performance standards for a modern car. Today a 60 hp (45 kW) engine would be seen as 'small' and a top speed of 125 kph as 'slow', despite a speed limit of 70 miles per hour (mph) (112 kph) on UK motorways. Why is all this power needed? Why don't we still have 'Petit Voitures' with 3.5 hp engines?

The key reason is speed. Firstly, the engine must be able to accelerate the car and enable it to climb hills. This is purely a matter of moving its mass. Then it must be able to deal with rolling friction and aerodynamic drag. These involve more subtle aspects of design.

Mass and kinetic energy

The words speed and velocity are usually used interchangeably but, to be precise, speed is a scalar quantity: it only has magnitude and we are not usually interested in whether this is in any particular direction. The more scientific term velocity is a vector quantity: it has a magnitude and a direction, say due east. In practice, speeds for cars are measured in miles per hour (mi h^{-1}) or kilometres per hour (km h^{-1}). The customary scientific unit is metres per second (m s^{-1}).

50 miles per hour = 80 kilometres per hour = 22.2 metres per second.

Every time a car is accelerated up to a particular speed, it gains *kinetic energy* due to its motion. This has to be supplied by the engine. When it is slowed down again, this kinetic energy normally has to be dissipated as *heat energy* in the brakes.

As described in Chapter 4, the kinetic energy, E_k, of an object in motion is proportional to its mass, m, and the square of its velocity, v.

$$E_k = 0.5\ mv^2$$

A mass of one kilogram travelling at a velocity of 1 m s^{-1} has a kinetic energy:

$$E_k = 0.5 \times 1 \times (1 \times 1) = 0.5 \text{ joules}$$

So a car with a mass of 1000 kg and a velocity of 30 m s^{-1} (approximately 110 km h^{-1} or 70 mi h^{-1}) would have a kinetic energy of:

$$E_k = 0.5 \times 1000 \times (30 \times 30) = 450\ 000 \text{ J} = 450 \text{ kJ}$$

The kinetic energy rises in proportion to the mass. Travelling at the same speed, a two-tonne vehicle would have twice the kinetic energy of a one-tonne vehicle. However, since it rises as the square of the speed, travelling twice as fast increases the kinetic energy by a factor of four.

If the car has been designed to reach this speed of 30 m s^{-1} in 20 seconds from a standing start, then the engine must deliver an average power to the wheels of:

$$\frac{450}{20} \text{ kJ s}^{-1} = 22.5 \text{ kJ s}^{-1} = 22.5 \text{ kW or 30 hp.}$$

This figure is critically dependent on the mass and the rate of acceleration. If the car is twice as heavy, then the average power required will rise to 45 kW. This would also be the power required if the mass stayed the same, but the car was only allowed 10 seconds to get up to speed.

Stopping the car poses an equally difficult problem. The kinetic energy has to be disposed of as heat in the brakes, and if anything it is desirable for cars to stop in the absolute minimum of time. Stopping the car in 20 seconds means converting 450 kJ of kinetic energy into heat in that time, requiring a heat dissipation rate of 22.5 kW. Doing so in 10 seconds would require 45 kW and, for an emergency stop in 5 seconds, the figure becomes 90 kW.

Put simply, the rapid acceleration of a heavy car needs a powerful engine. Stopping it needs very good brakes (something conspicuously lacking on early cars).

Climbing hills

Travelling on the flat is relatively easy, but what about climbing a hill? Here the engine must increase the *gravitational potential energy* of the car. As described in Box 4.1, the energy needed to move a mass, m, upwards through a height, H, against the gravitational pull of the Earth is mgH, where g is the acceleration due to gravity.

Suppose the one-tonne car has to climb a 1 in 20 hill, 100 metres high at a modest speed of 60 kph. The energy needed will be:

$$m \times g \times H = 1000 \times 9.81 \times 100 = 9.81 \times 10^5 \text{ J} = 981 \text{ kJ}$$

Doing this will involve travelling about 2000 m along the road. At 60 kph this would be covered in 2 minutes or 120 seconds. Thus the power required, the rate of doing work, will be:

$$\frac{981}{120} = 8.2 \text{ kJ s}^{-1} = 8.2 \text{ kW.}$$

This large power requirement is the reason that low-powered vehicles go up hills slowly (usually much to the annoyance of Mercedes owners).

Once again, mass is important. Doubling the weight of the car will double the required power to climb the hill, as will doubling the speed.

Descending a steep hill is another matter. Now the *gravitational potential energy* can be used to propel the car forwards, but it is still likely that large amounts of energy will have to be dissipated in the brakes. Disposing of 8 kW of steady heat production is no mean feat, equivalent to the output of two electric fires on each of four wheels. Unlike braking from travelling at high speed, which may only take a matter of seconds, this heat loss has to be sustained for a period of minutes. This is why it is usual to engage a low gear and use the resistance to motion of the engine to dissipate the heat.

Air resistance and rolling friction

Even when travelling on a flat road a car engine has to provide a large amount of power. At low speeds the energy produced by the engine is consumed approximately equally by: the accessories, such as the electric alternator and cooling fan; the drive train, such as the gearbox; aerodynamic drag; the rolling friction of the tyres (see Figure 8.5). At higher speeds, these last two factors dominate the energy consumption. Reducing them becomes very important for determining the engine power required to achieve a given speed, and overall fuel efficiency.

Aerodynamic drag is produced as the car moves forwards pushing against the air in its path. A certain amount of previously stationary air will be trapped by the front of the car, accelerated up to the same speed and pushed out of the way sideways. It will be given kinetic energy as a result, which ultimately has to come from the car engine.

Since it is a matter of kinetic energy, aerodynamic drag tends to vary with the square or even the cube of the velocity and becomes very important at high speeds. It can be reduced by streamlining the body. Here the developing car industry needed input from the infant aircraft industry.

The degree of success in streamlining is indicated by a car's **drag coefficient**. A cube, for example, has a high drag coefficient of over 1. An aerofoil shape like the wing of a plane can have a value of 0.05. A well-designed modern car will achieve a figure of about 0.3. Much of the pioneering work on car body streamlining was carried out by Paul Jaray in the wind tunnel of the German Zeppelin Airship Works between 1914 and 1923. His work led to the rounded shape of the Volkswagen Beetle, first produced as a prototype in the 1930s.

Rolling friction can be reduced by good design of wheel bearings, but mostly by the design of the tyres. Radial-ply tyres, introduced in 1949, have a lower rolling resistance than their cross-ply predecessors.

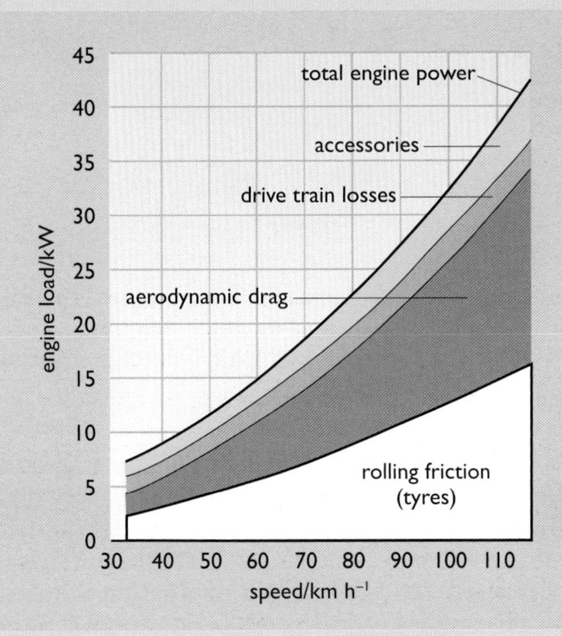

Figure 8.5 Power requirements and speed for a typical small car

Overall, a 'Petit Voiture' car engine of 3.5 hp is likely to be able to propel you to the supermarket at a leisurely 20 kph. If, however, you want to travel at 120 kilometres per hour *and* do so up hills *and* be able to accelerate to overtake the car in front, you will want something more powerful, more in the 100 hp/75 kW class.

The motorization of the USA

It was in the USA that sales of the petrol-engined motor car really took off, particularly in New York and California. Initially the USA imported large numbers of cars from Europe, particularly from France. Then, in 1908 Henry Ford sold his first Model T for US$850 (Figure 8.6). He believed in simple styling for mass sales. 'A customer can have a car in any color as long as it is black', he wrote (Ford, 1922). In 1913 he installed the first moving production line in one of his factories. Within two years it had cut assembly time per car from 12 hours to 93 minutes. Prices fell and sales soared. By 1923 sales of this model alone reached one million per year, but there was no shortage of competition from other manufacturers.

(a)

(b)

Figure 8.6 (a) Henry Ford (1863–1947) was born on a farm in Michigan. In 1896 he produced a petrol-engined quadricycle and went on to produce racing cars, even driving them himself. In 1903 he founded the Ford Motor Company, producing the Model T car in 1908 and introducing production line manufacture in 1913. The Ford company had a stormy history of labour relations into the 1940s because of his authoritarian management style; (b) The Model T Ford.

The effect on US society was dramatic. By 1923 Los Angeles and Salt Lake City had reached a level of car ownership of one to every three people. By 1925 motor vehicle production had become the largest industry in the USA and most cars sold were replacements, not first time purchases. In that year there were 153 cars per 1000 head of US population compared to only 13 in the UK. The 'car culture' of New York and California rapidly spread to the rest of the USA over the following decades. The continuing rise in car ownership since 1920 can be seen in Figure 8.7.

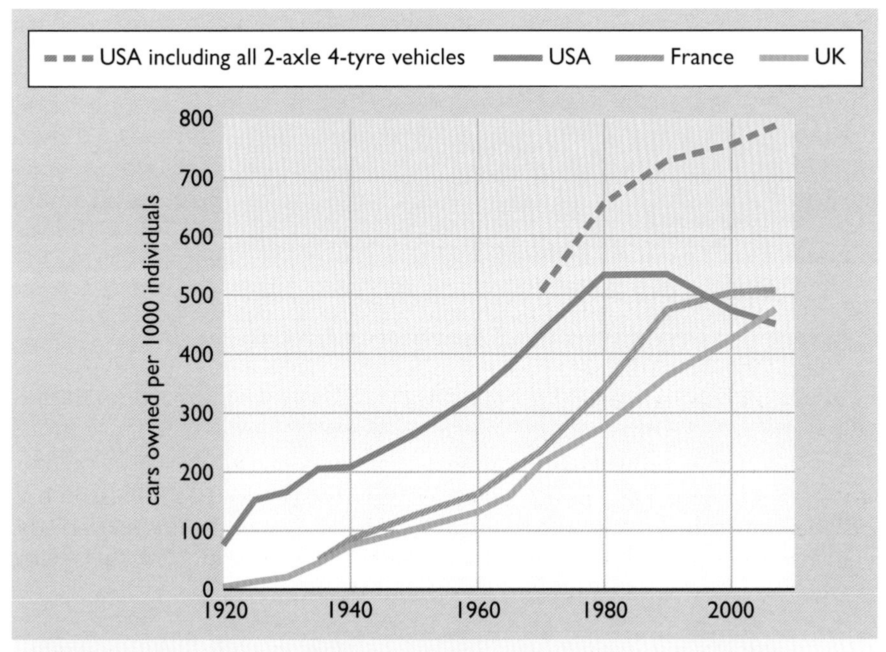

Figure 8.7 Relative car ownership for the USA, the UK and France 1920–2007 (sources: McShane, 1997; EC, 2000; EC, 2009; US BTS, 2010)

In the UK and France car ownership has increased following US figures but with a lag of about 35 years. Perhaps fortunately, it took until the 1970s for UK car ownership to reach that of the USA in the early 1930s. Recent figures might be interpreted as showing that US car ownership has been declining since about 1980. However, this may simply reflect a rather tight definition of a 'car'. Ownership of all vehicles with four tyres and two axles, including 'sports utility vehicles' or 'SUVs' continues to rise. In the USA there are now more than three vehicles to every four people. Elsewhere, car ownership may be very low. In India in 2004 it was under 10 cars per 1000 people. In China the figure for 2008 was only 29, but car production is so large that this number had doubled in only four years.

Aircraft petrol engines

The development of heavier-than-air flight required very light and powerful engines. The Wright brothers built their own lightweight 12 hp, four-cylinder petrol engine for their 1903 flight. New more powerful designs rapidly followed.

Aircraft piston engines have to deal with a particular problem not shared with cars. The higher they fly, the thinner the air becomes and this reduces the power available. The answer to this was to fit a **supercharger**, an engine-driven pump which compressed the air before feeding it into the engine. In this way engine performance could be maintained up to heights of 10 000 metres or more. An alternative way of compressing the air was to use a small turbine or **turbocharger** driven by the hot exhaust gases.

While supercharging an engine was initially developed for aircraft, it could also be used on the ground, allowing more air and fuel to be pumped through a conventional engine, increasing its power rating and its power to weight ratio. Turbochargers are now commonly fitted to many modern car engines.

Throughout the 1920s and 1930s, aircraft piston engines became larger and more powerful, reaching the peak of their design performance during World War II. The 12-cylinder Rolls-Royce Merlin engine which powered the Spitfire had a cylinder capacity of 27 litres and a maximum output of 1.48 MW at 3000 rpm. However, this level of performance required the use of high compression ratios and a very high grade of petrol.

Compression ratio and octane number

At the beginning of the 20th century petrol engines used low compression ratios of about 4:1. During the 1920s and 1930s, there was a continuous search for methods to improve their performance. As the years went by it became obvious that higher compression ratios allowed combustion at higher temperatures, producing more work from the same amount of fuel. Typical compression ratios in car engines in the 1930s were about 6:1 rising to 7.5:1 in the late 1940s and up to 9:1 or above today.

However, petrol must not spontaneously ignite in the engine; it must only do so when set off at the right time by the electric spark. If ignition is not synchronized with sparking, a clattering noise known as 'knocking' or 'pinking' will be produced. Worse still, it can lead to overheating and damage to the engine. The higher the compression ratio, the higher the grade of fuel required to prevent knocking. This anti-knocking capability is expressed as the '**octane rating**' of the fuel. Pure iso-octane (a hydrocarbon with eight carbon atoms in its molecule) has very good anti-knock capabilities and is given an octane rating of 100. Heptane (which has only seven carbon atoms) has very poor anti-knock properties and is given an octane rating of zero. A mixture of 80% iso-octane and 20% heptane would be assigned an octane rating of 80.

In practice commercial petrol consists of mixtures of a range of hydrocarbons, but it is characterized by its anti-knock performance or octane rating. There is practically no difference between the energy content per litre of high-octane and low-octane petrol. What is different is the ability for the fuel to be used in higher-compression engines with higher working temperatures and a higher thermodynamic efficiency. This is well worth doing. For example, raising the compression ratio of an engine from 7.5:1 to 10:1 can improve both output power and fuel efficiency by about 17%. Typically a compression ratio of 7.5:1 requires 85 octane fuel and one of 10:1 might require 100 octane fuel. In the UK modern 'premium' unleaded petrol has an octane rating of 95 and super-unleaded petrol has a rating of 98. In recent years careful design of cylinder heads and electronic engine management has allowed engines with compression ratios of over 10:1 to run on 95 octane petrol.

Lead additives

The petrol manufacturers have struggled throughout the 20th century to find ways of making 'high-octane' petrol out of low-grade crude oil. One way was to refine genuinely high-grade petrol, and if necessary, use catalytic cracking in the refinery to split long chain molecules into shorter ones, increasing the yield of volatile petrol. The other route was to find a chemical additive.

Tetraethyl lead was first introduced into petrol in 1916 in the USA by Thomas Midgley (who also has the dubious honour of inventing CFCs (chlorofluorocarbons) as refrigerants). He found that adding less than one gram of this lead compound per litre of petrol improved the octane rating by 10 to 15 points.

It also reduced engine wear, coating the hot exhaust parts in the cylinder head with a protective layer of lead oxide once it had been burned. This allowed cheaper engine components. Ultimately all the lead in the petrol would be discharged from the end of the exhaust pipe as a fine dust of lead oxide.

Although it was appreciated that lead was toxic, investigations had played down the possible environmental effects of it in the urban atmosphere. It was only in the 1960s that concern started to be expressed about possible brain damage, especially to children. During the 1980s the proportions of lead were steadily reduced and the octane rating held up by producing a better product at the oil refinery.

The general use of lead in petrol was banned in the EU and many other countries in 2000 and the UN has been seeking a worldwide ban. The introduction of totally lead-free petrol has been largely due to the need to remove other pollutants from the exhaust: the various oxides of nitrogen, collectively known as NO_x (see Section 8.4).

Ironically, the need to produce 'super' high-octane lead-free petrol has led to another pollution problem. Petrol companies initially included benzene as an octane-enhancer, but because of its toxicity its use is now severely restricted.

Alternative fuels

Although normally designed for use with petrol, spark ignition engines can use a wide range of other fuels.

Firstly, there are alternative fossil fuels which are of interest given a possible future shortage of conventional oil and petrol. *Liquefied petroleum gas* (LPG) and *compressed natural gas* (CNG) are widely used in road vehicles, and have a high octane rating. Uncompressed natural gas is used in spark ignition engines for electricity generation, particularly in combined heat and power (CHP) plants. *Methanol* (CH_3OH) can be produced from natural gas (as described in Chapter 7) and has been used as a high-octane racing fuel and a rocket fuel as well as more humble uses in bus fleets. However, its highly toxic nature poses some problems.

Secondly, there are biofuels which, if sustainably produced, could reduce global CO_2 emissions. For example, methanol has long been produced by the destructive distillation of wood (in a similar fashion to the coal gasification described in Section 5.4). Alternatively, it can be made from syngas produced by the gasification of biomass (using methods like those described in Section 7.10). *Ethanol* (C_2H_5OH) can be produced by the fermentation of sugars (it is of course the key ingredient of alcoholic drinks). It has a long history of use in car engines (Henry Ford in particular was very enthusiastic about its use) and is usually sold blended with petrol. A certain amount of petrol makes for easy starting of an engine in cold weather, and the ethanol has a conveniently high octane rating. A blend of 10% ethanol with 90% petrol is known as E10, and this is about the highest proportion that can be used in engines set up to run on conventional petrol. Beyond this, certain modifications are necessary. Since ethanol already chemically contains some oxygen, its combustion requires a different air/fuel ratio from that for conventional petrol (having the correct air/fuel ratio is essential for minimum pollution, as will be described a little later). The fuel system components may also need modification, as it can cause rotting of rubber hoses and oxidation of some metal parts. In many countries, Brazil for example, a blend of 85% ethanol with 15% petrol (E85) is on sale. In the UK, up to 5% ethanol is currently (2011) being blended with normal petrol and it is likely that this proportion will be increased in coming years. Another possible fuel is *bio-butanol* (C_4H_9OH), which can be produced from the fermentation of starch. This has closer properties to conventional petrol and it may prove easier to blend without requiring engine modifications.

Figure 8.8 Rudolf Diesel, German engineer (1858–1913). Diesel was a brilliant student, graduating with the highest-ever marks from the Munich Technical University. He patented his compression ignition engine in 1892, which used a blast of high-pressure air to inject the fuel into the cylinder. This was first exhibited at the Munich Exhibition of 1898. It has been widely used ever since and made him a millionaire. He mysteriously disappeared overboard from a ship in 1913.

8.3 The diesel or compression ignition engine

The petrol engine requires the air–fuel mixture to be ignited by a spark (or in the early days with a rapidly applied flame). The modern diesel engine sucks a full charge of air into a cylinder, where it is compressed and becomes very hot. At the top of the stroke, a small amount of fuel is injected and it spontaneously ignites, driving the engine through the rest of the cycle in the same manner as a petrol engine. In order to guarantee that the air–fuel mixture will be hot enough to ignite, the compression ratio used is higher than in a petrol engine, typically 14:1 to 22:1.

This form of engine has become synonymous with Rudolf Diesel (Figure 8.8) who patented an engine with a workable method of injecting fuel in 1892.

In a spark ignition engine, the air–fuel mixture burns rapidly. In a diesel engine it 'detonates' as a small explosion. This is why a diesel engine always produces more engine noise than the equivalent petrol engine. It is also difficult to achieve complete combustion, giving the engine a poor reputation for smoke emissions. The engine components, particularly the pistons, have to be made heavier and stronger in order to resist the rapid rise of pressure in the cylinder. Another early problem was that it was difficult to manufacture exhaust valves that would survive the high temperatures and pressures produced, so most early diesel engines adopted a two-stroke design. Even today, the largest diesel engines are two-strokes.

Injecting a measured amount of fuel into the cylinder against the high pressure of compressed air inside is difficult. Diesel used a high-pressure air blast. This required an expensive and bulky air pump and storage cylinder, which restricted the engine to stationary and marine use, and absorbed typically about 5% of the engine power. A basic design for a fuel injection pump was invented in 1910, but it presented many technical problems. This remains an expensive (and sometimes temperamental) part of a diesel engine.

Diesel did not achieve his original aim of running an engine on powdered coal dust, but his 1892 prototype, running on oil, achieved an efficiency of 26%, far higher than contemporary petrol or gas engines. Diesel engines running on oil, such as that shown in Figure 8.9, conveniently replaced small steam engines in many factories.

Note that the engine efficiencies for petrol and desel engines are usually quoted on the basis of the lower calorific value (LCV) of the fuel.

Figure 8.9 An early British 'oil engine' of the 1920s

A key feature of diesel engines is that they can be run on almost any grade of fuel as long as it can be pumped. Really thick oil might need preheating, but this could be done using the waste heat from the engine. They could even be adjusted to run on town gas by adjusting the compression ratio. Many modern engines used for electricity generation run on mixtures of natural gas and diesel fuel.

Diesel power for ships

Large diesel engines proved ideal for ships, especially since their efficiency could exceed 40%. They could easily be made more efficient than reciprocating steam engines or even steam turbines. Also fuel oil was far more convenient to handle than coal, the competing fuel. Oil could be mechanically pumped. Coal had a lower energy density and so took up more space. It also needed to be shovelled by hand, which required more manpower.

The difference in mechanical efficiency and the savings in wages were so great that it was economic to use diesel engines even though the fuel cost three or even four times as much per tonne. By 1926 over 5% of the world's shipping tonnage was powered by diesel and by 1937 the figure had exceeded 20%. Today, virtually all commercial shipping uses diesel engines.

Although as the 20th century progressed, virtually every other engine has got smaller, lighter and faster, large ships' diesel engines have remained stolidly slow and heavy. Currently, the world's largest production diesel engine is the Wärtsilä RTA-flex96C two-stroke used to power large container ships. It has 14 cylinders each with a piston nearly a metre in diameter and a stroke of 2.5 metres and weighs a mere 2300 tonnes (Wärtsilä, 2006). The maximum continuous power is nearly 109 000 hp (80 MW), at a stately speed of 102 rpm, typically connected to a large propeller nearly nine metres in diameter. A further 10 MW of electricity can be produced to run the ship by using waste heat to raise steam in a boiler to drive a steam turbine. The efficiency is over 50%.

Sadly, the fuel flexibility of large diesel engines has led to much of the world's shipping fleet using the cheapest low grade of high-sulfur oil. Although the sulfur dioxide produced by its combustion is mostly emitted far out at sea, it is a matter for international concern.

Diesel engines for road, rail and air

Small diesel engines for road vehicles did not appear until the 1920s. Even then, laws about smoke emissions meant that the two-stroke diesel engine was unacceptable. Road vehicles needed four-stroke engines and a small, reliable fuel injection pump. In 1924 the German company MAN introduced a 5 litre diesel engine for road vehicles at the Berlin Motor Show. In 1928 the British manufacturer Gardner introduced diesel engines for marine use and these were immediately experimentally fitted to buses. By 1934 most new trucks and buses in the UK were being ordered with diesel engines.

Diesel even took to the air. In Germany in the 1930s the Junkers company produced a successful two-stroke diesel aircraft engine which was used on transatlantic flights because of its good fuel efficiency. After World War II, the British English Electric company adapted the unusual weight-saving layout of this engine for their successful 'Deltic' railway locomotives produced in the early 1960s. These hauled express passenger trains at up to 160 kilometres per hour.

The replacement of steam locomotives with diesel power had a dramatic effect on the overall fuel efficiency of British Railways. In the mid-1950s the railway network accounted for about 5% of UK primary energy use, almost entirely as coal. Although many suburban passenger lines had been electrified, almost every other service used steam locomotives. These were designed for a high power to weight ratio, not fuel efficiency. An express steam locomotive on a long run might reach an efficiency of 11%. However, the figure for those used intermittently could be 5% or less, since they had to be fired up several hours before use and the boiler needed to be kept up to steam temperature even to move the smallest distance. Yet a diesel locomotive could have an efficiency of over 30% when running, and when it wasn't in use the engine was simply turned off. The potential fuel savings seemed enormous.

In 1955 British Railways announced a modernization plan involving the purchase of large numbers of diesel locomotives. This plan was implemented between 1957 and 1967 and included continuing expansion of the electrification of suburban railways. The effect was quite dramatic (see Figure 8.10). The total primary energy consumption of British Railways fell by almost three quarters. Although there had been a loss of freight and some passenger traffic to roads over this period, the bulk of this energy change was simply due to the substitution of one engine technology by another.

Since then, high-speed diesel-engine technology has continued to be developed, giving rise to the successful Intercity 125 units introduced in 1975 and now used both in the UK and Australia (see Figure 8.11). These use 2250 hp (1.7 MW) supercharged 12-cylinder four-stroke diesel engines at each end of a nine-coach train. The power is transmitted to the wheels using a diesel–electric transmission. Each engine, which has a thermal efficiency of about 40%, drives an electric generator which powers electric motors on the wheels. Although designed to operate at 125 mph (200 kph), they have reached 238 kph. The design has not been without its teething troubles, but careful study put many of the problems down to those of dissipating over 2 MW of waste heat from each engine! More recent UK diesel passenger train designs have turned to having a smaller separate engine under each carriage.

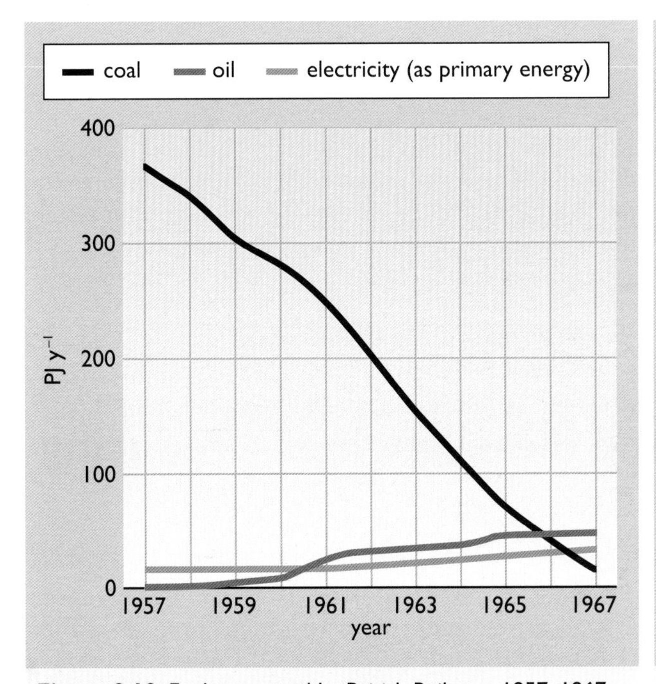

Figure 8.10 Fuel consumed by British Railways, 1957–1967 (sources: Ministry of Power, 1962; DTI, 1973)

Figure 8.11 Intercity 125 engine unit using a 2250 hp high-speed diesel engine

Back on the road, a small turbocharged diesel engine has been used for the modern 'Petit Voiture', the two-seater Smart (see Figure 8.12). This 800 cc engine can produce 41 hp (30 kW) at 4200 rpm and gives a fuel consumption of 3.4 litres per 100 km (over 80 miles to the gallon). The top speed of the car is electronically limited to 135 kph.

Figure 8.12 The Smart, manufactured by Daimler. The diesel version is capable of a fuel consumption of 3.4 litres per 100 km

DERV

Although the diesel engine can in theory run on a wide range of possible grades of fuel, concerns about clean combustion and the reliability of fuel injection pumps have made it necessary to tightly specify the fuel used for road vehicles, known as DERV (Diesel Engines for Road Vehicles). DERV is lighter than the heavy diesel oil used in ships, but heavier than the kerosene used in jet aircraft. It cannot be too thick and heavy or it may freeze in the vehicle tanks in winter. Nor can it be too light, otherwise it may not ignite properly under compression. This ability to ignite under pressure is characterized by a **cetane rating**, analogous to the octane rating of petrol. In practice modern DERV is almost identical to 'gas oil' used for domestic heating, but may contain additives to prevent freezing. Modern 'green diesel' (not to be confused with biodiesel below) also has the sulfur content refined out, both to stop potential emissions of sulfur dioxide and to allow the use of catalytic converters on engines.

Biodiesel

Biodiesel is being introduced in many countries, usually manufactured from energy crops such as oilseed rape or recycled vegetable oils. It has the advantage of being naturally low in sulfur content. The diesel engine has the potential to run on many different fuels, and there are many tales of neat cooking oil straight from the supermarket being used in cars. However, like ethanol, mentioned above, biodiesel can have a corrosive effect on some types of rubber pipework in the fuel system. Commercial biodiesel must also be compatible with fuel injection pumps designed for DERV and have the right properties to remain liquid in cold conditions. In practice commercial biodiesel is sold blended with DERV. A mixture of 5% biodiesel and 95% DERV is known as B5 and one with 20% biodiesel as B20. However, with suitable modifications to the fuel systems, higher proportions (up to B100) have been used in cars, trucks and railway locomotives.

8.4 Petrol and diesel engines – reducing pollution

When a fuel such as petrol, diesel or even coal is burned in an excess of air, the main combustion products are carbon dioxide (CO_2) and water vapour (H_2O). If there isn't enough air present, other products such as carbon monoxide (CO) and hydrogen (H_2) can be produced. Carbon monoxide is quite toxic; it combines with blood and inhibits the absorption of oxygen.

If there is sulfur in the fuel, then this will burn to form sulfur dioxide (SO_2). Diesel fuel, especially heavy fuel oil used in ships and power stations can contain considerable amounts of sulfur, 2% or more. Since sulfur dioxide is a serious contributor to problems of acid rain, it is desirable for refineries to remove as much sulfur from the fuel as possible.

The other serious pollutants are the oxides of nitrogen: nitrogen oxide (NO, commonly known as nitric oxide), nitrogen dioxide (NO_2) and dinitrogen oxide (N_2O, commonly known as nitrous oxide). In the mixture produced by car engines, typically 90% will be in the form of NO. Together they are referred to as NO_x. Like SO_2, these oxides are both an irritant to the respiratory system and a cause of acid rain. NO_x pollution can be particularly unpleasant in cities such as Los Angeles or Athens where there is plenty of sunshine. Under these conditions the NO component can combine with oxygen in the atmosphere by a photochemical reaction to produce more NO_2 and 'low-level' or 'tropospheric' ozone, O_3, giving a choking brown smog.

NO_x emissions can occur if there is any nitrogen contained in the fuel. Coal, for example, contains modest amounts in the form of ammonia compounds, which will burn to form NO. However, the most important mechanism for the formation of nitrogen oxides in engines is the 'thermal' one. Air is a mixture of gases and contains about 78% nitrogen. If the combustion temperature is hot enough, greater than 1500 °C, then this nitrogen will react with the oxygen in the air to produce nitrogen oxide.

$$N_2 + O_2 \rightarrow 2NO$$

As the mixture cools, then some of the nitrogen oxide may react with more oxygen to produce nitrogen dioxide (NO_2). Although 1500 °C may seem extraordinarily high, it is only necessary for a very small part of a flame to reach this temperature for the reaction to start. The reaction also requires a certain amount of time to take place. If the gases only reach 1500 °C for a few milliseconds, then the production of NO is much reduced.

This reaction is potentially reversible and in the catalytic converter, described below, NO can be persuaded to dissociate back into nitrogen and oxygen.

In many applications the formation of NO can be limited by very careful design of **low NO_x burners**. In these the combustion of the fuel, be it gas, oil or coal, is carefully regulated to eliminate 'hot-spots' where this reaction can take place. The general principle is to burn the fuel uniformly and keep the flame temperature down to that needed for the particular application. A domestic gas boiler, for example, only needs to heat water to 60 °C, so there is no need for a high flame temperature. A coal-fired power station needs to heat steam to 700 °C. This is a little more difficult. A gas turbine may need to run with a turbine inlet temperature of 1300 °C, which poses serious problems.

When burning fuel for an engine application there is a conflict between the need for a high temperature to give a high Carnot thermal efficiency (see Chapter 6), and avoiding high temperatures to keep the nitrogen oxide emissions down.

The situation is particularly difficult in petrol and diesel engines, where the combustion takes place in a relatively uncontrolled manner inside a cylinder and it has required many years of study and experimentation to reduce NO_x emissions to acceptable levels.

Emissions from petrol engines

Chemically, petrol consists almost entirely of carbon and hydrogen. When it is burned in air it should produce just carbon dioxide and water. Ideally, in a petrol engine the carburettor or fuel injection system should supply 14.7 parts by weight of air to one part of petrol. This is called the **stoichiometric ratio**. It is usually referred to by the Greek letter 'lambda' (λ). This 'perfect' mixture has a $\lambda = 1$ (Figure 8.13).

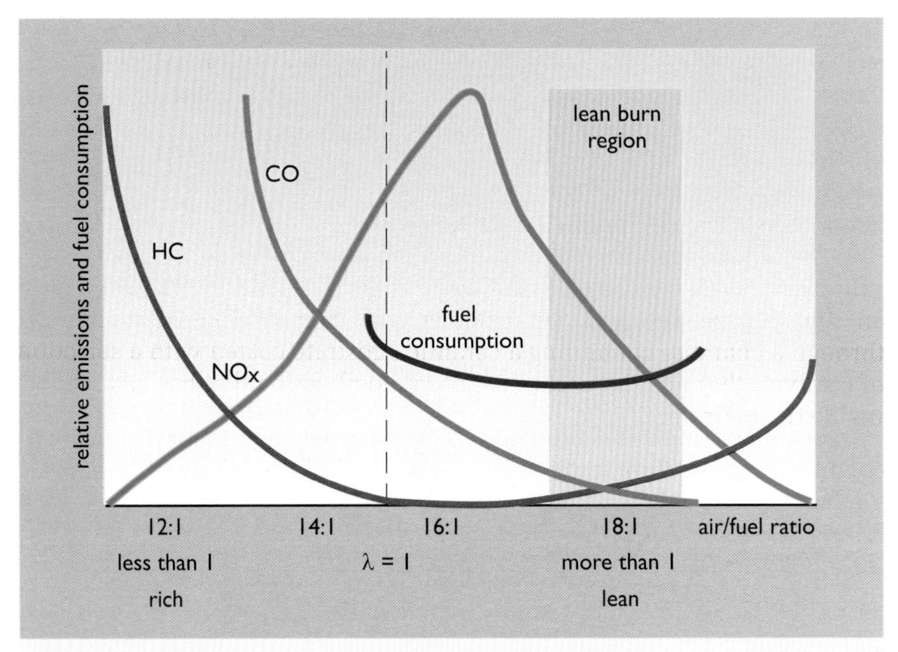

Figure 8.13 Air/fuel ratio in a petrol engine and relative emissions

If, though, there is a 'rich' mixture with too much fuel and not enough air, λ will be less than 1. There will not be enough air for the combustion process to proceed fully. Much of the carbon in the fuel will only be converted to carbon monoxide (CO) and not all the way to carbon dioxide (CO_2). Other products of incomplete combustion are unburned hydrocarbons (HC), ranging from unburned petrol down to a fine sooty carbon dust. Unburned benzene and partial combustion products such as 1,3 butadiene and aldehydes are particularly toxic.

It would seem, then, that the best thing would be to make sure that there was always a surplus of air in the air–fuel mixture, i.e. it would be 'lean'. Unfortunately lean mixtures burn hotter and usually produce more oxides of nitrogen (NO_X).

Since the 1970s, engine designers have wrestled the with problem of producing efficient petrol engines that minimize the emissions of the three major pollutants: unburned hydrocarbons, carbon monoxide and NO_X. They have come up with two basic commercial solutions; the lean-burn engine and the three-way catalytic converter.

The lean-burn engine

Changing the air/fuel ratio in a petrol engine radically alters the way the combustion process proceeds. As shown in Figure 8.13, if the mixture is too rich, there will be too much carbon monoxide and unburned hydrocarbons in the exhaust. Also, since the fuel is not all being burnt properly, the fuel efficiency will be poor as well.

If the mixture is made slightly weak (around 16:1) then there will be good fuel efficiency, but plenty of NO_x production. If the mixture is made weaker still (around 20:1), then the combustion can be made to proceed so rapidly that there isn't time for the NO_x to form. Achieving a balance between these requires extremely careful engine design, but the result is high efficiency, low HC and CO emissions, and NO_x emissions that are far lower than in normal engines.

The 3-way catalytic converter

The other alternative is to use a 'normal' engine, but to add what is literally an 'end of pipe' solution. The exhaust gases from the engine are passed through a chamber containing a ceramic substrate coated with a selection of catalysts that can complete the partial combustion that has taken place inside the cylinder.

A *catalyst* is a substance that can accelerate a chemical reaction without actually getting changed itself. The metal platinum is widely used in many chemical reactions, including the 'catalytic cracking' of petroleum in refineries. Inside the catalytic converter of a car exhaust three catalysts are used, each to convert a different pollutant, hence the term '3-way' catalytic converter. Platinum and palladium are used to help oxidize the unburned hydrocarbons and carbon monoxide to carbon dioxide, and rhodium is used to help convert the nitric oxides back to nitrogen and oxygen.

The main difficulty is that these reactions only work properly if the air–fuel mixture is very tightly controlled indeed, just slightly rich of $\lambda = 1$. The high level of accuracy required has meant that most manufacturers have abandoned the use of the carburettor, an essentially mechanical method of getting the right air/fuel ratio, in favour of fuel injection, usually electronically controlled. The best way to get the precise air/fuel ratio *entering* the engine is to measure the amount of oxygen in the exhaust gases *leaving* the engine and feed the result to a computerized engine management system controlling the fuel injection.

The other difficulty is that catalytic converters only work when they are hot. They only start working once they have been heated by the engine exhaust gases to more than a certain **lightoff temperature**, which can be 150 °C to 300 °C. This means that they aren't likely to be working effectively on short trips. Also, being suspended from the underside of the car, they are very vulnerable to damage, and, rather ironically, especially so from 'traffic calming' measures such as speed humps.

The use of catalytic converters also places more constraints on the petrol being used. Any attempt to burn leaded petrol in the car would result in the catalysts being poisoned with lead oxide and rendered completely useless in a matter of days. Thus the introduction of the catalytic converter has had to be accompanied by the simultaneous introduction of 'lead-free' petrol. Also, levels of sulfur in the petrol have to be minimized, otherwise the catalysts become slowly clogged with deposits of sulfur compounds and the catalytic converter may emit the characteristic 'rotten eggs' smell of hydrogen sulfide on warm-up.

Emissions from diesel engines

Unlike the petrol engine the diesel engine operates with a very wide range of air–fuel mixtures. Normally, at part throttle, the mixture is very weak. There is plenty of air to burn the fuel and emissions of carbon monoxide and unburned hydrocarbons can be very low.

However, at full power, the mixture can easily be far too rich, resulting in clouds of black sooty smoke (called particulates), all too familiar from trucks and buses. This pollution can, to a certain extent, be countered by careful engine management controls. Alternatively, various solutions based on exhaust filters are now being fitted to urban buses and trucks.

For diesel cars, 'oxidation' catalytic converters have become available to oxidize the remaining unburned hydrocarbons in the exhaust. These are essentially just '1-way' of the '3-way' type fitted to petrol cars. Their introduction has required the removal of sulfur from diesel fuel, since the catalyst would be poisoned by the sulfur dioxide produced in the exhaust. The manufacture of this 'Green' sulfur-free diesel requires more effort, expense (and energy use) at the refinery.

Generally, NO_x emissions from diesel-engine cars are higher than those from the equivalent petrol-engine version. The peak temperatures reached in the cylinders of diesel engines are higher than in spark engines, suggesting that they should produce more far more NO_x. The actual NO_x production is limited by the fact that the combustion takes place as a rapid detonation, limiting the time for chemical reaction to take place.

Obtaining best efficiency

The best way to reduce the emission of CO_2 from a petrol or diesel engine is to operate it at its maximum efficiency. Typically in a petrol-engined car only 24% (about one quarter) of the heat energy in the fuel will be delivered as work to the crankshaft. The figure for a car diesel engine is about 32%. One way to improve on this is to achieve a proper matching of the engine to its load.

Car engines are usually sized to meet rather inflated notions of driving performance. An ability to travel at 150 kilometres per hour may be appealing, but in practice the car may spend most of its time on short low-speed trips. Modern petrol car engines are usually designed to operate over a rotational speed range up to about 6000 rpm. For diesel engines this maximum speed is usually slightly lower, about 4000 rpm. However, best engine efficiency usually occurs at about a half of these rates of rotation.

The road speed of the car is matched to the rotational speed of the engine using a gearbox, which may be manually operated or automatic. In the past, cars were normally only manufactured with three- or four-speed gearboxes. The top gear was usually designed to give the absolute maximum speed on a flat road (essential for the advertising). However, considerations of energy efficiency suggest that the car should be able to cruise at a reasonable speed, say 100 kph, with the engine turning at its most efficient speed (around 2000 rpm for a diesel and 3000 rpm for a petrol engine). This requires an extra gear 'higher' than top gear. This is the reason that most modern cars are now supplied with five-speed (or even six-speed) gearboxes or an extra 'overdrive' gearbox. The lower engine speed also has the added benefits of reducing engine noise and extending engine life.

However, to get the absolute best efficiency from a petrol or diesel engine, it should also really be operated not just at its most efficient engine speed, but also at close to full power at that speed (i.e. with the accelerator pedal flat on the floor). One solution is to place both the engine and the gearbox under automatic computer control to continually optimize the gear ratio being used and the engine settings for best fuel economy. A further step is to use a hybrid petrol–electric drive system (see Chapter 9). This allows the petrol engine to be used at close to full power when required and also charge up a storage battery. When only small amounts of power are needed, the engine is switched off and the car runs on an electric motor using the stored energy from the battery.

It is also worth making sure that 'the maximum expansion of the gases is achieved after ignition' (as set out by de Rochas in 1862). Some recent petrol engine designs use the **Atkinson cycle**, a modification to the basic Otto cycle, in which the exhaust stroke is made effectively longer than the compression stroke by varying the inlet valve timing. This gives improved fuel efficiency and reduced exhaust noise.

8.5 The gas turbine

Chapter 6 has described how the invention of the steam turbine in the 1880s represented a step change in performance, compared with the steam engine, particularly in power to weight ratio. But why bother with the steam boiler? Why not an *internal combustion turbine engine*?

A Norwegian by the name of Egidius Elling patented a gas turbine as early as 1884, but the fundamental problem was that the metals of the day were not capable of withstanding the temperatures of the hot gases produced by a continuously burning flame. His solution was to use the flame to heat a water jacket, projecting a jet of steam and high temperature burned gases into a turbine. This held down the temperature of the mixture to a level that would not melt the turbine blades. In modern parlance this would be described as a **steam injected gas turbine** or STIG. Other European developers built experimental prototypes over the next 30 years, but it was difficult to compete with the proven performance of the competing diesel and steam turbines.

Two rival jet engines

The modern gas turbine grew out of the need for a high-performance aircraft engine. In the 1930s various inventors experimented with designs for an engine that could produce a jet of gas to propel an aircraft, rather than

BOX 8.3 Principles of the turbojet

The layout of the modern turbojet engine is little different from von Ohain's 1930s design. The key components are an axial compressor, a combustion chamber and a turbine (see Figure 8.14).

Axial compressor

The compressor carries out the function of the piston in a four-stroke engine, raising the pressure of the incoming air. However, it does this by using a succession of banks of turbine blades, progressively squeezing the air into a smaller volume. In a modern engine the compression ratio may be between 16:1 and 30:1. It has to be high, because at a jet aircraft's normal cruising altitude of about 10 000 metres, the air pressure is rather low and it is a matter of gathering enough air to burn a reasonable amount of fuel.

Combustion chamber

The combustion chamber is the heart of the engine where the fuel is burned, fed with highly compressed air. Here, the fuel has to be burned in an even manner in order to minimize NO_x emissions. This must be done all the way from minimum to maximum power output. Strangely, the first task of the combustion chamber is to reduce the velocity of the incoming air, so that the flame does not get blown out. The fuel, kerosene in the case of aircraft engines, is injected, burns and projects a jet of hot gases onwards to the turbine blades. In a modern turbine the gases leaving the combustion chamber can be up to 1300 °C, so the casings have to be made with similar attention to detail as the turbine blades themselves. Industrial gas turbines that have to meet stringent NO_x emission standards, are likely to have multiple banks of combustion chambers to cope with changing loads.

Turbine

The turbine extracts kinetic energy from the hot gas stream, converting it into work turning the main shaft through the engine to drive the compressor at the front. From the earliest days, finding ways of preventing the turbine blades from simply melting has been a major design problem. In the 1950s turbine inlet temperatures were only about 800 °C. Even when using the best nickel–chrome steels, this could only be tolerated by making the turbine blades hollow and pumping cooling air from the compressor through them. Since then, improvements in design and materials have led to a steady increase in inlet temperatures in successive models. The ability to withstand gas temperatures as high as 1300 °C is achieved by additionally coating the blade with a ceramic. Ceramics based on silicon carbide and silicon nitride can withstand higher temperatures than the metal beneath and act as a thin insulating layer.

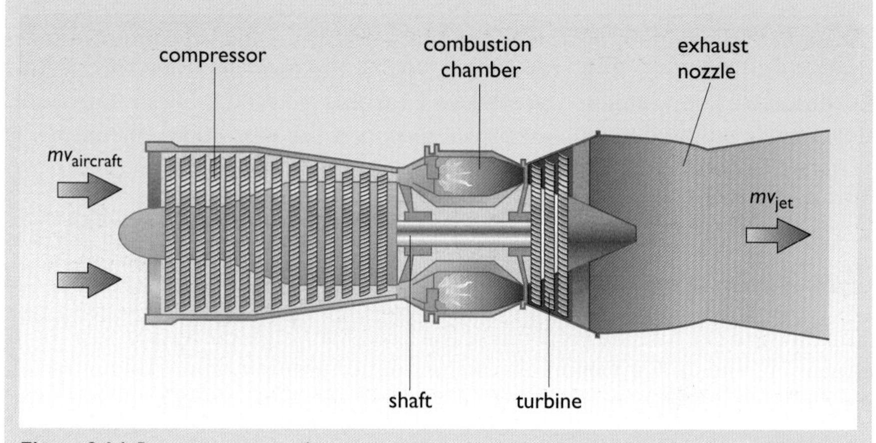

Figure 8.14 Basic components of a turbojet engine (source: redrawn from Rolls-Royce, 2002a)

relying on propellers driven by piston engines. Two inventors, on opposite sides in World War II, ended up producing working engines: Hans von Ohain (Figure 8.15) and Frank Whittle (Figure 8.16).

Their designs were basically similar. A continuous stream of air would be compressed by a turbine, fuel would be injected and burned, and finally the hot gases would be expanded through an output power turbine. Some of the work derived from this would be fed back to drive the compressor. The ejected hot gases produce thrust (see Box 8.3).

Figure 8.15 Hans von Ohain developed his jet engine in Germany in the late 1930s with the backing of Ernst Heinkel, the aircraft manufacturer. After the war he moved to the USA, eventually becoming Chief Scientist of the Aero Propulsion Laboratory at the Wright-Patterson US Air Force Base in 1975

Hans von Ohain conceived of his jet engine design in 1933 while studying for a doctorate at Göttingen University in Germany. By 1935 he had developed a working model to demonstrate his ideas and approached Ernst Heinkel, an aircraft manufacturer, for support. Heinkel saw it as a means to build the fastest aeroplane in the world and in 1936 offered him backing. By February 1937 von Ohain had tested a turbojet running on hydrogen fuel that produced over 100 kg of thrust, with the turbine spinning at 10 000 rpm. Heinkel urged von Ohain to press on to produce a full-size aircraft engine but using kerosene as the fuel. Work on an aircraft to use this engine, the He-178, began in early 1938 and it flew in August 1939. The world's first jet fighter, the Messerschmidt 262, made its first fully jet-powered flight in July 1942.

Figure 8.16 Frank Whittle developed his jet engine in England. He had to allow his patent on it to lapse because, as a Cambridge University student, he couldn't afford the £5 to renew it

In England a completely parallel development took place. Frank Whittle began experimenting with ideas for a turbojet engine while training as an RAF pilot. By 1930 he had designed and patented a jet aircraft engine, but could not get backers in an industrial depression. He could not even afford to renew his patent. In 1935, while studying for a degree in mechanical engineering at Cambridge University, he was approached by some ex-RAF officers who raised money for work on an experimental engine. They tested this in April 1937. The British Air Ministry began to take an interest and in 1939 gave Whittle a contract for a flight engine. The Gloster Aircraft Company built an experimental aircraft, which first flew powered by his jet engine in early 1941, demonstrating comparable speed with a Spitfire. In 1942 the Rolls-Royce company took over the manufacture of the jet engines and they have been making them ever since. Production engines were eventually installed in a twin-engine jet fighter, the Gloster Meteor, which first flew in March 1943.

Box 8.4 provides an explanation of propulsion and thrust.

BOX 8.4 Propulsion and thrust

Propulsion

We all spend time propelling ourselves, on foot or in vehicles, and in both cases we are making use of Newton's Laws of Motion in quite specific ways. To start moving at all, a forward force is needed (Newton's Second Law of Motion, see Chapter 4). But where does the force come from? When you walk, at every step you push backwards on the ground. And this is where **Newton's Third Law of Motion** enters. It states that if A pushes against B with a certain force, then B must inevitably be pushing against A with the same force in the opposite direction:

for every action there is an equal and opposite reaction.

So when you push backwards on the ground as you step, the ground pushes back, obeying Newton's Third Law, and this gets you moving. Similarly if you drive. The engine forces the wheels to rotate, so they push backwards on the ground where they make contact – and the ground pushes them forward, propelling the vehicle. *Friction* between foot or wheel and the ground is of course essential – as evident when trying to walk or drive on a sheet of ice. Railway locomotives are often rated by their **tractive effort**, the number of tonnes force they can exert on a train without the wheels slipping.

Thrust

How is motion achieved where there are no solid surfaces to provide frictional forces? A rocket can fly through outer space where there is nothing to push against at all. It does so by using **Newton's Second Law of Motion**; this is commonly stated as '*the force needed to accelerate a body is proportional to its mass multiplied by its acceleration*' (see Box 4.1). The combustion process in a rocket accelerates a stream of exhaust gases backwards. This is the *action*. The *reaction*, using Newton's Third Law, generates a forward force, the *thrust* of the engine pushing the rocket through empty space.

In the case of aircraft, the engines generate thrust by accelerating a continuous stream of air backwards, be it by using a propeller or a turbojet engine (as in Figure 8.14).

As explained in Chapter 4, the force (F) needed to accelerate a mass, m, with acceleration, a, is given by:

$$F = m \times a$$

To calculate the thrust, therefore, we need to know the mass of air and its acceleration. Assuming that the engine shown in Figure 8.14 is attached to an aircraft flying at a velocity $v_{aircraft}$, this is the velocity at which the air enters the compressor.

The hot exhaust gases are then ejected backwards at a higher velocity v_{jet}. Its average acceleration, a, is equal to the change in velocity divided by the time, t (in seconds), that it spends in the engine:

$$a = \frac{\left(v_{jet} - v_{aircraft}\right)}{t}$$

Taking a particular mass, m kilograms of air, (ignoring the small mass of the fuel actually burned); Newton's Second Law tells us that the force F (in newtons) needed to produce this acceleration is:

$$F = m \times a = m \times \frac{\left(v_{jet} - v_{aircraft}\right)}{t}$$

But m divided by t is the number of kilograms passing through per second: the **mass flow rate**. So it is a simple relationship:

thrust = mass flow rate × change in air velocity

Notice that only the *change* in the velocity of the air is involved here, so it doesn't matter whether the speed is measured relative to the ground or the aircraft, provided the same method is used for the initial and final speeds.

For a modern commercial aircraft engine the thrust is in the range 200–400 kN, approximately the weight of a mass of 20–40 tonnes.

Post-war developments

Even the hastily produced Messerschmidt 262 and Meteor could fly 20% faster than their piston-engined equivalents, and on low-grade kerosene rather than high-octane petrol. Despite this, the jet engine was of little practical use during World War II. Adolph Hitler failed to appreciate the potential of the Me 262 as a fighter plane until it was too late to deploy it.

The Whittle engine design was shared with the Americans and the Russians, who both immediately started their own development programmes. In Germany, Allied troops gathered up the various experimental and production machines as they advanced across the country and spirited them back home to study. Von Ohain ended up in the USA.

Figure 8.17 gives a good impression of the sheer power of the jet engine. It revolutionized aircraft design. Given the military pressures of the ensuing Cold War both the USA and the Soviet Union threw enormous amounts of money into development. By the end of the 1950s their military aircraft were regularly flying at almost three times the speed of sound.

Figure 8.17 Don't try this at home! A primitive gas turbine burning kerosene at the rate of one ton of fuel per hour being tested at Lutterworth Gas-turbine College in 1948.

In 1952, Britain unveiled the world's first jet airliner, the de Havilland Comet, designed to replace the piston-engined Douglas DC-4 Argonaut. The Comet had a cruising speed of 780 kph compared with the Argonaut's 460 kph, and a cruising altitude of 11 000 m compared with the Argonaut's 6100 m. The Comet's ability to fly at such high altitudes meant a smoothness of flight above bad weather which the Argonaut, and other piston-engined passenger aircraft of the day, couldn't match.

The Comet suffered embarrassing setbacks following a series of crashes due to metal fatigue in the bodywork, and in many ways the first commercially successful jet airliner was the Boeing 707 which entered service in 1958. This used four turbojet engines slung under the wings in a manner that made them easily removable. The body of the airliner could keep flying (and earning money) while exchange engines were being repaired or serviced. And even these needed less maintenance than their complex multi-cylindered piston engine predecessors.

Since 1958 the Boeing 707 has been succeeded by ever-larger designs, but essentially of the same layout. These have ushered in a whole new age of cheap mass air travel, with a matching growth in energy demand.

Modern jet engines

Early jet engines had simply concentrated on producing a narrow, high-speed jet of gas to push the aircraft forwards. However, for best fuel efficiency, it is better to produce a larger, slower, jet of air (see Box 8.5). This could be done by inserting more blades in the turbine to extract more of the kinetic energy of the gases leaving the engine, delivering it to the shaft running through its centre. In the prop-jet design developed in the 1960s this could then be used to drive a propeller at the front via a gearbox. However, the use of a large propeller limits the aircraft's top speed and is noisy.

BOX 8.5 Thrust and kinetic energy

Which is better – using a jet engine to produce a small amount of high velocity air or a larger amount of lower velocity air?

Thrust

Box 8.4 showed that the thrust delivered by a jet engine is equal to the mass flow rate (the number of kilograms of air passing through the engine per second) multiplied by the change in velocity of the air:

thrust = mass flow rate × change in air velocity

So, for example, accelerating 2 kg of air per second to a speed of 100 m s^{-1} should produce the same thrust as 1 kg of air per second accelerated to 200 m s^{-1}.

Energy

The jet engine turns the *heat energy* of the fuel into the *kinetic energy* of the moving hot exhaust air. Let us, for simplicity, consider a jet engine on the ground and, for calculation purposes, take the initial velocity of the input air as zero.

As described in Box 4.1 in Chapter 4, kinetic energy is proportional to the square of the velocity:

$$E_k = 0.5\ m\ v^2$$

Taking the first case above, projecting a mass of 2 kg of air backwards at 100 m s^{-1} will require:

$$E_k = 0.5 \times 2 \times 100^2 = 10\ 000 \text{ joules}$$

If there is a *continuous* mass flow rate of 2 kg s^{-1}, then this will require a *certain rate of conversion of energy* or *power*.

power = 10 000 joules s^{-1} = **10 000 watts**

In the second case, projecting a mass of 1 kg backwards at 200 m s^{-1} will require:

$E_k = 0.5 \times 1 \times 400^2 = 20\ 000$ joules

Again if this is repeated every second, then:

Power = 20 000 joules s^{-1} = **20 000 watts**

That is, twice the power requirement to produce the same thrust.

The function of the jet engine is to produce the maximum thrust for a given expenditure of energy. The figures above would suggest that it is better to produce a large amount of slower moving air rather than a small amount of high velocity air.

Modern engines for civil aircraft use a large **bypass fan** which drives air round the outside of the basic jet engine (see Figure 8.18). This slow bypass air surrounding the central high-speed engine exhaust also has the effect of making the engine much quieter. In a modern bypass engine design, about 80% of the thrust comes from the fan and only 20% from the hot jet engine exhaust.

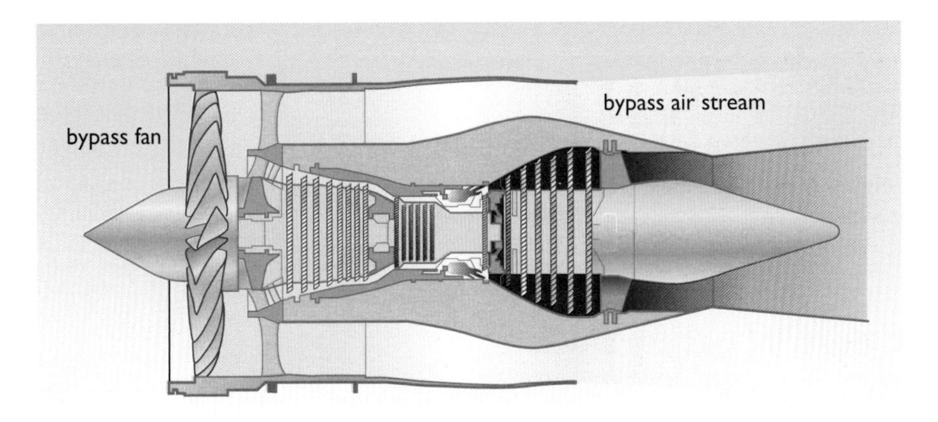

Figure 8.18 Rolls-Royce Trent using bypass fan (source: redrawn from Rolls-Royce, 2002b)

Turbofan engines are more efficient than turbojet engines, that is to say that for a given rate of fuel consumption, they produce more thrust, and that is what counts to push a plane through the sky. A modern turbofan such as the Rolls-Royce RB211-882 can produce 2.5 times more thrust per kg of fuel burnt than the 1943 German Me 262 engine, a result of both the turbofan design and steadily increased turbine inlet temperatures.

Aircraft fuel efficiency is not just a matter of cost; it is critical for the ability to carry out long-haul flights. For example, a Boeing 747-400 flying to New York from Tokyo needs to carry about 145 tonnes of fuel (nearly 40% of the total weight at take-off). During the flight, 125 tonnes will be burned with 20 tonnes kept in reserve. This is completely different from a car. A one-tonne car is likely to carry only 50 kg of fuel at maximum, but of course does have the option to stop and fill up again!

In practice, the engine design can be quite complex. It is desirable that the final stage of the compressor and first exhaust turbine run together at a high speed. This core jet engine then acts as a **gas generator** supplying kinetic energy to the downstream turbines which can turn more slowly on a separate concentric shaft. The Trent engine actually has three concentric shafts linking different sets of fans and turbines (see Figure 8.19). The final five banks of output turbines are used to drive the single bypass fan.

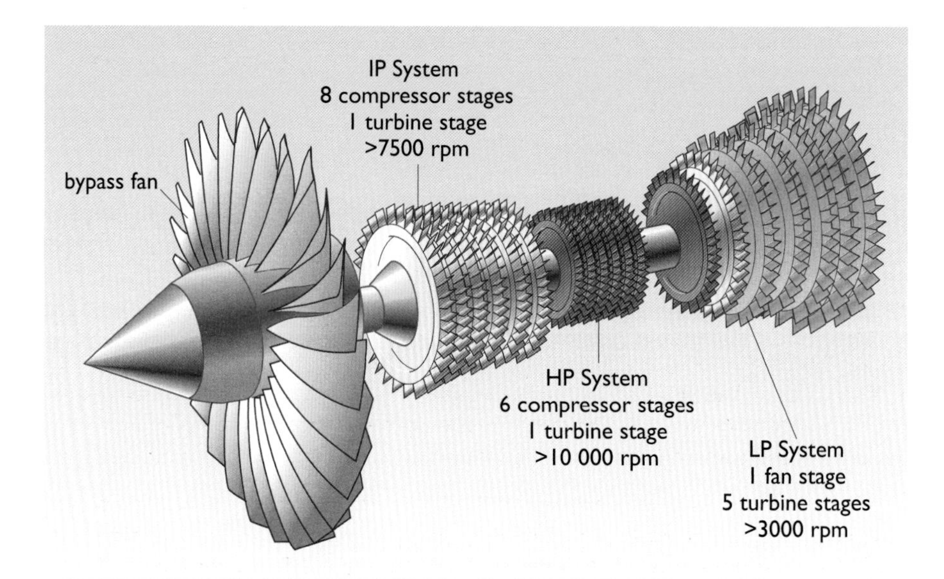

Figure 8.19 Different parts of the jet engine can be connected by concentric shafts running at different speeds. The Rolls-Royce Trent has three separate shafts (source: redrawn from Rolls-Royce, 2002c)

Industrial gas turbines

The ability to extract a large amount of the kinetic energy of the exhaust gases and deliver it to a shaft has meant that gas turbines could be used for purposes which had previously been the province of the diesel and steam turbine. In the 1960s, large gas turbines started to be used in naval ships, where high power to weight ratio is desirable, and then to drive generators in power stations to meet peak loads. The Rolls-Royce Trent engine is also sold in 'marine' and 'industrial' forms. Essentially these are much the same as the aircraft engine, but instead of driving the bypass fan, the shaft drives either a ship's propellers or machinery (usually an electric generator).

Gas turbines for cars?

One of the early questions was whether this technology could radically update car engine design. In 1950, the Rover company produced JET 1, a prototype gas turbine car. Although it showed that such a vehicle was technically feasible, the fuel efficiency was very poor.

Although small gas turbines can be produced with a high power to weight ratio, extracting energy from a turbine that may be spinning at 50 000 rpm to drive road wheels that may be going at 100 rpm is not easy. The result is something far bulkier than a conventional petrol engine.

The need to produce car engines with low carbon monoxide and NO_x emissions re-stimulated interest in gas turbines in the 1970s, but this faded away when it became clear that Otto engines fitted with catalytic converters could be made to meet the standards.

Improving power and efficiency

Military aircraft achieve extra thrust by using an afterburner – burning fuel downstream of the turbine. This is extremely noisy and cannot be sustained for long because it is very inefficient.

The simplest way to get more power output of a gas turbine, and make it more efficient, is to go back to the earliest designs and inject water, generating steam to expand through the turbine. This is regularly done in civil aircraft engines to produce the extra thrust for take-off. The increase in performance can be quite considerable. As an example, a sample 33.1 MW turbine without steam injection would have an efficiency of only 33%. In a steam injected (STIG) form it could produce 51.4 MW at 40% efficiency. However, to do this requires a continuous supply of high-quality water (exactly the same problem as for a steam railway locomotive). This isn't practical for aircraft and is only useful in industry if, for example, a factory requires large amounts of process steam.

In practice, the most popular approach in industry has been to use the waste heat from a gas turbine to raise steam in a boiler to drive a steam turbine; this combination was given the name Combined Cycle Gas Turbine or CCGT. This has enabled overall efficiencies of over 50% (on an LCV basis) to be achieved. This technology is described further in Chapter 9.

The fuel efficiency of aircraft engines could be improved by using higher combustion temperatures, which will require new turbine blade materials. However, higher temperatures are likely to mean higher NO_x emissions, which may be unacceptable. An alternative is to improve the amount of thrust produced per unit of fuel burned by increasing the bypass ratio. This may require a return to aircraft with propellors (or **open-rotor fans** as the new designs are called). These aircraft are likely to be slower and noisier than current jets. A more radical suggestion has been to use a central jet engine with extra bypass fans on each side driven by shafts from the main turbine.

In practice, it is likely that tackling the energy efficiency of air transport may require a complete redesign of the shape of aircraft which may take many years.

8.6 The Stirling engine

The petrol engine, the diesel engine and gas turbine are all internal combustion engines, but it would be wrong to think that these are naturally superior to external combustion engines like the steam engine. As was explained in Chapter 6, a high thermodynamic efficiency requires the use of high temperatures. The steam engine is limited in this respect by the physical properties of water. Throughout the 19th century it had a competitor which used air, and not steam, as its working fluid.

The first patent for a practical machine was taken out by the Reverend Robert Stirling in 1816; in 1818 he built a full-sized, 2 hp pumping engine. Much of the development was done by his younger brother, James. They wanted an engine which avoided the use of high-pressure steam, because they were concerned at the number of people killed in boiler explosions.

Principles

The Stirlings' early engines were mainly of the 'concentric piston' type. The essential principle is shown in Figure 8.20.

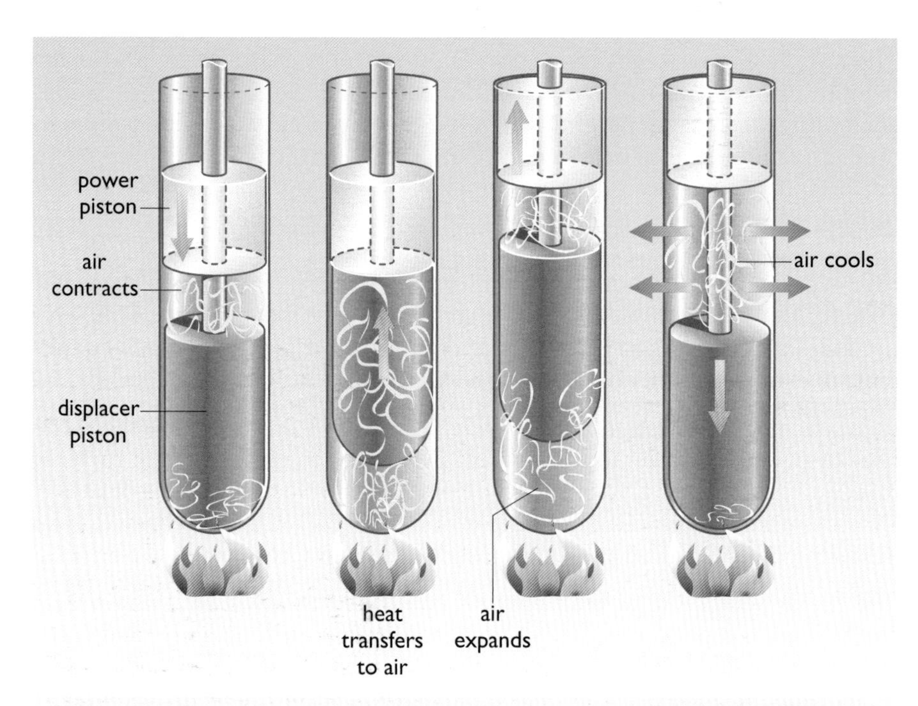

Figure 8.20 Principles of the operation of a Stirling engine (source: adapted from Ross, 1981)

If a gas is heated, it will expand. If it is then moved and cooled, it will contract. The concentric piston engine uses a cylinder heated at one end and cooled at the other. A loose-fitting displacer piston moves inside. Its main function is to shift the gas from the hot end to the cold end and back again. When the gas is at the hot end, it will expand, and when it is at the cold end, it will contract. A power piston which fits the cylinder tightly moves as the gas expands and contracts, extracting mechanical energy. The two pistons are driven 90° out of phase by a suitable linkage. A refinement is to make the gas flow through a heat-storing regenerator usually in the form of a mass of wire gauze between the hot end of the cylinder and the cold end. This improves the thermodynamic efficiency of the cycle.

The heat can be supplied from a range of different sources. Early ones used simple coal or wood fires. Gas burners proved very convenient for small-scale applications such as the table fan shown in Figure 8.21. Cooling was achieved simply by radiation to the surrounding air.

Figure 8.21 Before the electric fan came the gas-powered fan, driven by a small Stirling engine underneath. It is heated by a small gas burner fed from the normal domestic gas supply

Throughout the 19th century and early 20th century, Stirling engines competed with steam engines. The most successful were the smaller ones which could be used by ordinary people without the dangers of high-pressure boilers. Thousands were sold for farm water pumping and many were in use for 25 years or more. Unlike other engine designs, the Stirling engine can be scaled down to a very small size without serious loss of efficiency. It doesn't have the noise of the internal explosions of the petrol and diesel engines, and a well constructed one can run in almost complete silence. Other small-scale applications were dentists' drills, organ blowers and even gramophones! Many of these uses lost out to the small electric motor as widespread electrification spread across the world. Others succumbed to small, petrol engine designs developed in the 1920s.

The Philips engine

In 1937 the Dutch Philips company at Eindhoven became interested in portable lightweight generators for domestic valve radios. These needed to be silent, free from electrical interference and easy to use. They took Robert Stirling's basic concentric piston design and applied new high-temperature steels and auto-engine design. They also tried different working fluids instead of air, particularly helium and hydrogen.

Amazingly, development went on during World War II, without the knowledge of the occupying German forces. An almost silent 2.5 hp unit was field tested, hidden under a small cardboard box, powering a small open boat 50 miles around the canals of Holland.

In 1946, Philips were able to complete the development and produced a prototype engine-generator set of 200 watts output, which could run on kerosene or petrol. Compared to late nineteenth-century models they had increased speeds by a factor of ten, efficiency by a factor of 15 and power to weight ratio by a factor of 50.

Alas, by the early 1950s Philips realised that the spread of mains electricity and the development of the transistor was undermining much of the need for such a device and they abandoned large-scale production. There was, however, another market opportunity. The Stirling engine is a truly versatile engine. Instead of producing work from heat, it could act as a heat pump and perform as a refrigerator if driven by a motor. If the 'hot end' was held at room temperature, the 'cold end' of a single stage unit could reach temperatures sufficiently low to produce liquid air. Philips concentrated on this application and by 1963 they had achieved a temperature of 12 K, sufficiently low to demonstrate the effect of **superconductivity**: the complete disappearance of electrical resistance in some alloys.

An engine with a range of applications

The attractions of the Stirling engine have remained:

- It is quiet
- As an external combustion engine, it could potentially run on any fuel or even solar energy

- Unlike internal combustion engines, the combustion process can be carefully controlled.
- It is capable of reasonable efficiencies.

Since the 1950s there has been sporadic experimentation with Stirling vehicle engines (much by General Motors in the USA). This has shown that they can meet very tight emission standards without the use of catalytic converters. However, it must be said that the Stirling is not fully suited to direct-drive automotive applications. It has an inherent high thermal mass and is thus very sluggish in performance. For the moment, Otto cycle and diesel engines continue to rule for this application.

Given the right materials, a Stirling engine can be used to produce work from high temperature heat (500 °C or more). Conversion efficiencies of over 30% have been demonstrated in **dish stirling** solar power generation projects in Spain and the USA. These use a large concave mirror to focus sunlight onto a Stirling engine driving an electric generator (see Figure 8.22). Large-scale solar power generation projects are proposed using many hundreds of 25 kW Stirling engine modules. However, these projects will have to win out against rival schemes using large solar trough collectors and steam turbines, or the large-scale deployment of photovoltaic panels.

The relative silence of the Stirling engine makes it a good candidate for small-scale combined heat and power (CHP) generation units. In the UK trials have been carried out of a domestic micro-CHP unit. This is essentially a gas boiler which also generates electricity (see Figure 8.23). It is about the size of a domestic washing machine and uses a four-cylinder engine unit running on natural gas at the top, driving a generator below. The waste heat from the engine is fed into the normal domestic central heating system. Further trials of this and other such units are being carried out in other countries.

Figure 8.22 Boeing Solar Stirling Engine. The mirror concentrates sunlight to produce a very high temperature at the 'hot end' of the Stirling engine mounted on the arm.

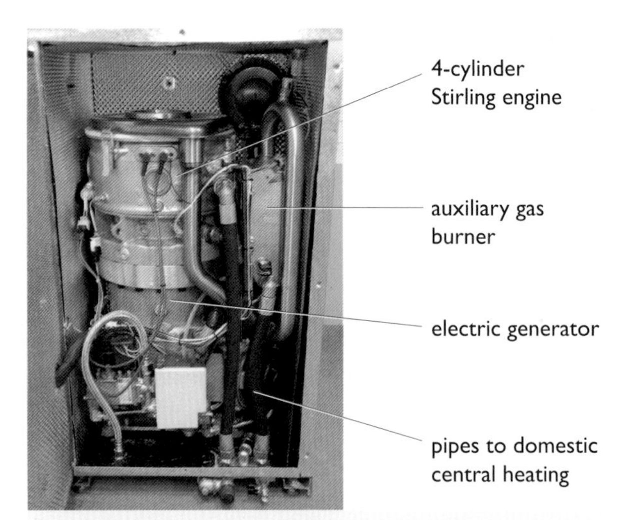

Figure 8.23 The interior of a gas-fuelled Whispergen domestic micro-CHP unit. It can generate 1 kW of mains electricity.

There are also intriguing possibilities for very low maintenance engines with virtually no moving parts. A **thermoacoustic** engine with a thermal efficiency of 30% has been described in the journal *Nature* (Backhaus and Swift, 1999). This could be described as consisting of just a loud noise in a carefully shaped container. The key function of the displacer piston in a Stirling engine is to move the working fluid from the hot end of the cylinder to the cold end and back again. But this can equally be achieved using a resonant sound wave. This raises the possibility of a whole range of future devices in which a Stirling engine turning heat into work could be coupled to a Stirling refrigerator to make a heat pump. The only moving parts might be a loudspeaker-like device to initially start the oscillation and possibly a diaphragm to separate the two halves. It is only by using modern computer modelling techniques that the design of such devices has become possible.

8.7 Which is the best engine?

This chapter has looked at a number of different engines developed from the end of the 19th century onwards. Chapter 6 has described the steam engines that preceded them. One might ask 'Which is the best engine?'

The answer is that they all have different applications, and one rather than another has been used at different times throughout history.

The reciprocating steam engine, fuelled first by wood and then coal, powered the Industrial Revolution. The first requirement is to raise steam in a boiler. Even in a small boiler this may take an hour or more. This might not be a problem in a factory but was not a good selling point for early steam cars! In railway locomotives, the thermal efficiency rarely exceeded 10%, since they were designed for high power to weight ratio rather than fuel efficiency. The reciprocating steam engine also has a high starting torque and a reasonable efficiency over a wide range of speeds, which means that it does not need the gearbox necessary for petrol and diesel engines. The best designs of triple expansion steam engines for factories and ships reached an efficiency of 25%. However, there were always considerable practical concerns about boiler explosions, and even today, steam boilers must be thoroughly inspected and even X-rayed to get insurance.

The steam turbine, invented in the 1880s, had a much better power to weight ratio than its reciprocating engine predecessor (see Figure 8.24). It immediately became the engine of choice for coal- and oil-fired steam ships, and for electricity generation. However, it lacked flexibility of operation and did not displace the reciprocating steam engine from the railways. The steam turbine still largely rules in the power station, fuelled by coal, oil or nuclear power. Single machines can be as large as 1 GW output and efficiencies can approach 40%.

Stationary internal combustion engines running on gas were introduced in the 1860s and proved popular as power plants for small factories well into the 20th century. They could be started and stopped quickly, without the inconvenience of raising steam in a boiler. Even today small reciprocating engines running on natural gas are used for CHP applications, with electricity generation efficiencies of 25–30%.

The petrol engine as the power unit for the motor car has changed the world since its introduction in the 1880s. There are now well over 500 million

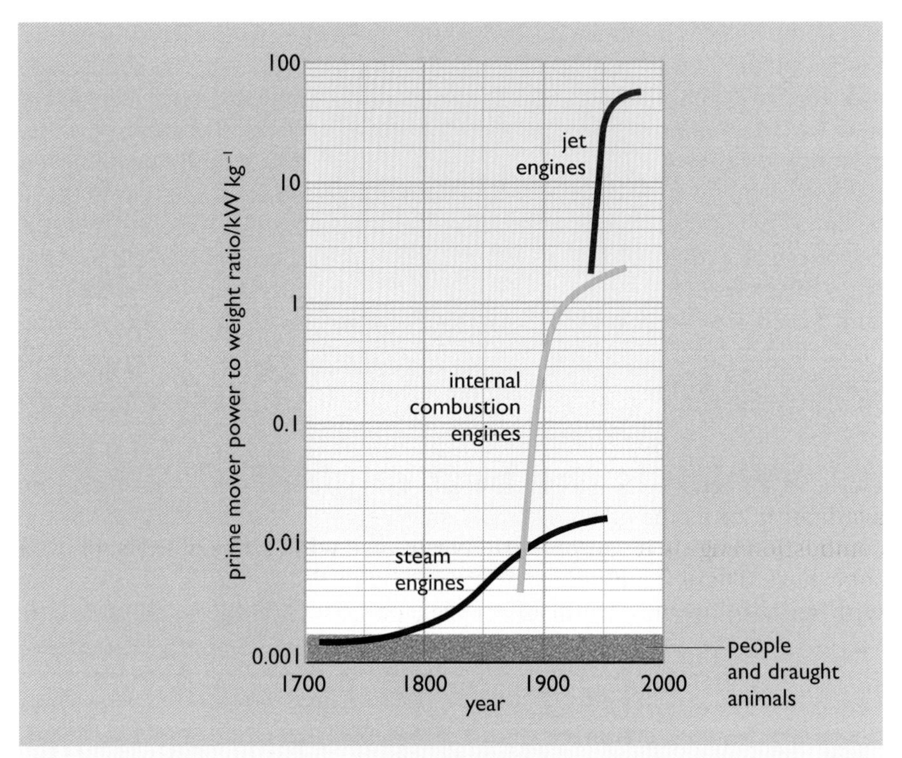

Figure 8.24 Power to weight ratio of different engines (source: adapted from Smil, 1994)

of them around the world, with typical thermal efficiencies of about 25%. Until the 1930s, the petrol engine was also the main power source for heavy vehicles such as trucks and buses. Until the 1950s it was also the power source for most aircraft. Petrol engines are most useful in the power range 1 kW to about 2 MW. Their widespread use has resulted in various forms of pollution. Key pollutants have been lead from the additives used to improve the octane rating, carbon monoxide from incomplete combustion and NO_x resulting from the high temperature combustion of the fuel.

The diesel or compression ignition engine, introduced in the 1890s, could run on a wider range and lower grade of fuel than petrol engines. Diesels are useful in the power range 20 kW to 80 MW and have a higher efficiency than the petrol engine. Large ships' diesel engines can achieve thermal efficiencies of 40–50%. Even when oil was far more expensive than coal, it could pay to use diesel engines because of their relatively high efficiency, the ease of starting and stopping, and the relative ease of handling the fuel. The development of a practical fuel injection pump in the late 1920s allowed the development of diesel buses, trucks and tractors, with fuel efficiencies of 30–40%, displacing their less efficient petrol predecessors. The replacement of steam locomotives by more efficient diesel-powered designs on British Railways in the 1960s gave a major reduction in overall energy use. Small turbocharged diesel engines are now highly competitive with petrol engines for cars and almost universally used in trucks and buses on the road, in railway locomotives and ships' engines. Diesel engines share some of the pollution problems of petrol engines, but especially suffer from incomplete combustion and a tendency to produce particulates (soot). The pursuit of clean combustion in road vehicles has required a tightly specified fuel, DERV. Biodiesel blends of fossil fuels and vegetable oils are being introduced.

The gas turbine, developed in the 1940s to run on kerosene, now rules the skies, having totally replaced the piston engine which required high-octane petrol. The modern turbofan design is more fuel-efficient than the original turbojet design used in the 1950s and there is still scope for further development. The industrial gas turbine is now widely used in power stations where natural gas is available, especially when coupled with a steam turbine; the total combined efficiency can reach 50%. Gas turbines are most useful in the power range 100 kW to 500 MW and have a very good power to weight ratio. Since they burn fuel at high temperatures, they may produce NO_x pollution.

The Stirling engine was widely used for small power applications before the availability of mains electricity and small electric motors. As an external combustion engine it can potentially run on a wide range of fuels, or even solar heat. There are also possibilities of very low maintenance engines with an absolute minimum of moving parts. Current devices are being built in the 1–20 kW output range, but efficient devices can be made down to powers of 5 W. Its key advantages are its relative silence compared to petrol and diesel engines and its wide possible range of fuels. These have made it of interest for new designs of solar power generation equipment and domestic micro-CHP units.

One of the major questions of the moment is 'Which is the best engine for a car?' For much of the 20th century, this meant 'Which engine will give the highest top speed?'. In the 1980s it changed to 'Which will give the lowest NO_x and CO emissions?'. Studies in the 1980s and 1990s commented favourably on steam engines, Stirling engines (and even small gas turbines) as alternatives to petrol and diesel engines. Since then, the development of catalytic converters and computer-controlled engine management systems has considerably improved the pollution performance of petrol and diesel engines.

Today, there is greater interest in overall fuel consumption and consequent CO_2 emissions. The battle is now between petrol and diesel engines (enhanced with computer control and hybrid electric drives) and new technologies such as fuel cells (see Chapter 14).

8.8 Summary

This chapter has looked at the development and design of three internal combustion engines: the spark ignition engine, the diesel engine and the gas turbine; and an external combustion engine, the Stirling engine.

In relation to the pros and cons of the performance of these different types of engine it has examined:

- the worldwide growth of car ownership
- the factors influencing the choice of engine size for a car
- the development of high-power petrol engines

- the need for high compression ratios for both power and fuel efficiency and the consequent need for high-octane petrol
- the need for the appropriate use of gearing between the engine and road wheels
- a few of the benefits and problems of alternative fuels for spark ignition engines
- the physics of the production of thrust and the importance of the bypass ratio in jet engine design, its overall fuel efficiency and some possibilities for increasing both that and power.

The chapter has also considered some of the very negative consequences and problems resulting from the use of these technologies, including the air pollution associated with petrol and diesel engines. These include lead oxide dust from the past use of lead additives in petrol, sulfur dioxide, carbon monoxide, NO_x and particulates. The production of the last three of these can be tackled by the use of 'lean burn' technology and catalytic converters.

Finally, it has compared the properties of the different types of engine discussed above with the steam engines described in Chapter 6. There is no clear 'best engine', rather there is a whole range of possibilities for different applications.

References

Backhaus, S. and Swift, G. W. (1999) 'A thermoacoustic Stirling heat engine', *Nature*, vol. 399, no. 6734, pp. 335–8.

DTI (1973) *UK Energy Statistics – 1973*, Department of Trade and Industry, London, HMSO.

EC (2000) *EU transport in figures – statistical pocketbook – 2000*, European Commission: Directorate-General for Energy and Transport (DG TREN), Belgium, Office for Official Publications of the European Communities.

EC (2009) *EU energy and transport in figures: statistical pocketbook 2009*, European Commission: Directorate-General for Energy and Transport (DG TREN), Belgium, Office for Official Publications of the European Communities; available at http://ec.europa.eu/energy/publications/statistics/doc/2009_energy_transport_figures.pdf (accessed 24 August 2010).

Ford, H. (1922) *My Life and Work*, Chapter IV, p. 71–2.

McShane, C. (1997) *The Automobile – A chronology of its antecedents, development and impact*, London and Chicago, Fitzroy Dearborn.

Ministry of Power (1962) *Ministry of Power Statistical Digest – 1962*, HMSO.

Rogers, G. F. C. and Mayhew, Y. R. (1980) *Engineering Thermodynamics, Work & Heat Transfer* (3rd edn), London, Longman.

Rolls-Royce (2002a) 'How a gas turbine works' [online], Rolls-Royce Group plc, http://www.rollsroyce.com/education/gasturbine/jet_engine_layout.htm (accessed 6 January 2003).

Rolls-Royce (2002b) 'How a gas turbine works' [online], Rolls-Royce Group plc, http://www.rollsroyce.com/education/gasturbine/different_jets.htm (accessed 6 January 2003).

Rolls-Royce (2002c) 'How a gas turbine works' [online], Rolls-Royce Group plc, http://www.rollsroyce.com/education/gasturbine/multiple_shafts.htm (accessed 6 January 2003).

Ross, A. (1981) *Stirling Engines*, Phoenix, AZ, Solar Engines.

Smil, V. (1994) *Energy in World History*, Boulder, CO, Westview Press.

US BTS (2010) 'National Transportation Statistics', Table 1.11 [online], US Bureau of Transportation Statistics, http://www.bts.gov/publications/national_transportation_statistics/#chapter_1 (accessed 10 July 2010).

Wärtsilä (2006) Wärtsilä RT-flex96C and Wärtsilä RTA96C Technology Review [online], http://www.Wartsila.com (accessed 27 July 2010).

Further reading

Booth, G. (1977) *The British Motor Bus – an illustrated history*, Ian Allan.

Loftin, L. K. Jnr (1998) *Quest for Performance: The Evolution of Modern Aircraft* [online], NASA History Office, http://www.hq.nasa.gov/office/pao/History/SP-468/cover.htm (accessed 28 July 2010).

Reader, G. T. and Hooper, C. (1983) *Stirling Engines*, London, E. & F. N. Spon.

Rolls-Royce (2007) 'Gas turbine technology: Introduction to a jet engine', VCOM13797 [online], http://www.rolls-royce.com/Images/gasturbines_tcm92-4977.pdf (accessed 23 August 2010).

Stone, R. (1999) *Introduction to Internal Combustion Engines* (3rd edn), Macmillan Press Ltd.

Strandh, S. (1989) *The History of the Machine*, London, Bracken Books.

Williams, R. H. and Larson, E. D. (1989) *Expanding Roles for Gas Turbines in Power Generation*, in Johansson, T. B. et al. (eds) *Electricity*, Sweden, Lund University Press.

Chapter 9

Electricity

By Bob Everett

9.1 Introduction

The previous chapters have looked at wood, coal, oil and natural gas as fuels. These are all *primary* fuels. Chapter 5 has also described 'town gas' made from coal, or more recently from oil. This is a *secondary* fuel, produced initially for lighting, but it soon found uses for cooking, driving internal combustion engines and eventually heating. It was fuel in a form that was *more convenient to use.*

Figure 9.1 King Steam and King Coal wonder what the Infant Electricity will grow into – an 1881 cartoon in the magazine *Punch*

Electricity is, in essence, another secondary fuel mostly produced from fossil fuels, though some, such as that from hydroelectricity and wind, is usually considered as *primary* electricity. Although we regard electricity as something essentially 'modern', it has been under continuous development since the beginning of the 19th century. By the 1880s a global electric telecommunications network was in place in the form of the telegraph, and mains electric lighting was just being launched. Electricity might have been regarded as an infant full of possibilities then (see Figure 9.1), but it still is so today. Many technologies such as the fluorescent lamp, the light-emitting diode (LED), microwaves and the fuel cell have taken 50–100 years to develop. This chapter celebrates some past inventors, many of whom were only in their 20s when they made their key developments. This innovation process is continuing and other new ideas, currently in the laboratory, are likely to emerge in the coming decades.

This chapter takes a historical look at the development of the current electricity system, focusing on the UK, but of course many of the topics are equally relevant worldwide.

It starts in the 19th century, by describing the basic technology of batteries and the development of the telegraph. It then describes early mains power generation, looking at some of the physics involved and the early days of electric lighting.

The next section follows the development of the humble incandescent electric light bulb and its more energy-efficient relatives, the fluorescent lamp and the light-emitting diode, through to the present day.

The chapter then turns to electric traction, as used in trams and trains, and also in electric cars, a technology that has revived after a long dormant period through most of the 20th century. Improved batteries, new magnetic materials and computer control are making these competitive with petrol and diesel cars. Hybrid petrol–electric and diesel–electric drives also have interesting possibilities.

The next section describes the world growth in electricity since the beginning of the 20th century. It looks at the expanding uses in telecommunications and computing, cooking and heating, refrigeration (including the devastating effects of refrigerants on the upper atmosphere) and the growing use of electric motors.

The following section looks at large-scale electricity generation and the development of today's enormous coal-fired steam turbine plants. It then looks briefly at hydroelectricity. Then it turns to the change in generation fuels in the UK away from coal since 1950, particularly the development of the combined cycle gas turbine (CCGT) and the growth of natural gas as a fuel. This fuel switching has had considerable benefits in cutting UK CO_2 emissions. The section ends by looking at the different possibilities for combined heat and power (CHP) generation.

Next, the chapter looks at the development of the UK National Grid, the long distance transmission of electricity and its local distribution and the recent development of undersea links between the UK and neighbouring countries.

The penultimate section describes the complexities of running an electricity network. It looks at the variation of the demand on the UK grid over the year and the basic workings of a power pool where there is competitive bidding between generators. This section also looks at the problems of peak demands and the variability of renewable energy supplies such as wind. Pumped storage and stronger grid links, particularly to hydropower resources, have a role to play in providing solutions. Finally the section turns to 'smart meters' and 'smart grids' aiming to provide the consumer with better information and further control options for dealing with peak demands.

The final section looks at electricity in the UK and five other countries around the world, all with different generation and use patterns.

9.2 Making electricity in the 19th century

Batteries and chemical electricity

Figure 9.2 The English chemist Sir Humphry Davy (1778–1829) was an early pioneer of electric batteries and arc lighting who also explored the use of electrolysis to isolate chemical elements. Most famous for his invention of the miner's safety lamp, he also discovered nitrous oxide, or 'laughing gas', the first chemical anaesthetic used

Box 4.7 described the basic properties of a battery. At the end of the 18th century it was observed that an electromotive force, a voltage, could be produced between plates or electrodes of two dissimilar metals, such as zinc and copper, if they were dipped in an **electrolyte** such as a dilute acid. Serious experimentation required a large high-voltage battery. In 1808 the chemist Humphry Davy (Figure 9.2), funded by the Royal Institution in London, constructed one with 2000 cells. Using this supply he demonstrated a basic **carbon arc lamp**, producing a bright spark 10 cm long when two carbon electrodes were withdrawn from each other. The electric current continued to flow through a stream of white-hot particles of carbon. It took another 60 years for this to become a commercial lighting proposition.

Davy also explored **electrolysis**, a key chemical application of electricity. He was able to produce pure metals without smelting by passing currents through dilute aqueous solutions of metallic salts. For very reactive metals that react with water, such as sodium, potassium or magnesium, he obtained the metals by passing currents through their molten salts.

Inventors explored the electrical and chemical properties of a wide range of materials during the 19th century, giving rise to a number of designs of non-rechargeable batteries or **primary cells**. In 1839 Sir William Grove developed a battery using zinc and platinum electrodes with solutions of sulfuric acid and nitric acid between them. It worked well and became a favourite power source for the expanding communications industry of the time – the telegraph. He also demonstrated a 'gas battery' using hydrogen and oxygen, or what would now be called a 'fuel cell' (see Chapter 14), but did not proceed with it because he could see no commercial application. In 1868 the French engineer Georges Leclanché developed a cell based on zinc and carbon. This was an immediate commercial success and 20 years later had been developed into the zinc–carbon 'dry' battery, the standard 'torch battery' for most of the 20th century. (More recent alkaline batteries are based on zinc and manganese hydroxide.)

Others were investigating the possibilities of **secondary cells**, i.e. rechargeable batteries, but the commercial need for these did not appear until the birth of mains electricity.

Magnetism and generators

The magnetic properties of electric currents proved equally fascinating. In 1820 Oersted showed that a pivoted permanent magnet would move in the presence of a wire carrying a current (the basis of what we would now call an ammeter). In the same year Ampère (Figure 9.3) founded the science of *electromagnetism*. He showed that when a wire carrying a current was placed parallel to a second conducting wire there was a force exerted between them. Later in the century the unit of current was defined in terms of this force and named the ampère (or amp for short). Ampère also experimented with passing currents through wires wound around a glass cylinder. Other experimenters found that when iron was used as the inner core it became an **electromagnet** and this could be used to attract or repel other magnets.

Figure 9.3 André-Marie Ampère (1775–1836), French physicist, mathematician and pioneer of electromagnetism

By 1831 Michael Faraday (see Chapter 4) had demonstrated that a bar magnet moving through a coil of wire would briefly produce a current. The basic connection between current, magnetism and force was now laid out. A current in a wire could produce a force on a magnet, and a coil moving in a magnetic field could produce a current in a wire. It was now obvious that mechanical energy could be converted directly into electrical energy and vice versa. By the end of 1831 Faraday had produced the first **dynamo**, a device capable of generating a continuous flow of direct current, and by 1834 instrument makers were selling the first commercial hand-driven rotary generators. They would have looked very similar to that shown in Figure 9.4, where a small coil of wire is rotated between the poles of a horseshoe permanent magnet made of iron. As the coil is rotated the changing magnetic effect induces a voltage in it. Two metal spring contacts pressing on metal 'sliprings' on the shaft are connected to the ends of the wire providing an entry and exit path for the current into an external circuit. The version shown will produce an **alternating current (AC)** flowing first one way and then the other. However, by using a more complicated arrangement of sliprings, called a commutator, it is possible to produce **direct current (DC)** that flows in one direction only.

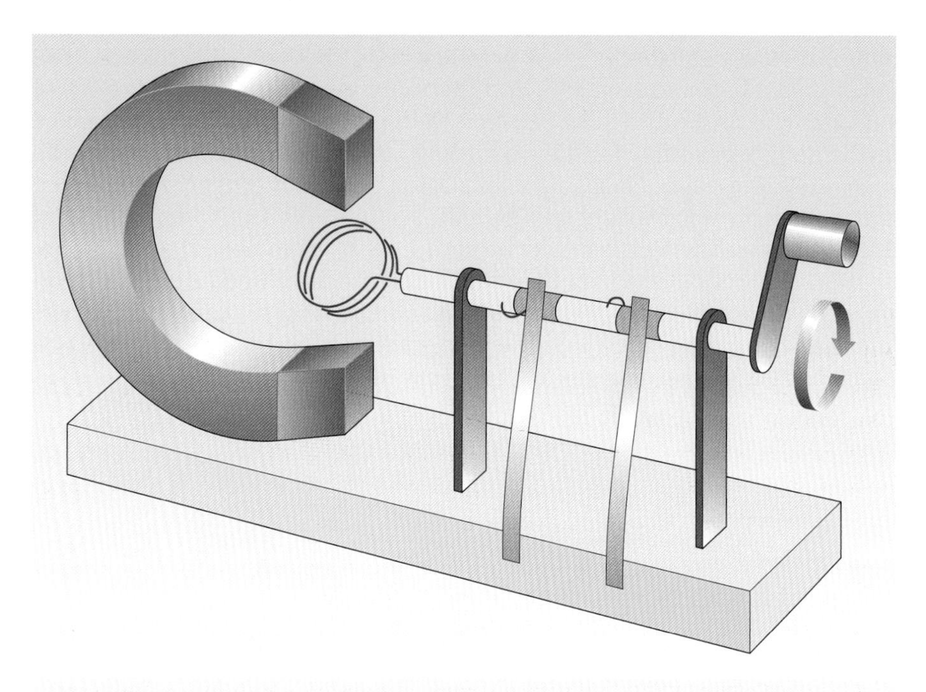

Figure 9.4 A diagram of a simple AC generator as might have been sold in the 1830s

Larger generators, with multiple magnets and whose coils were turned by waterwheels or steam engines, delivered a more substantial flow of current, and led to the first really practical uses of electricity. One of the first, developed in the 1840s, and which required very large direct currents at low voltages, was the electroplating of metals. This was extremely commercially desirable; a small amount of plated silver or gold could go a long way in creating the illusion of more expensive products.

The telegraph

Although the railways enabled a good postal service to be set up in the 1840s, there was an urgent need for something faster. Within a few years of Faraday's work with electricity and magnets, the telegraph – writing at a distance – had been invented by William Cooke and Charles Wheatstone. The railway companies in particular were very interested.

Figure 9.5 An early alphabetic signalling device. The diamond-shaped display board has the letters of the alphabet written on it. Messages were spelled out using a set of five meter needles mounted across the diamond

The principle was very simple. A switch could open or close a circuit formed with a battery and some wires. Somewhere else on that circuit, potentially many miles away if the wires were long enough, that changing switch position could be detected by deflecting a meter needle. The device shown in Figure 9.5 used a set of five needles to spell out the letters of the alphabet. It was first tried out in 1837 and its fame was assured after the police used it to telegraph ahead and arrest a suspected murderer seen boarding a London train at Slough in 1842. The telegraph soon came to be the key to the efficient scheduling and organization of the railways. Trains ran to timetables under the control of signalmen linked by telegraph.

However, the five-wire system was too complicated and unreliable. In 1837 the American Samuel Morse suggested using a single wire telegraph with a buzzer, or an electromagnet that might be connected so it inked a mark on a moving ribbon of paper. Morse invented a code, which bears his name to this day, consisting of 'dots and dashes' which were formed using short and long switch closures. In 1842 the US Congress gave Morse a US$30 000 grant for a telegraph to link Baltimore and Washington DC. It was a great success and spurred a rapid spread of telegraphy around the world. It also created a demand for batteries, high-quality copper wire, and operators. The young Thomas Edison (see below) learned about electricity by working as a telegraph operator.

The next major step was the laying of a telegraph cable on the floor of the Atlantic Ocean to join the USA and Europe. A cable laid in 1858 failed after a few weeks in operation and another laid in 1865 broke. Finally, a successful cable was laid in 1866 and the 1865 one was retrieved and repaired. Now a message could be sent across the Atlantic in a few minutes, instead of 11 days for the fastest crossing by steamship. By the 1870s there was a worldwide telegraph infrastructure.

In 1870 the telegraph was joined by the telephone, which allowed sounds to be transmitted at a distance. An Italian, Antonio Meucci, built a number of working prototypes between 1850 and 1862, though its invention is normally credited to a Scottish professor working in Boston, Massachusetts, Alexander Graham Bell. Like the long-distance telegraph network, the new urban telephone networks were all powered by batteries.

The rise of electric lighting and mains electricity

The use that really created a demand for mains electricity, as we know it today, was *lighting*. At this time it was mainly achieved using inefficient gas flames or oil lamps (see Chapters 3 and 5). Although carbon arc lamps had been demonstrated during the 1840s, it was hopeless trying to run them on batteries; they had to be coupled to steam-powered generators. Coastal lighthouses were an obvious application for the new brilliant light source and a number were installed in England, with steam generation plant, from 1856 onwards.

During the 1860s and 1870s considerable effort was devoted to improving the efficiency of dynamos. Dr Werner Siemens and Sir Charles Wheatstone looked at using *electromagnets* in generators to replace the heavy permanent magnets. These electromagnets could be run from the output of the generator itself. It might seem that such a machine could never be started – no current, no magnetic field. In practice the soft iron pole pieces held enough residual magnetism to get the process started. Once running they could supply their own magnetic field, i.e. they were *self-exciting*. These machines could be made lighter, cheaper and more efficient than those using the permanent magnet materials of the time.

The word '**generator**' has come to apply to any rotary device for making electricity, often including the power plant to drive it. The word '**dynamo**'

Figure 9.6 The Newcastle chemist Joseph Swan (1828–1914) made incandescent lamps but also had key expertise in vacuum pump technology

Figure 9.7 Thomas Edison (1847–1931) was a remarkable inventor and entrepreneur. In the 1870s he worked on automatic telegraph machines before producing a working incandescent light bulb in 1879. He set up an 'invention factory' at Menlo Park, New Jersey, employing a large number of researchers, which is why a large number of inventions are credited to his name. He promoted DC mains electricity supply and a number of US electricity utilities are still named after him

usually describes a small generator using a permanent magnet to supply its field and producing direct current. An **alternator** is a generator that produces alternating current. Its three-phase form, described later, is the normal generator used in modern power stations.

Carbon arc lamps give off a brilliant white light, excellent for lighthouses and searchlights, but they were just as labour intensive as oil lamps. The carbon rods slowly burnt away, needing constant replacement. Despite the technical difficulties, 30 000 spectators watched a floodlit football match in Sheffield in 1878. In the following year Blackpool started its seasonal 'illuminations' when the promenade and pier were lit by six powerful arc lamps. In the USA arc lamps mounted on 'moonlight towers' anything up to 90 m high were installed for lighting at main city street intersections.

By 1878 the combination of new generators and arc lamps was sufficiently well developed to compete with gas lighting. In many cases gas was cheaper, but the margin was narrow enough to encourage further development. While the gas companies worried about losing their street lighting contracts, they remained confident about hanging on to the domestic and office lighting market, for which the harsh arc lighting was totally unsuitable.

Then in December 1878 the Newcastle chemist Joseph Swan (Figure 9.6) demonstrated something far more saleable, an **incandescent lamp**, the 'light bulb' or 'glow lamp' as it was known at the time. He showed that a carbon fibre inside an evacuated glass globe would glow when carrying an electric current. The idea was not new – he had produced an example in 1848, but it did not last long enough to be useful. He started selling his lamps in 1879 at 25 shillings each (about £100 in modern terms).

At the same time the dynamic American inventor Thomas Edison (Figure 9.7) produced an almost identical lamp, which he proceeded to patent. A lawsuit followed. Ultimately, a commercial light bulb required both Edison's expertise in filaments and Swan's expertise in vacuum pump technology. In 1883 the two dropped their legal wrangles and formed the Edison and Swan United Electric Light Co. Ltd.

Glow lamps gave out a gentle light comparable to an oil lamp and were far more suitable for the home than the gas lamp (which smelt of sulfur) or the harsh arc lamp.

Indoor electric light immediately became a luxury status symbol (see Figures 9.8 and 9.9). In 1881 the Marquis of Salisbury had it installed in Hatfield House, and it was also put into the Savoy Theatre in London. Whereas conventional oil lamps (and even electric arc lamps) required continual attention, the incandescent lamps simply ran until their brittle filaments eventually burnt out or broke. They were, and still are, particularly sensitive to the supply voltage. If it is too low, they glow dimly and inefficiently. If it is too high they burn out very quickly. A 5% increase in voltage is sufficient to halve the life expectancy.

Thomas Edison proceeded with entrepreneurial zeal to set up generating systems. His first demonstration project was on the Holborn Viaduct in London in 1882, powered by the first public steam power station in the world. In New York, he formed the Edison Electric Illuminating Company and set up his Pearl Street generating station, installing six of his new 27 ton 'Jumbo' dynamos (see Figure 9.10) and laying cables in underground conduits. By the end of 1882, 192 buildings with over 4000 lamps had been connected. This was the first in a whole series of Edison supply companies in the USA.

The rush of developers to promote systems led to the same problems that the gas industry had encountered earlier in the century, involving rights to dig up the streets to lay cables. In 1882 the UK Parliament passed the *Electric Lighting Act* requiring that private electricity suppliers who wished to lay underground cables could only do so with the permission of the local authority.

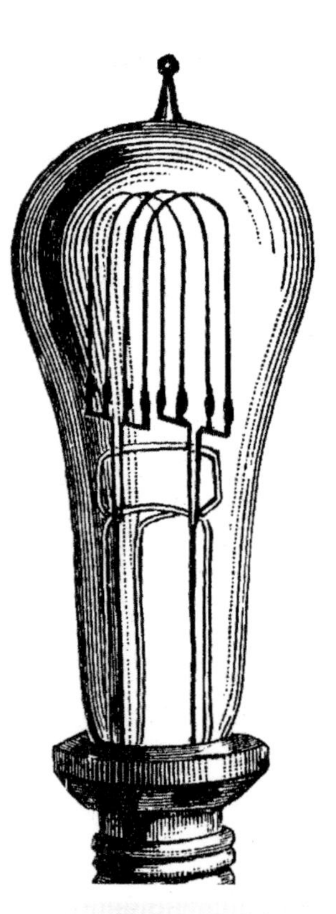

Figure 9.8 An early light bulb or 'glow lamp'

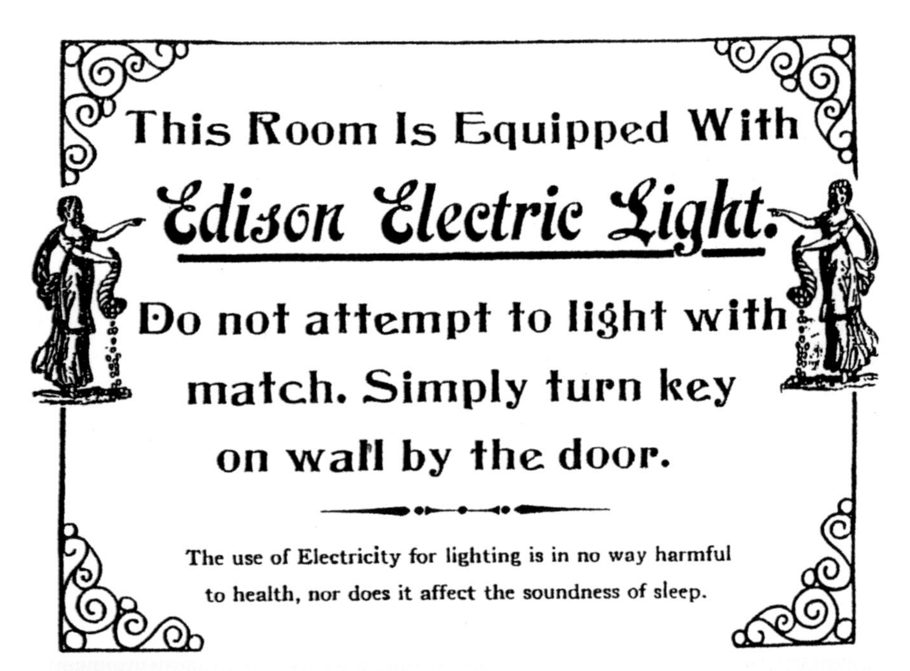

Figure 9.9 Early advertisement for electric light

Figure 9.10 Dynamo room of Edison's Electric Lighting Station, Pearl Street, New York. The enormous size and weight of the machines were a consequence of the poor magnetic materials of the day

AC or DC?

A battery produces a steady voltage and, given a constant load, a steady, unchanging direct current. A coil rotating between the poles of a magnet will produce an alternating current. Box 9.1 explains some of the details of AC.

BOX 9.1 Alternating current

Direct current flows continuously in one direction. In the example shown earlier in Box 4.5, it flowed from the positive to the negative terminals of a battery, through a load such as a torch bulb. An alternating current or voltage is one which goes through a complete cycle of changes periodically in time, reversing direction from positive to negative, as in Figure 9.11.

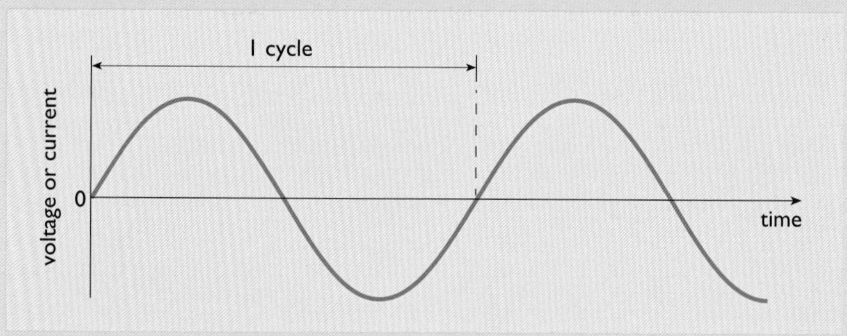

Figure 9.11 An alternating current (or voltage) waveform

The time for a complete cycle is called the **period**, normally denoted by the symbol T. Its inverse, the number of cycles per second, is called the **frequency**, usually denoted by the symbol f. The unit of frequency of one cycle per second is the hertz, named after Heinrich Hertz (see Chapter 4).

Thus:

$$f = \frac{1}{T} \text{ Hz}$$

The normal frequency of mains electricity in Europe is 50 Hz. In the USA it is 60 Hz.

Normal domestic mains electricity in Europe is AC delivered at a nominal 230 volts (230 V). Other values for the UK are stated elsewhere, but in practice it is not important because of wide permissible limits. A figure of 240 V was chosen in 1946 and later revised down to 234 volts. It has recently been altered to 230 V –6% +10% (i.e. it can legally be anywhere between 216 V and 253 V).

Obviously, the voltage is actually varying all the time, so the figure of 230 V is an average, a rather special one called the *root-mean-square* or r.m.s. voltage. If the voltage waveform is sinusoidal, as it is in Figure 9.11, then the r.m.s. voltage can be shown to be:

$$\frac{1}{\sqrt{2}} \text{ or } 0.71 \text{ times the peak voltage.}$$

Chapter 4 described the relationship between power, voltage and current for a DC circuit. As long as the load is a simple resistance (which it is for applications such as immersion heaters or incandescent light bulbs) the same holds for r.m.s. AC voltages and currents.

Power (watts) = voltage (volts) × current (amps)

Voltage (volts) = current (amps) × resistance (ohms) or $V = I \times R$

If the load is a resistance, R, then power, $P = \dfrac{V^2}{R} = I^2R$

But which was best for mains electricity? Arc lamps and incandescent bulbs could operate on either. Edison was a fervent supporter of DC supply provided by a steam-driven generator, but augmented with large rechargeable batteries, carrying out the same storage function that the gasholder did for town gas. This could keep the supply running if the generator had to be shut down for maintenance and could also cope with sudden peak loads. A DC supply was also convenient for charging other batteries, such as those in the battery–electric vehicles that started to appear in the 1890s. It was the preferred supply for variable speed electric motors such as those used on trams and early electric trains.

But the proponents of AC, such as George Westinghouse, had good arguments and eventually won the day. AC electricity could be generated at a high voltage for distribution and then easily changed in voltage down to that needed by the consumer using a transformer, a set of coils of wire wound round an iron core (see Box 9.2).

BOX 9.2 The transformer

The transformer is a device for converting the voltage in an AC electrical system, either 'stepping up' an input voltage to a higher value, or 'stepping down' to a lower one. It depends for its operation on the mutual interaction of changing electrical and magnetic fields, so it will only work with alternating currents.

A typical transformer consists of two coils of wire, wound in close proximity to each other around a 'core' usually made of thin layers of laminated sheets of a silicon–iron alloy that is easily magnetized (see Figure 9.12).

The first, 'primary', coil is connected to the input source, usually a mains supply. The 'secondary' coil is connected to the output – i.e. to the appliance or system requiring a higher or lower voltage for its operation. An alternating voltage applied to the primary coil induces an alternating magnetic field in the metal core. This field in turn induces an alternating voltage in the secondary coil. If the secondary coil has more turns of wire than the primary coil, the voltage induced in it will be higher than in the primary coil; and if it has fewer turns, the voltage will be lower. In a 'perfect' transformer, the ratio of voltages is the same as the ratio of number of turns. If the primary coil has N_p turns and the secondary has N_s turns, then the ratio of the primary voltage V_p to the secondary voltage V_s will be:

$$\frac{V_p}{V_s} = \frac{N_p}{N_s}$$

The law of energy conservation still applies, however, and the output power cannot exceed the input power. The voltage in the secondary coil may be higher than in the primary, but the current will be proportionately less; and if the voltage in the secondary coil is lower, the current in it will be correspondingly higher.

Transformers are very efficient devices. Typically only a few per cent of the input power is lost in the transformation process, mainly in heating the coils and in magnetizing the metal laminations. The energy losses in the laminations can be reduced by using modern silicon–iron alloys incorporating nickel and boron.

Transformers at low frequencies, such as 50 Hz, can be large and heavy. At higher frequencies they can be made smaller even though they carry the same amount of power. Modern electronic equipment, such as computers, TVs and low-energy light bulbs, uses **switch-mode power supplies**. They convert the 50 Hz to DC using semiconductor rectifiers and then convert this to high-frequency AC at about 30 kHz using high-voltage transistors. This is then fed to a much smaller transformer to change the voltages to those required.

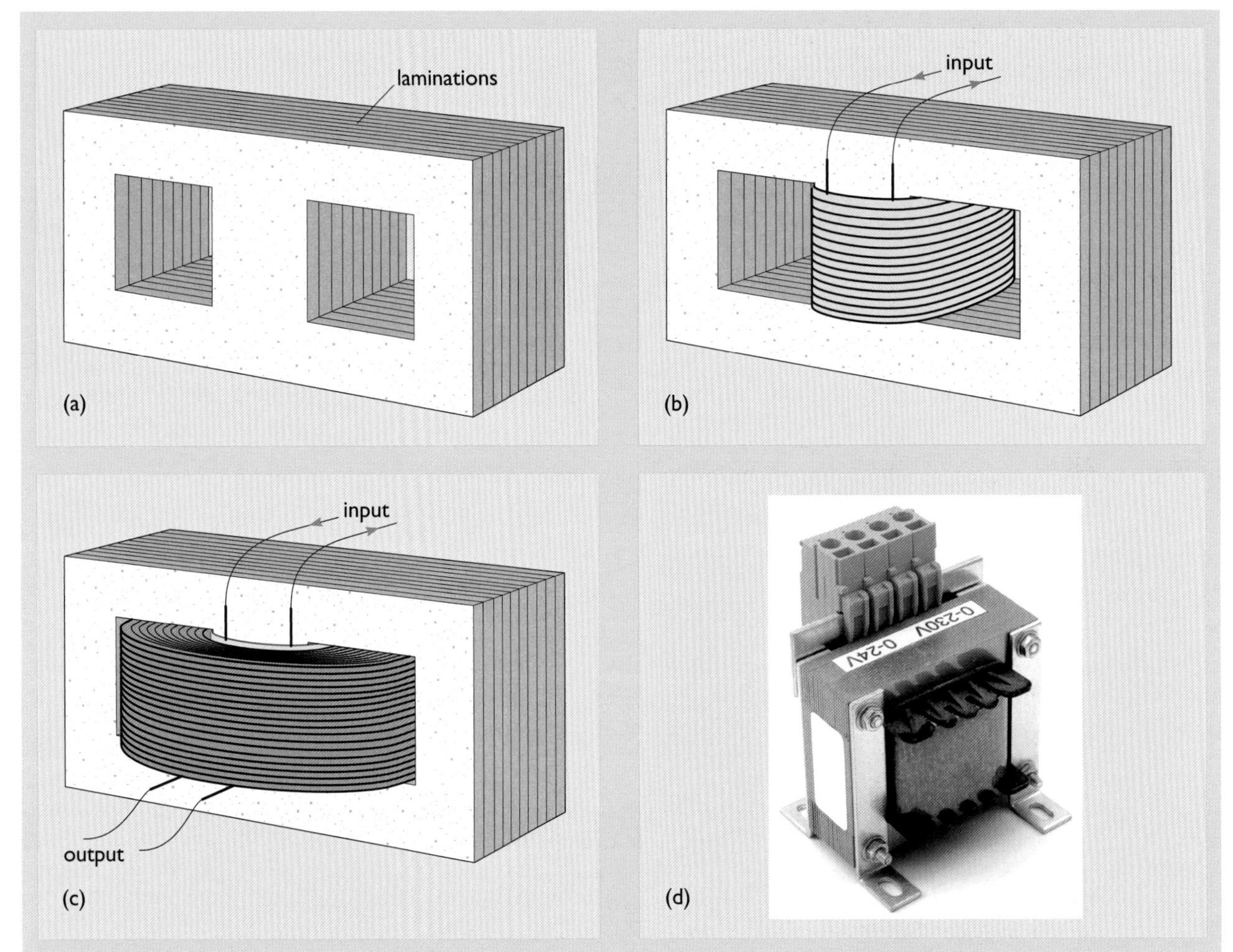

Figure 9.12 Construction of a transformer: (a) the basic core of laminations; (b) the input primary coil; (c) the output secondary coil; (d) a completed small transformer

Alternating current could be used for running both arc lamps and incandescent lamps, but it was initially not clear how it could be used directly to run electric motors. Then, in 1887, Nikola Tesla (Figure 9.13) showed how two separate AC supplies with the same frequency but displaced in phase could be used to produce a strong rotating magnetic field in a motor, dragging a magnet around with it. Although he started with two-phase electricity, he laid the foundations for the use of three-phase electricity, which has subsequently become the normal method of generation and distribution (see Boxes 9.3 and 9.4). Three-phase motors could be made very powerful indeed and over 90% efficient, eventually leading to their widespread use in factories to drive machinery, replacing direct drive from steam engines. AC motors could also be used to drive dynamos to generate DC electricity.

Figure 9.13 Nikola Tesla (1856–1943), a Serbian who emigrated to the USA, where he worked for the entrepreneur George Westinghouse. Tesla pioneered the use of AC electricity to drive motors and developed multiphase electrical distribution

BOX 9.3 Three-phase alternating current

A three-phase AC supply has three main conductors and a central neutral or star wire. The three separately generated alternating voltages between the star wire and each of the three conductors are identical but one-third of a cycle out of phase with each other (see Figure 9.14). Differences in electrical phase are expressed in degrees, taking a whole cycle as being 360°. Phase 2 is shifted by 120° relative to Phase 1, and Phase 3 is shifted by a further 120°.

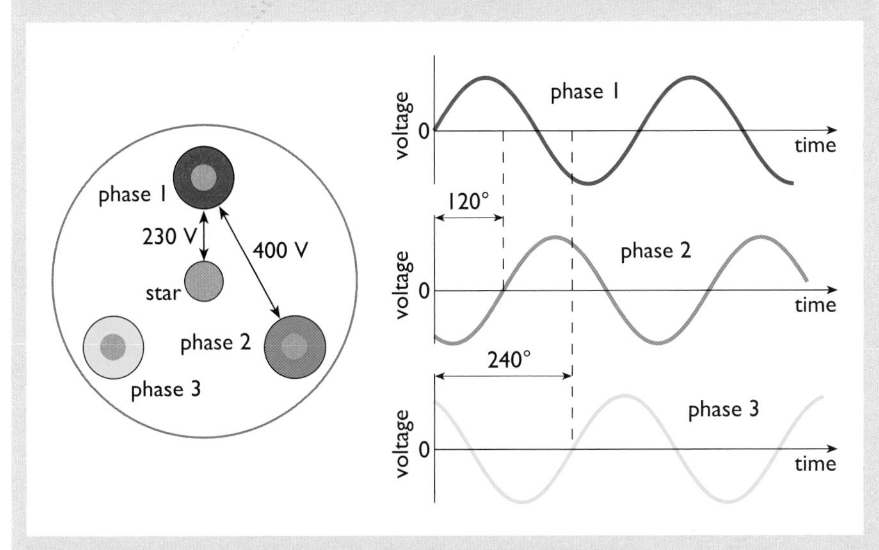

Figure 9.14 Voltage–time waveforms of three-phase electricity

This sounds complicated but has immense versatility. A typical three-phase distribution cable in a residential street has a potential difference of 230 V r.m.s. between the star wire and each conductor. It is normally described as being rated at 400 V AC because there is a difference of 400 V r.m.s between any one of the phase wires and another. A factory would use all three phases together in electric motors for machines. An individual house would tap off a 230 V single-phase supply between any one of the phase wires and the central star or neutral wire.

An electricity pylon normally carries two sets of three-phase wires hanging on long insulators, one set on each side, while the star wire runs along the top. If the currents are perfectly sinusoidal and the loads on all the phases are properly balanced, then theoretically all of the current is carried in the phase wires. The star wire can be made relatively thinner to carry any actual imbalance.

BOX 9.4 The three-phase alternator

Figure 9.4 showed a basic generator, with a coil of wire rotating between the poles of a magnet. The modern power station alternator, producing three-phase electricity, is based on a slightly different arrangement.

In its simplest form it can be thought of as a bar magnet (the **rotor**) which is rotated between a set of three fixed coils which together comprise the **stator** (see Figure 9.15). With the wiring connected in what is known as the 'star' connection, the inner end of each coil is connected to the neutral wire of the three-phase system and the outer ends form the three-phase wires. As the magnet is rotated, so a three-phase voltage is produced as shown in Figure 9.15.

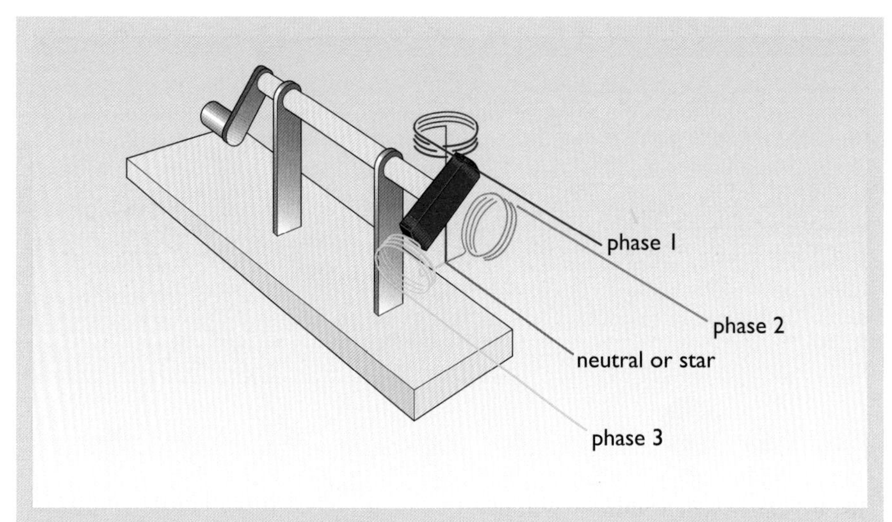

Figure 9.15 The three-phase alternator

In a more practical alternator, the bar magnet will be replaced by an electromagnet, a coil of wire carrying a large direct current. This is derived from a smaller separate DC generator called the **exciter**. The stator will take the form of a very complex set of windings mounted on a steel frame. The final generated power, which may run to currents of thousands of amps at voltages of 11 kV or more, emerges from the windings of the stator. Unlike the generator in Figure 9.4 this enormous current doesn't have to pass through a slipring.

Modern power station alternators can be very large indeed, up to a gigawatt in output, and are highly efficient (typically 98.5% at full load). Even so, dissipating the other 1.5% of heat losses may require pumping a coolant (water or even high-pressure hydrogen or helium) between the windings.

In 1888, Tesla went on to develop an *induction motor* that could be used on a single-phase AC supply. He showed that these could be made very small, just a fraction of a horsepower, and tolerably efficient (about 50%). This opened the door for a host of new electrical machine tools and domestic appliances, such as an electric table fan, first demonstrated in 1889.

Early power stations used separate generators to supply different loads. Experiments showed that it was safe to connect all the loads together and supply them from multiple three-phase AC generators connected together in parallel. This discovery was key to the development of the modern electricity system. However, whenever a new generator is connected to the system it must carry out a **synchronization** process. It is run up to exactly the right speed to produce electricity at the right frequency (50 Hz in the UK) and then finely adjusted so that its output will be produced at exactly the right *phase* with the National Grid. Only then are the switches thrown to connect its output. Today, this is done automatically under computer control.

High voltage or low voltage?

By the mid-1880s, the requirements of lighting and early DC motors had already set the normal mains supply voltage at between 100 V and 240 V.

This was not exactly ideal for filament lamps, but there were trade-offs to be made when it came to distribution wiring (see Box 9.5). Higher voltages allowed thinner wires to carry the same amount of power. They also allowed thinner pins to be used in plugs and sockets.

BOX 9.5 Choosing a lamp voltage

The current taken by a lamp will be equal to its power divided by the voltage:

$$I = \frac{P}{V}$$

A 100 W incandescent lamp designed for 230 V will therefore take:

$$\frac{100}{230} = 0.435\text{A}$$

Its resistance when running will be:

$$\frac{V}{I} = \frac{230}{0.435} = 529\ \Omega$$

Many modern lamps, however, are designed to run on 12 V. Examples are car headlamps and tungsten–halogen lamps run from the mains with transformers.

At 100 W a tungsten halogen lamp will take:

$$\frac{100}{12} = 8.3\ \text{A}$$

Its resistance needs to be:

$$\frac{V}{I} = \frac{12}{8.3} = 1.44\ \Omega$$

Thus a mains lamp must be designed to have a filament with a resistance (when hot) over 350 times larger than for a 12 V lamp of equivalent power. Low-voltage lamps have short thick filaments that are quite robust. Mains lamps have filaments that are long, thin and consequently fragile.

However, the 12 V lamp takes nearly 20 times more current than its mains counterpart. In practice, the supply wiring for 230 V can be made much thinner than that for an equivalent 12 V system. It is not just the wire that is affected, but also the size of the pins in plugs and sockets. UK domestic mains wiring is based on plugs and sockets capable of carrying 13 amps or a load of about 3 kW.

Early electrical wiring was not particularly safe and electric lighting can only have been slightly less of a fire risk than gas. Distribution voltages of up to 250 V, although dangerous, were rarely actually lethal. Over the years, safety standards have improved enormously, especially after the introduction of plastic PVC cable insulation in the 1950s. Although an electric circuit only requires two connections, a normal mains socket has *three* pins. The live pin carries the high voltage and the neutral pin acts as the current return and will in practice be at a voltage close to earth. The earth pin is what it suggests, a safety connection genuinely at **earth potential**: the voltage of the ground. Modern wiring regulations insist that all exposed metal surfaces on electrical equipment, as well as water taps in kitchens and bathrooms, are securely earthed. Bare wires at more than 35 V above earth potential must be kept safely out of reach.

The choice of distribution voltage is much more critical when it comes to transmitting power over long distances. Attempting to transmit any substantial amount of power at the normal local distribution voltages requires low-resistance cabling that would be impossibly thick. This severely limited the capabilities of early DC systems, as their economic radius of operation was put at about 2 km (Electricity Council, 1987).

The need to make efficient use of copper for wires was highlighted when in 1887 a French copper syndicate, seeing a boom in demand, cornered much of the world's supplies, forcing up the price. Today, domestic wiring still uses copper, but transmission lines are likely to use aluminium, more abundant but still expensive.

Today, distribution at voltages of 400 kV or more is considered normal. The early development of high-voltage distribution seems quite hair-raising by today's standards. One pioneer was Sebastian Ziani de Ferranti who in 1885 was brought in, at the age of only 21, to sort out the slightly chaotic Grosvenor Gallery AC system in London's prestigious Bond Street. He ensured that all of the distribution transformers worked in parallel across the mains, supplying a constant voltage, vital for the proper operation of incandescent lamps. He then turned to the problem of the company's rising demand (1 MW at the time). His solution was to build a new power station using the new, more efficient, steam turbine technology (see Chapter 6). The chosen site was 10 km away next to the Thames at Deptford, with power transmitted to a substation at the Grosvenor Gallery at 10 000 V. Ferranti was forced to develop his own high-voltage cable, insulated with wax-impregnated paper and surrounded by an outer iron casing. It was too thick to be made in coils and was made in 6 metre lengths, the maximum that could be carried through the streets by a horse and cart. The Board of Trade was understandably nervous of having this buried in the public streets. Ferranti arranged a demonstration of its safety. He had one of his assistants hold an uninsulated chisel while another hammered it into a live cable through the outer earthed casing (see Figure 9.16). Fortunately, the test was deemed successful and permission to lay the cable was granted. It remained in service for 40 years.

Figure 9.16 Sebastian Ziani de Ferranti looks on while testing his 10 000 V cable in 1890. His assistant is holding an uninsulated chisel as it is driven through the metal casing of the live cable

The technology for high-voltage transmission developed rapidly. In the USA, the Niagara Falls hydroelectric project was started in 1895 with three-phase transmission at 11 000 V over the 35 km to Buffalo. By the 1920s overhead lines reached voltages of 220 kV in the USA and underground cables were available for use up to 132 kV. This laid the technological basis for the construction of the UK National Grid in the 1930s (see Section 9.7).

Although high-voltage AC has been the standard for long-distance electrical transmission throughout most of the 20th century, high-voltage DC transmission is increasingly being used in new links. This is only possible because of the development since the 1970s of voltage converters using high-voltage and high-power semiconductors.

Simple metering and tariffs

Electricity was originally synonymous with 'electric lighting' and contracts were drawn up in terms of providing lighting to so many streets for a year. Entrepreneurs like Edison provided both the incandescent lamps and the electrical generation plant. If they improved the performance of either, it would be to their commercial benefit in providing the 'energy service' of illumination for a fixed price. This was fine as long as the consumers didn't want to use the electricity for anything else.

Forty years earlier, in the 1840s, gas companies had similarly charged for gas on the basis of the number of gas lamps in a house. As soon as other uses, such as gas cooking, were introduced it became necessary to develop proper gas meters and legislate that gas was to be sold by its energy content.

The same happened with electricity. It became a 'product' metered in UK Board of Trade 'units' of 1 kWh. Consumers were free to use it for any purpose. Whether they bought the most efficient light bulbs or the cheapest was up to them.

'Rotating disc' meters were introduced in 1889 and the basic design has lasted through to the present day. They are only now being replaced by electronic meters.

The actual way that a charge is made is known as a **tariff**. There are many different forms. Although we might think of 'buying electricity' as just a matter of kilowatt-hours, we are also paying for the transmission wires and the privilege of having a connection. The actual costs of generation only make up about a half of a UK domestic electricity bill today. The connection costs have to be borne whether or not the consumer takes any electricity at all.

So, a normal UK electricity bill consists of a **standing charge**, which is a fixed daily charge, and a **unit rate** of so many pence per kilowatt-hour. All or part of the standing charge may be disguised as a higher unit rate for the first few thousand kilowatt-hours per year. More complex tariffs are available, the most common being one that offers cheaper electricity at night. Industrial tariffs can be very complex, featuring not only cheap rates at night and during the summer, when demand is low, but compensating with very high prices during winter evenings, when demand can be very high. This subject will be revisited later under the heading of 'smart meter' and smart grids' (see Section 9.8).

9.3 The continuing development of electric lighting

Improving the incandescent light bulb

Chapter 5 described Welsbach's invention of the gas mantle and how it improved the efficacy of gas lighting by the use of phosphors in the 1880s and 1890s. This was a severe blow to early electric lighting entrepreneurs. In 1904 Welsbach redressed the balance by developing an electric incandescent lamp with a metal filament, using osmium. This could be raised to a higher temperature than carbon filaments, producing a brighter light and improving the lighting efficacy by about 50%. Today's incandescent lamps have tungsten filaments and are capable of running at even higher temperatures and efficacies. Tungsten is normally very brittle and it required considerable research to be able produce it in a wire form.

Then in 1913 Langmuir in the USA suggested filling the lamp with an inert gas such as argon rather than having a high vacuum. This inhibits the slow evaporation of the filaments allowing lamps to run hotter and brighter still. Finally, in 1934 came the 'coiled-coil' lamp, in which the filament is made as a fine coil and then coiled on itself. This reduces convection of heat in the inert gas. The everyday 'light bulb', also known as the general lighting service lamp (GLS), had almost reached its current state of performance. It remains almost unchanged today, over 70 years later, apart from improvements in its mass production.

Figure 9.17 A miniature tungsten–halogen lamp, introduced in 1960. It is about three centimetres high.

This is not to say that development has halted. In 1960 the miniature **tungsten–halogen lamp** was introduced (see Figure 9.17). This has a trace of a halogen element, such as iodine or bromine, added to the gas filling. This combines with any evaporated tungsten and makes sure it is precipitated back onto the filament allowing it to be run even hotter and brighter, increasing the efficacy by about 20%. In order to work, the glass bulb wall must be very hot, at least 250 °C, so the lamps are much smaller than conventional incandescent bulbs and are made of quartz glass to withstand the high temperature.

The fluorescent lamp

Chapter 3 introduced the terms *lumen* and *efficacy* for different light sources. Edison's original incandescent lamps had an efficacy of under 3 lumens per watt of electricity consumed. By 1920 the figure had been improved to almost 12 lumens per watt and they had re-asserted their price advantage over both gas and carbon arc lamps for street lighting. The dominance of the incandescent lamp wasn't challenged until the 1930s brought another contender – the fluorescent lamp.

In 1867 the French physicist Henri Becquerel showed that an electric current would flow through low-pressure mercury vapour inside a long sealed glass tube. In the late 1890s the inventor Peter Cooper Hewitt developed this idea to produce lamps that give off a bright bluish-green light. Although useless for domestic lighting, they were widely used by photographic studios. In those days of slow black and white film, they were interested in bright lights, but not too worried about colour rendering – the ability to show colours as the eye would see them under natural daylight (a topic to be discussed later in this section).

In 1933, the US General Electric Company marketed a new, brighter, mercury vapour lamp, the **high-intensity discharge lamp (HID)**. This had the high brightness of the arc lamp, but was sealed, did not give off any poisonous fumes, was maintenance-free and had a long life. This was a product to challenge the incandescent lamp for street lighting.

In Europe an alternative, the low-pressure sodium vapour lamp, with a slightly more appealing bright orange colour, was being introduced at the same time. The present generation of high-pressure mercury and sodium discharge lamps was developed in the 1960s, giving light over a wider range of the spectrum (a 'whiter' light) but they still retain their basic blue-green or orange tints. These are now familiar in modern street lights. **Metal halide lamps** are mercury lamps to which other metal vapours are added to modify the colour.

In fact low-pressure sodium lamps have the highest lighting efficacies, around 200 lumens per watt, but it is usual to sacrifice efficacy for better colour rendering.

The modern 'office' tubular fluorescent lamp also owes its existence to another early 20th century researcher, the Frenchman Georges Claude. He found that the newly discovered gas neon would glow a brilliant red in a discharge tube. He also perfected the art of bending the tubes to spell out words. The advertising industry, particularly in the USA, loved his red neon and blue-green mercury lamps and was prepared to pay for them (US$400 apiece in 1924). This specialist application made sure that fluorescent lamps became familiar items.

The problem was to produce a 'white' light. Becquerel's 1867 experiments showed that a mercury vapour arc also gave off ultraviolet light. If the tube was coated on the inside with a phosphor, such as zinc sulfide, this would be converted to visible light. Different phosphors could be used to adjust to precise colour. Getting a product to rival the incandescent lamp was not easy. Only in 1938, 60 years after Becquerel's demonstration, did the US General Electric Company produce a commercial tubular fluorescent lamp. This initial version had twice the efficacy of an incandescent lamp and twice the life expectancy.

The basic design has remained much the same since then (see Figure 9.18). It features a sealed glass tube, with metal end-caps. This contains a certain amount of argon gas and a small quantity of mercury, and is coated on the inside with a selection of phosphors. At each end of the tube is a small filament heater. Initially, when the lamp is off and the tube is cold, the mercury will exist as small droplets of liquid. When the lamp is turned on, mains voltage is applied between the two ends of the tube and a special starter circuit energizes the heaters. An arc forms through the argon, and as soon as this happens, the heaters are switched off. The liquid mercury now vaporizes and the mercury arc gives off ultraviolet light, which is converted to visible light by the various phosphors in the coating. In its basic design a large coil, or ballast, limits the current flowing through the tube.

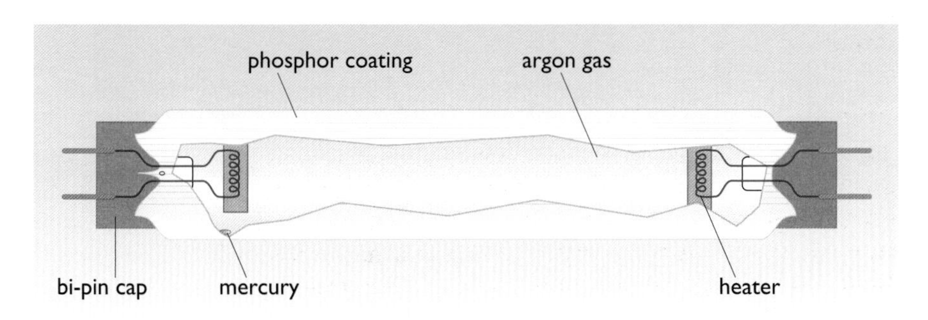

Figure 9.18 The basic fluorescent tube lamp

Since 1938, new design developments and phosphors have enormously improved its lighting efficacy, colour rendering and life expectancy. It has become the standard form of office lighting throughout the world.

Even so, by the 1970s the incandescent lamp was still stubbornly popular, although virtually unchanged in performance for over 30 years. What was needed was a 'compact fluorescent lamp' that could directly replace it. New techniques of bending narrow glass tubing enabled small folded fluorescent lamps to be produced. In 1980, Philips produced its SL series, larger than a normal incandescent bulb, but sufficiently small to fit in most normal light fittings. It was much heavier, with the ballast coil mounted in the base. It used only a quarter of the electricity of an equivalent incandescent lamp, and lasted much longer.

In the 1980s a new technological improvement, electronic ballasts, boosted the efficacy of both compact and full-sized fluorescent lamps. It had long been known that the higher the frequency of the AC applied to a fluorescent lamp, the more the mercury vapour became excited and the brighter the lamp became. But mains electricity only came at frequencies of 50 Hz and 60 Hz. It was not until the development of small, very cheap, high-voltage transistors in the late-1970s that this phenomenon could be exploited. These allowed the mains power to be converted to high-frequency AC at 35–40 kHz.

The result has been a lightweight compact fluorescent lamp (CFL) using only 20% of the electricity of a normal incandescent lamp and with eight times its life expectancy. Phosphors have been carefully chosen to match the colour balance of an incandescent lamp as closely as possible. The fact that these complex lamps can be sold for a price of £10 or less is an amazing feat of mass production. Figure 9.19 shows four different types of lamps available for domestic use.

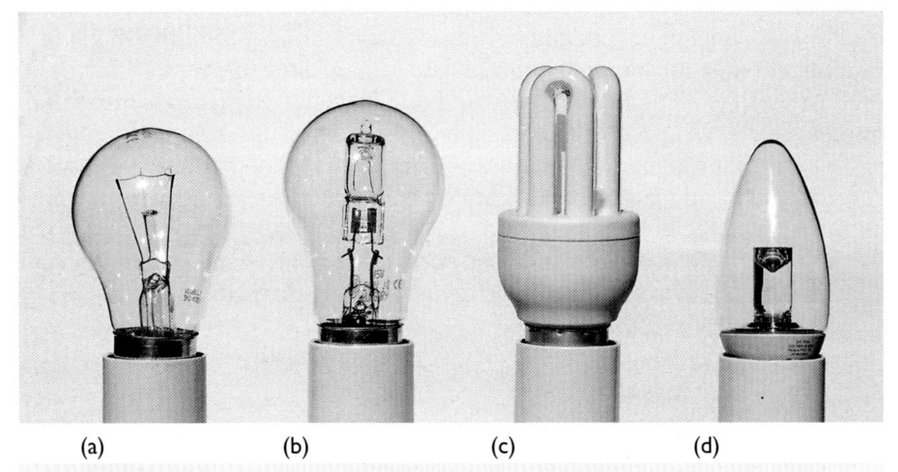

Figure 9.19 Types of lamp: (a) modern incandescent; (b) halogen replacement incandescent; (c) electronically ballasted CFL; (d) light-emitting diode

The light-emitting diode (LED)

It had been known since early in the 20th century that a current passed through the junction between dissimilar materials might produce light.

As with the fluorescent lamp, getting a commercial product has taken many decades. Practical designs have had to wait for the development of **semiconductor** materials. These are neither good conductors, nor complete insulators. In the mid-20th century it was discovered that tiny slices of semiconducting materials based on germanium and silicon would allow a flow of current only in one direction. This is the basic electrical property of an electronic device known as a **diode** or **rectifier**. This semiconductor technology was further developed to produce transistors and today's electronic devices.

Red LEDs based on the semiconductor gallium arsenide were introduced as small indicator lamps in the 1970s. Since then colours have progressed into orange, green, blue and even ultraviolet. Each colour has required the careful development of a specific semiconductor material. Power ratings have increased from milliwatts to watts and the overall brightness and size of LEDs are now sufficient for use in traffic signals, display boards, bicycle lamps and car indicators.

Modern white LEDs mostly use a basic blue indium gallium nitride semiconductor coated with a yellow-red phosphor based on cerium and yttrium. The luminous efficacies are competitive with incandescent lamps, but LEDs can last 20 times longer, drastically cutting down on maintenance costs for replacements. To date the LED substitute for a standard 60 watt incandescent light bulb has yet to appear, though at the time of writing (March 2011) 40 watt equivalent lamps are being manufactured.

Some of the elements used in the manufacture of phosphors for fluorescent lamps and LED semiconductors are not particularly common and there are concerns about their widespread availability, a topic that will be touched on in Section 9.4.

Efficacy and colour rendering index

Today there is a profusion of different electric light sources. Obviously the efficacy, the amount of light produced per watt of electricity, is important, but so is the lighting *quality*.

The incandescent lamp has been successful because it produces light over a wide spectrum – the colours of people, fabrics and paintings illuminated by it appear normal. A 'white' light can also be produced by a mixture of red, yellow, green and blue lights using phosphors. In an (old-fashioned) colour television using a cathode ray tube (CRT) they are excited by an electron beam. In a gas mantle the phosphors are excited by heat. In a fluorescent lamp they are excited by UV light and in a white LED red and yellow phosphors are excited by blue light. Different phosphor mixtures are used to produce 'cool' (bluish) or 'warm' (reddish) lighting. However, coloured objects lit by this 'mixed' light may not appear the same as under an incandescent lamp, and even worse, different people may perceive colours differently. The **colour rendering index (CRI)** is an index of lighting quality, with incandescent lamps getting the best ratings.

Table 9.1 gives a list of the most common modern lamp types, together with typical efficacies and CRIs. Generally the higher the power rating of the lamp, the better the efficacy.

Table 9.1 Lighting efficacies and colour rendering indices of modern electric lamps

Lamp type	Details	Lighting efficacy[1] /lumens watt^{-1}	Colour rendering index
General service lamp	The common incandescent light bulb	9–19	100
Tungsten halogen	Miniature incandescent lamps, often run on a 12V supply from a transformer	17–27	100
Fluorescent high-pressure mercury discharge lamp	Bluish white light, often using for shop lighting	40–60	85–95
Metal halide high-pressure mercury discharge lamp	Bluish white light, used for street lighting	75–95	~65
Compact fluorescent lamp (CFL)	Fluorescent replacement for normal incandescent light bulbs	70–75	75–85
Tri-phosphor tubular fluorescent (electronically ballasted)	Standard type for office and shop lighting	80–100	75–85
High-pressure sodium discharge lamp	Orange-white light, used for street lighting	75–125	~25
Low-pressure sodium discharge lamp	Pure orange light, used for street lighting	100–200	~5
White LED	Currently only available in low wattages	25–100	~85

[1] These values are lumens per watt of electricity consumed – the values in Table 3.1 of Chapter 3 are expressed in terms of primary energy

Source: manufacturers' data

Pollution problems

Energy-efficient lighting is not without its own pollution consequences. The presence of toxic mercury in so many lamps creates concerns about their safe disposal. Since 1980 manufacturers have progressively reduced the amounts in fluorescent tubes. Typically a modern CFL will contain 4 mg mercury (and less than 2 mg in the latest designs). A larger metal halide high-intensity discharge lamp may contain around 200 mg. This is a far cry from the hundreds of grams in Cooper Hewitt's original tubes. There has been concern that discarded lamps should be specially collected and treated as hazardous waste, but this has to be seen in the context of other large emissions of mercury into the environment, such as from coal-fired power stations.

Too many street lights may also be a bad thing. Many astronomers have complained of 'light pollution'. Today's city dwellers are no longer able to see and enjoy the stars because of the permanent haze of reflected orange light in the night sky.

Finally, concerns about global warming may mean that the standard incandescent lamp has reached the end of its days. Their sale is being phased out as an energy efficiency measure in many countries, though the more efficient tungsten halogen variety remains in use.

9.4 Electric traction

Just as electric lighting can be seen as a clean alternative to its messy oil and gas predecessors, so the use of electric motors to propel cars, buses and trains might seem more desirable than using steam, petrol or diesel engines. Over the past 120 years electric traction has fought it out with its competitors, winning some niche markets and losing others. However, electric drives and controls are increasingly being used in conjunction with diesel and petrol engines, improving their overall capabilities and minimizing pollution.

Electric trams and trains

Relying on lighting as the main outlet for electricity in the 1880s was not good economic sense. The competing gas works could be run steadily throughout the day, storing their output in gas holders. In the evenings, these had no trouble in meeting the peak lighting demands of their customers.

Electricity, on the other hand, had to be made exactly when it was needed. It was possible to store DC electricity in batteries, but these were extremely expensive. A large amount of electrical generation plant was thus required to meet only a few hours of actual lighting demand. Today we would say that this was operating with a low **capacity factor**. This is the ratio of the electricity produced over a given period to the maximum it could achieve running flat out continuously. A system with a 100% annual capacity factor would be one that ran flat out 24 hours a day, 365 days a year. (The term **load factor** is often used in this context though strictly speaking it refers to the demand, rather than the output of the particular generator supplying it.)

Obviously, to make the best return on the capital investment in a power station, a high capacity factor is desirable. What was needed were alternative daytime users for electricity, but they had to be willing to pay the extremely high unit prices of the day. Although Victorian steam-powered railways linked almost all the major towns and cities, the noisy, smoky, steam engine was not considered suitable for use on urban streets. People either walked or used horse-drawn buses or trams. The new tramways were only too happy to switch to electric traction and city authorities were only too glad to reduce their urban horse populations (see Box 9.6).

Electricity had plenty to offer. In 1879 Werner von Siemens demonstrated a miniature electric railway at an exhibition in Berlin. In 1883, Magnus Volk demonstrated his motorized horse tram on the seafront at Brighton, collecting electric current from a third rail laid along the track. It is still running today (see Figure 9.20).

Electric tram systems rapidly spread: Blackpool in 1885, Leeds in 1891 and Bradford in 1892. In the USA in Boston, Massachusetts, 9000 horses were made redundant by the electrification of the tram system. Electric traction also increased the speed and capacity of the trams. By the outbreak of World War I in 1914, the horse-drawn tram and bus had virtually disappeared from the major cities of Europe and the USA. In their place were the electric tram, the petrol bus introduced from about 1900 onwards, and new electric underground railways.

BOX 9.6 What was wrong with the horse?

Given the modern search for 'alternative' biofuels for road transport, one might ask why bus and tram companies abandoned motive power that ran on hay and oats in favour of very expensive electric traction.

The first problem was a 'horse shortage'. New railways encouraged the movement of goods and people, but once they arrived in cities, their final delivery involved horse transport. The demand was such that in 1873, a House of Commons Select Committee Enquiry was set up to see whether the country could actually breed enough horses to keep the cities running.

The second was the logistics of operation. Although a tram or bus normally used two horses, there had to be plenty of spares. An extra 'trace horse' would be used to get it up hills, the horses were changed four or five times a day and they were given one day's rest in every four. By 1898 London's remaining horse tram lines still required 14 000 horses to run their 1451 trams, nearly 10 per tram.

The third was that urban streets were covered in horse droppings. A horse could produce five tonnes of excrement and urine per year, a major health hazard requiring a whole army of 'crossing sweepers' to keep the pedestrian crossings clear.

An electric tram might require complicated wiring with a risk of electrocution, but it was easy to drive, clean, odourless and only required parking space for itself. The (not so clean) electricity generation plant (if actually owned by the tram company) could be placed conveniently out at the end of the line and the stables sold off to property developers. No 'modern' city of the 1890s could afford to be without an electric tram system.

As the petrol bus developed, the electric tram declined in popularity. They were accused of 'causing congestion'. From the 1920s onwards many tram systems were replaced by buses or rubber-tyred trolleybuses, and even the latter were phased out in the UK. However, trams and trolleybuses survived in many European cities, such as Amsterdam, Brussels, Vienna and Athens. But time moves on and by the 1990s attitudes to 'congestion' had changed. Towns were prepared to implement far-reaching plans to restrict car access and promote public transport. In south London, the tram tracks lifted in central Croydon in 1928 were put back again in 1997 and a new service installed with trams built in Vienna (see Figure 9.21).

Electric traction was also key to developing the use of railways in places where the steam engine could not venture, such as long tunnels. The technology of boring long deep tunnels had been perfected by 1870, but at that time the only way of hauling trains through them was with cables driven by steam engines at each end. The combination of reliable electric motors and mains electricity set off an explosion of underground railway building in the major cities of the world from 1900 onwards. It was no longer necessary to fight through city streets choked with traffic; it became possible to move rapidly beneath them. Building underground railways was an extremely expensive activity, but could be financed by property deals. Wherever the lines emerged into the fields on the outskirts of cities, the entrepreneurs were able to buy cheap farmland and immediately sell it to developers for housing, while promising rapid access to the city centres. Not to be outdone, the owners of some existing surface steam railways also adopted electric traction and joined the fray. The result was a growth

Figure 9.20 Volk's Railway, Brighton – the prototype electric horse tram system of 1883, which is still running, here photographed in 2002

Figure 9.21 A new tram in Croydon in 2002. The tram tracks that had been lifted in 1928 were replaced in 1997.

of suburban sprawl in cities like London and New York that continues to this day. The spread of other cities, such as Los Angeles, was initially due to the availability of rapid electric tram and trolleybus routes before these were replaced by the private motor car.

On main line railways electric traction enabled the use of really long tunnels such as the transalpine Simplon Tunnel from Switzerland to Italy, nearly 20 km long and opened in 1906. The Swiss pioneered the use of high-voltage AC supplies to trains through overhead wires. This has become essential to supply the latest generations of high-speed express trains, such as the Eurostar (see Figure 9.22) which operates through the Channel Tunnel from London to Brussels and Paris.

Figure 9.22 A Eurostar power car

As trains become faster their power requirements increase. The locomotive *Mallard*, which set the world speed record for a steam train of 202 kph in 1938, had a power rating of 2 MW. The Intercity 125 diesel trains described in Chapter 8 and capable of 238 kph have two 1.7 MW locomotives. The Eurostar, designed for running at 300 kph, requires up to 12 MW of electricity, about the average power requirement of a town of 10 000 people. This is delivered to the train through an overhead wire at 25 kV AC.

Currently only about 40% of the UK's main line system is electrified, and further electrification would be one way of mitigating the risks of peak oil discussed in Chapter 7.

Battery–electric vehicles

The battery–electric car is often seen as 'the car of the future', yet in 1899 a battery–electric car driven by the Belgian Camille Jenatzy was the first road vehicle to exceed 100 kph (see Figure 9.23). It used lead–acid batteries, weighed 1.5 tonnes and was powered by 50 kW of electric motors.

Battery–electric cars, buses and taxis were widely used during the first 20 years of the 20th century. They were slow, but in streets still clogged with horse-drawn traffic, this did not matter very much. As the century progressed and expectations of speed increased, the battery–electric vehicle was left carrying out the humbler duties of milk float, delivery truck and golf cart.

Slow but steady improvements in magnetic materials, control systems and batteries may have now changed the picture.

Early permanent magnets used soft iron. In the 1930s mixtures of iron with aluminium, nickel and cobalt, so-called 'alnico' magnets were developed. These gave a higher magnetic field for the same weight. More recently rare-earth permanent magnets have appeared using some of the lesser known metals. Samarium–cobalt magnets were introduced in Japan in the 1970s and neodymium–iron–boron magnets in Japan and the USA in the 1980s. These have allowed the construction of high-powered lightweight permanent magnet motors and generators. These are used in some electric and hybrid cars and also in modern wind turbine generators.

Modern high-powered car motors use AC, but of a variable frequency according to the desired speed, and produced from the DC battery supply using high-powered transistor switches.

The ability of a car motor to also function as a generator, plus the availability of modern electronic controls, have made the use of **regenerative braking** easier. This is the use of the electric motor during braking to convert some of the kinetic energy of the vehicle back into electrical energy to recharge the battery.

Battery technology has been the critical limiting factor. Even today few commercial electric cars have ranges of more than 100 km, yet this is what was being offered by Detroit Electric cars in 1910! The basic lead–acid battery has changed little since then and it is only in recent decades that there have been advances in other battery designs (see Box 9.7). The Tesla roadster (Figure 9.24) introduced in 2008 has a top speed of 200 kph and a range of nearly 400 km. It has a 215 kW (288 hp) variable speed AC motor driven by transistorized electronics. The lithium-ion battery consists of over 6000 small cells, similar to those used in laptop computers. It weighs 450 kg and holds 56 kWh, i.e. over 45 times as much energy as the standard lead–acid car battery shown in Box 4.7.

Figure 9.23 Camille Jenatzy's 1899 record-breaking battery–electric car, the 'Jamais Contente'

Figure 9.24 A Tesla battery–electric roadster introduced in 2008. It has a top speed of 200 kph and an advertised range of nearly 400 kilometres

BOX 9.7 Rechargeable batteries

The basic form of a rechargeable battery is a number of cells, each with a positive and negative plate with an electrolyte between them that is either a liquid or gel. A key factor is the **specific energy**, the number of watt-hours that can be stored per kilogram. Although these batteries can be recharged, depending on the type, the performance may deteriorate with the number of charge/discharge cycles. The replacement cost of the battery can be a significant part of the running cost of an electric car.

The **lead–acid battery**, invented in 1859, uses a positive plate of lead dioxide, a negative plate of lead, and sulfuric acid as the electrolyte. It is the standard rechargeable battery for starting petrol and diesel vehicles, and has been extensively used for traction applications. The principal problems are the limited life, which may be only a few hundred full recharging cycles, the expense and toxicity of lead and the consequent need for careful recycling and the relatively poor specific energy.

Other batteries are based on a positive plate of nickel hydroxide. The **nickel–iron (NiFe) battery**, patented in 1901, uses a negative plate of iron, and potassium hydroxide as the electrolyte. It was widely promoted by Thomas Edison for traction batteries.

The **nickel–cadmium (NiCad) battery**, patented in 1900, uses a negative plate of cadmium. Although widely used in rechargeable batteries, cadmium's toxicity creates disposal problems and it is being replaced by the nickel–metal-hydride battery.

The **nickel–metal-hydride (NiMH) battery** was developed during the 1970s and 1980s. Instead of iron or cadmium the negative plate is hydrogen, but this is held within a complex alloy of metals. The most common form uses nickel, cobalt and manganese/aluminium plus one or more of the 'rare earths': lanthanum, cerium, neodymium and praseodymium. The NiMH battery has about twice the specific energy of lead–acid batteries, but can last for several thousand recharge cycles.

The **lithium-ion (Li-ion) battery** essentially uses metallic lithium rather than hydrogen as the negative plate. They have been under development since the 1970s. They have about twice the specific energy of NiMH cells and can last for several thousand recharge cycles, but current production cells show a steady loss of performance with age.

Both NiMH and Li-ion cells are now mass-produced for computer batteries. Modern electric car batteries are essentially based on the same product, using hundreds or even thousands of small cells per car. Much research is going on into battery production quality control to produce a battery pack that can last for the whole life of a car (i.e. 150 000 km or more).

There are also other rival ways to store electrical energy. It is possible to store it in large capacitors, or ultra-capacitors, in the electric field between two charged plates. Since these do not involve any electro-chemistry, they could in principle have very long lives. These are being developed for automotive applications in China. Alternatively, electric energy can be stored as the kinetic energy of a flywheel, driven by a combined electric motor/generator. Although not designed for cars, such systems are in use in the UK in small railcars and would be suitable for trams.

Given the limited amount of energy stored in a battery, there remains the fundamental problem of performance. It is possible to travel a long distance slowly or a short distance at high speed, but the ability to travel a long distance at speed remains elusive. Current research is concentrating on the rapid charging of batteries. An alternative is to use easily replaceable battery packs.

If electric vehicles are going to replace more conventional petrol and diesel ones in the coming decades, there are serious questions that have to be asked about the availability of key chemical elements. This is a problem shared with new lighting and photovoltaic generation technologies (see Box 9.8).

BOX 9.8 Critical rare elements

Many electrical advances have depended on the properties of what may at first seem rather obscure elements. Some of them are what chemists call 'rare earths', i.e. the lanthanoids plus scandium and yttrium (see Figure 4.19). Although some of these elements are in fact not particularly rare, in many cases there are concerns about their global resources and availability (US DoE, 2010). Table 9.2 lists some examples and their sources.

Table 9.2 Key elements, their uses and sources

Element	Uses	Sources
	Batteries	
Cerium	Nickel–metal-hydride batteries, also phosphors in fluorescent lamps, LEDs and gas mantles	China, but other sources in Australia, Canada and the USA could be developed
Lanthanum	Nickel–metal-hydride batteries, also fluorescent lamp phosphors	China, but other sources in USA, Canada, Korea and Japan could be developed
Lithium	Lithium-ion batteries for mobile phones, computers and vehicles	Australia, Argentina, Bolivia, Chile, USA
	Lighting	
Europium	Red phosphor in fluorescent lamps, and CRTs	Found with yttrium (see below)
Indium	Semiconductor used in white LEDs and some PV panels	By-product of zinc refining in China, Canada, Korea and Japan
Terbium	Phosphor in fluorescent lamps, can also be used in permanent magnets	Mainly China, but other sources could be developed
Yttrium	Red phosphor in fluorescent lamps, CRTs, LEDs and gas mantles	China, but other sources in Australia, Canada and the USA may be developed
	Magnetic materials	
Dysprosium	An additive to neodymium–iron–boron magnets (see below) to improve high temperature performance	Mainly China but other sources in the USA could be developed
Neodymium	Used with iron and boron to make strong permanent magnets for motors and generators	Mainly China but other sources could be developed
Samarium	Used with cobalt to make strong permanent magnets for motors and generators	China, but other sources in Australia, Canada and the USA could be developed

Electric transmissions and hybrid electric drives

The flexibility and ease of control of the electric motor has led to its use in conjunction with other motive systems. As described in Chapter 8, modern diesel electric railway locomotives use electric transmission. The diesel engine drives a generator that supplies the electric power for the electric motors that drive the wheels. The controls ensure that the system produces the maximum tractive effort and acceleration without the wheels slipping on the rails. This is not new; an experimental steam–electric railway locomotive was built for the French railways in the 1890s.

A similar system for road vehicles, patented in 1905, was used in early petrol–electric buses. It was popular with drivers graduating from horse-drawn buses who could not cope with the complexities of a clutch and gearbox. Today this would be called a **series hybrid-electric drive**, one where the petrol engine drives the wheels *through* or *in series with* a generator and an electric motor. Today, the petrol–electric vehicle has reappeared in the form of cars such as the Toyota Prius (see Figure 9.25).

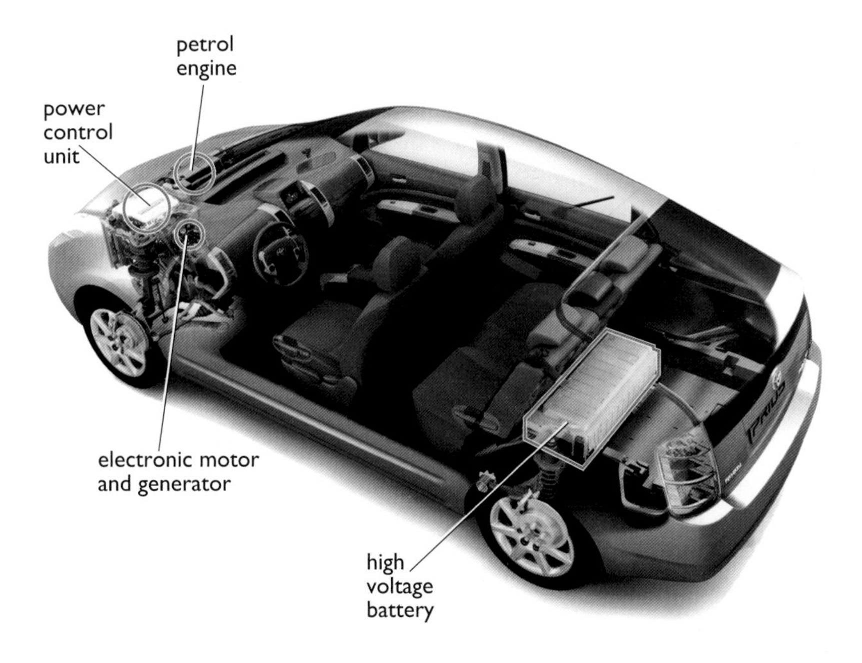

Figure 9.25 Cutaway section of a Toyota Prius

In its original 1997 form, when launched in Japan, the Prius had a 1.5 litre petrol engine and a 30 kW electric motor plus a modest rechargeable NiMH battery. In its current (2011) form the petrol engine has been enlarged to 1.8 litres and the electric motor doubled in size to 60 kW. Its road wheels are driven using a **parallel hybrid drive system**, i.e. *either* directly by the petrol engine *or* by the electric motor *or both*. A computerized control system optimizes the combination, making sure that the petrol engine is only ever used close to its peak efficiency.

When travelling slowly the petrol engine is cut out and the car travels under its electric motor and battery. At higher speeds, the petrol engine operates and recharges the battery. The motor is also used for regenerative braking when slowing down. In this case the computer-controlled system is designed to minimize fuel consumption and pollution. Its fuel consumption is quoted as 3.9 litres per 100 km (over 70 miles per UK gallon) in European driving conditions.

It must be said that this fuel performance can almost be matched by the best conventional diesel cars without the complexity and expense of hybrid drives. However, this combination of a petrol engine and a fairly large electric motor means that in principle most urban journeys could be made under electric power alone. A **plug-in hybrid** is essentially an electric car whose battery is normally charged using mains electricity, but which can also use its petrol engine if necessary on longer journeys. Unofficial conversion kits have been available for adding battery capacity to the Prius design and allowing it to be charged from the mains, and a new plug-in hybrid version is scheduled for release in 2012. The new Chevrolet Volt is specifically designed as a plug-in hybrid.

9.5 Expanding uses

On the world scale electricity consumption has expanded enormously, starting out with a prodigious growth rate of 30% per annum in the 1890s and, after 1930, continuing at a steady 7.5% per annum for 50 years (see Figure 9.26). In recent decades this world growth rate has declined and in many countries actually halted. World total generation in 2009 was just over 20 000 TWh. Over half of the growth in world electricity use between 2000 and 2009 was in China (BP, 2010).

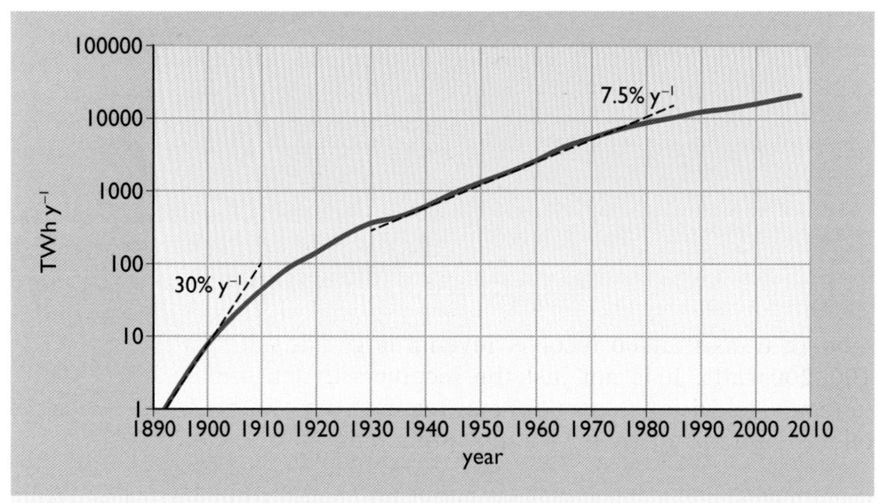

Figure 9.26 Growth in world electricity demand, 1890–2009. Note the logarithmic *y*-axis scale (sources: Romer, 1976; EIA, 2010a).

This increased demand has been the result of ever-expanding uses for electricity: in industry, telecommunications, and for a host of domestic applications. The following subsections chart the development of some of these.

Telecommunications and computers

The telegraph and the telephone relied upon wires for connection. In 1864 Maxwell published his paper predicting the existence of electromagnetic waves that could travel through space without needing wires (see Chapter 4). As mentioned earlier, in 1887 Heinrich Hertz built a crude microwave transmitter, which worked at about 5 GHz. He found that a powerful discharge in his transmitter circuit could make a spark jump across a gap in a receiving electric circuit some metres away. In order to get the best reception it was necessary to 'tune' the crude receiver to the frequency of his transmitter.

The young Italian engineer Guglielmo Marconi (Figure 9.27) successfully established a wireless link across the English Channel in 1899 when he was only 25, using somewhat lower radio frequencies than Hertz. Unlike the low-powered telegraph, which could be run from batteries, long-distance radio transmission required kilowatts of transmitter power and steam-driven generators. In 1901 Marconi demonstrated sending a wireless signal 2100 miles from Cornwall to St. John's Newfoundland and set up the first commercial transatlantic radio link in 1907. Public broadcasting of news and entertainment eventually followed with the British Broadcasting Company starting transmissions in 1922.

Figure 9.27 Guglielmo Marconi (1874–1937) was born in Italy, the son of an Italian country gentleman and an Irish mother. He began experimenting with radio on his father's country estate in 1895, before taking his equipment to the UK. He opened the first transatlantic commercial radio service in 1907. Later in the 1930s he developed the use of ultra-high-frequency (UHF) radio and radio navigation beacons

Really practical radio transmitters and receivers needed the development of thermionic valves. Looking somewhat like light bulbs, these allowed the amplification of minute electric signals. In receivers they needed a small amount of high-voltage electricity which could be supplied by dry batteries, but a large amount of current at low voltage to supply their heater filaments. For those listeners who did not have a mains supply this meant using a rechargeable battery that was taken every week to somewhere that did recharging (usually the corner shop). Understandably this added to the desire for access to mains electricity.

The introduction of the transistor in the 1950s made the fully battery-powered radio a reality, but by then it had been overtaken by an even more power-hungry device, the television receiver. UK TV transmissions had started in the 1930s, but were interrupted by World War II. In 1950 only 350 000 TV licences were sold in the UK. By 1954 the figure had risen to 3 million, largely as a result of the broadcasting of the Coronation of Queen Elizabeth II in 1953. A typical valve TV receiver required about 500 watts, while a modern transistorized receiver (even a large flat screen one) needs only 100–200 watts. It is not just the receivers that use electricity; a large terrestrial television transmitter may require a megawatt or more. Terrestrial television transmitters have a maximum range of about 100 km.

In 1945 Arthur C. Clarke suggested the possibility of broadcasting TV from a satellite in a **geostationary orbit**. A satellite 36 000 km out in space will circle the Earth in exactly 24 hours, and if it is above the equator this means that it stays exactly over the same position on the ground. This did not become commercially feasible until the 1980s. A single satellite can direct its transmission to receivers in a 'footprint' up to 1000 km across on the Earth's surface beneath it. At the frequencies used, above 10 GHz, electromagnetic waves can be focused to tight beams using the familiar dishes, allowing the use of very low power signals over long distances. This is essential because satellites are limited in transmission power, being reliant on limited supplies of electricity produced by large arrays of photovoltaic (PV) panels. (Clarke had originally suggested solar-powered steam engines!)

The availability of satellite TV has increased the market for receivers and the desire for access to electricity to the furthest corners of the most rural countries.

Since the 1990s, international communications have been transformed by the use of a combination of point-to-point microwave beams on the Earth's surface, and geostationary satellites and undersea cables using modulated light beams in optic fibres rather than electric signals in wires. A single telegraph wire of the 1890s could carry four simultaneous data channels. A microwave link of the 1990s could carry a thousand or more.

The late 1990s also saw the perfection of the lightweight mobile telephone, working as part of a cellular network of transmitters and receivers all linked together by microwave links. The handset can be small because it only needs to have a range of a few kilometres and a rechargeable battery is sufficient to power it. However, the thousands of cellular phone masts that now litter the countryside each contain a set of transmitters using 100 watts.

The development of the integrated circuit in the 1960s and 1970s enabled the manufacture of practical commercial computers, though still each the size of a large wardrobe. A university or military establishment would have been able to afford two or three. By the 1970s, such institutions began to use long distance telecommunication networks to link computers together. These links enabled users to send personal messages (e-mail), and protocols were set up to deal with their handling. Since then, computers have become faster, smaller and cheap enough to become a consumer item and the data links have expanded to become the modern internet. The IBM Personal Computer shown in Figure 9.28 only dates from the mid-1980s but would now perhaps be regarded as an 'antique'.

Figure 9.28 An early IBM PC 'portable' computer. The selling price with a 4.7 MHz processor, 256 kB of memory, one floppy disk drive and no hard disk was US$4225 in 1984 (about £6000 in 2011 equivalent)

Today, nearly all UK homes have a TV and many have more than one. In many European countries there are now more telephones than there are people. In the past, 'landline' telephones were passive instruments that sat unpowered, waiting to ring. Today they are likely to be mains-powered containing an answering machine and even a fax machine, continuously consuming 10 or more watts. Mobile phones have a permanently powered system of transmitters and receivers simply sitting waiting to be used. A more recent phenomenon has been the growth of internet servers, computers permanently switched on, consuming 50 or 100 watts, simply waiting to answer a request. These demands, though modest, all contribute to the rise in electricity consumption of the modern world.

Cooking and heating

Like town gas, electricity quickly advanced from being a lighting fuel to one used for cooking and specialist forms of heating. Electric equivalents of gas appliances appeared very rapidly. By 1890 the General Electric Company in the USA was selling electric irons, immersion heaters and an 'electric rapid cooking apparatus, which boiled a pint of water in 12 minutes' – an electric kettle (see Figure 9.29 for a 1921 version). In 1894 the City of London Electric Lighting Company staged an 'all-electric banquet' for 120 guests to launch their new rental scheme for electric cookers. Although the appliances became available, initially only the rich could afford the cost of the connection, house wiring and the high unit price of the electricity. By 1918 only about 6% of UK homes were wired and that was used almost exclusively for lighting. The price of electricity fell steeply in real terms during the 1920s and 1930s though it remained a far more expensive form of energy than coal or town gas (see Chapter 12).

Figure 9.29 An early electric kettle produced in 1921

Two decades of promotion had a significant effect. By 1939 about two-thirds of UK homes had an electricity supply. In those, almost all would have had electric lighting, 77% an iron, 40% a vacuum cleaner, 27% electric fires, 16% an electric kettle, 14% a cooker and less than 5% an electric water heater (Electricity Council, 1987). Town gas still remained the fuel of choice for cooking, while coal was preferred for heating.

After World War II the electricity grid was extended into rural areas of the UK making mains electricity and electric lighting almost universally available. New manufacturing techniques transformed the heavy cast-iron electric cooker into its white pressed-steel form of today. The competition between gas and electricity for cooking is still as strong as ever. Many fitted kitchens are sold with gas burners and an electric oven. They are also likely to include a relative newcomer, the microwave oven, which first appeared for sale in the USA in 1947. It was not until 1972 that a domestic version arrived in the UK. It heats food using a microwave transmitting valve, a cavity magnetron, originally developed for radar. Unlike a conventional cooker, which applies heat from the outside, this excites the water molecules inside, cooking food right through very quickly. This in turn has created a whole supermarket culture of pre-prepared 'instant' meals. This has, no doubt, contributed to the slow decline in energy use for cooking in UK homes since the 1970s.

Although electric fires were widely used as 'secondary' heating in bedrooms, from the earliest days, electricity was too expensive to tempt consumers away from coal, gas and oil as a main heating fuel in the UK until the 1960s. By this time not only had average unit prices fallen, but the industry was promoting cheaper 'off-peak' tariffs for use with storage heaters. These are essentially large blocks of special heat-resistant brick in an insulated case, which can be heated to a high temperature at night, and which then release the heat into the house steadily over the day. These initially sold well, providing a central heating system with a low capital cost, but could not compete against gas central heating once cheap North Sea gas appeared in the mid-1970s. By 2007 93% of the UK housing stock had central heating, but only about 8% used electricity for this purpose (DECC, 2010a).

Refrigeration

The idea of keeping food fresh by storing it at low temperatures in cold stores (instead of salting or drying it) was introduced in the 19th century. In the UK it initially used bulk ice imported from countries with a plentiful natural supply, such as Norway. In 1852 William Thomson (Lord Kelvin) showed that when a pressurized gas is allowed to expand it cools. This led to the development of many refrigeration processes during the 19th century. They were sufficiently good to enable the first shipment of frozen meat from Australia to London to be made in 1880 (a journey of six months).

Box 9.9 describes the principles of a modern domestic refrigerator, which uses electricity to pump heat from its interior to a condenser coil on the outside.

BOX 9.9 Heat pumps and refrigerators

Section 6.3 introduced the *heat engine* in which heat flowing from a high temperature to a low temperature can produce useful work. This process is reversible. A **heat pump** is a device for pumping heat from a lower temperature region and delivering it to a region at a higher temperature. However, this will require the input of energy from some source, usually in the form of mechanical work.

Examples are the refrigerator and the freezer, used for cooling, and the air conditioner, which can be used for either cooling or heating. A heat pump can also be specifically designed for heating a house.

The common domestic refrigerator is an insulated box with a heat pump attached. The temperature in the main compartment is typically around +5 °C, and −10 °C to −20 °C in the freezer compartment or dedicated freezer. The temperature outside is likely to be a normal room temperature of 15–20 °C. Since heat will naturally flow from the warm outside through the insulation into the cool interior, the heat pump has to balance this by rejecting an equal amount. It has to move heat from the low temperature interior to a higher temperature on the outside.

The heat pump uses two properties of liquids. First, when they evaporate, they absorb a large amount of energy, their **latent heat of vaporization**. For example, as described in Box 5.2, it requires 4.2 kJ of heat to raise 1 kg of water through 1 °C, but 2.26 MJ to convert water at 100 °C to vapour at the same temperature. The second property is that the boiling points of liquids vary according to the pressure. The higher the pressure, the higher the boiling point (see Box 6.5).

Water has too high a boiling point to be useful in refrigerators – they use a **refrigerant**, a liquid with a conveniently low boiling point, such as isobutane C_4H_{10}, which has a boiling point at atmospheric pressure of about −12 °C.

In a refrigerator, such as that shown in Figure 9.30, the refrigerant passes continuously through a cycle of processes. Leaving the cold interior as a low-pressure vapour, it enters the electrically driven **compressor**. This provides the required input energy needed to increase the pressure. This pressure increase raises the boiling point of the refrigerant to above room temperature, and the vapour condenses to a liquid as it passes through the **condenser** – a coil of pipes usually on the back of the appliance. The latent heat of vaporization which it gives up is rejected into the surroundings. The high-pressure liquid now passes through the **throttle**, which is a fine nozzle, into the **evaporator**, a network of pipes in the walls of the icebox of the fridge. In this low-pressure part of the system the boiling point of the refrigerant falls, and it evaporates, taking up heat from the interior.

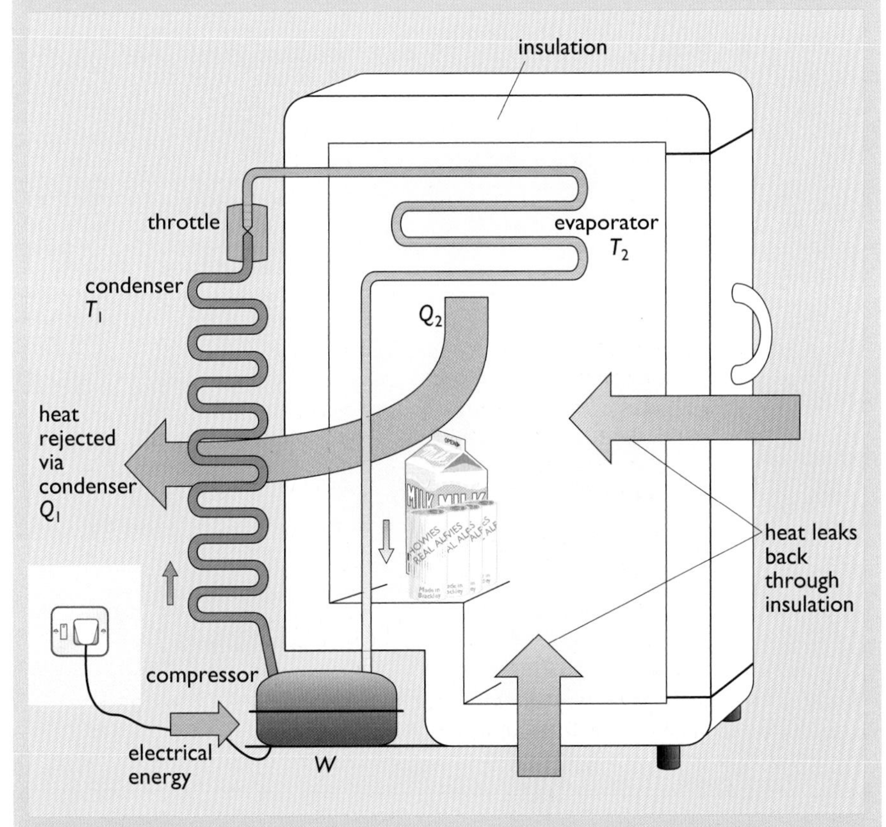

Figure 9.30 Simplified diagram of a domestic refrigerator

This entire sequence, taking in heat from a cooler region (at temperature T_2), using an input of mechanical work (W), and rejecting heat at a higher temperature T_1, is essentially the reverse of the sequence in a heat engine. It can be represented by reversing all the arrows in Figure 6.10.

The purpose of a heat pump described here differs from that of a heat engine and the measure of its performance reflects this. The aim here is to extract the maximum flow of heat (Q_2 in Figure 6.10) using the minimum rate of work, W. The **coefficient of performance** (COP) of a refrigeration or air conditioning unit is defined as the ratio of these two quantities:

$$\text{COP} = \frac{Q_2}{W}$$

Both Q_2 and W are, of course, quantities of power, and for most refrigerators W will in practice be the *electrical power* input needed to drive the compressor.

However, just as for the heat engine, Carnot's Law does limit the performance. Reasoning similar to that in Section 6.3 gives an expression for the maximum possible COP of a heat pump in terms of the Kelvin temperatures of the warm condenser (T_1) and the cool interior (T_2):

$$\text{COP}_{\text{max}} = \frac{T_2}{T_1 - T_2}$$

Taking a freezer as an example, with the exterior at room temperature (20 °C) and the interior at −15 °C, this theoretical maximum COP is about 7. In practice, values are very much lower, typically about 2.5, i.e. for every unit of work done by the compressor, 2.5 units of heat are removed from the interior.

An air-conditioning unit may be used to cool a whole building in a similar manner, pumping heat from an evaporator heat exchanger on the inside to a condenser on the outside. Many air-conditioners are reversible, providing cooling in summer and heating in winter. This is achieved by swapping over the functions of the two heat exchangers and thus the overall direction of flow of heat.

A heat pump can also be specifically designed to warm a building or to heat water. In this case the condenser is located inside the building, possibly connected to a central heating system, and the evaporator is located outside. In an **air-source heat pump** the evaporator is simply located in the surrounding air. A **ground-source heat pump** uses an evaporator heat exchanger buried in the soil. The operation of the heat pump makes the soil colder, resulting in heat flowing from the air down into the ground. Such ground-source heat pumps are now becoming more widely used. As with the refrigerator, one unit of mechanical work (usually provided by electricity) can be used to pump two or more units of heat into a building.

Although the domestic refrigerator is now regarded as an essential appliance, there was little demand for them when they were first produced in the USA in 1912. Shopping was normally done on a daily basis. Wealthier families might have an insulated 'icebox' to store food, but this was cooled by bulk ice delivered by a local supplier. The trade in 'natural ice' was replaced by bulk ice from steam-powered refrigeration plants as public health standards were increased. In 1926 the American General Electric company sold only 2000 (rather expensive) refrigerators, but by 1937 sales were up to nearly 3 million. By 1950 90% of US citizens living in towns and 80% in rural areas owned them (see Weightman, 2001). Other countries have followed suit. By 1975 nearly 80% of UK households owned a refrigerator. These have now been joined by freezers and fridge-freezer combinations. In 2009 domestic 'cold' appliances made up 4.4% of the UK's electricity demand. However, their total amount of electricity consumption has been falling gently since 1997 as better insulated refrigerators and freezers have been introduced.

The use of certain refrigerants has had a devastating effect on the Earth's upper atmosphere. Early refrigerators used sulfur dioxide or ammonia. These were unpopular because they are extremely toxic. Then in the 1930s a new class of chemicals, chlorofluorocarbons (CFCs) was developed. They appeared to be ideal for refrigerants as well as a host of other applications. However, in the 1970s it was discovered that these were damaging the Earth's ozone layer, which protects the Earth's surface from ultra-violet radiation from the Sun. Rapid international action took place to phase out their use, as described in Box 9.10. They have been replaced by less harmful alternatives.

BOX 9.10 CFCs and the ozone hole

Chlorofluorocarbons (CFCs) are hydrocarbons that have all their hydrogen atoms substituted by a combination of chlorine and fluorine. They appeared to be non-flammable and chemically inert. They were developed during the 1930s by Thomas Midgley, Jr. (who, as already mentioned, also has the dubious honour of inventing lead additives for petrol). He concentrated on CCl_2F_2, also known as CFC-12 or Freon-12, for use as a refrigerant (it has a boiling point at normal atmospheric pressure of about −30 °C). Other CFCs were used as propellants in aerosol cans and blowing agents for making foam plastics. By 1970 world production was about 600 000 tonnes per year (see Figure 9.31).

Then in 1974 two scientific papers were published. One suggested that the CFCs were reaching the stratosphere, 10 to 30 km above the Earth's surface, breaking up and releasing chlorine atoms (Molina and Rowland, 1974). The other paper said that chlorine atoms could be powerful ozone destroyers particularly at very low temperatures (Stolarski and Cicerone, 1974). Stratospheric ozone is vital in protecting the Earth's surface from ultra-violet radiation from the Sun, which can cause skin cancer in humans (ozone at ground level is a serious pollutant, as will be discussed in Chapter 13).

In 1984 measurements by the British Antarctic survey showed a 40% reduction in ozone over their survey site. Readings from a NASA satellite also showed that ozone levels had been dropping, but these had been ignored because it had been assumed that these were due to a faulty sensor! A large 'ozone hole' had been developing in the ozone layer over Antarctica. Its sheer size can be seen in the plot of satellite data in Figure 9.32. In 1987 a high-altitude research plane confirmed the presence of chlorine and low levels of ozone.

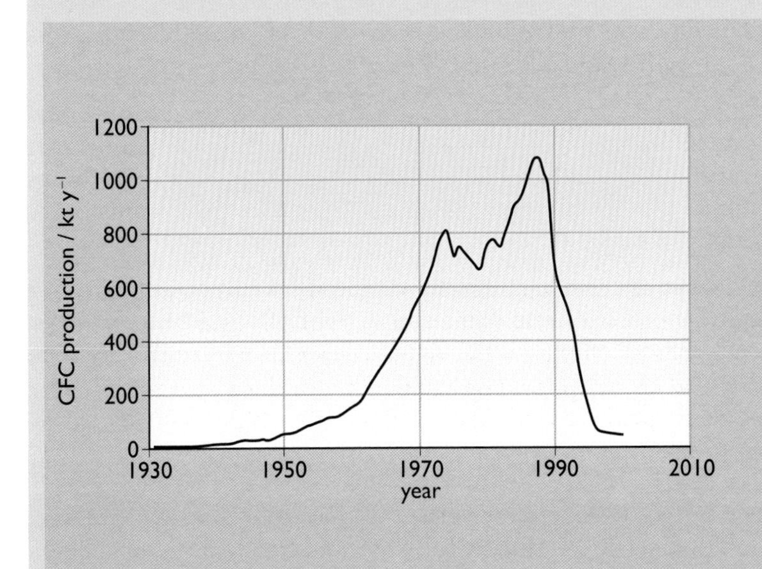

Figure 9.31 Sales of CFCs, 1930–2000 (source: adapted from Meadows et al., 2004)

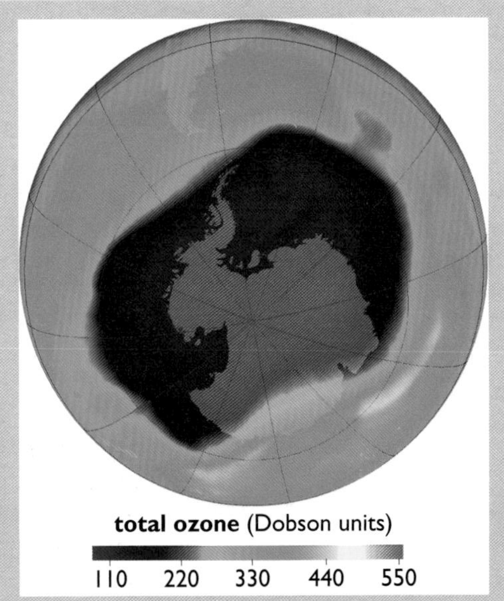

Figure 9.32 The largest observed ozone hole over the South Pole on 24 September 2006. The Dobson unit is a measure of the density of ozone in the upper atmosphere

The United Nations Environment Programme (UNEP) coordinated international action. In 1987 an international 'Protocol on Substances that Deplete the Ozone Layer' was signed in Montreal by representatives of 47 nations to restrict the use of CFCs. This was followed by further meetings and restrictions. By 2000 annual world production of CFC gases had fallen dramatically from its peak of 1 million tonnes.

Less harmful refrigerants, such as isobutane and hydrofluorocarbons (HFCs), which don't contain chlorine, have been substituted in new refrigerators. Programmes have also been put in place to ensure that the existing stock of refrigerators containing CFCs are collected at the end of their lives for special disposal.

The ozone hole appears to have reached its maximum size in 2006. A World Meteorological Organization/UNEP study concluded that the ozone layer would fully recover in approximately 2065.

CFCs have also been found to be extremely powerful greenhouse gases, so their phase-out has also been beneficial in reducing climate change.

Electric motors everywhere

The development of electric motors was a key step in the expanding use of electricity. Small motors allowed the development of labour-saving domestic appliances, such as the electric vacuum cleaner (1904), washing machine (1908) and dishwasher (1910). These immediately filled a need in middle- and upper-class homes brought about by the 'servant shortage' at the end of World War I (see Chapter 3).

The three-phase electric motor became the workhorse of large factories from the beginning of the 20th century. Before then many factories were powered by steam engines. There might be several separate steam engines, but this would be noisy and inefficient, or more likely, a single steam engine that was linked to many different machines through a complicated set of belts and shafts. Electric motors, though, could be fitted directly to individual machines and individually controlled, giving better productivity.

For the electricity companies, factories represented a welcome new market. They often ran day and night, with a more or less constant load, thus improving the overall load factor on the electricity supply system.

The trend to increased electrification of industrial processes has continued throughout the 20th century. The extreme is perhaps the fully automated car production line, where hundreds of electric motors power machines under automatic control that now carry out tasks that would once have been carried out by hand.

While electric motors can be extremely efficient on full load, many are only ever used on part load and therefore at poor efficiencies, and often for unnecessary reasons. Some of the worst offenders are the motors used on fans in office air conditioning systems.

Where electricity is used in the UK today

Back in 1920 nearly 70% of UK electricity was used in industry and over 10% was used for electric traction by railways and tram companies. There were fewer than a million connected consumers. Since then national electricity demand has increased tenfold. Although it has increased in all sectors of the economy, it has done so faster in the domestic and commercial sectors than in industry. Table 9.3 gives a breakdown of UK electricity use for the year 2009.

Table 9.3 Breakdown of UK electricity use in 2009

Energy industry use	7.8%
Losses	7.1%
Domestic	32.4%
Industry	25.9%
Commercial and public administration	23.6%
Transport (including associated buildings)	2.3%
Agriculture	1.0%

Source: DECC, 2010b

Over 100 years on from Edison and Swan, lighting remains an important end use for electricity. In 2008 it made up nearly 40% of electricity demand in the commercial sector and 13% in the domestic sector. Two-thirds of the domestic electricity use was for lights and appliances. Although electricity use for cooking, refrigerators and lighting has been falling, that for TVs and computers has been rising.

9.6 Large-scale generation

Competition vs economies of scale

In its early days, the generation of mains electricity was essentially a local affair. Power stations were small and, if they generated DC, limited in transmission distance by their distribution voltage. Yet today, most power stations are enormous and situated tens, or even hundreds, of kilometres from their customers. How has this come about?

The 1890s and the early years of the 20th century were a period of laissez-faire capitalism in many countries, including the UK and the USA. Government policy was that competition between electricity suppliers would give the maximum benefits to consumers. Where early utilities were municipally owned they were limited in size to their municipal boundaries. Given the rush to develop new systems, there was little incentive to standardize or for companies to combine to produce economies of scale. They were locked in competition with each other and also the gas companies. This picture was repeated in other countries, where cities were peppered with small power stations.

When World War I broke out in 1914, new munitions factories had to be rapidly constructed in the UK; electricity was the natural choice to power them. The military authorities, interested in standardized equipment, were appalled to find a profusion of different supply voltages and frequencies. By 1917, in London alone there were 70 separate companies with 50 different types of system and 20 different voltages. Most of the power stations were small and inefficient. The average efficiency of UK stations in 1920, virtually all coal-fired, was estimated to be under 10%.

In 1926 a UK government committee set up 'to review the national problem of the supply of electrical energy' recommended that a Central Electricity Board (CEB) be created and initially given the job of interconnecting the most efficient generating stations in England with a National Grid of high voltage transmission lines. The story of its construction is described in Section 9.7.

Although the power stations would remain in private hands, the publicly owned CEB would specify the actual level of generation of the selected stations from a central control room in London so as to achieve the lowest overall production costs. This conflict between ideas of independent electricity generation and enforced centralized control has remained a thorny problem right through to the present day. At the time opponents branded this central control as 'intolerable interference with the management of selected stations'. The press promoted it as implying promises of 'cheap and abundant supplies of electricity'. Eventually the centralizers won the day. The parliamentary bill setting up the CEB became law at the end of 1926.

The new National Grid standardized England on 50 Hz AC and forced the most inefficient stations to close. By the time its construction was completed in 1934 only 140 stations out of a total of 438 were left operating. The net result was that generation costs fell by 24% (Electricity Council, 1987). Not surprisingly this was seen as a triumph of central planning over piecemeal development. Further National Grid links were constructed with Wales and Scotland.

The nationalization of the industry in 1947 after World War II continued the process by combining 560 private and municipal electricity undertakings into a small number of Area Electricity Boards: 12 in England and two in the south of Scotland. It was at this time that 240 V AC was chosen as the 'standard' UK distribution voltage (more recently revised down to 230 V AC). The coordination of generation in England and Wales continued to be administered centrally. From 1957 onwards the state-owned Central Electricity Generating Board (CEGB) both owned and coordinated electricity generation in England and Wales. Scotland and Northern Ireland had separate arrangements. In the space of half a century, electricity generation changed from being a local activity through to a regional one and then to a state-controlled national one, a situation that continued until the privatization of the industry in 1989.

This centralization process opened the way for the development of larger generating units with higher efficiencies. During the 1950s, 60 MW was the 'standard size' for a coal-fired UK power station generator. By the 1970s single turbine generator sets of over 600 MW were being used in the UK, and over 1000 MW in the USA. A gigawatt generator is enough to supply the average requirements of about 2 million UK households. Drax Power Station in North Yorkshire was constructed and commissioned in two stages, with the whole station completed in 1986. The station has six 660 MW units giving it a total generation capacity of nearly 4 GW (see Figure 9.33). It is

Figure 9.33 The 4 GW Drax Power Station in North Yorkshire

the UK's largest coal-fired power station; in 2010 it burned nearly 10 million tonnes of coal and also co-fired nearly a further million tonnes of biomass fuel. Some of this biomass is UK-sourced and some imported but all is subject to the operating company's sustainability policy.

Efficiencies of coal-fired stations in England and Wales increased from about 17% in 1932 to 27% in 1960 and to 36% in 2000 though this rising trend has flattened out. Ultimately, the efficiency is limited by the chemical properties of steel. High temperature steam is highly corrosive and oxidizes conventional steels very quickly. Chapter 6 has described how a modern supercritical 660 MW turbine may use steam at 560 °C and 250 atmospheres pressure. This requires the use of expensive high-quality stainless steels in the boiler and turbine blades. Most large stations in the UK and USA built since the 1960s (including Drax) have used subcritical steam conditions and have limited temperatures to below 550 °C and pressures to 160 atmospheres. This gives a long station life with cheaper steels. It is a matter of a trade-off of energy efficiency against the capital cost of the plant.

It has become more convenient to concentrate these massive coal-fired power stations around coal mines rather than locating them near to the electricity consumers (see 'coal by wire' in Section 9.7). Large power stations also require large quantities of cooling water, so locations close to large rivers or the sea are favoured. These are also more convenient for the use of imported coal and liquefied natural gas.

This movement away from city centres has also been driven by pollution legislation. Until the 1990s rural stations could rely on tall chimneys to 'dilute and disperse' the SO_2 and particulates. Modern legislation insists on a more thorough clean-up, as described in Chapters 5, 13 and 14.

Hydroelectricity

Water power was an obvious early candidate for electricity generation. Although the plant is capital-intensive, the 'fuel' is essentially free and, worldwide, many large-scale schemes were brought into use in the early years of the 20th century; almost all of them are still in use. In the UK the bulk of hydroelectric power generation is in Scotland. The Galloway scheme brought 103 MW of generating capacity into operation in 1935. Since then a large number of other schemes have been added, bringing the UK's hydroelectric generating capacity to 1.4 GW. Unlike large thermal power stations, which may take hours to start up, hydro plants can produce electricity at short notice, often in only a few minutes. This makes them very useful for meeting peak demands. The UK has a further 2.7 GW of pumped hydro storage plant mainly for this purpose.

Developing hydroelectric sources has offered countries the opportunity of freeing themselves from reliance on imported fossil fuels. One of the earliest examples was in the Irish Republic which, after gaining its independence from Britain in 1921, did not want to be dependent on British coal. The Ardnacrusha scheme on the River Shannon was built by German contractors for one of the first nationalized electricity boards in the world. The investment represented 20% of the Irish government's budget

Figure 9.34 The 2 GW Hoover Dam on the Colorado river in the USA

for the year 1925. The original 70 MW was sufficient to supply 96% of the whole country's electricity demand in 1931. It is a measure of the growth of demand since then that today, even despite some expansion, it only supplies 2–3% of the Irish demand.

In the USA, truly enormous hydroelectricity projects could be undertaken. The Hoover Dam project (see Figure 9.34), which was started in 1932 on the border between Arizona and Nevada, has now reached an output capacity of over 2 GW, more than the *total* UK conventional hydroelectric capacity.

Today, the social and environmental effects of large hydroelectricity projects are highly controversial. The 18 GW Three Gorges Dam project in China is one such scheme and is described in Section 9.10.

New fuels

In 1950 coal provided 98% of the UK's electricity. Since then the dominance of coal has been challenged by oil, nuclear power and, more recently, natural gas. Figure 9.35 shows the evolution of the different generation fuels used in the UK.

As described in Chapter 7, during the 1950s and 1960s declining oil prices encouraged the construction of new oil-fired power stations. However, hopes that they would prove cheaper to operate than coal-fired ones were dashed by the oil price rises in the 1970s. Although there was a sharp surge in oil use in 1984 during the coal miners' strike, the use of oil as a major generation fuel has faded away.

Nuclear power stations started to appear in the UK during the 1960s; these are the subject of Chapters 10 and 11. The early Magnox stations were small compared to the gigawatt-scale coal-fired power stations being

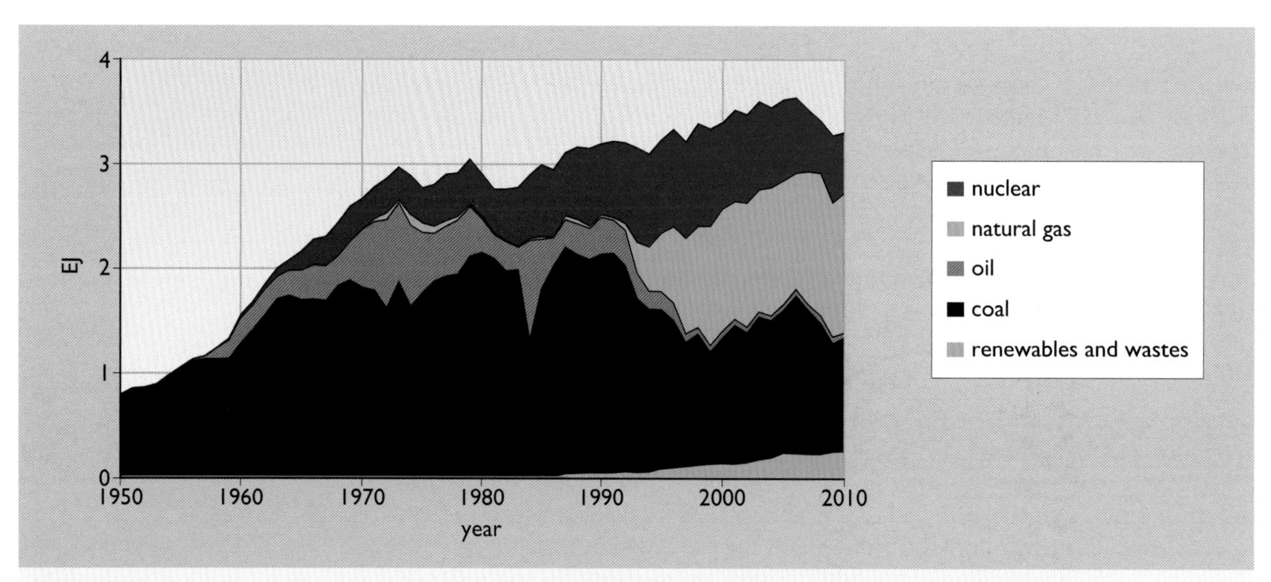

Figure 9.35 UK: fuels used for electricity generation 1950–2010 (sources: DECC, 2009; DECC, 2010b; DECC, 2011a)

built at the time, but the later advanced gas cooled (AGR) reactors and pressurized water reactors (PWR) are of the same scale, powering generators of 500 MW or larger. The PWR, in common with many other designs in use around the world, uses water to cool the reactor. Because of the added corrosion problems associated with radioactivity, and the obvious needs to maintain high levels of safety, steam temperatures are limited to below 400 °C. The generation efficiencies of these water-cooled reactors are thus slightly lower than for coal plants. As will be described in Chapter 10, the Magnox and AGR designs use CO_2 as a heat transfer medium in the reactor, employing it to raise steam in a heat exchanger away from the radiation problems. The AGRs thus have higher steam temperatures and generation efficiencies than PWRs.

The development of the CCGT and the UK dash for gas

Natural gas was not really available in quantity in the UK until the 1970s and was little used in UK power stations until the 1990s. However, its adoption has allowed the use of a new technology, the **combined cycle gas turbine** (CCGT) with a higher generation efficiency.

Chapter 7 has described the development of North Sea gas. When it was first introduced from 1970 onwards it was not clear quite how large the reserves would be. It seemed to make little sense to use it for electricity generation when there were plentiful supplies of coal and enthusiastic investment in nuclear power. A 1975 European Union Directive therefore restricted its use for large-scale electricity generation.

In the USA, on the other hand, large amounts of natural gas were available and there was a long history of using it for power generation with conventional boilers and steam turbines. In 1960 over 20% of US electricity was being generated from natural gas.

As described in Chapter 8, following the development of jet airliners in the late 1950s, gas turbines became mass-produced standardized items. Rather than propelling an aircraft they could be used to drive a shaft to turn a generator. The thermal efficiency of a simple 'open cycle' gas turbine is poor, under 30%. However, multi-megawatt engines were small, relatively cheap and could be run up to speed very quickly as 'peaking plant' to assist hydroelectric pumped storage schemes.

The exhaust gases of a gas turbine emerge at a temperature of 500 °C or more. This suggested to US engineers that they might be used with a conventional power plant as a **topping cycle**: an extra stage added at the 'hot' or 'top' end of a thermodynamic system. In 1963 a 27 MW gas turbine driving a generator was added to a conventional gas-fired plant in Oklahoma already using a 220 MW steam turbine. The hot exhaust of the gas turbine was directed to the existing steam boiler, supplementing the heat from the gas already being burned there.

Although there have been many such 'add-on' schemes to existing steam stations built since then, the CCGT station has been refined to a carefully integrated and optimized form with a high overall efficiency (see Box 9.11).

The Japanese Futtsu plant of Tokyo Electric Power, started in the mid-1980s, broke the trend to larger and larger conventional steam sets. It was built as a series of identical modules, gas turbines combined with steam turbines, each with an output of only 165 megawatts. As each module was completed it came into operation and began generating both electricity and cash flow in under three years from the start of construction. Replicating the design for successive modules cut the costs significantly. By 1997, the Futtsu plant had an output of 2000 MW, from 14 identical modules.

In the UK by the end of the 1970s, as can be seen in Figure 9.35, about 70% of the electricity was being generated from coal, mostly UK-produced and, to the eyes of the Thatcher Conservative government, excessively expensive

BOX 9.11 The Combined Cycle Gas Turbine (CCGT) power station

Chapter 6 has described the importance of using the highest possible temperatures in heat engines in order to obtain the maximum thermal efficiency. This is a consequence of Carnot's equation:

$$\text{Carnot efficiency} = 1 - \frac{T_{out}}{T_{in}}$$

where T_{in} and T_{out} are the inlet and outlet temperatures, respectively, expressed in degrees kelvin.

A conventional steam turbine can operate with outlet temperatures of 25 °C or less and inlet temperatures as high as 550–600 °C, yet, as described in Chapter 8, combustion temperatures in gas turbines can reach 1300 °C. This is limited only by the properties of the latest metal alloys and ceramics. Industrial turbines tend to use slightly lower temperatures to give a longer life. Designing a single heat engine that could operate between a very wide range of temperatures can be done by using a combined cycle, with two separate stages. Such a system is used at Shoreham power station in Sussex, built in 1999 (see Figure 9.36).

Figure 9.36 Shoreham CCGT power station near Brighton. Pollution regulations do not normally require such a high chimney for a gas-fired power station – Shoreham acts as a navigation landmark for coastal shipping.

The system is shown in Figure 9.37. Natural gas is burned in an industrial gas turbine at 1140 °C. The drive shaft, which spins at 3000 rpm, is connected both to the compressor and to one end of the generator, producing up to 240 MW of electricity. The exhaust gases from the gas turbine, which leave at about 630 °C are fed to a boiler, more formally known as a **heat recovery steam generator (HRSG)**, and used to raise steam at 540 °C. This is then fed to a conventional steam turbine connected to the opposite end of the generator via a clutch, producing a further 140 MW of electricity. Using the terminology described in Chapter 6, this is a *tandem-compound* system, with two turbines driving one generator. The station draws cooling water from a river estuary on one side of its site and discharges it into the sea on the other. This gives an average working temperature in the condenser of 22 °C.

The whole system is computer-controlled and can be operated by two staff. When starting from cold, the steam turbine is initially left disconnected from the generator. The gas turbine is run up to 3000 rpm using the generator as a motor and the gas burners are lit. It takes about 30 minutes to get the gas turbine stage fully operational and running at low power. Over the next hour steam starts to be produced in the boiler. This is initially used to clean out condensed water in the various stages of the steam turbine. Then the high-pressure steam is used to run this turbine up to 3000 rpm and the clutch connects it to the generator. It takes about 3 hours for the whole system to reach full power of 380 MW.

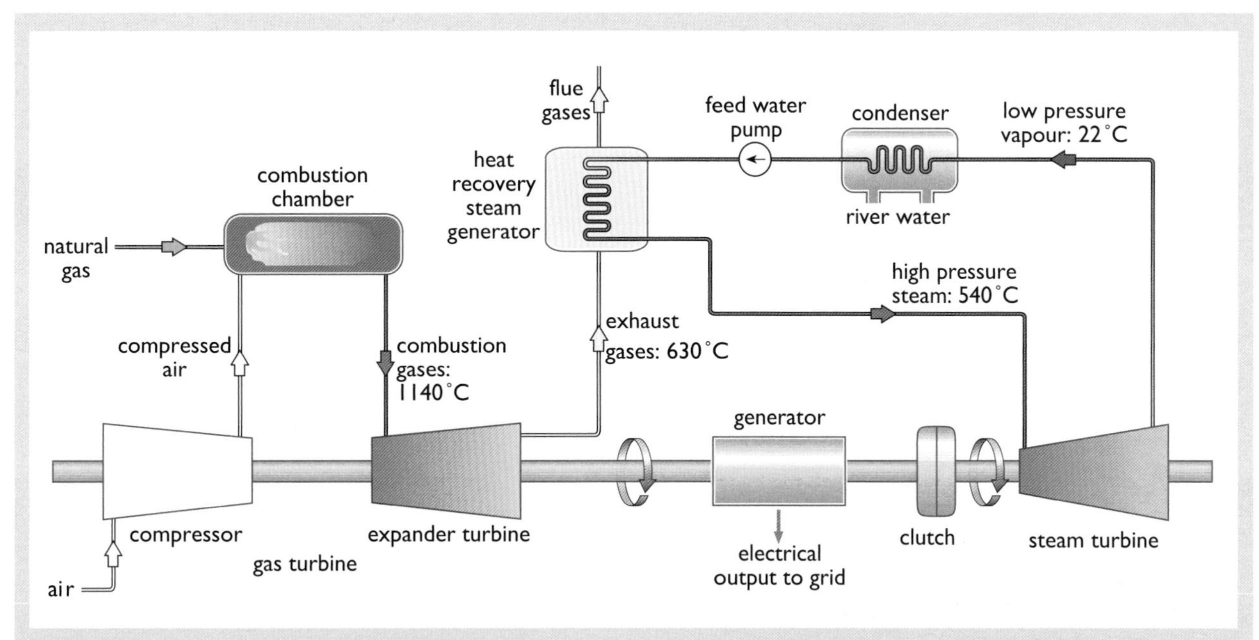

Figure 9.37 Schematic of the CCGT at Shoreham

The overall electricity generation efficiency is then over 50%.

Many CCGT plants are dual-fuel. They normally run on gas, but can also use light fuel oil.

As with conventional coal-fired power stations, NO_x emissions can be a problem, particularly given the high combustion temperature. They can be minimized by carefully controlling the fuel–air mixture and the speed of combustion.

compared to imported coal. There was a bitter confrontation between the government and the UK miners in 1984. Following this the government set about having the EU Directive withdrawn to allow the use of natural gas in power stations.

Between 1989 and 1992 the UK electricity industry was privatized. By this time the technology of CCGTs was well developed. Not only were they cheaper and quicker to build than coal or nuclear stations (a subject to be discussed in Chapter 12), but it had become possible to negotiate a fixed-price contract for 15 years' future supply of gas. This kind of contracting had not been done before.

Under the terms of the privatization, the old nationalized 'Area Boards' became Regional Electricity Companies (RECs) and were allowed to build their own power stations. Initially the RECs were monopoly suppliers to a sufficient number of small customers to guarantee sales of electricity from any moderate-sized power station that they might choose to build. Armed with a long-term gas supply contract and captive customers, all they had to do was approach merchant banks to supply the finance. Seeing an attractive and relatively risk-free investment, they were happy to oblige. The so-called 'Dash for Gas' was on. Between 1990 and 2010 the proportion of electricity produced from gas in the UK increased from almost zero to nearly 50%, largely at the expense of coal-fired generation (see Figure 9.38).

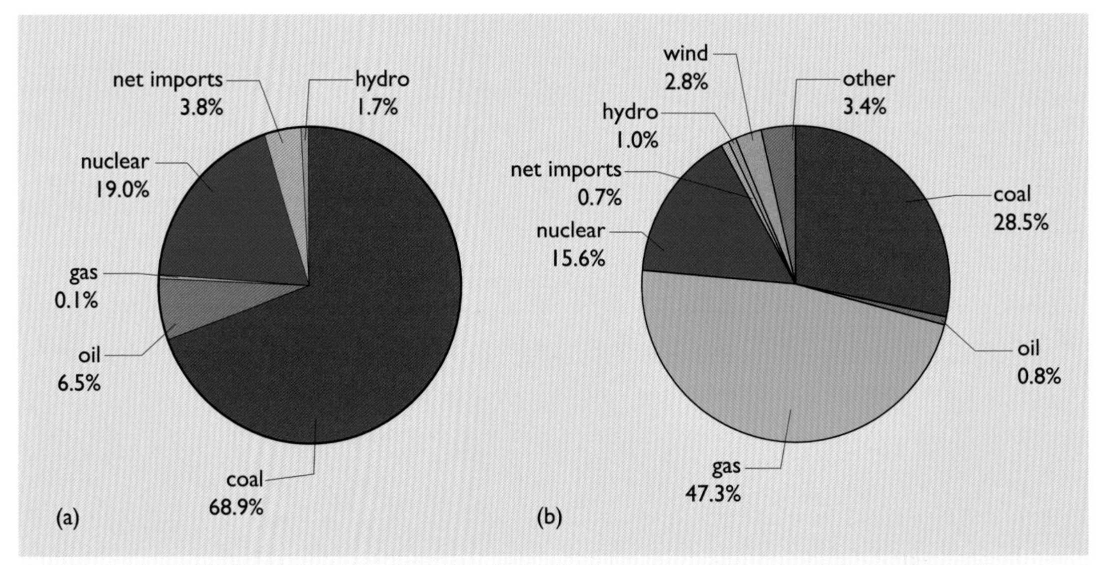

Figure 9.38 The Dash for Gas – UK electricity generation by fuel: (a) 1990; (b) 2010. Note that 'other' includes other renewable energy sources and energy from waste (sources: DECC, 2010c; DECC, 2011a).

The increase in the contribution from renewables and waste is also noticeable between 1990 and 2010. 2010 was a dry year and hydroelectric power generation was lower than in 1990, but there has been an increasing proportion of electricity from landfill gas and biomass and a rapidly increasing one of wind power, particularly as new offshore wind farms have been opened.

The environmental benefits of fuel switching

The progressive switch away from coal as a generation fuel has had a marked benefit in terms of CO_2 emissions. As pointed out in Chapter 1, when burned, oil and natural gas produce less CO_2 per unit of heat generated than coal. Nuclear power has very low CO_2 emissions and much of that is associated with uranium mining (discussed in Chapter 11). When used to generate electricity, gas has the benefit of increased generation efficiency compared to conventional coal stations. Table 9.4 shows the estimated performance of UK power stations in 2009. It shows that the emissions of gas-fired power stations per kWh generated were only about a half of those using coal.

Table 9.4 CO_2 emissions of different UK electricity generation technologies, 2009

Fuel	CO_2 emissions /g CO_2 kWh^{-1}
Coal	915
Oil	633
Gas	452
Nuclear[1]	16
UK average	466

[1] Estimate (SDC, 2006)
Source: DECC, 2010b

Looking back to 1950 the emissions of UK coal plants were nearly 1200 g CO_2 per kWh generated. Figure 9.39 shows the rise of UK electricity production since then and the rise, and fall, of the associated CO_2 emissions. If the UK had generated the 325 TWh it used in 2010 using the mix of technologies it used in 1950, then the CO_2 emissions would have been three times as large, over 450 Mt.

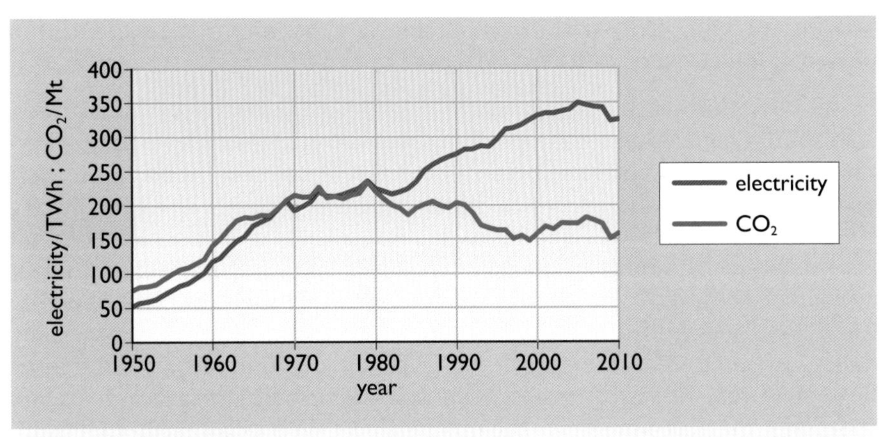

Figure 9.39 UK: electricity demand and CO_2 emissions from power stations, 1950–2010 (sources: DECC, 2009; DECC, 2010d; DECC 2011b)

Another environmental benefit has been the reduced sulfur dioxide emissions (as will be described in Chapter 13). Nuclear power plants do not emit SO_2 and as described in Chapter 7, sulfur is removed from natural gas at source. Since 1990 European environmental legislation has increasingly required the use of flue gas desulfurization in coal- and oil-fired power stations.

Combined Heat and Power (CHP) generation

Traditionally, in the UK, power stations have been seen as supplying only electricity, yet the generation process produces very large amounts of low temperature waste heat. As pointed out in Chapter 3, about a third of the UK's delivered energy is used for space and water heating, i.e. at final use temperatures of less than 60 °C. Combined heat and power (CHP or co-generation) plants not only produce electricity but also heat at a sufficiently high temperature to be useful, enabling them to achieve a high overall thermal efficiency.

At the end of 2009, UK CHP plant had a total electrical output of about 5500 MW (usually written as 5500 MWe). Most of this was industrial CHP in a relatively small number of large factories, chemical plants and oil refineries needing both electricity and heat in large quantities and at a range of temperatures (DECC, 2010c). These are sufficiently large to operate their own gas and steam turbine power stations. Power plants operated to produce electricity not for external sale are known as **autoproducers**.

Outside the large industrial plant CHP takes two forms. Small-scale CHP essentially takes the power station to the user, while large-scale CHP with community heating uses waste heat from new or existing large power stations.

Small-scale CHP

A small-scale CHP unit usually takes the form of a small reciprocating engine similar to a car or truck engine, but running on natural gas as a fuel

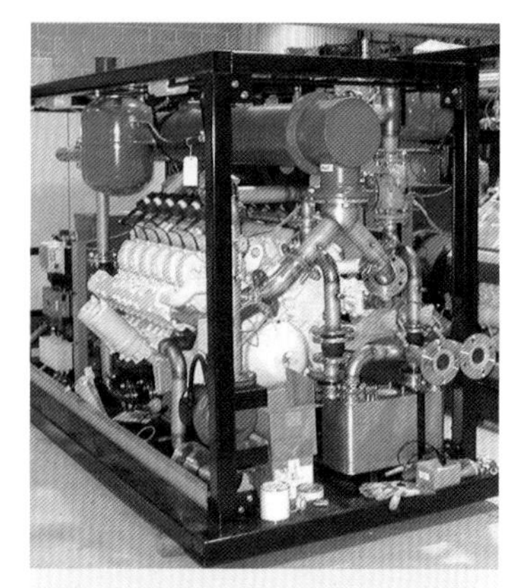

Figure 9.40 A small-scale CHP unit before installation; the gas-powered engine is at the front and the generator at the rear

(see Figure 9.40). This will drive an electrical generator typically with a power rating between 100 kW and 1 MW output.

In the UK there are hundreds of institutions, especially hospitals, community centres and large hotels, which have a sufficiently large year-round demand for both electricity and heat to warrant investing in their own CHP plant. Although a small-scale CHP unit may have an electricity generation efficiency of only about 30%, less than that of a conventional power station, the ability to use the waste heat makes it more energy efficient overall. This kind of comparison can be expressed as a flow chart showing the overall energy inputs and outputs of a system, known as a Sankey diagram (illustrated in Figure 9.41). The left diagram shows the energy flows for a coal-fired power station such as that used in the example in Box 3.1. The right diagram shows the energy flows for a small-scale CHP unit, including the use of waste heat. The actual amount of waste heat that can be used depends on the temperature at which it is required, but for typical heating applications the overall thermal efficiency can be over 80%. Such schemes have been extensively encouraged in the UK since the mid-1980s and are widely used in countries such as the Netherlands and Denmark.

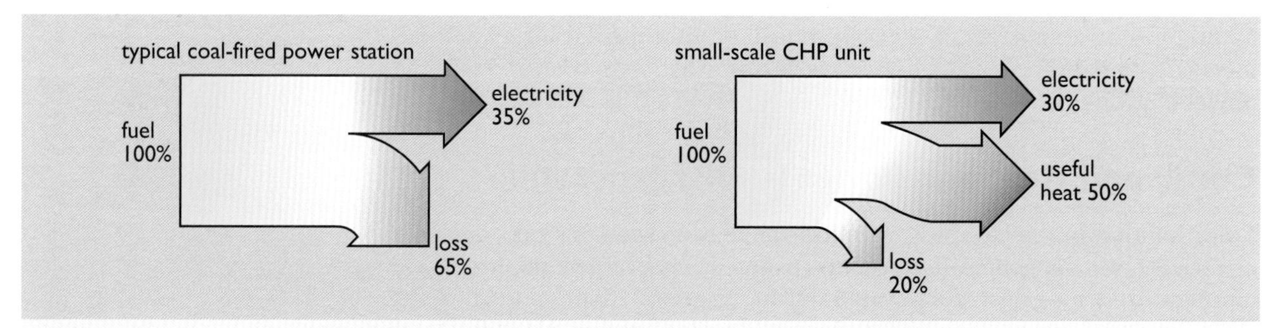

Figure 9.41 Sankey diagram comparing a conventional power station and a small-scale CHP unit

Over the past 20 years there have been repeated attempts to produce really small gas-fuelled CHP units suitable for individual homes, producing about 1 kW of electricity. Spark-ignition engines of this size are usually too small and unreliable, so development has concentrated on Stirling engine units which, as described in Chapter 8, are currently undergoing field trials.

Large-scale CHP with community heating

An alternative approach is to distribute waste heat as hot water from existing or specially adapted power stations via thermally insulated pipes to local buildings. Figure 6.23 has described the different steam turbine stages of a power station. The final low pressure (LP) turbine allows steam to be expanded to a low pressure vapour at 50 °C or less. Alternatively a **pass-out steam turbine** can be used. Depending on the heat load, steam can either be fully expanded through this to produce the maximum electricity, or at some point along it steam at typically 85–95 °C can be bled off for use as hot water. This does involve some loss in electricity generation efficiency, but this is made up for by the large amounts of heat that can be made available.

The centralized provision of water for heating is known as **district heating** or **community heating**. The heat sources may be central boilers, refuse incinerators or power stations. This has not been very popular in the past in the UK, because there was little tradition of using central heating until the 1970s and since then natural gas has been widely promoted for individual central heating systems. However, it is widely used in Scandinavia. In Helsinki in Finland 98% of heating is supplied by community heating from a variety of sources.

In the UK only about 2% of homes use community heating. A typical example is the Pimlico District Heating Undertaking, started in 1951 using heat from Battersea power station in central London. The power station was closed in 1982 but today the heat supply comes from a 3.1 MWe gas-fuelled CHP plant on the same site. It serves 3000 residential and 50 commercial customers. Today the most visible part of the scheme is a large hot water storage tank, the 'accumulator' (see Figure 9.42), built to take up any mismatch between the heat needs of the system and the electricity generation in the power station. It can store enough to keep the whole system running for more than a day. The Pimlico system is currently being linked to the nearby Whitehall CHP scheme, which serves mainly government buildings.

Figure 9.42 The Pimlico Accumulator stores heat for district heating originally supplied from Battersea power station, now from a gas-fuelled CHP plant

The potential for CHP with district heating in UK cities is enormous. A 2002 report (BRE, 2003) analysed heat loads in a large number of UK cities. It concluded that the economic potential was 18 GWe involving the connection of over 5 million homes. The potential for CHP in London alone was nearly 2.5 GWe.

Perhaps the most controversial possibility for CHP is that of using the waste heat from nuclear power stations. This has been implemented in many countries including Russia and Switzerland, and the heat mains can extend 20 km or more from the power station. This technology does, of course, provide heating with virtually no CO_2 emissions, but might face problems of public acceptance in the UK.

9.7 Transmission and distribution

The National Grid in the 1930s

Generating electricity is one thing; conveying it efficiently to the customer is another. The decision in 1926 to build the high-voltage National Grid was a bold one, seen by some as wholesale state interference in the free market of electricity supply. Yet it was essential for economies of scale in generation.

A primary decision was that only AC stations could be connected and they had to share a common frequency, chosen to be 50 Hz. It was also perhaps fortunate that the standardization took place so early. Electrical equipment has a very long life span. The last public DC supply in London was only disconnected in 1962. Many other countries still suffer from a variety of supply voltages and frequencies. The grid in Europe standardized on 50 Hz, while that in the USA uses 60 Hz. Japan, however, has a 50 Hz grid in the north-east and a 60 Hz system in the south-west. The modern global production of light bulbs has encouraged voltage standardization, but a typical laptop power supply is usually made to be extremely flexible and will accept AC at 50–60 Hz at anything between 100 and 240 volts. At higher power, it is possible to convert from AC to DC and from one frequency of AC to another by using a motor on one system coupled back-to-back to a generator on another. Today this kind of conversion can be done conveniently using high-voltage semiconductors.

Constructing the National Grid was not just a matter of engineering. Obtaining 'way-leave' permission for the pylons and overhead cables to be placed on private land was difficult in the late 1920s and still is. The Central Electricity Board employed a number of retired generals and admirals to convince aristocratic rural landowners (many of whom had little prospect of getting a mains supply themselves in the near future) of the benefits of the Grid.

By 1933, 4800 km of 132 kV transmission lines had been built. As well as enforcing the closure of inefficient generating stations, the National Grid gave greater flexibility of operation, especially if there was a failure in any one power station on the network. However, if too much went wrong all at once, there could be progressive collapse of a large portion of the system giving a wholesale blackout over a wide area for many hours. The CEBs response to one in 1934 was that this had been due to:

> a combination of circumstances that is not likely to recur, and there need be no apprehension of any such general failure in the future.
>
> (CEB quoted in Cochrane, 1985)

Since then system blackouts, although rare, have become larger, more disastrous and politically embarrassing as society has become more dependent on electricity. Widespread grid failures in the UK on Christmas Day 1960 were not very popular. Even less so was the blackout in 1977 that struck New York City for 25 hours, during which time there was widespread looting. In 1998 power to the city centre of Auckland, New Zealand was cut off for several *weeks* until large sections of a burned out high-voltage cable could be replaced.

A major concern today is exactly who is contractually responsible for keeping the lights on, and what to do to put them back on again when they go out. Starting up a collapsed network is known as a **black start** and requires considerable skill in system control and usually a good head of steam in a number of large coal-fired power stations.

The initial purpose of the National Grid was to interconnect power stations within particular *regions* of England and Wales, and to give local backup when needed. However, experiments in 1936 and 1937 showed that it was possible to run the whole *national* system connected together without disaster. By 1938 it was clear that the increasing demand in the south of England could only be met by supplying large amounts of electricity from the north. Starting in October 1938 all seven regions of the National Grid were permanently linked together, creating the biggest integrated electricity network in the world at that time.

Coal by wire

The long-distance transmission of electricity rather than the transport of coal has many implications. During World War II it freed up the railways for the transport of other essential materials. As UK electricity use rose after the war, large new power stations were sited near the mines in Wales and Yorkshire, while the growing industrial electricity demand was in the south of England.

A 1 GW coal-fired power station requires about 10 000 tonnes of coal a day to keep it going. That is ten large, slow, train-loads that were not particularly welcomed by those who wanted to upgrade the railway network for high-speed passenger and freight services. Transmitting the electricity via the National Grid meant that the coal trains only had to make the short journey between the mines and the power station, rather than the length and breadth of the country. The Grid could thus be thought of as a substitute for an entire railway distributing nothing but coal – 'coal by wire' as it was called.

One disadvantage is that removing the power stations from cities has meant that the power station waste heat is no longer available for district heating.

Figure 9.43 The 400 kV National Grid crosses the River Thames at Dartford

The Supergrid

Between 1926 and 1950 UK electricity demand increased by a factor of eight and the original National Grid could no longer cope. Work started on a completely new 'Supergrid' initially at 275 kV but upgradeable to 400 kV. The average height of the pylons was 42 metres, twice the height of the old 132 kV lines. They also used thicker wire. At 275 kV it was able to carry six times the power of a 132 kV line.

Once again negotiators had to go in search of way-leave permission for another 6400 km of lines. What had been difficult in the late 1920s was now compounded by new planning legislation. New Acts of Parliament had set up National Parks, Areas of Outstanding Natural Beauty and Sites of Special Scientific Interest. The Electricity Act of 1957 also put a duty on the electricity industry to have regard for the preservation of amenity, ranging from the 'natural beauty of the countryside' to 'objects of architectural or historic interest'. It is perhaps fortunate that the 1964 Supergrid crossing of the Thames at Dartford was not in an area noted for its scenic beauty since it required pylons 192 metres high (see Figure 9.43).

Putting cables underground can solve a lot of objections of visual intrusion, but costs can be 20 times those of overhead lines. The 275 kV Supergrid crosses north London almost invisibly under the towpath of the Grand Union Canal, whose water is used to cool the cable (see Figure 9.44).

The UK grid today

In the UK electricity is now mainly generated in very large stations as three-phase AC at a high voltage of 25 kV or 33 kV. It is then transformed upwards to the level of the Supergrid, 275 or 400 kV (see Figure 9.45). It may then travel hundreds of kilometres to a regional Grid Node, a major junction of the Supergrid and the older sections of the National Grid. These are usually

Figure 9.44 The 275 kV Supergrid cables run innocuously through north London under the towpath of the Grand Union Canal

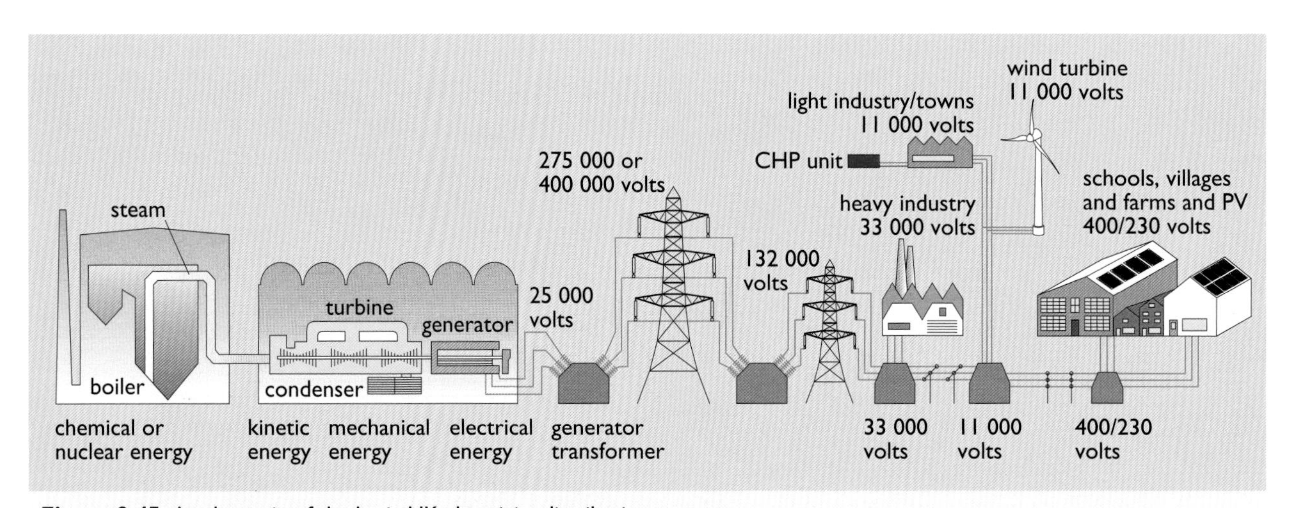

Figure 9.45 A schematic of the basic UK electricity distribution system

visible as a large collection of enormous transformers, where the voltage is reduced to 132 kV or 33 kV for distribution to local towns and cities, much of it over the route of the original 1930s Grid system. Large industrial users such as steel works are likely to take their supplies at this high voltage. For other users the power is further reduced in voltage to 11 kV or 6.6 kV. Small industrial users and others such as schools and hospitals take their power at this voltage, transforming it down to 400 V three-phase and 230 V single-phase in their own transformers. For domestic consumers the 11 kV supply is likely to be transformed down to 400 V three-phase. Individual consumers are given single-phase connections from this three-phase supply (see Box 9.3). In the UK these cables are usually buried out of sight, but in other countries, the wires very visibly run through the streets.

This is a picture of 'centralized generation', yet increasingly there is also **embedded generation**, smaller generators situated more locally to the loads. Solar photovoltaic panels or small CHP units may be installed in houses, schools or local leisure centres, connected to the grid at the 400/230 V level. Onshore wind turbines of around 1 MW output and industrial-scale

CHP units are likely to be connected at the 11 000 V level. Larger offshore wind farms may have high-voltage links running ashore at 11 000 or even 33 000 volts.

International links

Figure 9.46 shows the current (2011) extent of the UK and Irish grids. These are not just national, but are beginning to become part of an *international* system.

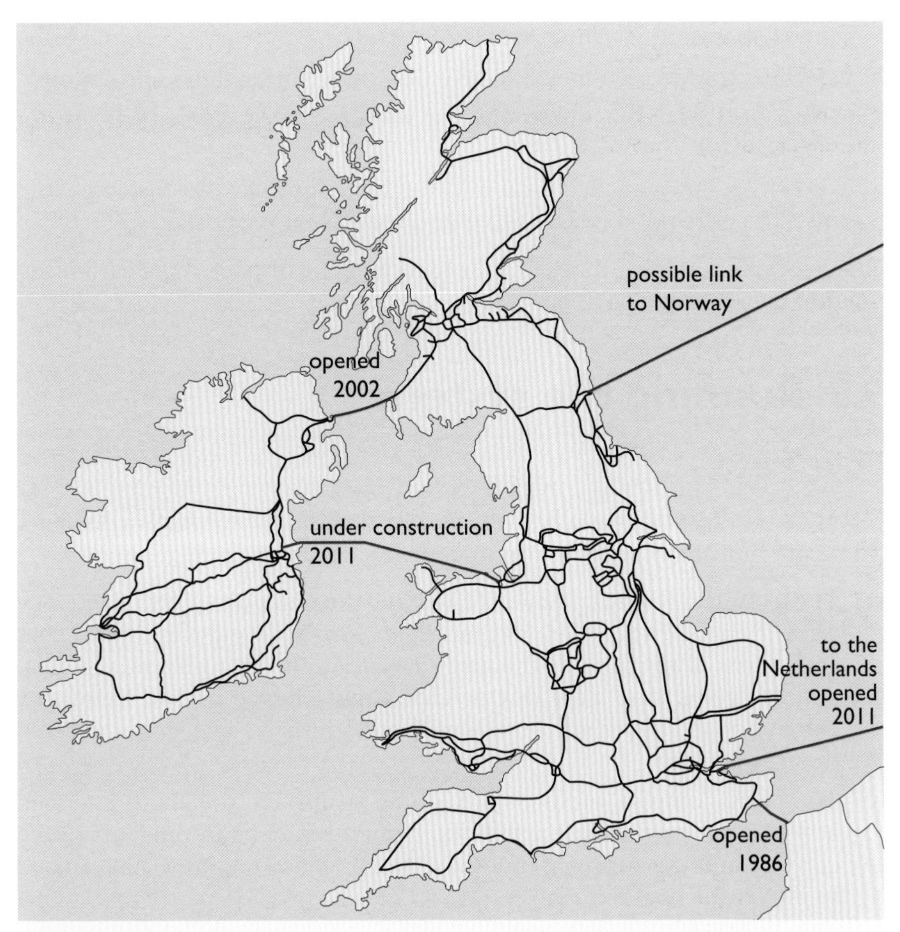

Figure 9.46 The National Grid in the UK and Ireland – 275 kV and 400 kV electrical supply system

In 1961 the British and French grids were linked together with a 160 MW cable under the English Channel. This was initially conceived to allow a mutual exchange of electricity since peak demands occurred at different times on the two sides of the Channel. Rather than attempt to synchronize the two grids the link was made in the form of high-voltage DC converted to and from AC at each end. Although the initial link was abandoned in 1982 after repeated damage from ships' anchors and trawler nets a replacement 2000 MW link was completed in 1986. This was laid in underwater trenches cut using a robot digger which crawled across the seabed on caterpillar tracks. Then the cables, each 50 km long and weighing 1700 tonnes, were laid in the trenches from a cable-laying ship and buried. Since then the link has largely operated as a conduit for importing French electricity into southern England, rather than on the basis of exchanging power at peak times. These imports typically make up about 3% of the UK electricity supply.

Since then other UK undersea links have been built or are under consideration:

- a 500 MW link between Scotland and Northern Ireland, opened in 2002
- a 1000 MW link between the Isle of Grain in the Thames estuary and Maasvlakte near Rotterdam in the Netherlands. Costing £500 million, it started operation in April 2011
- work has started on a new 500 MW link between England and Ireland
- a link between the north-east of England and Norway is under consideration
- extra links between Scotland and England through the Irish Sea and down the north-east coast of England have been proposed.

One reason for creating these links is to reduce the impact of the variability of wind generation (see Section 9.8).

9.8 Running the system

Four tasks

Mains electricity is essentially a commodity that has to be made on demand. There are four main tasks in running an electricity grid:

(1) The technical control problem of keeping the grid voltage and frequency within tightly specified limits, given multiple parallel connected generators. If the load on the grid rises, then the voltage tends to drop and the generators slow down. This slight change in grid frequency can be sensed anywhere on the network and used as a signal to boost the power input.

(2) The more complex problem of keeping supply and demand matched at all times, particularly when the demand may be in one part of the country and the supply in another. If CHP is used, there is also a need to match the heat demand.

(3) The need to 'keep the lights on' and to carry out a black start in the event of a grid failure.

(4) The overall task of optimizing resources in the supply of electricity (and heat where CHP is used).

The first three tasks are largely a matter of engineering, but the fourth has been challenging the minds of engineers, economists and politicians since the beginning of the 20th century.

What exactly is being optimized?

To a 'deep green' environmentalist such an optimization might mean supplying adequate 'energy services', such as lighting, heating, etc., using electrical systems, but involving appropriate levels of electricity conservation and distribution of waste heat, as well as actual electricity supply. The whole system would need to be run to minimize pollution, but doing so at a reasonable financial cost.

At the other extreme, perhaps, a dedicated free-marketeer might see this as all too complicated. Electricity is simply a 'product' and the aim is to maximize sales at minimum financial cost, while obeying any pollution regulations imposed by governments. It is something best 'left to the market'.

In the middle, there is a long history of state involvement, in the UK and elsewhere. A cheap and reliable supply of electricity is seen as something essential for the growth of an economy. Yet it may be something 'too important' to leave to the free market especially when fuel supplies, be they coal or oil or even gas, run short. Technologies such as nuclear power may require large amounts of finance that only the state is likely to be willing to provide. New technologies such as renewables may need subsidies for their introduction. Optimizing the running of such a system requires taking into account not just fuel costs, capital expenditure and the consequences of anti-pollution legislation, but also wider notions of national energy self-sufficiency and expectations of economic growth.

Ownership of the system

In practice, there are many possible mixtures of ownership and control of an electricity supply system. In its earliest days in the UK (and many other countries) local private and municipal companies supplied electricity within a few kilometres of their particular power station. After the completion of the original National Grid in 1934, the system in England, Wales and Scotland (i.e. Great Britain) came under state control even though most of the stations and distribution companies were privately owned.

After the 1947 nationalization and the 1957 Electricity Act, electricity supply in England and Wales became what is called a **vertically integrated system**. The generation of electricity was carried out by a state-owned monopoly, the Central Electricity Generating Board (CEGB), which sold it on to state-owned Area Boards, each of which had a monopoly relationship with its customers. Customers had no choice but to buy electricity from their local Area Board. A similar structure existed in Scotland. Although there was some independent electricity generation within large factories, technologies such as small-scale CHP were generally frowned upon.

By the 1980s there was considerable pressure within the European Union to treat electricity as a totally free market commodity, especially since there was considerable cross-border trade between different European countries. In the UK, the Conservative government pushed through the break-up (or 'unbundling') and privatization of the electricity industry in 1989. It was argued that it was bureaucratic and inefficient and that private companies could perform the role more flexibly at a lower cost.

For free marketeers, this privatization was seen as a role model for subsequent similar 'liberalizations' of state-controlled systems elsewhere in the world. The key ingredient was that most of the power stations became privately owned and were required to compete with each other. The actual mechanisms of this are complicated but central to the process is the notion of a centrally controlled **power pool**, an hour-by-hour competitive market in electricity. Electricity is bought in from all the competing power stations and then distributed through the National Grid to consumers. Exactly how the trading is arranged and who gets paid for what is governed by a set of **trading arrangements**. The present power pool in Great Britain is administered by the Office of Gas and Electricity Markets (OFGEM). The National Grid Company carries out the day-to-day running of the pool as system operator. Similar power pools exist in other countries and some are international. For example, the Nord Pool now covers Sweden, Norway, Denmark and Finland.

Although one aim of the 1989 UK privatization was to promote competition, since then many of the power generators and distribution companies have effectively been 'rebundled'. Mergers and takeovers have led to the industry being owned by a very few large (and mostly international) companies.

Balancing supply and demand

Different power stations have different operating characteristics. A modern 600 MW coal-fired plant may have a thermal efficiency of nearly 40%, but it may take eight hours or more to reach full power and efficiency when starting from cold. Also, if it is run at only part load, say producing only 300 MW, its thermal efficiency may suffer and reach only 35%. It is obviously best if this kind of station is run continuously at full power.

At the other extreme a small 30 MW 'open cycle' gas-turbine station may have a thermal efficiency of under 30%. However, such a station can be run up to full power in a matter of minutes. This kind of plant is best used for 'peaking' duties and may only be run for a few hundred hours a year.

In practice, demand for electricity varies widely, from hour to hour and from season to season. Figure 9.47 shows recent UK system data. On average, over a year, the load on the system is just under 40 GW, but on a summer

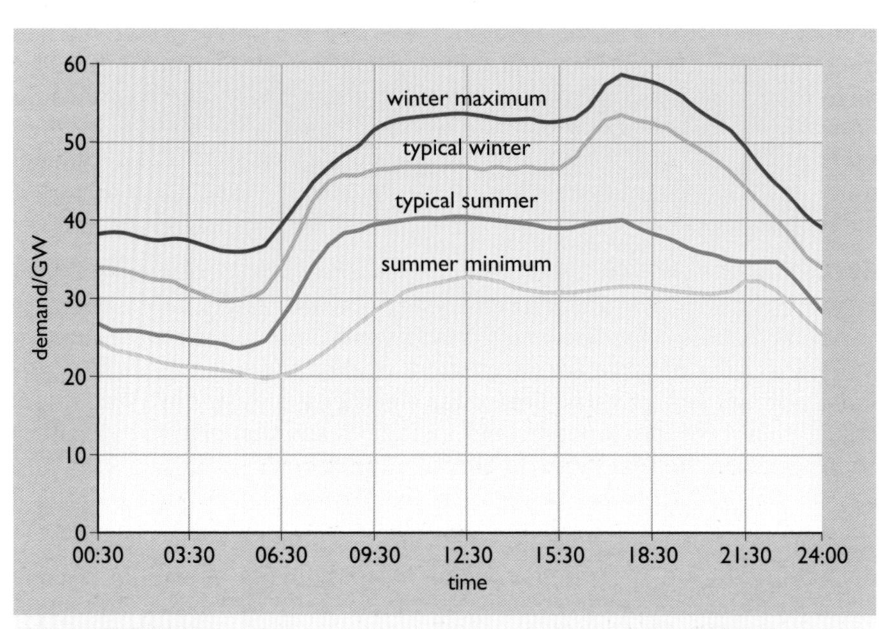

Figure 9.47 Variation of the electricity demand on the National Grid, 2009–10 (source: National Grid, 2010a)

night it can be 20 GW and it can rise to nearly 60 GW on a cold winter's evening. The demand can thus be thought of as a continuous 'base load' of around 20 GW, which essentially has to be produced for 365 days a year, with an additional variable demand on top.

Under the old nationalized CEGB system, power stations were classified according to a 'merit order'. Those with the lowest running costs, usually the nuclear stations and the largest and most efficient coal stations, were assigned to the top of this merit order. It was these that supplied the 'base load'. Next there were a large number of smaller and usually older coal-fired stations with lower thermal efficiencies. These were rated as 'middle merit order' and ran for most of the time in autumn, winter and spring, but shut down in summer. Finally, peaking plant, such as simple gas turbines, would be brought in to meet the mid-winter peak demands.

Under the current privatized power pool system the decision about which station runs at a given time is governed by a continuous process of competitive bidding under computer control. At any given time a computer model run by the National Grid Company estimates what the national demand will be in the following few hours and invites bids for the supply of electricity. Power station owners reply (or rather their computer programs do), the cheapest offers are accepted and the system is adjusted to bring appropriate stations online by remote control. The process has required the development of bidding strategies by power station owners to make sure that their particular commercial interests are maximized.

Overall, the system runs in a similar manner to that of pre-privatization days. The stations with the lowest running costs supply the base load. The lowest of all are likely to be renewable generators, such as wind power, which have zero fuel costs. As the demand rises, so stations with higher running costs bid into the pool and the electricity price rises. Above wind comes nuclear and then there is competition between coal- and gas-fired stations. In recent years gas-fuelled electricity has been cheaper than that from coal. Consequently gas-fired power stations have been operating almost all year round while coal has been used predominantly in the winter. Although hydro power has zero fuel costs its limited availability in the UK coupled with its high flexibility mean that it is normally kept for peaking use. At any time the competitive price is chosen to be sufficient to bring enough stations online to meet demand. The prices at times of peak demand can be very high.

The bidding process only applies to generators above a certain size. Small generators below about 1 MW are not constrained as to when they run, but are only paid a fixed rate for their electricity rather than the 'system price'.

Peak demands and pumped storage

Sudden surges in demand are a particular problem on the grid. As power station and grid link sizes have grown over the years, so have the potential problems arising from their failure. A 600 MW station might be generating at full power one minute, and the next a vital circuit-breaker could have tripped and it might be completely disconnected. There thus needs to be a temporary backup while other stations are brought online to cover the deficit.

The growth of radio and television has also produced problems by increasingly synchronizing the behaviour of large numbers of people. A mass rush for the electric kettle and the bathroom at the end of a popular TV show can produce an increase in national (and international) electricity demand of over 2 GW in a matter of minutes. Figure 9.48 shows the UK 'demand pickup' when the 1990 World Cup semi-final football match was transmitted.

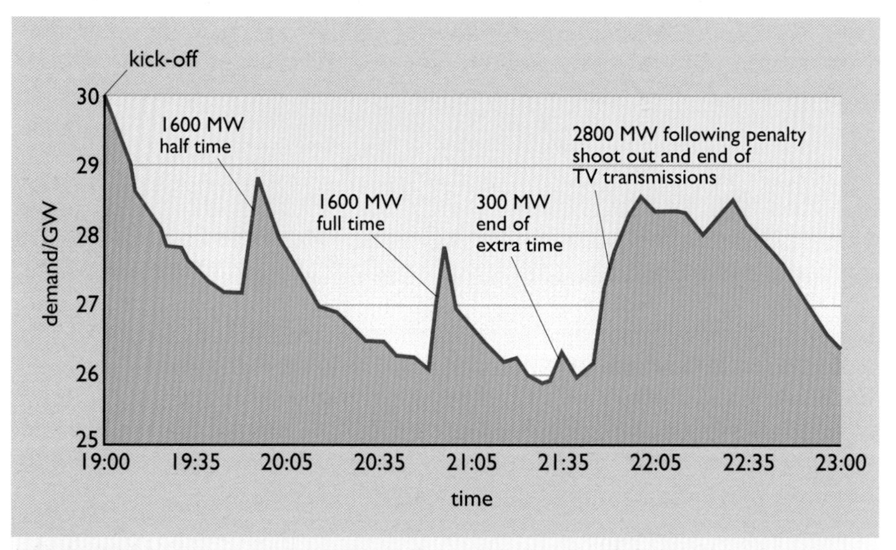

Figure 9.48 Variations in UK electricity demand during the 1990 World Cup semi-final (source: adapted from National Grid, 2010b)

To cope with these problems a number of **pumped storage** stations have been built. As described in Box 4.3, these are hydroelectric plants with very large turbines that can also act as pumps. At off-peak times surplus electricity from elsewhere in the grid is used to pump water up to a storage reservoir, usually high up on a mountain. When a surge in electricity demand occurs, the water is rapidly released back through the turbines, generating power in a matter of minutes. Although the output of these stations can be hundreds of megawatts, they are only designed to operate for short periods to cope while other more conventional plant is brought online. The UK currently has three pumped storage plants, two in Wales and one in Scotland. Their combined peak output power is over 2 GW or about 5% of the UK's typical winter electricity demand.

Coping with the variability of renewable energy

Renewable energy sources such as wind power, tidal power and PV do not produce electricity continuously at a steady rate. Their output is variable although to a certain extent it may be predictable (particularly in the case of tidal power). This has compounded the problem of matching supply and demand. It is likely that a future UK grid may have 30 GW or more of wind capacity on it. The question is always asked 'what do you do when the wind isn't blowing?'.

In Denmark, which in 2009 got 19% of its electricity from wind power, the answer is to use hydro power from Norway and Sweden via its strong international grid links. The increased use of wind power in the UK and Ireland is likely to require more pumped storage capacity and stronger grid links to Scotland's hydro resources, and possibly across the North Sea to those of Norway and Sweden.

Also, since weather systems (and their wind power) track across the British Isles from west to east, there is a good argument for stronger grid links between wind farms distributed from the far west of Ireland right across into Denmark and beyond. Although this doesn't eliminate the variability of wind power, it is likely to reduce it considerably.

Smart meters, smart grids and energy monitors

Currently, grid structures are 'dumb'. The consumer flicks a switch to turn on a light. It is for the 'system controller' to provide the electricity. What is needed is a two-way flow of information. There are many possibilities for 'smart' meters and an associated 'smart grid', i.e. one that has a parallel flow of *cost*, *demand* and *control* information. The following are some of the options:

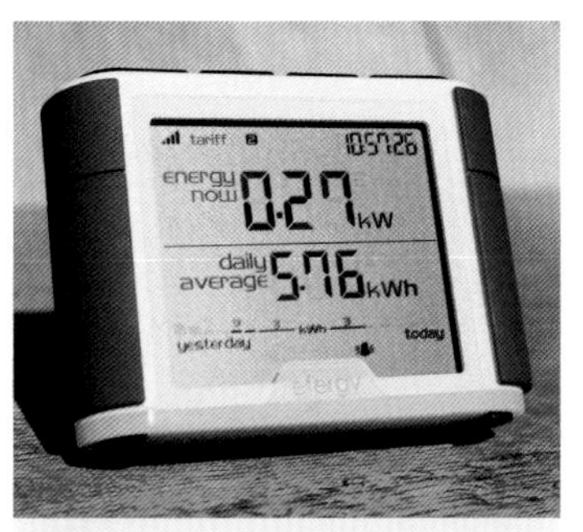

Figure 9.49 A domestic energy monitor radio linked to a current sensor on the electricity meter

- **Clear general energy use information** – if customers had clear instantaneous feedback on their electricity use, then they might be able to take action to reduce their consumption. Figure 9.49 shows a domestic energy monitor linked by radio to the electricity meter. It displays the instantaneous electricity consumption and can calculate daily averages including costs.
- **Remote meter reading** – this would mean an end to estimated billing.
- **Peak demand tariffs and information for all customers** – in the UK, the peak demand (as can be seen in Figure 9.47) occurs between 17:00 and 20:00 on winter weekdays. This is a time when offices and industry are closing for the day and domestic electricity consumers are arriving home and cooking meals. Although part of this demand is met from pumped storage plant, much of it has to be met from generation plant that would sit idle for most of the rest of the year. Experiments in California have shown an average peak demand reduction of 13% where customers were given warning of 'super-peak' prices. This is a matter of 'shifting demand' and delaying such activities as washing clothes to a time of lower prices.
- **Import–export metering for microgeneration** – the development of domestic micro-CHP and solar PV panels brings with it the need for two-way metering and billing for both imported and exported electricity.
- **Remote scheduling of loads** – not all electricity loads have to be met immediately. For example, tasks such as off-peak space and water heating can be delayed for a few hours. Other candidates in the wider community are water and sewage pumping (given adequate storage) and refrigeration in well-insulated refrigerators and cold stores.

The UK government has announced that it intends to start introducing smart meters in 2014 and complete their roll-out by 2019. Quite how many of the above features will be implemented has yet to be decided.

9.9 Electricity around the world

The 'litre of unleaded petrol' is now a globally traded standard product, and it might be tempting to think of the 'kilowatt-hour of electricity' in a similar way. Yet the manner of the generation of electricity and its use varies widely from country to country. Each of the sample countries discussed in this book uses different quantities of electricity and obtains it from different mixes of sources, as shown in Table 9.5. These have different pollution consequences and implications for sustainability. Also, the ownership and control is different in each country, with implications for investment in new generation technologies and in electricity conservation.

Table 9.5 Per capita electricity consumption and electricity generation by fuel, 2009

	Per capita consumption /MWh y^{-1}	Coal	Oil	Gas	Nuclear	Hydro, wind and other
UK	5.2	28%	1%	46%	18%[1]	7%
USA	13.1	45%	1%	24%	20%	10%
Denmark	6.3	49%	3%	19%	0%	30%
France[2]	7.7	4%	1%	5%	76%	14%
India[3]	0.6	69%	4%	9%	2%	15%
China[3]	2.4	78%	2%	1%	2%	16%
World[2]	3.0	41%	5%	21%	14%	19%

[1] Includes imports (mainly nuclear) from France
[2] 2008 data
[3] 2008 consumption and 2005 fuel breakdown
Sources: DECC, 2010c; EIA, 2010b; DEA, 2010a; DGEC, 2009; IEA, 2007 and 2010b.

UK

As can be seen from Table 9.5, per capita electricity consumption in the UK is slightly lower than that in Denmark and France, but far lower than that in the USA. The UK figure did not increase at all between 1995 and 2009 (see Figure 3.18). This is most likely due to campaigns to promote energy efficiency in appliances and lighting, and also strong competition from cheap natural gas for direct heating uses.

As described in Section 9.6, electricity is generated by a mix of coal, gas and nuclear plant, with a small amount of hydro power and other renewable sources. The use of gas has been growing rapidly at the expense of coal. However, as pointed out in Chapter 7, this has drained the UK's North Sea gas reserves all the quicker and increased concerns about the long-term supply of gas to the country.

The UK has a number of nuclear power stations, though their total output has been declining since 1998 as many of the early Magnox stations have been retired. An appreciable amount of electricity (typically about 3% of UK consumption) is imported from French nuclear stations via the Channel link.

The UK government has a policy of cutting national greenhouse gas emissions by 80% by 2050. Reducing CO_2 emissions from the electricity sector is a key element of that. It is suggested that this will require increased use of renewables, nuclear power and fossil fuel generation with carbon capture and storage. As can be seen from Figure 9.39, UK CO_2 emissions from power stations fell by about 30% between 1979 and 2010. This rate of decline will have to increase significantly if the government's targets for 2050 are to be met. This is discussed further in Chapter 14.

USA

On a per capita basis, US electricity demand is about twice that of the UK. Nearly half of US electricity is generated in coal-fired plant. Unlike the UK, where the use of coal has been falling, much of the growth in electricity demand in the USA has been met using coal, further contributing to the country's high CO_2 emissions. Much of the coal comes from mid-western states and is transported by rail.

Although 20% of the country's electricity comes from nuclear stations, many are approaching retirement, and there has not been much enthusiasm for constructing new ones. The USA has many large hydroelectricity schemes and is linked to others in Canada. Between them, the USA and Canada produce about a quarter of the world's hydroelectric power. The USA has considerable technical potential for renewably generated electricity from other sources, particularly from solar energy in the south-west and wind power in the central and northern states.

The US electricity industry is mainly organized on a state-by-state basis and consists largely of private utilities and generators, but with publicly appointed regulators with considerable powers. The California Energy Commission has been particularly forward-looking in promoting energy-efficient lighting and refrigerators since the 1970s.

The high dependence on electricity has created concerns about the reliability of supply, from both a technical and a commercial point of view. The weakness of the US electricity grid has been repeatedly criticized as, for example, contributing to grid failure in July 1977 that blacked out the whole of New York City. More recently, the need for stronger grid links has been emphasized in order to make use of the excellent wind resources in many US states.

France

In per capita terms, French electricity use is higher than that in the UK and Denmark. It also exports electricity to neighbouring countries. Demand had been growing strongly until 2000, partly because it has been promoted as a heating fuel. France does not have the supplies of natural gas available in the UK, or the extensive district heating schemes promoted in Denmark. Demand has flattened out and declined slightly since 2000.

Nuclear power and a modest amount of hydroelectricity are the principal sources. In environmental terms an average kilowatt-hour of French electricity involves very little CO_2 production but raises questions about nuclear safety and the disposal of nuclear waste. Although there is interest in expanding the use of wind power, it seems likely that the policy of heavy reliance on nuclear power will continue.

The electricity industry still consists largely of a vertically integrated monopoly system, Electricité de France (EdF), as set up in 1946. The French government has been very resistant to pressure from the European Commission in Brussels for any privatization. It was only in 2000 that the French parliament separated EdF into three parts: one for generation, one for grid management and one for distribution. Although EdF is now a private company, the French state still has a large shareholding.

Denmark

As can be seen from Table 9.5, Danish electricity demand on a per capita basis is similar to that in the UK. Danish national electricity consumption roughly doubled between 1975 and 2009, yet, as pointed out in Chapter 3, national primary energy consumption did not increase at all. The pursuit of energy efficiency in electricity generation has been a key element in this achievement.

In the early 1970s, the Danish electricity industry was almost totally dependent on oil for generation. Denmark's response to the price rises of 1973 was to adopt policies of security of supply, economic efficiency and environmental protection. Power stations under construction designed for oil were quickly switched to coal firing and there was an expansion in the use of CHP generation and district heating. In 2010, 53% of its electricity came from CHP (DEA, 2010b). The further reaches of the Copenhagen district heating system extend 35 km out from the centre of the city.

Figure 9.50 The twin Avedore CHP stations. Avedore 2 in the foreground is fuelled by gas and biomass. The older Avedore 1 runs on coal, but can also burn oil

The availability of natural gas from Denmark's North Sea sector from 1985 onwards has allowed the replacement of some coal-fired stations by CCGTs. The use of nuclear power had been considered in the 1970s, but proved politically unpopular and was ruled out in 1985.

Denmark is, of course, famous for its pioneering use of wind power. In 2009 wind supplied over 19% of the electricity demand and other renewables (mostly biomass CHP) brought the total to over 27%. In 2002 a new 570 MWe CHP plant, Avedore 2, was opened close to Copenhagen, fuelled by a mixture of gas and biomass (see Figure 9.50). Its opening has allowed the closure of three older coal-fired plants.

In other countries the extensive use of wind and CHP might be seen as excessively restrictive to the efficient operation of an electricity supply system. The Danish system avoids this through strong grid links with Sweden and Norway, which have very flexible hydroelectric power sources. There are benefits on both sides. In the wet years of 1989 and 1990, imports accounted for some 40% of Danish electricity consumption. In the dry year of 1996, gross exports from Denmark exceeded 50% of national electricity demand.

Organizationally, the Danish electricity industry has been traditionally based on small urban municipal systems and consumer cooperatives. It is these small organizations that have promoted many renewable and CHP schemes.

India

India is a developing country and the difficulties of its electricity industry are shared by many similar countries. As pointed out in earlier chapters, India has a rapidly growing population of about a billion people. Currently its per capita electricity consumption is only an eighth of that of the UK, with a general level of access to electricity roughly equivalent to that in the UK in the early 1940s. Reliable supplies of electricity are seen as essential for the development of industry in cities and also for the general policy of electrification of the whole country. About a fifth of electricity demand is used in agriculture, much of it for irrigation and it is heavily subsided (TERI, 2009). This contrasts with the UK, where only 1% of electricity demand is used in agriculture.

There is an extensive grid system used for long-distance transmission, particularly from the coal-mining areas in the east of the country to the major loads in the north-east. By 2007 over 80% of villages had some access to mains electricity. However, this does not necessarily mean that it extends into homes. It is estimated that in 2008 over 400 million Indians in rural areas did not have access to electricity (IEA, 2009).

Transmission and distribution losses are extraordinarily high, nearly 29% in 2006/7, compared to about 7.5% in the UK. However, much of this may be due to theft from local overhead distribution cables.

In 2008 the Indian electricity supply industry had about 140 GW of generation capacity (TERI, 2009) , but even this could not cope with demand and there are frequent power cuts in major cities. It is estimated that there is at least a 16% shortfall in generating capacity to supply the peak demand or, put another way, the country immediately needs another 20 GW of power plants to cope. This may sound a difficult enough problem, but if the country was to aim to reach current European per capita electricity use figures, then it would need to build at least another 1000 GW of generating plant, that is, 250 stations like the 4 GW one at Drax in the UK (see Figure 9.33).

The Indian industry is predominantly nationalized and controlled by power boards in each individual state. Even though domestic and agricultural electricity is heavily subsidized, most state utilities are loss-making and heavily in debt. According to a 2002 IEA report, of the total electricity generated only 55% is billed and 41% regularly paid for. The retail prices represent only 75% of the real average costs (IEA, 2002).

Coal is the major fuel used for electricity generation; however, much of it has a high ash content. This contributes to low generation efficiencies and high levels of pollution. The estimated emissions of Indian coal-fired plant in 2008 were nearly 1300 kg CO_2 per kWh (IEA, 2010a), 40% higher than for UK coal-fired stations. Most of the recent expansion has been through the use of coal or the indigenous, and limited, supplies of natural gas. The modest use of nuclear power is politically sensitive because of links to the production of nuclear weapons and recent tensions with neighbouring Pakistan. However, India has large reserves of thorium and there is interest in developing reactors to run on this (see Chapter 11). There is plenty of hydroelectric power in the mountainous northern states and there has been a rapid expansion in use of wind power in recent years. India boasted 13 GW of installed wind power as of December 2010, a tenfold increase since 2000.

China

Chinese electricity demand rose by a factor of six between 1990 and 2009 (BP, 2010). The per capita use in 2009 was about half of that in the UK. In 2006 over 70% of the Chinese electricity demand was in the industrial sector, a situation only paralleled in the UK before 1920. Agriculture used 3.5%, much of that, as in India, for irrigation (Ni, 2009). Unlike India, the rural level of access to electricity is high. It was estimated that in 2008 only 8 million people (out of 1.3 billion) lacked access to electricity (IEA, 2009).

As shown in Table 9.5, coal is the main generation fuel. As pointed out in earlier chapters, the continued growth of China's coal consumption raises serious questions about sustainability. Most of the coal mines are in the north of the country while the electrical load is in the expanding industrial cities of the south and east. Large amounts of coal are transported by road. There have been regular winter electricity shortages due to lack of coal.

Figure 9.51 The Three Gorges Dam under construction. This photo only shows approximately two-thirds of the total width of the dam.

Between 1990 and 2006 China opened nearly 400 GW of coal-fired power plant. Most of the recent plants have been large, high-efficiency supercritical and ultra-supercritical designs. Older, smaller and less efficient plants have been closed down. The overall CO_2 emissions for Chinese coal-fired plant in 2008 have been estimated at 900 g CO_2 per kWh (IEA, 2010a), far better than those in India (see above). Electrical transmission and distribution losses in 2006 were a creditable 6%, although the energy 'losses' in transporting the coal must be considerable.

China has considerable hydroelectric resources, particularly in the south-west where major rivers flow from the Himalayas.

The Three Gorges Dam (Figure 9.51) is the biggest hydro project in China and the largest dam in the world. It has involved damming the country's main river, the Yangtze, and the relocation of 2 million people. The project has 26 separate 700 MW generators with a total capacity of 18.2 GW. Fourteen units installed on the dam's left bank went into operation in September 2005, while the 12 plants on its right bank went into operation in October 2008.

9.10 Summary

This chapter has reviewed the historical development of electricity and described the relentless growth of the electricity industry during the 19th and 20th centuries, to the point where, at the start of the 21st century, it has become an essential ingredient of modern civilization.

One key element of this chapter is the continuing level of innovation. Section 9.2 celebrated some of the 19th century inventions leading to global telecommunications and electric lighting. Section 9.3 described the continuing development of different forms of more efficient electric light, to a level where the basic incandescent lamp, despite improvements, is now being phased out as too inefficient. Section 9.4 turned to electric traction and the long-dormant electric car, which could provide answers to a future shortage of oil for transport, that is, if sufficient supplies of 'rare earths' and other, similar key materials can be found.

Section 9.5 described the expanding uses of electricity throughout the 20th century and the devastating effect that one 'energy service', refrigeration, has had on the upper atmosphere.

Section 9.6 described the growth of the modern large coal-fired power station, the efficient CCGT and the environmental benefits of switching away from coal as a fuel. It also looked at ways to use the waste heat from power stations with CHP generation.

Section 9.7 looked at the growth of the National Grid and its current expansion with undersea links to neighbouring countries.

Section 9.8 described some of the complexities of operating a national grid system, including the problems of balancing supply and demand, coping with sudden peaks, or with the variability of renewable energy sources. It also described some of the possibilities for smart meters and smart grids, technologies likely to be developed in the next decade or so.

Finally, Section 9.9 looked at electricity in several countries, ranging from the UK, Denmark, France and the USA, where electricity demand seems to have reached a plateau, to India and China, with growing electricity use.

Electricity is *the* secondary fuel of today, clean and easy to use, and its use has brought many benefits, but it is not an unmixed blessing. Fossil-fuelled power stations are one of the leading causes of atmospheric pollution. The use of nuclear energy, the subject of Chapters 10 and 11, entails virtually no greenhouse gas emissions but, as we shall see, it brings with it other environmental and social concerns.

References

BP (2010) *BP Statistical Review of World Energy June 2010*, London, The British Petroleum Company; available at http://www.bp.com (accessed 11 April 2011).

BRE (2003) *The UK Potential for Community Heating with Combined Heat & Power*, Watford, Building Research Establishment, Client report number 211-533; available at http://www.energysavingtrust.org.uk (accessed 9 April 2011).

Cochrane, R. (1985) *Power to the People – The Story of the National Grid*, Sevenoaks, Newnes Books.

DEA (2010a) *Energy Statistics 1972–2009*, Copenhagen, Danish Energy Agency; available at http://www.ens.dk (accessed 25 March 2011).

DEA (2010b) *Danish Energy Policy 1970–2010*, Copenhagen, Danish Energy Agency; available at http://www.ens.dk (accessed 10 April 2011).

DECC (2009) *Historical electricity data: 1920 to 2008*, Department of Energy and Climate Change; available at http://www.decc.gov.uk (accessed 5 April 2011).

DECC (2010a) *Energy Consumption in the United Kingdom*: Domestic data tables, Department of Energy and Climate Change; available at http://www.decc.gov.uk (accessed 5 April 2011).

DECC (2010b) *Digest of UK energy statistics (DUKES)*, Department of Energy and Climate Change; available at http://www.decc.gov.uk (accessed 5 April 2011).

DECC (2010c) *UK Energy in Brief 2010*, Department of Energy and Climate Change; available at http://www.decc.gov.uk (accessed 20 March 2011).

DECC (2010d) '2009 Final UK Greenhouse Gas Emissions: data tables', Department of Energy and Climate Change; available at http://www.decc.gov.uk (accessed 10 April 2011).

DECC (2011a) *Energy Trends*, Department of Energy and Climate Change; available at http://www.decc.gov.uk (accessed 10 April 2011).

DECC (2011b) 2010 Provisional Greenhouse Gas Emissions, Department of Energy and Climate Change; available at http://www.decc.gov.uk (accessed 10 April 2011).

DGEC (2009) *L'électricité en France en 2008*, Direction générale de l'énergie et du climat; available at http://www.developpement-durable.gouv.fr/ (accessed 25 March 2010).

EIA (2010a) 'International Energy Statistics' [online], US Energy Information Administration, http://www.eia.doe.gov (accessed 25 March 2011).

EIA (2010b) *Annual Energy Review 2009*, Washington, DC, US Energy Information Administration, available at http://www.eia.doe.gov (accessed 25 March 2011).

Electricity Council (1987) *Electricity Supply in the UK – A Chronology*, London, Electricity Council.

IEA (2002) *Electricity in India*, Paris, International Energy Agency.

IEA (2007) *World Energy Outlook 2007*, International Energy Agency; available at http://www.iea.org (accessed 25 March 2011).

IEA (2009) 'The Electricity Access Database' [online], Paris, International Energy Agency, http://www.iea.org (accessed 10 April 2011).

IEA (2010a) *CO_2 Emissions from Fuel Combustion (2010 Edition)*, Paris, International Energy Agency; available at http://www.iea.org (accessed 10 April 2011).

IEA, (2010b) *Key World Energy Statistics, 2010*, Paris, International Energy Agency; available at http://www.iea.org (accessed 10 April 2011).

Meadows, D., Rander, J. and Meadows, D. (2004) *Limits to Growth: the 30-year Update*, London, Earthscan.

Molina, M. J. and Rowland, F. S. (1974) 'Stratospheric sink for chlorofluoromethanes: chlorine atomic catalysed destruction of ozone', *Nature*, vol. 249, p. 810.

National Grid (2010a) 'Seven Year Statement 2010' [online], http://www.nationalgrid.com/ (accessed 1 April 2011).

National Grid (2010b) 'National Grid powers up for World Cup 2010' [online], http://www.nationalgrid.com/uk/Media+Centre/WorldCup2010 (accessed 4 July 2011).

Ni, C. (2009) 'China Energy Primer' [online], Berkeley, CA, Lawrence Berkeley National Laboratory, http://china.lbl.gov (accessed 11 April 2011).

Romer, R. H. (1976) *Energy: An Introduction to Physics*, San Francisco, W. H. Freeman and Company.

SDC (2006) *The role of nuclear power in a low carbon economy*, Sustainable Development Commission; available at http://www.sd-commission.org.uk (accessed 9 April 2011).

Stolarski, R. S. and Cicerone, R. J. (1974) 'Stratospheric chlorine: a possible sink for ozone', *Canadian Journal of Chemistry*, vol. 52, pp. 1610–15.

TERI (2009) *TERI Energy Data Directory and Yearbook (TEDDY)*, New Delhi, The Energy and Resources Institute.

US DoE (2010) *Critical Materials Strategy*, US Department of Energy; available at http://www.energy.gov (accessed 26 March 2011).

Weightman, G. (2001) *The Frozen Water Trade*, London, HarperCollins.

Further reading

Byers, A. (1981) *Centenary of Service – A History of Electricity in the Home*, London, Electricity Council.

Patterson, W. (1999) *Transforming Electricity – The Coming Generation of Change*, Royal Institute of International Affairs, London, Earthscan.

Smith, G. (1980) *Storage Batteries*, London, Pitman.

Chapter 10

Nuclear power

By Janet Ramage

10.1 Introduction

Nuclear power is the subject of this and the next chapter. This one considers the present situation, the background science and the types of nuclear power plant currently in use. Chapter 11 then discusses possible new systems and the issues that might determine the future for nuclear power.

This chapter starts with a brief survey of the changing contribution of nuclear power to world electricity supplies over the past half century. A summary of the picture of atomic nuclei developed in Chapter 4 then leads to the first main topic: *radioactivity*. This was the earliest sub-nuclear effect to be detected, and is today undoubtedly the aspect of nuclear power that gives rise to the greatest public concern. It was also the tool that led to the discovery of nuclear fission, the central topic of the rest of the chapter, with accounts of present-day *nuclear fission* reactors and their merits and problems as power sources. The chapter also includes short introductions to the principles of *fast breeder* reactors and nuclear *fusion*, as a prelude to the more detailed discussion in Chapter 11.

Nuclear power has led to controversy since its inception. Before the general public had even heard of 'atomic energy', scientists in the 1940s were already divided, with some talking of 'electricity too cheap to meter' and others advocating no further development of the technology after World War II. In the event, the post-war period saw the emergence of new nuclear weapons and also the first nuclear power stations supplying electricity to the public. But the debate is by no means over. Supporters cite limited fossil fuel resources and concerns about climate change, while opponents raise issues such as safety, proliferation of nuclear weapons and the extent of the world's uranium resource. These issues are also discussed in Chapter 11.

Nuclear power worldwide

In the year 2009, nuclear power contributed to electricity generation in thirty different countries, with a world total of 435 reactors supplying about 2700 TWh a year from a total operational capacity of 375 GW. The USA accounted for about a third of the total output and France about a sixth. The only other countries contributing more than about three per cent of the total were Japan, the Russian Federation, South Korea, Germany, Canada,

Ukraine and the UK. In the latter, nuclear power contributed 69 TWh or about a fifth of the country's electricity production for the year (BP, 2010; DECC, 2009).

The first few nuclear plants started supplying power to national grid systems in the 1950s, but as Figure 10.1(a) shows, the rapid growth in nuclear capacity began over a decade later. From a mere 75 TWh in 1970, world annual nuclear output increased to over 700 TWh by 1980 – an average rise of about 25% a year. One country was largely responsible for this change: the USA, home to the world's first nuclear reactor, had the technology, and with fluctuating oil prices and increasing concern about resources, nuclear plants began to be seen as worthwhile investments.

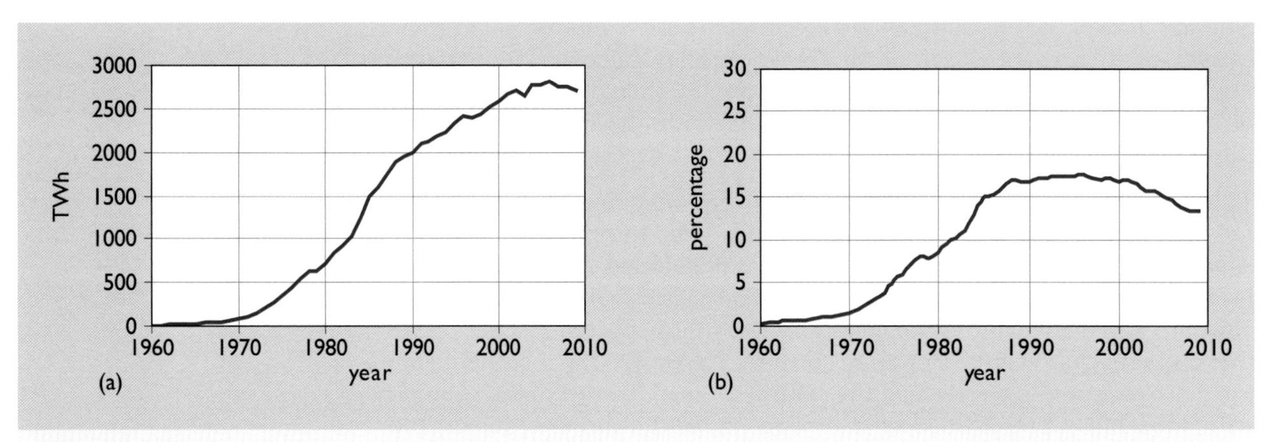

Figure 10.1 Nuclear contributions to world electricity generation: (a) annual nuclear output; (b) nuclear percentage of total (source: BP, 2010)

The worldwide increase received a further boost in the 1980s with the initiation of the French nuclear programme. As discussed in earlier chapters, France has only minimal fossil fuel resources, and in the 1970s a decision was reached at government level to develop nuclear energy as the dominant source for electricity. The reasoning may have been rather different, but the result was remarkably similar to the earlier growth in the USA. Between 1975 and 1985 the annual output of French nuclear plants rose from 18 to 224 TWh, a twelvefold increase over the ten-year period. Continuing growth, although less rapid, has meant that over the past few decades French nuclear output has continued to meet about three-quarters of the country's electricity demand.

World electricity consumption grew very rapidly throughout the 1970s and 1980s (see, for instance, Figure 9.26), but nuclear output was rising even faster and by 1990 accounted for about a sixth of total world electricity production (Figure 10.1(b)). Throughout the 1990s, nuclear output just kept pace with the ever-growing demand for electricity, but in the first decade of the present century its percentage contribution started to fall. Eventually, as new plant start-ups could no longer balance the shutdowns (Figure 10.2), the year 2007 saw for the first time an actual decrease in world nuclear output.

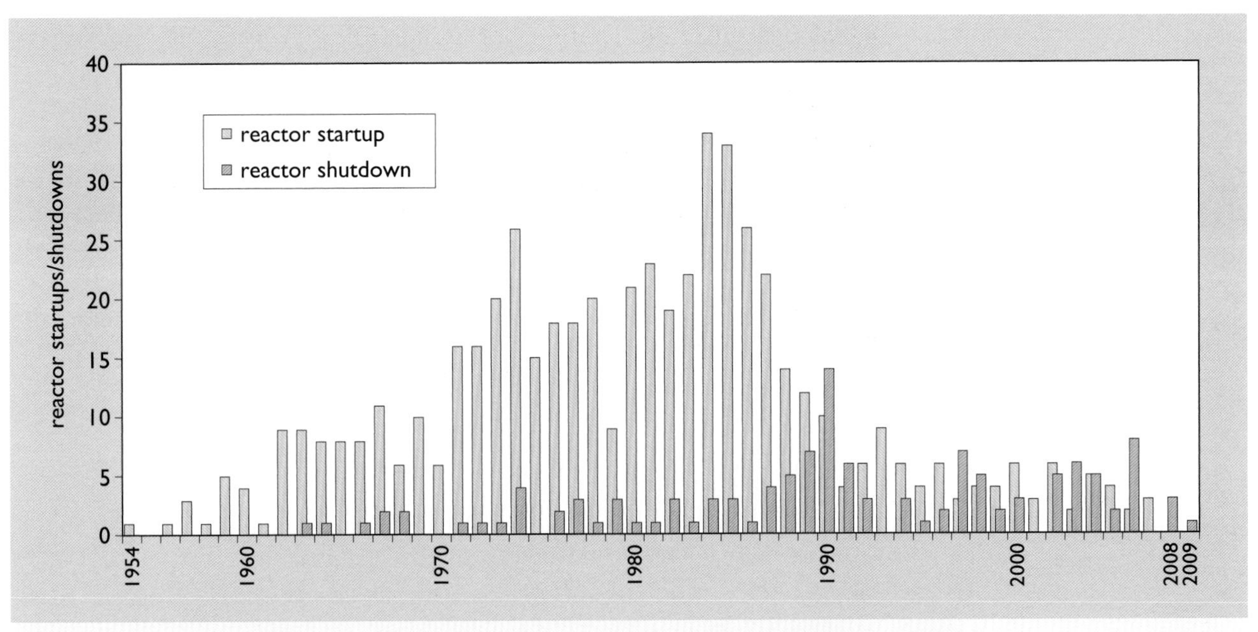

Figure 10.2 World reactor start-ups and shutdowns (source: redrawn from Schneider et al., 2009)

10.2 Nuclei: a brief summary

The account of atomic nuclei in Chapter 4 included the following points.

- The nucleus of any atom consists of two types of particle: protons and neutrons, collectively called nucleons. These are the 'heavy' particles in an atom, accounting for most of its mass.
- The main difference between a proton and a neutron is that the proton has positive electric charge whilst a neutron is electrically neutral.
- The number of protons in the nucleus, the atomic number, characterizes a chemical element.
- Nuclei of the same element may have different numbers of neutrons, creating isotopes of the element.
- The mass number of any isotope is equal to the total number of nucleons (protons plus neutrons) in the nucleus.
- The protons and neutrons in a nucleus are held together by the strong nuclear force acting between them at the tiny distances within the nucleus.

The actual *masses* of the sub-atomic particles did not enter in the accounts in Chapter 4, but they will be of considerable significance here. However, even the 'heavy' particles have masses that are tiny fractions of a kilogram. To avoid working with absurdly high (or low) powers of ten, a new, more appropriate, mass unit has been adopted for sub-nuclear particles. Similar reasoning applies to the energies of the particles. (The kinetic energy of a proton or neutron, even travelling at 100 million mph, would be about 1.6 millionths of a millionth of a joule.) Box 10.1 introduces the units used in nuclear calculations.

BOX 10.1 Units for sub-atomic masses and energies

The accepted unit for mass at sub-atomic level is the **unified mass unit (u)**, chosen to be roughly equal to the mass of one proton or neutron. Formally it is defined as one twelfth of the mass of a carbon-12 atom (because the carbon mass can be measured very accurately).

The unit for energies on the atomic scale is the **electron-volt (eV)**, defined as the energy gained by an electron (or proton) when it 'falls' through a potential difference of one volt.

In terms of the more familiar units for mass and energy, the values of these units are as follows:

one unified mass unit: $1\,u = 1.660 \times 10^{-27}$ kg

one electron-volt: $1\,eV = 1.602 \times 10^{-19}$ J

or one million electron-volts: $1\,MeV = 1.602 \times 10^{-13}$ J

This means that 1 kg of hydrogen contains about 6×10^{26} atoms.

To give some indication of the energy units, note that:

- 1 MeV (1 million electron-volts) is the kinetic energy of a high-speed proton or neutron travelling at about 30 million miles per hour
- the energies of the outer electrons of an atom, those involved in the chemical changes described in Section 4.4, are usually a few electron-volts
- the kinetic energy of the average atom of a gas at room temperature (see Section 4.3), also called its **thermal energy**, is about 0.025 eV – one fortieth of an electron-volt.

Mass-energy

Chapter 4 explained that mass and energy are now regarded as essentially two aspects of the same quantity, and that the mass-energy expressed as an energy (E) in joules and the same mass-energy expressed as a mass (m) in kilograms are related by Einstein's equation $E = mc^2$. Given that $c = 3 \times 10^8$ m s^{-1} this gives:

$$E(\text{J}) = m(\text{kg}) \times 9 \times 10^{16}$$

At the nuclear level, it is useful to have this relationship in terms of unified mass units (u) and millions of electron-volts (MeV) instead of kilograms and joules. Using the above values of the MeV and u, we find that the mass-energy of a particle expressed in MeV is 931 times its mass-energy expressed in unified mass units (u):

$$E(\text{MeV}) = m(\text{u}) \times 931$$

In other words, the mass of a single proton or neutron in energy terms is about 930 MeV, and the mass of an electron is about half an MeV. Box 10.2 shows how this might be used.

10.3 Radioactivity

A careful look at the relative atomic masses shown in Figure 4.19 reveals that the numbers of protons and neutrons tend to be equal for the lighter atoms. The nucleus of carbon, for instance, has 6 protons, so its relative atomic mass of about 12 means that it must have 6 neutrons. But as the nuclei become heavier, an excess of neutrons can be seen to develop. Mercury (Hg) for instance, with atomic number 80, has a relative atomic mass of about 200, so there must be 120 neutrons accompanying the 80 protons. We can see why this might be necessary: every proton repels every other proton due to their electric charge, so the forces needed to hold them all together increase steeply as their number rises. As neutrons have no electric charge and both protons and neutrons are attracted to each other by the strong nuclear force, any increase in the number of neutrons helps to maintain stability. Eventually, however, this is no longer possible, and beyond the element bismuth (83 protons and 126 neutrons) *there are no more stable nuclei*.

All isotopes of all the elements beyond bismuth are **radioactive**. Their nuclei spontaneously emit high-energy, electrically charged particles – which means of course that they change into nuclei of a different element (Figures 10.4 and 10.5 show examples). If this is so, how is it that we find uranium (atomic number 92) still existing in the Earth's crust? The answer is that it is decaying, but only very slowly indeed, as described later.

Radioactivity was discovered by Henri Becquerel in 1896, as an invisible 'radiation' that could fog photographic plates. This was fifteen years before even the idea of the nuclear atom, but several important features were gradually recognized:

- In many cases the 'radiation' decreases with time, over periods varying from seconds to years. In others it seems to show no detectable change during many years of observation.
- The type of 'radiation' and its intensity are characteristic of the emitting element. Physical or chemical changes – heating, compressing, even combining into different chemical compounds – have *no effect at all* on the radioactivity.

After Rutherford developed the idea of the nuclear atom, he was able to show that radioactivity is a *nuclear* process, and that there are three main types of radioactive emissions, easily distinguished by their penetrating powers (Figure 10.3).

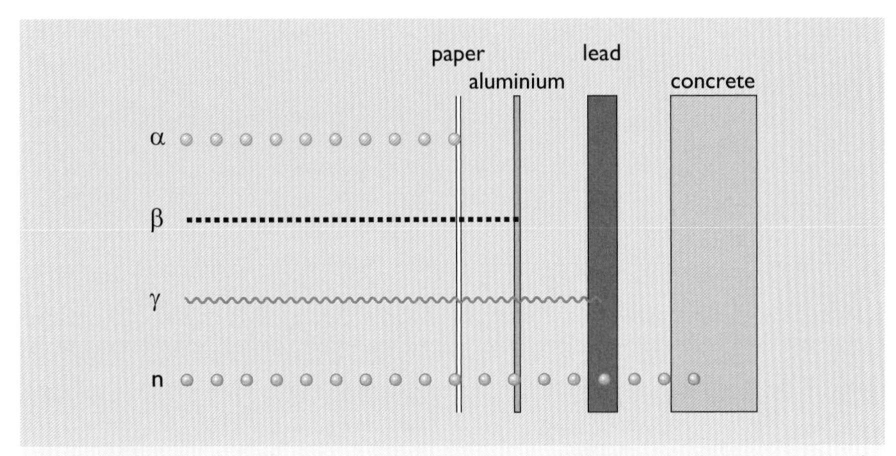

Figure 10.3 Penetrating powers of different types of radiation. Note that neutrons are not emitted in natural radioactivity, but are shown here for comparison because free neutrons play a major role in nuclear reactors

Alpha particles

Alpha particles (α-particles) are the least penetrating, being stopped by a sheet of paper, a couple of inches of air, or human skin. They can, however, present a serious health hazard if the radioactive substance enters the body – by inhalation or by ingestion in contaminated food or drink.

Rutherford identified α-particles as *helium nuclei*, consisting of two protons and two neutrons, and Figure 10.4 shows the process of **alpha emission** from the nucleus of Rn-222, an isotope of the radioactive gas **radon**, which has 86 protons. The loss of two protons means a *reduction of two* in the atomic number, so the result is a nucleus of the metallic element **polonium**, which has only 84 protons, two places lower in the periodic table (Figure 4.19). At the same time, the mass number is reduced by *four*, which determines the particular isotope of the new element, Po-218 in this example.

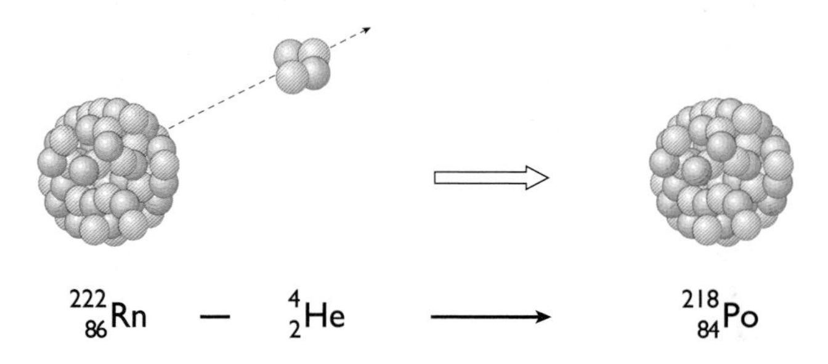

Figure 10.4 Alpha emission from a radon-222 nucleus results in a polonium-218 nucleus

The resulting isotope is often radioactive in turn (as in this case), so that an initially pure sample eventually contains a whole series of **daughter products**: different elements, each with its characteristic radioactive emissions. This multiplicity posed a problem for scientists trying to understand the processes, and remains a safety issue for those dealing with radioactive substances (see Section 10.7).

Beta particles

Beta particles (β-particles) are more penetrating than α-particles, but nevertheless are stopped by a thin sheet of metal or a few millimetres of almost any material. Historically, they were quickly identified as *electrons*, exactly the same as those already discovered a few years earlier (see Section 4.4) but with a great deal more energy. Normal β-particles from radioactive materials can penetrate into the skin, causing unpleasant burns but, like alpha particles, will only produce internal damage if the source is inhaled or ingested. However, the very energetic β-particles produced in a nuclear reactor can emit dangerously penetrating γ-rays as they decelerate.

An understanding of the origins of **β-emission** had to wait some twenty years for Rutherford's picture of an atom as a positive nucleus with surrounding electrons. But then the discovery that these particles were coming from the nucleus only deepened the mystery. How could a negative electron possibly come from a positive nucleus? And the development of the more detailed picture of a nucleus as a cluster of positive protons and neutral neutrons didn't help either.

By the 1920s, protons, neutrons and electrons were regarded as the basic indivisible particles of all matter. But this new evidence had to be explained. Emission of the negative beta particle changes the element to the next *higher* one in the periodic table, and, as Figure 10.5 shows, the loss of one very light electron does not change the mass number (131 in this case). In other words, whilst the atomic number, the number of protons, has *increased* by one, the total number of protons plus neutrons has not changed. So the number of neutrons must have *decreased* by one. Effectively, one neutron has turned into a proton by emitting an electron! This seemed to be the only explanation, but confirmation came only in 1932 when **free neutrons**, outside the nuclei of atoms, were first detected. Measurement showed that the mass of a neutron is indeed slightly greater than that of a proton, and, moreover, that a free neutron is *radioactive*, spontaneously turning into a proton by β-emission. If the masses are known, the energy released can be calculated (Box 10.2).

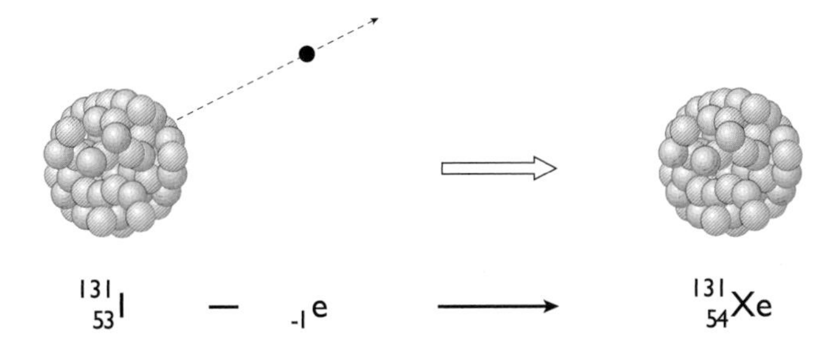

Figure 10.5 Emission of a beta particle (an electron) from an iodine-131 nucleus results in a xenon-131 nucleus

BOX 10.2 Energy from neutron decay

How much energy is released when a free neutron decays into a proton and an electron?

The calculation needs reasonably precise values for the masses of the three particles:

neutron mass	$M_n = 1.00867\,u$
proton mass	$M_p = 1.00728\,u$
electron mass	$m_e = 0.00055\,u$

The difference between the mass of the original neutron and the total mass of the resulting proton and electron is therefore:

$$1.00867 - (1.00728 + 0.00055) = 1.00867 - 1.00783 = 0.00084\,u.$$

If this spare mass appears in the form of energy, the relationship in Box 10.1 shows how much:

$$E(\text{MeV}) = m(\text{u}) \times 931 = 0.00084 \times 931 = \mathbf{0.78\,MeV}$$

If all of this became kinetic energy of the electron, it would be ejected as a penetrating β-particle travelling at about three-quarters of the speed of light.

Gamma radiation

Gamma rays (γ-rays) were discovered a few years after the other two types of radioactivity, mainly because they are much more penetrating and thus difficult to detect. When they were identified as very short wavelength electromagnetic waves, they were also called **γ-radiation**, whilst the photon view of radiation led to a third term: **γ-particles**. (All three of these names are currently acceptable.) Gamma radiation is essentially the same as high-energy X-rays (see Figure 4.15), and has similar effects on living matter. Gamma particles are stopped only by several centimetres of lead or steel, or a few feet of concrete, and together with neutrons they are the chief radiation hazard associated with nuclear technology.

Gamma emission doesn't change the numbers of protons or neutrons in the nucleus. It is a settling-down process by which a nucleus loses surplus energy, often following α- or β-emission, or fission. The nuclei before and after, differing only in their energy, are called **nuclear isomers.**

Radioactive decay and half-life

The theory of radioactivity developed by Rutherford and his younger colleague Frederick Soddy in 1903 is based on a conclusion drawn from their observations.

The rate at which particles are emitted is proportional to the number of radioactive atoms present at the time, and is not affected by any other factors.

As explained above, emitting an alpha or beta particle changes the nucleus to a different one, so it reduces the number of radioactive atoms of that type by one. The **rate of decay**, the rate at which the number of radioactive atoms decreases, is therefore proportional to the number present at any moment. As this number falls, the rate of decay also decreases, a situation that leads to the falling exponential shown in Figure 10.6. (This is the inverse of the *exponential growth* shown in Box 1.4.)

The rate of fall of the radioactivity is usually measured by specifying the **half-life**: the time taken for half of any sample to decay. The more rapid the decay, the shorter the half-life. Inspection of Figure 10.6 shows that the half-life is *the same no matter where you start.* Iodine-131 is a β-emitter with a half-life of 8.1 days, so half of any sample will decay in 8.1 days, half of the remainder will decay in the next 8.1, half of the new remainder in the next 8.1, and so on. It follows, of course, that the rate of emission of particles, the **activity** of the sample, is also becoming less at this rate. Box 10.3 describes the units of radioactivity and the half-lives of some isotopes.

Natural uranium consists mainly of three isotopes. The heaviest, **U-238**, accounts for over 99%. Most of the remainder, about 0.72% or one atom in 140, is **U-235**, and a tiny proportion, only 56 atoms in every million, is **U-234**. (Four more uranium isotopes have been detected in nuclear processes, including **U-233**, discussed in the next section.) All these isotopes are radioactive, but the long half-lives of the three main ones, 4500 million years for U-238, 700 million for U-235 and a quarter of a million for U-234, explain why we still find them on Earth. An interesting consequence of these half-lives is that a piece of uranium with today's proportions of U-238 and U-235 would have had *equal* amounts of these two isotopes about 6000 million years ago.

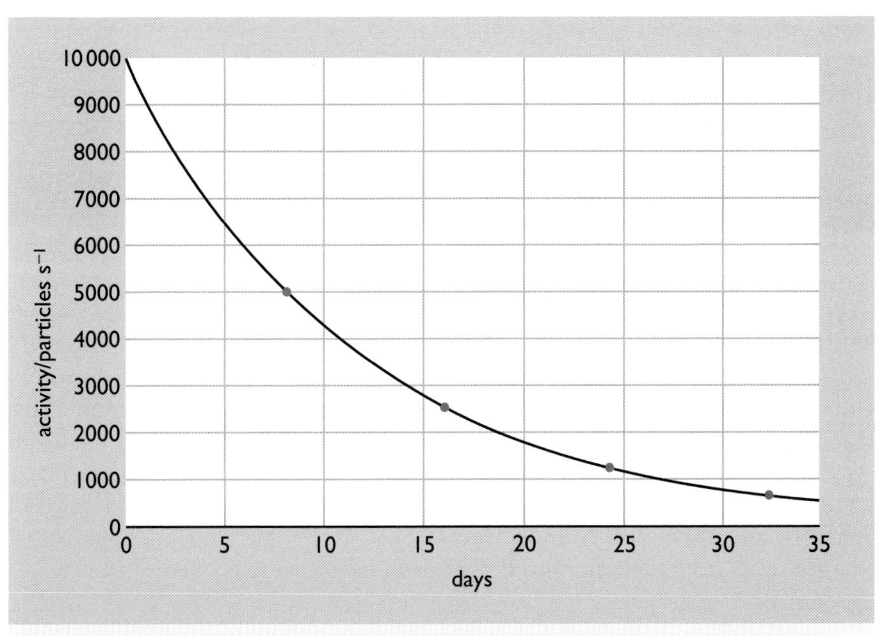

Figure 10.6 Radioactive decay curve of iodine-131 showing progressive halvings

BOX 10.3 Measuring radioactivity

A radioactive source can be characterized by the number of particles it emits per second. This is its *strength*, measured in **becquerels (Bq)**. A piece of material with an activity of 1 MBq is therefore emitting one million particles a second, and so on.

The becquerel replaced the earlier unit for the strength of a source, the curie (Ci). A one-curie source would emit 3.7×10^{10} particles a second, so the conversion is very roughly 1 Ci ≅ 40 GBq or 1 TBq ≅ 30 Ci.

Table 10.1 shows the half-lives and activities of a few **radioisotopes** (radioactive isotopes). Comparison of the two uranium isotopes shows, as one would expect, that greater activity leads to a shorter half-life. To put the data in context, note that:

- one gram is the mass of half a (dry) tea bag, and one microgram is a millionth of this – a speck of dust
- an activity of 10 000 Bq per cubic metre (undesirably high for continuous exposure) requires no more than two parts per thousand-million-million of iodine-131 in air
- the encapsulated sources considered safe for school laboratories have activities of perhaps 10 000 Bq.

So is 10 000 Bq a safe level of radioactivity or not? The first answer is that it isn't a 'level' but the strength of a particular source. And secondly, safety depends on *what* the source is and *where* it is. An encapsulated source used correctly in the laboratory adds a negligible amount to the natural radioactivity that we all receive continuously; but 10 000 Bq of plutonium in a soluble form entering the bloodstream would be a very different matter. In assessing safety – or potential danger – we do need data of the type in Table 10.1, but we also need information on the resulting doses and their effects. These topics are discussed in more detail in Chapter 13.

Table 10.1 Half-lives and activities

Isotope	Uranium-238	Uranium-235	Plutonium-239	Strontium-90	Iodine-131
[1]Particle	α	α	α	β	β
Half-life	4.5×10^9 years	7.0×10^8 years	24 000 years	28 years	8.1 days
[2]Activity of 1 g	12 000 Bq	79 000 Bq	2300 MBq	5.3 TBq	4600 TBq
[3]Mass for 10 000 Bq	0.81 g	0.13 g	4.3 μg	0.0019 μg	2.2 pg

[1] Gamma particles are also emitted in all cases except Strontium-90.

[2] This refers to one gram of the named isotope, and is the activity before it has produced any 'daughters'.

[3] Notice the units: 1 μg (microgram) is one millionth of a gram and 1 pg (picogram) is one millionth of a microgram or 10^{-12} g.

As mentioned above, the decay of a uranium isotope results in radioactive daughter products that in turn produce 'granddaughters' and so on, until the number of protons and neutrons is small enough to form a stable nucleus – in each case an isotope of lead. (Section 10.7 discusses some consequences of this sequence.)

An effect without a cause

What causes a radioactive nucleus to emit a particle at a particular moment? The extraordinary answer is that there is no immediate cause. Radioactivity, as Rutherford and Soddy recognized, is uniquely a *truly random* process. If a sample contains, say, a billion iodine-131 atoms, you can confidently predict that about half a billion will be left after eight days. But there is absolutely no way to predict *which* atoms will be left, nor when or whether a particular nucleus will decay. This is not a matter of inadequate apparatus, or insufficient knowledge. The radioactive decay of an individual nucleus is believed to be truly an effect without a cause, and the consequence is that we ourselves can neither cause nor prevent it.

Undesirable long-lived radioactive nuclei might in principle be converted individually into nuclei with shorter half-lives by bombardment with high-speed particles, but until such a system exists, we can only keep the material secure until enough half-lives have elapsed.

10.4 Nuclear fission

Experiments with neutrons

The discovery of radioactivity provided a completely new tool for the study of matter. Firing the particles into materials and observing their interactions with atoms or nuclei became one of the most fruitful experimental techniques of the twentieth century. Bombardment of ultra-thin gold foil with α-particles led Rutherford to the concept of the atomic nucleus (see Chapter 4), and it was his ex-student James Chadwick, bombarding the metal beryllium with α-particles, who in 1932 first identified free neutrons amongst the products.

Neutrons proved to be by far the most effective 'projectiles' for studying nuclei. Being relatively heavy, they pass easily through the cloud of electrons surrounding the nucleus, and being electrically neutral they are not deflected away by the positive nuclear charge. Enrico Fermi (Figure 10.7) in Rome, working his way through some sixty different elements, discovered that firing neutrons at target atoms often caused the target to become radioactive, emitting β-particles. This meant that elements one place higher in the periodic table were being produced (as discussed in Section 10.3), and Fermi eventually reached the ultimate target: uranium, the heaviest known element at that time. When bombardment again led to β-particles, he concluded that he must have produced elements with atomic numbers *greater* than 92, the so-called **transuranic elements**. As shown in Table 4.19, these are **actinoids**. But closer study showed another result: one that could accurately be called world shattering.

Fission

Fermi's experiments were soon repeated by others, including Otto Hahn and Lise Meitner in Berlin (Figures 10.8 and 10.9). Identifying the tiny amounts of the new materials produced by neutron irradiation of uranium was not easy, but eventually, in 1938, a brilliant chemical analysis established that *barium* was certainly one product (Hahn and Strassmann, 1939). How could adding a neutron to uranium result in this much *smaller* nucleus? Hahn

Figure 10.7
Enrico Fermi (1901–1954) and his brother Giulio were science prodigies, inseparable until Giulio died at the age of 15. Enrico took refuge in study, teaching himself from books found in the local flea market in Rome. At 17 he won a fellowship at Pisa, and by 1927 had made a major contribution: Fermi statistics – the basis for modern theories of metals and semiconductors. Turning to nuclear physics. he carried out the experiments that led ultimately to fission. All who knew him said Fermi was truly unique. His lectures were dazzling, a stream of new insights rather than standard approaches.

Figure 10.8
Otto Hahn (1879–1968) studied chemistry in Germany, and worked with Rutherford in Montreal before returning to Berlin. He distanced himself from the German nuclear weapons programme, and in the post-war years concentrated on the peaceful applications of radioisotopes and fission – including, in 1950, an analysis of the safety measures that would be needed if nuclear power stations were ever built.

Figure 10.9
Lise Meitner (1878–1968) studied physics in Vienna, but moved to Berlin in 1907. Barred as a woman from the main Department, she was allowed (as a special concession to Otto Hahn!) to use a spare room on the ground floor. By the 1930s she was Head of Physics. Being an Austrian citizen, she was allowed to continue despite her Jewish background, but the Anschluss changed that, and in 1938 she left, travelling illegally and escorted by a Dutch friend of Niels Bohr (Figure 10.11). Declining to join the Manhatten Project, she stayed in Sweden until retiring in 1960 to Cambridge (UK).

wrote to Meitner, now a refugee in Stockholm, 'perhaps you can suggest some fantastic explanation' (translated from Hahn, 1968).

Meitner was reading Hahn's letter when her nephew Otto Frisch arrived from Copenhagen to spend Christmas, and in the next couple of days, sitting in a small hotel, they found an explanation (Pais, 1991, p. 454, citing Frisch and Wheeler, 1967). The resulting paper (Meitner and Frisch, 1939) proposed a mechanism by which a heavy nucleus absorbing a relatively slow-moving neutron might become unstable and split into two lighter nuclei (Figure 10.10). They called the process kernspaltung: nuclear fission, and remarked that it should release a great deal of energy (Box 10.4).

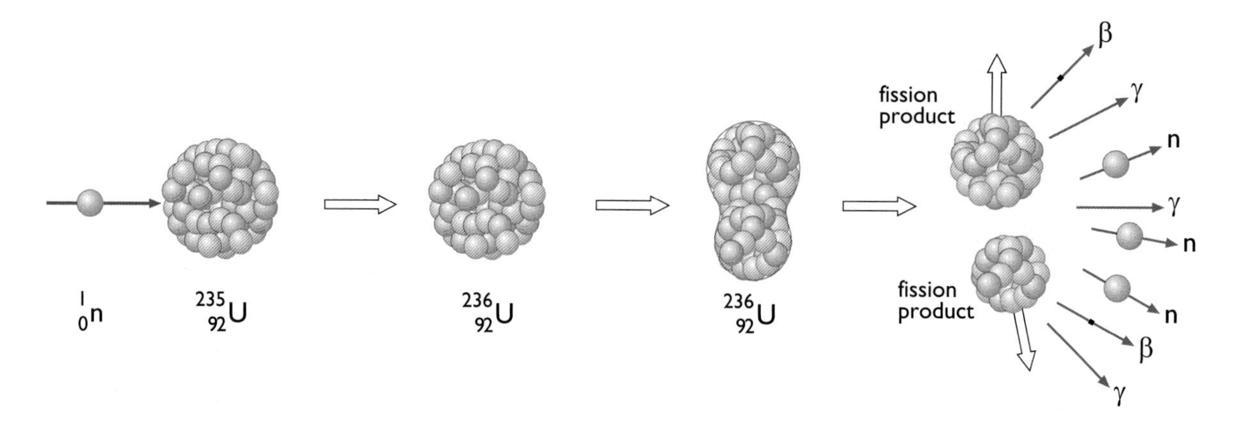

Figure 10.10 Fission of uranium-235. When a relatively slow-moving neutron is absorbed by a U-235 nucleus, the resulting unstable U-236 nucleus can take up a dumb-bell shape. Electrical repulsion between the two positively charged parts of this can then lead to fission, producing two new radioactive nuclei, the fission products. A few free neutrons (n) and gamma radiation (γ) are emitted in the process, and the fission products are likely to emit beta particles (β).

It didn't take long for the news to spread. In January 1939, Niels Bohr (Figure 10.11), in New York on his way to Princeton, met Fermi, a recently arrived émigré from fascist Italy. Two weeks later, at a physics conference in Washington, Bohr described the fission results, and Fermi suggested that free neutrons might be released in the process. Everyone knew that there would be surplus neutrons (see Box 10.5), but it did not necessarily follow that *free* neutrons would appear. The significance of Fermi's suggestion was immediately obvious to his audience. Neutrons cause fission, so a free neutron could cause a further fission, and more than one free neutron from each fission event could produce a divergent **chain reaction** (Figure 10.12). With only thousandths of a second between events, the result would be a very great deal of energy in a very short time – a nuclear explosion.

> This suggestion threw the meeting into an uproar while physicists who had facilities initiated calls to their laboratories to start the search for fission neutrons.
>
> (Manley, 1962)

Within a day, fission appeared for the first time in the news headlines. Within a year a hundred papers had been published in scientific journals. And during that year the Second World War broke out.

BOX 10.4 Energy from fission

Calculating the energy

It was not necessary for Meitner and Frisch to wait for measurements of the energy released in fission. The masses of individual atoms and the neutron were known quite accurately by the 1930s, so the total masses before and after the fission could be calculated by the method used in Box 10.2. The initial mass is that of the U-235 nucleus plus the initiating neutron. The final mass is the sum of the masses of the nuclei of the two fission products plus the masses of any spare free neutrons. If the final mass is less than the initial mass, the difference must be the mass-energy released in the process.

The fission of U-235 can lead to many outcomes, with different pairs of new nuclei and different numbers of free neutrons; but calculations showed that on average the final mass was less than the initial mass by slightly over one-fifth of an atomic mass unit – about 0.21 u.

As described in Box 10.1, $E(\text{MeV}) = m(\text{u}) \times 931$, so the energy released by the conversion of 0.21 u will be about **200 MeV per fission**.

This is only about 3.2×10^{-11} joules, which may seem very small, but the picture changes if we consider not one atom but one kilogram.

Each atom of U-235 has a mass of approximately 235 u.

Since $1\ \text{u} = 1.660 \times 10^{-27}$ kg (see Box 10.1):

Mass of one atom of U-235 = $235 \times 1.660 \times 10^{-27}\ \text{kg} = 390 \times 10^{-27}$ kg.

There are thus approximately 2.5×10^{24} atoms in one kilogram of U-235, so the energy released in its complete fission is:

$$E = 2.5 \times 10^{24} \times 3.2 \times 10^{-11}\ \text{J} = 80 \times 10^{12}\ \text{J} = \mathbf{80\ TJ\ per\ kg}$$

which is equal to the energy released in burning about 3000 tonnes of coal!

Note however that, as explained above, less than one per cent of natural uranium is U-235. Most of the rest, as we shall see, does not undergo fission.

Distributing the energy

Returning to the individual fission event shown in Figure 10.10, what happens to the energy that is released? The answer depends on the details of the process, but on average the energy is distributed as follows:

- More than four-fifths of the 200 MeV is carried off as kinetic energy of the two fission product nuclei. Collisions with surrounding atoms eventually redistribute this energy as heat.
- Just under 30 MeV is accounted for by gamma rays, beta particles, etc. Some of these are very penetrating and their energy may be 'lost' out into the surroundings.
- About 5 MeV becomes the kinetic energy of the free neutrons.

This third item was to become extremely important in practice. A neutron with kinetic energy of an MeV or so travels at a very high speed (see Box 10.1), and as discussed below, these **fast neutrons** produced in the fission process proved not to be ideal for running a nuclear reactor.

Figure 10.11 Niels Bohr (1885–1962) was born into a family of academics and bankers. He attended Copenhagen University, where he later established the world-famous Institute of Theoretical Physics. He applied the ideas of Planck and Einstein to Rutherford's nuclear atom, and the resulting 'Bohr atom' was a remarkable imaginative leap. He was in fact a remarkable man: an appallingly bad lecturer who was revered and loved by students and colleagues alike. For many theoretical physicists in the mid-twentieth century, Bohr was a greater influence even than Einstein, with whom he shared a mutual respect and affection.

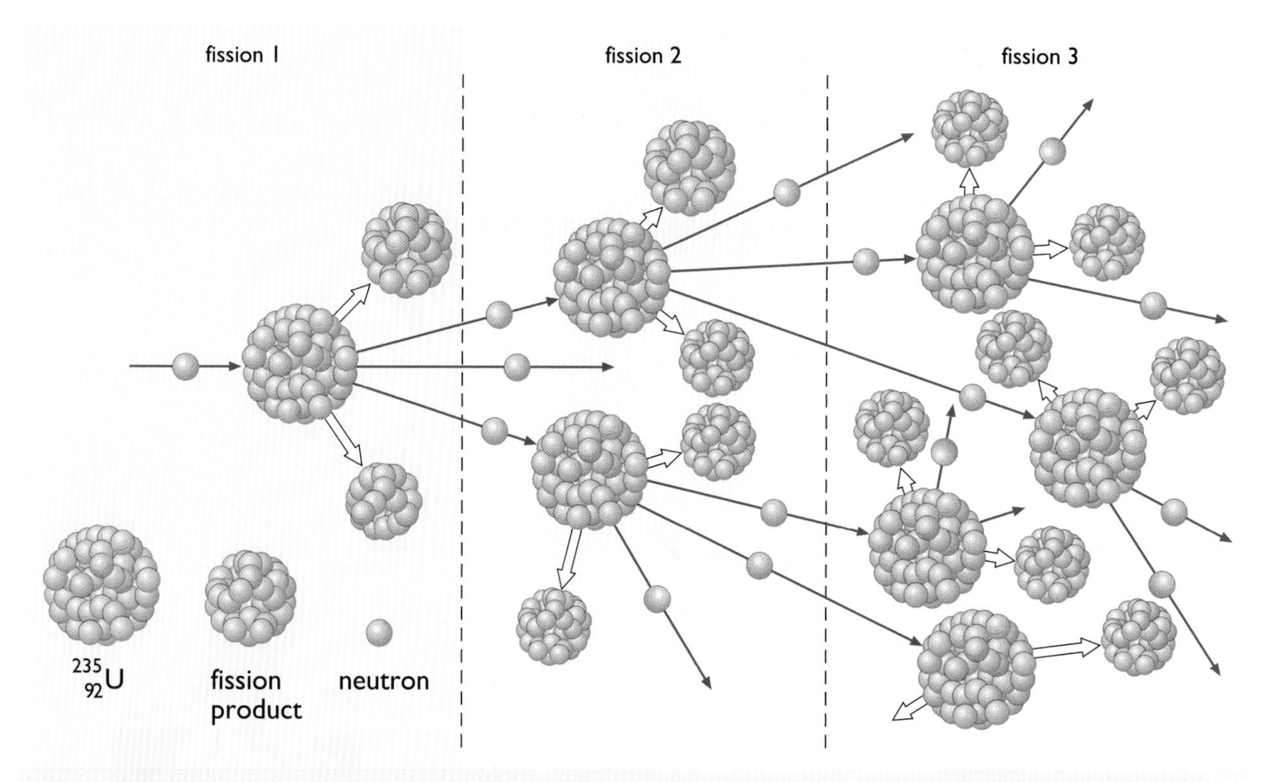

Figure 10.12 An explosive chain reaction. Suppose that a free neutron initiates the first fission (left-hand panel) and that two neutrons from this initiate new fissions (centre panel), and that on average two neutrons from each of these initiate further fissions (right-hand panel), and so on. Multiplying the successive generations in this way at intervals of a few milliseconds, the power output could reach megawatts in less than a second. (Note that this needs a very high concentration of U-235.)

BOX 10.5 **The surplus neutrons**

A uranium-235 nucleus undergoes fission into two new nuclei. If one of these products is barium, what is the other, and how many surplus neutrons should result from this process?

The atomic numbers of uranium and barium are 92 and 56 respectively. So the atomic number of the other element must be 92 minus 56, i.e. 36. This is the gas krypton.

Suppose now that the products are the heaviest stable isotopes of these two elements. The first column in Table 10.2 shows the mass numbers (protons plus neutrons) for these, and subtracting the numbers of protons gives the figures in the third column. Inspection shows that the original U-235 plus the initiating neutron provide *twelve* more neutrons than are needed by the two product elements.

Table 10.2 The spare neutrons in fission

Nucleus	Number of protons	Number of neutrons
U-235	92	143
Ba-138	56	82
Kr-86	36	50
Ba-138 + Kr-86	92	132

This calculation is misleading, however, in assuming that the product nuclei have their normal ratio of neutrons to protons. In practice they almost always start with too many neutrons, and this has two extremely important consequences. Firstly, it means that the number of *free* neutrons available for the chain reaction is not twelve but perhaps *two or three*. And secondly, it means that the new 'neutron-heavy' nuclei are likely to be *radioactive*. They are often β-emitters, because emission of a β-particle reduces the neutron–proton ratio (see Section 10.2). They also usually start with excess energy, so γ-emission is common.

Barium and krypton are not the only possible products of the fission of U-235. Other pairs of elements have been detected, but this example is typical in that the split is usually asymmetrical, with the atomic number of one product in the mid-fifties and the other in the mid-thirties. It is also typical in that the fission products are usually highly radioactive.

1939–1945: reactors and bombs

This book is concerned with the *controlled* use of energy. Explosions, whether chemical or nuclear, are outside its scope; but the development of nuclear weapons required the development of nuclear reactors, and these certainly are relevant.

The theoretical possibility of a fission weapon was evident to everyone after the Washington conference, but many scientists, including Bohr and Fermi, remained sceptical about its practicability. Bohr had established that it was the U-235 isotope that underwent fission, but, as already discussed, in natural uranium every U-235 atom is 'diluted' by about 140 atoms of U-238. Although on average two or three free neutrons are produced in each U-235 fission, many of these high-energy *fast neutrons* (see Box 10.4) are absorbed by the U-238 isotope or lost in other processes. In natural uranium, this leaves on average less than one neutron available to induce another fission. *So a chain reaction using fast neutrons in natural uranium is impossible.*

One solution would be to increase the proportion of the U-235 isotope; but producing this **enriched uranium** is not simple (see Section 10.7), and even if highly enriched uranium could be achieved, some free neutrons would still be lost in processes that did not lead to fission or by escaping out through the surface. This second point is very important, because the proportion that escape depends on the *size* of the piece of uranium. We can see this by considering two blocks of enriched uranium, identical in all respects except that block A is significantly smaller than block B. The greater size of B means that a fast neutron created by fission will on average spend a longer time in reaching the surface than a similar neutron in block A, and will therefore have a better chance of making a fission-inducing collision with a U-235 nucleus before it escapes.

In general then, if uranium with any particular concentration of U-235 is assembled into larger and larger blocks, a point may be reached where on average just *one* neutron from each fission causes another fission, sustaining a chain reaction. The mass of material that just reaches this point is called the **critical mass** for uranium with that degree of enrichment. There are of course other neutron 'losses', and in some cases the critical mass is never reached. In natural uranium, with less than one per cent U-235, the total neutron loss rate *always* exceeds the rate of production, no matter how large the block. So, as mentioned above, a chain reaction using fast neutrons in natural uranium is not possible.

There was, however, another option. Fermi had shown in the 1930s that *slowing down* the fast neutrons made them much more likely to initiate fission, increasing the chances of a chain reaction, and this was the method he used in the first reactor (see below). But the main problem at the time was the lack of the experimental data needed to assess the possibilities. The scientists in the USA persuaded Einstein to write to President Roosevelt (Einstein, 1939), to which he responded positively (President Roosevelt, 1939), and in February 1940 the sum of US$6000 (!) was made available for fission research.

Meanwhile, in England, Otto Frisch and Rudolf Peierls at Birmingham University were using the available data and inspired guesses to tackle the issue of critical mass. They concluded that for pure 100% U-235, about *one kilogram* would be enough.

> At that point we stared at each other and realized that an atomic bomb might after all be possible.
>
> (Pais, 1991, p. 493, quoting Frisch, 1979)

They presented their results in two memoranda (Frisch and Peierls, 1940a, 1940b), including the idea that an explosion could be achieved by rapidly bringing together two sub-critical masses of U-235. Their value for the critical mass proved rather optimistic, but even if ten times this was required (which proved to be the case), it would still be practicable.

By 1941, groups in several countries had projects for the construction of a controllable **nuclear reactor** (called at that time an atomic pile in the UK, uranium-pile in the USA and Uranbrenner in Germany). Everyone knew that a sustained chain reaction in natural uranium might be possible. And everyone knew that the others knew.

The first reactor

By 1941, Fermi was working on reactor development at the University of Chicago. In Britain, the Maud Committee, set up to consider the Frisch–Peierls conclusions, expressed confidence that a bomb was possible, and their report (Maud Report, 1941) was passed to the scientists in America. On 9 October the President made the decision to throw the huge resources of the USA into the development of an 'atomic bomb'. On 7 December, the Japanese attacked Pearl Harbor and the United States entered the war.

The bomb programme, now called the Manhattan Project, became a joint allied effort and most of the scientists working on it in the UK, including many refugees from Germany and occupied countries, crossed the Atlantic. The Chicago group and a Canadian–British–French team in Montreal worked on experimental reactors; the main site for weapons development was Los Alamos in New Mexico and there were reactors for plutonium production (see 'New fissile nuclei', below) elsewhere in the USA.

In Chicago on 2 December 1942, by carefully piling up lumps of uranium metal, surrounded by graphite to slow down the fast neutrons, and with strips of cadmium (a neutron absorber) to control the process, the Fermi team achieved criticality – the first controlled chain reaction (Figure 10.13). The Montreal reactor and plutonium-producing reactors elsewhere in the USA followed. The German project included a plant in occupied Norway producing heavy water, containing the hydrogen isotope deuterium, but they still had no operational reactor at the end of the war. Unknown to the Allies, it had been decided in 1942 to continue the nuclear work on a small scale only. The decision was made by the politicians, but the extent to which the German scientists, led by Werner Heisenberg, were actively pursuing the development of a weapon has remained the subject of debate ever since. (For more on graphite and heavy water, see Box 10.6.)

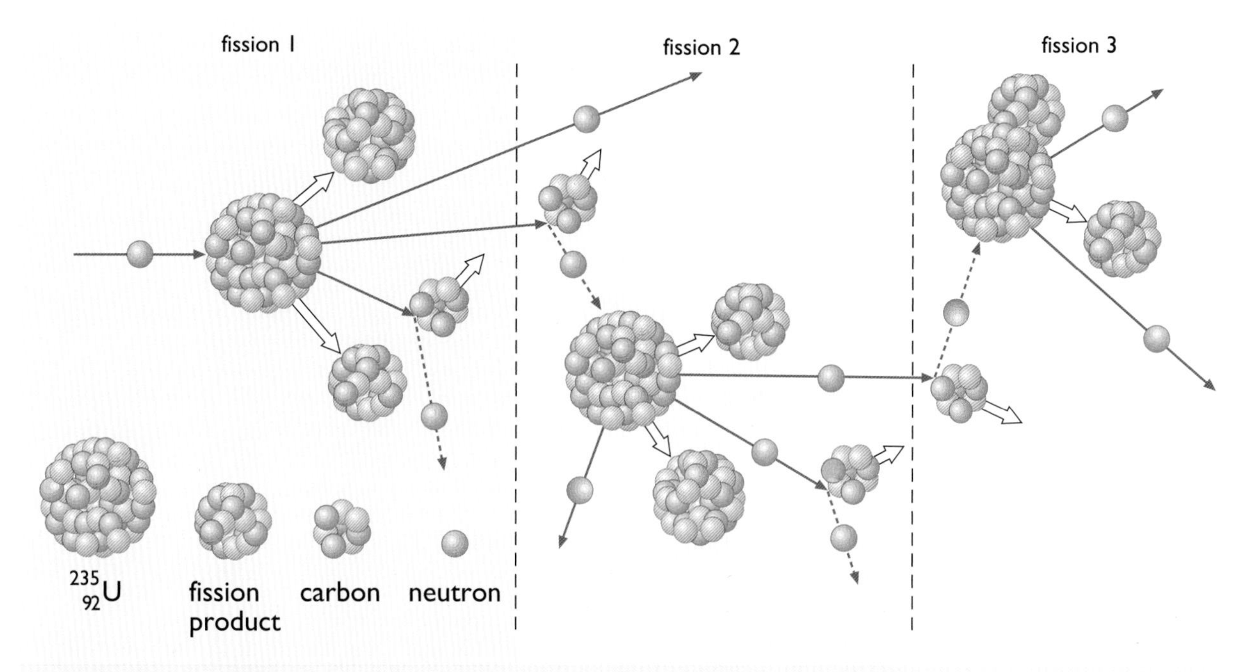

Figure 10.13 A controlled chain reaction. Some of the fast neutrons released in the first fission (left-hand panel) are slowed down in collisions with carbon nuclei. Just one of the resulting slow neutrons causes a second fission (centre panel), and just one from this causes another (right-hand panel), maintaining a steady chain reaction, even in natural uranium. (The many U-238 nuclei are not shown in the diagram.)

BOX 10.6 Moderators

The material that slows down the neutrons in a reactor to maintain the chain reaction is called the **moderator**. The main requirement for an efficient moderator is that it should rapidly reduce the speed of the neutrons without absorbing them. The best moderators have light atoms – for a simple reason. If a fast neutron collides with a stationary object of similar mass, such as a light nucleus, it will share its energy, reducing its speed; but if it collides with a much heavier nucleus it will be deflected without much change in speed. (Compare the result of a fast-moving ball bearing striking an identical stationary ball bearing, and the same ball bearing striking a stationary cannon ball.)

A common misunderstanding arising from its name is that the moderator in some way reduces the reactor output. In fact, by moderating the neutron speed, it increases the rate of fission.

The moderator atom is never completely stationary at the time a neutron collides with it. Like all atoms it will always have thermal energy (Box 10.1), and the resulting slow neutrons will also carry this residual energy. They are often referred to as **thermal neutrons**, and the reactors of virtually all currently operating nuclear power plants are of this type, called **thermal fission reactors.**

A number of different types of moderator have been used since the first reactor. Hydrogen should be ideal. Its nucleus, a single proton, has almost the same mass as a neutron, and it has the advantage of being readily available in the form of ordinary water (often referred to as **light water** in this context). Unfortunately, however, rather than just colliding and slowing down, a free neutron tends to *combine* with any available proton to form a new particle. The consequent neutron loss means that light water can only be used as a moderator if *enriched* uranium is used as the fuel.

The particle formed by the combination of a proton and a neutron is called a **deuteron**. Having just one proton, it must be a hydrogen nucleus, and the addition of one neutron means that it is the nucleus of the isotope *hydrogen-2*. Uniquely, the hydrogen isotopes and their nuclei are given individual names: hydrogen-2 is called **deuterium,** hydrogen-3 is **tritium** and its nucleus, consisting of one proton and *two* neutrons, is a **triton**.

Deuterium occurs naturally, accounting for about 1 atom in 7000 of natural hydrogen. It is an excellent moderator, fairly light, and with the great advantage that the colliding neutrons don't combine with deuterons. In the form of **heavy water** (D_2O), it can be an effective *liquid* moderator, and the low neutron loss means that the reactor can use natural uranium fuel, without the need for enrichment. Another advantage is that there is plenty of water in the world. However, extracting the heavy water involves separation of isotopes, and the one part in 7000 means that up to ten tonnes of water must be processed for every kilogram of heavy water.

The most easily available effective moderator is carbon. It can be obtained in very pure form as **graphite**, and although its nucleus is considerably heavier than a deuteron, its neutron absorption is low, so it can be used with natural uranium. As we have seen, this was Fermi's choice for the world's first reactor, but all the moderators described above were in use within a few years.

New fissile nuclei

Fermi's claim in the 1930s that he had produced transuranic elements by neutron bombardment of natural uranium proved correct. The quantities were far too small for chemical identification, but Hahn and Meitner had confirmed the result when they established the presence of a uranium isotope that emitted β-particles. As shown in Figure 10.5, β-emission increases the atomic number by one, so β-emission from uranium must lead to the element with atomic number 93. When this in turn was found to emit β-particles, it was evident that element 94 was also being produced.

Production of these two new elements in quantities just large enough for chemical analysis was eventually achieved in 1941, and they were given the names **neptunium** and **plutonium** (after the planets beyond Uranus).

The sequence leading to their production is illustrated in Figure 10.14. The first step is that a U-238 nucleus captures a neutron, thus becoming the

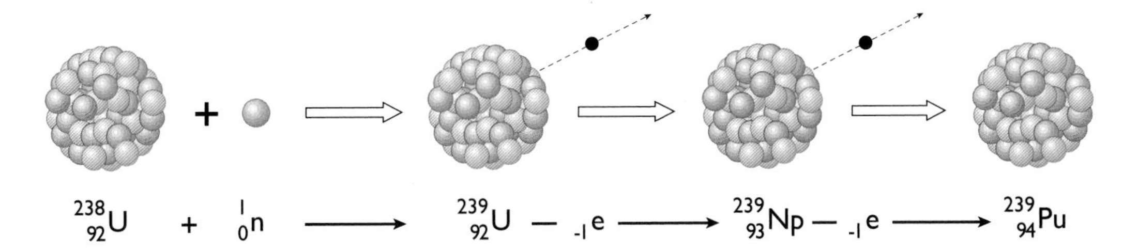

Figure 10.14 Production of plutonium-239 from uranium-238

heavier isotope **uranium-239**. This is radioactive, emitting beta particles (electrons). As we have seen, the result of beta emission is a change into an isotope of the next *highest* element, in this case **neptunium-239**. This is also a beta emitter, so the process is repeated, producing **plutonium-239**.

Both U-239 and Np-239 have relatively short half-lives (about 20 minutes and two days respectively) but Pu-239, although also radioactive, has a half-life of about 25 000 years. So bombardment of U-238 with neutrons results in a steady build-up of plutonium: a process called **breeding**. The discovery of this process leading to the production of plutonium was not announced to the world at large, and the reason for secrecy was obvious. As had been predicted on theoretical grounds, plutonium-239 *is fissile*.

A **fissile** nucleus is one that undergoes fission relatively easily – absorption of a slow neutron is sufficient. The only naturally occurring fissile nucleus is U-235. Plutonium does not exist in nature, except perhaps in minute quantities (Box 10.7), but is produced as a side effect in all uranium reactors. Isotopes like U-238, that can lead to fissile isotopes after absorption of a neutron are called **fertile**, and any reactor containing U-238 will necessarily produce some plutonium. However, *fast neutrons* are the most effective in initiating breeding, so the wartime plutonium-producing reactors had to maintain a fine balance between the slow neutrons required for a chain

BOX 10.7 A natural reactor

Some 2000 million years ago several natural nuclear reactors spontaneously achieved criticality. The evidence is still there at Oklo in Gabon, West Africa, in the form of stable isotopes resulting from millions of years of decay of the fission products.

How could a sustained chain reaction have happened in natural uranium? The first important point is that U-235 decays much faster than U-238. Although U-235 accounts for only one in every 140 atoms of natural uranium today, two billion years ago it was about one in 30 – in today's terms, about 3% enriched uranium. Secondly, the local ore was particularly rich in uranium; and thirdly, water was present in the area to act as moderator.

It is estimated that these reactors operated for about two million years and may have accumulated about seven tonnes of radioactive waste products, including plutonium. These of course decayed in turn, and after a few million years the remaining plutonium would perhaps have amounted to one or two atoms.

reaction in natural uranium and the fast neutrons needed for plutonium production.

Another potentially useful breeding process starts with the element **thorium**. Its atomic number is 90, two places lower than uranium in the periodic table, and it exists in nature as essentially a single isotope, thorium-232 (Th-232). This is similar to U-238 in several ways: it is radioactive, with a very long half-life, and more importantly, it is *fertile*. In a process exactly like that shown in Figure 10.14, a Th-232 nucleus captures a fast neutron and becomes Th-233, which emits a beta particle to become protactinium-233 (Pa-233), which again emits a beta particle and becomes **uranium-233**. And U-233, like Pu-239, is *fissile*. This route, breeding a new fissile uranium isotope from thorium, was already known in the 1940s, but has attracted only occasional interest over the years. It is described further in Chapter 11. (For more on plutonium and breeder reactors, see Box 10.11 and Section 10.8.)

Atomic bombs

With the discovery of plutonium, two fissile materials were available, and two potential routes to a nuclear weapon. Neither would be simple. The uranium method required isotope separation on a scale thousands of times greater than had ever been attempted, and the plutonium route required the construction of nuclear reactors on a large scale. In both cases there were still many unknowns. Nobody had yet attempted industrial-scale chemical and physical processing of intensely radioactive materials. And no one had ever worked with fissile material in quantities approaching the critical mass.

The decision was taken to proceed on both routes, and both reached the intended destination. On 16 July 1945 the first bomb, a plutonium device, was exploded at Alamagordo in New Mexico, on 6 August the first uranium device was exploded over Hiroshima and on 9 August a second plutonium bomb was dropped on Nagasaki.

'Swords into ploughshares'

Not long after the discovery of fission, and some years before the first reactor, it was suggested that controlled nuclear power might eventually fuel the boilers in power stations. This theme was taken up again in the post-war years (Figure 10.15). All reactors generate large amounts of heat energy – about 25 million kilowatt-hours for each new kilogram of plutonium in the war-time production plants, for instance – and this heat was already being used to generate electricity in the 1940s. The USA, while continuing plutonium production, also developed compact uranium reactors as marine propulsion systems. Needing no oxygen for combustion and with refuelling intervals of a year or more, these were particularly suitable for submarines.

However, it was Russia that in 1954 claimed the world's first nuclear power station, producing 30 MW of heat and 5 MW of electric power. The UK, without enrichment facilities until 1953, initially produced plutonium

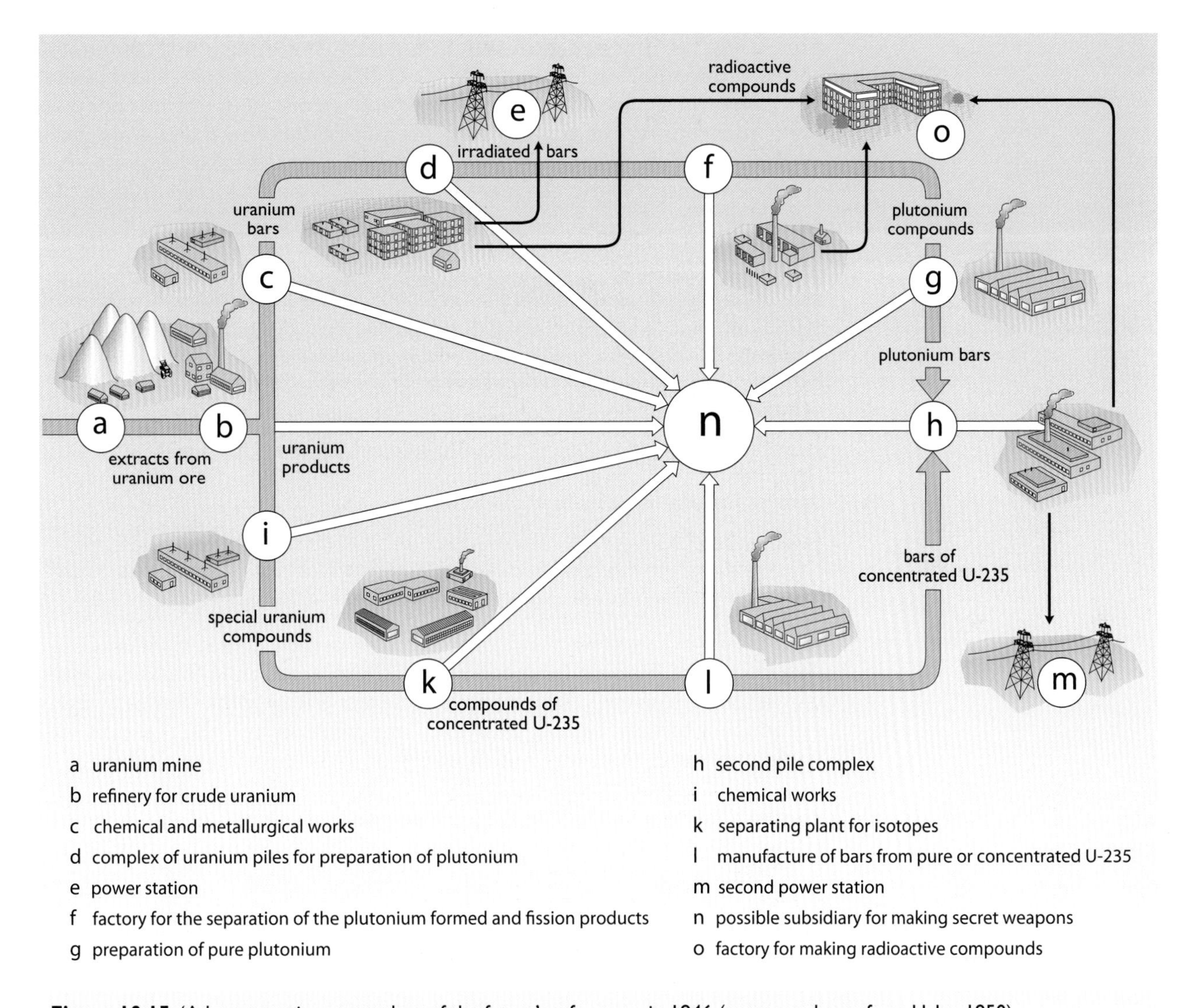

Figure 10.15 'A large atomic energy plant of the future' as foreseen in 1946 (source: redrawn from Hahn, 1950)

for its weapons in reactors of a different design, using natural uranium. The Calder Hall power station, inaugurated in 1956, was sited near two existing plutonium-producing reactors at Windscale (now Sellafield) in Cumbria, and its reactor was a development of their design. Operated by the United Kingdom Atomic Energy Authority (UKAEA), its main purpose was still plutonium production, but the UK was able to claim the world's first *large-scale* nuclear power station, supplying 90 MW of power to the grid.

10.5 Thermal fission reactors

A nuclear power station is in many respects similar to a coal-fired plant (see Chapter 6), but in this case the **steam generators** (also called boilers) producing steam for the turbo-generators use heat from a nuclear reactor.

This brings the further significant difference that although it is common to speak of '**burning**' nuclear fuel, this is not a chemical process, and does not produce carbon dioxide – a subject to which we return in the next chapter. Many different reactor designs have been considered in the seventy years since the first nuclear plants, and about half a dozen have come into use at various times. Nearly all have used thermal fission reactors (Boxes 10.6 and 10.8).

BOX 10.8 'Thermal'

It is worth noting the two uses of the word thermal. *All* nuclear power stations, like the fossil fuel plants, use hot steam to drive their turbines and are therefore *thermal power stations* in the sense introduced in earlier chapters.

At the time of writing (2010), virtually all operational nuclear power stations use thermal fission reactors using moderators (Box 10.6), so they are 'thermal power stations using thermal reactors'.

Fast neutron reactors (described in Section 10.8) do not require a moderator. The neutrons are thus not thermalized. Power stations using these reactors are therefore 'thermal power stations using fast neutron reactors'.

The basic requirement for a controlled chain reaction is that on average one neutron from each fission should interact with another fissile nucleus to induce a further fission. Less than one neutron and the reaction will fizzle out; more than one and it could quickly develop enough heat to melt the structure. There are therefore four essential components in any thermal fission reactor: the *fuel*, a *moderator*, the *coolant* and a *control system* (Figure 10.16). We'll look at each of these, the main safety factors and efficiency, before turning to a few specific reactor types currently in use.

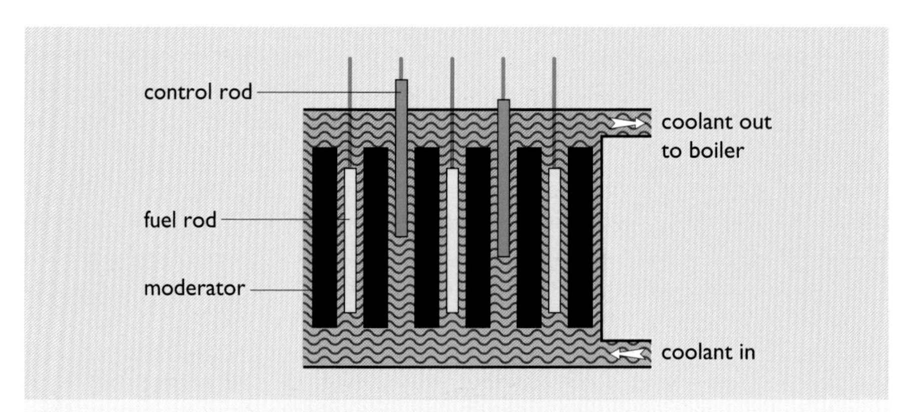

Figure 10.16 The essential features of a thermal fission reactor

The reactor

Fuel

A number of early power plants used *natural uranium* fuel, but most of these have reached the end of their productive life, and the remaining few, like more recent designs, now use enriched uranium, with U-235 content between 2% and 5%.

The two other known fissile nuclei mentioned in 'New fissile nuclei' above have both been used as minor components together with the uranium-235 in reactor fuels. Plutonium-239 is bred in any uranium reactor, as we have seen, but can also be added as a component of the fuel. Although plutonium does not exist in nature, there are today many potential sources. The **spent fuel** removed from a reactor inevitably still contains some of the plutonium bred during its use, and reprocessing this is one option. Then there are the many tonnes of plutonium potentially becoming available with the decommissioning of nuclear weapons. A further option is the use of thorium. As described above, a fast neutron reaction with this can breed the fissile isotope uranium-233. These possibilities are described further in Chapter 11.

Uranium reactor fuel is usually in the form of the oxide UO_2, a hard ceramic suitable for forming into pellets. If the fuel is to include plutonium, this can be in the form of plutonium oxide, PuO_2. In this case the fuel is called a **mixed oxide**, or **MOX**. The fuel pellets must be contained in a way that allows replacement of batches of spent fuel, and the **fuel rods**, the containers or cladding holding the fuel, are usually metal tubes. Steel, the obvious choice, is excluded in most cases due to its high neutron absorption, so other metallic alloys are used, as will be shown in the accounts of reactor types below.

A typical reactor might hold in the order of 100 tonnes of fuel, a third of this being replaced every year or eighteen months. The resulting heat output of some 2–3 GW would be sufficient for turbo-generators supplying 600–1000 MW of electric power. The fuels for specific reactors and fuel production are discussed in Sections 10.6 and 10.7.

Moderator

The function of the moderator in a nuclear reactor has been described in Box 10.6, and the fact that many present-day reactor types are identified by the moderator they use is an indicator of the importance of this choice. As will be shown in Section 10.6, the majority of the world's present nuclear plants have light water reactors, with ordinary water as moderator, and the main alternatives are heavy water reactors. The third moderator, graphite, has been used in reactors developed in the UK and in some Russian plants. For the reasons discussed in Box 10.6, both graphite and heavy water were originally chosen to allow the use of natural uranium, avoiding the need for enrichment; but as mentioned above, nearly all currently operating reactors, including these types, now use fuel with some degree of enrichment.

Coolant

The **coolant** is the heat exchange medium: the fluid – liquid or gas – that carries heat from the reactor core. (The name, rather inappropriate in a system whose main purpose is to *produce* heat, reflects its original function, which was to prevent overheating in reactors designed for plutonium production.) A significant advantage of a *liquid* moderator, whether heavy water or normal (light) water, is that it can also act as the coolant. Graphite is of course a solid, so a graphite-moderated reactor requires a different material as coolant. Some types, including those developed in the UK, use carbon dioxide gas.

Control system

Routine control of the heat output of a reactor is achieved by designing for a slightly *increasing* rate of fission but incorporating a neutron-absorbing material in the system. Cadmium and boron are good absorbers, and the reaction can be controlled by increasing or decreasing the amount of these present in the reactor core, either by adding absorber to the moderator or with **control rods** that can be moved in or out (or by some combination of these and other methods). Another aspect of control is the ability to deal with events such as an unplanned increase in the rate of fission, and all reactors need to include systems for rapid response to such emergencies.

Safety

Structures

Certain basic requirements determine the general form of all the reactors discussed here. The fuel must be distributed at the right density, surrounded by the moderator and in good thermal contact with the coolant. The coolant must stream freely past the hot fuel and for maximum plant efficiency must attain the highest possible temperature, which implies a high pressure. This has a major effect on the design because, as mentioned above, steel or other structural materials will absorb neutrons. The metals magnesium and zirconium are often used in reactors because of their low neutron capture. Only those reactors with very low neutron loss in the moderator and coolant can afford to have individual pressure tubes carrying the coolant. In most designs, the whole core is submerged in the flowing coolant in a single large pressure vessel.

It is worth considering for a moment the interior of the reactor core. The coolant must not occupy too much space, which means that it must be flowing very fast if it is to carry away the gigawatts of heat. So a very hot fluid streams at high speed and high pressure through narrow gaps and channels past the fuel rods, subjecting materials to mechanical stress as well as the combined effects of temperature and pressure, chemical attack and bombardment by sub-nuclear particles. The designer must ensure that even under these conditions fuel rods do not distort, control rods can move freely and nothing impedes the flow of coolant.

Shielding

The combined effect of radioactive fission products and the radioactivity induced in structural elements by neutron bombardment brings the radiation inside an operating reactor to about one trillion (10^{12}) times the level that a person could tolerate for even a short time. Without shielding, a 'safe' distance for working would be about eight kilometres away! So this is obviously an important aspect of reactor design. As shown in Figure 10.3, the two most penetrating components are neutrons and gamma radiation, so the shielding needs to be both a good neutron absorber and thick enough to stop the energetic gamma particles. It usually consists of a thick layer of concrete, and if the system also has a steel pressure vessel, this will absorb much of the gamma radiation.

Containment

The safe containment of the radioactive material itself is obviously important. The high pressure requires physically strong containment, and, as mentioned above, this is often achieved by surrounding the entire core by a strong pressure vessel. The fission products will include radioactive gases, so containing these and monitoring their emission from the plant is an important aspect of routine operation. Finally, the containment system should be designed to safeguard personnel in the plant and people in the surroundings in the case of 'non-routine events'.

Accidents

A main source of public concern about the safety of nuclear reactors is the potential for the release of very large quantities of radioactive substances, which can occur only if the containment fails. Major structural failure during normal operation is possible, but has never occurred, and the more likely event is that the pressure vessel is unable to withstand a sudden unplanned rise in core temperature and pressure, or an explosion resulting from this. Any explosion – nuclear or chemical – is the result of a runaway energy-producing chain reaction, multiplying so fast that the energy doesn't have time to escape by normal 'peaceful' means as heat and light. Instead, the energy density rises until the bonds holding the material together are broken and it blows apart. In the words of the unfortunate spokesperson after one accident, it is an 'energetic disassembly'. Obviously, a chemical explosion can only happen if suitable chemicals are present, particularly hydrogen. One possibility in reactors with water as a moderator or coolant is that the temperature rises to the point where the water turns into steam, which reacts with the zirconium alloy used as fuel cladding to produce hydrogen and zirconium oxide (i.e. a form of 'rusting'). The necessary conditions for such a chemical explosion might arise in a nuclear reactor as a result of either a sudden loss of coolant or a 'runaway' chain reaction, and we look briefly at each of these. The specific accidents mentioned are discussed in more detail in Chapters 11 and 13.

A major **loss of coolant** can have particularly serious consequences in a nuclear reactor. In a fossil-fuel plant, such failure would be dangerous, but at least the generation of heat energy could be stopped by cutting off the fuel or air supply. Not so in a nuclear reactor. Even if the chain reaction stops instantly, radioactive decay of the fission products continues, and the rate of energy release immediately after shutdown can be almost a tenth of full normal power – enough to cause the core temperature to rise at 100 °C per minute if the coolant is lost. The half-lives of many of the products are very short, and the rate may have fallen to a tenth of this level within an hour; but the heat produced during this short time is enough to melt the core and perhaps the floor of the reactor building as well – a 'China Syndrome' scenario (the fictional idea that the contents might continue to sink towards the diametrically opposite point on the globe!) The most serious known loss-of-coolant accidents were those at Three Mile Island (TMI) in Pennsylvania in 1979 and Fukushima in Japan in 2011. In both cases there was a partial core meltdown and some hydrogen was produced. At TMI, this gave rise to fears of a major explosion – the 'energetic disassembly' mentioned above. But this did not occur and the primary containment (reactor pressure vessel) remained intact, although the core was seriously damaged and the plant never reopened. At Fukushima there was a serious loss of coolant lasting for many days on a site with six reactors. It appears at the time of writing (June 2011) that meltdowns have occurred in three reactors and it is not clear if the reactor primary containments of all of these have remained intact. Leaking hydrogen has caused three serious chemical explosions, damaging the reactor buildings (see Chapter 11).

An explosion resulting from an uncontrolled divergent chain reaction should be virtually impossible in a reactor with only slightly enriched fuel. If the coolant is also the moderator, its loss means that no chain reaction at all can be maintained; and all reactors also have emergency systems for the injection of neutron absorbers. However, unforeseen combinations of events can lead even to this improbable occurrence, as at Chernobyl in 1986. A runaway chain reaction led to a steam explosion that breached the containment, exposing the core. The design of the Chernobyl reactor, with graphite moderator but water coolant, contributed to the accident, and it is claimed that this sequence of events would not be possible in most types of reactor. One design feature did however have unexpected merit. When the base of the vessel was blown downwards, molten fuel flowed out, leading to concerns that a self-sustaining chain reaction might re-start. This did not happen, and it was eventually discovered that the fissile material had been diluted by a simultaneous flow of large quantities of the sand that had formed part of the screening around the core; an interesting if somewhat unplanned 'safety system'.

It must be stressed that the events discussed here involved *steam and chemical explosions*. The fear that the fuel of a thermal reactor will somehow turn into an 'atomic bomb' is probably not well founded. A **nuclear explosion**, as shown in Figure 10.12, requires a high density of fissile nuclei in a quantity large enough to make the neutron loss negligible. The critical mass is about 10 kg for pure 100% U-235 and about 5 kg for pure Pu-239; but very much greater if the fissile material is dilute. The only way in which a thermal fission reactor with a few per cent of U-235 could become a nuclear bomb would be for an appreciable fraction of the fissile nuclei distributed throughout 100 tonnes of fuel to bring themselves miraculously into one small region.

Passive response

Following the events at Three Mile Island and Chernobyl, there has been increasing emphasis on the development of 'passive' response to failures. The aim is a system that closes down the reactor safely, not only without any action by the operators but without any requirement for electric power (currently needed to run pumps for emergency cooling, etc.). So, although no radically new reactor type has come into operation for several decades, recent years have seen increasing numbers of 'advanced' plants whose designers stress their improved safety. (See, for instance, **APWR** and **ABWR** in Section 10.6.)

Thermal efficiency

The study of coal-fired power stations in Chapter 6 showed that the heat-to-electricity efficiency of the plant depends critically on the temperature of the steam input to the turbines, and that the maximum permissible steam temperature and pressure is determined by the availability of high-temperature steel alloys. The same reasoning applies to nuclear plants, but with several important differences. In all types of nuclear plant except the boiling water reactors (see Section 10.6), the steam for the turbines is not produced directly in the reactor, but by the coolant carrying heat from the core to steam generators, and this of course involves some temperature drop and efficiency loss. But the really significant difference is that all the materials in the reactor core or its surroundings are subject to constant neutron bombardment, leading to embrittlement of metallic components and further limiting the maximum permissible temperatures.

Overall, the effect of these constraints is that the steam enters the turbines at significantly lower temperatures than the 500 °C of a modern supercritical coal-fired plant, and the resulting efficiencies are correspondingly lower. The earliest nuclear power plants, with steam temperatures only slightly above 300 °C, achieved little more than 20% heat-to-electricity efficiency, and although there has been considerable improvement, the current 35% or so of the best pressurized water reactors (see Section 10.6) falls short of modern coal-fired plants.

10.6 Types of thermal fission reactor

Since the first nuclear power stations were commissioned, about a dozen different types of thermal fission plant have been brought into use, but those currently operating (in 2010) fall into four main categories, essentially differing in their choice of moderator and coolant: *light water* for both, *heavy water* for both, *graphite moderator with gas coolant*, and *graphite moderator with light water coolant.* The great majority of the world's reactors fall into the first category, so the following brief summary of reactor types starts with **light water reactors (LWRs)**.

Light water reactors

Pressurized water reactors

The reactors in the majority of the world's nuclear plants use ordinary (light) water under high pressure as both moderator and coolant. The two main types are the **pressurized water reactors (PWRs)**, developed in the USA, and the **water-water-energy-reactor (VVER)** developed in Russia.

Figure 10.17 shows the main features of a PWR. The core of the reactor consists of about a hundred **fuel assemblies** held between top and bottom plates. Each fuel assembly is a cluster of a few hundred long thin fuel rods, zirconium alloy tubes a few metres long and a centimetre or so in diameter, packed with uranium oxide fuel pellets. The open core structure allows the water, acting as moderator and coolant, to flow freely at high pressure past

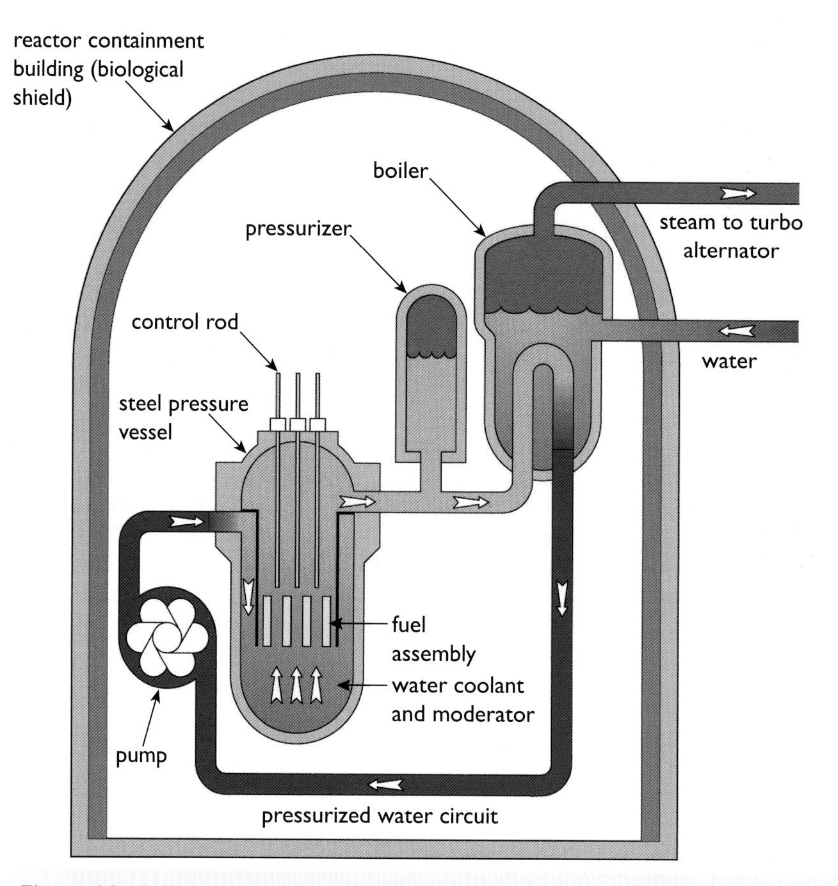

Figure 10.17 The main features of a pressurized water reactor (PWR) (source: adapted from RS and RA Eng, 2009)

the fuel rods. Heated as it passes through the core, it then gives up its heat in one or more steam generators (boilers) to produce steam for the power station turbines. Refuelling involves replacing perhaps a third of the fuel assemblies every year or so.

As in any thermal plant, the higher the steam temperature the greater the heat-to-electricity efficiency, but it is essential that the water coolant must remain liquid, and to achieve this at a temperature of 300 °C or more, it needs to be at a pressure of over 100 atmospheres. In a PWR, the core is contained in a pressure vessel surrounded by a primary concrete shield, and the entire system, including the steam generators, is enclosed in a concrete containment structure to prevent the escape of radioactive materials, with a steel lining to capture high-energy radiation. (The Russian VVER differs significantly in having several separate coolant circuits rather than a single pressure vessel.)

Maintaining the high pressure of the coolant water is very important. If one of the large pipes carrying heated water to the steam generators fractured, the water would at once flash to steam, losing its cooling effect. So a fast and effective emergency cooling system is needed, and the PWRs usually have several in place. (The failure in the Three Mile Island PWR was due in part to human intervention in the automatic safety systems.)

Design details of the PWR and VVER have undergone many changes over the years to increase efficiency, reduce cost, and to make the systems 'fail-safe'. Eventually, these changes have led to a new name for this type: **advanced pressurized water reactors (APWRs)**.

Boiling water reactors

Boiling water reactors (BWRs), also developed in the USA, come second to the PWRs in the number in use worldwide. Japan and a few other countries have generally preferred these to the PWR. The damaged reactors at Fukushima are BWRs.

The BWR is similar to the PWR in many respects, but, as the name suggests, the water that acts as coolant and moderator is allowed to boil into steam that drives the turbines (Figure 10.18). This has the disadvantage that the turbines are exposed to the potentially radioactive coolant, but not having separate steam generators reduces costs and heat losses. However, BWRs have generally operated at lower temperatures and pressures than PWRs, so their heat-to-electricity efficiencies are not very different.

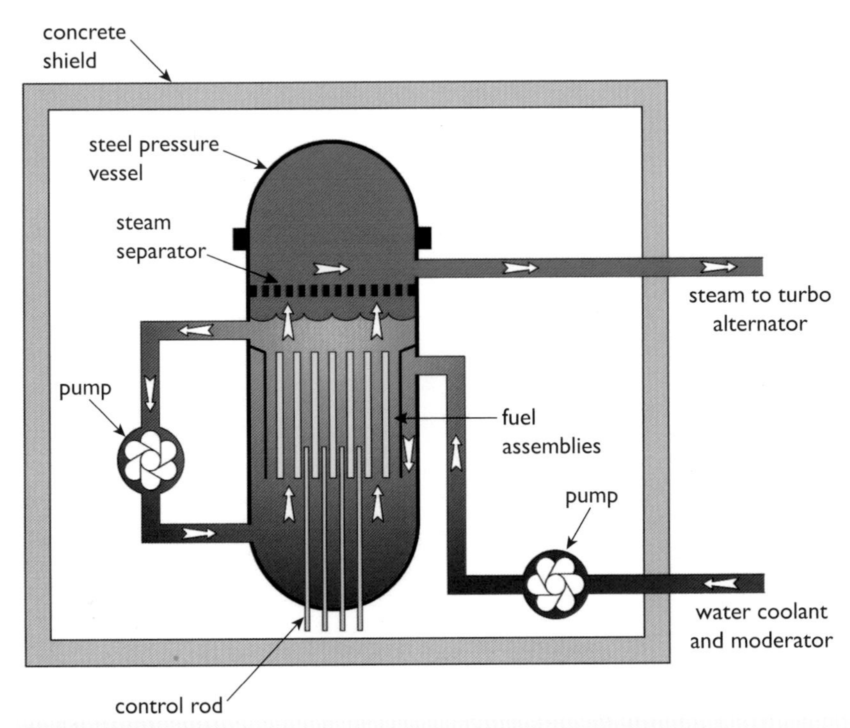

Figure 10.18 The main features of a boiling water reactor (BWR) (source: adapted from RS and RA Eng, 2009)

As mentioned above, design changes are often recognized by a change of name, and the **advanced boiling water reactor (ABWR)** is an example. It is still a large BWR (1356 MW electricity output), but with modified operating systems, structure and safety features, including passive responses to transients and accidents. At the time of writing, several ABWRs are operational, mainly in Japan.

Gas-cooled reactors

As mentioned earlier, the UK had no enrichment facility at the start of its nuclear programme, so the design of its reactors was governed by the need to achieve a chain reaction in natural (non-enriched) uranium. The original **Magnox** reactors (see Box 10.9) use graphite as moderator, carbon dioxide gas as coolant and a magnesium alloy (rather than neutron-absorbing steel) for the fuel cladding. As usual, the efficiency is limited by the maximum temperature and pressure that the materials and structure can withstand, and the best heat-to-electricity efficiency claimed for a Magnox plant is slightly over 30%.

The **advanced gas-cooled reactors (AGRs)** that followed also use graphite moderator and carbon dioxide coolant, but the uranium in their fuel is enriched to 2.3% U-235. As shown in Figure 10.19, the resulting structure is very different from a PWR or BWR. The main core is effectively solid graphite, with vertical channels extended by metal tubes up to the reactor pile cap to allow insertion of fuel elements and control rods from above. The two thousand or so fuel clusters are not long and thin as in the PWR, but short and chunky, stacked above each other. The carbon dioxide gas coolant is pumped through the core at a pressure of about 40 atmospheres, absorbing heat from the fuel and delivering it to the boilers.

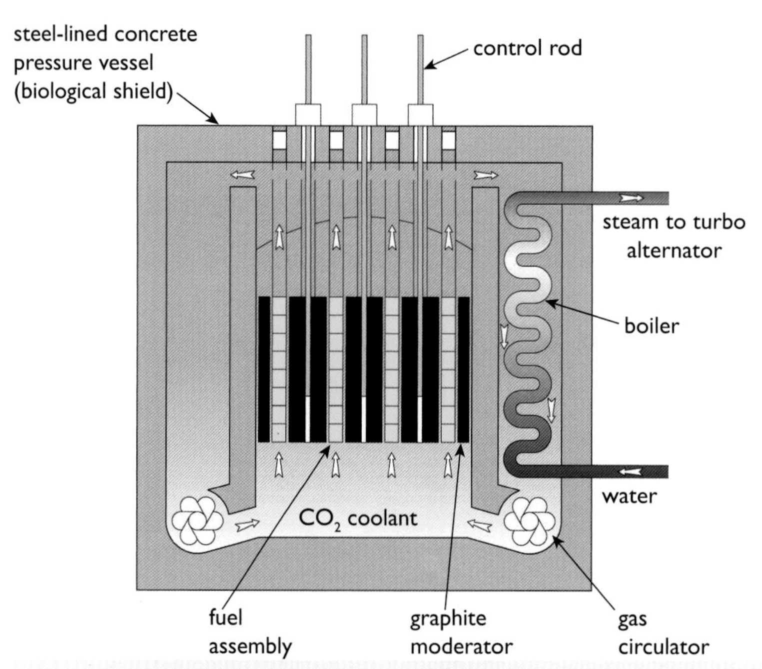

Figure 10.19 The main features of an advanced gas-cooled reactor (AGR) (source: adapted from RS and RA Eng, 2009)

The reactor is encased in a concrete pressure vessel which acts as the container for the carbon dioxide coolant gas and as a radiation shield. The walls, several metres thick, are pre-stressed by heavy steel wires, a system which was claimed by advocates of the AGR to be less likely to suffer catastrophic failure than the PWR. Other claimed safety advantages are the fact that the coolant is already a gas, and that the heavy graphite core could absorb heat if the cooling did fail. Nevertheless, as discussed in Box 10.9, the UK did not extend its AGR programme after the first group of plants.

BOX 10.9 Nuclear power in the UK

The four 23-MW generating sets of the UK's original Calder Hall power station used Magnox reactors, the name deriving from the magnesium alloy cladding of fuel elements. In 1959, Calder Hall was joined by the similar Chapelcross plant, also designed by the UK Atomic Energy Authority (UKAEA) primarily for plutonium production for military purposes. Both these plants, subsequently owned by British Nuclear Fuels Limited (BNFL), continued in operation into the present century, but are now closed.

In 1955 a programme was announced by the UK's Central Electricity Generating Board (CEGB) for the construction of nine civil Magnox power stations, with outputs ranging from 300 MW to over 1000 MW, and these were commissioned over the period from 1962 to 1971. In 1984 their operating lives were extended from the planned 25 years, and at the time of writing (2010) two remain in operation (Figure 10.20 and Table 10.3).

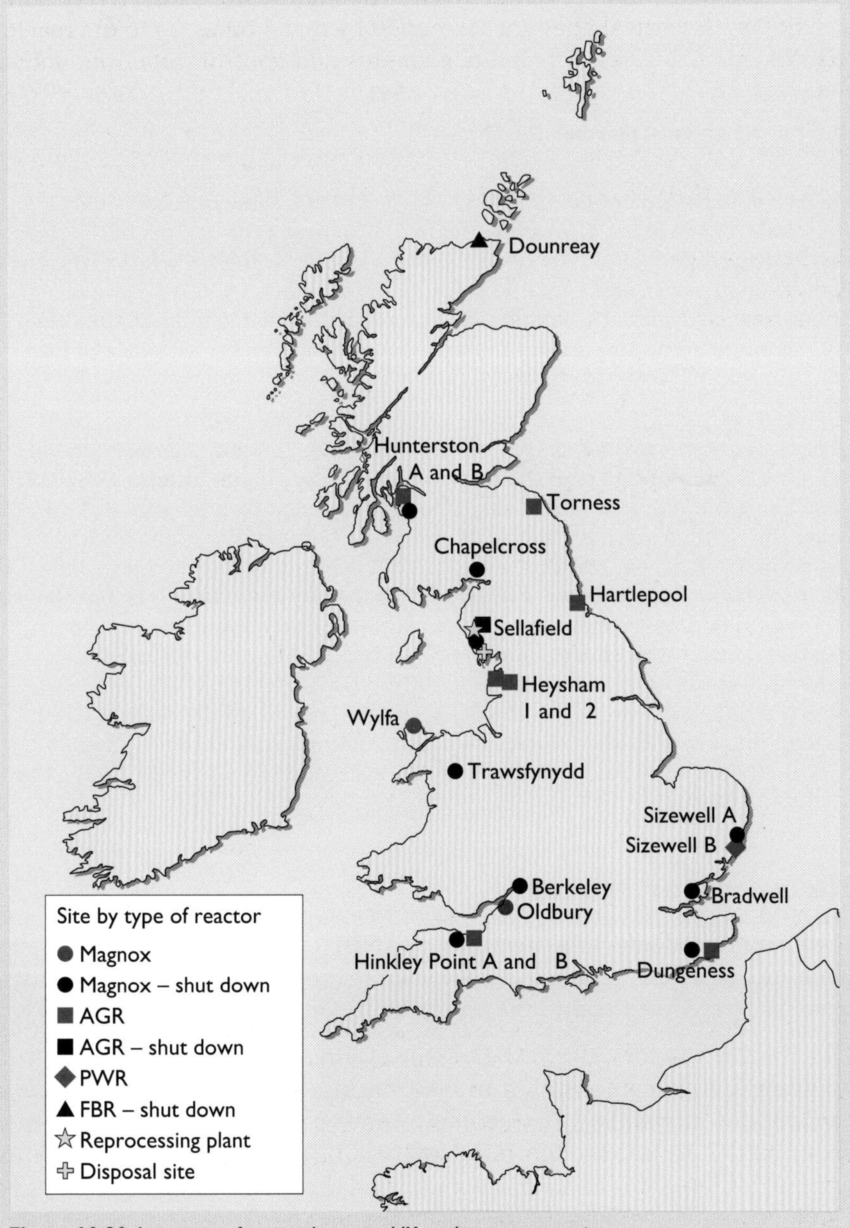

Figure 10.20 Locations of past and present UK nuclear power stations

Table 10.3 UK nuclear plants operating in 2010

Reactor and location	Start-up	Planned closure
Magnox		
Oldbury-on-Severn	1968	2010
Wylfa	1971	2012
AGR		
Hinkley Point B	1976	2016
Hunterston B	1977	2016
Hartlepool	1984	2014
Heysham 1	1984	2014
Dungeness B	1985	2018
Heysham 2	1988	2023
Torness	1989	2023
PWR		
Sizewell	1995	2035

For details of recent proposals, see Chapter 11.

Even before the first of the nine new Magnox plants came into use, their successors were being discussed, and in 1962 a prototype version of the AGR became operational. In 1964 the AGR was adopted for the UK's future nuclear power programme, and a total of seven twin-reactor (2 × 660 MW) power stations were eventually completed. At the time of writing (mid-2010), three of the fourteen reactors are temporarily out of operation for periods of a few months, but all seven plants are currently expected to continue in service over their planned lives.

There was, however, a potential rival to the AGR. In 1968 the prototype of another British design, the **steam-generating heavy water reactor (SGHWR)**, had come into operation, and the following decade saw a lengthy debate over the relative merits of the AGR and SGHWR for the next generation of UK nuclear power stations. The SGHWR eventually lost the battle, and in 1978 two more AGRs were ordered. But these were to be the last of the line. The AGR never managed to break into the LWR-dominated world market, and in 1981, long before the final two plants were operational, a planning application was submitted for the UK's first PWR. The site was one previously proposed for the SGHWR, at Sizewell on the Suffolk coast, where there was already a Magnox plant. After a long and bitterly fought public inquiry, the plan was approved, and the 1.2 GW Sizewell PWR came online in 1995.

Heavy water reactors

The **Canadian-deuterium–uranium (CANDU)** reactors, with some twenty plants operating in Canada and overseas, are the only other type of reactor to have made any inroad into the world dominance of the PWR and BWR.

Like the UK reactors, the CANDU was originally designed to use natural uranium fuel, but its solution to neutron loss was to use heavy water as coolant and moderator. The coolant, under pressure at a temperature of about 300 °C, flows through hundreds of horizontal metal tubes holding

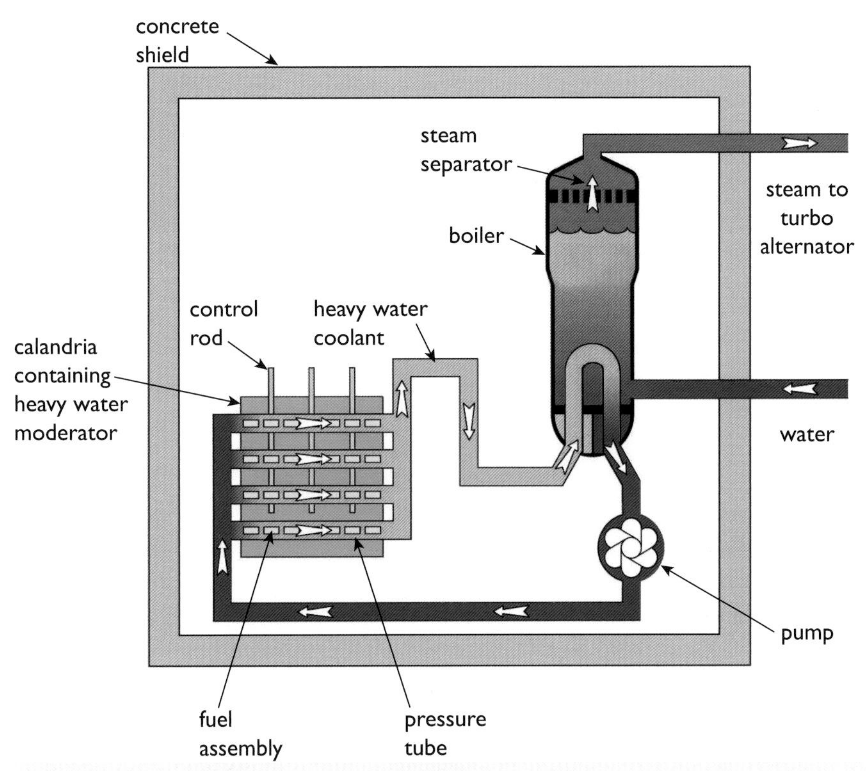

Figure 10.21 The main features of a CANDU reactor (source: adapted from RS and RA Eng, 2009)

the short fuel bundles (Figure 10.21). A major advantage, possible only because of the very low neutron absorption by heavy water, is that these tubes can be *double-walled*, insulating the hot, high-pressure coolant from the surrounding moderator, which does not need to be pressurized as it remains below 100 °C. The moderator fills a horizontal steel cylinder called a **calandria**, with the insulated tubes carrying the hot coolant in and out through the ends of the cylinder.

CANDUs have a relatively low thermal efficiency, but their high neutron economy allows the use of a range of different fuels. They have even used the spent fuel from PWRs, and have been considered by the USA for dealing with surplus weapons-grade plutonium. In recent years, as variants of the original CANDUs have come into use, a more general term has been adopted for reactors of this type: **pressurized heavy water reactors (PHWRs)**. India, which commissioned its first CANDU reactor over thirty years ago, now uses this term for all its reactors of this type, whilst Canada calls its newest version the **advanced CANDU reactor (ACR)**.

RBMK reactors

In addition to the VVER mentioned above, Russia developed other types of reactor, of which the main one was the **high power channel-type reactor (RBMK)**. Developed from early Russian plutonium-producing reactors, it uses graphite as moderator and light water coolant (Figure 10.22).

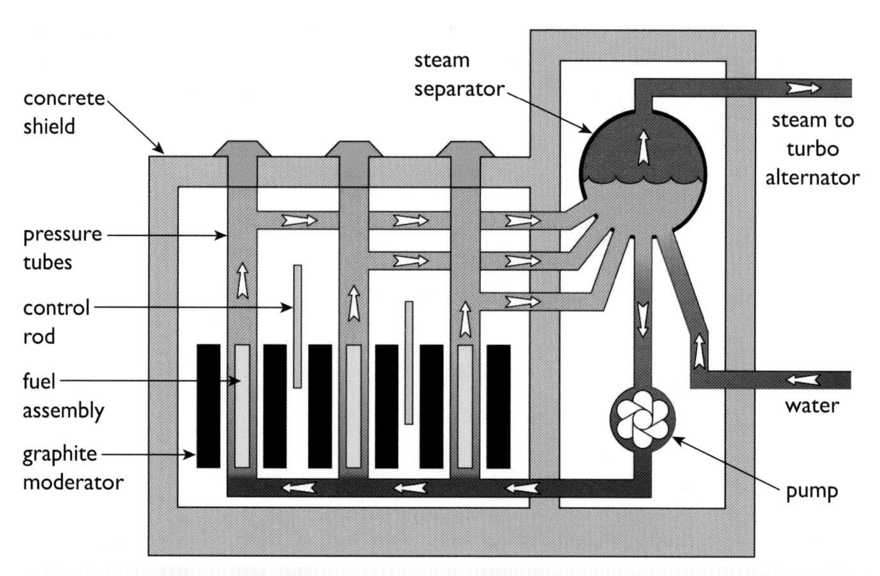

Figure 10.22 The main features of a Russian RBMK reactor (source: adapted from RS and RA Eng, 2009)

The RBMK, like early reactors elsewhere, was originally designed to use natural uranium, but subsequently changed to slightly enriched fuel. The Chernobyl reactors were of this type, and after the events in 1986 (see 'Accidents' in Section 10.5) a number of RBMKs in Russia and elsewhere were closed down. There are no plans to build more, but at the time of writing (2010) about a dozen are still operating.

Summary of reactor types

Table 10.4 shows the main types of thermal fission reactor in use worldwide in 2010. However, as, discussed, the last Magnox was commissioned in 1971 and the last AGR in 1985, and it is almost certain that Russia will build no more RBMKs.

Table 10.4 Types of thermal fission reactor in operation worldwide in 2010

Reactor	Fuel	Moderator	Coolant
PWR	Enriched uranium	Light water	Light water
VVER	Enriched uranium	Light water	Light water
BWR/ABWR	Enriched uranium	Light water	Light water/steam
Magnox/AGR	Natural uranium	Graphite	Carbon dioxide gas
CANDU/PWHR	Natural uranium	Heavy water	Heavy water
RBMK	Enriched uranium	Graphite	Light water/steam

As mentioned at the start of this chapter, during the past few years the overall rate of decommissioning of nuclear plants has exceeded the rate of construction. At the time of writing, in late 2010, over fifty new plants are under construction, forty of them accounted for by four countries: China, Russia, India and South Korea (WNA, 2010). When completed, the fifty plants should add a total of about 50 GW of electrical capacity to the 2010 total of 375 GW. But the latter figure is of course falling as more plants are decommissioned. Chapter 11 discusses these and other issues concerning the future of nuclear power.

10.7 Nuclear fuel cycles

To assess the full costs, the environmental effects or the social implications of any energy system, the entire sequence of events from the original primary energy to the final useful output needs to be examined. In the case of nuclear power, this means the complete fuel cycle outlined in Figure 10.23. Reactors have already been considered in some detail, so the topics here are the 'front end' and 'back end' of the cycle: *mining and extraction, enrichment and fuel fabrication*, and dealing with the *spent fuel*. The environmental and other effects of these processes will be discussed in detail in Chapters 11 and 13, so the following should be regarded as a short introduction.

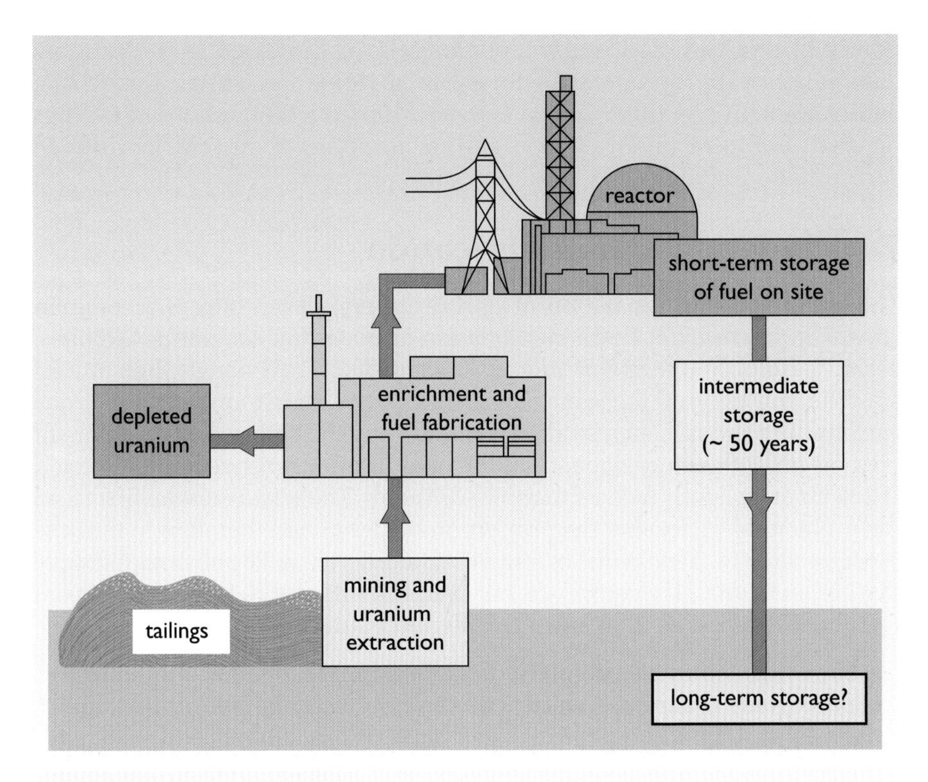

Figure 10.23 Nuclear fuel cycle

Mining and extraction

Uranium is a fairly common element, found in many types of rock and also in the oceans. As an element, it is a heavy metal nearly twice as dense as lead; but in nature it almost always occurs as a mixture of uranium oxides, represented chemically as U_3O_8. In customary mining terminology, a body of rock is defined as **uranium ore** if it contains an economically recoverable concentration of uranium. Currently this quantity varies from a maximum of a few per cent of U_3O_8 in the rock for a **high-grade ore**, down to a lower limit of about one part in a thousand. After extraction, the U_3O_8 is in the form of a compressed powder called '**yellowcake**'. Uranium itself is only mildly radioactive, so this powder does not present a major hazard and is the form normally transported to the next stage of processing. Typically, a nuclear plant generating 1000 MW of electric power requires about 200 tonnes of yellowcake per year, so the world's present nuclear power stations need a total annual supply of some 70 000 tonnes. Currently, only about two-thirds of this is 'new' uranium, the rest coming from stockpiles accumulated in earlier years. (It is estimated that over two million tonnes of uranium have been produced worldwide since 1945.)

The **tailings**, the residues after extraction of the uranium, do give rise to concern. As described in Section 10.3, the ore will have contained radioactive daughter products that have accumulated in the ground for millions of years. Moreover, one intermediate product is the radioactive gas radon (Figure 10.4), which has in the past had very serious consequences for workers underground (see Chapter 13). Initially in the form of a slurry, the tailings may have up to twenty times the radioactivity of the extracted uranium, and can also include chemically or biologically undesirable materials, so appropriate safe storage is obviously essential. Depending on the grade of ore, there can be between 100 and 1000 tonnes of tailings for each tonne of unenriched uranium, implying a current worldwide accumulation of up to fifty million tonnes per year.

Enrichment and fuel fabrication

The extraction and separation of useful minerals from ores is a common chemical process, but these methods are of no use in separating isotopes, because *isotopes are chemically indistinguishable.* Although their nuclei are different, they are the same element, with the same number of electrons surrounding the nucleus, so their chemistry is the same. Any enrichment method must therefore depend on the one difference between them: that a U-238 atom is slightly heavier than a U-235 atom. The scientists who first faced the problem of increasing the proportion of U-235 had therefore to think of processes in which *mass* makes a significant difference. The obvious example that comes to mind is a centrifuge, but practical difficulties (see below) meant that for several decades an entirely different method was used.

Gaseous diffusion was originally developed to produce highly enriched weapons-grade uranium. It starts with the conversion of U_3O_8 into uranium hexafluoride, UF_6. Known as **hex**, this uranium compound usefully becomes a gas when slightly heated. The method then uses the fact that the lighter hex molecules containing U-235 diffuse through a porous membrane slightly faster than the heavier molecules with U-238. The difference is extremely small, and a thousand or more repetitions of the diffusion process, consuming a great deal of energy, are necessary to increase the U-235 proportion to 3–4%. Many diffusion plants are still in operation worldwide, but centrifuging is now gradually taking over.

The principle of the **gas centrifuge** method has been well known for many years. When a fluid containing particles is spun at very high speed, the heavier particles move outwards faster than the lighter ones, leaving the fluid near the centre 'enriched' in light particles. However, the technology for efficient centrifugal separation of individual molecules with only a few per cent difference in mass was not easy, and this method has only in recent years become economically competitive with diffusion. The starting point is again conversion into hex, which is then spun at very high speeds to achieve the isotope separation. The major advantage over diffusion is that far fewer repetitions are needed, reducing the energy cost for a particular degree of enrichment by a factor of about fifty.

With either process, the enriched uranium is not the only end product. Inevitably, there must be a corresponding quantity of **depleted uranium** – the fraction left with less than the original proportion of U-235. The production of one tonne of 3.5% enriched uranium, for instance, could leave as much as six tonnes of depleted uranium with only a quarter of a per cent U-235. Depleted uranium is at present mainly stored for possible future fuel use (see Chapter 11), although some, in metallic form, has had (non-nuclear) military uses.

Spent fuel

No reactor fully burns all the U-235 in its fuel. The usual reason for replacing spent fuel is a build-up of fission products and actinoids, 'poisoning' the fuel and reducing the heat output of the reactor. Replacement details differ for different types of reactor, but typically a third of the fuel elements in a PWR are replaced each year or so, each one therefore remaining for about three years. Table 10.5 indicates possible contents of the spent fuel removed at the end of this period.

Table 10.5 The main constituents of 1000 kg of spent fuel from a thermal fission reactor

Content	Quantity/kg	Notes
U-235	7	Fission of U-235 will have contributed about two-thirds of the power output. Its remaining concentration in the spent fuel is similar to that in natural uranium.
U-238	940	The original 965 kg of U-238 in the fresh fuel will have been reduced by neutron absorption leading to plutonium and other actinoids.
Plutonium	9	More than half the plutonium produced from U-238 will have undergone fission, contributing about a third of the total power output.
Fission products	38	These lighter radioactive isotopes, with half-lives from fractions of a second to a few years, contribute over 99% of the initial radioactivity of the spent fuel.
Actinoids, U-236, etc.	6	The heavy radioactive isotopes, many with very long half-lives, contribute most of the radioactivity that remains after a few hundred years.

The data is typical for spent fuel from a PWR using enriched fuel containing 3.5% U-235 and 96.5% U-238, but the quantities can depend on the mode of operation – and could be significantly different for other types of reactor.

Initially the spent fuel is highly radioactive, and the energy transferred by the emitted particles to the surrounding atoms as heat is sufficient to melt solid materials in a few minutes. Any storage system must therefore provide both a radiation shield and an efficient means of heat extraction. After removal from the reactor, spent fuel is normally submerged under a few metres of water in a tank equipped with a cooling system, where it remains for at least a year. The activity will then have fallen to perhaps 10 000 MBq per gram of material. This is still almost a million times the activity of natural uranium, but it can now be moved to the next storage stage. Its subsequent treatment falls into two main categories: *direct disposal* and *reprocessing*.

Direct disposal

The direct disposal route involves leaving the spent fuel in tanks for 40 or 50 years, by which the time the radioactivity and heat production will have fallen to levels that in principle allow other, more compact forms of storage (Box 10.10).

BOX 10.10 Radioactive waste

The **radioactive waste** arising from the nuclear power industry or other civil uses of radioactive materials is usually classified as high-, intermediate- or low-level, depending on its activity.

High-level waste is responsible for all but a few per cent of the total radioactivity from nuclear waste. It is mainly either spent reactor fuel (Table 10.5) or the separated fission products and actinoids that result from reprocessing. In either case, the waste continues to require shielding and cooling and needs to be kept in monitored storage for perhaps 50 years. The proposed method for final disposal is to place the material in deep underground stores, but at present (2010) no country has agreed detailed plans for the type or the location of these, and it remains an issue for debate (see Chapter 11).

Intermediate-level waste also comes mainly from reactors, and accounts for rather more than a tenth of the total volume and rather less than a twentieth of the total radioactivity of all nuclear waste materials. Any substance in or near the reactor core is subjected to constant neutron bombardment, and many stable nuclei become radioactive after absorbing a neutron. This induced radioactivity means that components from the reactor core, and in particular items such as the fuel cladding removed if reprocessing is carried out, will be sufficiently radioactive to require shielding and safe storage. These are normally enclosed in concrete and placed underground.

Low-level waste, mainly materials and equipment that have become contaminated by radioisotopes, is an outcome of medical and research uses of radioactive materials, as well as the nuclear power industry. It accounts for nine-tenths of the volume of radioactive material to be disposed of annually, but contributes no more than 1% of the total activity. Nevertheless, it is usually regarded as material that should be buried or otherwise disposed of at special sites. In the UK most is sent to the Low Level Waste Repository at Drigg near Sellafield in Cumbria.

Reprocessing

As Table 10.5 shows, the spent fuel removed from a reactor can contain significant quantities of fissile U-235 and plutonium, and **reprocessing** is designed to extract these fissile materials for reuse. It involves chemical separation processes that were already used in the earliest reactors to obtain plutonium for the first nuclear weapons. In recent civil reprocessing, the spent fuel is moved from the reactor site after about a year to a reprocessing and fuel fabrication plant. There it must remain in cooling tanks for about another five years, until the activity has fallen enough to allow the three components, *uranium, plutonium* and *waste*, to be chemically separated. The fissile plutonium could in principle be used in reactor fuels and the uranium, now about 0.7% U-235, re-enriched for reuse. However, the future for reprocessing remains controversial, as will be discussed in Chapter 11.

10.8 Fast neutron reactors

The main difference between the thermal reactors discussed above and **fast neutron reactors (FNRs)** is that, as the name suggests, the sequence of fissions that is essential for a constant power output is maintained by the *fast neutrons* released in earlier fissions (Box 10.4). So one important practical difference between a thermal fission reactor and a fast neutron reactor is that the FNR does not need a means of slowing down the neutrons. In other words:

- A fast reactor *does not need a moderator.*
  However, as we have seen, fast neutrons are much less efficient than thermal neutrons in inducing further fissions. So a second important practical difference is the need for a high concentration of fissile nuclei to maintain the chain reaction. In other words...
- A fast reactor *needs much greater fuel enrichment.*
  In practice this means that the fuel, which might be either uranium or uranium together with another fissile material such as plutonium, needs a fissile content of between 10% and 30% compared with the 2–5% of a thermal reactor.

Since the first non-military fast reactor plant in the 1950s, a total of about twenty have operated at various times in eight different countries, accumulating only about 400 reactor-years of operation. (The total for thermal reactors is several hundred times this.) A 600 MW plant is currently operating in Russia. Many countries are currently considering FNRs in their future programmes, and according to the World Nuclear Association:

> They offer the prospect of vastly more efficient use of uranium resources, and the ability to burn [actinoids] which are otherwise the long-lived component of high-level nuclear wastes.
>
> (WNA, 2011)

One of the processes referred to by the WNA has already appeared above, in the earlier history of nuclear reactors. As described in Figure 10.14, fast neutrons can be effective in producing plutonium from the non-fissile U-238, and the prospect of 'breeding' this alternative fissile material has continued to attract attention during the seventy years since the first wartime plutonium-producing reactors.

The principle of the **fast breeder reactor (FBR)** is that the fast neutrons not only maintain the fission process but also breed plutonium from the U-238 in the fuel, which might initially be enriched uranium together with another fissile material such as Pu-239 as a minor constituent. It may seem odd to use plutonium in the fuel when the aim is to produce plutonium, but provided there are enough spare free neutrons for the breeding process, the quantity of plutonium should gradually build up, until a fuel cycle can be in principle be achieved in which more fissile material is extracted at each refuelling than was loaded in the fresh fuel.

Plants like this, where the fissile content of the extracted fuel is greater than that of the original fuel, are referred to as **breeders**. Other types, where the fissile content of the extracted fuel, is less than that of the original fuel, perhaps because the aim is to *consume* plutonium or undesirable actinoids, are called **burners**.

Liquid metal fast breeder reactors

The concentrated fuel of an FBR generates considerably more heat per cubic metre of core than the fuel in a thermal reactor. The coolant must be able to carry away this extra heat, should have reasonably heavy atoms so that it does *not* act as a moderator, and should not, of course, absorb neutrons. The majority of FBRs have used sodium, a metal that melts at about 100 °C and boils at about 900 °C, and is therefore liquid at the core temperature of about 600 °C without the need for high pressure. Sodium is chemically extremely active, igniting on contact with air and reacting violently with water, and its use has been the source of some of the problems suffered by FBRs.

Figure 10.24 shows in outline one form of **liquid metal fast breeder reactor (LMFBR)**. One essential structural difference from a thermal reactor is that the core in which fission is maintained is surrounded by a 'blanket' of U-238 in which the breeding occurs. The liquid sodium coolant circulates through the system and transfers heat to an intermediate circuit of the same liquid metal, which in turn provides steam for the turbo-generators.

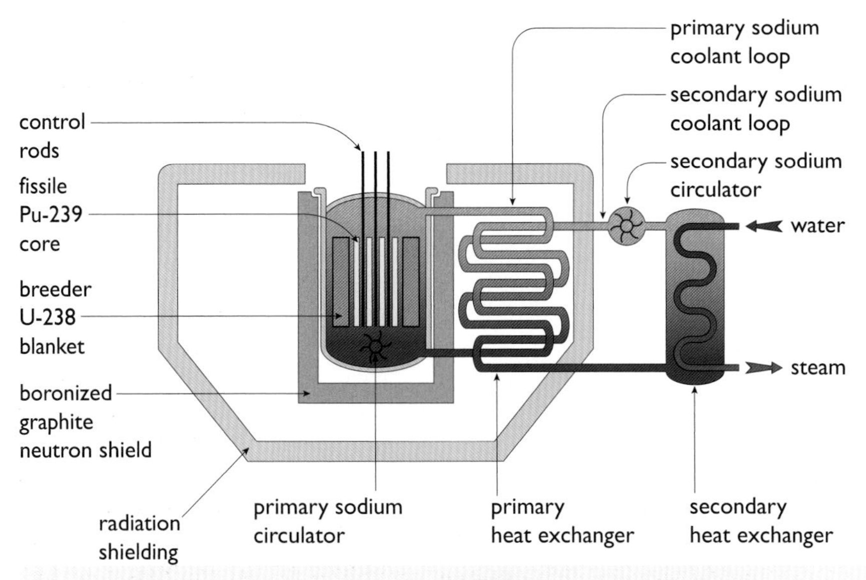

Figure 10.24 The main components of an LMFBR as constructed at Dounreay in Scotland (source: adapted from Austrian Institute for Applied Ecology, n.d.)

Safety

The main concerns about a breeder programme centre on its fuel, for three reasons: the concentration of fissile material, the necessity for reprocessing, and the central role of plutonium (Box 10.11). The concentrated fuel raises the question whether a runaway nuclear reaction could occur. A fully efficient 'atomic bomb' is not possible, and even a 'nuclear fizzle' would require a number of improbable conditions simultaneously. Nevertheless, there is some disagreement about probabilities, and the unlikelihood does seem to be of a different order from the virtual impossibility of a nuclear explosion in a thermal reactor. Similarly with reprocessing, where the significant difference is that whilst this is an option in a thermal reactor programme (see Section 10.7), it is a *necessity* in a breeder programme, where the main purpose is the extraction of plutonium.

BOX 10.11 The perils of plutonium

To see why plutonium causes concern we need to look more closely at its properties. One central fact is that with enough of it a bomb can be made. Not, it should be emphasized, easily. Plutonium is radioactive, fissile and toxic, and in unskilled hands is more likely to lead to unpleasant death than unlimited power. (Indeed, its extreme toxicity and the consequent blackmail power of a threat to distribute it is, perhaps, as great a reason for concern as the possibility of a bomb.)

It could be argued that bombs can also be made from highly enriched U-235, but the important difference is that enrichment requires complex and costly *isotope separation*. However, plutonium can be separated from uranium by simpler *chemical* means. The fissile Pu-239 produced in reactors does have 'diluting' isotopes, and true weapons-grade material should have less than 7% of the non-fissile Pu-240; but the plutonium from a breeder reactor, containing 80–90% Pu-239, could still be used in a weapon – if a rather unreliable one.

With a growing world surplus of plutonium, the concept has developed of using fast reactors as *burners*, consuming more plutonium than they produce (i.e. the reverse of a breeder). Many tonnes of weapons-grade plutonium are becoming 'available' through decommissioning of nuclear weapons by Russia and the USA. To put this in context, about eight kilograms of weapons-grade plutonium are required for a bomb, and the total quantity of plutonium in existence is thought to be over 1000 tonnes – a million kilograms. The merits and problems of dealing with plutonium are discussed further in Chapter 11.

10.9 Power from fusion

Nuclear fusion is the process that powers the stars, including the Sun. It is therefore the original source of almost all the energy that maintains the Earth's climate and its living matter. As the name suggests, it is the coming together of two lighter nuclei to form one heavier one. This is obviously the reverse of fission, so we might expect it to consume energy rather than produce it. If we attempted to fuse barium and krypton to create uranium, that would indeed be the case, but the result is very different for the lightest nuclei. We know this because the masses of all the particles concerned are known, and, as we saw in Box 10.2, if the total mass after a process is less than the total mass before, then the process releases energy. Suppose that a deuteron merges with a triton and the resulting particle then splits into an alpha particle and a free neutron (Figure 10.25). Box 10.12 shows the calculation, and the expected result that the energy released per kilogram of fuel should be considerably greater than for fission.

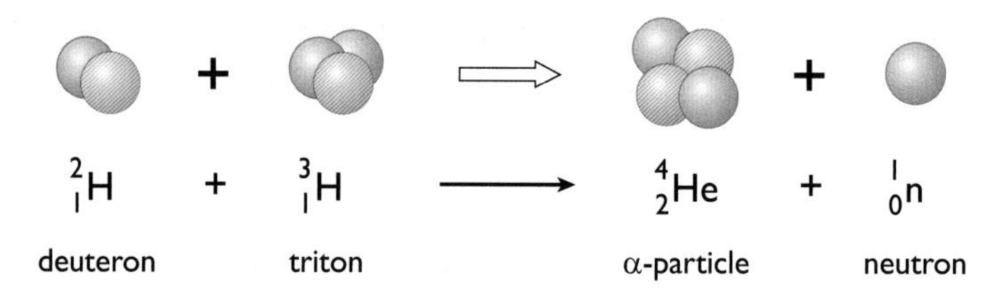

Figure 10.25 Fusion of the nuclei of deuterium and tritium producing a helium nucleus (α-particle) and a spare neutron

BOX 10.12 Energy from fusion

How much energy is released in the fusion of a deuteron and a triton?

The method in Box 10.2 can be used if the particle masses are known.

deuteron mass	2.0136 u
triton mass	3.0155 u
alpha particle mass	4.0015 u
neutron mass	1.0087 u

Simple arithmetic shows that if we start with a deuteron and a triton and end with an α-particle and a neutron, the total mass will have decreased by 0.0189 u.

If all this surplus mass appears as kinetic energy then, using the factors given in Box 10.1, the α-particle and the neutron will share a total of about:

$$0.0189 \times 931 = 17.6\,\text{MeV} = 17.6 \times 1.602 \times 10^{-13}\,\text{J} = \mathbf{2.82 \times 10^{-12}\ joules}$$

How much energy would be released in the fusion of 2 kg of deuterium with 3 kg of tritium?

As given in Box 10.1, 1 u = 1.660×10^{-27} kg. There are thus about 6×10^{26} deuterium atoms in two kilograms of deuterium, and the same number of tritium atoms in three kilograms of tritium.

So the energy released in the complete fusion of the five kilograms of the 'mixture' becomes:

$$6 \times 10^{26} \times 2.82 \times 10^{-12}\,\text{J} = 1.69 \times 10^{15}\,\text{J or } \mathbf{340\,TJ\ per\ kg}$$

Comparison with the corresponding figure in Box 10.4 shows that fusion should produce over four times as much energy as fission from each kilogram of fuel – but unfortunately the situation is not this simple (see main text).

There is, of course, the issue of how the two initial particles in Figure 10.25 are to be obtained. Deuterium, as discussed earlier, accounts for about one atom in 7000 of natural hydrogen, and can be extracted from water. But *tritium* is radioactive, a β-emitter with a half-life of only 12 years, and occurs only in very tiny quantities. So the tritium for the fusion reaction described in Box 10.12 must be produced independently. One possibility is the nuclear reaction shown in Figure 10.26: neutron bombardment of lithium-6, a fairly common isotope of lithium. This releases a few MeV of energy, but more significantly it raises a very interesting possibility. Figure 10.25 shows that the deuteron-triton reaction produces a spare neutron, and this is precisely what the lithium reaction needs in order to produce a new triton. In other words, there is the possibility of a form of chain reaction: neutron-triton-neutron-triton, etc., with continuous conversion of deuterium and lithium into helium – and continuous generation of energy.

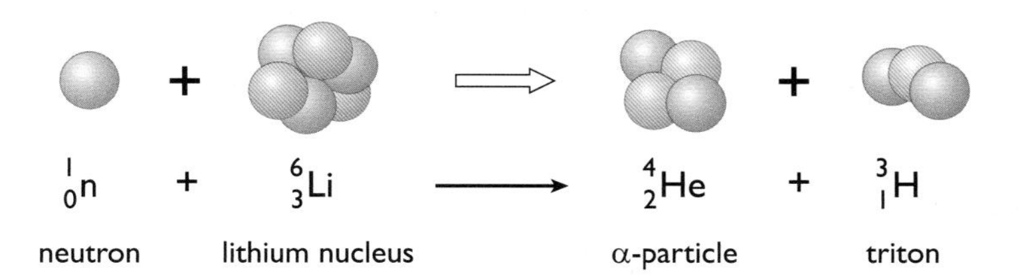

Figure 10.26 Collision of a neutron with a lithium nucleus producing a helium nucleus (α-particle) and a triton

There is however a serious problem: how to persuade a deuteron and a triton to merge. We know that the strong nuclear force holds together the closely packed protons and neutrons in a nucleus. But the deuteron and triton both have positive electric charges, so initially there is an electrical force pushing them apart, and this becomes stronger the closer they approach each other. To put it another way, they have an enormous energy mountain to climb before the strong nuclear force can take over and fuse them. One way to overcome this barrier is speed. If the two particles approach each other with very high kinetic energies, they may penetrate the barrier and fuse, finally releasing energy.

Chapter 4 explained that the kinetic energies of the atoms of a gas increase as the temperature rises. So could a high temperature raise the kinetic energies of the deuteron and triton enough to allow fusion? The answer is yes, in principle; but the temperatures required for this **thermonuclear fusion** are millions of degrees, and under such conditions not only is any material completely vaporized, but even the electrons are stripped from their atoms – a state of matter called a **plasma**. And even if such a plasma could be produced, how could it be *contained*? This has been the major problem in the attempts to achieve controlled fusion over the past half century. In the core of the Sun, at a temperature of 15 million degrees Celsius, enormous gravitational forces hold the plasma. On Earth, other means are needed to contain the high-temperature mass of charged particles, and some of the systems that have been tried will be considered in the next chapter.

10.10 Summary

The essentials

This chapter has extended the accounts of types of energy and the nature of matter in Chapter 4. Its content may be summarized as follows:

- an introduction to radioactivity, the inevitable accompaniment to nuclear processes

- an account of thermal fission reactors: their modes of operation, essential components and safety aspects, with brief descriptions of specific reactors in present-day use
- short introductions to other nuclear options that will be treated in more detail in the next chapter.

In conclusion

In looking at the uses of nuclear energy, we notice one major difference from earlier chapters. There, we have seen a multiplicity of systems using different types of energy for a range of different purposes. Here we find that almost all present-day operational nuclear-fuelled systems are versions of just one type: a *thermal fission reactor*, used for one main purpose: the *generation of electricity*.

This raises interesting questions. Are there fundamental reasons for this limited civil use of nuclear energy? Is it an intrinsic feature of the resource, or a consequence of its association with weapons issues, or just the result of a series of financial or political decisions? The next chapter will address these more complex issues – not only the possibilities of new types of nuclear plant, but the many other questions that arise when we consider the future for nuclear power.

References

Austrian Institute for Applied Ecology (n.d.) 'Reactor Types': 'Fast breeder Reactors (FBR)' [online], http://www.ecology.at/nni/index.php?p=type&t=fbr (accessed 23 June 2010).

AJ Software & Multimedia (2008) 'Library': 'Historical Documents': 'The Nuclear Age Begins' [online], http://www.atomicarchive.com/Docs/Begin (accessed 24 June 2010).

BP (2010) *BP Statistical Review of World Energy*, London, The British Petroleum Company; available at http://www.bp.com (accessed June 2010).

DECC (2009) *Digest of UK energy statistics (DUKES)*, Department of Energy and Climate Change; available at http://decc.gov.uk/en (accessed June 2010).

Einstein, A. (1939) Letter to President Roosevelt; available at AJ Software & Multimedia (2008).

Frisch, O. R. and Wheeler, J. A. (1967) 'The discovery of fission', *Physics Today*, vol. 20, no. 11, pp.43–52; also available at http://www.physicstoday.org (accessed July 2010).

Frisch, O. R. (1979) *What little I remember*, Cambridge, Cambridge University Press.

Frisch, O. R. and Peierls, M. (1940a) 'On the Construction of a "Super-bomb" based on a Nuclear Chain Reaction in Uranium', memorandum; available at AJ Software & Multimedia (2008).

Frisch, O. R. and Peierls, M. (1940b) 'On the Properties of a Radioactive "Super-bomb"', memorandum; available at AJ Software & Multimedia (2008).

Hahn, O. (1950) *New Atoms*, New York, Elsevier pp. 82–83.

Hahn, O. (1968) *Mein Leben*, Munich, Bruckmann K G, p.151.

Hahn, O. and Strassmann, F. (1939) 'Concerning the existence of alkaline earth metals resulting from neutron irradiation of uranium', *Naturwissenschaften*, vol. 27, no. 11.

Manley, J. H. (1962) 'Atomic Energy', *Encyclopædia Britannica*, London, vol. 2, p. 648.

MAUD Report (1941) 'On the Use of Uranium for a Bomb'; available at AJ Software & Multimedia (2008).

Meitner, L. and Frisch, O. R. (1939) 'Disintegration of Uranium by Neutrons: A New Type of Nuclear Reaction', *Nature*, vol. 143, no. 3615, pp. 239–240; also available at AJ Software & Multimedia (2008).

Pais, A. (1991) *Niels Bohr's Times*, Oxford, Oxford University Press.

President Roosevelt (1939) Response to Dr Einstein; available at AJ Software & Multimedia (2008).

RS and RA Eng (2009) *Nuclear energy: the future climate*, London, The Royal Society and The Royal Academy of Engineering.

Schneider, M., Thomas, S., Froggatt, A. and Koplow, D. (2009) *The world nuclear industry status report 2009*, Commissioned by the German Federal Ministry of Environment, Nature Conservation and Reactor Safety; available at http://www.bmu.de/english/nuclearsafety/downloads/doc/44832.php (accessed June 2010).

WNA (2010) 'Reactor Database' [online], World Nuclear Association, http://world-nuclear.org/info/reactors.htm (accessed 24 June 2010).

WNA (2011) 'Fast Neutron Reactors' [online], World Nuclear Association, http://world-nuclear.org/info/inf98.htm (accessed 4 July 2011).

Chapter 11

The future of nuclear power

By Dave Elliott, Bob Everett and Janet Ramage

11.1 Introduction

The previous chapter looked at the physics of nuclear power systems. This chapter looks at various issues concerning its immediate and long term future.

It starts with a brief résumé of the global state of the industry and a description of the different 'generations' of reactor design. It moves on to consider the various pressures on nuclear reactor operators that have led to the decline in reactor construction since the 1990s. This is followed by a consideration of the available uranium resource, the energy implications of using low-grade ores and possible solutions for dealing with a shortfall. The next topics relate to nuclear safety, reviewing the implications of four major nuclear accidents, including that at Fukushima, and the safety needs for future reactor designs. This is followed by nuclear wastes disposal and reactor decommissioning and the need to store waste out of the human environment for thousands of years. It then gives a brief description of the factors determining the cost of nuclear electricity and some technical possibilities for reducing costs through improved reactor design.

The final sections look beyond the basic simple 'burner' reactor and consider the costs and benefits of reprocessing nuclear fuel, including a look at the nuclear proliferation issues involved in the use of uranium enrichment and the separation of plutonium in reprocessing. It then moves on to consider the future possibilities of 'fast breeder reactors', the breeding of uranium from thorium and fusion power.

The final topic is the recent resurgence of interest in nuclear power as a low-carbon technology in a world of possible energy shortages.

11.2 The current state of play

As described in the previous chapter, at the end of 2009 there were over 400 power reactors in operation in 30 countries around the world supplying about 14% of global electricity demand. This represents a saving in global CO_2 emissions of about 2.5 billion tonnes compared to conventional coal-fired generation. As shown in Figure 11.1, there are several countries where nuclear power provides a third or more of the total electricity supply.

Over 60% of the reactors used around the word are pressurized water reactors (PWRs), as first developed in the USA. A further 20% are boiling water reactors (BWRs). At the time of writing (2011) a total of 66 reactors were under construction: 27 of these are in China and 11 in Russia; 55 of these reactors are PWRs. (IAEA, 2011).

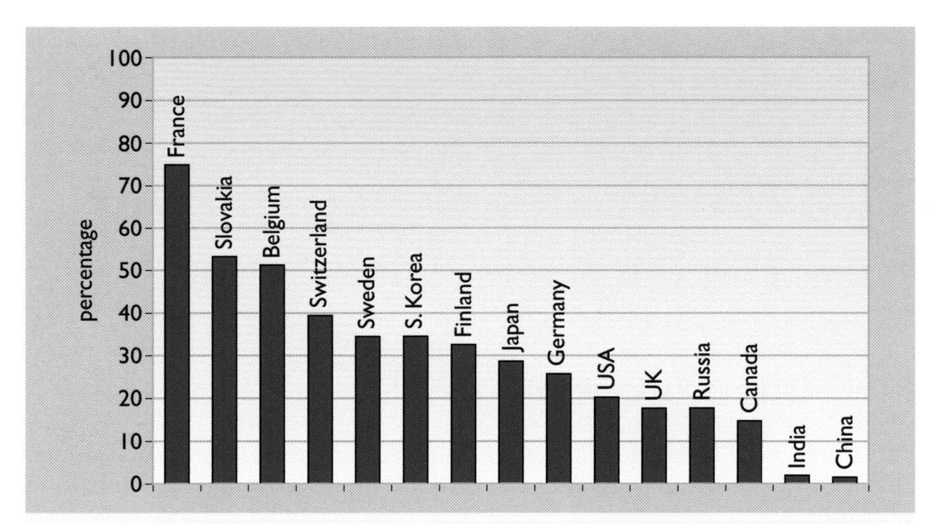

Figure 11.1 Percentage of electricity produced by nuclear power in selected countries, 2009 (source: WNA, 2010a)

In describing past, present and future reactor designs, the nuclear industry has developed a system of four 'Generations' (see Figure 11.2).

- Generation I reactors are those built in the 1950s and 60s and include the UK Magnox design.
- Most of today's operational reactors are Generation II, the basic types described in the previous chapter.
- Generation III designs are mostly 'evolutionary' improvements on the Generation II reactors for deployment in the immediate future, aiming to reduce costs, increase safety margins and reduce the volume of waste produced.

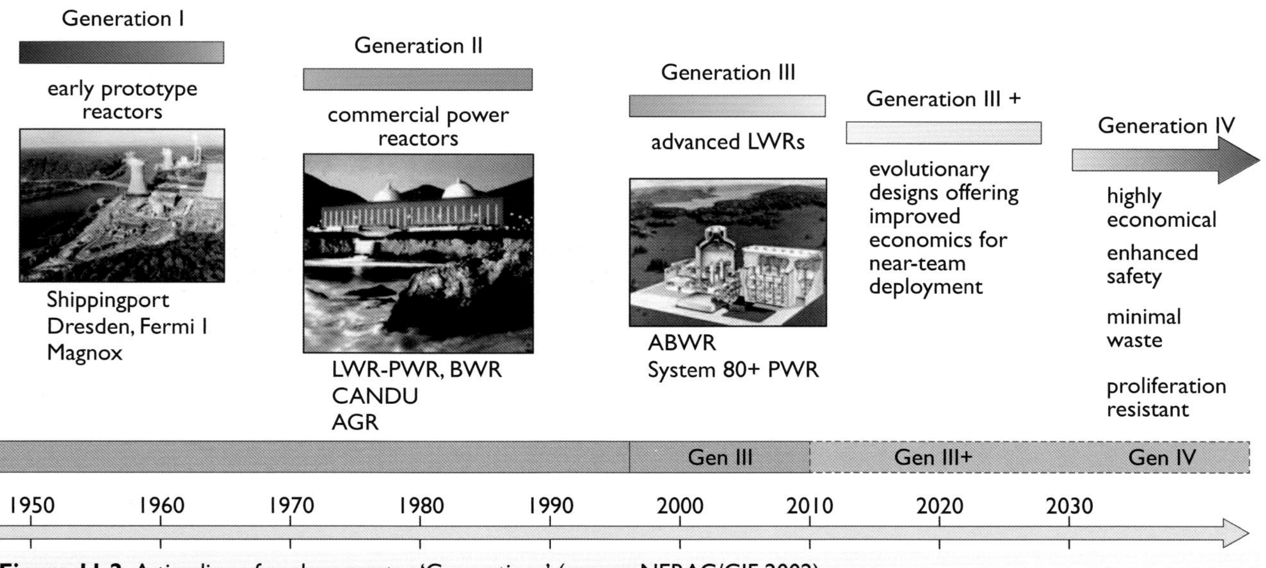

Figure 11.2 A timeline of nuclear reactor 'Generations' (source: NERAC/GIF, 2002)

- Generation IV designs represent more advanced reactor designs for deployment after 2030. Some of these involve radical changes in reactor design or fuel handling.

Some of the Generation III and IV designs are described later in the chapter.

In addition to power plants, the expansion of the nuclear industry over the last forty years or so has involved the development of uranium mining, fuel fabrication and enrichment facilities in various locations around the world, backed up by interim waste storage facilities for the used fuel and waste. Some countries (mainly the UK and France) also have reprocessing plants to extract plutonium from the used fuel, and some also have plans for long-term waste repositories. Nuclear power has become a major international industry.

11.3 Reasons for the decline in reactor construction

The previous chapter has noted the decline in world nuclear reactor construction in the late 1990s. As pointed out above, new reactors are still being built, but at the time of writing (2011) only two were in Europe and one in the USA. Why is this?

In the 1950s the early nuclear power stations were inseparable from military needs to produce plutonium for weapons and prototype power plants for nuclear submarines. Indeed the widely used PWR design is derived from one of those designs. Today, nuclear power is seen as a fully commercial, rather than a semi-military activity. But the industry is at the centre of a set of conflicting pressures, illustrated in Figure 11.3. Two of these, the need for low-carbon electricity supplies and issues of national energy security, are in its favour. Many other pressures are not so favourable.

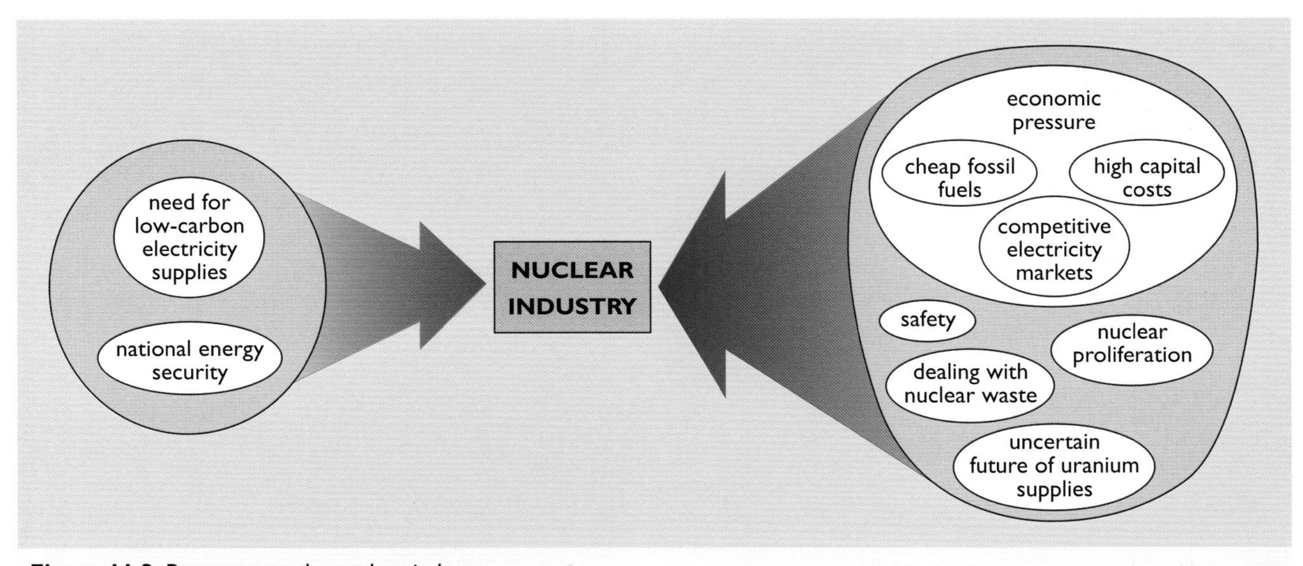

Figure 11.3 Pressures on the nuclear industry

Favourable pressures

National energy security

As pointed out in Chapter 2, this has been a particularly important factor in France, which has little in the way of fossil fuel resources. It was a serious consideration more widely in Europe and the USA during the 1970s when oil prices peaked. The development of the resources of Alaskan oil and North Sea oil and gas in the late 1970s contributed to the collapse of the oil price in 1986. A possible world of energy shortages gave way to one of global availability of cheap fossil fuels throughout the 1990s. However, rising oil prices and, in Europe, the decline of North Sea gas production have made this a topic of interest once again.

Need for low-carbon energy sources

The topic of climate change became a matter of international discussion in the 1990s, leading to the Kyoto Protocol in 1997, which set targets for reducing carbon emissions. Since then there has been interest in nuclear power as a possible 'low carbon' electricity source.

Unfavourable pressures

Economic pressures

Free markets and privatized electricity systems

In the UK (and France) nuclear electricity was developed by state-owned electricity utilities. However, the 1980s marked the rise of 'free marketeers'. In 1979 Margaret Thatcher became prime minister in the UK and in 1981 Ronald Reagan became president of the USA, both champions of free markets. They held that governments should not attempt to 'pick winners'; the choice of electricity generation fuels and technologies should be left to the competitive marketplace. The UK government pressed ahead with the privatization of the state-owned electricity industry in 1989 (as described in Chapter 9), including its nuclear stations. This was supported by some environmental groups who felt that it would expose many of the 'hidden' costs of nuclear power. Despite the rhetoric, the government had to withdraw the older Magnox stations from privatization at the last minute as unsaleable and keep them in public ownership. In France, the utility Electricité de France (EDF) is still largely state-owned, despite pressure from the European Commission. The Commission has also been particularly keen on a pan-European competitive electricity market. This has made it very difficult for countries such as France to argue the case for national subsidies for nuclear power.

Capital costs

The Sizewell B PWR plant (Figure 11.4) was under construction at the time of electricity privatization, but since then no new reactors have been built in the UK. Nuclear reactors are highly capital intensive. They also have a poor record of being built to budget. The cost of Sizewell B increased from £1.7bn to £3.7bn during construction. Capital repayments can represent

Figure 11.4 Sizewell B PWR in Suffolk

70% of the total cost of the electricity generated. Plants therefore need a long operational lifetime over which this investment is repaid in order to be economic. Whilst in the past power plants could be state funded, the new free-market doctrine asked that they should be assessed on the basis of 'commercial rates of return'.

The problems of reducing the capital costs of future reactor designs are discussed later in this chapter.

Cheap fossil-fuel based electricity

As pointed out in Chapter 9, the privatization of the UK electricity industry conveniently coincided with the withdrawal of an EU directive prohibiting the use of natural gas for large scale electricity generation. The availability of this and cheap CCGT technology gave rise to the 'dash for gas'. In addition, cheap imported coal cut the costs of coal-fired electricity generation. Faced with this combination, nuclear electricity was deemed by the UK government to be uneconomic in the mid-1990s.

Safety

The fundamental need to have a radioactive core makes the inspection and maintenance of nuclear power stations and waste processing facilities extremely difficult. It means that repairs that might only take days or weeks on a fossil-fuelled power station may require shutdowns of months. Per kilowatt-hour of electricity generated, operation and maintenance costs for nuclear plants are typically over twice those of gas-fired power stations.

Although the coal, oil and gas industries have serious accidents, the nuclear accidents at Three Mile Island in the USA in 1979, Chernobyl in Ukraine in 1986 and Fukushima in Japan have deeply influenced the public perception of nuclear power. They have also spurred the industry to seek better reactor designs. Both of these topics are discussed later.

Nuclear waste

The radioactive waste from nuclear power generation can be extremely long-lived. Some may require isolation from the human environment for a period of over 100 000 years. At the time of writing (2011) there is still no agreed policy for the disposal of nuclear wastes in the UK or in the USA. Power station operators are seeking to minimize the *volume* of waste produced by using 'enhanced burn-up' (see Section 11.5), although this does not necessarily reduce the *total amount of radioactivity* produced.

Future uranium supplies

At present world uranium mining does not supply enough ore to run the world's current stock (or fleet) of reactors. The difference is made up mainly from the large military uranium stocks left over from the cold war. If the world's nuclear reactor fleet is to be expanded, then a larger uranium supply will be needed. There are conflicting claims as to whether or not this will be possible.

Nuclear proliferation and terrorism

Although nuclear power is a 'commercial' enterprise, the possibility of diverting power station uranium or plutonium to weapons manufacture remains. The consequences of a terrorist attack on a nuclear power plant, enrichment facility or reprocessing plant are also a matter for concern. Currently, the vast majority of the world's reactors run on a simple 'open' fuel cycle (as shown in Figure 10.23). The future development of a 'closed' fuel cycle' (i.e. with reprocessing of spent fuel) and fast breeder reactors are critically dependent on resolving these issues.

11.4 The uranium resource

Availability of uranium ore

Figure 11.5 'Yellowcake' uranium oxide

Typically, fuelling a 1 GW PWR reactor requires an annual supply of about 200 tonnes of unenriched uranium per year (expressed in terms of the actual content of uranium metal) which has to come from a finite resource of uranium. The difficult relationship between quoted figures of resources and reserves was introduced in Chapter 5 in the context of coal and discussed further in Chapter 7. Uranium has yet another set of terminology, but only expressed in different *grades of resource*.

Uranium is potentially available at a wide range of concentrations. The ratio of the mass of uranium (expressed in terms of the 'yellowcake' oxide U_3O_8 – shown in Figure 11.5) to the mass of ore is called the **ore grade**. The highest grade of ore currently being mined is nearly 18% in Canada and the lowest being mined on a large scale is 0.029% in Namibia (Kreush et al., 2006).

The current average ore grade of world production is about 0.2%–0.3% (Harvey, 2010). As with other minerals, there are both underground and open-cast mines.

Estimates of the world uranium resource are published by the Nuclear Energy Agency and International Energy Agency in their *Red Book* (OECD, 2010). The prime uranium *resources* are classified as:

reasonably assured resources (RAR) – those thought to exist with sufficient confidence that mining operations can proceed

inferred resources (IR) – those that require further direct measurements before making an investment decision.

As might be expected, the lower the grade of ore, the more expensive the uranium is likely to be to produce, so the resource estimate is further broken down by cost: under US$40 per kg of uranium metal, under US$80 per kg and under US$130 per kg. In 2009, the *Red Book* introduced a higher price band of under US$260 per kg.

Table 11.1 shows 2009 figures for the world's uranium resource (RAR plus inferred resources available at a cost of less than US$260 per kg of uranium) and world uranium production. Roughly two thirds of the world's resource and production is in just five countries.

Table 11.1 World uranium: identified recoverable resources and production, 2009

Country	Recoverable resources of uranium /1000 tonnes U	Percentage of world total resources	Production /tonnes
Australia	1679	27%	7892
Kazakhstan	832	13%	14 020
Russia	566	9%	3564[1]
Canada	545	9%	10 173
USA	472	7%	1453
South Africa	300	5%	563
Namibia	284	5%	4626
Brazil	279	4%	345
Niger	276	4%	3243
Ukraine	224	4%	840[1]
Uzbekistan	115	2%	2429
China	171	3%	750[1]
India	80	1%	290[1]
Other	487	8%	584
World total	6306	100%	50 772

[1] Uranium Institute/Word Nuclear Association estimate

Note: Totals may not add due to rounding

Sources: OECD, 2010; WNA, 2010c

What about the balance of supply and demand? Unlike the fossil fuels where world annual production meets annual demand (as described in Chapter 2), world uranium consumption (estimated as likely to be 68 000 tonnes in 2010) is considerably larger than the current 'primary' uranium supply, that from mining production, which was just under 51 000 tonnes in 2009 (WNA, 2010c).

This is only possible because of supplies of uranium from stocks left over from the cold war. The left-hand side of Figure 11.6 shows world uranium supply from 1950 to 2005 and the fuel demand from the world's reactors.

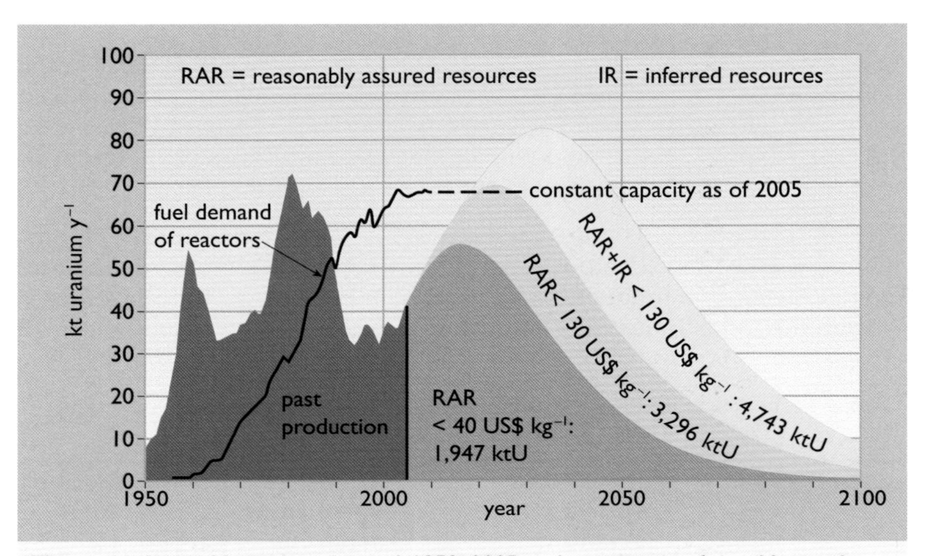

Figure 11.6 World uranium: demand, 1950–2005, and projections of possible supply from known sources (source: adapted from EWG, 2006)

During the 1950s and 60s a large amount of uranium mining was carried out to produce nuclear weapons and feed the new nuclear power industry. Mining costs were not a great consideration. The end of the cold war era left both the USA and Russia with large stocks of uranium. The world's reactor fleet has been partly running on these ever since and world uranium mining output dropped to around 35 million tonnes per year in the 1990s. In 2000 the world market price was under US$50 per kilogram.

What are the future prospects? A study by the German Energy Watch Group using resource data from the 2005 *Red Book* has produced projections of possible future world uranium supply. These are shown in the right-hand side of Figure 11.6; the left-most (orange) area includes only RARs costing less than US$40 per kilogram and the middle (yellow) area only those costing between US$40 and US$130 per kilogram. In the language of oil reserves, these might be equivalent to 'proven reserves'. The right-most (blue) area includes 'inferred' resources available at less than US$130 per kilogram.

It is clear that in the short term, as the existing uranium stocks are used up, world mining production will need to increase considerably just to keep the existing reactors running. The recent renewed interest in nuclear power has seen a surge in uranium prices which reached over US$200 per kilogram in 2007.

In the longer term, new RARs of uranium will need to be discovered and, most importantly, brought to market quickly at an acceptable price. It may be hard to persuade investors to put money into new nuclear plants with life expectancies of 50 years if there is no guarantee in the form of 'proven reserves' of a fuel supply for them. The nuclear industry is naturally keen to point out that although the low demand for mined uranium over the past 20 years has discouraged exploration, the known resource has doubled over this period. In addition to the figures given in Table 11.1, the 2009 *Red Book* quotes a further 10 million tonnes of 'undiscovered' uranium resources.

Also, since the present costs of the yellowcake ore represent under 1% of the final cost of nuclear electricity, the industry could well afford to pay far more for its uranium.

Energy return over energy invested

This topic, the concept of overall 'energy balance' of a technology, has been introduced in Chapter 7, in the context of Canadian tar sands.

The size of the uranium resource base is highly dependent on the ore grade. In Australia, which has the world's largest resources, 90% are of a grade of 0.06% or less. Similarly in Kazakhstan most of the ore has a concentration far below 0.1% (EWG, 2006).

One attraction of nuclear power is that an enormous amount of energy can be produced from a small amount of fuel. However, with an ore grade of 0.1%, 1000 tonnes of rock have to be processed to produce just one tonne of yellowcake. Even more rock may have to be removed to gain access to the ore in the mine. In an open-cast mine this may amount to a further 5000 tonnes. This all requires energy. It also has serious waste implications, which will be discussed later (see Section 11.7).

An alternative approach is to dissolve the uranium by pumping sulfuric acid into the ground. An acid solution containing uranium (and other associated minerals) can then be pumped back to the surface. This is known as **in-situ leaching (ISL)**. This method also has its own energy (and environmental) implications.

Figure 11.7 Centrifuges filled with uranium hexafluoride gas are used to 'enrich' the level of the isotope U-235; the gas is spun rapidly in a rotating tube inside the vertical gold-coloured casing

Once the yellowcake has left the mine, the uranium has to be processed, enriched and fabricated into suitable fuel for the power station. Chapter 10 described the two main methods of enrichment, gaseous diffusion and the use of centrifuges (as shown in Figure 11.7). The diffusion method is highly electricity intensive and could require an input equivalent to 3–4% of the final electricity output of a reactor. The centrifuge method is much more energy efficient and may only require 0.1% of the final electricity output.

The overall 'energy return over energy invested' (EROEI) for nuclear power production is a hotly debated topic. The invested energy must include the significant energy input in constructing and operating a nuclear power station. If high grade ores are used, the total energy used in preparing the uranium fuel for use may be only a small fraction of the total life cycle energy use of a power station. But if the ore grade declines, it can become significant.

Recent estimates (Harvey, 2010) suggest that at the current world average ore grade of 0.2–0.3% the EROEI for a 1 GW nuclear power plant is about 17–19. For an ore grade of 0.01%, the EROEI drops to 5.6 for underground mining and to 3.2 for open-pit mining. It could be as low as 2 or as high as 10 if ISL is used. As will be described in Chapter 13, the EROEI values for electricity from fossil fuels may also be low, though those for renewable energy sources can be much higher.

The energy used in mining also has implications for the overall carbon emissions of nuclear power. At present mining operations are likely to use fossil fuels, particularly oil, but in principle there is no reason why they should not use renewable energy or even nuclear electricity.

11.5 Solutions to uranium fuel limitations

In the short term there are a range of solutions to a possible shortage of uranium:

- making full use of 'secondary' uranium sources
- using reprocessed plutonium in the form of Mixed Oxide Fuel (MOX) in existing reactors
- recycling spent PWR fuel into CANDU reactors
- breeding more plutonium in existing reactors by using 'enhanced burn-up'

In the longer term, there are options for new reactor designs:

- they could be made more fuel-efficient by increasing the operating temperatures
- there could be a move to a 'closed' fuel cycle using reprocessed plutonium in new purpose-built reactors
- 'fast breeder reactors' might be developed specifically to use up the large stocks of U-238 available
- this technology could be taken further to breed uranium from the large potential resources of thorium.

In the very long term, fusion power might possibly be brought into production.

'Secondary' sources of uranium

There are several 'secondary' uranium sources:

- tail stocks of depleted uranium from previous enrichment
- weapons-grade, highly enriched uranium
- uranium recovered by reprocessing from existing reactors.

Re-enrichment of existing 'depleted uranium' tails

As described in Chapter 10, natural uranium contains about 0.7% U-235 (the rest being U-238). Most modern reactors have been designed to use uranium fuel enriched to 3.5% U-235 or more. The production of one tonne of 3.5% enriched uranium could leave a further seven tonnes of depleted uranium 'tails' at the enrichment plant containing mainly U-238 and only 0.3% U-235. There are large world stocks of these tails (not to be confused with waste 'tailings' at the mines). These are mostly in the form of canisters of uranium hexafluoride (the UK has 25 000 tonnes and the USA has over half a million tonnes).

Although it would be desirable to separate as much of the U-235 from the U-238 as possible, this has to be tempered by considerations of what is cost effective. As pointed out above, advances in enrichment technology, particularly in the design of gas centrifuges, mean that it is now considered economic to produce more enriched uranium from these stocks. This will, of course, still leave large quantities of depleted uranium, but with an even lower U-235 content (around 0.15%).

Highly enriched uranium (HEU)

HEU is uranium that has been very highly enriched (typically 98% U-235) for use in atomic weapons. It can thus be blended with depleted uranium to give a U-235 concentration suitable for conventional reactors. The end of the cold war and international agreements to reduce the numbers of nuclear warheads have made large quantities of HEU available for commercial use. In 1993, the USA and Russia concluded an agreement whereby Russia would supply 500 tonnes of HEU per year for 20 years, the equivalent of 153 000 tonnes of natural uranium.

Uranium recovered from reprocessing

Reprocessing spent reactor fuel was introduced in the previous chapter and will be revisited later in this chapter. Its function has been the separation of plutonium for other uses (particularly nuclear weapons), but it also separates the remaining uranium from the fission products, which are treated as waste. In 2007 the UK had about 30 000 tonnes of depleted uranium from the reprocessing of spent fuel from Magnox reactors (DECC, 2008). These are fuelled with unenriched uranium, so the recovered uranium contains only about 0.4% U-235. There is a further 5 000 tonnes of uranium recovered from AGRs. These have been fuelled with enriched uranium and their recovered uranium has a higher proportion of U-235 (around 0.9%). In principle, this uranium could be sent for re-enrichment. However, in practice, it is contaminated with other uranium isotopes, particularly U-236. Although, as shown in Figure 10.10, this can split into fission products in a chain reaction it does not *always* do so. It may instead emit gamma radiation and remain as a relatively stable isotope which is a strong neutron absorber. Its presence makes the use of recovered uranium less desirable than freshly mined uranium. If uranium prices were to rise, this view might change.

Using mixed oxide fuel (MOX)

As described in Chapter 10, normal reactor fuel consists of an oxide of natural or enriched uranium, but an alternative fuel with similar properties can be made by mixing plutonium oxide with uranium oxide made from depleted uranium tails (i.e the rejected material from the enrichment plant). This is known as mixed oxide fuel or MOX. Its properties are not identical to uranium oxide and in practice reactors only use a small proportion at any one time. As of January 2009, approximately 6% of the world's reactors were licensed to use MOX.

The plutonium may come from military stocks. In June 2000 the USA and Russia agreed to dispose of 34 tonnes each of weapons-grade plutonium by 2014. An alternative source is from the reprocessed spent fuel. In 2007, the UK had a stock of 100 tonnes of plutonium which potentially could be used for MOX fuel.

Recycling PWR fuel into CANDUs

CANDU reactors, described in the previous chapter, were originally designed to run on natural uranium containing only 0.7% U-235. A proposed DUPIC process (Direct Use of PWR fuel In CANDU) aims to recycle PWR spent fuel which may have a U-235 content of 1% into these reactors. The fuel could then produce more energy without the need for reprocessing, although the radioactive spent fuel will need repackaging and transportation. It is claimed that this dual fuel cycle could reduce a country's uranium requirements by 30% (WNA, 2010b). Experiments started in a Chinese CANDU reactor in early 2010 (WNN, 2010).

'Enhanced burn-up' in existing light water reactors

Nuclear power station operators are under pressure to maximize the energy output from a given amount of uranium fuel, minimize operational downtime for refuelling and minimize the volume of nuclear waste produced per kilowatt-hour of electricity generated.

All three of these objectives can be achieved to some degree by increasing the reactor **burn-up**. This is a measure of the total heat produced by a tonne of uranium in reactor fuel (see Box 11.1).

Early reactor designs such as the Magnox, using unenriched fuel with a U-235 concentration of 0.7%, have burn-ups of around 5 GWd tU^{-1}. Modern reactors using uranium enriched to 3.5% are likely to operate with burn-ups of 30–40 GWd tU^{-1}.

What about the amount of uranium mined? A high burn-up requires enriched uranium. At first sight it might be thought that a progression from 5 GWd tU^{-1} to 40 GWd tU^{-1} is a spectacular improvement. If the reactors have the same electrical efficiency, it does indeed represent an eightfold reduction in the volume of spent fuel emerging from the reactor. There is also likely to be an underlying modest improvement in the amount of unenriched uranium required. A study by Massachusetts Institute of Technology (MIT, 2003) concluded that a future improvement of burn-up from 50 GWd tU^{-1} to 100 GWd tU^{-1} would halve the spent fuel production, and give a 6.5% reduction in natural (unenriched) uranium input.

A high burn-up may not be without costs. A burn-up of 100 GWd tU^{-1} would require the use of uranium enriched to almost 10% U-235. This implies more energy use at the enrichment plant and then a harsher radiation

BOX 11.1 Burn-up

The burn-up of a particular nuclear reactor is a measure of the useful heat output produced per tonne of fuel consumed. Formally it is defined in terms of two quantities:

(1) The total heat output per year. This is expressed in the slightly curious unit of 'gigawatt-days' (GWd) where 1 gigawatt-day = 24 gigawatt-hours.

(2) The annual quantity of fuel 'consumed'. This is the number of tonnes of uranium loaded per year (tU), or where MOX fuel including plutonium is used the 'tonnes of initial heavy metal (tHM)'. (This is obviously less than the actual added load which includes the fuel cladding, the oxygen in oxide fuel, etc.)

The burn-up is equal to the first of these quantities divided by the second. For a uranium reactor it has units of GWd tU^{-1}. Burn-up only relates the heat output of the reactor to the fuel input. It is not the same as the overall electricity generation efficiency. As in any power station, the electrical output will be some fraction of this. Assuming, say an efficiency of 33%, then this means that a burn-up of 30 GWd tU^{-1} will produce only 10 gigawatt-days (or 240 GWh) of electricity per tonne of uranium.

The operation of most present nuclear plants involves the removal of a certain fraction of the reactor fuel at regular intervals. For example, an increasingly common sequence is to remove one-third every eighteen months, in which case each element will spend four and a half years in the reactor. But regardless of the particular regime, the quantities used in the burn-up are always per year.

What determines the burn-up of a reactor?

It is important to realize that any given burn-up necessarily implies a certain degree of enrichment. Obviously if the number of fissile atoms in each tonne of fuel is increased, fewer tonnes of fuel are needed for a given heat output, so the burn-up rises. (It should also be borne in mind that greater enrichment means mining more ore for each final tonne of fuel and correspondingly leaving more wastes and depleted uranium.)

A second factor is the amount of time that the fuel spends in the reactor. The longer the time, the more uranium is likely to be used up. Also, as pointed out in Chapter 10, energy from the fission of the plutonium that is always bred in the reactor can be significant. With each batch of fuel spending a few years in the reactor, plutonium contributes slightly less than half the total energy produced. However, the rate of breeding of plutonium decreases after a few years, so there comes a time where it is most economic to replace the fuel.

environment within the reactor, which may affect component life. Also, although it reduces the volume of the spent fuel produced, it has the disadvantage that it is more radioactive. According to the UK nuclear waste agency Nirex, fuel from a PWR running with a burn-up of 55 GWd tU^{-1} would be around 50% more radioactive than that from one running at 33 GWd tU^{-1} making long-term disposal harder (Edwards, 2008).

11.6 Nuclear power and safety

No energy technology is completely safe. The question of whether or not nuclear power is 'safer' than other energy technologies is hotly debated and will be discussed further in Chapter 13.

Although accidents involving the exposure of workers to radioactivity may happen within nuclear plants, the main safety issues concern the possible release of radioactivity into the environment. This could happen as a result of:

- An accident due to the malfunction of some part of a nuclear plant (including those making and reprocessing fuel). This could be due to poor design, construction errors, human errors, or, as at Fukushima, an earthquake followed by a tsunami (Box 11.2).
- A deliberate terrorist or military attack on a plant. This could include a 'cyber-attack' on the computer control systems.

Although an accident involving a major release of radioactivity has a low probability, the consequences are likely to be widespread and long lasting. The main concerns are those fission products which may enter the food chain and ultimately be absorbed by humans. The radiation from these can then give rise to cancers, which may take many years to develop. Table 11.2 gives details of three of these fission products (see Chapter 10 for more on half-lives).

Table 11.2 Fission products that may be absorbed by the human body

Isotope	Half-life	Properties
Iodine-131	8.5 days	becomes concentrated in cow's milk and is absorbed in the human thyroid gland; may cause thyroid cancers
Caesium-137	30 years	has similar chemical properties to potassium and becomes distributed throughout the body
Strontium-90	29 years	has similar chemical properties to calcium and becomes concentrated in the bones

The very short half-life of I-131 means that it is only a serious matter of concern immediately after an accident. The long half-lives of Cs-137 and Sr-90 mean that land may remain contaminated for many decades.

The public perception of nuclear safety in the UK and the USA has been influenced by a number of nuclear accidents, particularly the Windscale fire of 1957, the US Three Mile Island accident of 1979 and the Chernobyl accident in Ukraine in 1986. The accident at Fukushima in early 2011 very quickly had a significant effect on public opinion in Germany and Italy (see Box 11.2).

Understandably, these accidents, particularly Chernobyl, led to the widespread cancellation of nuclear power station projects, the winding down of nuclear research programmes and the closure of many experimental or prototype reactors. Many countries, notably Germany, set timetables to completely phase out nuclear power, turning instead to electricity generation from natural gas and renewables. Prior to 2011, Germany had

BOX 11.2 Four significant nuclear accidents

1957: Windscale, Cumbria, UK

In October 1957, a fire broke out in the graphite moderator of a UK military plutonium-producing reactor at Windscale (now renamed Sellafield). It started essentially because the basic physical processes surrounding the absorption of energy by graphite moderators were poorly understood at the time. The reactor was *air cooled*, providing a source of oxygen for the graphite fire. The chimneys of the plant had been fitted with filters to catch any stray radioactive material, but these were overwhelmed by the intensity of the fire. There was a large release of radioactivity over the surrounding countryside. Two million litres of milk were thrown away and crops were deliberately destroyed over an area of 800 km^2 in order to prevent radioactivity entering the food chain. Iodine pills were issued to the local populace to flush out any absorbed I-131 from their bodies.

1979: Three Mile Island, Pennsylvania, USA

The failure of a pump in this PWR led to a loss of coolant in the reactor and a partial melting of the fuel in the reactor core. The problem was made worse by the confused response of the reactor operators. There was no major release of radioactivity, but a large amount of hydrogen built up in the reactor and a chemical explosion could have resulted. At one time the mass evacuation of over half a million people seemed imminent. Commercially, the reactor had to be written off as unrepairable.

1986: Chernobyl, Ukraine

Until Fukushima in 2011 this was the most serious accident to date. Ironically, it occurred when the reactor was being tested at low power in order to check the safety systems. A rapid total meltdown of the core led to destruction of much of the reactor building (see Figure 11.8). About 6% of the core's radioactive content was released into the atmosphere (Read, 1993; NEA, 1995). A radioactive plume spread right across north-western Europe reaching northern Finland and the UK. The incident caused the immediate deaths of three power station workers but the total number of premature deaths as a result of radiation exposure could eventually be up to 40 000 (Medvedev, 1990). The need to reduce human exposure to radiation led to the evacuation of the nearby city of Pripyat (population 50 000) and the setting up of a 30 km exclusion zone around the reactor. This is still (2011) in force and Pripyat is still unoccupied. Farming restrictions were put in place in many north-western European countries. Some, relating to the radiation monitoring of hill farm sheep, are still in force in the UK. The neighbouring state of Belarus put the economic cost of the accident to them at US$235 billion over 30 years (IAEA, 2006). This is more than the value of a whole year's global nuclear electricity output.

Figure 11.8 The damaged Chernobyl reactor in 1986

Some of the above figures on health impacts are disputed, which is perhaps not surprising given that radiation exposure may undermine immune systems, leading to potentially fatal illnesses that might not be seen as being linked to the exposure.

A '10 year after' review suggested that around 2 500 of the 200 000 'liquidators' who were brought in to clean up the plant might develop cancers, while a further 2500 people from the immediate area might also develop cancers. A follow-up UN assessment in 2000 identified 1800 cases of thyroid cancer in children, although these were seen as treatable, e.g. by surgical removal, and it was claimed that, overall, there was 'no evidence of a major public health impact' (UNSCEAR, 2000). However, reports of illnesses and deaths have continued, and longer term it has been estimated that there might be up to a further 8000 cases of cancer, although it has also been claimed that some of the health effects that had been attributed to the accident might have had psychosomatic causes or be due to the stress resulting from over-zealous relocation of people out of the contaminated area (UNDP/UNICEF, 2002).

2011: Fukushima Daiichi, Japan

On 11 March 2011 a serious earthquake hit north-eastern Japan. The epicentre was deep below the seabed approximately 70 km off the coast, giving rise to a tsunami or tidal wave which devasted coastal towns. The combined effects of the earthquake and tsunami caused over 15 000 deaths. The earthquake affected many nuclear power stations across the area. They were all equipped with sensors which immediately triggered a shutdown. Two reactor sites, Fukushima Daiichi and Fukushima Daini, were damaged by the tsunami. At Daini, the four reactors were successfully shut down; however, at Fukushima Daiichi (subsequently referred to as Fukushima) there were serious problems.

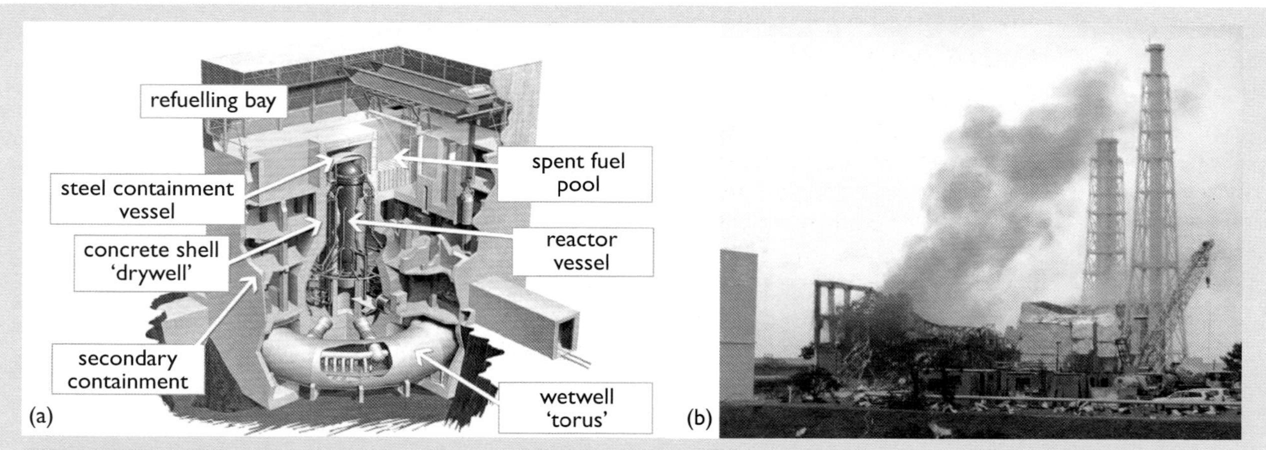

Figure 11.9 Fukushima Daiichi: (a) a section through a BWR as used at the site; (b) damage to the reactor buildings by hydrogen explosions

This site houses six boiling water reactors (BWRs) commissioned during the 1980s (i.e. Generation II reactors) and owned by the Tokyo Electric Power Company (TEPCO). Figure 10.18 in Chapter 10 has shown the key elements of the BWR design and Figure 11.9(a) shows the basic layout of a reactor building. It is designed so that fuel can be removed from the reactor and stored in fuel pools to the side of the reactor. (Figure 11.11, later in this chapter, shows a UK example of such a fuel pool.)

Cooling is a vital function in both a nuclear reactor and its associated fuel pools. When the control rods are inserted into a nuclear reactor to shut it down, the chain reaction stops, but the fission products continue to produce heat. It is thus essential that the reactor cooling system continues to work, even though the turbines may no longer be generating electricity. Reactors are thus designed to have multiple cooling systems and multiple electricity supplies to operate them. At Fukushima the site power supply consists of two separate links from the site to the Japanese grid, 12 emergency diesel generators and large emergency rechargeable batteries.

At the time of the earthquake, only three of the Fukushima reactors were operating. The earthquake severely damaged all of the links to the Japanese grid. As designed, the emergency diesel generators started and maintained cooling. Then the unforeseen happened. Although the site had been designed to withstand a tsunami 5.9 metres high, the one that struck 40 minutes after the earthquake was estimated to be over 10 metres high. The wave swamped the buildings of reactors 1–4, damaging switchgear and pumps and putting all but one of the diesel generators out of action. Some cooling was maintained for about a day using internal emergency systems and battery power. Replacement generators could not be brought in because of the inland devastation caused by the earthquake. The power station operators were powerless to prevent serious damage to the reactors.

Without cooling, the temperature inside the reactors rose, boiling off the water and exposing the zirconium fuel cladding to steam. As mentioned in Chapter 10 zirconium reacts with steam to produce hydrogen. Water levels also dropped in the spent fuel pools. The three reactor cores melted, possibly having melted right through the bottom of the reactor vessel in all three cases (Government of Japan, 2011).

In the succeeding days after the earthquake there were three explosions from leaking hydrogen, destroying the roofs of three of the reactor buildings (see Figure 11.9(b)) and spreading radioactivity, particularly caesium-137 and iodine-131, in the immediate vicinity of the site. Fortunately, for most of this time the wind was blowing out to sea. The Japanese government initially declared a 20 km exclusion zone to be evacuated around the site, but have since extended this in a north-westerly direction. This has involved moving 200 000 people.

TEPCO initially had to resort to using fire engines and building site concrete pumps to pump emergency cooling water. However, due to the extent of the damage to the reactors and buildings, substantial volumes of water emerged from the bottom of the structures heavily contaminated with radioactivity, hindering access to the site and flowing out into the sea.

At the time of writing (June 2011) the full extent of the damage or the total release of radiation is still not fully clear. Electric power has been restored to the site together with cooling to the reactors and spent fuel ponds. The leaking radioactive water from the site is being gathered into storage tanks. The levels of radioactivity in the damaged buildings are such that workers can only stay in parts of them for a few hours before accumulating a whole year's permitted radiation dose. Although the accident was initally rated as being far less serious than Chernobyl, it has now been classified as being on a similar scale.

reconsidered nuclear power, but the Fukushima accident caused it to return to its previous non-nuclear policy.

The response of the nuclear industry to Chernobyl has been to stress safety in new Generation III and IV designs. A basic key element is **secondary containment**, a building structure that is sufficiently strong to contain any radioactive material if the primary containment of the reactor vessel were to fail. The Three Mile Island PWR had secondary containment, whereas the Chernobyl design did not. The Fukushima reactors are an early BWR design and its secondary containment has been criticised for its poor performance. For maximum safety secondary containment would need to be sufficiently strong to resist an internal explosion, as happened at Chernobyl and could have happened at Three Mile Island.

New designs also feature 'passive safety' elements which are not reliant on complex emergency systems. For example, emergency cooling could be achieved by using a large, high-level tank of water and boron salts (boron is a strong neutron absorber) using gravity rather than relying on electric pumps powered by batteries or diesel generators. This is a feature that could have been useful in preventing meltdown at Fukushima.

The overall probability of an accident that proceeds all the way to damage to the core for current US reactors has been put at about 1 in every 10 000 years of reactor operation (Sailor et al., 2000). For a fleet of 400 reactors (approximately the current global total) this would mean that damage to a reactor core could be expected once every 25 years. As it happens, this is indeed the elapsed time between the Chernobyl and Fukushima accidents. It has been suggested that a goal should be set to reduce the frequency to 1 in 100 000 reactor-years of operation (MIT, 2003).

The need to increase safety may conflict with other trends in new reactor designs:

- An extension of design life to 60 years.
- The use of higher temperatures and steam pressures or the use of coolants such as sodium in order to increase thermal efficiency.
- The use of higher levels of enrichment and fast neutrons in order to increase burn-up and reduce the volume of spent fuel.

There is also the difficult question of who should insure nuclear power stations and the rest of the nuclear infrastructure against accidents. While governments in both the UK and USA have been keen to see nuclear power in the private sector, the major burden of insurance has been left with the state. It remains a largely unspecified 'extra' cost to nuclear electricity not considered in many cost comparisons of rival technologies.

11.7 Waste disposal and decommissioning

Waste and the nuclear fuel cycle

A 1 GW coal-fired power station may consume more than 3.5 million tonnes of coal a year and produce 1.5 million tonnes of waste in the form of ash to be disposed of. In the UK pulverized fuel ash (PFA) normally finds a home *within* the human environment by being turned into lightweight concrete for building.

The volumes of fuel and waste in the fuel cycle for a 1 GW nuclear plant are more modest, but being radioactive, they need to be *excluded* from the human environment, in some cases for an extremely long time. The radioactive parts of the power plant at the end of its life are a significant part of the total nuclear waste produced.

Figure 11.10 summarizes the normal 'open' fuel cycle (i.e. without reprocessing) giving an estimate of the quantities of waste produced at each stage. It also shows the flow of uranium through the cycle from the 216.8 tonnes of unenriched uranium metal in the original mine ore through to the 25.4 tonnes of enriched uranium entering the reactor. Note that there are small losses of uranium throughout the cycle.

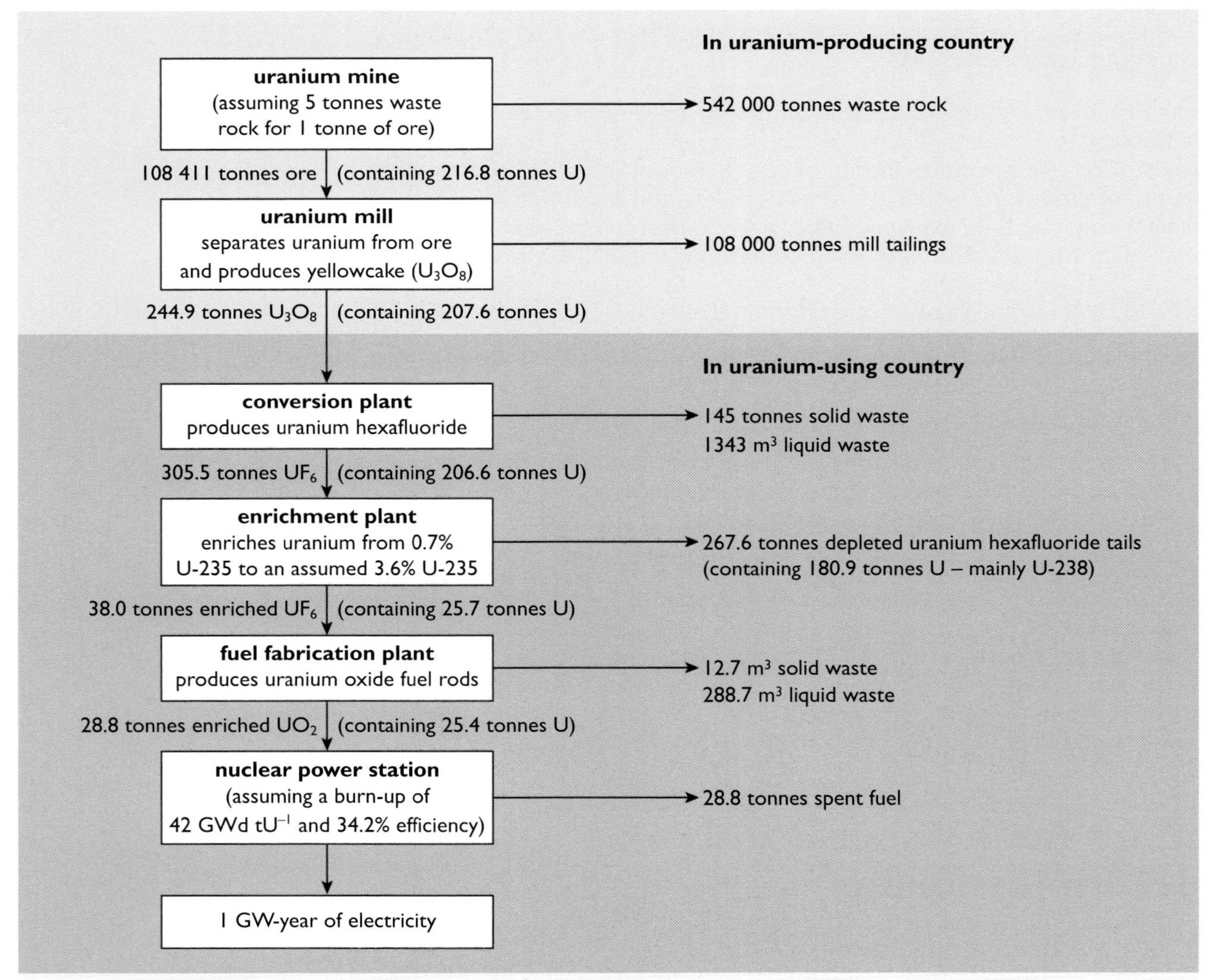

Figure 11.10 Mass flows for the production of 1 GW year of electricity in a normal 'open' fuel cycle (source: adapted from Harvey, 2010)

It can be seen that the greatest volume of wastes, the mine waste rock and the mill tailings, remain at the mine (the uranium mill is likely to be located at the mine to minimize the transport costs).

The remaining steps in the fuel cycle are likely to take place in the user country. At each successive step the wastes are likely to be more radioactive. The wastes from the conversion, enrichment and fuel fabrication are classified as 'low' or 'intermediate' wastes. At present in the UK these are packaged and kept in surface dry storage.

The depleted uranium tails are not classified as 'waste' because they are potentially a 'fuel' for future fast breeder reactors. They are packaged and stored as uranium hexafluoride. In 2007, the UK had a stock of 25 000 tonnes.

The bulk of the radioactivity comes from the spent fuel. This is classified as 'high-level' waste. Many of the fission products in this have short half-lives so in the UK it is normally kept for several years at the reactor site. The radioactive decay process continues to produce heat, so storage is in water in large cooling tanks.

After removal from the reactor, the level of radioactivity in the spent fuel falls rapidly. This is due to the short half-lives of many of the fission products. The overall radioactivity will have fallen by a factor of 80 after a year. However, that still leaves other fission products and actinoids such as plutonium with longer half-lives. Thus by year ten it will have only decreased by a further factor of six. Waiting for the radioactivity to fall by yet a further factor of six may require over 100 years, far longer than the life of the reactor. At some point in time, highly radioactive material has to be transported from the reactor site. At present in the UK, after an initial cooling period, high-level wastes are sent to Sellafield where they are again stored in cooling tanks (see Figure 11.11) with some then being sent on to a 'vitrification' plant where they are converted into a stable glassified form. They then remain in surface dry storage.

Figure 11.11 Spent fuel cooling tanks at Sellafield

(a)

(b)

(c)

Figure 11.12 A deliberate high-speed train crash in the UK in 1984 into a nuclear waste container in order to demonstrate its strength (it survived intact)

The movement of nuclear wastes by rail (or road) has long been seen as a major potential accident hazard and a terrorist target. In 1984 the UK nuclear industry commissioned a high speed train crash with a nuclear waste container to publicly demonstrate its strength (see Figure 11.12). The experiment was considered successful.

Long-term waste storage

The solution to the problem of the long-term storage of nuclear waste remains unresolved. Any storage site would need to keep nuclear materials in a stable form but out of the human environment for many thousands of years. Deep underground repositories could be the answer. These would have to be sited in known stable geological rock strata. There must also be no possibility that moving ground water could carry radioactivity away from the site.

Figure 11.13 shows a scheme as proposed in Sweden. The nuclear waste would be packaged in cast iron containers sealed in copper canisters. These would be stored in rows of vertical holes embedded in bentonite clay in tunnels 500 metres below ground. The repository would need to be able to dissipate the heat generated from the continuing radioactive decay. In neighbouring Finland tunnelling work has started on a deep repository which, if licensed, could start accepting waste in the 2020s. There are many more uncertainties in the USA, where the proposed repository project at Yucca mountain in Nevada has been abandoned. In the UK, the current situation is that communities have been invited to come forward as hosts for a possible deep geological repository, with Sellafield in Cumbria being a possible option, but a functioning system is not expected to be available until 2040.

It is the long half-lives of the actinoids which create the requirement for a cooling-off period of thousands of years. For example plutonium-239 has a half life of 24 000 years. This has led to suggestions for separation of the actinoids in a reprocessing plant from the other fission products such as strontium and caesium which only have half-lives of about 30 years. The fission products might then only need to cool off for a period of hundreds of years, while the actinoids could be dealt with separately, possibly being 'transmuted' into non-radioactive elements by being bombarded with neutrons in a special reactor. This is still an area for future research.

Plant decommissioning

At the end of its life a nuclear power station needs to be decommissioned and dismantled. Many of its components, particularly the reactor pressure vessel together with its surrounding pipework, will have become highly radioactive.

Currently UK industry policy is to remove all the non-radioactive components, but to leave the reactor core within the existing weatherproofed building in a so-called 'safe store'. It can then 'cool off' for a period of around 100 years before final dismantling.

The lack of a policy on long-term waste storage and the apparent willingness of the UK industry to postpone decommissioning is not popular with environmentalists. They see this as simply leaving today's problems for future generations to solve.

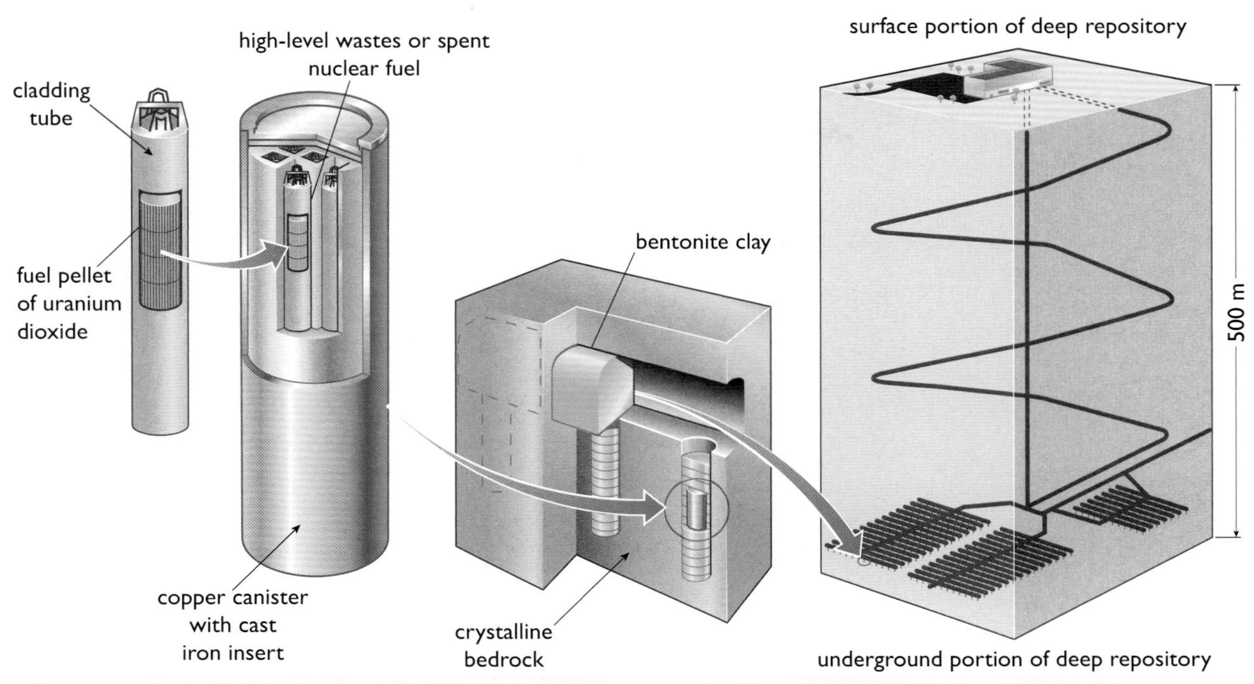

Figure 11.13 A proposed underground nuclear waste repository in Sweden (source: adapted from SKB, 2010)

11.8 **Generation costs and the prospects for cheaper nuclear power**

The factors determining generation cost

Given the ability to produce large amounts of energy from a small amount of relatively cheap fuel it might be expected that the economics of nuclear power might be quite favourable. In practice they are dominated by the complex and capital intensive nature of the nuclear power station itself.

The key factors affecting the overall cost of electricity from a power station (be it nuclear, or any other sort) are:

- The capital repayment costs, i.e. the costs of the loan to build the plant. These are in turn influenced by:

  the actual capital cost of construction

  the interest rate charged on the loans for construction

  the construction time, which may be several years

  the working life of the station and thus the amount of time available to repay the loans. This may be 40 years or more.

- The fuel costs including waste management
- The thermal efficiency of the plant
- The operating and maintenance costs
- The decommissioning costs.

Figure 11.14 shows the relative contributions of these to a 1994 estimate of the electricity cost for a proposed UK twin PWR plant, Sizewell C. The breakdown is broadly similar to more recent UK estimates. These figures and those for a rival gas-fuelled CCGT plant will be discussed in more detail in the next chapter.

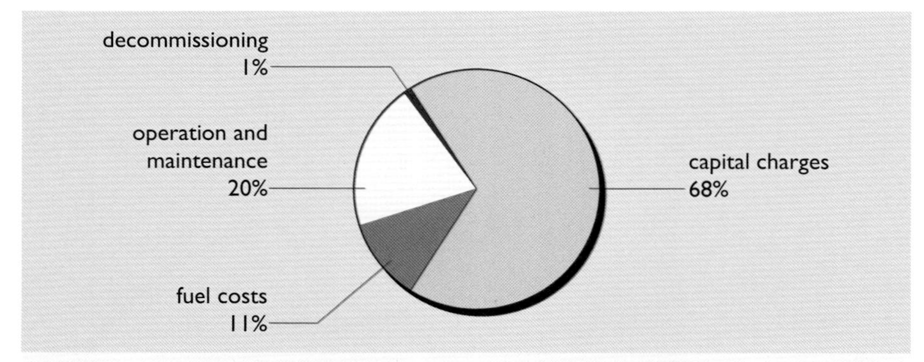

Figure 11.14 Electricity cost breakdown for a proposed UK PWR, 1994 (source: based on Sadnicki, 1994)

As can be seen from Figure 11.14 the capital repayment charges make up the major part of the overall electricity cost. The capital cost for power plant is normally quoted per kilowatt of generating capacity. Costs for nuclear power stations are higher than for basic coal-fired plant and several times more expensive than gas-fired CCGT plant. Nuclear operation and maintenance costs (quoted per kilowatt-hour of electricity generated) are typically over twice those for gas-fired CCGT plant. Note, also, that decommissioning only accounts for a small fraction of the final electricity cost, even though the actual decommissioning capital costs are over 10% of the capital construction cost. This is a consequence of the use of 'discounted cash flow' a (slightly difficult) topic discussed in the next chapter.

Any broad comparison of different technologies requires a careful specification of the factors listed above. Many renewable energy technologies such as wind power or hydroelectricity, for example, have high capital costs but zero fuel costs.

However, making realistic price comparisons is fraught with difficulties. Nuclear power station construction in many countries has a long history of building delays and budget overruns. It is thus necessary to take a very careful view of any quoted estimates for costs. Enthusiastic manufacturers are likely to quote costs assuming construction of ten or more identical reactors. Cautious investors and governments are more likely to insist on 'First Of A Kind' (FOAK) prices, particularly for new designs. The difference between these costs may be a factor of two or more.

At the time of writing (2011) nuclear power is seen by the UK government as being cost-competitive with other technologies, particularly if a *carbon price* is included. This is a notional sum which covers the estimated damage to the environment caused by the emissions of CO_2 from a plant. It may be expressed in terms of CO_2 or its carbon content. As noted in Chapter 5, 44 grams of CO_2 contains 12 grams or carbon, so £1 per tonne of CO_2 is equivalent to £3.67 per tonne of carbon. The subject of the 'cost of carbon' is discussed in Chapter 13. It is likely that carbon prices will rise in coming years. For the purposes of comparing technologies the UK government has assumed a carbon price on a rising scale from £14.10 per tonne of CO_2 in 2010 to £70 per tonne in 2030 and continuing upwards. This carbon price has been included for fossil-fuelled electricity generation technologies in the cost-comparison table shown in Table 11.3, where the CO_2 emissions of nuclear and wind power have been assumed to be zero.

Table 11.3 Estimated costs of electricity in the UK from different sources assuming construction in 2009 (2009 figures)

	Nuclear (FOAK) /p kWh^{-1}	**Offshore wind (FOAK) /p kWh^{-1}**	**Onshore wind /p kWh^{-1}**	**Gas CCGT /p kWh^{-1}**	**Coal (without CCS) /p kWh^{-1}**
Without carbon price	9.90	16.09	9.39	6.52	6.42
Carbon price	0.0	0.0	0.0	1.51	4.03
Total	9.90	16.09	9.39	8.03	10.45

Source: Mott Macdonald, 2010

A 2008 study for the California Energy Commission (Sourcewatch, 2010) estimated the cost of nuclear power at 15.3 US¢ per kWh, more expensive than supercritical coal-fired generation, gas CCGTs, wind power or even concentrating solar thermal electricity.

The 2003 MIT study quoted in Section 11.5 estimated the cost of nuclear electricity as 6.7 US¢ per kWh. It pointed out that a carbon price of US$100 per tonne of carbon (about £17 per tonne of CO_2) would be needed to make it cost-effective against coal. However, a price of over US$200 per tonne of carbon (£34 per tonne of CO_2) would be needed to make it cost-competitive against gas CCGTs.

At the time of writing US$1.60 ≈ £1.00

What could be done to improve the competitiveness of nuclear power? Table 11.4 lists some possible incremental improvements.

Table 11.4 Comparative US electricity costs (US$2002)

Case	**Cost/US¢ kWh^{-1}**
Nuclear (light water reactor)	6.7
+ reduce construction cost 25%	5.5
+ reduce construction time from 5 years to 4 years	5.3
+ reduce operation and maintenance costs	5.1
Pulverized coal plant	4.2
CCGT (moderate gas prices)	4.1

Source: MIT, 2003

Achieving these reduced construction costs, construction times and operation and maintenance costs may require changes in the design of future reactors.

Reducing nuclear costs

Nuclear power stations are usually large (500 MW or more) and relatively few are built. Each one tends to be constructed on site with little prefabrication. By contrast the 150 MW gas turbine of a CCGT plant is factory assembled and then moved to the power station site by ship and road. The average construction time for recent US nuclear reactors has been about six years. A CCGT plant can be constructed in under two.

There are several design routes that could reduce costs:

- adopting modular construction to allow reactor elements to be factory assembled
- simplifying the design to reduce the number of components
- simplifying the fuel handling to reduce operation and maintenance costs
- increasing operating temperatures to give a higher thermal efficiency.

Many new designs have been proposed as part of the Generation III and Generation IV studies. Below are a few examples.

The Westinghouse Advanced Passive AP1000

This is a Generation III evolutionary development of previous PWR designs. It tackles some safety issues and attempts to reduce construction costs.

It features:

- a steel secondary containment vessel
- a passive gravity-fed emergency water cooling system
- a simplified design with 35% fewer pumps and 85% less control cable compared to previous Generation II PWR designs
- a simplified safety system with fewer safety-related pumps and pipework
- a physically smaller size
- modular construction

and it is claimed that it could take only three years to construct.

The pebble-bed high temperature reactor

This is a Generation IV reactor involving a radically different design. It has the possibility of simplified fuel handling, a small reactor unit size (100 MW) and a higher thermal efficiency.

The key element is the fuel packaging. This takes the form of billiard ball sized 'pebbles' with a hard silica casing. Each pebble contains thousands of tiny coated particles, each with a small kernel of uranium oxide surrounded by a layer of graphite forming the moderator (see Figure 11.15).

The pebbles are sized and designed so that bringing them together in a hopper in the reactor achieves a critical mass and starts the chain reaction. The reactor is designed to operate at a high temperature (900 °C) and is cooled by pumping high-pressure helium gas through the stack of pebbles. The hot gas then drives a gas turbine and a second stage steam turbine as in a CCGT, giving a thermal efficiency of 40% or more.

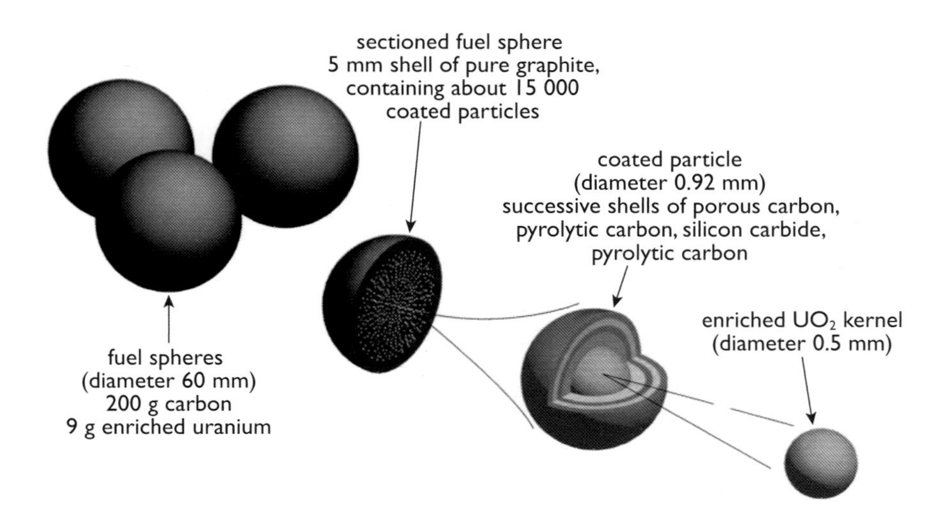

Figure 11.15 Fuel 'pebbles' as used in the pebble-bed reactor (source: adapted from PBMR, 2010)

At the end of the fuel cycle the spent fuel should ideally remain safely encapsulated within the pebbles for disposal without reprocessing.

Although this is a Generation IV design, it is not new. One prototype, the Dragon reactor, started operation in the UK in 1965, and another, the AVR, in Germany in 1967. More recently the South African electricity utility Eskom has been developing a 100 MW design, although work on that has now been halted due to budget cuts.

Other high temperature reactors

The considerations of Carnot efficiency described in Chapter 6 suggest that nuclear power plants should attempt to produce heat at as high a temperature as possible in order to increase the electricity generation efficiency. In Chapter 10, it was pointed out that current PWR designs only use steam temperatures of around 300 °C, even though modern coal-fired power stations operate at 500 °C or more. A number of high-temperature reactor designs have been proposed. A major problem is finding a suitable coolant. Carbon dioxide has been used successfully in the UK AGRs and helium in prototypes such as the pebble-bed reactor. What could be used to improve the thermal efficiency of proven conventional PWR designs?

Water is cheap and convenient but, as pointed out in Chapter 6, it requires very high pressures to contain it at high temperatures. High temperature steam is also highly corrosive to steel and requires the use of special alloys. This has considerable safety implications for nuclear power stations.

Any pipework used within the reactor core must, in addition, be relatively transparent to neutrons in order not to inhibit the fission process, and it must last for the whole reactor life (25–60 years) without suffering from metal fatigue. This is a rather difficult design brief, but, as pointed out in Chapter 10, alloys using the metal zirconium are of interest.

An alternative approach is to extract heat from the reactor core using liquid metals or molten salt. These can be molten at the reactor temperature without requiring pressurized pipework. They can carry heat from the reactor to an external heat exchanger where high-pressure steam can be raised without worries of neutron damage to the high-pressure pipework.

Sodium or a sodium/potassium mixture are possible choices for liquid metals. However, these react violently with water and have severe safety implications. Molten lead is a possible choice for future reactor designs. Obviously these metals or salt are solid at room temperature and the pipework needs an auxiliary heating system to melt the metal or salt before reactor operation can commence.

11.9 Beyond the basic open fuel cycle

Closing the fuel cycle – reprocessing

Unlike a coal-fired power station, where the fuel is totally burned, the fuel in a nuclear reactor has to be removed when it can no longer sustain a satisfactory rate of nuclear fission. It still contains potentially useful amounts of U-235 and plutonium (bred from the uranium during the period of operation in the reactor).

Should the spent fuel just be thrown away, or is it worth recovering the remaining uranium and plutonium for use in MOX? This recycling is known as 'closing the fuel cycle'. In the past, the plutonium was highly valued for atomic weapons. In today's commercial environment, its only value is as potential nuclear fuel. In the longer term, reprocessing facilities will also be needed for any future development of fast breeder reactors (see below).

Since plutonium and uranium are different elements, they can be separated chemically in a reprocessing plant using a process called **PUREX** (Plutonium Uranium Recovery by EXtraction). This is not easy, since the spent fuel, even after cooling for a year or more, is still very highly radioactive.

The UK has a **Thermal Oxide Reprocessing Plant (THORP)** and a MOX fabrication plant at Sellafield in Cumbria (see Figure 11.16). Similar plants exist in the USA, France and Russia.

Figure 11.16 Sellafield reprocessing plant and MOX fabrication facility

What are the savings in uranium? The study by Massachusetts Institute of Technology (MIT, 2003) has considered the (rather large) uranium requirements of a possible future global fleet consisting of 1500 GWe of nuclear reactors. It estimates that conventional reactors using a conventional once-through fuel cycle might require 306 000 tonnes of natural uranium (expressed in terms of the uranium metal) per year. That is 204 tonnes of uranium for 1 GW-year of electricity, comparable with the figures quoted in Figure 11.10.

It has also considered a 'closed' fuel cycle in which the plutonium is recycled and blended with depleted uranium to produce MOX. This is shown in Figure 11.17 with quantities scaled for 1 GW of electrical generating capacity. In this cycle the spent MOX is not recycled again because it contains too many higher actinoids such as americium bred from plutonium. This process, which requires the recycling of 220 kg of plutonium per gigawatt-year, would only require 171.6 tonnes of natural uranium, a saving of 16% compared with a simple open fuel cycle. It would also use up 3 tonnes of otherwise unwanted depleted uranium.

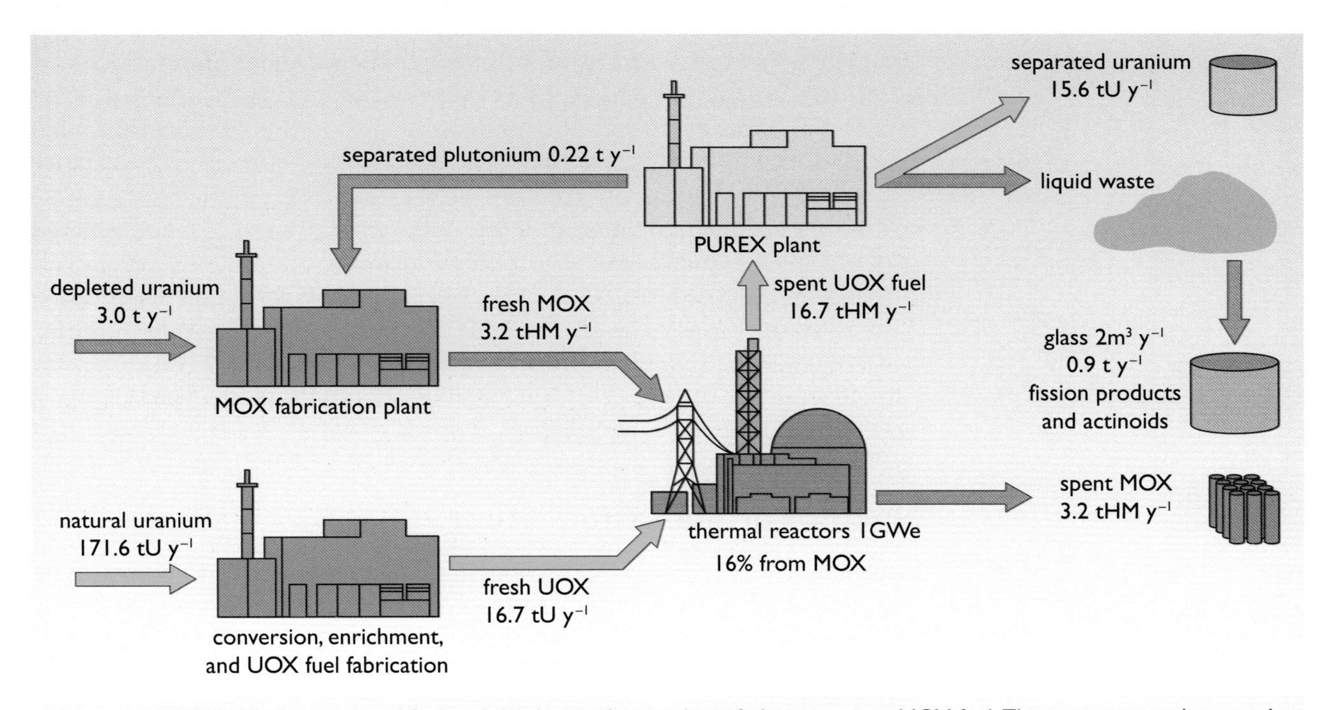

Figure 11.17 Possible future closed fuel cycle with a single recycling of plutonium into MOX fuel. The amounts are the annual quantities required for 1 GW of reactor output (source: adapted from MIT, 2003).

In this cycle the reprocessing PUREX plants also produce 15.6 tonnes of separated uranium, containing a small amount of U-235. In principle, this too could be recycled, but, as pointed out above, in practice it also contains small amounts of U-236 which is a strong neutron absorber. It is thus most likely to be sent to storage, for possible future use if the uranium price were to become high enough.

Alternatives to the PUREX system are under development at the Argonne National Laboratory in the USA. These are designed to separate out *all* of the actinoids (uranium, plutonium, americium, etc.) for recycling into new reactor fuel. Since the remaining fission products have relatively short half-lives, these processes could result in a reduction in the long-term radioactivity of any nuclear waste (i.e. it would decay significantly in hundreds of years rather than thousands).

Although reprocessing saves on uranium mining, it comes at a cost:

- The reprocessing and MOX fabrication plants are expensive to build and operate. Because of this there are only a few such facilities in the world and it has recently been announced that the UK Sellafield MOX plant is to be closed.
- The spent reactor fuel is highly radioactive and needs to be left to cool in large water tanks for several *years* before reprocessing can start.
- There are serious transportation problems:

  radioactive spent fuel must be moved from the power stations to the reprocessing plant

  plutonium must be moved between the reprocessing plant and the MOX fabrication plant (it helps if these are on the same site)

  MOX fuel rods need to be sent back to the power stations.
- The movement of the fuel and its reprocessing exposes workers to extra radiation hazards. Nearly 80% of the collective radiation dose associated with the complete nuclear fuel cycle comes from reprocessing activities (UNSCEAR, 1993).
- It increases the total amount of low-level wastes, such as contaminated clothing and chemicals used in the processing.
- It has its own safety issues. The Sellafield THORP plant was closed in 2005 for three years after a leak went undetected for nine months, allowing a solution containing 20 tonnes of uranium and 160 kilograms of plutonium to flow into the secondary containment. The operating company was fined £500 000.

Nuclear proliferation issues

Nuclear reactors were originally operated to produce plutonium for atomic weapons. A major concern about the continued used of nuclear power is the possibility of diverting materials for the manufacture of new weapons. There are two key points where this might happen, at enrichment plants and in the reprocessing fuel cycle.

Uranium enrichment plants

Early reactors such as the UK Magnox plants and the Canadian CANDUs do not require enriched uranium. Later reactors have used uranium enriched to 3–5% U-235 and future fast neutron reactors may require enrichment levels of 10–20%. Any concentration over 20% is classified as 'weapons useable'. A practical atomic bomb would only require about 25 kg of uranium, but it would need to be enriched to a level of 90% or more (Mark et al., n.d.). At least seven countries worldwide possess enrichment facilities. There have been suggestions that uranium enrichment should be restricted to just a few sites and brought under global international control as a non-proliferation measure. This also supposes that there will be long-term political stability at the proposed sites.

Reprocessing plants

Reprocessing potentially makes plutonium available for weapons manufacture. As little as 5 kg of plutonium could be used to make a crude atomic bomb, yet as is obvious from Figure 11.17, the reprocessing fuel cycle for a single 1 GW reactor involves many times that amount. Although producing highly enriched uranium is a highly energy intensive process, plutonium can be separated from spent fuel, or even stolen MOX fuel, *chemically*.

The simplest way to reduce this risk is to abandon reprocessing. This may leave the world's nuclear industry excessively reliant on limited global uranium resources.

There are alternative technical possibilities. It may be possible to develop new reprocessing methods, so that the highly radioactive minor actinoids remain with the recycled plutonium. This would make subsequent reactor fuel handling more difficult, but would make plutonium theft a dangerous exercise. Another is the development in the USA of an **integral fast reactor (IFR)**, in which reprocessing and fuel recycling are integrated into the reactor plant. These technical possibilities may be decades away from commercialization.

As with enrichment, there may be a need to establish a secure fuel cycle with reprocessing restricted to a few global sites under international control.

Plutonium production in fast breeder reactors

The possibility of converting unwanted depleted uranium, consisting of 99.7% U-238, into useful reactor fuel is very attractive, and Section 10.8 has described a basic process whereby plutonium might be bred by surrounding the core of a reactor with a blanket of depleted uranium.

This breeding process requires a strong flux of 'fast' neutrons, hence the name 'fast breeder'. In practice the process is slow and it can take years of irradiation to produce a reasonable amount of plutonium.

The breeder blanket would then be sent to the reprocessing plant where the plutonium would be extracted and turned into MOX fuel. An alternative is to make sure that the breeding takes place within the existing charge of reactor fuel, i.e. equivalent to achieving a very high burn-up rate. The ultimate goal would be a reactor that could breed as much new fissile plutonium as it consumed in U-235. It could then, in theory, be refuelled simply with depleted uranium.

The high neutron flux necessary requires a fast reactor running on uranium enriched to 15–20% U-235 rather than only 3.5% in a conventional PWR. Also, instead of water to convey heat from the reactor, the designs have used a liquid metal, sodium (or a sodium/potassium mixture). Although sodium can potentially withstand high temperatures, it is highly reactive, burning in air and reacting violently with water. These features have created concerns for safety and, in practice, operational difficulties, sodium fires and a history of poor reliability.

In the UK, the Dounreay fast reactor, illustrated in Figure 10.24, was first started in 1959 and was operational until 1977, producing 14 MW of electricity. Several other countries have constructed fast reactors. In France the 1.2 GW Superphénix became operational in 1985 but was closed in 1998 because of 'excessive costs'. In Russia a 600 MW fast reactor, the B-600, has been operational since 1980. A 500 MW prototype fast breeder reactor is under construction at Kalpakkam in southern India and may be completed in 2011.

Breeding uranium from thorium

As mentioned in the previous chapter, thorium-232 is *fertile*. Although it is not extensively mined, there may be three times as much thorium available in the world as uranium. Also, naturally occurring thorium is almost totally made up of the Th-232 isotope.

If a Th-232 nucleus is bombarded with neutrons in a fast breeder reactor (most likely fuelled by uranium) it can capture a fast neutron to become Th-233, which then emits a beta particle to become protactinium-233 (Pa-233), which again emits a beta particle and becomes uranium-233 (see Figure 11.18).

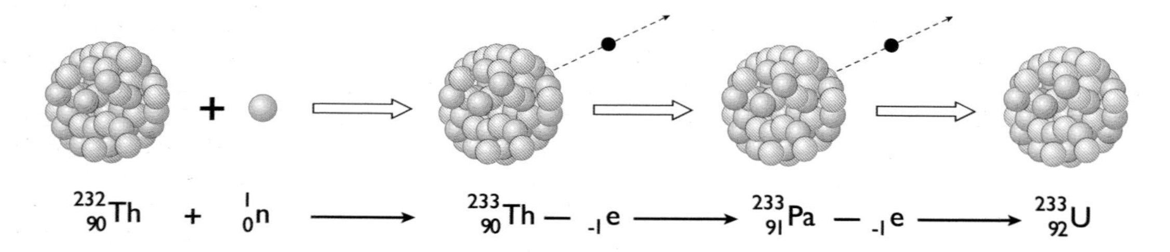

Figure 11.18 Production of U-233 from Th-232

U-233, like Pu-239 produced from U-238, is fissile. It can be chemically separated from the thorium in a processing plant and then used as fuel in a reactor.

The use of thorium and U-233 was extensively explored in the 1960s and 70s, particularly in a reactor at Shippingport in the USA. The breeding reaction is slow because the intermediate product protactinium is a strong neutron absorber. Also, other reactions produce U-232, the decay products of which produce intense gamma radiation, making reactor shielding difficult. Although it may be possible to produce nuclear weapons using U-233, the gamma ray emissions from the U-232 might make this nuclear fuel an undesirable commodity for any terrorist to steal.

This cycle is of particular interest in India, which has a quarter of the world's thorium reserves, and it may be used in their new prototype fast breeder reactor mentioned above.

Nuclear fusion

Nuclear fusion has been the subject of research for more than half a century, but the only device to produce net output power has been the hydrogen bomb, first exploded in 1952. The basic physics of the enormous potential production of energy of the deuterium–tritium (D–T) reaction and a method for producing tritium have been described in the previous chapter. The controlled production of fusion power has remained elusive.

It is easy to see in general why there are problems. Two nuclei can only fuse if they come close enough for the short-range nuclear force of attraction to overcome their intense electrical repulsion. To overcome this electrical barrier the nuclei must approach each other at extremely high speeds, and several different methods have been investigated in the attempts to achieve this.

One way to achieve very high speeds is to raise the particles to a very high temperature. This is the principle of the thermonuclear approach. Calculations show that the deuterium–tritium reaction needs a high particle density at a temperature of *several million degrees Celsius.* At this temperature all atoms are stripped apart and what may have started as a gas becomes a *plasma* of hot nuclei and electrons. A major problem is to design a container that will not instantly melt.

Much of the fusion research effort of the past half-century has been concentrated on magnetic containment – using the force that acts on a charged particle moving in a magnetic field to keep the plasma compressed. In the **Tokamak** arrangement, devised in the 1950s by the Russians Andrei Sakharov (see Figure 11.19) and Igor Tamm, the particles move in a doughnut-shaped ring (the torus) surrounded by large coils carrying an electric current (Figure 11.20). These coils produce a magnetic field along the torus, and the electrically charged particles spiral around perpendicular to the direction of this field. A current flowing in the plasma heats it, and additional energy may be supplied by fast-moving deuterons and tritons injected into the system. This principle is used in the Joint European Torus (JET) in England (see Figure 11.21) and the Tokamak Fusion Test Reactor (TFTR) at Princeton in the USA. Both of these have achieved fusion for extremely short periods of time, but neither has yet produced more energy from fusion than the energy input needed to run the system.

Figure 11.19 Andrei Sakharov (1921–1989) was born in Moscow, the son of a physics lecturer. He was drawn into work on the hydrogen bomb. He suggested a 'layer cake' bomb of alternate heavy and light elements – the fission reaction in the heavy elements would set off the fusion reaction in the light ones. In 1950 he proposed his Tokamak idea for controlled nuclear fusion, but continued to work on ever larger hydrogen bombs. In the 1960s he started expressing reservations about nuclear testing, eventually urging an end to the cold war. Despite being the 'father of the H-bomb', in 1975 he was awarded the Nobel Peace Prize. The citation described him as 'the conscience of mankind'.

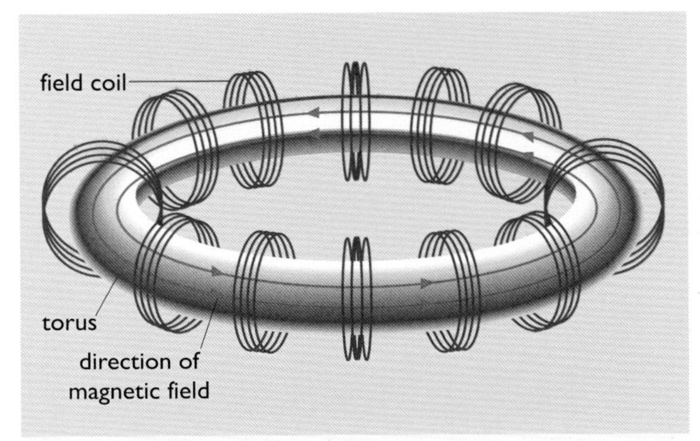

Figure 11.20 The Tokamak principle

Figure 11.21 The Joint European Torus (JET) at the Culham Science Centre in the UK

A more advanced Tokamak machine, the International Thermonuclear Experimental Reactor (ITER), is planned, at a cost now put at around 18 billion euros. It is being funded by the USA, Russia, the EU (including the UK), China, India, Japan and South Korea. It is under construction at Cadarache in the Provence area of France and is scheduled for completion by 2018. The expectation is that it will be able to produce pulses of 500 MW. If that is achieved, a larger more advanced plant is envisaged as the next step, possibly, if all goes well, leading on to a commercial system by 2050.

An alternative approach has been a micro version of the method used in the hydrogen bomb. In the bomb, the D–T core is essentially surrounded by an independent fission (i.e. uranium or plutonium) bomb. When this explodes, it compresses the core, achieving the conditions necessary for fusion. The aim of the **inertial confinement method** is to achieve this implosion in a more controllable way. The National Ignition Facility at the Lawrence Livermore National Laboratory in California has been a leader in this approach, using an intense blast of radiation from an array of powerful lasers to explode the surface coating of a sphere containing the D–T fuel. As with the Tokamak systems, inertial confinement has yet to produce more energy than it consumes.

Even if controlled fusion were eventually achieved in these devices, there are still many other problems to be addressed:

- Could the containment systems survive the intense neutron bombardment produced by a continuous fusion reaction for years on end?
- What are the waste implications of replacing radioactive system components?
- How will energy in the neutron flux be collected and used to produce useful electricity?
- Where would a long-term supply of tritium come from? Although, as described in Chapter 10, it can be produced from lithium, this element may be in considerable demand for the production of batteries for electric cars.
- What would happen if there was a leak of radioactive tritium into the local water supply?

The UK Atomic Energy Authority has suggested that fusion might supply up to about 20% of total global electricity by 2100 (UKAEA, 2007). That is about the level of contribution from renewables, including hydro, at present.

11.10 A nuclear renaissance? – conflicting views

Is nuclear power a viable option? In the 1990s after Chernobyl and Three Mile Island, the answer in the USA and Europe seemed to be 'no'. However, in recent years there has been a 'nuclear renaissance'. France and Finland have started new nuclear projects, based on the European pressurized water reactor (EPR) (see Box 11.3).

BOX 11.3 The European pressurized water reactor (EPR™)

This is a Generation III+ 'evolutionary' version of the conventional pressurized water reactor, developed by a French/German consortium including the French company AREVA. Units are currently under construction at Olkiluoto in Finland (Figure 11.22), Flamanville in France and Taishan in China. It is rated at 1.65 GW and will have an estimated electrical efficiency of 36–37% (AREVA, 2009).

Figure 11.22 An EPR reactor pressure vessel arriving at Olkiluoto, June 2010

It has been designed to operate with uranium enriched to up to 5% U-235, reprocessed uranium or MOX fuel (in variable proportions and up to 100%) with a service life of 60 years.

It has four parallel cooling loops to four separate steam generators. The secondary containment is a double-walled concrete shell and has been designed to withstand external incidents such as plane crashes, as well as a possible internal core meltdown. It also has a large low-level tank of borated water to provide emergency cooling.

In the UK, the mid-1990s view that nuclear power was uneconomic compared to gas and coal-fuelled electricity has changed with considerations of giving credit for its low CO_2 emissions (as shown in Table 11.3). In 2008, the UK government announced a major programme for the replacement of nuclear power stations being closed, in order to cut national CO_2 emissions and in the interests of energy supply diversity (BERR, 2008). At the time of writing (2011) eight possible sites for new plants have been identified.

Around the world there are currently 66 reactors under construction in 14 different countries. Many others have expressed interest in nuclear power programmes.

Is there enough uranium to go round? In 2006 the German Energy Watch Group expressed doubts about future supplies (see Figure 11.6). Certainly any expansion of conventional nuclear power will require an expansion of global uranium mining to cope (and appropriate investment). The nuclear study by MIT (MIT, 2003) and another by the UK's Sustainable Development Commission (SDC, 2006) did not see major problems in uranium supply. Besides, it is estimated that the UK's existing stocks of plutonium, depleted uranium tails and recovered uranium from reprocessing could fuel three 1 GW PWRs for 60 years.

To reprocess or not to reprocess? Only a minority of EU countries (principally France) favour reprocessing. In its 2008 Nuclear White Paper the UK government decided that the spent fuel from the new reactor programme currently planned in the UK would *not* be reprocessed (BERR, 2008). This is despite the existence of a ready-built reprocessing plant at Sellafield. The MIT study also advised the US government against reprocessing, pointing out that it 'presents unwarranted proliferation risks'. The UK's SDC also commented on the increased risk of nuclear proliferation that any global expansion of nuclear power would bring.

The MIT study advised the US government to pursue a policy of large-scale construction of new conventional reactors using an open, once-through fuel cycle. On the other hand the SDC advised the UK government that 'there is no justification for bringing forward plans for a new nuclear power programme, at this time, and that any such proposal would be incompatible with the Government's own Sustainable Development Strategy' (SDC, 2006).

From a commercial viewpoint, any 'renaissance' in Europe is likely to be dependent on whether or nor the new French and Finnish EPR reactors can be completed without excessive delays or budget overruns. At the time of writing (June 2011), that looks uncertain. There are also the commercial consequences of Fukushima to consider. Three reactors have been damaged beyond repair and the Japanese government has announced that the remaining three on the site will also be closed. The share price of the operating company, the Tokyo Electric Power Company (TEPCO), fell by 80% between early March and June 2011.

Elsewhere in Asia the situation is different, and it could be that a new expanding programme of large-scale construction will unfold in China and South Korea.

In the longer term it is possible that the focus for nuclear power could shift from electricity generation projects to plants designed mainly for producing heat for industrial processes, or possibly for the production of high-added-value transport fuels like hydrogen or liquid synfuel.

The changing energy market may also lead to a switch from large gigawatt-scaled plants to smaller units, which can be deployed more easily, rapidly and widely in countries without large-scale grid connections. There are already prototype systems being developed in the USA and Japan.

It is too early to say what the full effects of the Fukushima accident will be on public opinion of nuclear power and government policies. As mentioned earlier, Germany has reverted to its post-Chernobyl policy of a nuclear phase-out and a referendum in Italy has proved strongly anti-nuclear.

11.11 Summary

The debate about nuclear power is a complex one and there are strong and polarized views about it. Some of the issues are technical, others are political and of a form not shared by other energy technologies.

This chapter has described the following basic points:

- Nuclear power is a major global energy option. Currently uranium-fuelled burner reactors provide reliable electricity generation on a large scale without directly producing any of the CO_2, SO_2 and NO_x emissions associated with fossil-fuelled power plants.
- Considerations of national energy security and an effective response to climate change are key drivers for a 'nuclear revival'.
- Many of the currently operating nuclear power stations were ordered and built during the 'oil crisis' years of the 1970s. Since then the availability of cheap fossil fuels and a culture of 'free market competition' in electricity generation has not acted in favour of nuclear power. The capital costs of power station construction make up a major proportion of the cost of nuclear electricity. Reducing these construction costs, extending station working lives and improving thermal efficiency are major goals for new reactor designs.
- There are concerns about the future available supplies of uranium. Currently world uranium mining does not supply all of the world's consumption. The mismatch is made up from 'secondary' uranium sources, of which countries such as the UK and the USA have reasonable stocks, although these may be more inconvenient and expensive to use than freshly mined uranium.
- In the short term (i.e. over the next decade) any expansion of nuclear power using conventional reactors will require a major expansion in uranium mining. In the longer term new reserves of uranium will need to be discovered and brought to the market. While there may be potentially large resources of uranium available, if these are of a low ore grade they could be expensive and both energy and CO_2 intensive to mine.
- No energy technology is accident-free. However, the public perception of nuclear power has been adversely affected by major nuclear accidents such as those at Chernobyl and Fukushima. Increasing reactor safety is a major goal for new reactor designs.
- The long-term solution for dealing with nuclear wastes remains unresolved. This is particularly difficult because it involves timescales in excess of 100 000 years. Power station operators and new reactor designs are striving to reduce the *volume* of spent fuel by using 'enhanced burn-up'. This does not significantly reduce the *total radioactive content* of the waste. Deep geological storage of wastes may provide a solution but it has yet to be put into practice.
- There are serious political concerns about the possible diversion of nuclear materials for weapons production. These focus on the production of enriched uranium, necessary for current and future 'high-burn-up' reactor designs, and the use of a 'closed' fuel cycle involving the reprocessing of spent fuel to extract plutonium for recycling. Reprocessing could potentially extend uranium supplies and is a proven technology.

However, it has many disadvantages and its widespread commercial future use is not clear.

- There are further *technical* options for extending uranium supplies based on the breeding of plutonium from the large available stocks of U-238 using fast breeder reactors (FBRs). However, the use of these is critically dependent on the *political* acceptability of reprocessing. Although prototype FBRs have been built by several countries, their commercial viability remains unproven.
- The breeding of fissile U-233 from thorium is another technical option, tested at the prototype level. World thorium reserves are estimated to be larger than those of uranium. However, the commercial viability is as yet unproven.
- In the longer term, fusion energy from hydrogen could become a significant power source, though a working prototype may still be decades away from operation.

References

AREVA (2009) 'The path of greatest certainty', AREVA, http://www.areva.com (accessed 29 October 2010).

BERR (2008) *Meeting the Energy Challenge: A White Paper on Nuclear Power*. Department of Business, Enterprise and Regulatory Reform, London.

DECC (2008) 'Energy Markets Outlook Report', Department of Energy and Climate Change, London, The Stationery Office; available at http://www.decc.gov.uk (accessed 27 October 2010).

Edwards, R. (2008) 'Nuclear super-fuel too hot to handle', *New Scientist*, vol. 198, no. 2651, pp. 8–9.

EWG (2006) *Uranium Resources and Nuclear Energy*, EWG-Paper No. 1/06, Energy Watch Group; available at http://www.lbst.de/publications/studies__e/2006/EWG-paper_1-06_Uranium-Resources-Nuclear-Energy_03DEC2006.pdf (accessed 13 August 2010).

Government of Japan (2011) 'Report of the Japanese Government to the IAEA Ministerial Conference on Nuclear Safety: The Accident at TEPCO's Fukushima Nuclear Power Stations' [online], http://www.iaea.org/newscenter/focus/fukushima/japan-report (accessed 28 June 2011).

Harvey, L. D. D. (2010) *Carbon Free Energy Supply,* London, Earthscan.

IAEA (2006) *Chernobyl's Legacy: Health, Environmental and Socio-Economic Impacts and Recommendations to the Governments of Belarus, the Russian Federation and Ukraine (second revised edition),* Chernobyl Forum 2003–2005, Paris, International Atomic Energy Agency; available at http://www.iaea.org/Publications/Booklets/Chernobyl/chernobyl.pdf (accessed 5 January 2011).

IAEA (2011) 'Power Reactor Information System' [online], International Atomic Energy Agency, http://www.iaea.org/programmes/a2/index.html (accessed 23 August 2011).

Kreush, J., Neumann, W., Appel, D. and Diehl, P. (2006) *Nuclear fuel cycle*, Nuclear Issues Paper No.3., Berlin, Heinrich Böll Foundation, http://www.boell.de (accessed 18 June 2011).

Mark, C., Taylor, T., Eyster, E., Maraman, W. and Wechsler, J. (n.d.) 'Can Terrorists Build Nuclear Weapons? [online]', Washington, DC, Nuclear Control Institute, http://www.nci.org (accessed 30 June 2011).

Medvedev, Z. (1990) *The Legacy of Chernobyl*, Oxford, Basil Blackwell.

MIT (2003) *The Future of Nuclear Power: An Interdisciplinary MIT Study*, Boston, MA, Massachusetts Institute of Technology; available at http://mit.edu/nuclearpower (accessed 5 January 2011).

Mott MacDonald (2010) *UK Electricity Generation Costs Update*, Brighton, Mott MacDonald; available at http://www.decc.gov.uk (accessed 19 March 2011).

NEA (1995) *Chernobyl: Ten Years On – Radiological and Health Impact*, Nuclear Energy Agency, Paris, OECD.

NERAC/GIF (2002) *A Technology Roadmap for Generation IV Nuclear Energy Systems*, US DOE Nuclear Energy Research Advisory Committee and the Generation IV International Forum; available at http://gif.inel.gov/roadmap (accessed 27 October 2010).

OECD (2010) *Uranium 2009: Resources, Production and Demand (Red Book)*, OECD Nuclear Energy Agency and International Atomic Energy Agency, OECD Publishing.

PBMR (2010) Photo gallery: fuel composition, Pebble Bed Modular Reactor (Pty) Limited [online], http://www.pbmr.com/index.asp?Content=213 (accessed 6 January 2011).

Read, P. R. (1993) *Ablaze: the story of Chernobyl,* London, Secker and Warburg.

Sadnicki, M. J. (1994) Nuclear Review Background Paper, Sizewell B and Sizewell C, Hoskyns Group plc.

Sailor, E. C., Bodansky, D., Braun, C., Fetter, S. and van der Zwaan, B. (2000) 'A nuclear solution to climate change?', *Science*, vol. 288, pp. 1177–8.

Sourcewatch (2010) 'Comparative electrical generation costs' [online], Center for Media and Democracy, Madison, WI; available at http://www.sourcewatch.org/index.php?title=Comparative_electrical_generation_costs (accessed 4 November 2010).

SDC (2006) *The Role of Nuclear Power in a Low Carbon Economy*, SDC position paper, London, Sustainable Development Commission; available at http://www.sd-commission.org.uk/publications.php?id=344 (accessed 5 January 2011).

SKB (2010) 'Our final method of disposal [online], Stockholm, Swedish Nuclear Fuel And Waste Management Company, http://www.skb.se/Templates/Standard____24109.aspx (accessed 6 January 2011).

UNDP/UNICEF (2002) *The Human Consequences of the Chernobyl Nuclear Accident*, United Nations Development programme and UN Children's Fund.

UNSCEAR (1993) *Report to the General Assembly*, New York, United Nations Scientific Committee on the Effects of Atomic Radiation.

UNSCEAR (2000) *Report on the Effects of Atomic Radiation*, New York, United Nations Scientific Committee on the Effects of Atomic Radiation.

UKAEA (2007) *Fusion: A Clean Energy Future*, UK Atomic Energy Authority brochure JG07.246, Harwell.

WNA (2010a) 'Nuclear Share Figures 1999–2009' [online], London, World Nuclear Association, http://www.world-nuclear.org/info/nshare.html (accessed 5 January 2011).

WNA (2010b) 'Processing of Used Nuclear Fuel' [online], London, World Nuclear Association, http://www.world-nuclear.org/info/inf69.html (accessed 5 January 2011).

WNA (2010c) 'Uranium production figures 1999–2009' [online], London, World Nuclear Association, http://www.world-nuclear.org/info/uprod.html (accessed 5 January 2011).

WNN (2010) 'Chinese Candu reactor trials uranium reuse' [online], London, World Nuclear News, http://www.world-nuclear-news.org/ENF-Chinese_reactor_trials_Candu_fuel_reuse-2403101.html (accessed 5 January 2011).

Chapter 12

Costing energy

By Bob Everett

12.1 Introduction

As you progress through this book, you will realize that there is a wide range of possible ways of providing society's requirements for energy services, and also of conserving energy. Which methods are actually used in any particular place is a matter of local circumstances, the perceived cost of the energy supply, and of attitudes to investment in energy projects. These topics are discussed in this chapter. It must be said that some of these concepts are not easy to grasp and the mathematical equations may appear daunting at first sight. Try not to be put off by this as these are complex issues and you will not be alone if you find some of the ideas difficult to follow at first.

First, this chapter looks at recent energy prices in the UK and compares them with prices in a number of other countries. Next, it looks at how these prices have changed over the years in 'real' terms and in relation to earnings. This requires some explanation of the 'retail price index' and how inflation affects the perceived value of money when used as a unit of account. This leads on to the question of the affordability of energy and the topic of fuel poverty.

The following section then looks at the problems of investing money in projects both to *supply* energy and to *save* it. If the 'project' is something simple like a low-energy light bulb with a lifetime of a few years, then it may only be necessary to consider the time taken to get our invested money back – the 'payback time'. If, however, something more substantial is being developed, like a power station, then money will have to be borrowed from investors and financial institutions, which will want to charge interest. The money borrowed will have to be repaid over the working life of the station, which may be 30 years or more.

This kind of analysis involves understanding 'real' interest rates and 'discounted cash flow' calculations. As examples, the economics of a proposed nuclear power station and a rival combined-cycle gas turbine scheme are compared. This raises additional questions of how financial risk can be included in the calculations, and how high interest rates and short investment lifetimes may predispose investors to one technology rather than another.

The chapter then goes on to explore how different people and organizations may have different financial outlooks, and looks at the difficult question of what constitutes an equitable discount rate in order to make sure that the present generation is not getting energy benefits at the expense of future generations.

The final section looks at some of the real-world complications arising from the basic free market outlook on energy, including: the perceived need for countries to subsidize their own fuel industries in order to maintain security and diversity of supply; external costs due to pollution (dealt with in more detail in Chapter 13); and the problems of using subsidies to encourage new technologies.

12.2 Energy prices today

The previous chapters have described how different countries have different magnitudes and patterns of energy use. They also have different energy prices and taxation policies, and varying levels of subsidy between the different sectors of the economy. This section starts with the prices that usually provoke the most complaints from consumers – those of petrol and diesel. Box 12.1 provides some useful information on converting energy prices.

BOX 12.1 Converting energy prices – a summary

Different forms of energy are traditionally measured in different units.

Electricity has always been sold by the kilowatt-hour (kWh). Gas has usually been sold by volume in cubic metres or cubic feet, but is priced by its energy content, now normally expressed in kWh on gas bills in the UK. Typically one cubic metre of natural gas as normally supplied in the UK contains about 39 MJ of energy. (Remember that 3.6 MJ = 1 kWh.)

Petroleum is still sold by volume. Crude oil, which varies in energy content depending on its source, is sold by the barrel (159 litres), containing roughly 5.7 GJ of energy. Petrol and diesel fuel (DERV) for road vehicles, and domestic heating oil, are all sold by the litre which contains roughly 10 kWh of energy.

Coal is sold by the tonne. That for domestic heating is likely to contain about 28 GJ per tonne. Power station coal may have a lower energy content, typically 25 GJ per tonne in the UK.

Although most of the prices in this book are quoted in pounds and pence, oil is traditionally priced in US dollars. In early 2011 approximate exchange rates were £1.00 = US$1.60 = €1.15.

So, when reading this chapter remember that:

£1 per GJ equals precisely 0.36p kWh^{-1}

US$100 per barrel roughly equals US$17.50 per GJ, 40p per litre, £11 per GJ or 4p kWh^{-1}.

Petrol and diesel fuel

Crude oil is a globally traded commodity, so there is a 'world price'. This has been extremely volatile in recent years, the price of a barrel of oil having risen to over US$140 in July 2008, collapsed to US$35 by the end of 2008 and risen again to US$115 by the time of writing (March 2011). The actual cost of transporting a barrel of oil from the Persian Gulf is only about US$3, so at present Middle Eastern oil is potentially available across the world at much the same price.

As pointed out in Chapter 7, petrol requires careful refining and blending, and it too is a globally traded 'standard' commodity. Figure 12.1 shows petrol prices in a number of different countries in January 2011. The basic untaxed fuel price (where figures have been available) is similar, about 50p per litre. This is perhaps not surprising, given that it is all likely to have been refined from crude oil priced at about US$100 a barrel or 40p per litre.

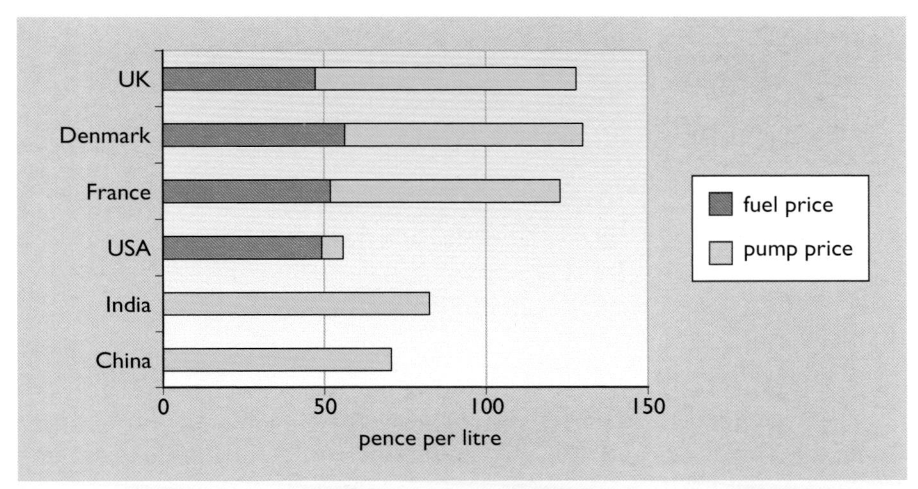

Figure 12.1 Comparative petrol prices, January 2011; note that untaxed fuel price data are not available for India and China (sources: DECC, 2011a; EIA, 2011a; EIA, 2011b; Hindustan Petrol, 2011; Independent, 2011)

However, what is different is the level of tax charged by the different governments. In January 2011 in the UK petrol cost about £1.28 per litre of which over 60% was tax. This petrol price was almost the highest in Europe, third only to Greece, the Netherlands and Denmark. Almost all European countries have high tax rates on petrol. In contrast, in the USA petrol sold for only 56p per litre, of which only 12% was tax.

The UK is unusual in having little difference between the price of petrol and road diesel fuel. In many other countries, diesel is cheaper than petrol. This is in part to encourage the adoption of more efficient diesel engines, and in part to avoid penalizing commercial transport excessively.

Domestic energy prices

In the UK domestic sector, heating oil is sold with a low rate of tax and its 2009 price was similar to those in France and the United States, about 50p per litre or 5p per kWh (see Figure 12.2). In Denmark there is a higher rate of tax to encourage the use of alternative district heating or biomass heating.

In the UK natural gas for the domestic market has a low rate of tax and is similar in price to that in France, around 4.5p per kWh. Again, prices are higher in Denmark, over 7.0p per kWh because of their taxation policy. This has caused considerable friction with the European Commission in Brussels since it conflicts with their policy of 'harmonization of energy prices'. The Danish government successfully argued that the taxes had been applied to protect the environment, something which was also a key policy

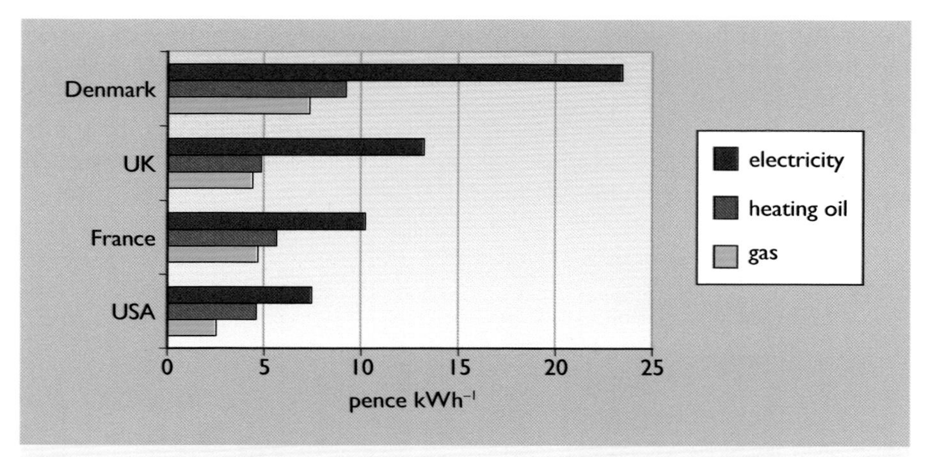

Figure 12.2 Comparative domestic fuel prices, 2009 (source: IEA, 2010a)

objective of the European Union. US natural gas prices are considerably lower than European ones.

Average UK domestic electricity prices in 2009 were about 13p per kWh, higher than in France and the USA but far lower than the highly taxed rate of over 23p per kWh in Denmark.

It is worth remembering that the two heating fuels, oil and gas, have to be burned in devices such as boilers with a limited efficiency, usually between 60% and 90%. This means that the *useful* energy price will be higher than the *delivered* energy price. For example, if gas at 4.5p per kWh is burned in a boiler with an efficiency of 60%, then the useful heat energy output will cost 4.5/0.60 = 7.5p per kWh. Even worse, burning coal at 4.0p per kWh in an open fire with an efficiency of 25% gives a useful energy price of 4.0/0.25 = 16p per kWh. It would be cheaper to use electricity at 13p per kWh and an electric fire with an end-use efficiency of virtually 100%.

It should also be noted that figures quoted for domestic electricity are only average prices. At night, off-peak electricity for heating purposes is sold at a considerably lower rate, currently about 5p per kWh in the UK (as compared to the average cost of 13p).

Therefore, it is perhaps not surprising that there is an increasing use of electricity for heating, particularly in France and the USA, where in the recent past heating fuels have usually been coal and oil. In France, these more traditional fuels are being displaced by nuclear-generated electricity. However, in the USA it is increasingly coal-fired electricity that is being used. The resulting CO_2 emissions pose considerable environmental problems.

The figures above are for mains electricity. Yet electricity is also regularly purchased in another form – batteries. A 'D-cell' battery for a torch or radio costs about £1 in the UK, yet it only holds about 6 watt-hours of electrical energy. This is equivalent to over £160 per kWh. The energy from small long-life batteries for watches and cameras can be 50 times more expensive than this, reaching a million pounds a gigajoule! For these applications it is not the *quantity* of energy that is important, but the *quality* – a reliable, portable supply available when needed. This, it seems, is an energy service that has an enormous value.

Industrial energy prices

Industrial energy prices are usually far lower than those for the domestic consumer. They also tend to be far closer to the world average prices for globally traded commodities such as oil and coal. Figure 12.3 shows the trends in UK industrial energy prices between 1998 and 2009. Note that these exclude value added tax (VAT) and the Climate Change Levy (a carbon tax), introduced in 2001, which currently adds 0.164p per kWh to gas and coal prices and 0.47p per kWh to electricity prices.

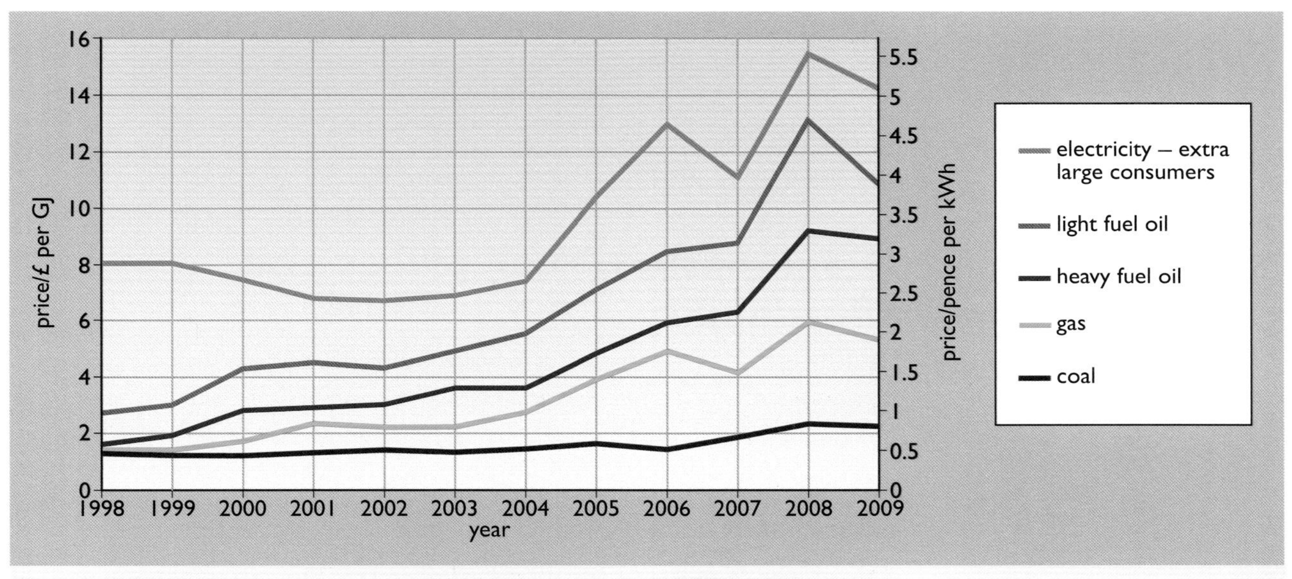

Figure 12.3 UK: recent trends in industrial energy prices (source: DECC, 2011a)

Perhaps the most remarkable feature of this chart is how the prices of coal, gas and heavy fuel oil, which were all just over £1 per GJ in 1998, have diverged. By 2009 UK coal prices had increased slightly to about £2 per GJ (0.72p per kWh). This price was largely determined by imports from countries such as Russia and Colombia. UK production only supplied about a third of UK coal use.

In 1998 gas and oil prices were low. North Sea gas was selling at about £1.40 per GJ (0.5p per kWh). Light fuel oil for office heating sold at just under £3 per GJ and heavy fuel oil (used in power stations) sold for £1.60 per GJ. Since then the prices of oil and gas have risen dramatically. By 2009, the price of gas had risen by almost a factor of four to over £5 per GJ and heavy fuel oil by a factor of more than five to nearly £9 per GJ. The price of light heating oil also quadrupled over the same period.

Somewhat strangely, despite rising gas prices, UK electricity prices to large consumers actually fell, from £8.10 per GJ (2.9p per kWh) in 1998 to £6.70 per GJ (2.4p per kWh) in 2002. This is likely to have been due to fierce competition in the electricity market which cut profits to the point where many suppliers, and notably the nuclear industry, were experiencing financial difficulties. Since then prices have risen to about £14 per GJ (5p per kWh) in 2009. The rising price differential between gas and coal has meant that gas-fuelled electricity has become less economic and there has been a continued use of coal-fired electricity generation.

In Europe and the USA, there is a basic assumption that all electricity should, within reason, have the same price to all consumers (though large industrial ones do get it more cheaply than small domestic consumers). This is not so in India, where there is a general policy of subsidizing electricity differently for different consumers. In 1999–2000 the average electricity price was about 3p per kWh. Domestic consumers only paid about 80% of this, but electricity for agriculture and irrigation was almost given away. Over 100 GWh of electricity was sold at an average price of 0.4p per kWh. This, of course, required a massive subsidy cash flow of over £5 billion per year. Some of this money was recovered by charging far higher prices of the industrial and commercial sector, and particularly the railways (IEA, 2002). There is a major problem of theft of electricity and non-payment of bills. As a result, in 1999–2000 the total revenue from the sale of electricity only covered 74% of the cost of generating it.

12.3 Inflation, 'real' prices and affordability

Section 12.2 has expressed energy prices in terms of pounds and US dollars. These different currencies are freely exchangeable for each other (and euros) and can be used to purchase energy, or other goods or services. Here money is being used for its principal purpose, as a medium of exchange, without which we would have to resort to barter. Later in this chapter, Section 12.4 looks at a second use of money, as a *store of value over time.*

A third use of money is as a *unit of account.* This allows the 'value' of things that might otherwise seem difficult to describe to be discussed, for example, costs of pollution or even the loss of a human life, topics that will be looked at in Chapter 13.

Currently (March 2011), in London in the UK, domestic electricity costs 13.8p per kWh. Twenty-five years ago it cost 5.7p per kWh. Obviously the price of electricity has risen, but what does it mean in 'real' terms? This requires some thought about the variation of the value of money over time.

The value of money

Since around 1920, the pound sterling, like almost all world currencies, has suffered from inflation. This is best thought of as a 'disease of money' which progressively decreases its purchasing power in real terms from year to year. Money is normally thought of as the benchmark against which to assess the value of goods and services. But in reality it is the other way round: money is the medium through which things are exchanged and their value discussed.

In order to use money as a unit of account in different years, the crucial question is how much of it is required in order to buy the same goods and services in different years. In the UK, as in most other industrial economies, government statisticians regularly assess prices and compile statistical analyses of their year-by-year changes. The **retail price index** (RPI) is an economic indicator that specifically addresses the prices of a representative mixture of typical household goods (a hypothetical 'basket of goods' as shown in Figure 12.4) at a point in time. (The UK has recently changed to a *consumer price index* that specifically excludes any interest payments such as on mortgages, but this doesn't really affect the basic process described here.)

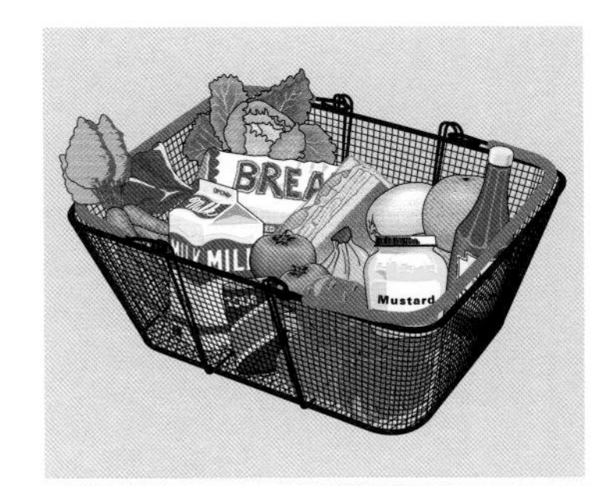

Figure 12.4 A sample batch of everyday household items. It is these (and a few others) that enable the 'real' value of money to be measured

Indices such as these are usually expressed as being equal to 100 in a specific base year. Figure 12.5(a) shows a plot of the UK retail price index from 1950 to 2010 with 2010 as the base year. From the figure it can be seen that goods that would have cost £100 to purchase in that year could have been purchased for £30 in 1980 and only £3.75 (3 pounds and 15 shillings in 'old' money, pre-decimalization) in 1950.

The chart is shown plotted with a logarithmic *y*-axis, so an equal percentage growth over time should show as a straight line. It is obvious that the line appears to be steeper in the 1970s and 80s.

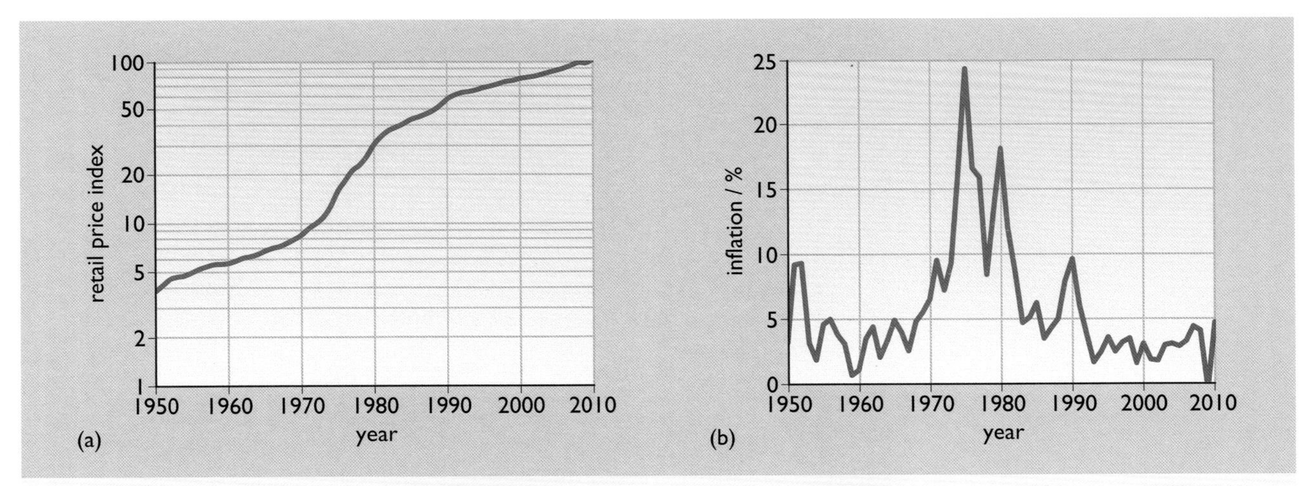

Figure 12.5 UK: economic indicators: (a) changes in retail price index (year 2010 = 100); (b) annual inflation rate 1950–2010 (source: ONS, 2011a)

Inflation is usually described as being *the rate of change of the retail price index per year.* Mathematically speaking, this is the slope of the RPI–time chart, and is shown in Figure 12.5(b). The period of high inflation between 1973 and 1984 can be clearly seen.

There has been a 27-fold increase in prices in the UK since 1950, but for most people this has not involved social hardship because average UK wages have risen even faster – in money terms by a factor of nearly 80 over the same period. This increase in the ratio of earning power over prices has been matched in virtually all industrialized nations. Looking back further in history, there have been times, such as in the economic depression following World War I, when inflation levels have fluctuated widely, and have even been negative, with both prices and wages falling.

Although price cuts are always popular, wage cuts are certainly not, and can provoke social unrest.

If inflation gets out of control, as it did in Germany in the early 1920s (see Box 12.2) it can become extremely difficult to know how to carry out economic analysis at all. Low inflation is always a prime economic political objective for governments.

BOX 12.2 Hyperinflation

In 1922 and 1923 inflation in Germany reached staggering proportions, peaking at 2500% per month. At the beginning of 1922 the mark was worth US$2.38. Eventually, in November 1923, the exchange rate was stabilized at 4 200 000 000 000 marks to the US dollar. The money was so worthless that it required thick bundles of banknotes to buy even basic groceries (see Figure 12.6). Yet life went on and workers were paid in similar thick bundles of notes. The 'real' value of the goods did not change, just the value of the money used as a medium of exchange. Traders and banks used the US dollar ('real money') as a unit of account to keep track of what was going on. The mark became useless as a 'store of value' and many people's life savings were wiped out.

Figure 12.6 This 1922 German shopkeeper had to keep his money in a tea chest because there wasn't room for it in the cash register

Thus, to be strictly accurate when quoting prices, the 'year of valuation' when it is being used should also be stated. For example, in 1971 on average a dozen eggs cost 22 (new) pence – that is £0.22 in 1971 pounds or £(1971). Between 1971 and 2000 the RPI rose by a factor of 8.4, so it can be said that those eggs, valued at 22p in 1971, were worth $8.4 \times$ £0.22 = £1.85 in pounds of the year 2000 or £(2000). In fact, in the year 2000 a dozen eggs actually cost on average £1.72, so in 'real' terms eggs had become slightly cheaper in relation to the other items in the statisticians' reference batch of commodities (bread, potatoes, etc.).

An egg is an egg, whether the year is 1971 or 2000, but its value is being measured in different money.

At the beginning of Chapter 1 of this book, Figure 1.1 showed a chart of world oil production and prices, where these are expressed in US$(2009). The actual annual average dollar prices of the day have been converted into year 2009 values by 'adjusting them for inflation' using a dollar retail price index. Converting the value of the oil into money of a particular standard year shows historically exactly how serious was the sevenfold increase in real price between 1973 and 1979, and how, by 1998, prices had almost fallen back to their 'real' pre-1972 levels.

To take another example, Figure 12.7 shows the variation in 'real' UK petrol and diesel prices between 1990 and 2009, in terms of both 'pump prices' and the basic fuel cost excluding taxes and duty.

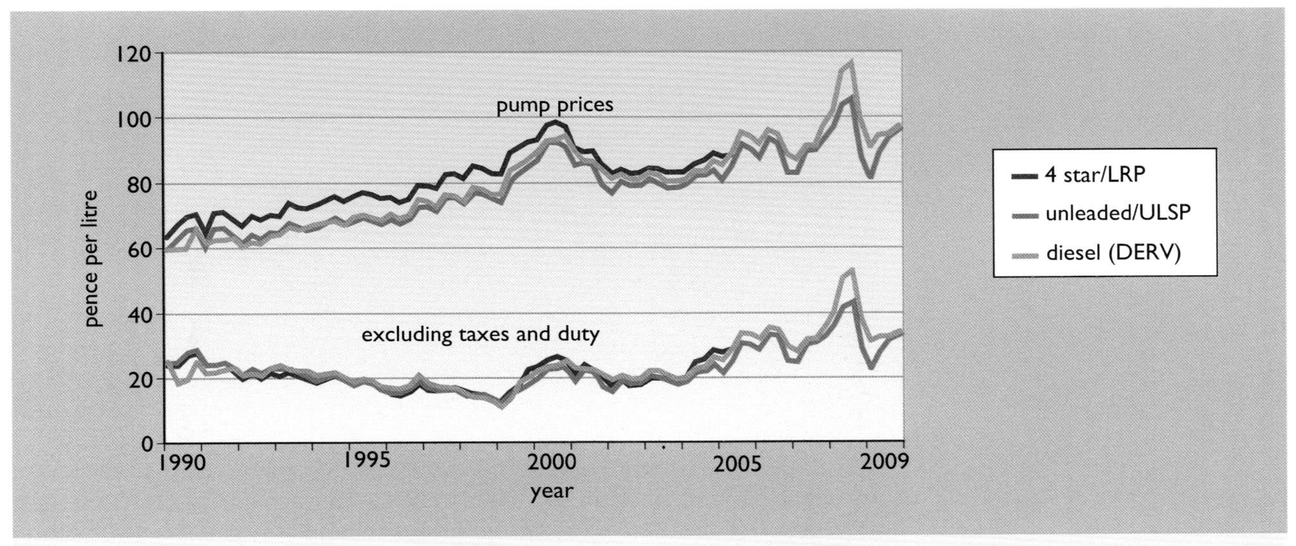

Figure 12.7 UK: 'real' petrol prices, 1990–2009, i.e. adjusted for inflation, expressed in £(2005). LRP = lead replacement petrol; ULSP = ultra-low sulfur petrol. (source: DECC, 2010)

As will be appreciated from looking at Figure 1.1, 1990 was a year when world oil prices had fallen from their peak in 1979 but still had further to fall. Between 1980 and 1982 real petrol prices had been about £(2005)1.00 per litre but had fallen to just over 60p per litre in 1990. Given increasing environmental concern about pollution, in 1993 the UK government introduced a progressively increasing fuel tax whose effect can be seen in the chart. The progressive increase was eventually halted after protests in 2000, but by then pump prices were almost the same in real terms as they had been in the early 1980s.

The large tax component in UK road fuels (and those in other EU countries) has the effect of masking the price volatility of oil. Between 1999 and 2009 the real untaxed price of petrol varied by a factor of four, yet the pump price only varied by 40%. If the world price goes up, then the consumer response is usually to ask the government to cut the fuel tax.

In a similar fashion, it is possible to look back further in time at how domestic energy prices in Great Britain have changed over the whole course of the 20th century. The values, expressed in Figure 12.8 in £(2010) are for Great Britain, rather than for the whole UK, because natural gas was not available in Northern Ireland until about 1990. Note that, unlike vehicle fuel, these domestic fuels have always carried a low rate of tax.

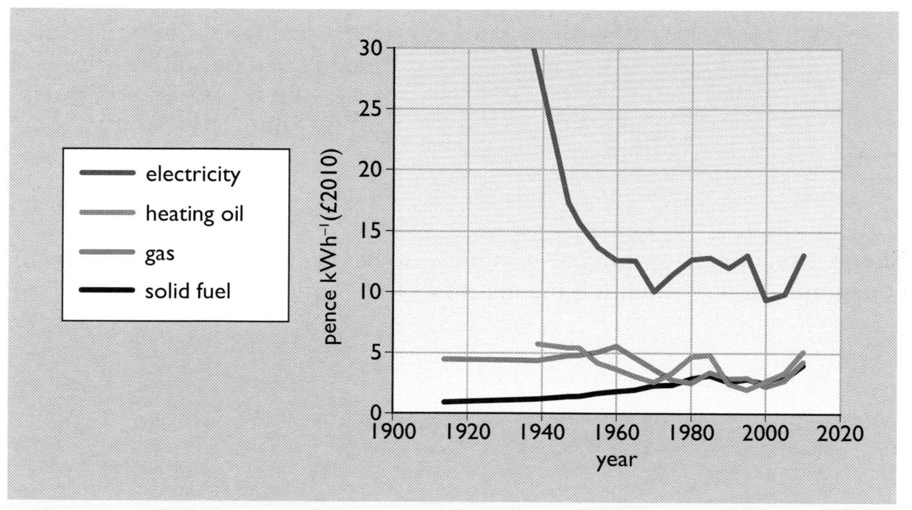

Figure 12.8 GB: real domestic fuel prices, 1914–2010, expressed in 2010 pence per kWh (source: Evans and Herring, 1989; DECC, 2011b)

The figure shows quite clearly why solid fuel was the preferred option in Great Britain over the first half of the 20th century – seen in today's money, it was cheap! Town gas, the convenience fuel made from coal by a quite labour-intensive process, sold for about five times the price of solid fuel, and electricity was a very expensive luxury fuel. In 1885 the Brighton Electric Light Company was selling electricity for 1 shilling (5p) per kWh. This is equivalent to £4.60 per kWh in 2010 money! Fortunately for their customers, the company had cut its prices by a factor of 12 by 1898. Even so, electricity prices only creep onto the scale at the top of Figure 12.8 in the 1930s. Since then, continuous improvements in technology have brought the price of electricity down, falling to below 15p per kWh in 2010 equivalent terms in the 1950s. The fall between 1995 and 2000 was partly a consequence of the extension of electricity privatization to the domestic sector with a continuous price war between suppliers, and partly due to the increased use of cheap North Sea gas for electricity generation.

Oil for heating was not a major fuel in Britain before World War II, so its price only enters the chart in 1939. Its price fell during the 1950s and by the 1960s it had become cheaper than town gas. Sales of oil-fired central heating expanded. The electricity industry also started marketing electric central heating using 'off-peak' storage heaters and cheap night-time electricity. However, as described in Chapter 7, in the 1970s the situation changed dramatically. Oil prices rose, although the availability of North Sea oil to some extent cushioned the UK consumer against the enormous jump in world prices in the 1970s. In addition, new supplies of North Sea gas were being brought into Britain, selling at half the price of town gas and competing heavily with coal. The scene was set for a massive boom in sales of gas central heating and a continuing decline in the use of oil and coal.

Affordability and fuel poverty

Although the RPI allows the value of goods from one year to another to be compared, it says little about their affordability. As described above, the monetary price of eggs rose by a factor of eight between 1971 and 2000. However, average wages increased by a factor of almost 15 over the same period.

In 1971 the average UK gross weekly wage was £28.70. This sum would have bought $(28.70 \div 0.22) \times 12 = 1565$ eggs. By 2000, the average weekly wage had risen to £420, which would have bought $(420 \div 1.72) \times 12 = 2930$ eggs, almost twice as many. Put another way, eggs were twice as affordable to an 'average person' in 2000 as they were in 1971.

Energy use, like everything else, is likely to be determined by the ability of consumers to buy it, which means it has to be judged in terms of earning power. Just as government statisticians compile a retail price index, they also compile a similar one of *average weekly earnings*. This can be used to show the change in relative affordability of energy. Figure 12.9 shows energy prices for Great Britain adjusted by the average earning power in each year and expressed in 2010 pence per kWh.

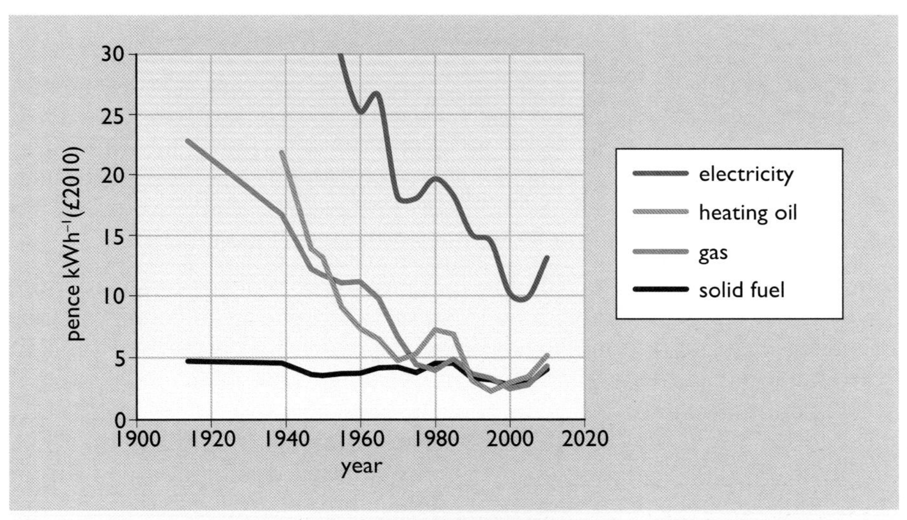

Figure 12.9 GB: energy prices relative to earnings, 1914–2010 (sources: updated from Evans and Herring, 1989; DECC, 2011b; ONS, 2011b)

The declining curves for heating oil, gas and electricity show how they have become more affordable over the last century. A week's average wages in 2010 would have bought more than twice as many kilowatt-hours of electricity as a week's wages in 1955. However, the recent price rises since 2000 suggest that the downward trend may not be totally irreversible. The line for coal shows an almost constant price between 1914 and about 1985 (when large-scale imports started). This may reflect the labour costs involved. The coal price largely followed the earnings of the UK miners, which in turn kept pace with general earnings.

How much did families actually spend on energy? The percentage of UK household expenditure on fuel and power peaked in the 1960s at about

6% (see Figure 12.10). By 2000 it had fallen to only 3.3% but it has risen since then. However, it should be noted that this figure *does not* include expenditure on transport and fuel for the motor car!

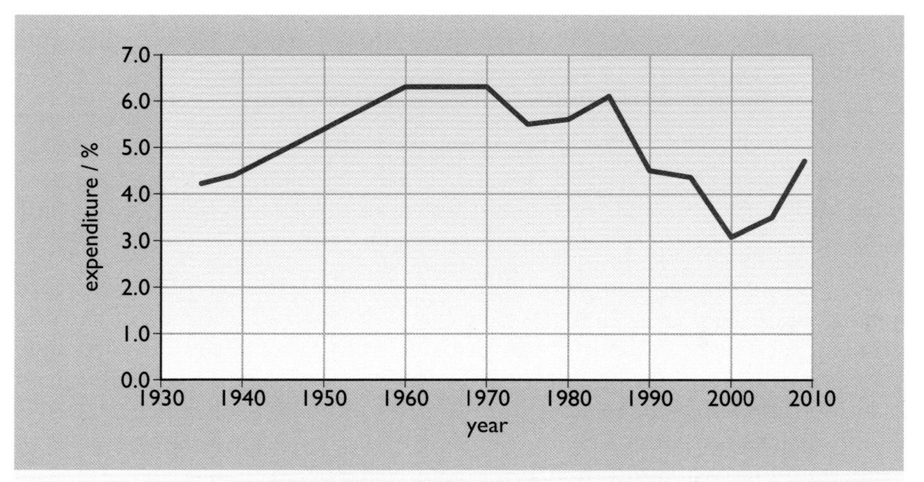

Figure 12.10 UK: percentage of household expenditure on fuel and power, 1935–2009 (sources: Stone, 1966; DECC, 2010; ONS, 2010)

It is all very well to say that energy has become more affordable to the 'average worker', but what about the poorer sections of the community? The UK *Family Expenditure Survey* divides the population according to income into tenths or 'deciles'. In 2000, for the poorest tenth of the population, 'fuel and power' accounted for 6.5% of the household expenditure; electricity accounted for 3.6% of the total expenditure. Food accounted for a further 21%.

A household is defined as being in **fuel poverty** if it needs to spend more than 10% of its income on energy in order to achieve a satisfactory degree of comfort. This is defined as having a minimum temperature of 21 °C in the living room and 18 °C elsewhere in the house. It has been estimated that in 2003 approximately 2 million households in the UK (i.e. about 1 in 12) suffered from fuel poverty, down from an estimated 6.5 million in 1996. However, by 2007 rising fuel prices had increased this figure to 4 million households. About a third of these comprised households of a single person aged over 60 (DECC, 2010).

Fuel poverty creates a difficult energy pricing problem. It is easy to say that energy, and especially electricity, is too cheap and that higher prices would encourage conservation, but raising prices with environmental taxes is likely to hit the poor hardest. The most satisfactory solution is to improve the insulation standards and heating systems of the housing stock, so that the 'satisfactory degree of comfort' can be achieved with an affordable amount of energy.

If fuel poverty is a problem in the UK, it is even worse in India. The country has a rapidly rising population and an acknowledged serious energy shortage. Wood-fuel for cooking is undoubtedly being gathered on an unsustainable basis in many areas and electricity shortages cause frequent blackouts. According to a 1993/4 survey (NSSO, 1996) the annual average expenditure in India on fuel and lighting amounted to 7.4% of household income in rural areas and 6.6% in urban areas. At first sight these figures appear similar to those for the UK, but it must be remembered that these are proportions of a far lower income. In Chapter 3 it was pointed out how expensive and inefficient fuel-based lighting is compared to modern electric lighting with fluorescent lamps – yet according to another 1993/4 survey, 62% of the Indian rural population and 16% of the urban population still used kerosene for lighting at that time. The Indian government has an overall policy of extending the electricity grid to as many villages as possible, and statistics suggest that 81% of villages (though not individual homes) had some sort of connection by 2007 (TERI, 2009).

The question is sometimes asked 'how can the poor afford electric light?', and the answer is 'with great difficulty' – but more cost-effective light may be obtained by spending money on electricity rather than on kerosene.

12.4 Investing in energy

This section describes the *textbook theory* of investment in energy, where energy is seen as a simple, tradeable commodity in a perfect investment marketplace. However, there are many complications with this view, some of which will be looked at in the next section.

Price and cost

This chapter has so far described the price of energy today and over the past century. The next question is, how much would it cost individuals or organizations to produce their own energy or to invest in energy saving?

At first sight the two words 'price' and 'cost' would seem to mean the same thing.

- A **price**, though, is something *determined in the marketplace*. Conventional economic theory says that if there is a shortage of a particular freely traded commodity then the market price will rise so that supply matches demand.
- A **cost**, at least as used in this chapter, assumes that an energy producer (a private individual, a company, or perhaps a government department) can make some *surplus or profit* out of a particular investment. Normally a cost figure would include such things as materials consumed, labour and also repayments on invested capital at rates that would be competitive with other projects. The cost is therefore *a measure of the minimum amount for which the producer must sell energy in order for its production to make a profit.* Put simply, for the purposes of this chapter: price = cost + profit.

If something is being sold at a price equal to the cost of producing it, then there is no profit. However, 'profit' is a word which should be used carefully, not least because profits tend to be taxable.

The distinction between cost and price is demonstrated clearly in the case of crude oil from the wells of Persian Gulf states. Here the cost consists of:

- a large historic capital investment in exploration and drilling until an oil field has actually been discovered, followed by other investments in setting up the wells, pipelines and port facilities to ship it out.
- The *marginal cost* operational cost of manpower and energy to actually pump out the oil and load it onto tankers.

At the time of writing (2011) the world crude oil *price* is over US$100 per barrel. This is the amount of money that consuming countries are prepared to pay for the limited supply of oil available. Yet the marginal cost of pumping another barrel from the existing wells in the Gulf States is likely to be less than US$2. The remaining US$98 is available to pay off the capital cost of the original drilling and exploration (much of which took place decades ago, and has probably been paid off completely by now), for profit, and for investment in finding the next oil well. As pointed out in Chapter 7, this is becoming increasingly difficult and expensive. Ultimately, the world's finite resources of oil can only be sold once, and the revenue will have to be sufficient to fund more sustainable alternatives for the producing nations.

Balancing investment against cash flow

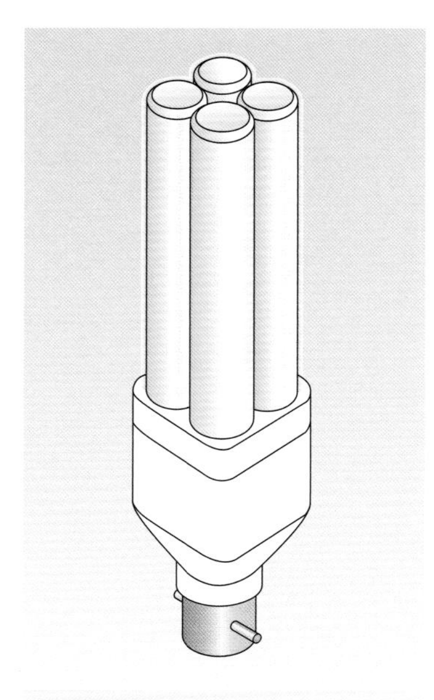

Figure 12.11 A compact fluorescent lamp

The Middle East is fortunate to still have large energy reserves. Elsewhere, practical energy economics is largely a matter of carefully balancing a capital investment in a project against the flow of money from that project over a long period of time. The financial return on investment can be expressed in a number of ways, of which the simplest is the payback time. Exactly the same financial analysis can be applied whether it is a matter of investing money to *produce* energy or to *save* energy.

A simple example is the investment involved in using a compact fluorescent lamp (CFL), like the one shown in Figure 12.11, rather than two possible alternatives: a conventional incandescent lamp, or a slightly more efficient halogen one as shown in Figure 9.19.

Currently (2011) in the UK a good-quality, 20-watt, electronically ballasted CFL costs about £10 in the shops (although they are also widely available for far less, typically £1, under various subsidy schemes). A conventional 100-watt incandescent bulb of equivalent light output costs only 50p (or rather *did* cost 50p, since their sale is being phased out in the UK and many other countries). A third alternative is a 70-watt halogen lamp with the same light output costing £3. Clearly the incandescent lamp is the cheapest, so why 'waste' money on buying the CFL or a halogen lamp?

The answer is that CFLs use much less electricity and last much longer. There is thus a balance to be struck between an investment in capital equipment (the CFL) and the energy savings in kilowatt-hours spread over the life of the lamp. Depending on the level of use, the extra capital expenditure can be recovered in a year or less. Given this short time frame, inflation and interest on money can be ignored. The calculation is given in Box 12.3.

A CFL is a low-cost investment and its practical life is relatively short. Other energy investments require a longer time-perspective, but the same considerations will apply whether it is a matter of investment in a power station, a wind turbine, or cavity wall insulation in a house.

A combined-cycle gas turbine (CCGT) power station is an example. It requires a large capital investment at the beginning of the project, a continuing supply of gas fuel, a certain expenditure on operational staff and occasional expenditure on maintenance. It can be built within two

BOX 12.3 Calculating the payback time on using a compact fluorescent lamp

An electronically ballasted CFL typically uses only 20% of the electricity of a 100-watt incandescent bulb of equivalent light output or only 30% of that of a 70-watt halogen replacement. In addition, the CFL will last 8000 hours against only 1000 hours for an incandescent bulb or 2000 hours for a halogen lamp.

In order to calculate the payback time it is necessary to look at the expenditure over the entire life of a CFL. Currently UK electricity costs, on average, about 13p per kWh.

20-watt compact fluorescent lamp

In its 8000-hour life a 20-watt CFL will consume:

$$8000\,\text{h} \times 20\,\text{W} = 160\,000 \text{ watt-hours} = 160\,\text{kWh of electricity.}$$

At 13p per kWh, this will cost: 160 × 13p or £20.80

So, including the £10 cost of the lamp, the total cost over 8000 hours = **£30.80.**

70-watt halogen lamp

In 8000 hours a 70-watt halogen lamp will consume:

$$8000\,\text{h} \times 70\,\text{W} = 560\,000 \text{ watt-hours} = 560\,\text{kWh of electricity.}$$

At 13p per kWh, this will cost: 560 × 13p or £72.80

Each halogen lamp only last 2000 hours, so over the period of 8000 hours, 4 lamps will be needed. So, including the £12 cost of the four lamps, the total cost over 8000 hours = **£84.80.**

100-watt incandescent lamp

In 8000 hours a 100-watt incandescent lamp will consume:

$$8000\,\text{h} \times 100\,\text{W} = 800\,000 \text{ watt-hours} = 800\,\text{kWh of electricity.}$$

At 13p per kWh this will cost: 800 × 13p or £104.00

Each incandescent lamp lasts 1000 hours, so over 8000 hours 8 lamps at 50p each will be needed. This adds another £4.00, bringing the total to **£108.00.**

Overall profit

So the overall profit in using a compact fluorescent lamp compared to:

(a) the halogen lamp is £84.80 – £30.80 = **£54.00**, and

(b) the incandescent lamp is £108.00 – £30.80 = **£77.20.**

Both of these sums are many times the initial cost of the CFL.

Payback time

The figures above look good, but how long will it take to get the investment back?

A simple way to work this out is to plot a comparative expenditure–time chart for the three options (see Figure 12.12).

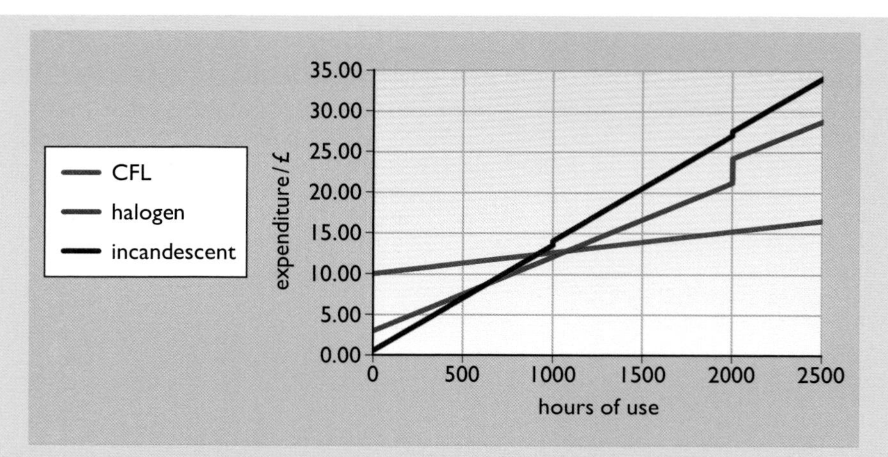

Figure 12.12 Expenditure–time chart for CFL, halogen and incandescent lamps

The 20-watt CFL initially costs £10 and will consume 20 kWh in every 1000 hours of use. This amount of electricity will cost 20 × 13p or £2.60. So after 1000 hours of use, the total expenditure will have been £10 + £2.60 = £12.60 and after 2000 hours it will have been £10 + £2.60 + £2.60 = £15.20. This is plotted in Figure 12.12 as the red line.

The 70-watt halogen lamp initially costs £3. In each 1000 hours it will consume 70 kWh of electricity, costing 70 × 13p or £9.10p. After 1000 hours of use, the total expenditure will have been £12.10 and after 2000 hours £21.20. This is plotted in Figure 12.12 as the blue line.

The 100-watt incandescent lamp costs 50p and in each 1000 hours it will consume 100 kWh of electricity, worth £13.00. After the first 1000 hours the total expenditure will have been £13.50. It will then be time to buy a new lamp. In the next 1000 hours it will consume a further £13.00 in electricity, so after 2000 hours the total expenditure will have been £26.50. This is plotted in Figure 12.12 as the black line.

When the three plotted options are compared, it can be seen that after about 900 hours of continuous use, the costs of using the CFL have become less than those of the incandescent lamp. After about 1200 hours they are lower than using the alternative halogen lamp. These numbers are equivalent to 2.5 and 3.3 hours use per day respectively for a year.

years and once constructed it should be able to produce electricity for 20 years or more.

These costs need to be turned into a **levelised cost** for the electricity from such a plant. How might this be done? As a first attempt, the capital cost of the plant can be '**annuitized**', i.e. spread out over a number of years by dividing it by the lifetime of the plant. The annual fuel, operation and maintenance costs can then be added to give the total annual cost. Finally, this value can be divided by the annual electricity output to give the cost in pence per kWh. Wear and tear on this kind of equipment is often proportional to the number of hours run, so operation and maintenance (O & M) costs are often expressed as a certain sum per kWh of electricity produced. Box 12.4 shows a sample calculation.

This approach seems reasonable enough if the money for capital investment (£320 million) happens to be available. However, the capital expenditure

BOX 12.4 A combined-cycle gas turbine power station

A proposed gas-fuelled 400-MW combined-cycle gas turbine (CCGT) has a capital cost of £800 per kW. It is expected to run for 7500 hours per year at full power and has a design lifetime of 20 years. Its overall electrical generation efficiency is 50%. Operating and maintenance (O & M) costs have been estimated at 0.6p per kWh of electricity generated. Gas is available at £5.56 per GJ or 2.0p per kWh of gas over the lifetime of the plant. (The approximate sample capital and O & M costs have been taken from Mott MacDonald, 2010.)

Calculate the cost of the electricity in pence per kWh ignoring any capital interest charges.

Solution

If the station unit runs for 7500 hours per year, in each year it will produce:

$$7500\,\text{h} \times 400\,\text{MW} = 3\,000\,000\,\text{MWh of electricity}$$

Total capital cost is: $£800 \times 400\,000 = £320$ million

Capital cost spread over 20 years is: $\dfrac{£320\,000\,000}{20} = £16$ million per year

Annuitized capital cost per kWh of electricity produced is:

$$\frac{£16\text{ million}}{3\,000\,000\,000\text{ kWh}} = 0.53\text{p kWh}^{-1}$$

Gas cost per kWh of electricity generated is: $\dfrac{2.00\text{p}}{50\%} = 4.00\text{p kWh}^{-1}$

O & M costs $= 0.60\text{p kWh}^{-1}$

Total cost per kWh $= \mathbf{5.13p\ kWh^{-1}}$

Note that the cost of the fuel makes up over three-quarters of the final electricity cost.

has to be made now, whereas the benefits in electricity only arise in the future (and some of it will be in 20 years' time, which is a long time to wait). What happens if the money has to be borrowed from a bank or raised from shareholders? They might be prepared to supply £320 million but will want to receive a steady stream of interest payments for the privilege. This makes the calculation more complex, and this example will be revisited and expanded upon in the next subsection.

Discounted cash flow analysis

Interest rates and discount rates

In practice, a pound tomorrow is not worth the same as a pound today. An important concept is the **time value of money**; there are several separate factors involved and it is important to understand them. They are:

(1) The effect of **inflation**, already discussed, which progressively erodes the capacity of money for purchasing real goods.

(2) The normal **time preference for money**. Given a choice, most people would generally rather have a pound today than a pound in the future. Put another way, we would need to be offered a pound plus some additional sum, say $x\%$, next year to forego the use of one pound today. This preference will exist even when the inflation rate is zero.

(3) The **opportunity cost** of a potential investment. Money can be lent out and interest charged. This means that we can forgo the use of a pound today in order to have a pound plus an additional sum in the future. There are always likely to be a number of different opportunities for investing money. If we spend money now rather than invest it, or use it in a poor investment rather than a better one, then an economist would say that we have incurred a cost. This is known as the opportunity cost.

Different individuals' time preferences for money will vary according to their financial circumstances, but in general it is these that determine interest rates as quoted by banks. Someone opening a savings account at a bank is giving up the use of a pound today, but expects to be able to withdraw somewhat more than a pound at some time in the future. The theory assumes that they will only open the account if the rate of interest offered is sufficiently good to overcome the attractions of spending the money now.

As an example, £100 is invested at a 10% rate of interest per annum. In this case, the investor would expect to be able to withdraw £110 in one year's time. Suppose that this £110 is needed to settle some bill that is expected to arise at that time. By investing only £100 now a larger bill can be paid in the future. If this bill is due quite a long time in the future then the effect of the investment can be very marked. After ten years' investment at 10% interest, £100 would have a **future value** of £260. Put another way, the **present value** of £260 in ten years' time at an interest rate of 10% is only £100.

In this case, future payments are being *discounted*, effectively saying that sums of money in the future can be expressed in terms of smaller sums today. This concept leads to a technique of economic appraisal known as **discounted cash flow (DCF)** analysis. This allows a series of bills at various times in the future to be expressed as a single lump sum in the present. For example, there may be three impending bills: £100 today, £110 in one year's time and £260 in ten years' time. As shown above, the separate 'present values' of each of these is £100; they can be added together to have a **net present value (NPV)** of £300. This is the total amount of money available today to settle all three bills, given that some of the money can be invested at an interest rate of 10%.

If the interest rate were lower, say only 5%, then the net present value would be different. A net present value calculation is a convenient tool for dealing with payments over long periods of time.

In practice the terms 'discount rate' and 'interest rate' tend to be used interchangeably. They are not necessarily quite the same thing, however. The discount rate is literally the rate at which the value of future income or expenditure is 'dis-counted'. It may be similar to a bank's quoted interest rate, but with an extra allowance to cover project risks. This will be discussed later in this section.

We can now go back and look at the power station example from Box 12.4 through the eyes of a potential investor – would it be better to invest in this plant or in something else? The value of investing in this project must be competitive with the **opportunity cost**, the value of the next best opportunity foregone in financing a given investment. If money can be safely invested to produce a 5% return, in say, chemicals or electronics, or even just a bank savings account, then why bother with gas-fired power stations unless these can produce better returns?

In practice, for a new investment, the interest rate is likely to be something 'given' by a bank, and this value in turn is likely to have been politically and economically influenced by the government treasury.

The effects of inflation must also be allowed for. It is very important to realize that discounting and inflation are completely separate things. Inflation describes how prices rise with time, eroding the 'real' value of the very money that is the basis of the calculation. Discounting, in contrast, describes the fact that because of uncertainty about the future, people usually prefer to have money today rather than the promise of money tomorrow; and that if we do have money now it can always be invested in order to make even more money later.

In order to adjust for the effects of inflation a 'real' interest rate needs to be calculated, rather than the purely monetary one. This is simple enough:

real interest rate = monetary interest rate – rate of inflation

It is worth noting that since both 'monetary' interest rates and inflation rates vary with time, so can the 'real' interest rate, which is the difference between the two.

Historically, bank interest rates in the UK (and other countries) have varied widely from year to year. Figure 12.13 shows how the official Bank of England rate has decreased from a high value of over 15% in 1980 to only 0.5% in 2010. Actual large commercial loans are likely to be made at rates higher than this. The chart also shows the 'real' interest rate, net of inflation. Although during the 1970s real interest rates were actually negative, since about 1982 the rate has been positive but has been declining since the early 1990s.

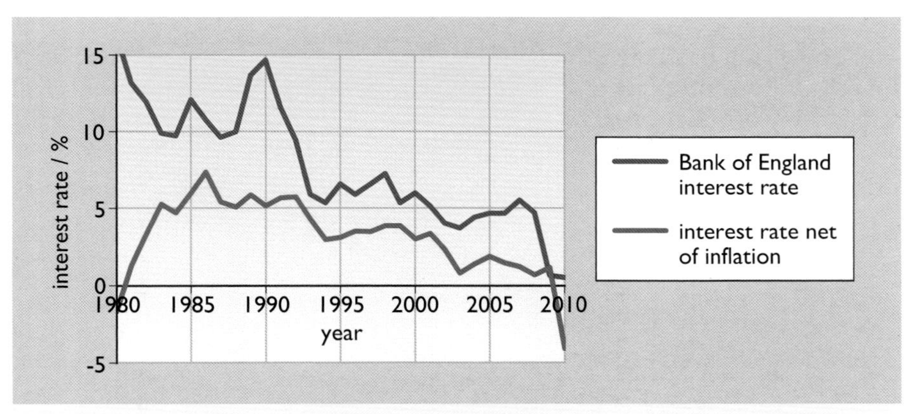

Figure 12.13 UK: Bank of England interest rate and real interest rate net of inflation, 1980–2010 (sources: Bank of England, 2011; ONS, 2011a)

Project lifetime

Generally if we borrow money, we agree to pay it back over a certain number of years. For example, mortgage payments for a house might be spread over 15 or 25 years. Financial institutions are unwilling to consider loans spread over longer periods than this because of the essential uncertainty about the future. Yet the working life of a nuclear power station might be 40 years or more and that of a hydroelectric station in excess of a hundred. Although this does not often create problems of analysis for most energy projects, it is worth remembering that the capital repayment time (to pay off a loan) may be shorter than the physical working lifetime of the scheme.

Basic discounting formulae

Although the actual choice of interest rates and repayment times for discounting is fraught with complications, its mathematical application is relatively straightforward. Although the formulae below may look daunting, they are standard functions in computer spreadsheet packages.

Present value

The present value of a future sum of money is an important quantity. Its general formula is fairly easy to explain.

Given a discount rate of 10%, then a sum of £100 today (i.e. its present value) is equivalent to a future value of:

£100 × (1 + 0.1) in one year's time

£100 × (1 + 0.1) × (1 + 0.1) in two years' time, or

£100 × $(1 + 0.1)^n$ in n years' time

In general for a discount rate r, the value of a sum V_p in the present is equivalent to a sum V_n in n years' time, where:

$$V_p = \frac{V_n}{(1+r)^n}$$

Net present value

The above is just the present value of one sum of money at some time n years in the future. If a whole sequence of sums of money spread over a long time are included, then the total present value of all of these is the net present value (NPV).

If these sums of money, spread over n years, are denoted as V_1, V_2, V_3,..., V_n, then:

$$\text{NPV} = \frac{V_1}{(1+r)} + \frac{V_2}{(1+r)^2} + \frac{V_3}{(1+r)^3} + \ldots + \frac{V_n}{(1+r)^n}$$

Or

$$\text{NPV} = \sum_{i=1}^{n} \frac{V_i}{(1+r)^i}$$

This has become a standard function in computer spreadsheets.

Annuitization

If all of the repayments on a loan are equal, of value A, then that is an **annuity** continuing for a fixed number of years to repay an initial capital sum and all its interest. The present value of this annuity is:

$$V_p = A \times \left(\frac{1}{(1+r)} + \frac{1}{(1+r)^2} + \frac{1}{(1+r)^3} + \ldots + \frac{1}{(1+r)^{n-1}} + \frac{1}{(1+r)^n} \right)$$

Inverting this (after a little bit of mathematical manipulation) A can be given as the annuitized value of a present capital payment V_P:

$$A = V_p \times \frac{r}{1-(1+r)^{-n}}$$

Putting it more simply, if an amount V_P is borrowed from the bank now, a sum A must be paid back every year for n years to pay off the loan.

Values of A as a function of r and n can be tabulated as shown in Table 12.1. This is sometimes called a **capital recovery factor**. Although nowadays project costings (Box 12.5) are worked out on computer spreadsheets, in the past banks and finance houses used simple books of pre-calculated tables such as these, and they are still useful to give quick answers.

BOX 12.5 Annuitization: a simple method for costing

The concept of capital expenditure having an annuitized value depending on the discount rate and the project's financial lifetime is extremely useful. Table 12.1 shows the annuitized values for every £1000 of capital for various different real discount rates and project lifetimes. Using the table, it is easy to calculate the annuitized cost of a given capital sum over a given period. To this must be added the annual running cost (fuel, maintenance, etc.). When this is done, the levelized cost of energy is given by the simple formula:

$$\text{cost of energy} = \frac{\text{annuitized capital cost} + \text{average running costs}}{\text{annual average energy produced}}$$

This is sufficiently accurate for many purposes, though not in the case where annual costs are highly variable, when it is necessary to use a full NPV calculation.

Table 12.1 Annuitized value of capital costs (annual cost in £ per £1000 of capital) for various discount rates and capital repayment periods

Capital repayment period/years[1]	Discount rate/%						
	0	2	5	8	10	12	15
5	200	212	231	250	264	277	298
10	100	111	130	149	163	177	199
15	67	78	96	117	131	147	171
20	50	61	80	102	117	134	160
25	40	51	71	94	110	127	155
30	33	45	65	89	106	124	152
35	29	40	61	86	104	122	151
40	25	37	58	84	102	121	151
45	22	34	56	83	101	121	150
50	20	32	55	82	101	120	150
55	18	30	54	81	101	120	150
60	17	29	53	81	100	120	150

[1] This is not necessarily equal to the total physical lifetime of the project.

The example of the power station from Box 12.4 can now be revisited, this time taking into account discount rates to show that the cost of 5.13p per kWh calculated previously is now an underestimate – see Box 12.6. It also considers a simple wind turbine project.

BOX 12.6 Simple calculations using discounting

CCGT

As an example of discounting the cost per kWh of electricity from the power station considered in Box 12.4 can be recalculated, assuming a discount rate of 10% per year and using the figures in Table 12.1. Note that this involves ignoring the (short) construction time.

First the annuitized value of the capital cost is needed. From Table 12.1 the annuitized value of £1000 over 20 years at 10% per year = £117 per year.

Therefore, for a borrowed sum of £320 million the annual repayments will be:

$$320\,000 \times £117 = £37.44 \text{ million}$$

This is far higher than the annual repayment figure of £16 million used in Box 12.4. (This figure can be derived from Table 12.1 using the annuitized value of £1000 over 20 years at 0% per year = £50 per year, i.e. 320 000 × £50 = £16 million)

Annuitized capital cost per kWh of electricity produced is:

$$\frac{£37.44 \text{ million}}{3\,000\,000\,000 \text{ kWh}} = 1.25\text{p kWh}^{-1}$$

Gas cost per kWh of electricity generated is:

$$\frac{2.00\text{p}}{50\%} = 4.00\text{p kWh}^{-1}$$

O & M costs = 0.60p kWh^{-1}

Total cost per kWh = **5.85p kWh^{-1}**

This is somewhat higher than the earlier figure of 5.13p kWh^{-1}.

Onshore wind turbine

The same procedure can be used to calculate the cost of electricity from a proposed 1 MW (1000 kW) wind turbine. This has a capital cost of £1500 per kW and a life expectancy of 25 years. It operates with a capacity factor of 30% (a factor which is determined by the wind speed at the site) and its operation and maintenance costs are 1.5 pence per kWh. Its fuel costs are, of course, zero.

Repeating the calculation above:

Total capital cost is: 1000 × 1500 = £1.5 million

From Table 12.1 the annuitized value of £1000 over 25 years at 10% per year = £110 per year.

Annual capital repayments is: 1500 × £110 = £165 000

Annual electricity generation is: 1000 × 30% × 365 × 24 = 2.63 million kWh

Annuitized capital cost per kWh of electricity generated is:

$$\frac{£165\,000}{2\,630\,000\ \text{kWh}} = 6.27\text{p kWh}^{-1}$$

O & M costs = 1.50p kWh^{-1}

Total cost per kWh **= 7.77p kWh^{-1}**

This procedure can be used for other capital-intensive renewable energy technologies involving a large initial investment and a short construction time.

A discounted cash flow calculation in detail

A full calculation for a major project will require considerable detail. In Boxes 12.4 and 12.6, two major cost categories were used: the capital cost (those occurring at the start of a project), and running costs (those recurring throughout the project lifetime) consisting of O & M and fuel costs.

In practice, a scheme is more accurately described by a list of costs or a 'cost breakdown'. The two major cost categories still apply – capital costs and running costs. However, there may be other costs over the whole project lifetime. A hydroelectric dam will have a large initial capital cost. It may last for over a century but is likely to have new turbines fitted every 30 years or so. This adds extra complexity to the calculation. There is also an extra category, decommissioning and disposal costs, which occur at the end of a project. In the CCGT and wind turbine examples above these have been considered minimal; however, in the case of a nuclear power station they involve a large financial expenditure spread over a very long time.

Box 12.7 describes a detailed discounted cash flow calculation for a large (2.6 GW) nuclear power station, Sizewell C, as proposed in 1994. This involved cash flows spanning a period of over 100 years. The box compares the costings with those its main rival technology, the gas-fired CCGT, and looks at critical effect that the choice of discount rate has on the results.

In 1995 the UK government made a decision, on the basis of this comparison, to back CCGTs rather than nuclear power. Many CCGT plants have since been built with a good track record of rapid construction and proven performance. However, the economic argument may have now swung the other way. Natural gas prices have increased dramatically since the 1990s, as shown earlier in Figure 12.3. The decline in UK North Sea gas production described in Chapter 7 also raises questions about the long term availability of gas. Also increasing concern about climate change has meant that a 'carbon price' (to be discussed in Chapter 13) is now likely to be added to the CCGT costs.

Uncertainties remain over the practical construction costs and timescales of new large nuclear power stations, such as the Finnish EPR described in Chapter 11. If a major accident occurs (such as at Fukushima in Japan) reactors may have to be written off before they have achieved their full working life. It may also incur large damage liabilities for the operating company.

BOX 12.7 Case study: Sizewell C – a proposed nuclear power station

In 1994, the UK government conducted a review into the future of nuclear power (DTI, 1995). As part of this, the privatized UK company Nuclear Electric proposed the construction of a giant 2.6-GW twin pressurized water reactor, based upon the existing 1.2-GW Sizewell B reactor and to be known as **Sizewell C.** The costs below are based on the company's submission to the review (Nuclear Electric, 1994) and a detailed critique (Sadnicki, 1994) prepared for a client in the coal industry. It was estimated that the power station would take six years to build, followed by a 40-year operational life, after which the station would have to be decommissioned. Decommissioning would be done in a number of stages: Stage 1 would involve removing the fuel; Stage 2 would then involve the dismantling of all the non-radioactive parts. The plant would then be left for 40 years to allow the high-level radioactive material inside the reactor to decay away, before Stage 3, the final dismantling stage, more than a century after the work started on construction. The costs involved are summarized in Table 12.2. They are expressed in £(1993) where £1(1993) ≈ £1.60(2010).

Table 12.2 Sizewell C nuclear power station cost breakdown

Item		Period/years	Amount
Construction costs		7	£3450 million total
Running costs	Fuel supply	40	0.3p kWh^{-1}
	Fuel disposal	40	0.1p kWh^{-1}
	Operation and maintenance (O & M)	40	0.7p kWh^{-1}
	Total running costs	40	1.1p kWh^{-1}
Decommissioning and disposal costs	Stage 1 – de-fuelling	5	£50 million
	Stage 2 – removal of outside shield	5	£160 million
	Surveillance during 'cooling down' period	40	£1 million per year
	Stage 3 - back to green-field site	10	£190 million
	Total decommissioning and disposal costs		£440 million

Sources: Nuclear electric, 1994; Sadnicki, 1994.

There is obviously an enormous expenditure involved. The total construction, running and decommissioning costs amount to over £12 billion spread over a period of more than a century. Balancing this is a large output of electricity. Nuclear Electric's proposal assumed that the station would operate at a load factor of 85% (i.e. it would run at full power for 85% of each year), producing approximately 19 TWh of electricity per year. These flows of cash and electricity are shown in Figure 12.14.

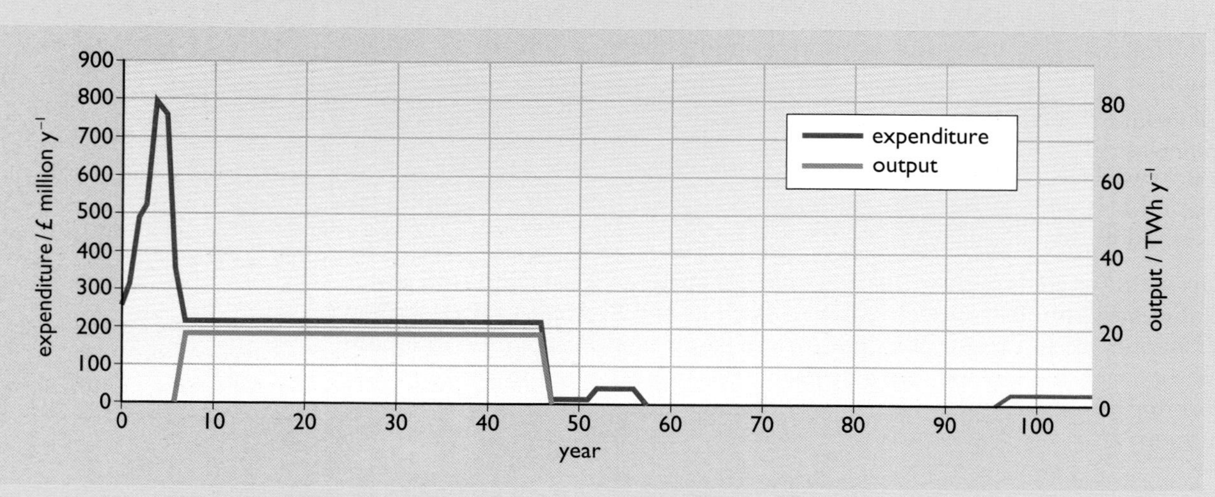

Figure 12.14 Sizewell C: assumed expenditure £(1993) and electrical output (source: Sadnicki, 1994)

The aim of the calculation is to produce a single levelized cost of the electricity per kWh. The first step is to calculate the net present value of every year's financial expenditure. This requires making an assumption about the discount rate, or rather discount rates, involved.

Addressing the costs involved in decommissioning in particular has been a difficult problem. These costs amount to a total sum of £440 million to be spent over a period of 60 years, starting over 40 years into the future. Nuclear Electric argued that this should be dealt with by setting up a pot of money of only £270 million on the last day of operation of the station in its 46th year. This could be put into a 'safe investment' which would then earn sufficient interest to pay off the total £440 million. Such a long-term investment could only be expected to earn 2% interest.

The concept of decommissioning costs over such a timescale raises difficult questions about (a) the reliability of the use of money as a 'store of value' and (b) 'intergenerational equity', i.e. issues of whether or not economic benefits enjoyed by this generation are being taken at the expense of future generations. These topics are discussed later in this section.

Assume for the moment that decommissioning does indeed cost £270 million. This and all the remaining expenditure of construction and operation in the calculation are discounted back to year zero, the time when work first started, using the 'commercial' interest rate. The value of this was a subject of considerable debate at the time, with suggestions in the range 8–12%. Taking a figure of 10%, a single figure for the net present value of all the costs can be calculated as £3.36 billion. Note that this is far less than the undiscounted figure of £12 billion.

The next step is to calculate the net present value (in kWh) of every year's electricity generation, again discounted back to year zero. This amounts to 96.3 billion kWh. It may seem odd to discount electricity generation in kWh, but this has an equivalent 'real' monetary value that can only be assigned when the calculation is complete.

The cost of electricity in pence per kWh can then be calculated as:

$$\frac{\text{NPV of capital costs, fuel, O \& M and decommissioning (£)} \times 100}{\text{NPV of electricity (kWh)}}$$

$$= \frac{\text{£3 360 000 000} \times 100}{\text{96 300 000 000 kWh}} = 3.49\text{p per kWh}$$

We can also break the answer down into its different components to show how the initial capital cost dominates the final electricity cost as shown in Table 12.3 and in Figure 11.14.

Table 12.3 Breakdown of electricity cost[1]

Component	Cost/p kWh^{-1}	Percentage
Capital charges	2.38	68%
Fuel costs	0.40	11%
Operation and maintenance	0.70	20%
Decommissioning	0.01	<1%
Total	3.49	100%

[1] In 1993 prices.

Having arrived at this answer it is worth noting that it can be expressed in various other ways:

- *Net Present Value of the whole project*: Using the discount rate of 10% and this electricity price, then the NPV of all the costs is equal to the NPV of all the benefits (i.e. the electricity).

  The NPV of the project as a whole = NPV (benefits) – NPV (costs) = 0.

  If the electricity could be sold at more than 3.49p per kWh, or alternatively the costs could be reduced, then this project NPV would become greater than zero and there would be some room for profit. Generally where this kind of analysis is carried out, projects with overall NPVs greater than zero are worth doing and the larger the NPV the better.
- Percentage **Internal Rate of Return (IRR%)**: This is the value that the discount rate must take in order to make the overall project NPV equal to zero, i.e. in this case 10%. It is worth borrowing money at this rate to carry out the project. However, if the cost of borrowing were higher than this, then the project would become uneconomic.
- **Benefit–Cost Ratio (B/C)**: This is the ratio of the net present value of all the benefits to the net present value of all the costs. At a discount rate of 10% and an electricity price of 3.49p per kWh this equals 1. Generally projects with a B/C ratio greater than 1 are worthwhile considering, and the higher the B/C, the better.

The debate surrounding the 1994 UK government review on nuclear power concerned the relative commercial merits of nuclear electricity and those of its chief rival, the combined-cycle gas turbine. The key financial factors are given in Table 12.4.

Table 12.4 Financial characteristics of different electricity generation technologies

Technology	Construction time	Capital costs £(1993)	Operation and fuel costs	Operational lifetime
Sizewell C	7 years	£1340 kW^{-1}	1.1p kWh^{-1}	40 years
CCGT	2 years	£450 kW^{-1}	1.7p kWh^{-1}	20 years

Source: Sadnicki, 1994; DTI, 1995.

The nuclear power station detailed in Table 12.4 has high capital costs and a long construction time, but relatively low running costs over a long operating lifetime. The competing CCGT could be built quickly and cheaply, has minimal decommissioning and dismantling costs, but has higher running costs over a shorter life. Which is best?

Setting aside the many objections as to whether Nuclear Electric's cost and performance estimates were correct, the decision rests largely on attitudes to finance.

Ignoring discounted cash flow (effectively having a discount rate of 0%) means the investor is prepared to wait patiently for the returns on a particular energy investment. The higher the discount rate the less the investor is prepared to wait. Anyone wanting quick returns may be unwilling to invest in projects that require a long construction time.

The electricity cost calculation above for Sizewell C can be repeated with different discount rates and the results compared with similar calculations for a CCGT station. The results are plotted against discount rate in Figure 12.15.

Because the cost of nuclear electricity is dominated by the large capital costs of the initial construction, it is very dependent on the discount rate, whereas the cost of electricity from CCGT generation is dominated by the price of the gas. In this example, if a discount rate of 0% (i.e. a long return time) is acceptable, then nuclear power is cheaper than gas. If a discount rate of 10% is chosen then the position is reversed. So the choice as to which technology is most economic is largely dependent on the cost of borrowing money.

The 1994 government review also prompted debate on whether or not the financial markets could actually put up almost £3.5 billion for a nuclear power station at all, as this would require a considerable degree of faith in the success of one project. A further problem was whether or not private investors would be prepared to lend money over the full 40-year life of the station. If they insisted on the loans being paid off in 15 or 20 years, then the cost of the electricity would have to go up to 5p per kWh to meet the costs.

A CCGT project, on the other hand, was considered to be a far more attractive prospect for investors. An entire 400-MW power station could be bought for a modest £180 million and its assumed operating life of 20 years was more in line with normal expectations of loan repayments. The conclusion of the UK government's nuclear review was that the Sizewell C station was the best option that the nuclear industry could offer in the short term, but that it could not compete with new gas-fired stations. Sizewell C itself was not built, but since then, many new CCGT stations have been constructed.

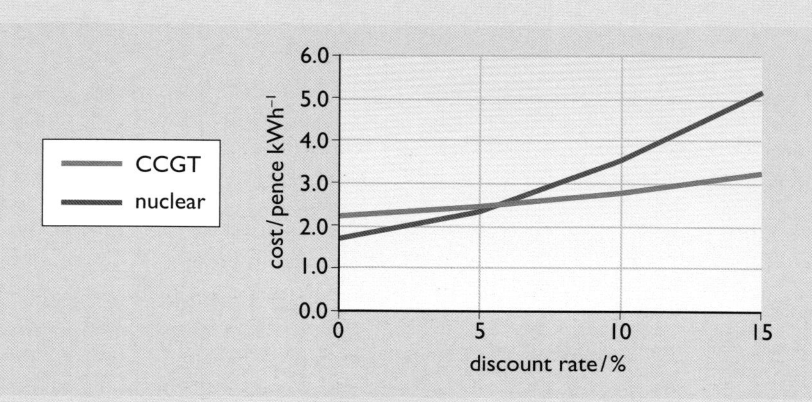

Figure 12.15 Comparison of electricity costs with discount rate for CCGT and nuclear plants

Dealing with normal financial risk

The calculations above have assumed that there is a single 'commercial' interest rate for the project, plus, of course, a very low interest rate for the 'safe storage' of the decommissioning money. In practice, financial institutions have to safeguard their investors, so they only like to lend money on low-risk investments. If there is any possibility that a project may fail or not produce the promised returns, then there has to be an extra allowance to cover this risk. Often the borrowing may be split into two parts. A bank may supply a part of the money at a normal interest rate. This is called **debt**. The remainder is raised by selling shares to shareholders, with a prospect of likely good returns, but no guarantees. This is called **equity**. The 'debt' always gets paid out first and the shareholders carry the project risks. Depending on the level of risk, shareholders may want potential returns of 20% or more.

Some types of energy project always carry some risk, but others may be seen as sufficiently risk-free to be paid for by debt alone. The financial risk on such projects can be estimated by sensitivity analyses. If the cost per kilowatt-hour of energy generation can be reduced to a simple discounted cash flow equation, the most likely values can be put in to give a 'best estimate' of the energy costs. It can then be analysed to establish which factors have most influence. For example, the cost of electricity from a CCGT project could be expressed in terms of its capital cost, discount rate, gas fuel costs, operating and maintenance costs and the load factor. Each particular value can be varied around a central estimate (marked 100% on the *x*-axis) to give a 'spider diagram' such as shown in Figure 12.16.

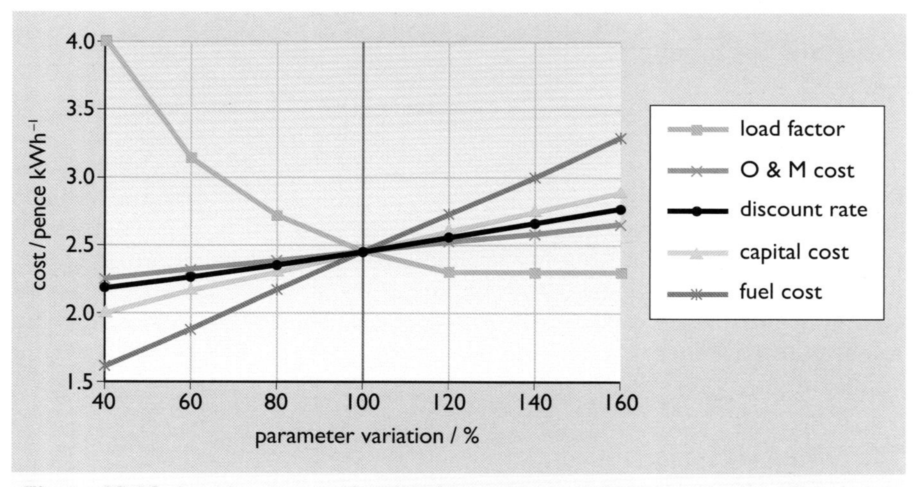

Figure 12.16 A spider diagram showing the sensitivity of electricity cost to different parameters for the CCGT plant described in Box 12.7. Costs are in £ (1993).

This diagram shows that two parameters, the fuel costs and the load factor, are more important than the others. Variations in these factors make a bigger difference to the overall electricity cost than variations in others. In a competitive electricity market a margin of 0.5p per kWh or less may make the difference between commercial success and failure. A rise in gas prices of about 35% would force up the electricity cost by this amount, as would a reduction of about 30% in the running hours. The worst possible case scenario might be if a rise in gas prices were to make the station unable to compete with others in the marketplace, automatically reducing the load factor. A wise management would thus make sure that it had a firm contract for long-term fuel supply at a good price and a ready market for all its electricity.

In the extreme case, a rise in fuel prices could lead to the mothballing of the plant, a fate which has befallen the giant Grain oil-fired power station on the Thames estuary, built in 1979 (Figure 12.17).

Figure 12.17 An unwanted investment – the 1.4 GW Grain oil-fired power station on the Thames estuary built in 1979 and here photographed in 2001, has been little used for want of cheap oil to burn. A gas-fuelled power station has now been built next to it.

What are acceptable discount rates and investment lifetimes?

Conventional textbook economics assumes that all investors have free access to information and free access to credit. Anything less than this is seen as a 'distortion of the market'. Discount rates are based on opportunity cost and bank interest rates, as discussed. However, in practice different individuals and organizations have different discount rates and time perspectives. It is worth looking at the time preference for money more carefully. Often, discount rates are high and time perspectives short. This can lead to an economic mismatch between the judgements made by consumers who might conserve energy, and large utilities and corporations interested in supplying it.

The private individual

It must be said that for domestic consumers energy investments are not a very exciting prospect. Often the attractions of spending money *now* will outweigh the attractions of savings in the future. Attractions for expenditure like food or holidays compete with investments such as compact fluorescent lamps and cavity wall insulation. Even though these two technologies have short payback times, they have required much government promotion in the UK in recent years. Conversely, sales of double glazing, with a much poorer economic justification, have been surprisingly high. This would appear to be because double glazing is extremely 'visible' and adds to the perceived value of a house in a way that 'invisible' cavity wall insulation does not.

Many private individuals, especially those on low incomes and in rented accommodation, have high perceived discount rates and short investment time horizons. At the extreme, those living on state benefits or low pensions may have no access to credit. Given the choice of a pound today or a pound tomorrow, they will always say they need it now. Their discount rate is infinitely large. For those living in rented accommodation, investment is also limited by the length of their lease or the probable time when they will move on. Yet it is these very people who are likely to suffer most from fuel poverty and need energy savings.

That said, even the poorest individuals may save money to provide an income in old age (some people would continue to do this even if their savings earned zero interest), or to provide for their children's education. The UK state pension system does not actually have any accumulated cash reserve; today's pensioners are paid for by today's workers. However, private pensions are different. An individual pays into a fund over their working life, building up a 'store of value' to be repaid in their old age. The contents of private pension funds make up an enormous body of money, potentially available for investment in energy projects.

Commercial companies

Many commercial companies may find investment in energy financially difficult. In the extreme they may treat energy costs as 'running costs' to be paid out of a particular annual budget, or the accounting system may simply not have an 'energy investment' budget. For this reason companies may only consider energy projects with a payback time of less than one year. They may also be in rented offices, with no incentive to invest in capital equipment that lasts longer than their lease.

If they are in a very competitive market, companies may also feel that they need to plough as much as money as possible into their latest product. Anything less than this could lead to commercial failure. This can lead to very high perceived discount rates and short financial time horizons.

Public sector organizations

In its 2003 *Green Book*, the UK Treasury suggests that public sector organizations should make a real return of 6% on investments, though this point is qualified with a long discussion on the proper evaluation of risk and 'optimism bias' (HM Treasury, 2003). This figure has changed slightly over the years, with the 2003 figure being lower than the 8% quoted at the time of the 1994 government nuclear review, but higher than an earlier figure of 5% used in the 1970s. Public sector bodies are also encouraged to enter into public–private partnerships (PPP), however, in which case higher commercial interest rates will apply.

Intergenerational equity

Very-long-term investments, such as those involved in paying for the decommissioning of nuclear power stations, may span one or more generations. This raises the question of whether or not today's generation is paying its full share of the costs. This concept of fairness across generations is known as **intergenerational equity**.

It may seem strange that in the Sizewell C calculation above the £440 million of decommissioning costs amount to nearly 4% of the total £12 billion undiscounted lifetime costs of the station, but only feature as less than 1% of the electricity cost. This is because the decommissioning fund is only set up at the end of the life of the station, more than 45 years after the start of construction. The 10% discount rate used in the main financial calculation heavily 'dis-counts' this future expenditure, because, as pointed out at the beginning of this section, a pound tomorrow is not considered to be worth the same as a pound today. It can be argued that this discounting process also undervalues other long-term elements that have not been entered into the equation: for example, the benefits of avoided CO_2 emissions, which may last for a hundred years or more; and the disbenefits of the nuclear waste, which could last for thousands of years.

Section 12.3 looked at how the value of money is measured against real goods, such as a dozen eggs. Money is used as a unit of account for the 'real' value of the eggs. A dozen 1971 eggs have the same food value as a dozen eggs in the year 2000, yet they are assigned different values in the money of their respective years, 22p in 1971 and £1.72 in 2000. Likewise, a derelict nuclear power station requiring £440 million of dismantling work must be considered as a 'real', unpleasant object, with real impacts on its surroundings. And yet, at a discount rate of 10%, on paper the strange magic of DCF can discount a 'future' derelict power station down to a negative asset of under £1 million in present day values. This gives the distinct impression that there must be something wrong with such an accounting system.

Mathematically speaking, discounted cash flow sets up a value function, expressed in money, and the economic analyst aims to maximize it. However, this may not always give very acceptable answers. For instance, in 1973 the mathematician Colin Clarke took investment in the whaling industry as an example. He demonstrated that using DCF it could be considered economically preferable for a fishing fleet to hunt blue whales to extinction rather than let the population of the species produce a sustainable annual catch. All that was required was for the discount rate used to be more than twice the breeding rate of the whale population. If the natural rate of population increase was 5% per year and the whaling company used a discount rate of over 10%, then the ruthless pursuit of profit maximization would lead to the extinction of the species (Clarke, 1973).

The notion of a discount rate can be justified on the basis of our time preference for money – whether or not we want 'jam today' or 'jam tomorrow' over a normal time period of financial investment, say up to 30 years. This seems to be a morally acceptable question. Dealing with longer timescales is an altogether more difficult problem. It is less a problem of finance and more one of utilitarian philosophy and the ethics of discounting the future at all.

Large parts of the UK are still littered with coal mining and other waste from previous generations. Projects such as the nuclear power station in Box 12.7 are seen 'looking into the future' and heavily discounting both future benefits and disbenefits.

The oil shale waste tip shown in Figure 7.29 conversely can be seen from the present looking back into history. Its continued existence poses the question: 'if lighting oil was in such demand in the 19th century, couldn't they have afforded just a little bit extra to clean up the mess left behind?'. Equally one might feel grateful for the technological advances that James Young pioneered and say that a waste tip is a small price to pay.

Turning back to the future there is the question of whether or not future generations would tolerate today's standards of waste management. Given the progressive tightening of legislation on particulate smogs, acid rain, petrol containing lead, and general aspects of health and safety, etc., it seems unlikely that they will. It is possible that in the future society will insist on a far more thorough and expensive clean-up of today's technologies than is budgeted for at present.

A conventional short-term economic analysis may only consider the maximization of benefits to a small group of investors in the present generation. However, many economists and philosophers believe that the analysis should really deal with society as a whole and include the welfare of future generations. This has been the basis of the *Stern Review on the Economics of Climate Change* commissioned by the UK Treasury (Stern, 2006).

These considerations could imply that the real discount rate applied should be very low. In his mathematical theorem about blue whale fishing, Clarke showed that the use of a zero discount rate would lead to the maximum sustainable catch. Another study of the philosophical, rather than mathematical, basis of discounting also came down in favour of a rate close to zero (Broome, 1992).

Others have argued that, hopefully, economic growth and technological innovation should make future generations wealthier in real terms. The 2003 Treasury *Green Book* suggests a discount rate of 3.5% for time periods of 0–30 years, falling to only 1% for periods of over 300 years. More recent Treasury advice (Lowe, 2008) has suggested figures of 3% for 0–30 years falling to only 0.86% for more than 300 years.

Such estimates also have to take into account the finite possibility of catastrophe; for example, an economic collapse akin to the inflation in Germany in the 1920s could wipe out the 'store of value' in any potential nuclear decommissioning fund, leaving future generations to foot the bill out of their own resources.

However, Broome does warn that if future climate change were to cause the world economy to start contracting, then an appropriate discount rate might become negative. This would imply that it would pay to invest now to stave off future disaster. He adds, though, that 'it is not plausible that any of this will be reflected in people's present expectations of interest rates'.

Discounted cash flow – summary

Discounted cash flow analysis is a simple technique for taking account of the time value of money in cost calculations. It was originally developed for the comparison of the relative economics of different projects where some form of borrowing has to be made. The difficulties in its use are mainly those of choosing an appropriate discount rate (particularly one taking account of inflation) and an investment lifetime, which may often be shorter than the actual operational lifetime of the project. So when a particular figure of *x* pence per kWh has been produced, it is vital that all the key assumptions are clearly stated. It is all too easy to produce low figures using unrealistically low discount rates, long project lifetimes and inflated performance estimates.

It must be said that the basic concept of the time value of money is difficult to grasp. Even when a discounted cash flow calculation has been explained in detail it often provokes the exasperated question 'Can't you just tell me in simple terms what the payback time is?'

Those readers who have that feeling can take comfort from this quote from Lord Peston during the debate on electricity privatization in the UK parliament in 1989:

> When I was a junior economist in the Treasury [...] I remember preparing a paper which was to go to Ministers. It contained a formula. The people who looked after me asked me whether my formula had to go in [...] I said 'Yes, my formula must go in'. They were horrified at the idea that a document going to Ministers would contain a formula. It was a discounted cash flow formula, as I was one of the first to think that discounted cash flow might apply to government matters. [...] the old Ministry of Power had not the slightest idea of what discounted cash flow or compound interest were.
>
> (Lord Peston quoted in Hansard, 1989)

This is not to say that at the time Peston is referring to (probably the late 1940s) Treasury officials were particularly ignorant. The notion of maximizing return on investment was only introduced in the Dupont and General Motors Corporations in the USA in the 1920s. Before then, the accounting systems in use did not distinguish clearly enough between interest charges and general operating costs to allow this sort of analysis to be carried out at all.

12.5 Real-world complications

In the UK at present, energy is treated as a freely traded free-market commodity. This is in line with the European Union's programme on the harmonization of energy markets. There is a 'world price' for oil and a similar one for coal. There has also been legislation towards the 'liberalization' of electricity markets, with the aim that eventually all consumers, industrial or domestic, should have the choice of being able to buy their electricity from anywhere, including from another country.

This view is considerably different from that prevailing 40 years ago, and even now is only grudgingly accepted by many European countries for a whole range of reasons, some of which are described below.

Energy security and diversity of supply

It is easy to argue that energy is just a tradeable commodity if a country has plenty of it or reliable access to someone else's supplies. It is less easy to do so if it hasn't. During the 20th century, two world wars gave Europeans ample experience of severe food and fuel shortages. Since World War II, politicians in the European Union have been very careful to ensure that the Common Agricultural Policy keeps the larders full, even if it does require an extensive subsidy system. In India, a country with a long history of droughts and famines, subsidized electricity is almost given away to the agricultural sector.

Fuel shortages are extremely unpopular and are likely to provoke rapid political action. Current International Energy Agency policies insist that each member state has 90 days' worth of reserves of fuel supplies, but even so, quite minor petrol shortages can provoke 'panic buying'. Any wise government will make sure that it has adequate energy supplies to hand.

Since the 19th century, industrial countries such as the UK and Germany have manufactured goods for export. Until recently, the factories have been run on coal (either directly, or on coal-generated electricity). It has thus been in the interests of their governments to 'promote' indigenous coal production to provide reliable energy supplies for home industry and ensure full employment. In wartime, the need to run the energy system flat out became even more important.

In other countries, such as France, Japan, the Soviet Union, India and, more recently, China, the rapid expansion of heavy industry and the electricity industries has been seen as essential to 'modernize' and compete. The overall result has been that these countries developed extensive subsidy systems to their coal-based industrial infrastructure. This in turn created a political power base of large numbers of miners and industrial workers interested in maintaining the status quo. In Germany, for example, there was a law saying that only German coal could be burnt in power stations, even though this required a large subsidy, the *kohlepfennig*.

However, this was increasingly difficult to justify against a background of falling world coal prices from large pits in Australia, South Africa and South America. In Germany by 1995 the coal subsidy was running at nearly £70 per tonne, or over £5 billion per year. This had been reduced to about £2 billion per year by 2008.

Figure 12.18 The political price of coal. The president of the National Union of Miners, Arthur Scargill, is arrested in 1984 during the bitter miners' strike over UK coal prices

In the UK, between 1980 and 1990 government grants to the loss-making British Coal company totalled almost £15 billion. The choice between subsidized home-produced coal or cheaper imported coal resulted in a showdown in 1984 between the then-Conservative government and the miners. There was a long and bitter miners' strike (see Figure 12.18). The government won the day, in part by running oil-fired power stations flat out to keep the lights on. Eventually the government paid out large sums in redundancy and retraining allowances to miners, equivalent to many years' subsidy payments. The subsidy payments were totally phased out by 2008, but, as noted in Chapter 5, UK coal production has fallen dramatically.

More recently the decline in UK North Sea oil and gas production described in Chapter 7 has created tensions between the UK government and oil and gas companies. A number of difficult (and unanswered) questions have arisen: What government 'encouragement' should they be given to drill for new UK supplies? What level of tax should they pay, particularly given high world oil and gas prices? And if they are given 'tax breaks' isn't this a form of subsidy?

More surprisingly, subsidies can also be set up as part of a programme to implement 'free markets'. In 1989, the UK government started the break-up and privatization of the state-owned Central Electricity Generating Board (CEGB). The aim was to reduce electricity costs by promoting competition within the industry. However, this exposed the true costs of the existing nuclear power plants, in particular the older Magnox ones, which had to be hastily withdrawn from the privatization and remained in state hands. A tax system, the **Non-Fossil Fuel Obligation** (**NFFO**), levied on fossil-fuel generated electricity, had to be set up to subsidize these nuclear plants. Once in place, though, a small proportion of the funds were used to support new renewable energy projects. However, the sums paid out to renewable energy were in terms of millions of pounds, rather than the billions paid out to the nuclear industry (or spent on coal subsidies).

Externality costs of pollution and disaster

A further question is whether or not all of the costs of a particular energy technology are genuinely included in the economic calculations. There may be other **externality costs**. These include costs resulting from air pollution, noise pollution and those resulting from the risk of major accidents in mines or power stations; although the effects of these are described in Chapter 13, assigning a monetary value to such things can be difficult and the subject of much debate. However, over the past 60 years, detailed analysis has resulted in a whole range of 'clean-up' programmes being implemented. For example, fitting flue gas desulfurization (FGD) equipment to coal-fired power stations in the 1990s, as described in Chapter 5, added about 0.5p per kWh to their generation cost. However, this was far less than the estimated cost of acid rain damage to trees, buildings and human health, estimated at the time at 4–5p per kWh (Pearce et al., 1992). As will be described in Chapter 13, the costs associated with climate change are still a matter of fierce debate but an initial attempt to internalize the resulting costs has been made with the UK Climate Change Levy. This carbon tax, introduced in 2001 and increased in 2009, adds 0.47p per kWh to business electricity bills and 0.164p per kWh to business bills for gas.

Conflicts between supply and demand

At the time of the privatization of the UK electricity industry in 1989, it was held that the government didn't really need an energy policy, because it could all be left to the free market. However, energy utilities are likely to have been set up in such a way that their profits increase with their energy sales. They are also likely to have better access to funds and information to expand energy supply than their consumers do to invest in energy conservation. The result is likely to be excessive energy use and increased resulting pollution. The concept of **least-cost planning** in energy has been widely used, particularly in California since the 1970s. This analyses the provision of the end energy service from the viewpoint of the consumer, rather than just the energy supplier. It aims to make sure that the consumer gets the best energy service for minimum cost. This has resulted in programmes of 'market intervention' to decouple energy utility profits from sales and to promote energy efficiency. More recently, starting in 2002, the UK has introduced obligations on the electricity and gas utilities, the **Energy Efficiency Commitment**, now called the **Carbon Emission Reduction Target (CERT)**, for them to assist their customers in cutting energy use. This is done through widespread subsidies on energy-efficient lighting, loft insulation and cavity wall insulation.

Where energy utilities are state-owned, the provision of cheap fossil-fuelled energy can be seen as part of the 'beneficence' of the state, whether this was in the old Soviet Union or modern-day Venezuela, where in 2009 petrol was being sold for 2p per litre. This may be popular with consumers but is ultimately financially unsustainable.

There is a wide range of subsidies, both explicit and implicit, that artificially lower the price of fossil fuels and fossil-fuel generated electricity. A joint study by the IEA/OECD/World Bank estimated that in 2009 global fossil fuel subsidies were US$312 billion per year (OECD, 2010; IEA, 2010b) (see Figure 12.19). This represents about 0.6% of global GDP. These subsidies are particularly large in energy-producing countries such as Iran, Saudi Arabia and Russia. The study argues that phasing out these subsidies could enhance energy security by producing a 5% reduction in global energy consumption, reduce greenhouse gas emissions by 2 $GtCO_2$ and bolster economic growth. This will be discussed further in Chapter 13.

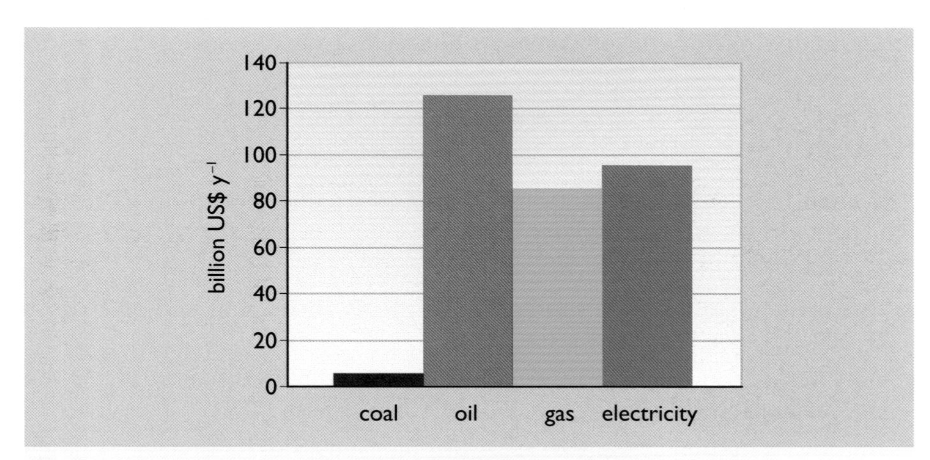

Figure 12.19 Global fossil fuel subsidies in 2009 (source: OECD, 2010)

New technologies and economies of scale

Chapter 9 has described how UK government intervention in the electricity 'free market' in the 1920s led to the introduction of the National Grid, allowing the use of larger, more efficient, power stations and resulting in lower electricity costs. Other new energy technologies may have high costs simply because of low volume of production. If only mass production and consumption could be encouraged, then the technology would be considered economic.

Solar electricity generation from photovoltaic (PV) panels is a good example. A 1999 study for Dutch Greenpeace (KPMG, 1999) concluded that PV electricity could be competitive with that generated from fossil fuels if the demand for panels could be raised to a sufficient level to support large production plants. For the required economies of scale, each plant would need to produce panels with 500 MW of peak output (or roughly five square kilometres) per year, approximately three times the 1999 world demand for PV panels. Since then PV has been heavily subsidized in countries such as Germany and Spain. Ten years on, in 2009, world PV production had reached 10 700 MW of peak output and costs have fallen dramatically.

Supporting renewable energy

Since the 1990s UK (and much of European) government policy has shifted from one of 'implementing free markets' where 'governments should not pick winners' to matters of depleting fossil fuels and energy security, climate change and active support for renewable energy.

The UK's NFFO mechanism for the support of renewable energy was withdrawn in 2001 and replaced by a **Renewables Obligation** on electrical utilities to buy a steadily increasing proportion of their electricity from renewable sources. If they do not meet this obligation then suppliers have to pay what is effectively a fine, currently 3.7p for every kWh of renewable energy that they fail to purchase.

This system has been criticized as excessively complicated. An alternative **Feed in Tariff** approach, where electricity from renewable (and CHP) supplies can be paid clearly stated high prices has been used successfully in Germany and Spain for some years and has now been introduced in the UK.

Reasons for subsidies

Although fossil-fuelled energy can be seen as a commodity to be traded (and priced) in global markets there are many reasons why individual governments might wish to subsidize its use:

- Ensuring adequate food production through cheap fuel and electricity to farmers, particularly for irrigation.
- Alleviating fuel poverty by providing cheap fuel and electricity to the poorer sections of society, particularly the elderly.
- Improving national energy security through subsidies for domestic energy production. This may also create employment and cut energy imports.
- Redistributing (state-owned) national resource wealth, for example in oil-rich countries, by making petrol available very cheaply.

However, these subsidies may end up being wasteful, reducing investment in energy efficiency and leading to excessive energy use and resulting pollution. Fuel poverty, for example could equally be tackled with 'economic' fuel prices, but with subsidies for low-energy lighting and home insulation.

There are also good reasons for taxing energy use and various forms of 'market intervention':

- Raising basic national tax revenue and providing the income streams for the more worthy of the purposes above.
- Discouraging consumption to reduce national energy imports and improve the balance of payments.

- Raising the effective fuel price to encourage investment in energy conservation.
- Raising revenue for investment in cleaner and more sustainable energy technologies.

12.6 Summary

Energy is obviously a vital commodity for the modern industrial society. As such its prices are subject to a host of economic and political pressures. The most obvious of these in recent years has been the political control exerted by OPEC over oil prices.

Section 12.2 has described how the same basic fuels may be sold for considerably different prices in different countries, largely as a result of their tax policies. It has also described the recent rises in global oil and gas prices and their divergence from coal prices, which have remained relatively low.

Although in many countries energy prices are much the same to different consumers, this is not universal. In India electricity is provided at very low prices indeed for agricultural purposes.

Section 12.3 turned to the subjects of inflation, 'real prices' and the affordability of energy. Despite the continual grumbles about the high costs of energy, and the increase in prices over the past decade, it must be said that the 'real' price of energy is almost at an all-time low in relation to the earning power of workers in the UK. However, rising fuel prices are likely to disadvantage the poorer sections of the community, whether these are the elderly in the UK or the rural poor in India.

Section 12.4 discussed the complex topic of investing in energy, starting with the distinction between a cost, which is something that can usually be calculated, and a price, which is something determined in a competitive marketplace.

It described a simple calculation on the payback time for a compact fluorescent lamp and then turned to a simple calculation of the levelized generation cost for a combined cycle gas turbine generation plant (CCGT).

It then moved on to more detailed projects where money would have to be borrowed and loans paid back over time. With these it is vital to appreciate the processes of inflation and of discounted cash flow calculations. The CCGT calculation was revisited to show the effect of having to pay interest on a loan by using an annuitization table. Then a full discounted cash flow calculation for a nuclear power station was described and compared with a rival CCGT plant.

The section continued by describing how financial risk can be dealt with by splitting loans into debt and equity components with different effective interest rates.

It then described how different investors may have different perspectives on acceptable discount rates and project lifetimes. These should always be clearly stated in any particular calculation.

The section ended with a discussion of intergenerational equity and the extent to which discounting, particularly over long time periods, is financially and philosophically acceptable.

Section 12.5 considered many real-world complications in the costing and trading of energy. There are obviously conflicts between national energy security and global free trade in energy. The removal of subsidies for coal mining in the UK led to a bitter miners' strike in 1984. Similar political conflict has taken place in other countries.

The external costs of energy need to be taken into account even though they may be poorly quantified.

There may be conflicts between supply and demand. Energy utilities may have a vested interest in increasing energy use and better access to funds and information than their consumers do for alternative energy conservation measures.

New technologies, particularly renewable energy ones, may need financial support to achieve the economies of scale to become competitive with existing energy technologies.

Finally, although this chapter has discussed 'real' prices, it is important to remember that money is only used as a medium of exchange, a unit of account and, somewhat imperfectly, a store of value. It is used to facilitate the provision of energy services such as warm homes, cooked dinners, illumination and transport. These are the 'real' end products.

References

Bank of England (2011) 'Statistical Interactive Database' [online], Bank of England, http://www.bankofengland.co.uk (accessed 10 March 2011).

Broome, J. (1992) *Counting the Cost of Global Warming*, Report to the Economic and Social Research Council, Whitstable, White Horse Press.

Clarke, C. (1973) 'The economics of over-exploitation', *Science*, vol. 181, pp. 630–4.

DECC (2010) *UK Energy in Brief 2010* Department of Energy and Climate Change; available at http://www.decc.gov.uk (accessed 10 March 2011).

DECC (2011a) *Premium Unleaded Petrol/Diesel Prices in the EU* (spreadsheet), Department of Energy and Climate Change; available at http://www.decc.gov.uk (accessed 14 March 2011).

DECC (2011b) *Quarterly Energy Prices*, Department of Energy and Climate Change; available at http://www.decc.gov.uk (accessed 14 March 2011).

DTI (1995) *The Prospects for Nuclear Power in the UK – Conclusions of the Government's Nuclear Review*, HMSO.

EIA (2011a) 'Weekly Retail Premium Gasoline Prices (Including Taxes) [online]', US Energy Information Administration, http://www.eia.doe.gov (accessed 14 March 2011).

EIA (2011b) *Petroleum Marketing Monthly*, US Energy Information Administration, March; available at http://www.eia.doe.gov (accessed 14 March 2011).

Evans, R. D. and Herring, H. (1989) *Energy Use and Energy Efficiency in the UK Domestic Sector up to the Year 2010*, Energy Efficiency Office, HMSO.

Hindustan Petrol (2011) *Revision of Petrol Price effective 15th/16th January 2011* [online], http://www.hindustanpetroleum.com/Upload/En/UPdf/MS_16012011.pdf (accessed 12 March 2011).

HM Treasury (2003) *The Green Book: Appraisal and Evaluation in Central Government*, London, TSO; available at http://www.hm-treasury.gov.uk/data_greenbook_index.htm (accessed 12 March 2011). See especially Annex 4 on risk and uncertainty and Annex 6 on discount rate.

IEA (2002) *Electricity in India*, Paris, International Energy Agency; available at http://www.iea.org (accessed 13 March 2011).

IEA (2010a) *Key World Energy Statistics*, Paris, International Energy Agency; available at http://www.iea.org (accessed 6 March 2011).

IEA (2010b) *World Energy Outlook 2010*, Paris, International Energy Agency.

Independent (2011) 'China raises petrol, diesel prices' [online], *Independent*, 22 February 2011, http://www.independent.co.uk/life-style/china-raises-petrol-diesel-prices-2221947.html (accessed 14 March 2011).

KPMG (1999) *Solar Energy: From Perennial Promise to Competitive Alternative*, Hoofddorp, Netherlands, Bureau voor Economische Argumentatie.

Lowe, J. (2008) *Intergenerational wealth transfers and social discounting: Supplementary Green Book guidance*, London, HM Treasury; available at http://www.hm-treasury.gov.uk (accessed 12 March 2011).

Mott MacDonald (2010) *UK Electricity Generation Costs Update*, Brighton, Mott MacDonald; available at http://www.decc.gov.uk (accessed 7 March 2011).

Nuclear Electric (1994) 'Further nuclear construction in the UK', Volume 1 of a submission to the government's *Review of Nuclear Energy* (DTI, 1995).

NSSO (1996) *Sarvekshana,* vol. XIX-3, no. 66, National Statistical Survey Office, Government of India.

OECD (2010) *The Scope of Fossil-Fuel Subsidies in 2009 and a Roadmap for Phasing Out Fossil-Fuel Subsidies*, IEA, OECD and World Bank joint report prepared for the G-20 Summit, Seoul (Republic of Korea), 11–12 November 2010; available at http://www.oecd.org/dataoecd/8/43/46575783.pdf (accessed 4 April 2011).

ONS (2010) *Family Spending: A Report on the 2009 Living Costs and Food Survey*, UK Office for National Statistics; available at http://www.statistics.gov.uk (accessed 7 March 2011).

ONS (2011a) *Retail Price Index*, UK Office for National Statistics; available at http://www.statistics.gov.uk (accessed 7 March 2011).

ONS (2011b) *Average Weekly Earning Index*, UK Office for National Statistics; available at http://www.statistics.gov.uk (accessed 7 March 2011).

Pearce, D., Bann, C. and Georgiou, S. (1992) *The Social Costs of Fuel Cycles*, Report to Department of Trade and Industry, London, HMSO.

Peston, M. (1989) *Hansard* [online] 16 May 1989, col. 1108, http://hansard.millbanksystems.com/lords/1989/may/16/electricity-bill (accessed 12 March 2011).

Sadnicki, M. J. (1994) Nuclear Review Background Paper, Sizewell B and Sizewell C, Hoskyns Group plc.

Stern, N. (2006) *Stern Review on the Economics of Climate Change,* London, HM Treasury, http://webarchive.nationalarchives.gov.uk/+http://www.hm-treasury.gov.uk/sternreview_index.htm (accessed 14 March 2011).

Stone, R. (ed.) (1966) *The Measurement of Consumers' Expenditure and Behaviour in the UK, 1920–1938*, New York, Cambridge University Press.

TERI (2009) *TERI Energy Data Directory and Yearbook (TEDDY) 2009*, New Delhi, The Energy and Resources Institute.

Chapter 13

Penalties: assessing the environmental and health impacts of energy use

By Stephen Peake, Bob Everett, Godfrey Boyle and Janet Ramage

13.1 Introduction

The preceding chapters on fossil and nuclear fuels will have left you in no doubt that our use of these fuels, whilst conferring numerous benefits, also incurs very substantial penalties. These include adverse impacts on the Earth's ecosystems and climate, together with deleterious effects on the health of humans and many of the other species with which we share the planet. These effects are not only detrimental in themselves, in ways that can often be quantified, but they in turn create substantial additional monetary costs to society over and above the simple market prices of the fuels themselves.

This chapter reviews the predominant penalties of energy use, primarily in terms of the major causes for concern (climate change, air pollution, accident risk, etc.). It focuses mainly on the environmental and health impacts of fossil and nuclear fuels; those of hydroelectricity, wind power and the other renewable energy sources are also briefly mentioned.

It starts by classifying and identifying the impacts of energy use and then goes on to examine various ways of assessing and comparing these across different fuel chains, with a focus on various electricity generating systems.

It should be stressed at the outset that there is no comprehensive assessment method enabling all of the widely varying and qualitatively differing impacts involved in our use of energy to be compared together and 'objectively'. The chapter then outlines methods of monetizing the cost of environmental- and health-related 'energy externalities' before considering some of the major causes for concern including climate change, air pollution, and accidents and risks, the latter concentrating on nuclear energy. Finally, it looks at energy return over energy invested (EROEI) for different systems.

13.2 Classifying the impacts of energy use

As well as delivering innumerable social benefits most, if not all, forms of energy, exploited through whatever technology, also deliver at least some form of negative social and environmental impact. Planning more sustainable energy systems frequently involves weighing up the relative costs and benefits of alternative policies (e.g. regulations, taxes, subsidies) towards different energy sources, fuels and technologies. Policymakers (economists in particular) require investments in anything other than least-cost (market cost) energy technologies to be supported by evidence of the climate, human health, ecological or national security benefits that justify higher up-front capital or operational expenditures.

There are various criteria that can be used to analyse and classify the impacts of energy use, including:

- scale
- source
- public concern.

By scale

The range of scales associated with energy-related impacts (Holdren and Smith, 2000) include:

- global scale – for example, emissions of carbon dioxide and other greenhouse gases produced by fossil fuel combustion can contribute to global climate change
- regional scale – for example, acid rain caused by the emission of sulfur dioxide (SO_2) and oxides of nitrogen (NO_x) from power stations can have adverse effects a thousand miles away from the emission sources
- community scale – for example, emissions of air pollutants from vehicle engines in urban areas
- workplace scale – for example, the health hazards entailed in coal mining, or in oil and gas extraction
- household scale – for example, the emissions of smoke and other pollutants from wood burning.

By source

Table 13.1 lists the main environmental and social concerns associated with the principal energy sources. It illustrates the extremely wide variety of quantitatively and qualitatively different impacts that humanity's use of energy has on the environment and on society.

Table 13.1 Environmental and social concerns associated with the principal energy sources

Source	Causes of concern
Oil	Global warming, air pollution by vehicles, acid rain, oil spills, oil rig accidents, land despoilation (for example oil sands mining).
Natural gas	Global warming produced by CO_2 from combustion and methane from pipe leakage, methane explosions, gas rig accidents.
Coal	Global warming produced by CO_2 produced from combustion and methane leakage from mines, acid rain, environmental spoliation by opencast mining, land subsidence due to deep mining, spoil heaps, ground water pollution, mining accidents, health effects on miners.
Nuclear power	Radioactivity (routine release, risk of accident, waste disposal), misuse of fissile material, nuclear proliferation, land pollution by mine tailings, health effects on uranium miners.
Biomass	Effect on landscape, habitats, and biodiversity; ground water pollution due to fertilizers; use of scarce water; competition with food production.
Hydroelectricity	Effects of the construction of dams: displacement of populations, effect on rivers and ground water, visual intrusion and risk of accident, downstream effects on agriculture, methane emissions from submerged biomass.
Wind power	Visual intrusion of turbines in sensitive landscapes, noise, bird and bat strikes, and interference with TV reception, communications and aircraft radar.
Tidal barrages	Destruction of wildlife habitat, reduced dispersal of effluents
Geothermal energy	Release of polluting gases (SO_2, H_2S, etc.), ground water pollution by chemicals including heavy metals, seismic effects.
Solar energy	Occupation of large land areas (in the case of centralized concentrating solar and PV plant), use of toxic materials such as cadmium in the manufacture of some PV cells, visual intrusion in rural and urban environments.

By public concern

Another way to classify the impacts of energy use is to consider the contribution of energy sources to some of the most important public concerns, as shown in Table 13.2. This obviously lists many pollutants and problems, some of which will be described in more detail later in this chapter.

Table 13.2 Principal causes of concern relating to energy use and impact pathways

Impact category	Impact	Pollutant/cause
Direct impacts on human health: These are mainly from fossil fuels, traditional biomass and nuclear power. The majority of health impacts are related to poor air quality and to a lesser extent cancers caused by heavy metals or radioactive elements in the food chain.	**Mortality (premature death)**	
	Reduction in life expectancy	Particulates, SO_2
	Cancers	Benzene; 1,3 butadiene; diesel particulates; heavy metals (Hg, As, Cd, Pb, Se); radionuclides; dioxins High levels of electric and magnetic fields.
	Fatality risk from traffic and workplace and nuclear accidents	Accident risk
	Morbidity (ill health)	
	Cancer risk (non-fatal)	As for fatal cancers (above)
	Respiratory hospital admissions	Particulates, low-level ozone, SO_2
	Restricted activity days	Particulates, low-level ozone
	Congestive heart failure	Particulates, carbon monoxide
	Cerebrovascular hospital admissions; cases of chronic bronchitis; coughs in children/asthmatics; lower respiratory problems	Particulates
	Asthma attacks; symptom days	Low-level ozone
	Loss of amenity, impact on health; heart attacks; angina pectoris; hypertension; sleep disturbance	Noise
	Risk of injuries from traffic and workplace accidents	Accident risk
Global, regional and local climate change: The main culprits are fossil fuels and energy-related land use change involving biomass which can both lead to increases in greenhouse gas emissions.	The impacts of climate change are widespread. Rapid global environmental change threatens many species and ecosystems on Earth, including human social and economic systems Related impacts: Increased death rates as a result of extreme weather, vector-borne diseases (i.e. transmitted by an intermediary such as mosquitoes) such as malaria and dengue fever, bad sanitation, cholera from polluted water, increased air pollution and smog, aeroallergens, respiratory disorders	Greenhouse gases: carbon dioxide (CO_2), methane (CH_4), nitrous oxide (N_2O), halocarbons

Table 13.2 (*Continued*)

Impact category	Impact	Pollutant/cause
Loss of biodiversity/ ecosystem damages from fossil fuel combustion and unsustainable biomass use such as deforestation	Ocean acidification (e.g. destruction of coral reefs)	CO_2, SO_2
	Acidification of lakes	
	Direct forest impacts	Acid rain damage caused by SO_2 and NO_x
		Deforestation in the process of coal mining or oil exploration
	Climate change induced changes in range of species	Greenhouse gases
	Regional climate change stress/collapse of unique ecosystems e.g. collapse and release of additional greenhouse gases from Amazon (Lenton et al., 2008), upland peat bogs in Europe and permafrost zones of the northern hemisphere across Europe and Asia.	Greenhouse gases
Damage to food production due to the combustion of fossil fuels	Reduced yields for crops such as wheat, barley, rye, oats, potato, sugar beet and sunflower seed.	SO_2, NO_x, low-level ozone
	Ocean acidification threatens marine harvests of many fish species.	CO_2
Damage to buildings produced by the combustion of fossil fuels	Accelerated rusting of galvanized steel, corrosion of limestone, sandstone, mortar, rendering, paint, and glass.	SO_2, NO_x
	Blackening of buildings.	Particulates
Water pollution	Contamination of oceans and freshwater sources from energy extraction, processing and transportation	Marine oil spills (Figure 13.1)
		Mining effluents such as water used to wash coal, water used for enhanced oil recovery, leachate from uranium mining, etc.
Land contamination	Solid wastes are produced from various fuel chains including coal combustion, oil extraction and the nuclear fuel cycle.	See for example the 19th century oil shale mining waste tip in Figure 7.29
Land cover/footprint/ community displacement	Land-intensive sources including large hydro, forestry, large-scale wind.	
Amenity loss	Several energy sources contribute to noise over various stages of the fuel chain.	Road transport, rail and aircraft noise.
Visibility/aesthetic impact	Fossil fuel exploration, mining, the construction of power stations, electricity pylons, wind farms and biomass farming.	

Source: adapted from CEC, 1995a.

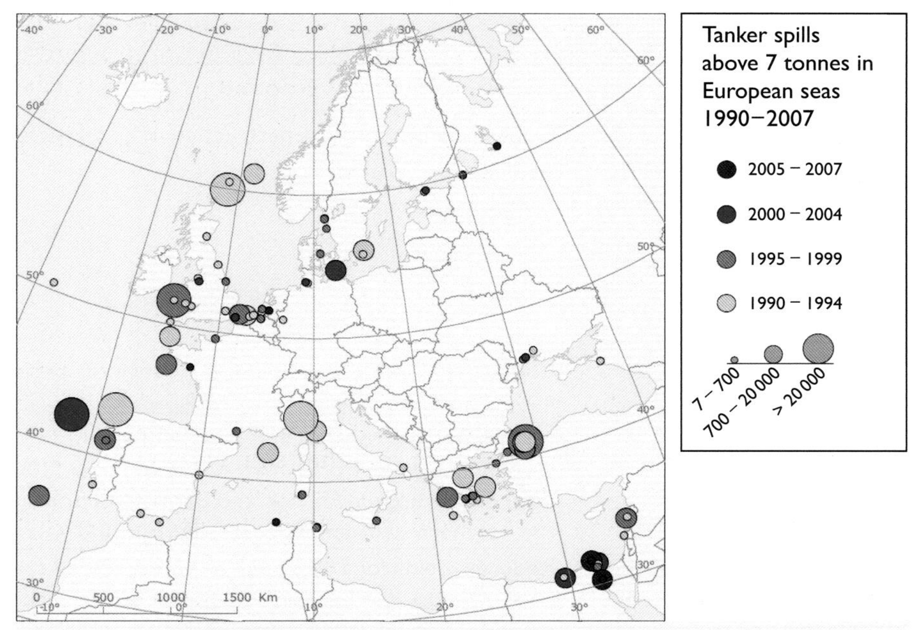

Figure 13.1 Large oil spills (>7 tonnes) from tankers in European waters 1990–2007 (source: EEA, 2008)

13.3 The full financial costs of energy use

This section provides an overview of some of the methods used to estimate those energy costs that are not fully reflected in market prices for one reason or another. It also looks at the effect of subsidies (already touched on in Chapter 12) and on the use of 'shadow prices' for the damage done by pollutants in assessing appropriate levels of abatement or choices of technology.

Energy systems are continually evolving, shaped by a variety of forces (scientific, technological, economic, political). The evolution is partly the result of billions of individual decisions based on individual preferences (small car versus large car; gas versus electric cooker) and in part planned by policymakers through a myriad of policy interventions (regulations, taxes, research and development, etc). The evolution of energy systems in turn demands that policymakers assess and monitor the trade-offs between the economic, environmental and social dimensions of sustainable energy. New scientific information and large-scale accidents occasionally force the public and planners to revisit and reconsider what trade-offs can be accepted and at what price.

Despite conflicting protestations about the 'true' costs and benefits of energy use from growthists, peakists and environmentalists (Chapter 1), there are many methodological uncertainties involved in establishing both the benefits and costs of energy use. Although this is a worldwide problem, the costs of energy use show themselves on the regional, local and even domestic scale.

There are two ways in which we do not always pay (or see) the full cost of different energy sources or technologies. Firstly, there may be numerous *energy subsidies* that governments provide for particular fuels, technologies or markets. Secondly, there are the environmental and health *externalities* (Box 13.2) associated with the use of different fuels over their life cycles.

Energy subsidies

Energy subsidies are a key way in which governments seek to pursue their political, economic, social and environmental objectives. What exactly is a subsidy? As noted at the beginning of Chapter 12, petrol is a globally traded commodity. In the USA, for example, it is sold at a 'world price' with a small extra amount of tax. In European countries, particularly Denmark, petrol is sold with a high level of tax. However, in other countries, particularly oil-producing ones, it may be sold at *less* than the world price; in this case, the difference between the sale price and the world price can be regarded as a 'subsidy'.

In 2009 global fossil fuel subsidies were estimated to be around US$300 billion per year (see Figure 12.19) or about 0.6% of global GDP. This is a significant fraction of some estimates of the costs of substantially decarbonizing the global economy.

Nearly a half of the subsidies were in the energy exporting nations of Iran, Saudi Arabia and Russia (IEA, 2010a). In some ways this can be seen as 'the beneficence of the state' sharing out the 'natural wealth of the country' to its inhabitants. It has a parallel in the long tradition that UK miners always received free coal for their homes.

Many of the world's most advanced economies (e.g. the G20 nations) are committed to reducing fossil fuel subsidies. A joint study by the IEA, OECD and World Bank argues that phasing out these subsidies would deliver a triple win: enhancing energy security, reducing greenhouse gas emissions and bolstering economic growth (OECD, 2010). The authors of the report argue that phasing them out would mean that governments could make more economically efficient investments (e.g. reducing poverty, improving standards of health and education). Cutting these subsidies would lead to a reduction in world annual carbon dioxide emissions of about 2 billion tonnes. However, the report points out the necessity to protect the basic energy service needs of the poor. This may require significant changes in social and energy pricing policy within countries. For example, the country with the fourth largest volume of subsidies is India, an energy-importing nation. As pointed out in Chapter 3, kerosene and LPG are used by the poor for lighting, in part because of the low level of electrification of the country. In the electricity sector, described in Chapter 9, a fifth of national generation is used by the agricultural sector and is heavily subsidized. Increasing these energy prices to a 'world level' may require the provision of alternative and sustainable ways of providing illumination for the poor and guaranteeing agricultural productivity.

From time to time governments argue in favour of the use of subsidies to reduce the impacts of energy use on the environment and human health.

Many countries (both developed and developing) spend significant sums on support for 'clean' energy technologies such as renewables, nuclear and more recently carbon capture and storage. There are obvious conflicts between these subsidies and those supporting fossil fuels. According to the IEA/OECD/World Bank report, fossil fuel subsidies have been shown to encourage waste, increase price volatility, increase fuel tampering and smuggling, and undermine the competitiveness of renewables and more efficient energy technologies. In summary, energy subsidies are therefore a policy tool that can both undermine as well as support the pursuit of sustainable energy systems.

The external costs of energy use

An **externality** can be defined as 'The costs and benefits which arise when the social or economic activities of one group of people have an impact on another, and when the first group fail to fully account for their impact' (CEC, 1995a, p. 4). There may be social, economic and environmental externalities associated with any economic activity. Interest in attempting to quantify environmental externalities dates to the 1960s, in part triggered by the search for economically optimal solutions to air pollution.

Almost all energy sources have negative impacts at some stage in their 'life cycle'. Those impacts vary from the very local (noise in the ears of the operator of the digger, or soot in the lungs of the miner) to the regional (acid rain, changes in regional climate) to the global (global warming, ocean acidification).

Chapter 12 looked at the monetary costs of energy, measured in conventional costing terms. These exclude the additional energy externalities or 'environmental adders' not fully reflected in the conventional market prices. In this book the term **energy externalities** is used to include the social externalities (welfare losses due to nuisances such as noise or visual intrusion), human health externalities, and ecological externalities associated with the different life cycles of different energy sources/fuels.

Many of these externalities can usually be dealt with in some way. It may involve better standards of health and safety at a mine, fitting some form of pollution abatement equipment or making an appropriate choice between different technologies. In order to assess the appropriate response, the particular externality first needs to be valued and expressed in financial terms. This may not be particularly easy and requires the use of appropriate units. The visual nuisance of a wind turbine is a matter of its size and location and is not directly related to the amount of electricity it produces. However, the effects of acid rain or global warming are likely to depend on the quantities of pollutants produced. While SO_2, NO_x and particulate emissions are usually discussed in terms of tonnes, the emissions of CO_2 are often described either in terms of tonnes of CO_2 or tonnes of carbon. Emissions of other greenhouse gases, such as methane, are usually described in terms of their carbon dioxide equivalent (CO_2e). The conversions between these conventions are described in Box 13.1.

BOX 13.1 Carbon conversions

Carbon dioxide emissions are a major contributor to climate change. However, as will be described later in the chapter, they only form part of a larger global *carbon cycle*, in which carbon may exist in various chemical forms. It may exist as almost pure carbon in coal, carbohydrates in wood, or carbonates in the shells of sea creatures. It is thus often appropriate to express CO_2 emissions in terms of just their carbon content.

As explained in Box 5.3, the atomic weight of carbon is 12 and that of oxygen is 16. The combustion of 12 tonnes of carbon will thus produce 44 tonnes of CO_2:

$$\begin{array}{ccccc} C & + & O_2 & \rightarrow & CO_2 \\ 12 & & 2\times16 & & 44 \end{array}$$

The emission of 44 tonnes of CO_2 can thus be taken as equivalent to the emission of 12 tonnes of carbon, the conversion involving a factor of $44/12 = 3.67$.

In emission terms: 1 tonne C $\equiv$ 3.67 tonnes CO_2

A *carbon price* may be expressed either in £ per tonne of carbon or £ per tonne of CO_2. A price of £1 per tonne of CO_2 (£1 tCO_2^{-1}) is thus equivalent to £3.67 per tonne of carbon (£3.67 tC^{-1}).

Carbon dioxide is only one of a number of greenhouse gases. As will be described later, in Section 13.4, they have different direct global warming potentials (DGWP). The emission of a tonne of methane is likely to cause 25 times as much warming as a tonne of CO_2. In order to put these on an equal footing for assessment purposes, the emissions are expressed in **carbon dioxide equivalent (CO_2e)**.

In emission terms: 1 tonne methane $\equiv$ 25 tCO_2e

Quantities of different greenhouse gases emitted on an annual basis by a country, for example, may be added together into a single carbon dioxide equivalent (CO_2e) using their DGWPs.

Greenhouse gas equivalent accounting is also used to describe a more comprehensive measure of concentrations of greenhouse gases in the atmosphere than the CO_2 concentration alone. For example, as shown in Figure 1.20(a), in 2005 the CO_2 concentration in the atmosphere was 379 ppmv. However, when the concentrations of five other greenhouse gases are taken into account the total *carbon dioxide equivalent* (CO_2e) concentration was around 450 ppmv. When targets for future greenhouse gas concentrations are presented or discussed, it is always worth checking the units carefully.

The damage done by the emission of a tonne of pollutant can be expressed as a **shadow price**, a notional sum of money that can be entered into financial calculations in order to make investment decisions. The use of a shadow price for carbon, a *carbon price*, has already been mentioned in Chapter 11 in the context of the overall economics of nuclear power. Putting a price on the damage done by pollution is obviously a difficult task. Box 13.2 discusses how such external costs can be valued and their importance in determining the optimum financial balance between the costs of the damage done by pollution and the costs of pollution abatement.

BOX 13.2 How different external costs are monetized

The 'optimum' level of pollution – setting a balance

Pollution, up to a point, is generally considered tolerable in the world of economics. It could be argued that to reduce the environmental impacts associated with energy systems to zero would also require reducing economic activity to near zero. An economic balance has to be made between the damage done by pollution and the costs of abatement. Hence some level of pollution is economically acceptable and there is an 'economic optimum level of pollution'.

In order to understand what an optimum level of impacts and pollution means we need to understand the difference between **marginal abatement costs** and **marginal damage costs**.

Often, the cost of abating the environmental impacts of the first few units of pollution is cheaper than the next (the so-called 'low-hanging fruit') and so on until the last few units are very expensive. An illustration of this principle is shown in Figure 13.2, which shows a marginal abatement cost curve for greenhouse gas emissions for the global economy. This curve includes measures to reduce methane and nitrous oxide emissions from agriculture. These savings have been expressed in terms of their equivalent global warming potential in carbon dioxide equivalent, CO_2e (as described in Box 13.1). World greenhouse gas (GHG) emissions in 2009 were about 50 Gt CO_2e.

Figure 13.2 looks forward to a (relatively high growth) 'business as usual' forecast of 70 Gt CO_2e for 2030. It then demonstrates the costs of saving up to 38 Gt CO_2e. Each vertical bar represents a different opportunity to reduce GHG emissions. The width of each bar represents the annual reduction potential of GHG emissions for each abatement measure in 2030 if immediate action is taken. The height of each bar represents the global average cost in Euros of avoiding one tonne of CO_2 equivalent for each measure; it is a weighted average over regions and time. The curve focuses on the 'technical opportunities' for abatement (i.e. 'without a material impact on the lifestyle of consumers') up to a value of €60/t CO_2e^{-1} using technologies that are either available today or 'offer a high degree of certainty about their potential in a 2030 time horizon' (McKinsey & Company, 2009, p. 9).

The graph is ordered left to right from the lowest-cost abatement opportunities to the highest-cost. At the left, bars below the line indicate net financial *savings* over the lifetime of the measures. Indeed it suggests that the emissions of some 11 Gt CO_2e of greenhouse gases can be abated at a net profit! Many of these options are in the field of 'energy efficiency'. In the centre of the diagram there are a range of other low-cost options, many of which involve land-use management. At the right are a range of higher cost technological options, most of which involve energy supply, for example nuclear power, renewables and fossil fuels with carbon capture and storage (CCS).

This and other abatement or resource–cost curves should be viewed with caution, since there are obviously considerable simplifications involved. However, the overall shape of the curve and the broad ordering of options within it are helpful.

The optimal level of abatement is the point at which there is a balance between the additional cost of reducing emissions by one tonne and the marginal damage done by the emission of another tonne. Although some pollution is still being produced, in the language of environmental economics it is said that the cost of that pollution has been properly financially accounted for, it is no longer an externality and it is said to have been 'internalized'. Externalities are properly only considered to be those external costs over and above the optimum level.

Where it can be demonstrated that energy externalities exist, it means that in terms of economic theory that the 'prices are not right' and in turn that possibly the 'invisible hand of the market will not do an optimal job' (CEC, 1995b, p. 4). Policymakers then have a range of options at their disposal to try to bring the system back into balance. These include '**financial adders**' such as pollution taxes or regulations such as cap and trade schemes (e.g. the European Emissions Trading Scheme – which is described later).

Economic valuation of impacts

Valuing the environmental impacts associated with energy systems requires the assessment of all sorts of impacts over all the various stages of a fuel chain. It is an ambitious methodological challenge. It also includes ethically difficult problems such as the economic valuation of loss of human life. However, in order to carry out cost–benefit analysis some kind of financial valuation must be done.

In general there are three stages to the valuation of impacts (CEC, 1995b, p. 42):

Stage 1: quantify the impacts in physical terms – e.g. tonnes of crop damaged, deaths caused, amount of architectural damage due to acid rain, etc.

Stage 2: Use market prices to value impacts where these are available (e.g. food prices, cost of repairing stonework).

Stage 3: Where there are no market prices for impacts (e.g. increased risk of illness or death or loss of recreational value) use one of a number of indirect impact valuation methods such as:

(a) ask people in the form of a questionnaire how much they are willing to pay to avoid the impact

(b) analyse how certain impacts may affect related markets (e.g. the price difference between similar houses that are near and slightly further away from a busy road)

(c) analyse how much people are prepared to spend on alternatives to avoid the particular impact.

Health impacts can be assessed using the value of lost earnings. The valuations must, of course, include the loss of life. The **value of a prevented fatality (VPF)**, also called the **value of a statistical life (VSL)** can be thought of as the 'willingness to pay' to avoid an anonymous premature death. Policy decisions may be based on values between €1 and 5 million (CEC, 2005). This can be extended to ideas of the **value of a life year (VOLY)**, about €50 000 (£1 ≈ €1.15). Health effects may be expressed in terms of **years of life lost (YOLL)** – an example of its use is given later in the chapter.

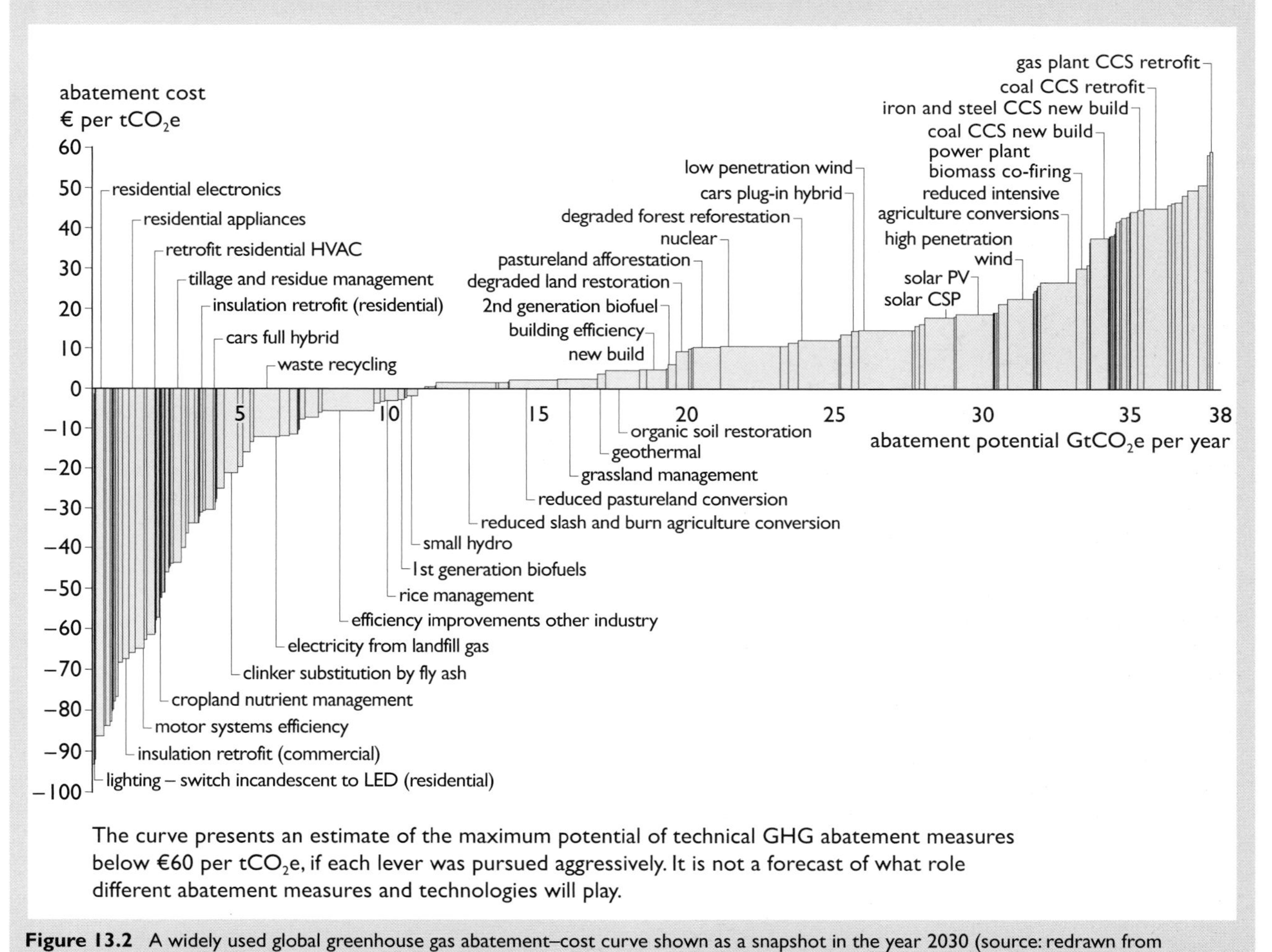

Figure 13.2 A widely used global greenhouse gas abatement–cost curve shown as a snapshot in the year 2030 (source: redrawn from McKinsey and Company, 2009)

Fuel chain analysis

Determining the optimum pollution balance is likely to require the study of more than just, for example, an individual power station. It should really proceed all the way from the original energy source right through to the final *energy service*, i.e. the full fuel chain. It should also cover the whole *life cycle* of the individual elements, be they power stations or batteries for electric cars.

Policymakers in general are interested in two types of quantitative information about different fuel chains:

(1) the magnitude of different kinds of impacts (e.g. increase in cancer following exposure to radiation or loss of biodiversity as a result of acidification)
(2) the quantification in monetary terms of these impacts.

Fuel chain analysis and its related life cycle analysis are key tools for answering the first question above and identifying at what stage in a fuel chain there are likely to be significant externalities. Life-cycle methodologies and database tools have improved in the last two decades. Fuel chain analysis (Box 13.3), once the preserve of publicly funded academic analysis (e.g. CEC, 1995b) is now done on a commercial basis, often for compliance with legislation.

A diagram of energy flows in an economy or an understanding of the production of enriched uranium for a nuclear reactor (Figure 11.9) involves the concept of a fuel chain. Sometimes the technique is used to analyse just a part of a chain. When it is used to capture all or nearly all of the chain it can be referred to as a **full fuel chain analysis**.

A really full assessment should include the construction and disposal of the components within it. This is **life-cycle analysis**, also sometimes referred to as 'cradle to grave' or even 'cradle to cradle' (where recycling may be involved). The analysis of the reprocessing of spent nuclear fuel discussed in Section 11.9 is an example of a fuel cycle involving recycling.

No fuel chain is immune from causing some form of negative external effect. Fuel chain analysis can be used to account for and assess different quantities including:

- materials: it may require 100 000 tonnes of uranium ore to fuel a 1 GW nuclear power station for a year (see Figure 11.9)
- energy: the energy return over energy invested (EROEI) of Alberta oil sands is estimated to be about 7 (see Table 7.8)
- impacts: on average, each kWh of electricity generated from coal in the UK in 2009 involved the emission of 915 g CO_2 (see Table 9.4)
- externalities: the estimated damage cost of emitting a tonne of SO_2 in 2008 at low level in western Europe (the Netherlands) is €12 400 or £10 800 (Bennink et al., 2010).

From pollutant quantities to monetary estimates: the impact–pathway approach

Fuel chain analysis can expose at what stages of a chain pollutants are emitted and in what quantities, but how can their total effect be valued?

A common methodology used for this is the **impact–pathway approach** (CEC, 2005). Figure 13.4 shows the approach applied to sulfur dioxide emissions from European industrial energy processes and electricity generation. It starts by measuring emissions, then estimating the dispersion

BOX 13.3 Fuel chain analysis

This is vital to assess the full energy, economic, social and environmental costs associated with different systems. Figure 13.3 shows some example fuel chains.

Figure 13.3(a) shows a chain from a coal mine to the production of a kWh of electricity from a power station. At each stage along the chain, pollutants may be emitted. Those from the power station itself will be discussed later in this section. However, there are obviously other costs that will need to be taken into account, for example mining accidents, noise nuisance involved in coal transport, etc. In this example the end product is taken to be electricity, but the chain can be extended on to the final *energy service* which might be illumination as shown in Figure 3.1. The final end product there is the quantity of useful light produced by a lamp and the overall efficiency calculation carried out in Box 3.1 is an example of a fuel chain analysis.

Figure 13.3(b) shows two alternative fuel chains which can both potentially produce a transportation energy service expressed in units of passenger-kms. This fuel chain might be called a 'well to wheels' approach. The first uses an electric car using electricity from a coal-fired power station. The second uses a conventional petrol car using an internal combustion engine.

A really full assessment may also need to take into account the various costs associated with construction and disposal including, for example, the power station used in both parts of Figure 13.3 and the batteries used in the electric car in part (b). Such a 'full fuel chain analysis' can also be called a *life-cycle analysis*.

Fuel chain and life-cycle analysis is important in determining the overall useful energy output from a process. The concept of Energy Return Over Energy Invested (EROEI) has been mentioned in Chapters 7 and 11 and will be revisited later in this chapter.

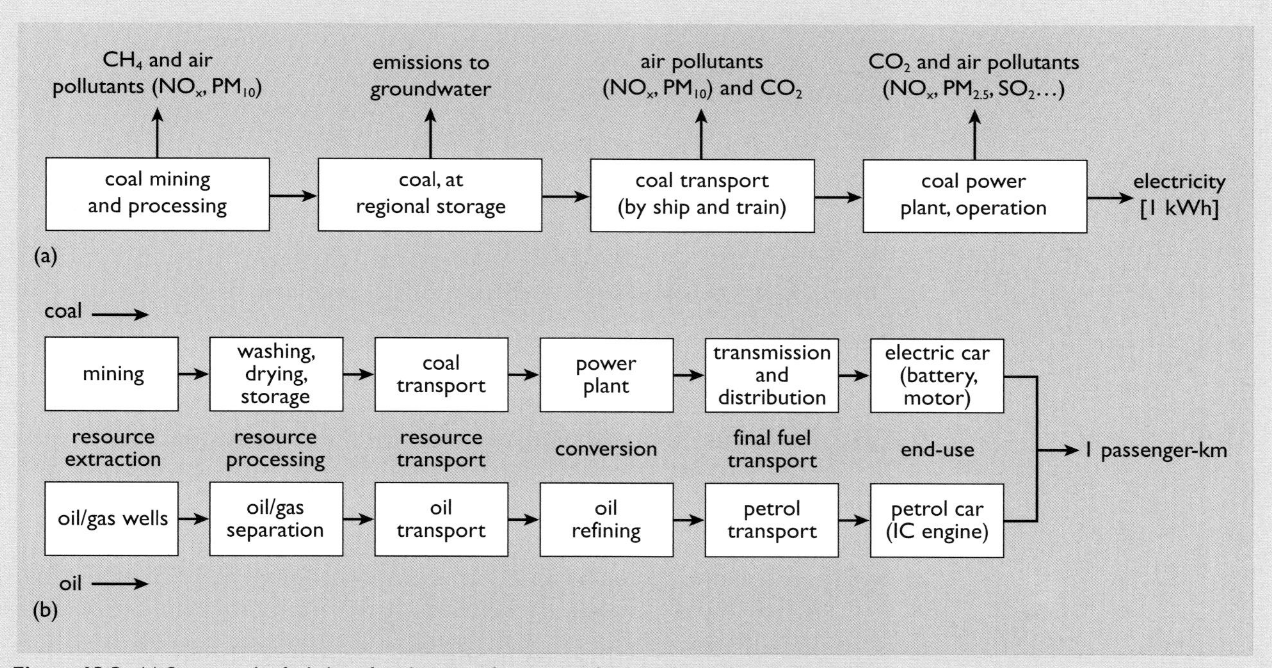

Figure 13.3: (a) Stages in the fuel chain for electricity from a coal-fired power station showing impacts at each stage; (b) various stages in two passenger transport fuel chains (sources: (a) Bauer, 2008; (b) Lazarus et al., 1995)

of pollutants in the environment and the subsequent increase in ambient concentrations. Then the impacts on, for example, crop yields or health are evaluated. The methodology finishes with an assessment of the resulting cost. This phase is perhaps the most difficult. Some direct damage costs will be clear, but in other cases various techniques need to be used to try to attach values – as described in Box 13.2.

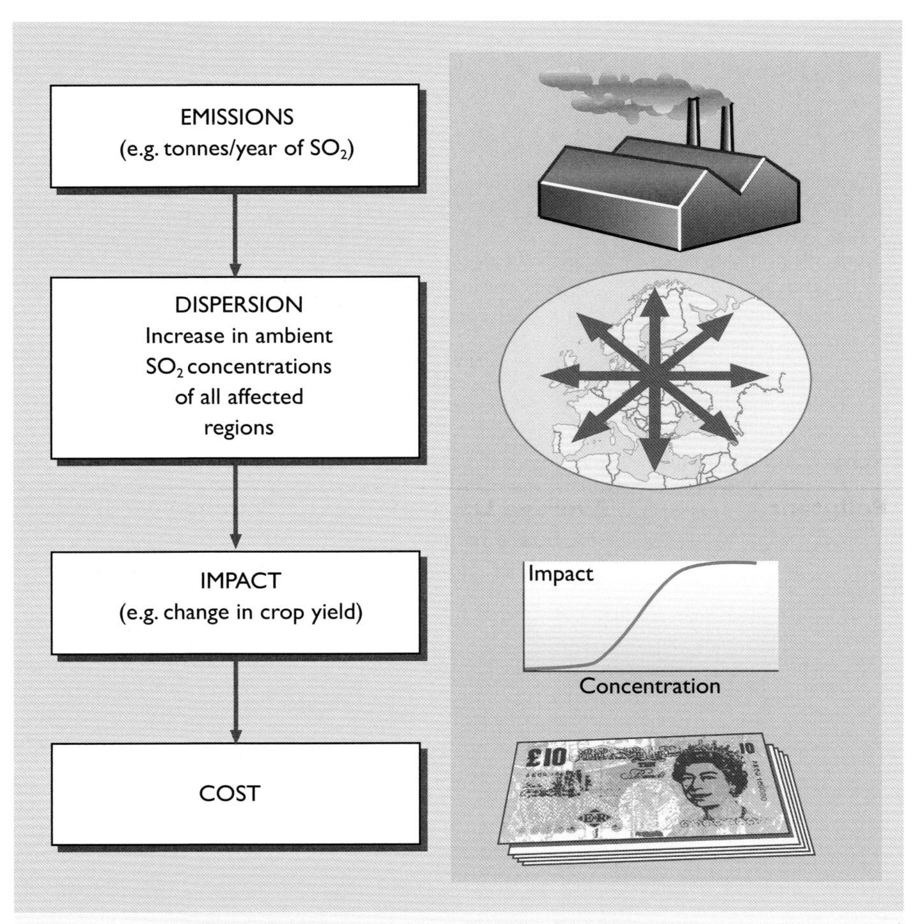

Figure 13.4 The impact–pathway approach for SO_2 emissions

Obviously damage values will vary according to the precise location and circumstances of the emissions. Pollutant emissions in town centres are likely to do more damage than emissions in the countryside. However, the acid rain pollution produced by SO_2 and NO_x may show its effects up to a thousand kilometres away from the point of emission.

The pollution damage from coal-fired power stations and heavy industry has been a matter of concern in both Europe and the USA since the 1970s. While assessment has initially focused on acid emissions and particulates, similar methods can be used for pollution problems related to climate change.

Non-climate-related pollution

A recent comprehensive study, *The Hidden Costs of Energy*, has been produced by the US National Research Council of the National Academies (NRC, 2010). It analysed the US energy system as it was in 2005 using impact–pathway analysis. Although it did not consider all sectors in detail it identified US$120 billion of non-climate-related external costs. This amount is equivalent to over 1% of US GDP. Of this damage about a half was due to the use of coal-fired electricity plants (which provide about a half of US electricity). The NRC study considered the emissions and valued the damage from over 400 US coal-fired power plants. They included four main airborne pollutants: SO_2, NO_x and two grades of particulate matter, $PM_{2.5}$ and PM_{10}; the finer details of the effects of each of these pollutants are described in Section 13.5.

The researchers produced valuation estimates of the damage produced by the emission of one tonne of each of these pollutants, i.e. the marginal damage cost, as described in Box 13.2. The exact figures varied from plant to plant, but Table 13.3 shows average values. The table also shows higher damage estimates from a recent similar Dutch study for power stations assuming that the pollutants are emitted at low level in the Netherlands (Bennink et al., 2010).

Table 13.3 Estimates of marginal pollutant damage in the USA and Netherlands

Pollutant	Average US damage costs in 2007 /£ t^{-1}	Damage assuming emission in the Netherlands in 2008 /£ t^{-1}
SO_2	3570	10 800
NO_x	980	7520
$PM_{2.5}$	5840	26 000
PM_{10}	280	1710

Note: converted to £ using £1 = US$1.60 = 1.15 euros
Sources: NRC, 2010; Bennink et al., 2010.

While the precise values in these tables may be questioned, they are obviously significant sums of money. As pointed out in Section 5.6, a large power station with no pollution abatement equipment can produce up to 20 tonnes of SO_2 per hour. The annual damage costs associated with this can easily run into hundreds of millions of pounds. As will be described later in this section, the damage costs can far exceed the commercial value of the electricity produced.

These considerations have led to regulations, in Europe, the USA and elsewhere insisting on the fitting of proper pollution abatement equipment to power stations, as also described in Chapter 5.

Climate-related damage: the social cost of carbon

Impact–pathway assessment has been used to assess the pollutants listed in Table 13.3 but the effects of greenhouse gases are obviously on a global scale and have an impact over a long period of time. How can the damage cost of emitting a tonne of CO_2 (or any other greenhouse gas) be calculated?

The 2006 Stern Review on climate change commissioned by the UK Treasury has already been mentioned in Chapter 12. According to the review:

> '[climate change] ... has profound implications for economic growth and development. All in all, it must be regarded as market failure on the greatest scale the world has seen.'
>
> (Stern, 2006, p. 25)

The reason for the scale of this market failure is probably the enormous magnitude of the external costs associated with the emissions of CO_2 and other greenhouse gases. Yet, as with the effects of SO_2 and particulates, a proper economic assessment of pollution abatement requires a *valuation* of the marginal damage done by the emission of a tonne of CO_2. This is often given the name **social cost of carbon (SCC)**. Since this 'cost' is principally related to global warming, the SCC can equally apply to other greenhouse gases as long as they are appropriately weighted by their direct global warming potentials (DGWP). This term has been described in Box 13.1 and will be revisited later in the chapter. The SCC may be quoted in terms of £ per tonne of carbon dioxide equivalent, or just in £ per tonne of carbon. As described in Box 13.1 a carbon price of £1 tCO_2^{-1} is equivalent to £3.67 tC^{-1}. Where mixtures of greenhouse gases are involved the appropriate measure is tonnes of *carbon dioxide equivalent* or CO_2e.

SO_2 and particulates have relatively short lives in the atmosphere. A tonne emitted today is likely to have inflicted its damage within a year or two. If the pollution ceases, then environmental recovery may follow quite quickly.

Carbon dioxide and other greenhouse gases, however, once emitted stay in the atmosphere for a long time before they are broken down or reabsorbed by natural processes. Carbon dioxide in particular is part of the 'carbon cycle' described in the next section. It has a long **residence time** in the atmosphere of about 100 years. The effects of global warming (rising temperatures, rising sea levels, etc.) are on an even longer timescale of a thousand years or more.

Any assessment of the global warming damage done by a tonne of CO_2 has to take into account its effects over a long period of time. This raises the issues of 'discounting the future' described at the end of Chapter 12.

The SCC thus has to be an estimate of the present discounted value of the additional social costs (also sometimes called 'marginal social damage') that an extra tonne of carbon released now would impose on current and future society right across the globe (Hope and Newbery, 2008; Watkiss, 2011).

In principle, there are three basic steps involved in calculating the SCC:

(1) Estimate the short-term costs of all the various impacts of climate change plus the costs of any adaptations society makes to help minimize/cope with those impacts. For example, the impacts might include rising food prices and the adaptation costs of better coastal defences. Obviously, this process involves considerable uncertainties about the values to use.

(2) Sum these costs to all societies right across the globe. This raises questions of 'global equity'. Should damages be treated equally in different economies that have different levels of average incomes?

(3) Sum these costs over time. This step raises philosophical questions of how far into the future it is appropriate to look and an appropriate discount rate to use.

The tools that are used to estimate the SCC are called **integrated assessment models (IAMs)** (Figure 13.5). They are integrated in the sense that they include a complex climate model and an economic model. The first is used to model how greenhouse gas emissions affect atmospheric concentrations, radiative forcing and in turn climate impacts over all regions, extending into the future. The economic model is used to estimate discounted 'present values' of the costs of climate damage and compare these with the costs of abatement.

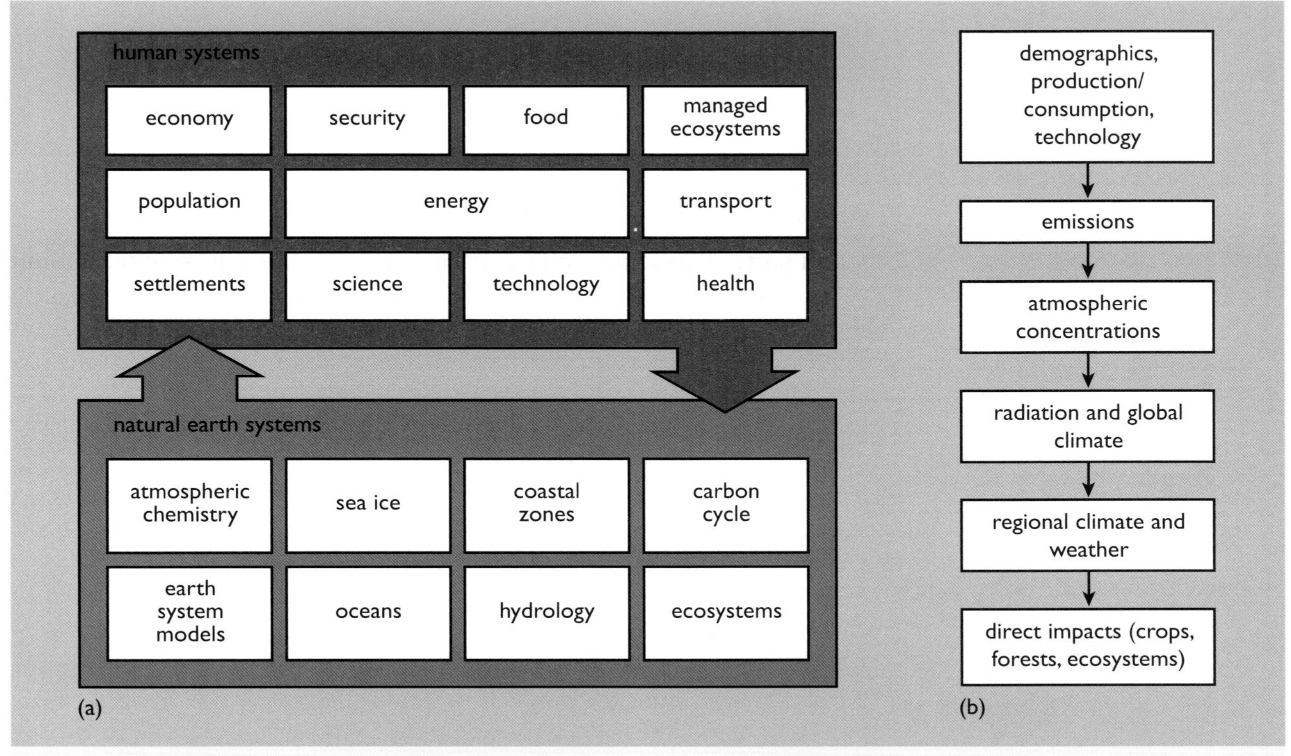

Figure 13.5 Two different examples of integrated assessment models, which show the interactions between natural and human systems: (a) based primarily on the detailed components; (b) using the key stages involved (sources: (a) Edmonds and Calvin 2010; (b) Hope, 2005)

There is a wide range of published values for the SCC in the literature, varying from a few US dollars to hundreds of dollars per tonne of CO_2. The mean value, taking into account economic and scientific uncertainty, is believed to be in the range US\$50–100 tCO_2^{-1} (£30–60 tCO_2^{-1}). The actual value depends on the future climate scenario. The SCC is higher for 'growthist' views of future global CO_2 emissions and lower for low emission scenarios (Hope, 2011).

The values for the SCC are particularly sensitive to *scientific* assumptions such as climate sensitivity as well as, for example, *philosophical* and *economic* assumptions about appropriate discount rates. There are obviously many other uncertainties about a wide range of other scientific and economic parameters.

Market-based carbon pricing: 'cap and trade'

The range of uncertainty surrounding the SCC makes it difficult for governments to justify any particular value for some form of market intervention such as a carbon tax. It can always be argued that the value is either too high or too low.

'**Cap and trade**' is an alternative approach to dealing with pollution abatement. The concept arose in the USA in the 1970s as a result of a search for 'least cost solutions' to the problem of acid rain. A scheme was implemented by the US Environmental Protection Agency (EPA) in 1990. It only applied to large plants emitting SO_2 such as power stations and heavy industry. The aim was to allow their owners flexibility in complying with progressive reductions in SO_2 emissions (Tietenberg, 2006).

Each year an overall cap was set on the *total* SO_2 emissions for the participating plants. This was progressively reduced from year to year. Each plant was given an allocation for its annual emissions (effectively a licence to pollute). Plant operators could reduce their emissions below what was expected of them by fitting abatement equipment or burning more expensive low sulfur coal. In this way they could create an 'emission credit'. This could then be sold to the owners of other plants who otherwise would not meet their targets. The penalty for emitting more than the given allocation and failing to buy the appropriate emission credits was set sufficiently high, US$2000 per ton, to encourage compliance.

The value of the emission credits in the trading marketplace effectively gives a *revealed price* for the SO_2 (about US$200 per ton at auctions between 1999 and 2004). It allows power plant owners to make commercial decisions about whether or not to invest in abatement technology, or even to switch to a completely different one, burning gas rather than coal. Between 1990 and 2008 US SO_2 emissions from electricity generation fell by a factor of two (EIA, 2009).

Given its usefulness in reducing acid rain in the USA, US negotiators ensured that emissions trading was incorporated in the 1997 international Kyoto Protocol relating to emissions of greenhouse gases (GHGs). The Protocol allows nations, or groups of nations, to establish GHG trading schemes on the same basis. The European Union's Emissions Trading Scheme (EU ETS) is a Europe-wide cap and trade scheme, introduced in its first phase in 2005. It covers electricity generation and the main energy-intensive industries. In the UK, for example, these account for around 50% of national CO_2 emissions. Each EU member state is set an overall allowance, a cap on the total emissions from all the installations covered by the system. This is then broken down into allowances for individual installations. As in the US SO_2 scheme those that emit less CO_2 than their allowance can sell their 'emission credit' to others that have exceeded their allowed limits.

The current EU ETS Phase II (2008–2012) builds on the lessons from the first phase, and has broadened to cover CO_2 emissions from other industries including flaring from offshore oil and gas production and petrochemicals. Phase III of the EU ETS (2013–2020) will bring new innovations into the scheme such as auctioning of credits.

At the time of writing (July 2011), the traded price for a permit to emit a tonne of CO_2 in the scheme is about €12 tCO_2e^{-1} (£11 tCO_2e^{-1}).

There are obviously other 'carbon prices' in use for taxation and policy evaluation purposes. Box 13.4 lists the main examples.

BOX 13.4 What is the price of a tonne of carbon dioxide?

There are at least five alternative methods to conceive of the shadow price of a pollutant such as carbon dioxide:

(1) A social *cost* of carbon. This is a valuation of the marginal damage done by the emission of an extra tonne of CO_2 or its greenhouse gas equivalent. The most likely value of this is between £30 and £60 tCO_2^{-1}

(2) The price may be considered as that which someone is willing to pay to buy emission credits in a greenhouse gas trading scheme. At the time of writing (July 2011) in the EU Emissions Trading Scheme this was about €12 tCO_2e^{-1} (£11 tCO_2e^{-1}). For sectors covered by the EU ETS, UK government policy appraisal now uses estimates of future trading prices with starting values of £21 tCO_2e^{-1} (2009 prices) with a range from £12 to £26 tCO_2e^{-1} (Watkiss and Hope, forthcoming).

(3) The price may be seen as the amount of a specific pollution tax. For example electricity, gas and heating fuel bills for UK businesses carry a *climate change levy* set by the UK government. The current (2011) tax rate for electricity is 0.47 p kWh^{-1}. Given that the average specific CO_2 emissions from electricity generation are just under 0.5 kg CO_2 kWh^{-1} (see Table 9.4) this can be seen as equivalent to a carbon price of about £10 tCO_2^{-1}.

(4) The price may be valued as the *penalty price* charged by an energy regulator for failing to meet some climate-related target. For example UK electricity suppliers are obliged to supply a rising percentage of electricity from renewable sources. The scheme is regulated by the Office of Gas and Electricity Markets (OFGEM). If the suppliers fail to meet their targets they are charged a 'buy-out' or penalty price. This is currently 3.7 p kWh^{-1} (OFGEM, 2010). Again using the average UK specific emission figure, this can be seen as equivalent to £75 tCO_2^{-1}.

(5) The price may be interpreted as that implied by a given reductions target. A regulator might look at a curve such as Figure 13.2 and decide that in order to achieve an emission reduction of X tonnes per year, a price of £Y CO_2e^{-1} will be required. Current UK government policy for sectors outside the ETS is based on a price of £51 tCO_2e^{-1} (with a range of £25 to £76). A target that *reduces* over time may require a progressively *increasing* carbon price. For example, the UK Government has agreed a target of an 80% reduction in national greenhouse gas emissions from 1990 levels by 2050. This will require a steady reductions of emissions year on year (a topic that will be discussed in the next chapter). In order to achieve this a carbon shadow price is suggested for analysis purposes starting at £14.10 tCO_2e^{-1} in 2010, rising to £16.30 in 2020, £70 in 2030 and £135 in 2040. This is used in a comparison of electricity generation technologies later in this chapter.

What does all this mean to the average person?

A carbon tax of £50 tCO_2^{-1} is roughly equivalent to the following:

2.5p kWh^{-1} on the price of gas-fuelled electricity

1p kWh^{-1} on natural gas for heating

12p on a litre of petrol or diesel. This is about a 10% increase on UK July 2011 pump prices and roughly equivalent to an increase of US$30 a barrel in world oil price.

Using damage costs to compare electricity generating systems

So far this section has concentrated on the pricing of pollution and external costs. How are these prices used to influence investment in technology? First we consider the pollution costs of electricity from conventional coal-fired power stations and then the use of a carbon price in comparisons of 'high carbon' and 'low carbon' electricity generation technologies.

The external costs of coal-fired electricity

Figure 13.4(a) has set out a chain for assessment including mining, and transport of the coal. In practice 80% or more of the environmental damage arises from the combustion of the fuel and the emission of SO_2, NO_x and particulates. The environmental effects of these have been summarized in Table 13.2 but will be described in detail later in this chapter. Table 13.3 has given some financial estimates of the marginal damage done by the emission of an extra tonne of each of these. These values can be used to determine the overall damage costs of a power station, identify which are the most serious pollutants and assess the cost–benefit of fitting pollution abatement.

Today's electricity systems contain a mixture of power plants of different vintages and grades of cleanliness. Box 13.5 assesses the external costs of a 'dirty' power station with little pollution abatement. The figures are typical of what may have been considered 'normal' in both the USA and Europe in the past.

The sample calculation in Box 13.5 shows that the external costs from SO_2, NO_x and particulates can exceed the commercial electricity price. This kind of analysis, carried out in the 1980s, was key to legislation in Europe and the USA which insisted on pollution abatement being fitted to new and existing coal-fired power stations. Flue gas desulfurization (FGD) can reduce SO_2 emissions by 90%, and electrostatic precipitators can cut particulate emissions by over 95%. There are various techniques to cut NO_x emissions by 60% or more. If the above calculation with US prices is repeated using emission factors for a more modern conventional power station design, for example a 1995 study for a new UK power station at West Burton in Yorkshire (CEC, 1995c), then the external electricity cost falls to only 0.5 p kWh^{-1}. The NRC study in the USA mentioned earlier showed a wide range of externality costs between the best and worst power plants, but the estimate for a national average for non-climate-related pollution was 3.2 US cents per kWh, equivalent to 2 p kWh^{-1} (NRC, 2010). However, they

BOX 13.5 External costs of a 'dirty' coal-fired power station

As described in Chapter 5, modern coal-fired power stations are fitted with various systems to reduce the amount of pollution they produce. However it is worth carrying out a sample calculation using one of 'yesterday's power plants' using today's environmental damage estimates:

A 660 MW coal-fired power station built in 1970 has been fitted with little in the way of pollution abatement systems. Every hour it emits 10 tonnes of SO_2, 2.5 tonnes of NO_x, 0.5 tonnes of $PM_{2.5}$, 0.5 tonnes PM_{10} and 650 tonnes of CO_2.

The values are typical of the worst-performing 5% of US plants in 2005 (NRC, 2010).

(1) *Use the US values in Table 13.3 to calculate the non-climate-related external damage costs per kWh of electricity generated.*

(2) *Using a carbon price of £50 per tonne of CO_2e calculate the climate related damages per kWh.*

In each hour the power station will produce 660 000 kWh. It is convenient to express the pollutant emissions as a specific emission factor in grams per kWh. For example the SO_2 the factor is:

$$\frac{10.0 \times 1000000}{660000} = 15.2 \text{ g kWh}^{-1}$$

The pollutant costs shown earlier in Table 13.3 in £ per tonne can be converted to pence per gram by dividing by a factor of 10 000. The total for each pollutant can then be calculated:

Table 13.4 Damage costs for coal-fuelled electricity

Pollutant	Specific emission factor /g kWh^{-1}	Damage cost /p g^{-1}	Total cost /p kWh^{-1}
SO_2	15.2	0.357	5.43
NO_x	3.8	0.098	0.37
$PM_{2.5}$	0.8	0.584	0.47
PM_{10}	0.8	0.028	0.02

The total non-climate-related damage costs are thus **6.3 p kWh^{-1}**

If the calculation is repeated using the Dutch figures in Table 13.3, then the damage costs are over 21 p kWh^{-1}.

Since the damages have been expressed in 2007 money this should perhaps be compared with a UK large-scale electricity price for that year (see Figure 12.3) of about 4 p kWh^{-1}. It is thus possible for the external costs to be larger than the 'commercial' costs. It is also obvious that the most serious pollutants are the acid emissions and the fine particulates.

The climate-related damages can be calculated in the same way, for example by using an assumed carbon price of £50 t CO_2e^{-1}

Table 13.5 CO_2 damage costs for coal-fuelled electricity

Pollutant	Specific emission factor /g kWh^{-1}	Damage cost /p g^{-1}	Total cost /p kWh^{-1}
CO_2	985	0.005	**4.9**

This is obviously a significant cost and, as will be described later, is sufficient to affect the choices made about electricity generation technologies.

calculated that climate change costs could well add a *further* 3 cents per kWh assuming a carbon price of US$30 per ton of CO_2e (about £18 tCO_2e^{-1}).

Obviously, fitting pollution abatement equipment involves capital and operational costs and slightly reduces the overall generation efficiency. The use of flue gas desulfurization involves a whole extra chain of mining for the limestone used. In the competitive electricity market in the UK these extra costs have helped tip the balance away from coal in favour of natural gas-fuelled generation. Since natural gas is a low-sulfur fuel the non-climate-related external costs of gas-fuelled generation are far lower than for coal, the NRC estimate being only 0.16 US cents per kWh, equivalent to 0.1 p kWh^{-1}. These costs relate mainly to NO_x emissions and fine particulates.

If the emissions from a coal-fired power station are reduced, then the significance of other parts of the fuel cycle become relatively more important. There are a whole range of issues. Assigning a 'damage cost' to them is likely to be a matter of local or regional circumstances:

- Mining illness and accidents, a topic which will be discussed later in the chapter.
- Coal transportation. As pointed out in Chapter 5, coal is likely to be transported thousands of kilometres by train, ship, or even, in China, by road. This requires consumption of energy, involves the emission of pollutants and creates a strain on the rail and road networks of the particular countries. As pointed out in Chapter 9, an alternative is to site the power stations close to the mines and use 'coal by wire', long-distance transmission grids instead. This alternative has its own problems of visual intrusion of pylons in the landscape.
- Environmental damage due to mining, a topic that will be discussed later in the chapter.

Comparing the overall costs of different electricity generation technologies

Electricity can be produced from many different sources, using a range of technologies. The direct production costs of conventional nuclear, CCGT and wind generators have been discussed in Chapters 11 and 12. The key components are:

- capital costs including those of decommissioning
- operation and maintenance costs
- fuel costs including those of waste.

There may also be CO_2 disposal costs for future power plants using carbon capture and storage.

There are also the external costs to be considered:

- those related to climate, i.e. a carbon price
- other atmospheric pollution, such as the SO_2, NO_x and particulate damage discussed above
- water pollution due to leaching of heavy metals from mines and ash
- accidents in the fuel chain
- those related to land use, transportation and mining

- costs, and benefits, related to security of supply in terms of fuel mix diversification
- costs and benefits related to increased flexibility of the energy system, to facilitate greater use of intermittent sources like wind.

Figure 13.6 provides a schematic breakdown of total production and external costs for a hypothetical power plant.

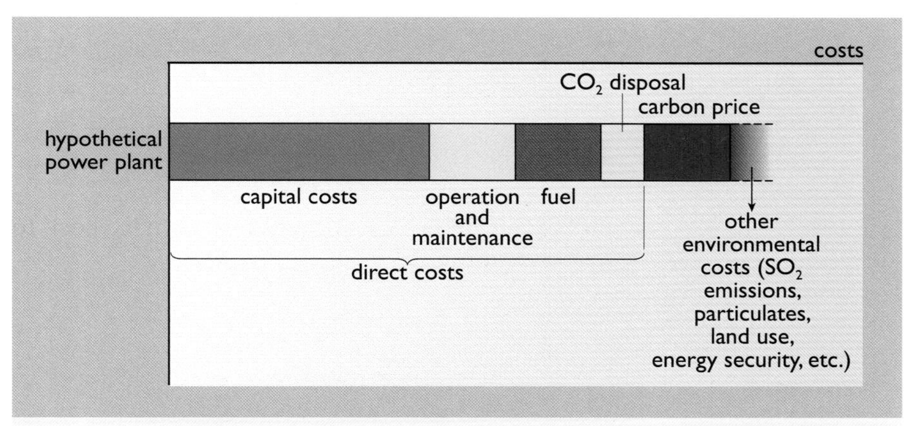

Figure 13.6 Schematic breakdown of total production costs for a hypothetical power plant (adapted from Bennink et al., 2010)

The full external costs of different technologies are rarely compared in a single study. However, climate related costs are increasingly being taken into account. As pointed out in Chapter 9, different electricity generation technologies involve the emission of different amounts of CO_2 in order to produce a kilowatt-hour of electricity. Any thorough analysis should extend this to the total *greenhouse gas* emissions involved, including the full fuel cycle.

Figure 13.7 shows a range of estimates for the specific greenhouse gas emissions for electricity generation technologies in Europe expressed in g CO_2e kWh^{-1}. This includes estimates for future technologies such as carbon capture and storage (CCS). Obviously there is an enormous range between the worst coal-and lignite-fired power stations and the 'low-carbon' technologies, renewables and nuclear. The use of waste heat from power stations for district heating can reduce the effective emission factors of fossil-fuelled plant. The wide range of values for oil-fired electricity reflects the wide size range of plant in use. Generally, smaller generation plant have lower efficiencies and higher specific emissions. The figure also shows the average 2009 performance of UK electricity generation in terms of power plant CO_2 emissions alone, as given in Table 9.4.

Figure 13.8 shows a cost comparison of different electricity generation technologies for the UK, assuming construction in 2009 and including a rising carbon price as described as the fifth option in Box 13.4. For simplicity the small greenhouse gas emissions involved in nuclear, wind power or hydro have been ignored. It is also obvious, comparing this chart with Figure 13.6, that a whole range of other environmental costs have been ignored.

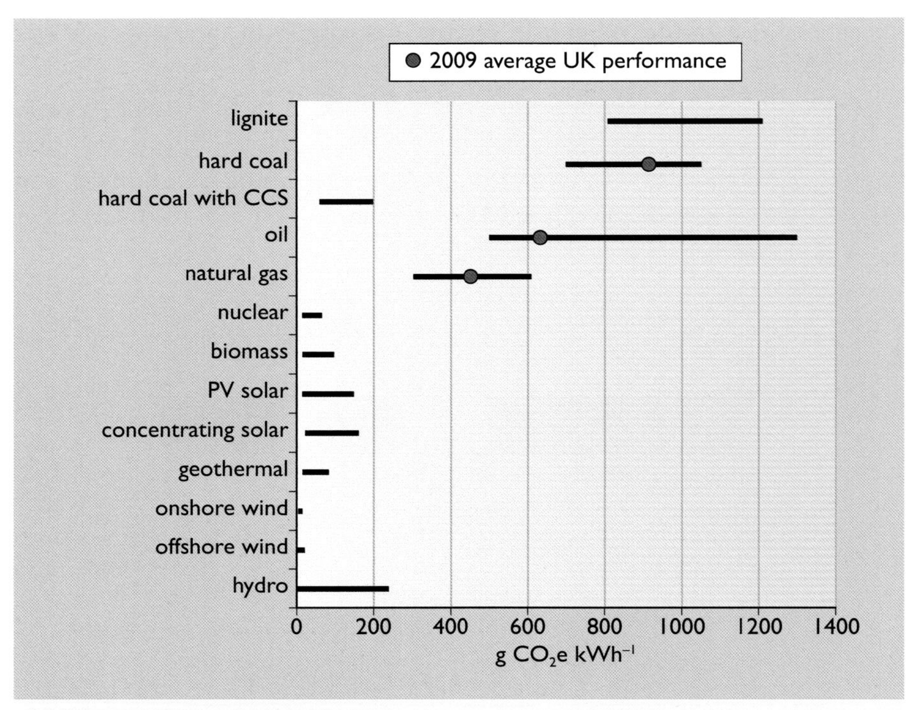

Figure 13.7 Specific greenhouse gas emissions for a variety of electricity generation technologies in Europe; the 2009 average performance figures for UK plant are for CO_2 emissions alone (sources: EEA, 2011; DECC, 2010)

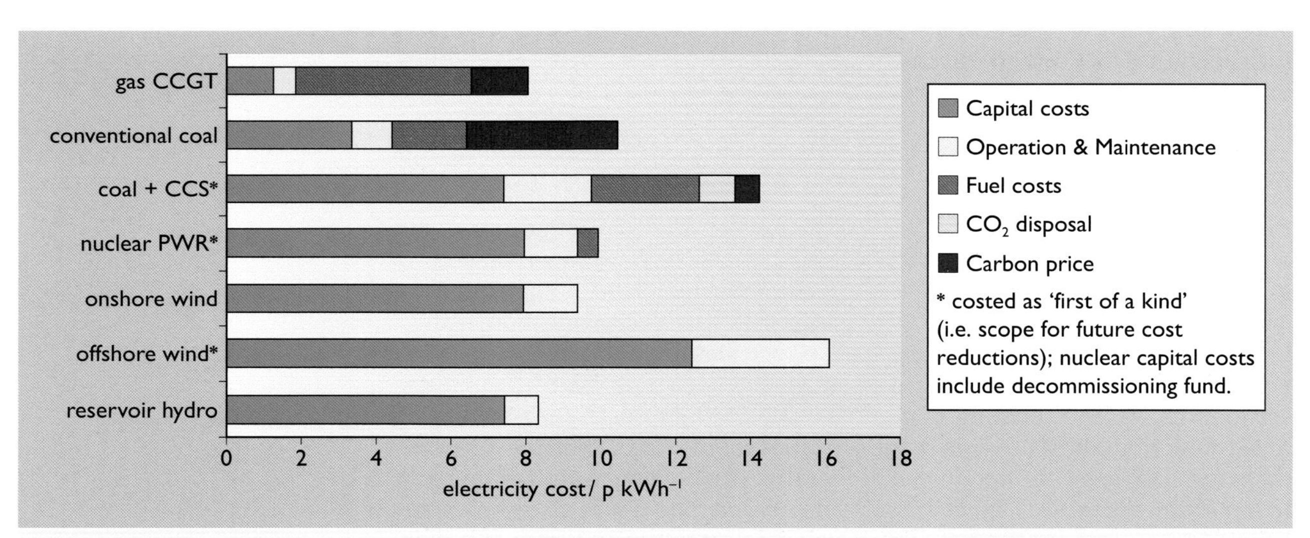

Figure 13.8 Cost breakdown of electricity from different generation technologies (in pence per kWh) assuming construction in 2009, using a discount rate of 10%, including a rising carbon price of £14.10 tCO_2e^{-1} in 2010 to £16.30 in 2020, £70 in 2030 and £135 in 2040. 'Coal + CCS' refers to post-combustion carbon capture and storage. The values for gas CCGT and onshore wind turbines are not intended to be the same as those given in Box 12.6 (source: Mott MacDonald, 2010).

The figure shows that the carbon price makes up only a fraction of the overall costs for most technologies, with the exception of conventional coal technology. Its inclusion obviously changes the overall balance of technology choice, particularly between conventional coal and lower carbon alternatives. The calculations assume a rising carbon price with time. If the evaluation is repeated for, say, 2023, then the carbon price for conventional coal will have risen to over 10p kWh^{-1}, dominating its overall costs and making coal with CCS a cheaper alternative.

13.4 Climate change

The human perturbation of the carbon cycle

Probably the most important global-scale impact of humanity's fossil fuel use is global warming resulting from the emissions of greenhouse gases (GHGs) such as carbon dioxide (CO_2), methane (CH_4) and nitrous oxide (N_2O). As described in Chapter 1, these human-induced or 'anthropogenic' GHG emissions are now considered to be the principal cause of a process that has already led to a rise in the Earth's mean surface temperature of around 0.8 °C since the Industrial Revolution. In this section we consider some of the effects and consequences of this climate change in more detail. Since this is a matter of international concern, the emissions of most greenhouse gases are covered by the **Kyoto Protocol** of the United Nations Framework Convention on Climate Change (UNFCCC). This was initially adopted at a meeting in Japan in 1997 and aims to stabilize the concentrations of GHGs in the atmosphere at levels that would not cause 'dangerous anthropogenic climate change'. In recent years this has been interpreted as restricting the long-term rise to not more than 2 °C above the pre-industrial level (i.e. about 1.2 °C above the current level).

Of the GHGs, CO_2 has the largest overall contribution and nearly all of the emissions are from the combustion of fossil fuels. Figure 13.9 shows the rapid increase in CO_2 emissions from fossil fuel combustion since 1950. There is also a small extra contribution (about 5% not shown) from cement manufacture, principally by the elimination of CO_2 from limestone ($CaCO_3$); this is not a combustion process.

Not all of the emitted carbon dioxide remains in the atmosphere. As shown in Table 13.6, between 2000 and 2005 the total anthropogenic emissions from fossil fuels and cement manufacture were about 7 GtC y^{-1}. Of this about 2 GtC y^{-1} has been absorbed by the oceans, dissolving in the surface layers. About a further 1 GtC y^{-1} is absorbed on land, for example by photosynthesis in growing trees and vegetation. This is part of the overall global carbon cycle described in Box 13.6. Since the overall take-up by the sea and land is less than the amount emitted by humanity, the quantity remaining in the atmosphere has been rising at the rate of about 4 GtC y^{-1}.

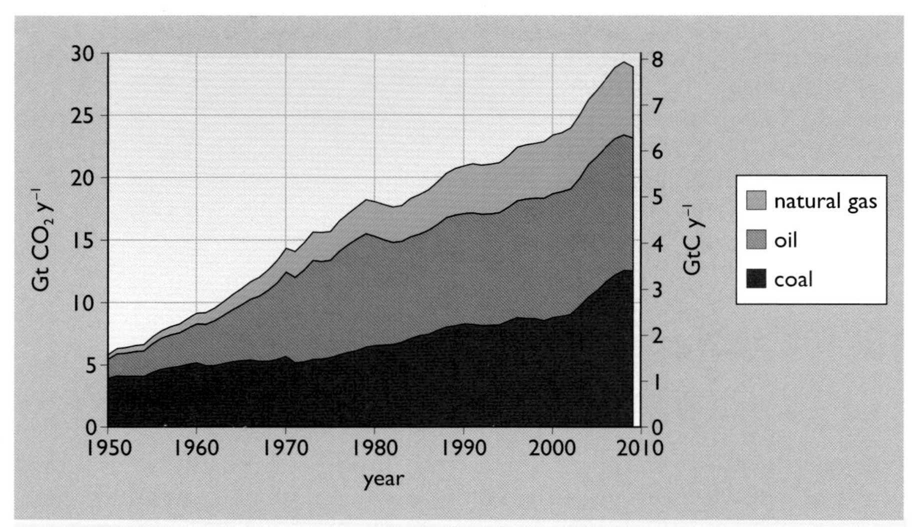

Figure 13.9 Global carbon dioxide emissions from fossil fuel combustion 1950–2009 (sources: Boden et al., 2010, IEA, 2010b, BP, 2010)

Table 13.6 Net flows of carbon to and from the atmosphere, 1980–2005

Flows of carbon[1]	1980s /GtC y^{-1}	1990s /GtC y^{-1}	2000–2005 /GtC y^{-1}
Atmospheric increase	3.3 ± 0.1	3.2 ± 0.1	4.1 ± 0.1
Emissions (fossil fuel, cement manufacture)	5.4 ± 0.3	6.4 ± 0.4	7.2 ± 0.3
Flows from atmosphere to oceans	(1.8 ± 0.8)	(2.2 ± 0.4)	(2.2 ± 0.5)
Flows from atmosphere to land	(0.3 ± 0.9)	(1.0 ± 0.6)	(0.9 ± 0.6)

[1] Positive values are flows to the atmosphere; negative values (in brackets) represent uptake from the atmosphere.
Source: IPCC, 2007a, Table TS.1, p. 26.

As a result, atmospheric CO_2 concentrations have increased from about 285 parts per million by volume (ppmv) at the end of the pre-industrial era, around 1850, to 390 ppmv in 2010 – an increase of over 30% in 150 years (Figure 1.20(a)). Humans are now increasing CO_2 concentrations in the atmosphere by approximately 2 ppmv per year.

So, in total, how much carbon dioxide remains in the atmosphere and how much has been absorbed by the oceans or terrestrial ecosystems? The IPCC estimates that total emissions of CO_2 from fossil-fuel burning and cement production from 1850 to 1998 were approximately 270 ± 30 GtC (IPCC, 2000, Table 2.1). In addition, approximately 136 ± 55 GtC has been emitted as a result of land-use change (agriculture and urbanization), and mainly from forest ecosystems, for example by removing and burning the standing stocks of biomass in timber (Table 13.7).

BOX 13.6 The carbon cycle

Carbon dioxide is emitted into the atmosphere by a number of *natural* processes, for example the respiration of plants and animals, and the process of the decay of organic material. It is also taken up from the atmosphere by other natural processes: photosynthesis in plants and algae on land and phytoplankton in the sea. When these die they may decay and their carbon content will remain in the soil on land or sink to the bottom of the sea. If this carbon remains buried, over a period of millions of years it will slowly form the fossil fuels (coal, oil and gas). CO_2 dissolved in seawater may be used to build the calcium carbonate shells of small sea creatures. When these die, these sink to the bottom of the ocean and eventually build up as layers of carbonate rocks.

These processes constitute the *natural carbon cycle* which has, of course, been operating for millions of years. These are shown at the left and right of Figure 13.10. Carbon cycles through both the biosphere (living plants and animals) and the geosphere (the Earth and its rocks). The carbon cycle is thus described as a **biogeochemical process**.

The increasing human emissions of CO_2 from the combustion of fossil fuels have added an extra factor. This flow is larger than the take-up of CO_2 by the sea and land and has led to a steadily increasing CO_2 concentration in the atmosphere.

What would happen if anthropogenic CO_2 emissions ceased tomorrow? The natural take-up processes should slowly reduce the atmospheric concentration back towards its pre-industrial level. However, a tonne of CO_2 emitted today is likely to remain in the atmosphere for a hundred years. This is known as its **mean residence time**.

It must be said that the overall net *flows* of carbon involved are very small compared to the overall global *stocks* of carbon in the sea, soil, fossil fuel resources and standing vegetation. However, humanity's intervention is a significant change in what appears to be a finely balanced system with multiple feedbacks.

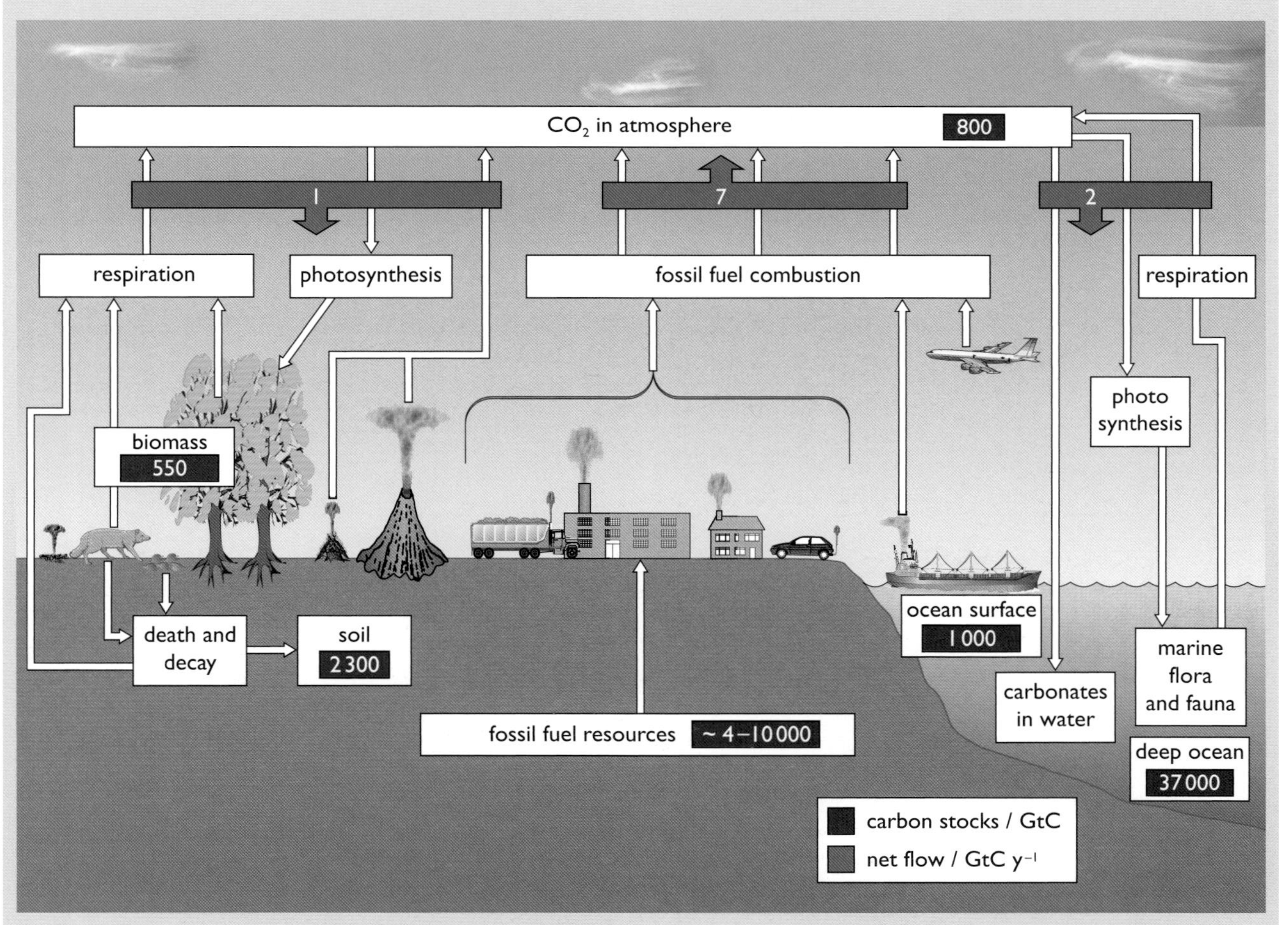

Figure 13.10 The carbon cycle (flow figures are 2000–2005 averages)

Table 13.7 Changes in global carbon stocks between fossil sources, land, atmosphere, oceans and forests, 1850–1998

Global carbon stock	Carbon mass/GtC
1 Total emissions of CO_2 to atmosphere from fossil-fuel burning and cement production	270 ± 30
2 Emissions as a result of land-use change	136 ± 55
3 Increase in carbon dioxide in atmosphere	176 ± 10
Remainder (1 + 2 – 3) is what is estimated to have been taken up by oceans and forests	230 ± 95

Source: IPCC, 2000, p. 4.

Other greenhouse gases

The other anthropogenic greenhouse gases, though individually less important, are together estimated to increase the global warming effect created by anthropogenic CO_2 emissions by about 56%. The two most important are methane (CH_4) and nitrous oxide (N_2O). It has been estimated that by 2007 their atmospheric concentrations had increased by about 150% and 18% respectively since 1850. A significant proportion of their increased emissions is estimated to be due to energy-related activities.

As mentioned earlier, the relative warming effect of greenhouse gases can be expressed in terms of their direct global warming potentials (DGWPs). These are expressed relative to that of CO_2. Table 13.8 shows the range of global warming potentials of those greenhouse gases regulated under the Kyoto Protocol. Since different gases have different residence times in the atmosphere, the figures have been calculated using the warming effect over one hundred years.

Table 13.8 Direct global warming potential for greenhouse gases regulated under the Kyoto Protocol

Greenhouse gas	DGWP for 100-year time horizon
Carbon dioxide (CO_2)	1
Methane (CH_4)	25
Nitrous oxide (N_2O)	298
Hydrofluorocarbons (HFCs)	100–14 800 depending on the gas
Perfluorocarbons (PFCs)	7390–12 300 depending on the gas
Sulfur hexafluoride (SF_6)	22 800

Source: IPCC, 2007a.

The first three gases in Table 13.9 occur naturally. The last three are almost completely man-made and used as refrigerants or in industrial processes. In addition, there are the halocarbons, which include the widely used man-made refrigerants chlorofluorocarbons (CFCs). As described in Chapter 9, their chlorine content causes depletion of the Earth's protective ozone layer in the upper atmosphere – the stratosphere. Halocarbons are also potent greenhouse gases, CFC-12 (the most commonly used in the past) having a DGWP of 6600. The phase-out of CFCs has been covered by the Montreal Protocol of 1987. This has had the fortunate side-effect of reducing their contribution to global warming. It has led to their replacement with alternative non-ozone depleting refrigerants such as HFCs, which do not contain chlorine, though these need to be carefully chosen in order to have a low DGWP.

Factors in global warming

Overall there are many factors affecting the total atmospheric radiative forcing or 'warming' effect. Figure 13.11 shows the relative magnitudes of these, ranked by their level of scientific understanding. The role of the main greenhouse gases is obviously the most important, but there are others.

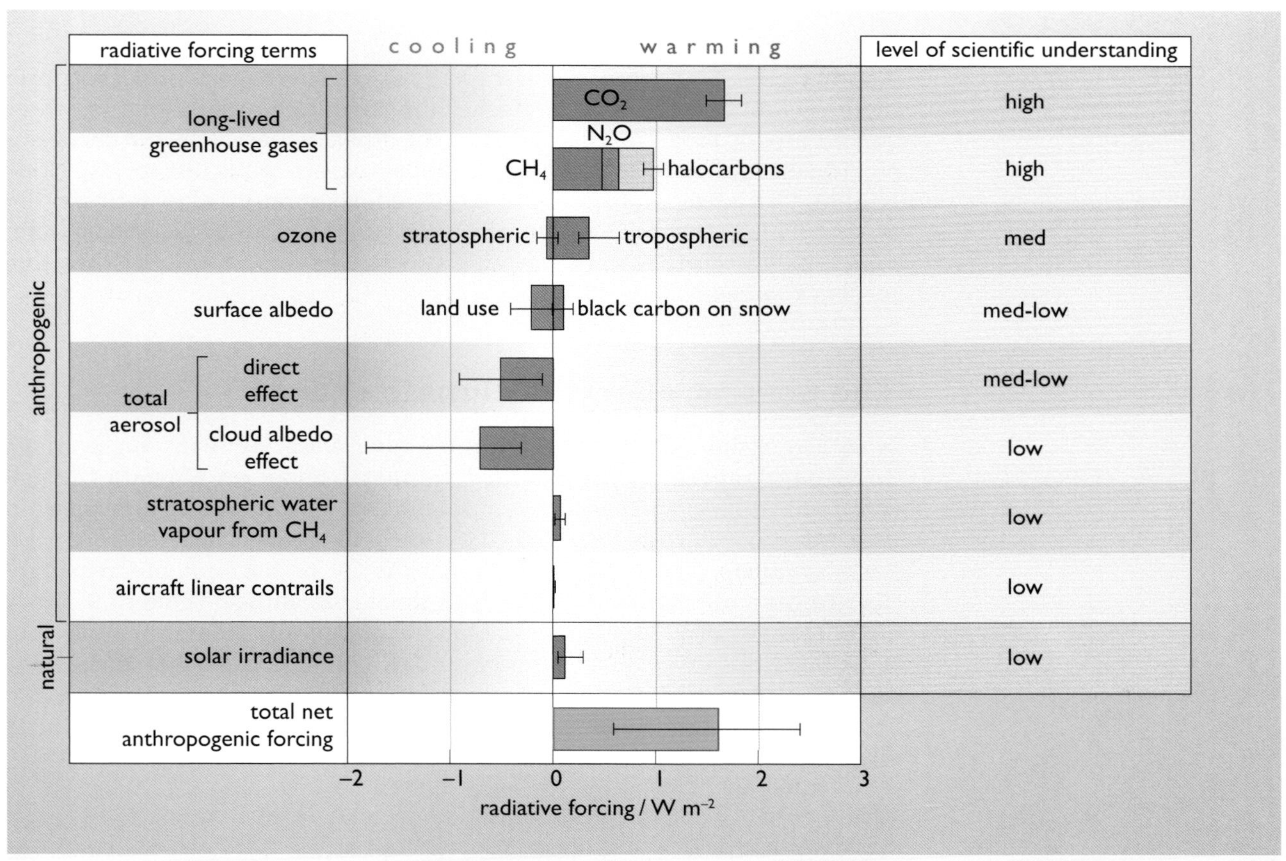

Figure 13.11 The radiative forcing effects of various factors, arranged in order of level of scientific understanding. Positive radiative forcing effects are shown in orange and negative effects in blue. The black horizontal lines indicate the range of estimates for these forcings (source: adapted from IPCC, 2007a)

Ozone is another important greenhouse gas. It is found in the lower atmosphere – the troposphere – as well as in the stratosphere. It is formed in the troposphere through the interaction of oxygen, nitrogen oxides and certain hydrocarbons. It is estimated that the warming effect due to formation of ozone in the troposphere is greater than the cooling effect of ozone depletion (caused by halocarbons) in the stratosphere (see Figure 13.11).

In contrast to the global warming effects of the above gases, there is the *cooling* effect of sulfate aerosols and suspended fine particles that are generated in the atmosphere by burning fossil and biomass fuels. These are periodically also produced in very large quantities by volcanic eruptions and can then cause a measurable drop in global temperature. These aerosols tend to reflect incoming solar radiation (i.e. they have a high 'albedo') but absorb little outgoing infrared radiation. In addition they seed the formation of clouds, which also reflect a large amount of incoming solar radiation. At present, as shown in Figure 13.11, the total cooling effect of aerosols offsets almost half of the warming effect of the other greenhouse gases. However, this arguably beneficial influence is likely to diminish in coming decades as SO_2 and particulate emissions are reduced by non-climate-related pollution regulations.

Another greenhouse gas, water vapour, also plays a very important role in the 'natural' greenhouse effect. Human-induced emissions of water vapour are too small to directly affect the global water cycle, but there is a significant indirect effect. Increased global temperatures due to emissions of other GHGs are likely to cause increased evaporation of water from the oceans. This in turn will lead to increased water vapour in the atmosphere, which then leads to further warming and further evaporation – a 'positive feedback' effect.

The present effects of climate change

Is anthropogenic climate change and global warming really happening? Successive reports of the Intergovernmental Panel on Climate Change (IPCC) have been increasingly confident that the increased atmospheric concentrations of anthropogenic GHGs are having a significant effect on the Earth's climate. The 'signals' of climate change are becoming increasingly discernible amid the 'noise' of natural climate variability. Table 13.9 shows a wide range of key atmospheric, climatic and biophysical indicators that have been observed during the twentieth century. Note that this is not an exhaustive list. It includes both changes attributable to anthropogenic climate change and those that may be caused either by that or by natural variations.

While many indicators are changing in a direction consistent with a warming world it is important to note that at the time of publication of IPCC's AR4 in 2007, some had *not* been observed to change, including:

- diurnal temperature range (the variation in temperature from the daily high to the nightly low)
- the thermohaline circulation of the global ocean (described in Chapter 14)

Table 13.9 Observed changes in climate according to the IPCC's Fourth Assessment Report (AR4)[1]

Indicator	Observed changes	
Concentration indicators	**Year(s)**	**Level**
Atmospheric concentration of CO_2	1000–1750 2000 2010	280 ppmv 368 ppmv 388 ppmv – a rise of about 2 ppmv per year over last decade (a)
Atmospheric concentration of CH_4	1000–1750 2000 2010	700 ppbv 1750 ppbv 1870 ppbv.(a)
Atmospheric concentration of N_2O	1000–1750 2000 2010	270 ppbv 316 ppbv 323 ppbv (a)
Tropospheric concentration of O_3	1750–2000	Increased by 35 ± 15%, varies with region
Stratospheric concentration of O_3	1970 –2000	Decreased, varies with altitude and latitude
Atmospheric concentration of HFCs, PFCs, and SF_6	1947–2007	Increased globally
Temperature/weather indicators		
Global mean surface temperature	1906–2005: increased by 0.74 °C, with 1956–2005 trend nearly twice that for 1906–2005; land areas have warmed faster than the oceans. The warming between 1977 and 2007 was widespread over the globe, and greatest at higher northern latitudes. January 2000 to December 2009: warmest decade on record. (b)	
Northern hemisphere surface temperature	Average northern hemisphere temperatures during the second half of the 20th century were *very likely* higher than during any other 50-year period in the last 500 years and likely the highest in at least the past 1300 years.	
Hot days/heat index, cold/frost days	Changes in extremes of temperature are consistent with warming. Observations show widespread reductions in the number of frost days in mid-latitude regions, increases in the number of warm extremes (warmest 10% of days or nights) and a reduction in the number of daily cold extremes (coldest 10% of days or nights) Heatwaves have increased in duration beginning in the latter half of the 20th century.	
Precipitation	Long-term trends in precipitation amounts from 1900 to 2005 have been observed in many large regions. Significantly increased precipitation has been observed in the eastern parts of North and South America, northern Europe and northern and central Asia. Drying has been observed in the Sahel, the Mediterranean, southern Africa and parts of southern Asia.	
Heavy precipitation events (i.e. storms and flash floods)	Substantial increases in heavy precipitation events have been observed.	
Frequency and severity of drought	More intense and longer droughts have been observed over wider areas, particularly in the tropics and subtropics since the 1970s.	

(Continued)

Table 13.9 *(Continued)*

Indicator	Observed changes
Biological and physical indicators	
Ocean warming/Global mean sea level	Observations since 1961 show that the average temperature of the global ocean has increased to depths of at least 3000 m and that the ocean has been absorbing more than 80% of the heat added to the climate system. Such warming causes seawater to expand, contributing to sea-level rise. Global mean sea level rose in the range 12–22 cm during the 20th century. Global average sea level rose at an average annual rate of 1.8 mm (1.3 to 2.3 mm) from 1961 to 2003. Losses from the ice sheets of Greenland and Antarctica have probably contributed to sea-level rise from 1993 to 2003.
Ocean acidification	Decrease of pH of 0.1 units from pre-industrial times.
Duration of ice cover of rivers and lakes	On average, the general trend in northern hemisphere river and lake ice over the past 150 years indicates that the freeze-up date has become later at an average rate of 5.8 ± 1.9 days per century, while the breakup date has occurred earlier, at a rate of 6.5 ± 1.4 days per century.
Arctic sea-ice extent and thickness	Annual average Arctic sea-ice extent has shrunk by about 2.7 ± 0.6% per decade since 1978 based upon satellite observations. The decline in summer extent is larger than in winter extent, with the summer minimum declining at a rate of about 7.4 ± 2.4% per decade.
Glaciers and ice caps	During the 20th century, glaciers and ice caps have experienced widespread mass losses and have contributed to sea-level rise. The Greenland and Antarctic ice sheets taken together have very likely contributed to the sea level rise of the past decade. It is very likely that the Greenland Ice Sheet shrank from 1993 to 2003, with thickening in central regions more than offset by increased melting in coastal regions. Average Arctic temperatures increased at almost twice the global average rate in the past 100 years. Arctic temperatures have a high decadal variability, and there was also a warm period from 1925 to 1945.
Snow cover	Snow cover has decreased in most regions, especially in spring.
Permafrost	Permafrost and seasonally frozen ground in most regions display large changes in recent decades. Temperatures at the top of the permafrost layer have generally increased since the 1980s in the Arctic (by up to 3 °C). The maximum area covered by seasonally frozen ground has decreased by about 7% in the northern hemisphere since 1900, with a decrease in spring of up to 15%.
Atmospheric circulation patterns	Changes in large-scale atmospheric circulation are apparent (e.g. El Niño[1] and Atlantic oscillation patterns).
Timing of spring/plant and animal ranges/breeding, flowering, and migration	Altered timing of spring events for a broad range of species and locations. Spring arrives earlier by 2.3–5.3 days/decade in the last 30 years. The vast majority of studies of terrestrial biological systems reveal notable impacts of global warming over the last three to five decades, which are consistent across plant and animal taxa: earlier spring and summer phenology and longer growing seasons in mid and high latitudes, production range expansions at higher elevations and latitudes, some evidence for population declines at lower elevations or latitudinal limits to species ranges, and vulnerability of species with restricted ranges, leading to local extinctions. (c)
Coral reef bleaching (i.e. die-back)	Increasing evidence for climate change impacts on coral reefs.

Table 13.9 *(Continued)*

Indicator	Observed changes
Economic indicators	
Weather-related economic losses	Global losses reveal rapidly rising costs due to extreme weather events since the 1970s that may be driven in part by climate change (but mostly due to changes in exposure to risk). For specific regions and perils, including extreme floods on the some of the largest rivers, there is evidence of an increase in occurrence. (c)

Notes:
[1] 'El Niño' is a warm water current that periodically flows along the coast of Ecuador and Peru, disrupting the local fishery. It is associated with fluctuations in the circulation of the Indian and Pacific oceans, called the 'Southern Oscillation'. During El Niño events the prevailing trade winds weaken and the equatorial counter-current strengthens, causing warm surface waters in the Indonesian area to overlie the cold waters of the Peru current. Such events have climatic effects throughout the Pacific region and in many other parts of the world. (IPCC, 2007a).
Sources: unless otherwise indicated IPCC, 2007a; (a) Blasing, 2011; (b) NASA, 2010; (c) IPCC, 2007c, pp. 99, 104.

- small-scale phenomena such as tornadoes, hail, lightning and dust storms
- the extent of Antarctic (as distinct from Arctic) sea ice.

The observed changes in global weather are consistent with the predictions made by sophisticated computer models of the world's climate. These models calculate the effects of increasing greenhouse gas concentrations, taking into account the effects of suspended fine particles, variations in solar activity, positive feedback effects and numerous other factors. Such models predict further increases in global mean temperature during the 21st century of between 1.1–6.4 °C (depending on the assumptions regarding future emission levels and the climate's sensitivity to them) together with a rise in mean sea levels of between 20 and 50 cm (IPCC, 2007a).

The future consequences of climate change

A rise in global mean temperature of a few degrees Celsius, coupled with a rise in sea level of a few tens of centimetres by the end of the 21st century, might at first sight seem relatively unproblematic. But these rising average levels do not reflect the increases in *extremes* of temperature, sea levels, rainfall and other weather-related phenomena that are likely to accompany them.

Increasing mean temperature means increasing heat energy within the planet's weather system and this is likely to manifest itself in an increasing frequency of more extreme weather events such as hurricanes and storms. Model projections also suggest that future temperature increases will be greater on land than in the oceans; greater inland than in coastal regions; and greater at high latitudes than at lower latitudes.

Increased temperatures will lead to increased evaporation of water, which will in turn lead to increased precipitation (rain and snow). It is estimated that a temperature increase of one degree Celsius will lead to about a 2.5% increase in precipitation. Again, the increases in precipitation are likely to result in more extreme events, such as flooding. But paradoxically, despite such increases in rainfall, increased evaporation in summer is likely to lead to droughts in some regions.

Increased evaporation means that many regions are likely to become more humid. This, coupled with increases in temperature, will lead to significant increases in the so-called 'heat index' (an indicator of discomfort) and thus to additional heat-related illnesses and deaths – though in some regions these could be at least partially offset by decreases in winter deaths due to cold.

Melting ice is another major consequence of global warming. Mountain glaciers worldwide appear to be retreating, and there is considerable evidence of ice melting in the Arctic, but this has little impact on global sea levels since Arctic ice is mostly floating on water. However, much of the ice in the Antarctic is land-based. Very substantial global sea-level rises would result if the supports that anchor the enormous West Antarctic Ice Sheet to land were to melt, causing the sheet gradually to slide into the sea. This, it is estimated, could cause global sea levels to rise by several *metres* over the next 1000 years (IPCC, 2007a).

However, the various phenomena are not fully understood and there is considerable uncertainty as to many (though not all) of the consequences of climate change (both natural and anthropogenic). This is due in great measure to the non-linear nature of the response of the climate systems to changes in temperature and other inputs.

Some of the *ecological* consequences of global warming can be predicted with reasonable confidence; others are highly uncertain. Some species are unlikely to be able to adapt quickly enough to increasing temperatures and may decline or become extinct in some regions. Increasing temperatures are likely to lead to pests and vector-borne diseases (such as malaria, which is transmitted by mosquitoes) that were once confined to the tropics becoming prevalent in previously temperate regions. Warmer winters in many regions are likely to mean that more pests survive winter than in the past.

Figure 13.12 shows an IPCC summary (IPCC, 2007b) of the likely consequences of climate change during the 21st century, for human health, ecosystems, agriculture and water resources, if current climate policies remain unchanged and global warming continues. As the global mean surface temperature increases, the risks to different systems (water, ecological, food, coastal, those impacting on human health) increase, precisely because they are interdependent. Collectively, they comprise the complexity of what we call 'the climate system'.

As mentioned earlier, recent policy discussions have focused on limiting the global temperature rise to less than 2 °C above pre-industrial times. This is likely to require a cut of 50% or more in global GHG emissions by 2050. Some campaigners (including the government of the Maldives – see Figure 1.21) are pressing for a limit of 1.5 °C. This would require an almost immediate and very rapid reduction in emissions.

Global warming and climate change are perhaps the most challenging and profound of the many impacts of human energy use on the planet. The next chapter discusses some measures that could be taken to reduce the

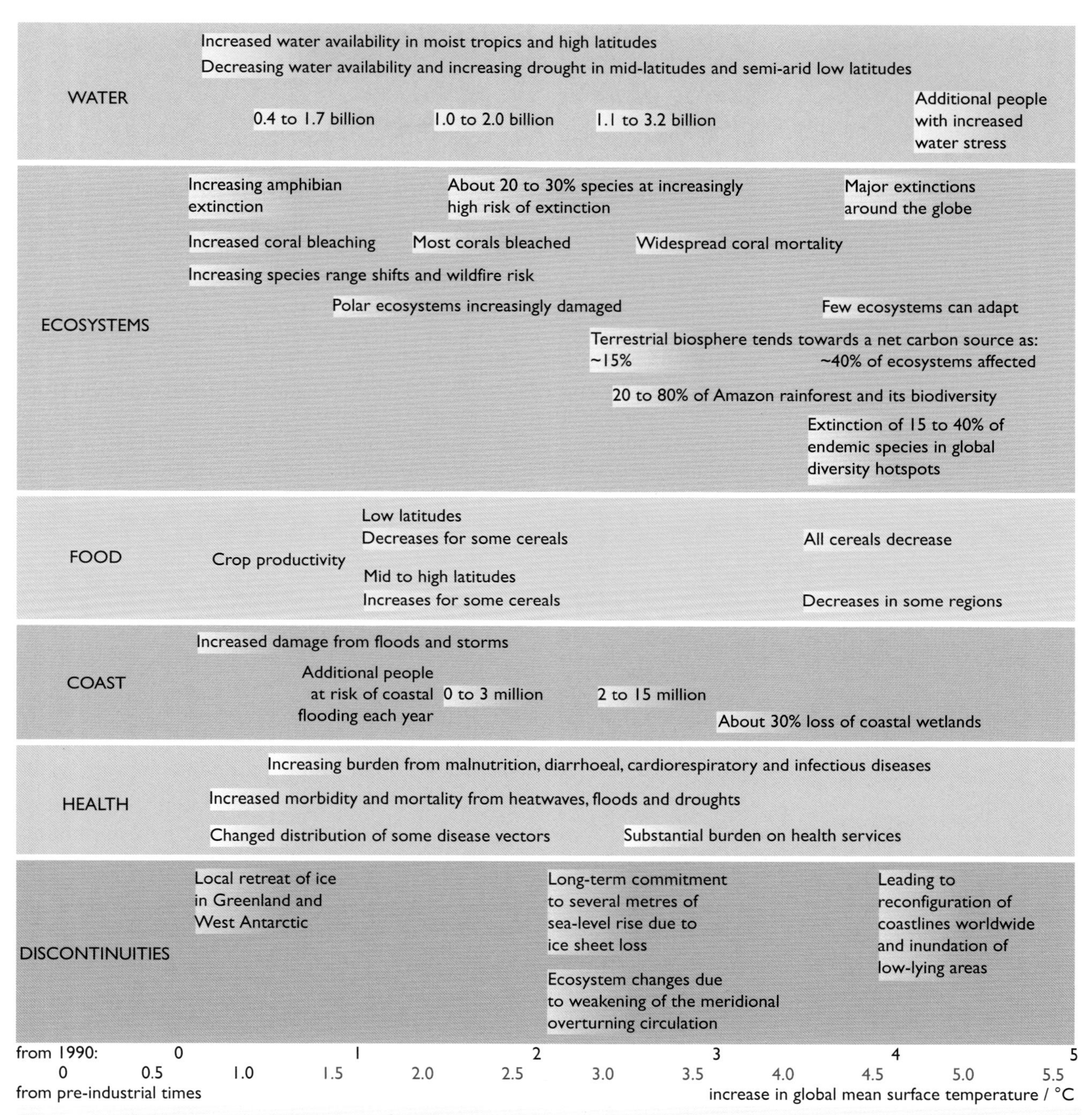

Figure 13.12 Summary of projected impacts with increase in Global Mean Surface Temperature; the left-hand side of items in the figure indicates onset of that impact (source: adapted from IPCC, 2007b)

emissions of GHGs and sequester the CO_2 from fossil fuel combustion. However, even if the world is successful in stabilizing atmospheric carbon *concentrations* during the 21st century, it may take a hundred years or more for the global mean *temperature* to stabilize. It is then likely that sea levels will still continue to rise gradually for hundreds of years thereafter as the oceans slowly warm and physically expand (IPCC, 2007a).

13.5 Air-quality-related impacts

Forms of airborne pollution

The previous section has described the effects of the main greenhouse gases. There are, of course, many other atmospheric pollutants whose effects have been summarized in Table 13.2. These can appear in different forms:

- A gas, consisting of individual atoms or molecules – like the oxygen and nitrogen of the air itself. Gaseous atmospheric pollutants include SO_2, NO_x and carbon monoxide. The term vapour is often used in this context. Strictly, a **vapour** is a gas at a temperature where it can be liquefied under pressure. This is the case for water at normal temperatures (see Box 6.5), so we are quite correct to talk of 'water vapour' in the air. The common use of 'vapour' for other airborne substances that we can see, such as a coloured gas or a mist, is not strictly correct.
- An **aerosol,** consisting of tiny droplets or particles as little as one micron (a thousandth of a millimetre) in diameter, but nevertheless each containing millions of atoms much larger than the individual free atoms or molecules of a gas or vapour.
- Particulate matter (PM or particulates). These are solid particles that are small enough to float in air. Only a few microns in size, they are individually visible as the motes we see in a beam of sunlight. At higher concentrations they appear as smoke from chimneys.

In industrial areas 'smog', a mixture of smoke and fog, is likely to consist of particulates and sulfur dioxide. This is likely to be a winter phenomenon. In modern cities, particularly those with large numbers of cars, a haze of brown **photochemical smog** is likely to occur in summer. This is produced by chemical reactions between NO_x and other pollutants and air in the presence of bright sunlight to produce ozone and nitric acid.

A summary of the sources and properties of the key atmospheric pollutants in relation to energy use is provided in Box 13.7.

BOX 13.7 Key energy-related atmospheric pollutants

Carbon dioxide (CO_2)

This is a major greenhouse gas produced by the combustion of carbon-based fuels.

Methane (CH_4)

A powerful greenhouse gas produced by the decay or digestion of organic material. It can be emitted from agriculture or as a by-product of fossil fuel extraction. Since natural gas is almost pure methane leaking gas transmission pipes are a major source of concern.

Chlorofluorocarbons (CFCs) and hydrofluorocarbons (HFCs)

Extremely powerful greenhouse gases artificially produced originally as refrigerants. CFCs have been phased out because of damage caused by their chlorine content to the ozone layer (see Chapter 9). However, some replacement HFC refrigerants, although not damaging to the ozone layer, are still powerful greenhouse gases. They can be released into the atmosphere from production plants, when refrigerators are serviced and when they are finally dismantled.

Sulphur dioxide (SO_2)

SO_2 is the a by-product of combustion of fuels containing sulfur. However, the amounts vary greatly. Biomass fuels such as wood are usually naturally low in sulfur. Natural gas normally has the sulfur removed at source, as do petrol and DERV (see Chapter 7). Coal and heavy fuel oil can contain up to 5% sulfur, so today the main producers of SO_2 are coal- and oil-fired power stations and international shipping which may run on cheap, low-grade fuel oil.

As a gas, sulfur dioxide is acidic and can affect health and vegetation. It affects the lining of the nose, throat and airways of the lungs. However, the main polluting effects result from the conversion of SO_2 into sulfuric acid (H_2SO_4) in the atmosphere. If it remains in the air, this can produce an aerosol of sulfuric acid droplets which we experience as an unpleasant haze. The acid can also react with fine particulates to form sulfate aerosols. SO_2 is the main contributor to acid rain but, somewhat perversely, sulfate aerosols high in the atmosphere have a global *cooling* effect.

Nitrogen oxides (NO_x)

NO_x is mainly emitted as a by-product of combustion. It consists of three oxides of nitrogen: N_2O, known as nitrous oxide or dinitrogen oxide, which is, as mentioned above, a powerful greenhouse gas, and two acidic gases, NO, nitric oxide and NO_2, nitrogen dioxide. NO_2 is the more powerful acid gas.

These gases may result from the combustion of fuels which contain nitrogen, for example coal or biofuels such as wood. More commonly, NO and NO_2 are the produced by oxidation of the nitrogen in the air by the high temperature combustion of any fuel, but particularly in petrol and diesel engines. N_2O is also produced in agriculture by the breakdown of nitrogenous fertilizers.

In the presence of bright sunlight NO and NO_2 can react with oxygen and other pollutants to produce low-level ozone (see below). They are thus known as **ozone precursors**. In the atmosphere, NO_2 dissolves in water vapour and oxidizes to form nitric acid (HNO_3), visible as a brown haze in modern urban summer smog. NO and NO_2 can affect human health and vegetation and provide a further contribution to acid rain. Their effects are therefore similar to those of SO_2. Since a major source is internal combustion engines, which are often in close proximity to people, the immediate effects of NO_x on the respiratory system are more obvious. In particular, nitrogen dioxide (NO_2) has both acute and chronic effects on airways and lung function.

Particulate matter (PM)

The sources of PM are diverse, and include road traffic, domestic heating (coal and wood fuels), construction work, quarrying and industrial activities. They can also be produced by chemical reactions in the atmosphere as a result of other forms of air pollution. The actual particles can take many different forms including soot, dust, grit, sea salt and biological particles. Fine particles can be carried into the lungs and can be responsible for causing premature deaths among those with pre-existing lung and heart disease. The most commonly used classification is based on the size of particles. PM_{10}s are particles 10 μm (microns or thousandths of a millimetre) in diameter or smaller. **$PM_{2.5}$s** (less than 2.5 μm in diameter) are small enough to penetrate deep into the lungs. These may be particularly dangerous if they are contaminated with acids from other pollutants.

Carbon monoxide (CO)

CO is mainly emitted as a by-product of the incomplete combustion of any fuel containing carbon and is also formed by the oxidation of hydrocarbons and other organic compounds. Within the human body it prevents the transport of oxygen by binding to the haemoglobin in the blood, effectively starving the body of oxygen. A badly adjusted combustion system such as a gas fire can produce enough CO to kill a person in a matter of hours in an ill-ventilated room. By far the major producer of carbon monoxide is the internal combustion engine, although the toxic effect is of course less dramatic in the open air. Before the introduction of natural gas, carbon monoxide poisoning was more common because it was a major component of town gas (see Chapter 5).

VOCs (volatile organic compounds)

VOCs are a range of chemical compounds of different kinds and from different sources and include benzene, 1,3-butadiene and polycyclic aromatic hydrocarbons (PAHs). They are hydrocarbons, mainly resulting from incomplete combustion in internal combustion engines, and vapours from solvents and similar materials used in industry. Many of these are carcinogenic, and they contribute to the 'chemical' smog of urban areas such as Los Angeles, Mexico City and Bogotá. Their major environmental impact is as a precursor to the formation of ground level ozone.

Ground or low-level ozone (O_3)

O_3 is a gas whose molecules consist of three oxygen atoms. In the lower atmosphere, the troposphere, ozone occurs naturally but is also a by-product rather than a direct product of our use of motor vehicles. It is generated at or near ground level in a series of complex interactions, energized by sunlight, between oxygen and two other precursor pollutants: NO_x and VOCs. It is itself a pollutant, chemically active, and damaging to the health of crops, trees and human health. It can cause irritation to the eyes and nose and very high levels can cause damage to the airway lining. O_3 can persist for several days and spread over large areas. The low-level ozone concentration is often used as a general indicator of the level of atmospheric pollution. In the upper atmosphere O_3 is not a pollutant but in fact plays a vital role in shielding the Earth from harmful UV radiation (see Chapter 9).

Heavy metals

Heavy metals such as lead (Pb) and mercury (Hg) have various health effects, the most important of which relate to deterioration of the immune system, the metabolic system and the nervous system. Some are known or suspected to be carcinogenic. Until recently lead additives were used as an octane-enhancer in petrol (see Chapter 8). Lead from vehicle exhausts can accumulate in the brain and can impair the cognitive development of children. Leaded petrol has now been phased out in nearly all countries in the world. Mercury may be emitted as vapour from coal-fired power stations, but can be intercepted if they are fitted with flue gas scrubbers as described in Chapter 5.

Dioxins and furans

These are related to VOCs and take the form of polycyclic aromatic hydrocarbons which include chlorine. They are produced in small quantities in many combustion processes and as by-products in some industrial processes, particularly where aromatic hydrocarbons and chlorine are present. Dioxins are highly toxic, even in very small quantities. They can be present in both the flue gases and the ash produced by Municipal Solid Waste (MSW) incinerators where waste plastics are burned.

'Clean air' has been a major concern in Europe and the USA for over 50 years and this has led to significant reductions in pollution levels. However, air pollution from traditional fuels in developing countries remains a serious problem.

Wood burning in developing countries

The IEA (2006) estimates that in developing countries, especially in rural areas, 2.5 billion people rely on biomass, such as fuelwood, charcoal, agricultural waste and animal dung, to meet their energy needs for cooking. It further suggests that in many countries, these resources account for over 90% of household energy consumption.

Often, these fires are located indoors and emit smoke and other pollutants that are particularly harmful to health, especially at the high concentrations found in confined spaces. For example, an estimated 70% of Indian households in 2004 used biomass as their main cooking fuel. The adverse health effects of this included increased incidence of acute respiratory infection, chronic respiratory disease, tuberculosis, lung cancer, heart disease and adverse pregnancy outcomes. Overall, it is calculated that some 6 to 9% of premature deaths of Indian women and children (the groups most exposed to indoor pollution) at that time were attributable to biomass fuel use in households (Smith et al., 2000, cited in Holdren and Smith, 2000). And even before it is burned, the gathering and cutting of wood is usually a laborious and sometimes a hazardous business, which often results in additional adverse effects on human health.

Furthermore, domestic cooking fires seldom achieve complete combustion of the wood fuel and so emit significant quantities of greenhouse gases. Figure 13.13 shows total greenhouse gas emissions (including CO_2, CH_4, non-methane hydrocarbons and N_2O) arising from the burning of renewable sources such as wood, dung and crop residues. It is expressed in terms of the equivalent number of grams of CO_2e per megajoule of useful cooking energy delivered (i.e. taking into account the efficiency of the stove). It is clear from the figure that the greenhouse gas emissions from these renewables, under typical (non-optimum) conditions, can be higher than those for non-renewable sources such as kerosene or liquefied petroleum gas – although greenhouse emissions from the use of one renewable source, biogas, are much lower.

Considerations of smoke pollution and the non-sustainable gathering of biomass have led the Indian government in particular to promote the use of LPG and kerosene for cooking and lighting. In China the use of biogas from the anaerobic digestion of wastes has been heavily promoted in rural areas.

The effects of acid rain

The problem of **acid rain** is common to both developing and developed countries. As described in Box 13.5, emissions of SO_2 and NO_x lead to the production of sulfuric and nitric acids which contribute to acidification. Measurements in clouds downwind of major pollution sources have detected water drops with the acidity of vinegar (i.e. a pH of 3.0 or less), and when these are incorporated into rain or snow the result is acid precipitation.

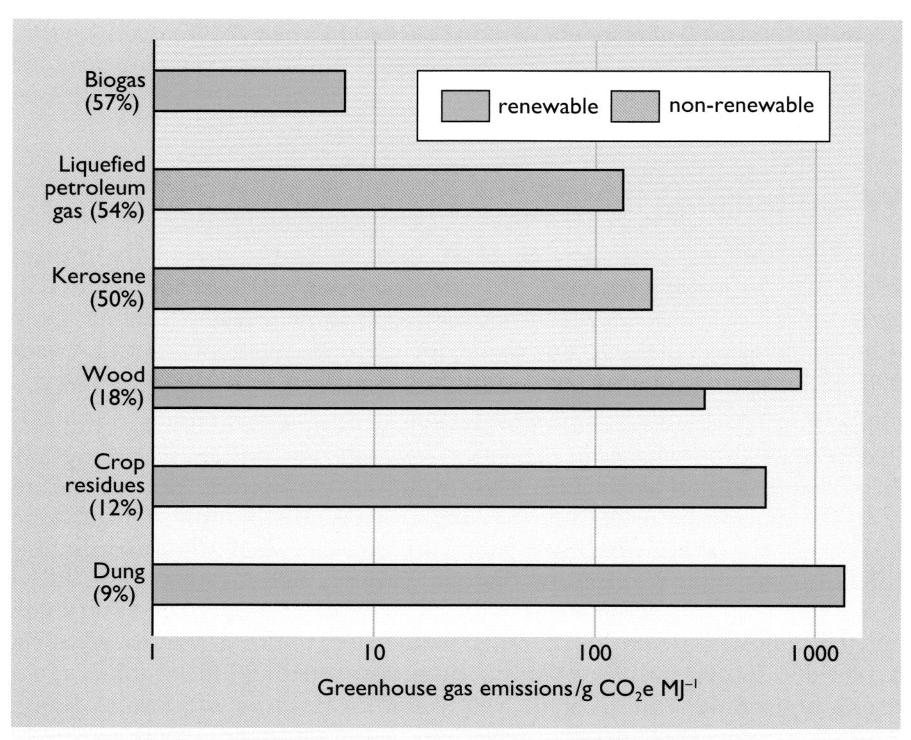

Figure 13.13 Greenhouse gas emissions from household fuels in India. Includes warming from all greenhouse gases emitted, weighted by stove distribution in India. Percentages shown in parentheses are average energy efficiencies of stoves used for fuel combustion (source: UNDP, 2000; Smith et al., 2000)

More commonly called acid rain, this has damaging effects on vegetation, lakes and fish, buildings and structures. It can cause respiratory diseases in humans, especially the vulnerable.

As early as the 1960s, it was found that some lakes in Scandinavia and the eastern USA were so acidic that they could no longer support life, and the late 1970s saw growing signs of an even more widespread potential disaster: *waldsterben*, or the death of the forests. By the mid-1980s, along a swathe from Scandinavia through the Low Countries, Germany and Poland and into the Czech Republic, up to half of all forest trees were reported to be damaged in varying degrees; about one in twenty seriously so. Similar effects were being observed in the eastern regions of North America.

As techniques for tracing pollutants through the atmosphere improved, it became generally recognized that the change from acid gases to acidic aerosols takes place in long plumes which can stretch over a thousand miles downwind from the source. For North America and western Europe, this means eastwards.

By the 1980s, however, acid rain levels were already falling in most developed countries, the effect of increasingly stringent limits to permitted emissions levels of SO_2 and NO_x. The fact that the deleterious effects appeared when SO_2 emissions were already falling led to two rather different interpretations: that these were delayed effects of the acid rain of earlier years, or that acid rain was irrelevant. However, no other single cause of the *waldsterben* has been found, and the current view of many experts is that acid rain entering the soil initiates a process that ultimately weakens the trees, making them more vulnerable to other stresses such as those caused by the periods of low rainfall during the 1980s. Pollution can impact on forest productivity in numerous and sometimes indirect ways (Godish, 1997, pp. 108–12). Acid will tend to remove some essential nutrients from soils (particularly K, Ca, Mg) and may mobilize metals such as aluminium to toxic levels, which can damage roots. Adding nitrogen, the main nutrient of plants, may create an imbalance in resources and make trees more vulnerable to diseases and frost.

In Asia the problem of acid precipitation is increasing (see, for example, Jing and Yining, 2011). China, the world's biggest coal consumer, is experiencing many of the problems that Europe encountered thirty or so years ago. But there are success stories. In the industrial city of Chongqing, a few years ago metal and concrete structures were corroding so fast that rust removal and repainting were needed every one or two years, and in coniferous forests of the region up to half the trees were dying. Chongqing's air quality has improved substantially in recent years. Although Chinese national SO_2 emissions doubled between 1986 and 2005 the ambient SO_2 levels in Chongqing fell by more than a factor of six (LBNL, 2009).

Particulates

In the late nineteenth and early twentieth centuries industrial London was famous for its 'peasoup' fogs when mixtures of smoke and SO_2 would reduce visibility to a few metres. The great London smog of December 1952 which lasted for four days was critical for obtaining effective pollution legislation. It mainly affected central London and the smoke density at County Hall, opposite the Houses of Parliament, reached almost 4500 $\mu g\ m^{-3}$ on two successive days. SO_2 densities also rose to over 1000 $\mu g\ m^{-3}$ (Brimblecombe, 1987). The death rate, particularly amongst children under one and people over 45 soared, mainly due to heart and lung disease. In the week after the smog the death rate for the adminstrative County of London rose by a factor of three. Overall it is believed to have caused at least 4000 deaths. The high (and rapid) death rate is believed to be due to a 'cocktail effect' of the *combination* of particulates and SO_2. A subsequent enquiry pointed the finger of blame at domestic coal fires. In 1956 the Clean Air Act was rushed through Parliament giving local authorities the power to insist on the use of low-sulfur 'smokeless fuels' such as coke and anthracite. The introduction of low-sulfur North Sea gas in London in the 1970s has dramatically reduced the levels of both particulates and SO_2. Average levels in 2000 were well over ten times lower than in 1950 (GLA, 2002).

In the 1952 smog a high level of pollution produced a large number of deaths that were highly visible in the mortality statistics. Does it follow that a smaller level of pollution over a long period of time would produce a lower level of deaths? This question, whether the results obtained from the effects at high concentrations still apply at levels hundreds of times smaller, has been the subject of much debate. The current view, based on more recent measurements correlating health effects and particulate concentration, is that the effect is proportional to the exposure down to the lowest levels: there is no 'safe threshold' below which the pollutant has no effect at all (WHO, 1999).

In Europe pollution standards have been steadily tightened up since the 1950s. The presence of fine particulates ($PM_{2.5}$) are carefully monitored and estimates of years of life lost (YOLL) made (see Box 13.2). Figure 13.14 shows a map of the concentration of YOLL due to $PM_{2.5}$ pollution illustrating that it is concentrated in urban areas and industrial countries.

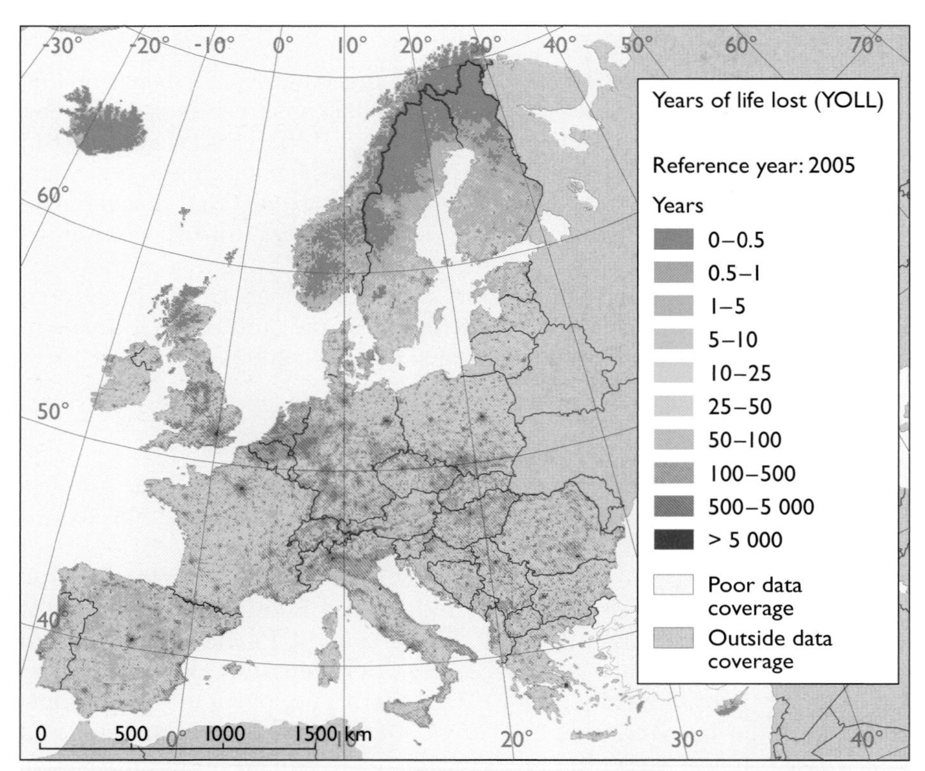

Figure 13.14 Years of life lost due to $PM_{2.5}$ pollution in European countries, 2005 (source: EEA, 2005)

Progress in reducing air pollution

Developed countries

Concerns over the environmental effects of air pollution have led to major reductions in the emissions of the main gaseous pollutants. This is as a result of legislation enforced at the local level (such as the UK's Clean Air Act) and at the wider European and US scale to tackle the effects of acid rain. In the 27 EU-associated states SO_2 emissions fell by a factor of three between 1990 and 2005 (EEA, 2008). In the USA SO_2 emissions from electricity generation (the main emitter) fell by a factor of two between 1990 and 2008.

In the UK, as shown in Figure 13.15(a), emissions of SO_2 fell by nearly 90% between 1990 and 2009. This is mainly due to progressively reducing EU emission targets for 'large combustion plant' which include power stations and heavy industry. However, it is part of a longer trend which can be seen in the chart from 1970 onwards which is related to the declining use of coal and its replacement for heating by low-sulfur natural gas.

This trend can also be seen in the declining UK emissions of particulates shown in Figure 13.15(b). The significant reduction in 'residential' particulate emissions is due to the switch from coal-fired domestic heating to natural gas. (The large dip obvious in 1984 is due to the temporary replacement of coal-fired electricity with oil-fired during the miners' strike of that year.)

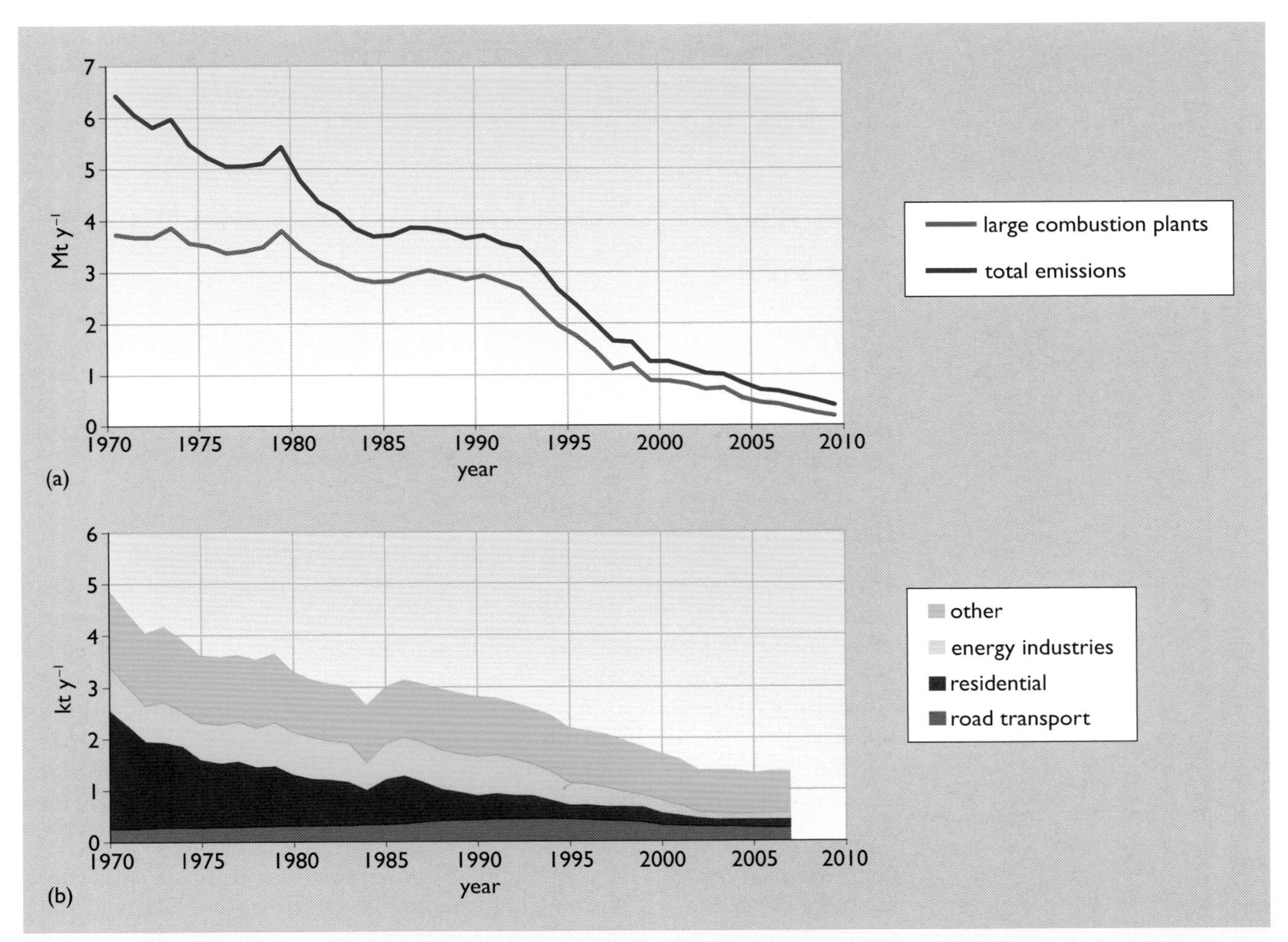

Figure 13.15 UK pollutant emissions: (a) sulfur dioxide, 1970–2009; (b) particulates (PM_{10}), 1970–2007. Note that the large combustion plants are those in power stations and heavy industry (sources: (a) DEFRA, 2011; (b) DEFRA, 2008).

Developing countries

In the urban areas of developing countries, emissions of most of the pollutants discussed in the text far exceed those found in similar areas of the developed world. For example, particulate (PM) levels in many cities are extremely high. The principal causes include poorly tuned and maintained engines, particularly diesels and the polluting two-stroke engines often used in motorcycles and motorized 'rickshaws'; and the uncontrolled burning of solid fuels (wood and coal) and refuse. The effects of such emissions on the environment and human health are profound. In China, it is estimated that there are some 170 000 to 290 000 premature deaths per year from urban air pollution (Holdren and Smith, 2000).

Some economists point to the fact that in those countries we now call 'developed', levels of pollution also initially rose during the early stages of development, but later fell as higher levels of affluence gave rise to higher environmental standards, resulting in the introduction of cleaner technologies and practices. The implication is that increasing living standards will in themselves, given time, lead to reductions in environmental impact. However, it remains to be seen whether rising affluence alone will be sufficient to enable the polluted cities of the developing world eventually to reduce their emissions. Other, more active approaches may prove necessary.

13.6 Nuclear power: impacts of routine operation

As in the case of the energy systems discussed above, the routine operation of nuclear plants has some deleterious effects for the population at large and also specifically for workers in the industry. This section is concerned with the nature and consequences of these effects and also provides the background for the discussion of 'non-routine events' in Section 13.7.

Public impacts

Exposure of the public to radioactivity can arise within each stage of the nuclear fuel cycle, from the extraction of uranium ore to the final disposal of spent fuel. In order to assess the significance of this exposure, it is useful to compare it with the **natural background radiation** to which we are all subjected. This arises from charged particles or electromagnetic radiation reaching the Earth from outer space, from radioactive rocks and from naturally occurring radioisotopes in our food. It varies over the surface of the Earth, with an average annual dose per person of about 3 mSv (see Box 13.8).

Over the past few decades, the increasing use of medical X-rays, and more recently CT scans, has led to a significant rise in the *average background exposure* in many countries. The US is the extreme case, with this average rising from 3.6 mSv per person in 1980 to 6.2 mSv per person in 2006 (*Scientific American*, 2011). For the UK at the time of writing (2011), the average background is about 4 mSv per person per year. With a population of about 60 million, this means an annual total population dose of about 240 000 person-sieverts per year. The data in Box 13.8 then suggests that background radiation might be responsible for about 2400 cancers per year – rather less than 1% of the UK total diagnosed annually.

Estimates of the average individual dose due to the routine operation of nuclear power plants vary widely. The World Nuclear Association (WNA, 2011a) suggests that the world average annual dose per person received by the public from nuclear power is 0.0002 mSv per year, which is some fifteen

BOX 13.8 The effects of radioactivity on health

The basic units of radioactivity have been described in Box 10.3. But what effect does it have on the human body? As the high-speed particles or penetrating radiation from a source of radioactivity pass through any material, living or otherwise, they tear apart the atoms or molecules, and the energy dissipated in doing this is a measure of the damage they cause. The term **ionizing radiation** indicates the type of biological damage that can occur, and this is about ten times greater for energy delivered by heavy particles – alphas or neutrons – than by betas, gammas or X-rays. It also depends on other factors such as the type of tissue and the rate at which the energy is delivered. The unit that takes these factors into account is the **sievert** (Sv). In simplified form, the relationship between the sievert and the energy deposited in tissue is as follows:

- Beta particles, gamma rays and X-rays deliver a radiation dose of one sievert in depositing 1J (one joule) of energy per kilogram of tissue.
- Alpha particles and neutrons deliver a radiation dose of one sievert in depositing *one-tenth* of a joule of energy per kilogram of tissue.

How large a dose is 1 sievert? What effect does it have? Data comes from three main sources: laboratory experiments on tissue samples or on animals; studies of those unfortunate groups of people who have been exposed to high radiation levels; epidemiological studies of larger populations at low doses. To summarize the often-controversial results of many studies over many years may mean dangerous over-simplification. Nevertheless, it is perhaps useful to give some idea of the orders of magnitude.

- **Single large doses in a short period**: For a single person a dose of 10 Sv or more almost certainly means death within hours or days. Doses of 1–10 Sv lead to radiation sickness and disability for weeks or months, and can be fatal. Below 1 Sv the symptoms decrease, until at about 0.1 Sv there may be no immediately obvious effects.
- **Long-term effects of lower doses**: Smaller doses can only be statistically analysed over a large number of people. They are often expressed in terms of **person-sieverts**: 100 person-sieverts could mean 100 people each receiving a dose of 1 Sv or 20 000 people each receiving 5 mSv (millisieverts), and so on. Over a wide range of doses and population sizes, the long-term rate of induced cancers appears to be 1–2 per 100 person-sieverts. The total number of genetic effects down to about the tenth generation of descendants is thought to be of the same order.

Radioactive discharges and exposure to radioactivity

Figures such as those above provide the basis for international standards which are used by many countries in legislation governing permitted exposures and permitted discharges from nuclear plants into the air or water. The latter must of course take into account the nature of radioactive isotopes that are released. Are they airborne? What is their half-life? By what routes, if any, can they enter the human food chain?

One example of a fission product is iodine-131 (I-131). Although its half-life is only 8 days, this radioisotope is particularly hazardous. Emitted as a gas, it condenses on vegetation and is easily taken up by grazing cattle, entering their milk. This provides a short route into the human body. The iodine then concentrates in the thyroid gland, where its beta and gamma emissions mean an increased dose to the surrounding tissue.

> Estimates of routine I-131 emissions from a nuclear power station tend to fall in the range 1–10 million becquerels (MBq) per day (i.e. a few nanograms of this isotope). To convert this figure into health effects requires data on weather, population distribution, etc., and a more detailed analysis than is possible here; but the general conclusion is that the annual population dose from this source is unlikely to reach even 1 person-sievert. This implies that, on average, there would not even be one induced cancer over the normal lifetime of the plant. The situation can be very different if a significant fraction of the total I-131 in a reactor is released in an accident, as described in the next section.

thousand times smaller than the average yearly dose from background radiation. (For the UK, for instance, this figure implies national annual total emissions due to nuclear power of about 12 person-sieverts – a negligible contribution to the nation's background total of 240 000 person-sieverts.)

The WNA total has been questioned, but perhaps the more interesting issue is whether such a world average is relevant. It has been argued that even national averages may be inappropriate when the emissions arise from a relatively small number of plants, and that exposure levels are likely to be higher near nuclear facilities. There has been research into possible 'clusters' of cancer occurrences around such facilities but there are conflicting opinions on the evidence.

The effects of mine tailings and of spent fuel are also the subjects of controversy. As explained in Chapter 10, uranium mining can produce very large volumes of radioactive tailings. The operation of nuclear power plants can also produce large quantities of low-level waste. Both of these require suitable storage to reduce the public exposure to radiation. Some of the radioisotopes have very long half-lives, and although the activity may be low, the cumulative dose to the public continues to rise, and can reach a very high level after a long period. But this in turn raises the question of whether it is legitimate to 'count' a dose that accumulates over a period much greater than any one human lifetime.

Similar reasoning could be applied to the long-lived isotopes in spent fuel, but there is also particular concern with the immediate or shorter-term effects. It is generally agreed that the public exposure due to a *reprocessing* plant is greater than from the routine operation of a power station. The 'cluster' argument may also have more force in this case, but the relatively small number of civil reprocessing plants means that reliable data is still rather sparse. The final issue, the ultimate disposal of spent fuel, has been discussed in earlier chapters, where we have seen that it remains unresolved.

Workplace impacts

Risks to *workers* in the nuclear industry can also occur throughout the nuclear fuel cycle (Figure 10.18), but the hazards of uranium mining in particular have long been known. The high incidence of lung cancer in workers in the Joachimstal mine on the Czech–German border was recognized in the nineteenth century, when the main products of the mine were silver and other metals. With the discovery at the mine of uranium-containing pitchblende and the identification by Marie Curie of an associated radioactive gas (later to be named radon), this became one of the earliest recognized cases of radioactivity-related industrial disease. Numerous subsequent studies have continued to identify a relationship between uranium mining and lung cancer (WISE, 2011).

The maximum annual dose allowed for radiation workers is in the region of 20 mSv per year – several times the average background discussed above – but it is claimed that in practice doses are usually kept well below this level. However, as the discussion of accidents below reveals, circumstances may change the rules. Moreover, the presence of highly radioactive materials also means that minor workplace incidents whose effect might be negligible in other industries can have serious consequences.

13.7 Accidents and risks across fuel chains

The chapter so far has considered a variety of environmental and health impacts linked to energy systems in the course of their normal life cycle. Exposure to workplace hazards and accidents in the energy industries is significant both for workers and in some circumstances for the public. The energy industries are no exception to the rule that fatal accidents are a feature of life. Mining accidents lead to tens of thousands of deaths a year worldwide. Oil rigs have experienced serious fires, liquefied natural gas tanks have exploded, dams have burst – sometimes killing hundreds in a single incident. Air disasters bring similar numbers of fatalities. Worldwide we kill ourselves and others in our hundreds of thousands on the roads every year. This section considers the impacts of energy systems in terms of two non-routine events: accidents and catastrophes. It is useful to distinguish between the two, as explained in Box 13.9.

BOX 13.9 Types of data

It is important to be aware of the types of data that form the basis for the statistics. The three aspects of any energy system determining its effects on human health or the environment are:

(1) **Routine operation**. The construction, normal operation and decommissioning of any energy system usually has some deleterious effects. Ideally these should be predictable and measurable, and they are often subject to regulation.

(2) **Accidents**. These are a feature of all activities, and most of the energy industries are sufficiently well established to have accumulated data on 'normal' accident rates and their consequences.

(3) **Catastrophes**. These are large-scale, individually unpredictable, and rare.

The first two categories should supply the data needed to estimate the effects of an existing or proposed system, or to compare the effects of different systems, such as those used to generate electric power.

Whether the consequences of catastrophes should be similarly included in the statistics is a subject of debate. The data they provide is certainly different: instead of averages over periods of time and many similar systems, the only available information might come from a single event, which may or may not be repeated in the future.

Risk assessment

One way to deal with this situation is to assess the risk from potential catastrophes – those that have not yet occurred as well as those that have. The results of such assessments can then take their place with the other data in assessing the environmental or health effects of a system. **Risk assessment** is a two-stage process, requiring estimates of the *probable consequences* of each event and of the *probability that it will occur*. As will be discussed, it relies heavily on statistical reasoning.

Biomass harvesting and forestry

As already noted, the manual gathering and harvesting of traditional biomass sources for domestic use in developing countries can be a hazardous occupation. Even when wood fuel is harvested on a commercial basis, taking advantage of mechanization in such forms as the chainsaw, working conditions among forestry workers in many countries are still often very dangerous and unhealthy. Fully mechanized forestry as practised in some developed countries is considerably less hazardous, but still involves risks to safety. Globally, the annual mortality rate for forestry workers is estimated to lie between 32 000 and 160 000, though by no means all of these are related to energy production (Holdren and Smith, 2000). Nevertheless, although the totals are high and the consequences for individuals may indeed be catastrophic, the scale of individual events in this field probably places them in the 'accidents' category of Box 13.9.

Hydro and wind power

Hydro power is a substantial contributor to world electricity supplies and the contribution from wind power is growing fast.

Workplace hazards in the hydroelectricity industry mainly occur when large dams are under construction, and are likely to involve the same levels of fatalities and injuries to workers as those on other large construction projects.

The collapse of a dam can be a major catastrophe for the local communities. During the twentieth century, some 200 dam failures outside China are thought to have resulted in the deaths of more than ten thousand people. And within China, in one year alone, 1975, it is estimated that almost a quarter of a million people perished in a series of hydroelectric dam failures (Sullivan, 1995).

In comparison with hydroelectricity, wind power is a relatively new technology. By the end of 2010 there had been approximately 40 operator deaths, most of them due to falls and injury by blades. Overall the death rate appears to be low, approximately 0.15 deaths per TWh of electricity generated (Gipe, 2011).

Coal, oil and gas

The extraction of coal from the ground, whether by surface (opencast) or underground mining, and the underground extraction of oil and gas from beneath the Earth's surface, entail substantial risks of death or injury to the workers in the industries involved – even in countries like the UK where safety considerations are today given a very high priority.

Table 13.10 gives some recent data on fatalities and injuries for the offshore oil and gas industry and deep coal mines. There is a continuing annual incidence of a few fatal injuries and several hundred serious but non-fatal injuries in Britain's energy-related industries.

Table 13.10 Annual average number of fatalities and injuries in UK offshore oil and gas installations and deep mining

	UK offshore oil and gas installations April 1995–March 2010	**UK deep mining April 2006–March 2011**
Fatalities	1.7	2.0
Over 3 day injuries	186	223
Total injuries	238	263

Sources: HSE, 2010; HSE, 2011.

Although the *average* death rate on UK offshore oil and gas rigs since 1995 has been low, there has been the occasional major *catastrophe*, for example in 1988 when 168 people died following an explosion on the *Piper Alpha* gas rig. As mentioned in Chapter 7, this led to an upgrading of safety standards on offshore oil and gas rigs, and perhaps has assisted in maintaining a good safety standard since then.

Worldwide death rates for coal mining are very roughly estimated by Holdren and Smith (2000) to lie somewhere between 30 000 and 150 000 per year. In the USA the mining death rate has dropped dramatically from over 1000 a year in the 1940s to about 65 per year between 2001 and 2007 (MSHA, 2011).

Chinese coal mining has had a particularly bad safety record, leading to the closure of many small mines. After peaking in 2002 at 7000, annual deaths from coal mine accidents in China in 2010 have fallen to approximately 2430 people (AFP, 2011).

Nuclear power: a question of risk

The nuclear power industry has experienced very few serious accidents resulting in immediate or near-immediate fatalities – certainly fewer than either coal mines or oil installations. Nevertheless, it is the thought of a catastrophic accident that is at the centre of much of the public concern about nuclear power. Is there a rational justification for this?

The difference between 'catastrophes' and other industrial accidents (as categorized in Box 13.9) is not that the latter are less serious – any fatality is of course serious – but that the catastrophes are rare. Being rare, they don't provide sufficient 'historic' data to calculate annual averages of the type shown above for offshore oil and gas or for mining. So if they are to be taken into account in assessing the desirability of new plants, questions about the past health or environmental effects per year or per kilowatt-hour must be replaced by those about the probabilities of repetitions in the future.

Assessing the actual probabilities of specific accidents or predicting consequences is outside the scope of this text, but it is useful to look at how, generally, probabilities are obtained. Box 13.10 describes a simple scenario with a selection of possible responses.

BOX 13.10 Not in my back yard

It is proposed to build a nuclear power station near your home. When you express your concern about the possibility of a serious accident, you are informed that the chances are one in ten million per year of an accident serious enough to cause perhaps ten immediate deaths and 10 000 ultimate deaths from cancer. How do you react?

(a) That would be terrible. I'd rather live without electricity.

(b) So what? In 'ten million years' I'll be dead anyway.

(c) £5000 to a penny that the reactor will survive ten years.

(d) How do they know?

The remainder of this section initially adopts response D, asking how such probabilities are obtained and comparing results of one analytical study with the little available data, before briefly considering the implications of the other responses.

Probabilities and consequences

Probabilities are not certainties. The chances are 1 in 6 that a rolled die will show a 2; but this doesn't mean that every sixth throw will be a 2, nor even that a 2 will turn up once in every six throws. It is perfectly possible to get no 2s at all in a dozen throws, or to throw half a dozen in succession. In this sense, response 'B' shows a poor appreciation of probabilities, whilst someone responding C understands what they mean (and evidently hopes to make a profit).

How is a figure obtained for the probability of a certain type of accident? One method is to use past history where it is available: 'How frequently has this type of switch failed?' The probability of different accidents can then be assessed – or reduced at the design stage – by accumulation of such information on all the component parts of a system. On the whole, this approach has proved sufficiently reliable to be used routinely for many years by designers of complex systems throughout industry.

Risk assessment like this has moved from the design office to the public domain largely with the rising concern about reactor accidents; indeed it could be seen as a counter to response A. As people became aware of the appalling consequences of the worst conceivable accident, it became necessary to point out (or to find out, according to your view) just how small the probability was that such an accident would occur. A rational person, it is argued, will reach decisions on a basis of probability multiplied by consequences. If you accept a situation with a one in a thousand chance of 5 fatalities a year, then you should accept a one in a million chance of 5000 fatalities, and so on. And similar reasoning should determine your choice between alternatives.

Accidents in complex systems are caused by combinations of circumstances, so we need combinations of probabilities to assess their likelihood. If one in every 500 switches is faulty, and one in every 200 indicator lights, then the probability that a switch and its indicator light both fail should be one in 500 times 200, or one in 100 000. Knowing individual probabilities means it is possible to calculate overall probabilities for complex sets of events – like a switch failing to open and a valve failing to close and … .

The reasoning can run either way. In order for a reactor core to melt, the following and the following and the following must all happen, and the combined probability will be the likelihood of that conjunction. Or you can ask what happens if X fails, and in addition Y fails, and so on, and calculate the probabilities and consequences of all possible accidents. (This is the basis of 'hazop', widely used to develop safer processes and procedures in other industries.)

There are complications of course. If an electrical surge caused both the switch and the indicator light to fail, the result would be the unfortunate consequence of what is known as **common mode failure**. Allowing for common mode failures is essential in assessing risk, and eliminating them is an important aspect of design. The best-known example of what not to do is probably the 1975 boiling water reactor accident at Brown's Ferry in Alabama, USA, where a relatively small fire caused by technicians using a candle to test for leaks in one air duct affected hundreds of cables and put out all the carefully planned emergency cooling systems.

Two other difficulties are 'unknowns' and people. There can be parts of the system for which probabilities based on past history are not available. Risks must then be assessed by calculation, by analogy with similar cases, and where possible by experiment. The behaviour of the massive steel pressure vessel of a pressurized water reactor under the stresses which might result from an accident, and the behaviour of emergency cooling water as it hits the hot core, are examples where there is little history to go on, and this has led to controversy over both.

The accident at Fukushima was largely a consequence of, and error in, designing for an 'unknown', in that case a major earthquake accompanied by a tsunami. The nuclear site had been designed to withstand a tsunami 5.9 metres high. The one that arrived was over 10 metres high.

People present similar problems. Brown's Ferry is only one of many cases where human ingenuity eventually prevented serious consequences, but the incident at Three Mile Island (TMI) seems initially to have been the result of a determined effort by all present to counter the actions of the safety systems. The Chernobyl catastrophe was also due in part to deliberate over-riding of the safety systems in order to carry out experimental low power operation. It should perhaps be added that the TMI engineers were in a control room with, at one moment, *over a hundred* alarms sounding and lights flashing. Like those at Chernobyl or Fukushima who put themselves at severe risk, they were heroes to be there at all. Any risk assessment that ignores the human factor will obviously be of little value, yet to give figures for the chances that people will behave in certain ways clearly introduces another spectrum of uncertainties.

Despite all these difficulties, estimates are made. The Rasmussen Report (Rasmussen, 1974), also known as Wash-1400, produced a detailed analysis of the probabilities and consequences of accidents in light-water reactors (PWRs and BWRs). It was criticized by the nuclear industry for its 'conservative' assumptions, and by opponents of nuclear power for its omissions. Nevertheless it may be worth looking at a few of its results, if only to obtain an idea of its assessments of the orders of magnitude of the risks and to compare these with actual events.

Table 13.11 shows simplified versions of the Rasmussen probability-times-consequences conclusions for just three different potential accidents:

(1) a core meltdown with little or no release of radioactivity from the plant

(2) a meltdown with an explosion violent enough to breach the containment, releasing large amounts of radioactivity, but in circumstances where the exposed population is very small

(3) a meltdown with an explosion violent enough to breach the containment, releasing large amounts of radioactivity, in the worst conditions with, for example, the reactor in a densely populated area and weather conditions that maximize the effect, exposing a population of 10 million or so.

The second column in the table shows the assumed ranges of the activity released in each case, whilst the third allows for further uncertainties about the worker and population doses that would result from each event. The

Table 13.11 Estimates of accident probabilities and consequences

Example	Activity released/ TBq	Total dose to population /person-sieverts	Consequent number of deaths		Probable number of events per million reactor-years	Predicted number of deaths per thousand reactor-years
			Immediate	Delayed		
1. Meltdown *without* major breach of containment	0–500	0–100	0	0–10	10–100	0–1
2. Meltdown with breach of containment under *average* conditions	500 000–5 million	0.1–1 million	1–10	1000–10 000	0.1–10	0.1–100
3. Meltdown with breach of containment under *worst* conditions	500 000–5 million	1–10 million	1000–10 000	10 000–100 000	0.001–0.01	0.001–1

See the main text for further explanation and comment on these figures
Source: Ramage, 1997, adapted from Rasmussen, 1974.

next two columns show the deaths expected from these doses. (column 5 follows from column 3 and the data in Box 13.8, with the assumption that all induced cancers lead to premature death.) The figures in the sixth column are determined by combining the probabilities of many events to give a range of probabilities for each accident, and the final column shows the result of multiplying probabilities by consequences. It should be noted that this is just part of a very much larger set of conclusions – and that this analysis is for essentially one type of reactor.

The evidence

Fortunately, the number of serious reactor accidents has been far too small for any statistically valid test of these estimates, but it may be worth looking briefly at some of the accidents already mentioned in earlier chapters – noting however that not all involved a light-water reactor.

The first is the 1957 fire in the graphite moderator of the Windscale reactor (see Box 11.2). The fire overwhelmed the safety filters for the cooling air and, as described in Box 11.2, there was a release of radioactive gases including some 1000 TBq of iodine-131. This puts the event in the first row of Table 13.11 (and perhaps a little beyond it in seriousness). The accident was initially claimed not to have caused a single death, but if the relationship between activity release and the consequences indicated by the first example in Table 13.11 is at all correct, this seems statistically unlikely. In the late 1980s the official view changed, with the statement that the event probably led to a few tens of deaths from cancer – far too small a number, of course, to be detected in the mortality statistics, but now at least consistent with Rasmussen (see also Arnold, 1992 and McSorley, 1990).

Iodine-131 features largely in concerns about reactors. As discussed in Section 13.6, the routine I-131 emissions from a nuclear plant, a few million becquerels per day, are unlikely to cause detectable harm. However, the total I-131 content of a reactor can be *several million terabecquerels*, over a million, million times these routine amounts; and it is the possibility of an appreciable fraction of this escaping that causes concern.

At Three Mile Island, a combination of improbable events – a switch failing to open, a valve failing to close, misinterpretation of meter readings – resulted in a chemical explosion within the reactor. To the credit of the designers, the containment was not breached, but about 1 TBq of I-131 was released, less than a millionth of the total content. This puts the event in the first example in Table 13.11, making it improbable that there was even one resulting death.

The most serious accident to date for which we have reasonably full details was at Chernobyl in 1986 (see Chapters 10 and 11). The 'lid' of the reactor was blown out and fell back tilted, leaving the interior exposed, and about 6% of the core content was released. The 31 people who received lethal radiation doses in the first few days included pilots who flew helicopters low over the open reactor in an attempt to drop neutron absorbers into the core. The radioactivity released was of the order of the largest release shown in Table 13.11 (example 3), and the estimates of delayed deaths quoted in Box 11.2 range up to 40 000.

At Fukushima, *three* reactors melted down, probably with some breach of containment in each case. In April 2011, the Japanese Nuclear and Industrial Safety Agency (NISA) estimated that there had been a release of 130 000 TBq of I-131 and a further 6000 TBq of caesium-137 (NISA, 2011). Fortunately, during most of these releases the wind was blowing out to sea, i.e. it was not a 'worst case'. So, assuming that the NISA figures are correct, the incident might best be represented by the second example in Table 13.11, but with a lower activity release than suggested there. It is worth noting, however, that the Japanese authorities increased the maximum annual dose allowed for radiation workers (Section 13.7) to 250 mSv for the Fukushima clean-up workers.

Worldwide there have now been a total of five reactor meltdowns in approximately 14 000 reactor-years' operation of conventional power generating reactors (WNA, 2011b). This suggest an average of 350 events per million reactor years or considerably higher than any of the assumed frequencies in column 6 of Table 13.11. As mentioned in Chapter 11, for current US reactors the overall probability of an accident resulting in damage to the core has been estimated at about 1 in every 10 000 years of reactor operation, or 100 per million reactor-years (Sailor et al., 2000). It has been suggested that a goal should be set to reduce the frequency to fewer than 10 per million reactor-years of operation (MIT, 2003).

The responses to risk

Having asked 'How do they know?', is it now easier to assess the other three responses offered in Box 13.10? When all the sums are done, how should we respond to probabilities? Is one of the first three responses the 'correct' one?

- If the rational response is to compare probability-times-consequences for different alternatives, then as far as routine emissions and reactor accidents are concerned, the figures – if even remotely reliable – place nuclear power amongst the safest energy sources, and the gambler who chooses response C is on good ground.
- But perhaps this response is not appropriate for potentially large-scale catastrophes if they are sufficiently improbable? It has been argued that, as individuals or societies, we normally don't bother at all about risks whose chances are less than perhaps one in a million a year; that we do in practice adopt the light-hearted approach and choose response B.
- On the other hand, there are those who argue for response A: that if the consequences are sufficiently awful we shouldn't accept the risk at all, no matter how low the probability.

The answer may not be a simple one. It is arguable that the degree of risk we accept depends on the degree of control that we have as individuals. (Despite the accident statistics, we may prefer driving to flying.) And if the risk is imposed, our criteria may be even more severe: we may reject an involuntary and imposed risk but accept an appreciably higher one that is voluntary.

To dismiss any of these views as 'not rational' seems neither justifiable nor particularly useful.

13.8 Energy return over energy invested (EROEI)

This is a topic that has already been discussed in relation to unconventional sources of oil in Chapter 7 and nuclear power in Chapter 11. As pointed out earlier in this chapter, EROEI analysis is a form of fuel chain analysis and in its fullest form incorporates life cycle analysis.

In its simplest form:

$$\text{EROEI} = \frac{\text{useful energy output from an energy system}}{\text{the energy consumed in running it}}$$

However, the term is used slightly differently depending on the precise definitions of the terms 'useful energy output' and 'energy inputs' but the overall picture is the same. For example, considering the extraction of coal from a mine:

$$\text{EROEI for coal at mine gate} = \frac{\text{energy in coal produced} - \text{energy consumed in production}}{\text{energy consumed in production}}$$

In its fullest sense the 'energy consumed in production' includes a proportion of everything required for the construction and operation of the mine, including the manufacture of all its machinery.

For coal mining in the USA it has been suggested that the EROEI is about 80 (Cleveland, 2005), i.e. 1 GJ of energy must be expended to obtain 80 GJ of saleable energy.

The EROEI for US oil and gas in the 1930s was over 100, but by 1970 this figure for US produced oil had fallen to about 30, and by 2005 to between 11 and 18 (Murphy and Hall, 2010).

EROEI of electricity generation

However, there are also transport energy costs to be taken into account. As pointed out in Chapter 5, coal is commonly transported thousands of kilometres by rail, sea or road to the final user, often a power station.

$$\text{EROEI for coal at power station gate} = \frac{\text{energy content of coal available for generation} - \text{energy consumed}}{\text{energy consumed}}$$

The *energy consumed* is the sum of *all* the energy consumed in getting it to the power station gate: the energy used in setting up and operating the mine plus that used in transporting the coal plus that of setting up the appropriate transport infrastructure.

The long-distance transportation of coal can consume 5% of the energy content. That for natural gas, either by pipeline or as liquefied natural gas, can consume 10% or more.

If the end point of the analysis is the electricity *output* from the power station, then the generation efficiency and the energy consumed in the building and decommissioning of the power station must be included.

For renewable energy technologies such as wind power, there are no fuel costs. However, there is energy consumed in constructing, operating and decommissioning a wind turbine:

$$\text{EROEI} = \frac{\text{lifetime electricity production} - \text{energy consumed over life}}{\text{energy consumed over life}}$$

For nuclear power, as discussed in Chapter 11, there are possibly large energy inputs involved in mining uranium, particularly if low-grade ores are used.

When all these factors are taken into account, EROEI estimates for fossil-fuelled electricity generation can be rather low, and certainly lower than for some rival technologies, as shown in Table 13.12.

Table 13.12 Comparison of energy return over energy invested (EROEI) for various electricity sources, from two authors

Electricity source	EROEI	
	Heinberg	Harvey
Coal, including 480 km transportation by rail and river	Not given	5 at 32% efficiency 6.7 at 42% efficiency
Natural gas, including 4 000 km transportation by pipeline	Not given	2.2 at 54% efficiency
Nuclear	1.1–15	16–18 at 0.2% grade ore 3–5 at 0.01% grade ore
Wind	18	(Germany) 40–80 (20-year lifespan)
Concentrated solar thermal	1.6	8–40 (20-year lifespan)
Solar PV	3.75–10	(Central Europe) 8–25 (25-year lifespan)
Tidal	About 6	Not given
Wave	15	Not given
Hydro	11–267	Not given

Sources: Heinberg, 2009, p. 55; Harvey, 2010, p. 478 (citing Spath et al., 1999, and Spath and Mann, 2000).

This comparison gives favourable values of EROEI for nuclear power, as long as high-quality grades of ore can be used. Many renewable technologies have favourable values: those for wind are in the range 18–80, i.e. wind turbines produce about 18–80 times as much energy over their lifetime as is required to construct and operate them.

Liquid fuels

The EROEI of oil and possible alternative liquid fuels is also a major concern. As described in Chapter 7, liquid vehicle fuels can be produced from a wide range of sources. Interest in 'alternative fuels' has been stimulated both by 'peak oil' and the needs to combat climate change. Table 13.13 shows some estimates of EROEI for oil and different liquid fuels.

Table 13.13 EROEI for liquid fuels

Fuel	EROEI
'Conventional' oil	19
Tar sands	5.2–5.8
Oil shale	1.5–4
Ethanol	0.5–8
Biodiesel	1.9–9

Source: Harvey, 2009.

From a purely 'peak oil' point of view, EROEI may not be the prime concern. Liquid fuels are a premium product and it may well be considered commercially acceptable to expend 1 GJ of energy in extracting 3 GJ of oil from oil shale, giving an EROEI of 2. EROEI is more likely to be considered important from a climate change point of view, since the 'energy invested' is likely to involve extra CO_2 emissions.

Thus although EROEI analysis is a very useful way of comparing energy systems, a full life-cycle analysis of greenhouse gas emissions should be carried out as well, for example as has already been shown for electricity production in Figure 13.7.

Future EROEI for fossil fuels

As pointed out above, the EROEI values for fossil fuels have declined over the past century. They are likely to continue to decline over the next few decades. This is due to multiple factors:

- As the best fossil reserves are depleted, they are likely to be replaced by those requiring a higher energy investment (for example oil wells located in less accessible regions or under the sea).
- Reserves located close to the consumers are likely to be replaced by those further away and requiring longer transportation (for example, the replacement of UK North Sea gas with liquefied natural gas brought from the Persian Gulf).
- Tightening emission standards are likely to mean lower overall energy efficiencies for power plants. For example, the use of flue gas desulfurization reduces the generation efficiency by about 1%. The use of carbon capture and storage (CCS) may require up to 40% more coal to produce the same amount of electricity.

13.9 Land requirements

Another dimension that needs to be taken into account in the assessment of the sustainability of different energy systems is their land footprint.

For example, coal mining using mountain-top removal (MTR) in the Appalachian mountains in the eastern USA has been particularly controversial. This is a heavily forested mountain chain and highly regarded as a tourist attraction, yet it contains high-quality coal. The mountain tops are removed into the adjacent valley bottoms in order to expose the coal seams, which can then be surface-mined. This has serious implications for the biodiversity of the forest, pollution of mountain streams, health impacts on local communities and loss of tourism (see Epstein et al., 2011).

Hydroelectric schemes can also have a significant impact. During the 20th century the construction of large dams has flooded about half a million square kilometres of land (roughly the total area of Spain), and is estimated to have led to the displacement of some 30 to 60 million people. In China alone, according to World Bank estimates, 10 million people were displaced by reservoirs in the period from 1950 to 1989, and more than one million people were displaced by the 18 GW Three Gorges Dam, the world's largest hydroelectric project (Figure 13.16).

Figure 13.16 A 2006 view of the Three Gorges Dam region, which was completed in 2009. The photograph covers a distance of approximately 50 km from left to right (source: NASA images, 2006)

Such schemes may also not be entirely 'climate friendly'. The rotting of vegetation on the flooded land may produce considerable amounts of methane. It has been claimed that in some circumstances hydro schemes can produce more greenhouse gases than a coal-fired plant of the same generating capacity (WCD, 2000).

What of wind power? Wind farms need to cover a large area in order to intercept sufficient wind power to make a significant contribution to national electricity supplies. In Europe, large-scale development is proceeding offshore and it is estimated that a total wind-farm area of about 1200 km^2 would be needed to provide 10% of the UK's electricity demand. The turbine towers, of course, only occupy a tiny fraction of this area.

13.10 Summary

This chapter has assessed the environmental and human health dimensions of the sustainability of energy systems. It has considered key pollutants, their sources and their impacts across a range of scales from local impacts to global ones.

In many countries there are subsidies which encourage the use of fossil fuels. These are in conflict with the needs to cut global greenhouse gas emissions, and with subsidies in other countries that are promoting cleaner technologies.

The development of policies to promote sustainable energy systems requires that monetary values are placed on their full environmental and health impacts. While such impacts are undesirable, in practice their costs have to be balanced against those of abatement. When this balance is achieved it may be said that an economic 'optimum' level of pollution has been reached.

Fuel chain analysis is a useful tool in assessing the critical points in a fuel chain where pollutants are emitted and environment damage is done. Impact–pathway assessment is an important technique for placing financial values on the various environmental and health impacts of energy systems. The values may be expressed in terms of a 'shadow price', for example the damage done by the emission of a tonne of SO_2. Calculating damage done by the emission of a tonne of CO_2, the 'social cost of carbon' is not particularly easy. Policymakers have thus turned to other methods of valuation such as 'emissions trading'. Setting a 'carbon price' is very important in making decisions about investment in future technologies, for example the relative economics of coal-fired electricity generation, nuclear power, renewables or CCS.

The chapter has described the wide-ranging effects of climate change, both as it affects the world at present and as it is likely to do in the future. It has also described the detailed effects of those pollutants not directly involved in climate change. Environmental standards have improved considerably over the past 40 years and there have been significant reductions in the emissions of SO_2, NO_x and particulates in Europe and the USA.

Accidents (ranging from fatal to relatively minor) can and do occur across many of the fuel chains and technologies encountered in this book. Those related to nuclear power are of particular concern. Section 13.7 has discussed the nature of 'risk' and 'risk assessment' when assessing catastrophes such as Chernobyl and Fukushima.

Beyond the headlines of pollution and accidents there are other considerations to be taken into account when comparing different energy sources and technologies. These include EROEI and land requirements.

All energy technologies have some negative environmental and human health impacts. The task that politicians and policymakers face as they plan the transition to a lower carbon and safer future is to weigh up the trade-offs between the various impacts of the range of technologies currently available or likely to be available in the near future.

References

AFP (2011) 'China says over 2,400 dead in coal mines in 2010'; http://latestchina.com/article/?rid=30962 (accessed 19 July 2011).

Arnold, L. (1992) *Windscale 1957: Anatomy of a Nuclear Accident*, London, Macmillan.

Bauer (2008) *Life Cycle Assessment of Fossil and Biomass Power Generation Chains: An analysis carried out for ALSTOM Power Services*, Switzerland, Paul Scherrer Institut; available at http://gabe.web.psi.ch/research/lca (accesssed 10 July 2011).

Bennink, D., Rooijers, F., Croezen, H., de Jong, F. and Markowska, A. (2010) *VME Energy Transition Strategy: External Costs and Benefits of Electricity Generation*, Delft, CE Delft; available at http://www.cedelft.eu (accessed 11 July 2011).

Blasing, T. J. (2011) 'Recent Greenhouse Gas Concentrations', Oak Ridge, TN, Carbon Dioxide Information Analysis Center; http://cdiac.ornl.gov/pns/current_ghg.html (accessed 30 March 2011).

Boden, T. A., G. Marland, and R.J. Andres (2010) *Global, Regional, and National Fossil-Fuel CO_2 Emissions* [online], Carbon Dioxide Information Analysis Center, Oak Ridge National Laboratory, U.S. Department of Energy, Oak Ridge, TN, http://cdiac.ornl.gov/trends/emis/tre_glob.html (accessed 11 December 2010).

BP (2010) *BP Statistical Review of World Energy*, London, The British Petroleum Company; available at http://www.bp.com (accessed 14 December 2010).

Brimblecombe, P. (1987) *The Big Smoke: A History of Air Pollution in London Since Medieval Times*, London, Methuen & Co. Ltd.

CEC (1995a) *ExternE: Externalities of Energy, ii: Methodology*, Brussels; available at http://www.externe.info (accessed 21 July 2011).

CEC (1995b) *ExternE: Externalities of Energy, i: Summary*, EUR 16520 EN Brussels http://www.externe.info (accessed 21 July 2011).

CEC (1995c) *ExternE: Externalities of Energy, iii: Coal and Lignite*, EUR 16522 EN Brussels [online], http://www.externe.info (accessed 21 July 2011).

CEC (2005) *ExternE, Externalities of Energy, 2005 Methodology Update,* EUR 21951; available at http://www.externe.info (accessed 15 July 2011).

Cleveland, C. J. (2005) 'Net energy from the extraction of oil and gas in the United States', *Energy*, vol. 30, pp. 769–82.

DECC (2010) Digest of United Kingdom energy statistics (DUKES), Department of Energy and Climate Change; available at http://www.decc.gov.uk (accessed 5 April 2011).

DEFRA (2008) Emissions of Particulates (PM10), by source: 1970–2007 [online], http://archive.defra.gov.uk/evidence/statistics/environment/airqual/aqemparts.htm (accessed 19 July 2011).

DEFRA (2011) Emissions of Sulphur Dioxide [online], http://www.defra.gov.uk/statistics/environment/air-quality/aqfg20-aqemsox (accessed 19 July 2011).

Edmonds and Calvin (2010) Overview of Integrated Assessment Models, Improving the Assessment and Valuation of Climate Change Impacts for Policy and Regulatory Analysis November 18, 2010, National Center for Environmental Economics, Washington, DC. (http://yosemite.epa.gov/ee/epa/eerm.nsf/vwRepNumLookup/EE-0564?OpenDocument).

EEA (2005) 'Years of life lost (YOLL) in EEA countries due to $PM_{2.5}$ pollution, 2005 [online]', Copenhagen, European Environment Agency, http://www.eea.europa.eu/data-and-maps/figures/years-of-life-lost-yoll (accessed 23 March 2011).

EEA (2008) *Energy and Environment Report 2008*, Copenhagen, European Environment Agency; available at http://www.eea.europa.eu (accessed 9 July 2011).

EEA (2011) 'LCA emissions of energy technologies for energy production' [online], Copenhagen, European Environment Agency, http://www.eea.europa.eu/data-and-maps/figures/lca-emissions-of-energy-technologies (accessed 9 July 2011).

EIA (2009) *Annual Energy Review 2009*, US Energy Information Administration; available at http://www.eia.doe.gov (accessed 15 July 2011).

Epstein, P. R., Buonocore, J. J., Eckerle, K., Hendryx, M., Stout III, B. M., Heinberg, R., Clapp, R. W., May, B., Reinhart, N. L., Ahern, M. M., Doshi, S. K. and Glustrom, L. (2011) 'Full cost accounting for the life cycle of coal', *Annals of the New York Academy of Sciences*, vol. 1219, no. 1, pp. 73–98.

Gipe, P. (2011) Wind Energy – The Breath of Life or the Kiss of Death: Contemporary Wind Mortality Rates http://www.wind-works.org/articles/BreathLife.html (accessed 21 June 2011).

GLA (2002) *50 years on: The struggle for air quality in London since the Great Smog of December 1952*, London, Greater London Authority; available at http://static.london.gov.uk/mayor/environment/air_quality/docs/50_years_on.pdf (accessed 18 July 2011).

Godish,T. (1997) *Air Quality*, 3rd Edition, Boca Raton, FL, Lewis Publishers.

Harvey, L. D. D. (2010) *Carbon-Free Energy Supply: Energy and the New Reality 2*, London, Earthscan.

Heinberg, R. (2009) 'Searching for a miracle: "net energy limits" & the fate of industrial society'. Post Carbon Institute and International Forum on Globalization, False Solution Series, no. 4; available at http://www.postcarbon.org/report/44377-searching-for-a-miracle (accessed 31 March 2011).

Holdren, J. P. and Smith, K. R. (2000) 'Energy, the Environment and Health' in Goldemberg, J. (ed) (2000) *World Energy Assessment: Energy and the Challenge of Sustainability*, United Nations Development Programme,

United Nations Department of Economic & Social Affairs and World Energy Council, New York, NY, pp. 61–110.

Hope, C. (2005) 'Integrated Assessment Models' in Helm, D. (ed) *Climate Change Policy*, Oxford, Oxford University Press.

Hope, C. (2011) *The Social Cost of CO_2 from the PAGE09 Model*, University of Cambridge, Cambridge Judge Business School Working Paper 5/2011; available at http://www.jbs.cam.ac.uk/research/working_papers/2011/wp1105.pdf (accessed 15 July 2011).

Hope, C. and Newbery, D. (2008) 'Calculating the social cost of carbon' in Grubb, M., Jamasb, T. and Pollitt, M.G. (eds.), *Delivering a low carbon electricity system*, Cambridge, Cambridge University Press, pp. 31–63.

HSE (2010) *Offshore injury, ill health and incident statistics 2009/2010*, London, Health and Safety Executive; available at http://www.hse.gov.uk/offshore/statistics/hsr0910.pdf (accessed 19 July 2011).

HSE (2011) *Accident, dangerous occurrence and disease statistics in deep mines* (spreadsheet), HM Inspectorate of Mines; available at http://www.hse.gov.uk/mining/statistics.htm (accessed 19 July 2011).

IEA (2006) 'Energy for Cooking in Developing Countries' in id., *World Energy Outlook*, Paris, International Energy Agency, pp. 419–22.

IEA (2010a) *World Energy Outlook 2010*, Paris, International Energy Agency.

IEA (2010b) *CO_2 Emissions from Fuel Combustion 2010: Highlights (spreadsheet), Paris,* International Energy Agency; available at http://www.iea.org (accessed 11 July 2011).

IPCC (2000) *Land Use, Land Use Change and Forestry*, IPCC Special Report, Cambridge, Cambridge University Press.

IPCC (2007a) *Climate Change 2007: The Physical Science Basis: Technical Summary*, Intergovernmental Panel on Climate Change, Cambridge, Cambridge University Press.

IPCC, (2007b) *Climate Change 2007: Synthesis Report An Assessment of the Intergovernmental Panel on Climate Change*, Intergovernmental Panel on Climate Change, Cambridge, Cambridge University Press.

IPCC (2007c) *Climate Change 2007: Impacts, Adaptation and Vulnerability*, Intergovernmental Panel on Climate Change, Cambridge, Cambridge University Press.

Jing, L. and Yining, P. (2011) 'A hard rain is falling as acid erodes beauty' [online], Beijing, *China Daily*, http://www.chinadaily.com.cn/usa/2011-01/12/content_11833224.htm (accessed 27 July 2011).

Lazarus, M., Heaps, C. and Hill, D. (1995) *The SEI/UNEP Fuel Chain Project : Methods, Issues and Case Studies in Developing Countries*, Boston, MA, Stockholm Environment Institute, Boston, http://www.energycommunity.org/documents/fcaproj.pdf (accessed 11 July 2011).

LBNL (2009) *China Energy Databook*, Berkeley, CA, Lawrence Berkeley National Laboratory; available only as CD from http://china.lbl.gov/databook.

Lenton, T. M., Held, H., Kriegler, E. Hall, J. W., Lucht, W., Rahmstorf, S. and Schellnhuber, H. J. (2008) 'Tipping elements in the Earth's climate system', *Proceedings of the National Academy Science*, vol. 105, no. 6, pp. 1786–93.

MIT (2003) *The future of nuclear power: an interdisciplinary MIT study*, Boston, MA, Massachusetts Institute of Technology; available at http://mit.edu/nuclearpower (accessed 5 January 2011).

McKinsey & Company (2009) 'Pathways to a low carbon economy', Global Greenhouse Gas abatement curve Version 2; available at http://www.mckinsey.com/en/Client_Service/Sustainability/Latest_thinking/Pathways_to_a_low_carbon_economy.aspx (accessed 12 September 2011).

McSorley, J. (1990) *Living in the Shadow*, London, Pan Books.

Mott MacDonald (2010) *UK Electricity Generation Costs Update*, Brighton, Mott MacDonald, Case 1, pp. 87, 97; available at http://www.decc.gov.uk (accessed 21 July 2011).

MSHA (2011) 'Injury Trends in Mining' [online], Mine Safety and Health Administration, Arlington, VA, http://www.msha.gov/mshainfo/factsheets/mshafct2.htm (accessed 4 April 2011).

Murphy D. J. and Hall, C. A. S. (2010) 'Year in review – EROI or energy return on (energy) invested', *Annals of the New York Academy of Sciences*, vol. 1185, no. 1, pp. 102–18.

NASA (2010) '2009: Second Warmest Year on Record; End of Warmest Decade' [online], NASA, Goddard Institute for Space Studies, http://www.giss.nasa.gov/research/news/20100121 (accessed 4 April 2011).

NASA images (2006) 'Rise of the Three Gorges Dam', http://www.nasaimages.org/luna/servlet/detail/NSVS~3~3~9657~109657:Rise-of-the-Three-Gorges-Dam (accessed 23 March 2011).

NISA (2011) *News Release: INES (the International Nuclear and Radiological Event Scale) Rating on the Events in Fukushima Dai-ichi Nuclear Power Station by the Tohoku District – off the Pacific Ocean Earthquake*, April 12th 2011, Nuclear and Industrial Safety Agency; available at http://www.nisa.meti.go.jp/english/files/en20110412-4.pdf (accessed 21 July 2011).

NRC (2010) The Hidden Costs of Energy, National Research Council, National Research Council Academies Press, Washington; available at http://www.nap.edu (accessed 16 July 2011).

OECD (2010) *The Scope of Fossil-Fuel Subsidies in 2009 and a Roadmap for Phasing Out Fossil-Fuel Subsidies*, IEA, OECD and World Bank joint report prepared for the G-20 Summit, Seoul (Republic of Korea), 11–12 November 2010; available at http://www.oecd.org/dataoecd/8/43/46575783.pdf (accessed 4 April 2011).

OFGEM (2010) *The Renewables Obligation buy-out price and mutualisation ceiling 2010–11– information note*, Office of Gas and Electricity Markets; available at http://www.ofgem.gov.uk (accessed 17 July 2011).

Ramage, J. R. (1997) *Energy: A Guidebook*, Oxford, Oxford University Press.

Rasmussen, N. (1974) *Reactor Safety Study: An Assessment of Accident Risks in US Commercial Nuclear Power Plants* (WASH-400) Washington DC, US Nuclear Regulatory Commission.

Sailor, E. C., Bodansky, D., Braun, C., Fetter, S. and van der Zwaan, B. (2000) 'A nuclear solution to climate change?', *Science*, vol. 288, pp. 1177–8.

Scientific American (2011) 'Exposed: Medical imaging delivers big doses of radiation', May; available at http://www.scientificamerican.com/article.cfm?id=exposed-graphic-science (accessed 19 July 2011).

Smith, K. R., Uma R., Kishore, V. V. N., Zhang, J., Joshi, V. and Khalil, M. A. K. (2000) 'Greenhouse Implications of Household Stoves: An Analysis for India', *Annual Review of Energy and Environment*, vol. 25, pp. 741–63.

Spath, P. L., Mann, M. K. and Kerr, D. R. (1999) *Life Cycle Assessment of Coal-fired Power Production*, NREL/TP-570-25119, National Renewable Energy Laboratory, Golden, Colorado; available at http://www.nrel.gov/docs/fy99osti/25119.pdf (accessed 21 July 2011).

Spath, P. L. and Mann, M. K. (2000) *Life Cycle Assessment of a Natural Gas Combined-Cycle Power Generation System*, NREL/TP-570-27715, National Renewable Energy Laboratory, Golden, Colorado; available at http://www.nrel.gov/docs/fy00osti/27715.pdf (accessed 21 July 2011).

Stern, N. (2006) *The Economics of Climate Change: The Stern Review*, Cambridge, Cambridge University Press; available at http://www.hm-treasury.gov.uk/d/Chapter_6_Economic_modelling_of_climate-change_impacts.pdf (accessed 15 July 2011).

Sullivan, L. R. (1995) 'The Three Gorges Project: Dammed if they Do?', *Current History*, vol. 94, no. 593, p. 266.

Tietenberg, T. H. (2006) *Emissions Trading: Principles and Practice*, Washington, RFF Press.

UNDP (2000) *World Energy Assessment: energy and the challenge of sustainability*, New York, NY, United Nations Development Programme.

Watkiss, P. and Hope, C. (forthcoming) *Using the Social Cost of Carbon in Regulatory Deliberations*, submitted to Wiley Interdisciplinary Reviews Climate Change.

Watkiss, P. (2011) 'Aggregate economic measures of climate change damages: explaining the differences and implications', *Wiley Interdisciplinary Reviews: Climate Change*, vol. 2, pp. 356–72, doi: 10.1002/wcc.111.

WCD (2000) Dams and Development: A New Framework, The Report of the World Commission on Dams, London, Earthscan, p. 77.

WHO (1999) *Provisional Global Air Quality Guidelines*, Geneva, World Health Organization.

WISE (2011) 'World Information Service on Energy, Health Hazards for Uranium Mine and Mill Workers – Science Issues [online]', http://www.wise-uranium.org/uhm.html (accessed 19 July 2011).

WNA (2011a) 'Nuclear Radiation and Health Effects' [online], World Nuclear Association, http://www.world-nuclear.org/info/inf05.html (accessed 14 July 2011).

WNA (2011b) 'Cumulative reactor years of operation' [online], World Nuclear Association, http://www.world-nuclear.org/uploadedImages/org/info/cumulative_reactor_years.png (accessed 19 July 2011).

Further reading

Finer, M., Jenkins, C. N., Pimm, S. L., Keane, B. and Ross, C. (2008) 'Oil and Gas Projects in the Western Amazon: Threats to Wilderness, Biodiversity, and Indigenous Peoples', *PLoS ONE*, vol. 3, no. 8.

POST (2011) *Postnote Update No.383: Carbon Footprint of Electricity Generation*, London, Parliamentary Office of Science and Technology; available at http://www.parliament.uk/post (accessed 16 July 2011).

Chapter 14

Remedies: towards a sustainable energy future

By Godfrey Boyle, Stephen Peake and Bob Everett

14.1 Introduction

This final chapter considers how the sustainability problems of our present uses of fossil fuels might be mitigated. Obviously this will require technological and socio-economic changes. But as the previous chapters have described, change and innovation in the field of energy are a continuous and ongoing process. Figure 14.1 shows a set of timelines for environmental issues and various technologies discussed in this book. If a transition to a low carbon world within the next few decades seems revolutionary, then it is only necessary to look at how many other past energy revolutions have happened on a similar timescale.

This chapter focuses on five approaches to improving the sustainability of energy use: 'cleaning-up' fossil fuel combustion, geo-engineering, making more use of hydrogen as an energy carrier, reducing consumption and exploiting renewable energy resources. The topic of the future use of nuclear power has already been discussed in Chapter 11.

Cleaning up fossil fuel combustion is a twofold problem. Firstly, this is a matter of dealing with pollutants such as particulates, SO_2 and NO_x. Most of the techniques for doing this have already been described in earlier chapters. Secondly, there is the problem of reducing CO_2 emissions into the atmosphere; Section 14.2 considers some of the options.

The next section discusses some geo-engineering proposals – large-scale systems aimed at preventing global warming.

The 'hydrogen economy' (Section 14.4) is a vision of a possible future based on hydrogen as an energy carrier, possibly with the large-scale use of fuel cells. The hydrogen might be derived from fossil fuels, with capture and sequestration of the associated carbon at source, or from renewable energy sources.

Reducing energy consumption might seem an intractable problem in a world of ever-increasing demand; Section 14.5 considers some of the possibilities, concentrating on energy for heating buildings and energy use in transport.

Renewable energy resources have considerable potential to improve the sustainability of future energy use; the main resources are described briefly in the penultimate section.

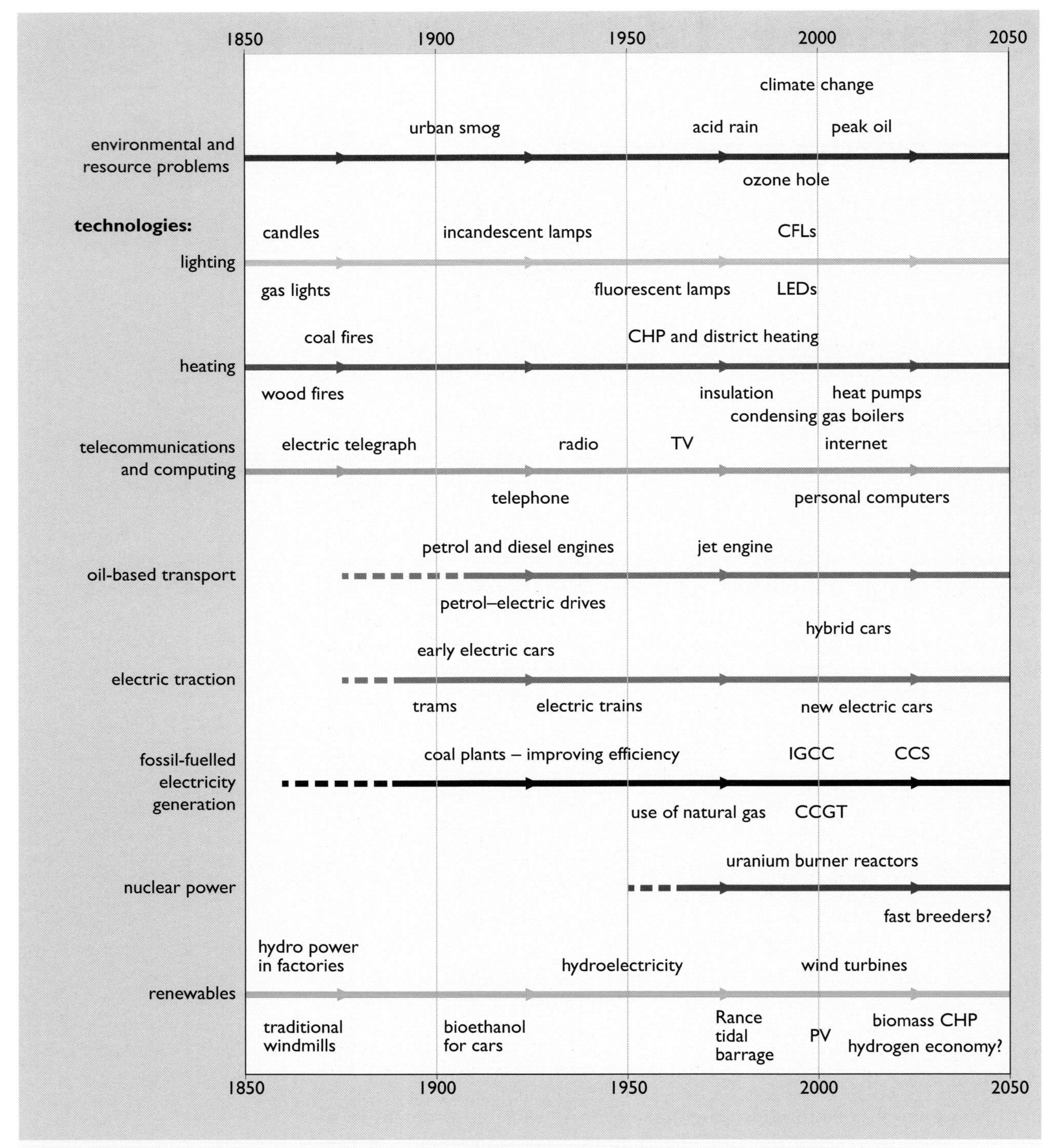

Figure 14.1 Timeline of environmental and resource problems and key energy technology developments since 1850

Finally, to end both this chapter and this book, the concluding section looks at some scenarios for future UK and world energy use aimed at a long-term stabilization of the rising global temperature.

14.2 Cleaning up fossil fuels

As described in the previous chapter the burning of fossil fuels brings with it various pollution consequences. Serious concerns about carbon dioxide emissions and climate change have only really arisen in the last forty years. Concerns about other pollutants have a longer history. The sources and detailed pollution effects of particulates, SO_2 and NO_x have been described in the previous chapter. However, a whole range of techniques have been used to combat these problems, many of which have already been described in earlier chapters.

Particulates

Smoky urban chimneys and diesel trucks belching black smoke are today considered as unacceptable. The urban smogs of the late nineteenth and early twentieth centuries eventually led to enforceable 'clean air' legislation in many countries. In the UK a major step was the 1956 *Clean Air Act*. As described in Sections 5.4 and 13.5, this promoted 'smokeless' fuels such as anthracite and coke. In practice, as shown in Figure 13.15(b), the particulate emissions of domestic heating in the UK have since decreased enormously, not so much from using cleaner coal-based fuels, but because of the switch to cleaner burning natural gas (see Chapter 7).

Particulate emissions from transport remain a serious problem in many cities around the world. The diesel engine is a major culprit. As described in Section 8.4, emissions from these can be cut by proper engine tuning, the use of exhaust filters and oxidation catalysts.

Fly ash from coal-fired power stations is also a serious problem, particularly where pulverized fuel combustion is used. As described in Section 5.6, particulate emissions can be minimized by the use of cyclones, bag filters and electrostatic precipitators.

Sulfur dioxide

One way to prevent the emission of SO_2 from fossil fuels is to remove the sulfur before they are burned. As described in Section 5.4, the removal of as much sulfur as possible was essential, even in the early nineteenth century, when coal was converted to make town gas and coke. Where the coke was sold on as a 'smokeless fuel' it was also sulfur-free. The separated sulfur was processed and sold as fertilizer.

In the transport sector, as described in Chapters 7 and 8, progressively tightening clean air legislation has required low-sulfur petrol and diesel. This has become particularly important where catalytic converters have been introduced to reduce NO_x emissions. Any sulfur in the fuel risks damaging the catalysts.

Figure 14.2 Sulfur removed from Athabasca tar sands sitting on the quayside at Vancouver in Canada

Sulfur is also removed from natural gas at source. This is in part to prevent corrosion damage to transmission pipework. The 'sulfur' smell of town gas was useful in detecting leaks. A substitute smell is deliberately added to natural gas for the same purpose.

Today over 90% of world sulfur production is 'recovered' from oil and gas (see Figure 14.2). The fertilizer industry is a major consumer.

In power stations fuelled by coal or heavy fuel oil, the sulfur dioxide can be removed during or after combustion using flue gas desulfurization or FGD. As described in Section 5.7, this can be done by absorbing it in limestone (calcium carbonate) producing calcium sulfate (gypsum). The limestone can be mixed with the fuel, in which case the gypsum emerges with the ash. Alternatively, the gases can be washed with a slurry of limestone. The resulting gypsum may be of good enough quality to be sold for plasterboard manufacture.

Nitrogen oxides

These mainly result when the combustion temperature is greater than 1500 °C, sufficiently high for the nitrogen in the air to combine with any oxygen present. It is only necessary for a small part of a flame to reach this temperature for this reaction to take place. This is a problem for the combustion of all fossil fuels.

In gas boilers and gas-fired power stations NO_x formation can be minimized with careful burner design, keeping the combustion temperature down and tightly controlling the fuel–air mixture so that there is only the minimum necessary oxygen present to burn the fuel. This is particularly difficult in CCGT plant where the combustion temperature may need to be 1200 °C.

In coal-fired power stations careful furnace design can keep the combustion temperature down (see Section 5.7). There may also be organic nitrogen compounds in the coal itself which will produce NO_x when burned even at low temperatures. Selective catalytic reduction (SCR), involving the injection of ammonia into the flue gases, can be used to convert this NO_x back to nitrogen. An alternative is 'gas reburn' where a small amount of natural gas is injected into the flue gas stream, initially burning by combining with the oxygen from the NO_x reducing it back to nitrogen.

NO_x emissions from road vehicles have been particularly problematic, especially since they are often emitted in towns and cities. As described in Section 8.4, lean burn engine designs and catalytic converters have been introduced to reduce emissions from petrol engines. Selective catalytic reduction is being introduced in the USA for large diesel trucks.

Carbon dioxide

There remains the extremely serious problem of carbon dioxide emissions, which are unavoidable when carbon-based fuels undergo combustion. It is likely that worldwide emissions of CO_2 will need to be reduced by around 80% well before the end of the twenty-first century if there is a reasonable chance of limiting global warming to 2 °C above its pre-industrial levels.

These emissions can be reduced through the improvement of energy efficiency in end use (a topic to be discussed later). They can also be cut with improved efficiency in energy conversion, particularly electricity generation and by switching from high carbon fuels to lower carbon ones. However, the next major step to decouple carbon dioxide emissions from fossil fuel combustion involves the use of carbon capture and storage technologies on a large scale. This possibly represents the next stage in the evolution of fossil fuel heat and power systems.

Nuclear power is of course a low-carbon technology and its future possibilities (including resource limitations) have been discussed in Chapters 10 and 11. In the transition to a future low carbon economy, fossil-based heat and power systems will also need to be integrated into an overall energy system that includes a growing proportion of renewable energy technology. This involves many problems that can only be touched on in this book.

Fossil fuel switching

When burned coal, oil and natural gas all produce different amounts of CO_2 per unit of heat produced. Coal typically produces 60% more CO_2 per MJ than natural gas (see Table 1.3). Also, as pointed out above, coal and heavy fuel oil may contain sulfur. There are considerable environmental benefits in switching from coal and oil to natural gas. It is a clean burning fuel. The sulfur content is usually removed at source and the product as sold is about 98% methane. When burned it produces virtually no SO_2 and various techniques have been developed to reduce NO_x emissions.

Natural gas can also be used with a high efficiency. In the home, as described in Box 7.6, it can be burned in condensing gas boilers, which can have an efficiency of over 90% and low NO_x emissions.

In the electricity sector gas can be burned at a high efficiency in combined cycle gas turbines (CCGTs). These have been described in Section 9.6. This section has also described the overall CO_2 emission benefits seen in the UK electricity sector as it has moved from coal-based generation to increased amounts of nuclear- and gas-fuelled generation.

Although the CO_2 emissions of gas CCGT plant are quite low, there are further possibilities of using them with carbon capture and storage, reducing the emissions even further.

In the transport sector, there are benefits to be had in switching from petrol and diesel to liquefied petroleum gas (LPG) and compressed natural gas (CNG). These can produce lower CO_2 emissions and lower levels of particulates.

As pointed out in Section 9.4, there has been a resurgence of interest in the use of electricity in transport. This has taken the form of:

- new electric tram systems, electrification of existing rail lines and new electric high speed rail links
- petrol–electric hybrid cars including batteries to improve their fuel efficiency
- 'plug-in hybrids' – cars with a petrol engine and a battery chiefly charged from mains electricity
- electric cars charged only from the mains.

Of course, switching transportation energy use 'to electricity' raises the question of how the electricity is to be generated. At present over 40% of the world's electricity is generated from coal. Given concerns about the future availability of oil and gas, the question has to be asked, 'how can coal be made more sustainable?'

'Clean coal'

This is a term that has been used to cover a range of technologies that attempt to address some of the pollution problems of coal use, particularly for electricity generation. Sometimes 'clean coal' refers only to flue gas desulfurization and low NO_x burners, which are now 'standard' equipment for today's power stations in Europe and the USA. These are, of course, technologies that are vital for those regions of the world that still suffer from acid emissions and of particular interest given the high sulfur content of many of the world's current coal reserves.

For the purposes of this book, clean coal technologies are those that go beyond the 'basic' supercritical power station design described in Chapter 5. A key element is to reduce the CO_2 emissions. One way to do this is by increasing the overall generation efficiency of coal power stations, another is to use carbon capture and storage, described a little later in this chapter.

Chapter 6 has explained the importance of combustion temperature in determining plant efficiency. There has been a slow increase over the years in steam pressure and temperature, from 'subcritical' plants (circa 165 bar pressure and 560 °C peak temperature) in the 1980s, through 'supercritical' ones to 'ultrasupercritical' ones today (with pressures above 240 bar and temperatures around 600 °C). This has required the development of special steel alloys to cope with the high-pressure steam in the boilers.

Chapter 5 also described a pressurized fluidized bed combustion (PFBC) plant which has raised the combustion temperature to 900 °C, allowing a stream of hot combustion gases to drive a gas turbine as well as providing lower temperature heat to drive a steam turbine. The *Integrated Gasification Combined Cycle (IGCC)* power plant is an alternative approach to driving a gas turbine using coal. This uses a large coal gasifier (as described in Section 7.10) to produce a stream of 'syngas', a mixture of carbon monoxide and hydrogen, which can be burned (at a high temperature) in a CCGT (see Box 14.1).

Pilot IGCC plants have been built in the USA and Spain. Their design involves a large amount of 'industrial chemistry'. Because of their complexity they are capital-intensive and the cost of electricity from them is currently (2011) considered higher than that from conventional pulverized coal plant. The need to use oxygen rather than air creates a significant energy penalty, resulting in an overall generation efficiency that may be no better than a PFBC plant.

However, the chemical processing could be extended to extract carbon dioxide and send it for either enhanced oil recovery or sequestration.

Carbon capture and storage (CCS)

Although fossil fuel combustion always *produces* carbon dioxide, the CO_2 does not necessarily have to be allowed to *escape* into the atmosphere.

BOX 14.1 The Integrated Gasification Combined Cycle (IGCC) power plant

The **Integrated Gasification Combined Cycle (IGCC)** power plant is one approach to reducing emissions from coal-fired electricity generation. In its simplest form it can be thought of as a coal gasifier (seen at the left in Figure 14.3) providing syngas to burn in a CCGT power station (at the right). The gasifier is fed with oxygen rather than air in order to prevent the gas stream being diluted with unwanted nitrogen. Liquid oxygen is separated from air by cooling it to –183 °C.

Immediately after the gasifier is a cooler followed by a washing stage which allows the removal of dust, ungasified tars and ammonia produced from the nitrogen content of the coal. Then follows a sulfur removal stage which ensures that the final combustion of the gas produces the minimum amount of SO_2. The result is a clean mixture of hydrogen, carbon monoxide and a small amount of methane, which is burned in the gas turbine at a high temperature, about 1200 °C. The waste heat from the turbine (at about 600 °C) is fed to a heat recovery steam boiler, raising steam to drive a steam turbine. This is supplemented by steam produced by the initial gas cooling stage.

Although the generation efficiency of the CCGT is high, about 50%, the overall efficiency of the whole plant is lower, about 38% (MIT, 2007). The air separation unit is particularly energy intensive, consuming about 10% of the gross electricity generation.

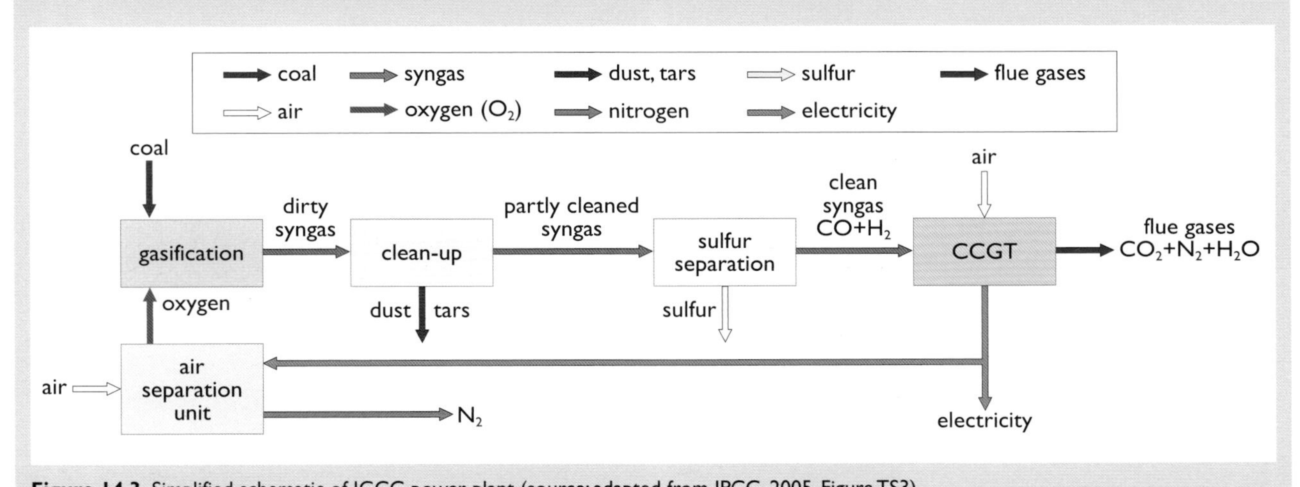

Figure 14.3 Simplified schematic of IGCC power plant (source: adapted from IPCC, 2005, Figure TS3)

It can be captured and **sequestered**, i.e. stored in *isolation* from the atmosphere, most likely deep underground.

CCS involves three stages: capturing the CO_2, its transport and its final storage.

Capturing the CO_2

Carbon capture is best applied to large stationary sources such as power stations and industrial plants. In 2009 approximately 30% of UK CO_2 emissions were from power stations and the proportion for the USA is very similar. The prime interest is in reducing the emissions from coal-fired power stations, where its use represents the most ambitious form of 'clean coal' technology. There are two main possibilities: **post-combustion capture** and **pre-combustion capture**. Post-combustion capture might be simplified by burning coal in oxygen rather than air, a process known as **oxyfuel combustion**. These alternatives are illustrated in Figure 14.4. The first three are coal-based, while the fourth uses natural gas.

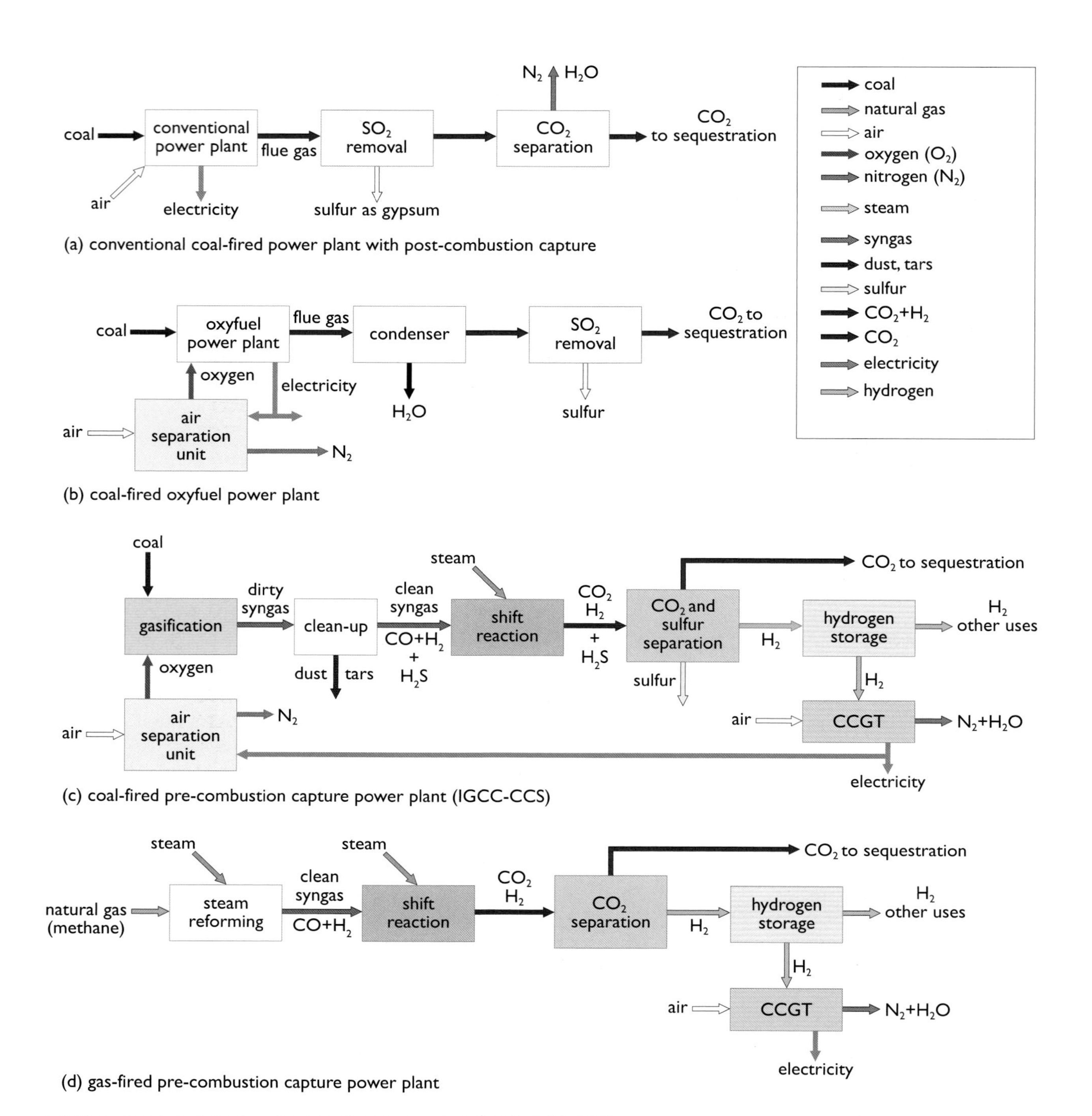

Figure 14.4 Schematic diagrams of electricity generation plants with carbon capture: (a) basic coal-fired post-combustion capture; (b) coal-fired oxyfuel; (c) coal-fired pre-combustion capture IGCC-CCS; (d) gas-fired pre-combustion capture CCGT (source: adapted from IPCC, 2005, p. 26)

Post-combustion capture

In its simplest form, shown in Figure 14.4(a), post-combustion capture involves burning the coal in a relatively conventional power plant. The SO_2 in the flue gases can be extracted using flue gas desulfurization (FGD), as described in Section 5.6. Finally, the CO_2 is captured from the flue gases using one of a number of possible chemical processes as described in Box 14.2.

BOX 14.2 Separating the CO_2

Carbon dioxide is already commonly extracted from streams of syngas in chemical plants and also 'acid' natural gas (in order to reduce the CO_2 concentration to make it saleable). Pilot tests are currently (2011) being carried out on scrubbing CO_2 from the flue gases of conventional power stations (for example Longannet in Scotland). The aim is to extract 90% or more of the CO_2.

There are a number of technical options of which the first two are those currently preferred (IPCC, 2005):

- The gas stream can be washed with ethanolamines, organic liquids related to ammonia. These chemically react with the CO_2. The resulting compound can be pumped away and can be reheated in order to recover the CO_2 and regenerate the original liquid for re-use.
- The CO_2 can be absorbed by methanol or an organic solvent called Selexol. Again the liquid can be pumped away and heated to recover the CO_2 and the solvent.
- Gas separation membranes can be used which are more permeable, for example, to hydrogen than carbon dioxide.
- The CO_2 can be absorbed in solid sorbent pellets under pressure. Like the liquids, these can be removed and then the CO_2 can be recovered by releasing the pressure.

However, as pointed out in Chapter 5, the flue gases of a conventional power plant consist mainly of nitrogen. The CO_2 makes up only about a quarter of the total mass flow, about 650 tonnes per hour for a 660 MW plant. The flue gases are also at around atmospheric pressure. A very large volume of gas thus has to be processed to extract a relatively small proportion.

This problem can possibly be reduced by *oxyfuel combustion*, which basically burns the coal in pure oxygen (though in practice in order to limit the furnace temperature it is a mixture of oxygen with some recycled flue gases). The resulting final flue gas stream consists mainly of CO_2, water vapour and sulfur dioxide. The SO_2 can be extracted by flue gas desulfurisation and the water vapour can be condensed out by cooling the gases, leaving a relatively pure stream of CO_2. The CO_2 separation stage is thus eliminated, but at the cost of having to provide an oxygen supply and some complex recycling of the flue gases. At present (2011) this technology is still in the development stage, with pilot plants having been built in Scotland and Germany. Full-size demonstration plants may be constructed in Germany in the near future.

Pre-combustion capture

An alternative to post-combustion capture is to remove the CO_2 before combustion. Section 7.10 has described how coal is gasified at the Great Plains Syngas plant in North Dakota and converted to a stream of methane and carbon dioxide. The CO_2 is then piped to Canada to be used for enhanced oil recovery.

Although synthetic natural gas (i.e. methane) is the main product of that particular plant, the overall gasification process is similar to that shown in Figure 14.3. In that basic IGCC design, the CCGT power station section at the right is fed with clean 'syngas', hydrogen and carbon monoxide. However, it only requires a little more 'industrial chemistry' to produce the system illustrated in Figure 14.4(c). The syngas can be converted into a stream of hydrogen and carbon dioxide using the *water gas shift reaction*, also described in Section 7.10:

$$\begin{array}{ccccccccc} CO & + & H_2O & \rightarrow & CO_2 & + & H_2 \\ \text{carbon monoxide} & + & \text{water} & \rightarrow & \text{carbon dioxide} & + & \text{hydrogen} \end{array}$$

Any sulfur present as hydrogen sulfide in the gas stream can be removed chemically and converted to elemental sulfur. Then, most importantly, the CO_2 and hydrogen can be separated. The fact that the gas stream is likely to be under pressure and that the CO_2 makes up over 80% of it by mass simplifies the task. The CO_2 can then be piped away for sequestration, leaving an important fuel, relatively pure hydrogen. This can be stored, piped away for other uses (as described in the 'hydrogen economy' later in this chapter) or burned immediately in a CCGT plant to generate electricity. Unlike the natural gas fuelled CCGT plant described in Chapter 9, which produces CO_2, the combustion products of this one are mainly nitrogen and water vapour. Such a power plant would be known as an **Integrated Gasification Combined Cycle with Carbon Capture and Storage** or **IGCC-CCS.**

Although avoiding the large CO_2 emissions from coal combustion has considerable attractions, as shown in Figure 14.4(d), pre-combustion capture using natural gas is technically simpler. Natural gas can be converted to syngas using steam reforming. This, as described in Section 7.10, is the first step used in gas-to-liquids (GTL) plants. The syngas can then be converted to hydrogen and CO_2 as in the coal-based process of Figure 14.4(c). The hydrogen can be burned in a CCGT, but, unlike the coal-based process, there is no need to sacrifice some of the electricity to drive an oxygen-producing air separation unit.

Transporting the CO_2

All of the systems described above produce streams of CO_2 which will be at a rate of several hundred tonnes per hour for a large power station. This CO_2 then has to be transported under pressure to its final storage site, which may be hundreds of kilometres away. This can be done by pipeline, or by ship. Since CO_2 is highly soluble in water that can then become acidic, it has to be thoroughly dried before transport in order to minimize corrosion damage. The CO_2 needs to be compressed to 10–40 atmospheres pressure in a pipeline or to 100 atmospheres or more for ship transport. Ultimately, it is likely to be compressed to several hundred atmospheres pressure for injection deep underground. This compression carries a significant energy penalty. Overall a power plant using CCS may consume up to 40% more fuel in order to produce the same amount of electricity as a conventional one.

Final CO_2 storage

There are many possible locations for sequestering the CO_2. One is in the many large, deep, saline aquifers (i.e. porous rock layers containing salty water) that lie beneath the Earth's surface. These are similar to the reservoir rock for conventional oil and gas. Indeed, as mentioned above, CO_2 is already pumped into many of these for enhanced oil recovery (EOR). Depleted oil and gas reservoirs are also good candidates for use, as are deep, un-mineable coal seams.

The Norwegian company Statoil has since 1996 been successfully sequestering some 1 million tonnes of CO_2 per year in a saline aquifer, the Utsira formation, 800–1000 m beneath its Sleipner gas production platform in the North Sea (Figure 14.5). The gas in this field is 'acid', i.e. it has a high CO_2 content, which must be separated out in order to make the gas saleable. Sequestration is used rather than release into the atmosphere in order to avoid payment of Norway's national carbon emission tax.

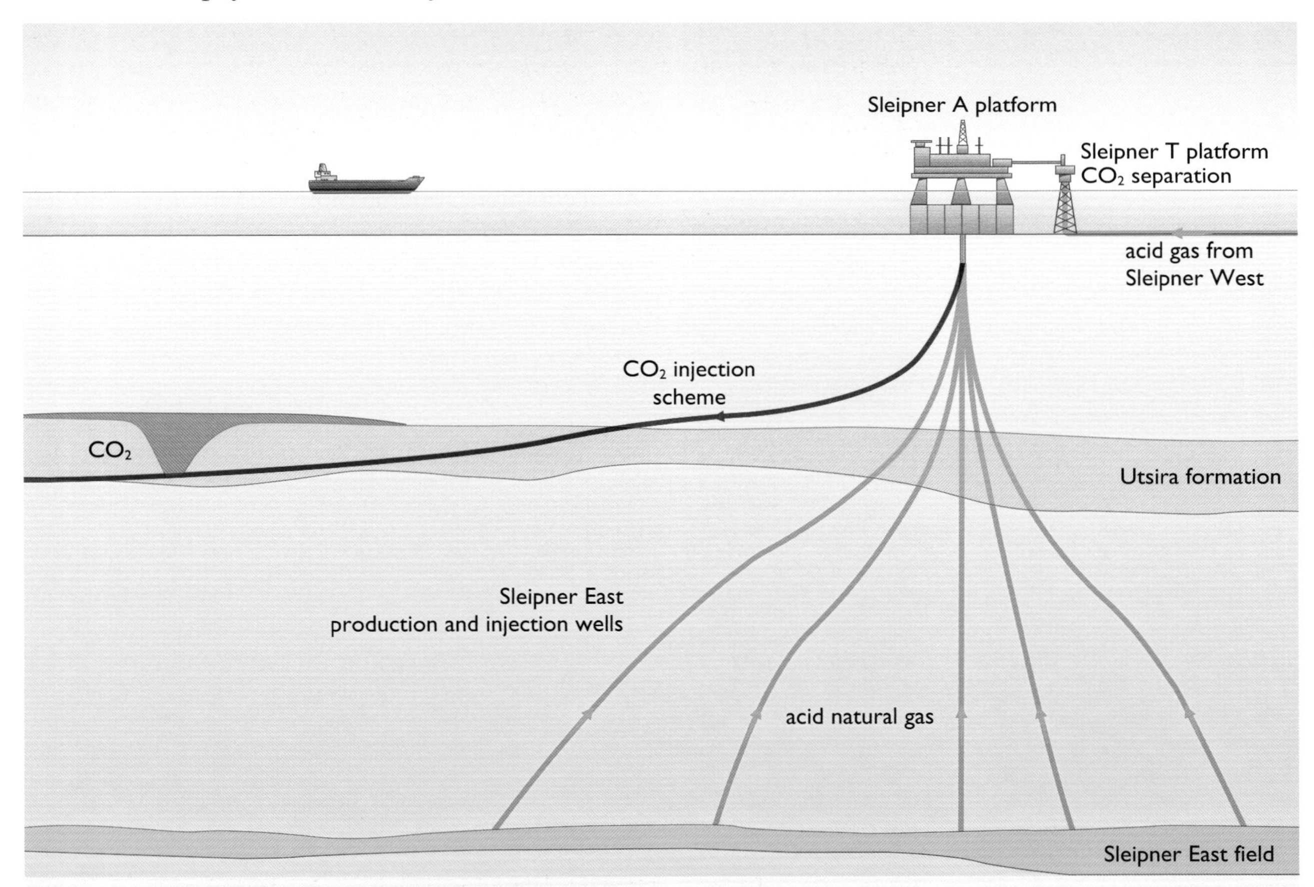

Figure 14.5 Norwegian Statoil's Sleipner field CO_2 sequestration project

Alternatively the CO_2 can be deposited in deep coal seams, and this may be part of a process of enhanced recovery of coal bed methane.

CO_2 storage potential

The above options sound promising, but could a sizeable proportion of the CO_2 produced from the world's fossil fuel combustion be sequestered in such ways? In 2009 global CO_2 emissions were nearly 30 Gt (see Figure 13.9). The IPCC has estimated that the amount of CO_2 being captured and sequestered could be up to 5 Gt per year by 2020 and possibly over 30 Gt by 2050 (IPCC, 2005).

There are many large saline aquifers around the globe. The storage capacity available depends on the type, but Statoil have calculated that the Utsira formation alone might be used to store some 1 Gt of CO_2 per year for the next 600 years. This is roughly equivalent to the current total of CO_2 emissions from all of the EU's electric power plants.

Estimates of the capacities of different geological reservoir types show that oil and gas fields could store between 675 and 900 $GtCO_2$, un-mineable coal seams 3–200 $GtCO_2$ and deep saline formations 1000–10 000 $GtCO_2$.

Another possibility is the injection of CO_2 deep into the world's oceans, a topic that will be discussed later in this chapter.

A working system?

Is there a complete working large-scale power plant incorporating a carbon sequestration and storage system anywhere in the world? The answer is that most of the key components described above exist but have not yet been brought together into a complete system. Table 14.1 summarizes their status.

Table 14.1 Development status of CCS technologies in 2005

Component	Technology	Status
Capture	Industrial CO_2 capture (e.g. natural gas production)	Mature technology
	Post-combustion capture	Economically feasible
	Pre-combustion capture	Economically feasible
	Oxyfuel	Demonstration phase
CO_2 transportation	Pipeline	Mature technology
	Shipping compressed or liquid CO_2	Economically feasible
Storage	Enhanced oil or gas recovery	Mature technology
	Depleted oil or gas fields	Economically feasible
	Saline aquifers	Economically feasible
	Enhanced coal bed methane recovery	Demonstration phase

Source: IPCC, 2005.

Costs

It will be appreciated from Figure 14.4 that the options for electricity generation using CCS are complex. Some of them involve a lot of 'industrial chemistry'. The capital costs are thus higher than for conventional plant. There are also energy penalties associated with the CO_2 capture itself: the production of oxygen rather than air for combustion, the compression or liquefaction of the CO_2 for transport, and the final high pressure CO_2 compression for insertion into the underground reservoirs. This means that in order to produce the same amount of electricity a CCS plant will require more fuel (up to 40% more) than a conventional one. Thus, although these technologies have the potential to solve a climate problem, this may be at the cost of worsening that of fossil fuel depletion.

Table 14.2 shows some estimates of levelized electricity costs from conventional and CCS electricity generation technologies, including some already shown in Figure 13.8 (see Chapter 12 for details of the term 'levelized costs').

Table 14.2 Levelized costs of electricity generation technologies for a 2009 construction start

Technology	Capital costs[1] /p kWh^{-1}	Operating costs /p kWh^{-1}	Fuel costs /p kWh^{-1}	CO_2 transport /p kWh^{-1}	Total /p kWh^{-1}
Conventional coal	3.34	1.08	1.99	–	6.41
Post-combustion capture coal + CCS[2]	7.41	2.33	2.87	0.96	13.57
Coal IGCC	6.17	1.31	2.03	–	9.51
Coal IGCC + CCS[2]	8.20	2.23	2.83	0.95	14.21
Gas CCGT[3]	1.24	0.60	4.69	–	6.53
Gas CCGT + CCS[2]	2.98	1.13	6.50	0.43	11.04

[1] Capital costs calculated using 10% discount rate.
[2] CCS plants included in this table, are 'first of a kind', i.e. not mass-produced.
[3] Gas CCGT plant figure is not intended to agree with that calculated in Box 12.6.
Source: Mott MacDonald, 2010.

It is difficult to say how realistic these cost estimates are until some demonstration plants are built. It would appear that producing electricity from fossil fuels and sequestering the CO_2 will involve a cost penalty of between 5p kWh^{-1} and 7p kWh^{-1}. Section 13.3 has discussed the importance of a carbon price in comparing such technologies. The figures above represent a cost of over £50 to save the emission of a tonne of CO_2. Other estimates of €38–58 tCO_2e^{-1} (£33–50 tCO_2e^{-1}) have been given earlier in Figure 13.2.

14.3 Geo-engineering options

There are other technological 'remedies' for the problem of climate change which can be classified as 'geo-engineering'. There are two main methods: carbon dioxide removal (CDR) and solar radiation management (SRM) (Royal Society, 2009).

Carbon dioxide removal (CDR)

The CCS process described in the previous section sequesters carbon dioxide underground to prevent its release into the atmosphere. CDR methods involve locking up carbon in vegetation (particularly forests), the soil, or in the oceans. As described in Box 13.6, the carbon dioxide in the Earth's atmosphere is part of an overall global cycle. While CO_2 is being emitted into the atmosphere by the combustion of fossil fuels, it is being taken out again by photosynthesis, mainly in plants on land and ocean phytoplankton, and also by dissolving in the waters of the world's oceans. Enhancing these removal processes offers possible ways in which the atmospheric concentration of CO_2 might be reduced.

Land-based methods

Reforestation

When plants grow they absorb CO_2 from the air. Much of this is incorporated into the cells of the plant itself. In the case of a large tree, this may amount to several tonnes of carbon which will remained locked up in its wood as carbohydrates for its whole life. Globally, as shown in Figure 13.10, there may be 550 Gt of carbon contained in the world's biomass. While much publicity is given to the recent deforestation and burning of areas such as the Amazon rainforest, it should be remembered that Europe and much of the eastern USA was once covered with thick forest. This has been cleared for farmland over the past two millennia. However, there is a certain scope for reforestation.

For example, in Scotland by 1900 tree-felling had reduced the level of tree cover to only 5% of the land area. By 2006 programmes of tree planting (not necessarily for carbon sequestration) had increased this to 17% and the Scottish government has a policy of increasing this to 25% by the second half of the 21st century (Scottish Executive, 2006).

It has to be stressed that the CO_2 is only 'sequestered' in the tree or its wood. If the tree dies and rots, most of its carbon content will eventually be converted back into CO_2 which will find its way back into the atmosphere. Similarly, if the tree is burned for fuel its carbon will (normally) be converted back into CO_2. Thus, once a new forest is established, its tree stock must be continuously maintained. Cut wood could be used as a building material, but again the carbon is only effectively sequestered for as long as the building is standing and the wood is intact.

In order to sequester significant quantities of carbon, afforestation would have to be carried out on a very large scale. Estimating the potential contribution that forestry could make to CO_2 removal is fraught with uncertainty. The IPCC has suggested that about 1.3 $GtCO_2$ per year could be sequestered at a cost of under US\$20 tCO_2e^{-1}. This represents about 4% of current global CO_2 emissions. This potential could be as high as 14 $GtCO_2$ per year at costs of up to US\$100 tCO_2^{-1} (IPCC, 2007).

Clearly, although carbon sequestration using forests can play a useful role, it is unlikely to provide a complete solution.

Bioenergy with carbon capture and storage

The coal gasification and sequestration processes described in the previous section could be equally applied to wood or other biomass. Indeed, biomass has the benefit being a naturally low sulfur fuel. Hydrogen and electricity could be produced and a stream of CO_2 sent for underground sequestration. If the biomass was sustainably grown, then this could result in a net extraction of CO_2 from the atmosphere. Electricity could be generated with *negative* CO_2 emissions.

Biochar

As shown in Figure 13.10, there is an estimated 2300 Gt of carbon contained in the world's soil. When vegetation is buried not all of it rots; some of it starts the long process of a return to coal. A certain amount of carbon could be sequestered by the simple process of burying (or just ploughing in) crop wastes. Solid carbon for burial can also be artificially produced by the low temperature pyrolysis of biomass (i.e. the charcoal manufacturing process described at the beginning of Chapter 5). The use of biochar for carbon sequestration seems highly controversial. On the one hand, it is argued that it is also an overall 'soil improver' providing, for example, better water retention, and that there are very large areas of poor farmland that would benefit. On the other, its use may conflict with maximizing the food and biomass yield from a particular area of farmland.

There are, of course, more general uses of biomass and biofuels, which will be discussed later in this chapter, but these are not considered part of 'geo-engineering'.

Carbon sequestration in oceans

The world's ocean depths contain a vast store of CO_2, about 37 000 Gt, or nearly 50 times as much as the atmosphere (see Figure 13.10). There are two natural processes that, together, gradually remove CO_2 from the surface of the oceans and deposit it at greater depths.

The first is the continuous dissolving of CO_2 in water; indeed, the *surfaces* of the world oceans are already saturated with CO_2. However, some of this is slowly removed to the ocean depths. In the Atlantic Ocean, warm surface waters flow as the Gulf Stream from the Caribbean towards northern Europe. When it reaches Arctic latitudes it cools and sinks to the bottom of the ocean, carrying its dissolved CO_2 content with it before flowing southwards. This flow is part of a global **thermohaline circulation**, also known as the '**Great Ocean Conveyor Belt**'. Cold, CO_2-rich water from the depths of the North Atlantic is 'conveyed' southwards to Antarctic regions before eventually surfacing in the Indian Ocean and the equatorial Pacific (Figure 14.6). There the CO_2 eventually escapes, but the residence time is about 1000 years (Royal Society, 2009).

The second process depends on the fact that, at the surface of the world's oceans, microscopic organisms called phytoplankton grow by photosynthesis, harnessing the energy of sunlight and absorbing the CO_2 dissolved near the surface. Phytoplankton are then grazed by other small organisms, zooplankton, which in turn are consumed by marine animals, mainly fish. Eventually, when these die, a proportion (about 30%) of the carbon in their remains sinks to the deep ocean, where it remains for as much as 1000 years before being eventually converted back into CO_2 by the respiration of deep-sea bacteria and returning to the surface waters. This process is sometimes termed 'the biological pump'.

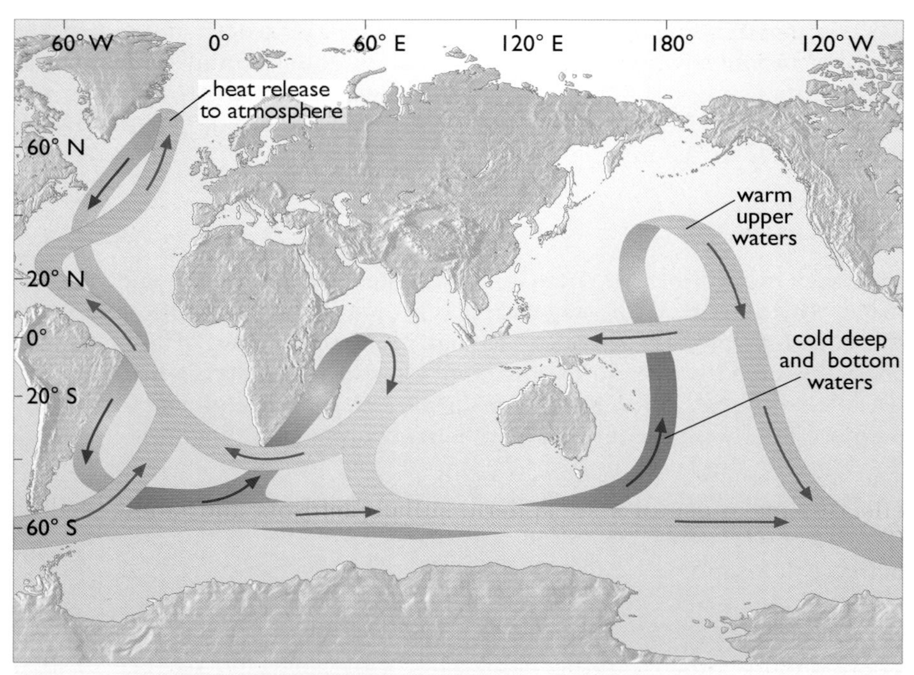

Figure 14.6 The 'Great Ocean Conveyor Belt'

Several geo-engineering methods for accelerating this photosynthetic process have been proposed. One involves the addition of general nutrients, such as nitrates and phosphates, to large areas of ocean in order to increase the growth of phytoplankton and the other organisms that depend on them. More specifically, iron could be added to those areas of the ocean that are known to be lacking in this essential micronutrient. This in turn would increase the amounts of CO_2 eventually deposited from their remains in the ocean depths.

These approaches need to be taken with care, since there are areas of the world's oceans that already suffer from *excess* nitrogen and phosphate run-off from land-based agriculture. This has given rise to *dead zones* of oxygen-depleted water resulting from the over-stimulation of phytoplankton growth.

The direct injection of CO_2 deep into the ocean is another possibility. However, there are concerns that all these approaches could upset the ecological balance of the oceans, and a great deal of additional research would be required before any large-scale projects could be undertaken – though small-scale pilot projects to investigate the potential of these techniques are likely (Royal Society, 2009).

Solar radiation management (SRM)

As explained in Box 1.6, the Earth sits in a radiation balance between the incoming solar radiation and the outgoing infrared radiation. An increase in global temperature could, in theory, be reduced by reflecting more radiation back into space. Reflecting just 2% would be sufficient to offset the warming that may happen as a result of a doubling of atmospheric

CO_2 levels from their pre-industrial levels. However, even reflecting 2% would require very large projects.

Section 13.4 has explained that not all of humanity's actions result in global warming. Some give rise to *cooling*. As shown in Figure 13.11, the most significant of these are due to sulfate aerosols resulting from SO_2 emissions. When suspended in the atmosphere these can directly reflect solar radiation back into space. This effect can be amplified as they seed the formation of clouds, reflecting even more radiation.

The large-scale emission of aerosols into the atmosphere can be quite significant. The 1991 eruption of the volcano Mount Pinatubo in the Philippines projected so much dust and sulfur dioxide into the upper atmosphere that it resulted in a temporary decrease in global temperature of 0.5 °C.

The solar radiation management methods that are most likely to be cost-effective make use of stratospheric sulfate aerosols and cloud seeding (Royal Society, 2009). Proposed schemes involve the continuous injection of many thousands of tonnes of sulfate particles into the upper atmosphere, equivalent to a permanent major volcanic eruption. This may well be in conflict with efforts to avoid 'acid rain'. Somewhat perversely, the adoption of flue gas desulfurization has had the effect of clearing the skies over many industrial areas, allowing in more solar radiation.

Another possibility is the creation of clouds in otherwise cloud-free areas by the forced evaporation of water. This could be done by spraying seawater into the air using the power of offshore or floating wind turbines on so-called 'cloudships'.

Other more extreme suggestions include constructing mirrors in space to shield the Earth from the Sun, though this does raise the question of the CO_2 emissions and expense of the launch vehicles.

14.4 Fuel cells and a fossil-fuel-based hydrogen economy

The fuel cell: energy conversion without combustion

The combustion of fossil fuels has been described in Chapters 5 and 8. This inevitably involves the production of a range of pollutants. Although these can be 'cleaned-up' to a considerable extent, this involves additional costs and low-level pollutant emissions still remain. However, there is one energy conversion device that generates electric power without combustion: the fuel cell, invented in the early 1840s by Sir William Grove (Figure 14.7).

In Grove's apparatus (Figure 14.8), supplies of hydrogen and oxygen were combined in a container where they reacted continuously, with the help of a platinum catalyst, to produce a small but steady flow of electric current. Grove's work was regarded by most as merely a scientific curiosity until the 1950s when fuel cell technology was further developed by the chemist Roger Bacon in Cambridge. In the 1960s fuel cells were successfully used to generate electricity, heat and drinking water for the US *Gemini* and *Apollo* spacecraft, but they were very expensive. Although still costly, in

Figure 14.7 Sir William Grove (1811–1896), born in Swansea in Wales, was both a lawyer and a scientist. In 1839 he invented a battery that used zinc and platinum electrodes, widely used for the new telegraph networks (see Chapter 9). In the early 1840s he invented a 'gas battery' using oxygen and hydrogen, what would today be called a 'fuel cell'. He was one of the founders of the Chemical Society but concentrated on the legal profession after 1853, becoming a High Court judge in 1880

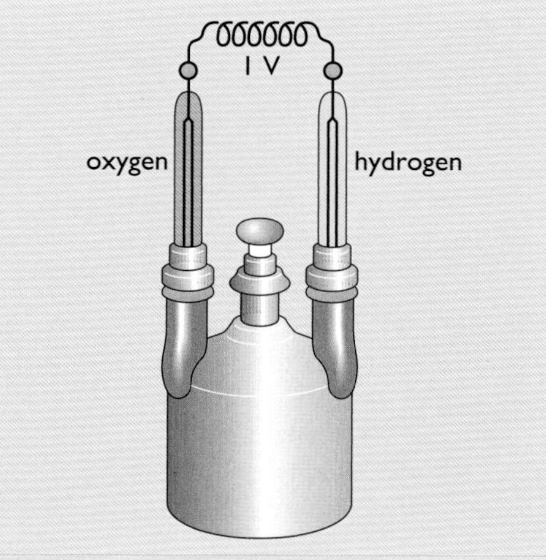

Figure 14.8 Grove's original apparatus

recent years there have been substantial falls in fuel cell prices and major improvements in performance, trends that lead many to believe they could soon become competitive with more conventional energy conversion systems. The basic principles of fuel cells are described in Box 14.3.

BOX 14.3 Fuel cells: the basic principles

How do **fuel cells** work? Many people will be familiar with school experiments in which water is split into its constituents, hydrogen and oxygen, by the process of 'electrolysis' – passing an electric current between two electrodes immersed in water. Fuel cells operate in a manner that is essentially the reverse of electrolysis – by combining, rather than splitting, hydrogen and oxygen. This process generates an electric current, water – and some 'waste' heat.

Fuel cells use any two reactants that are respectively oxidizing (i.e. a source of oxygen) and reducing (i.e. readily combining with oxygen), but the most common reactants are hydrogen and oxygen (or air, which is approximately 20% oxygen).

The principal characteristics of fuel cells and their mode of operation are shown in Figure 14.9. As can be seen, most fuel cells consist of two electrodes, an 'anode' (the node from which electrons leave the device) and a 'cathode' (the node through which electrons re-enter the device). These can be made from a variety of electrically conducting materials. The electrodes are usually coated with platinum, or a platinum group metal such as palladium or ruthenium, which acts as a catalyst. **Catalysts** are substances that increase the rate of a chemical reaction without themselves undergoing any permanent chemical change. Between these is placed an 'electrolyte', of which again there are a variety of types.

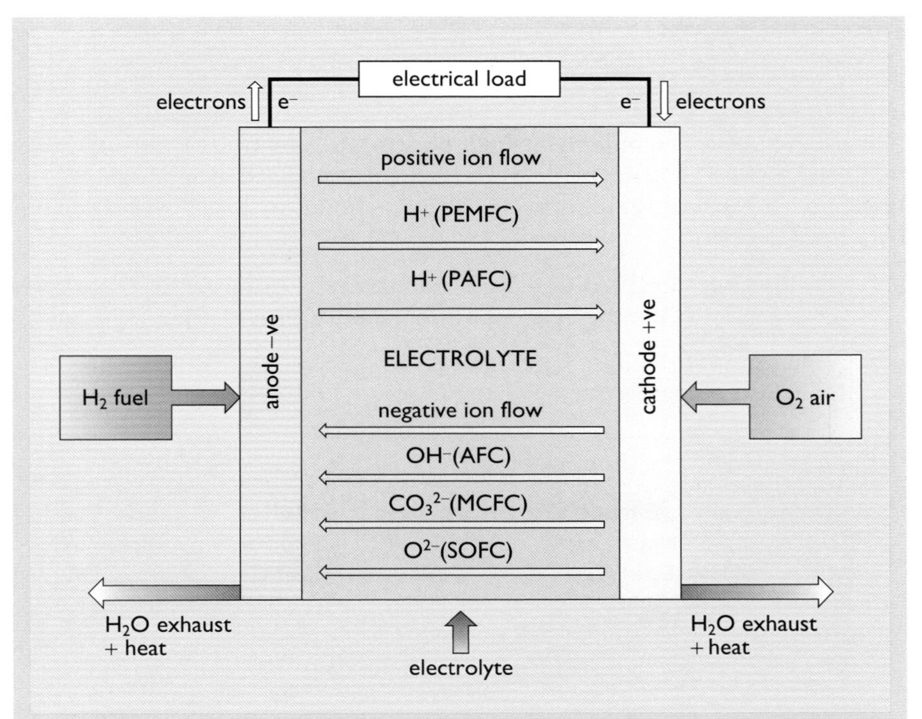

Figure 14.9 Principles of fuel cell operation. For key to abbreviations see Table 14.3 (source: Laughton, 2002)

Normally, hydrogen is the fuel fed to the anode, while oxygen (from air) is supplied to the cathode. Both the anode and the cathode are porous, allowing the gases to flow through them. With the aid of the catalysts present on the surface of the electrodes, the hydrogen splits into hydrogen ions (i.e. protons, shown as H+ in the diagram) and electrons (shown as e^- in the diagram). The electrons flow away from the anode into an external electrical circuit where they can be made to deliver useful energy. Meanwhile, the hydrogen ions flow through the electrolyte to the cathode where (again with the aid of a catalyst) they combine with the oxygen supplied to the cathode and the incoming electrons from the external electrical circuit to form water vapour. Depending on the type of cell, some 30–60% of the energy content of the input fuel is converted to electricity; the rest appears as heat, but this can often be used, either for space or water heating or to provide energy for the 'reformers' that may be required to convert, say, natural gas into the pure hydrogen required by the fuel cell.

The key advantage of the fuel cell as an energy converter is that electricity is produced *directly*. This means that its efficiency can be higher than the limits set by Carnot for heat engines (see Section 6.3). It also means that there are no emissions of the gaseous pollutants that are associated with combustion processes, such as SO_2, NO_x or particulates. If pure hydrogen is the fuel, there are no CO_2 emissions and the only other emission, apart from some 'waste' heat (which can be put to good use), is water vapour.

One difficulty of present fuel cell designs is the poisoning of the catalysts by impurities, particularly sulfur, in the fuel. For many types extremely pure hydrogen is necessary. This is normally provided by the electrolysis of water or the steam reforming of natural gas.

Fuel cell types

The science and technology of the fuel cell have advanced enormously since Grove's day. There is now a wide variety of fuel cell types, each using different arrangements of fuel, electrolyte, electrodes, etc. One way of classifying fuel cells is by the type of electrolyte used. A summary of the main fuel types and their characteristics is shown in Table 14.3 followed by a brief description of each.

(1) **Alkaline Fuel Cell (AFC)** – This was the type used in the US *Apollo* spacecraft and later the Space Shuttle, where it provides both electricity

Table 14.3 Classification of fuel cells

Characteristics	Fuel cell type					
	Alkaline Fuel Cell (AFC)	**Proton Exchange Membrane Fuel Cell (PEMFC)**	**Direct Methanol Fuel Cell (DMFC)**	**Phosphoric Acid Fuel Cell (PAFC)**	**Molten Carbonate Fuel Cell (MCFC)**	**Solid Oxide Fuel Cell (SOFC)**
Electrolyte	Aqueous potassium hydroxide (30–40%)	Sulfonated organic polymer (hydrated during operation)	Sulfonated organic polymer (hydrated during operation)	Phosphoric acid	Molten lithium/ sodium/ potassium carbonate	Yttria-stabilized zirconia
Operating temperature	60–90 °C	70–100 °C	90 °C	150–220 °C	600–700 °C	650–1000 °C
Charge carrier	OH^-	H^+	H^+	H^+	CO_3^{2-}	O^{2-}
Anode	Nickel (Ni) or platinum group metal	Platinum (Pt)	Platinum-ruthenium (Pt, Ru)	Platinum (Pt)	Nickel/ chromium oxide	Nickel/yttria–stabilized zirconia
Cathode	Platinum (Pt) or lithiated NiO	Platinum (Pt)	Platinum-ruthenium (Pt, Ru)	Platinum (Pt)	Nickel oxide NiO)	Strontium (Sr) doped lanthanum manganite
Co-generation heat	Low temperature	Low temperature	Low temperature	Temperature acceptable for many applications	High temperature	High temperature
Electrical efficiency, %	60	40–45	30–35	40–45	50–60	50–60
Fuel sources	H_2 (removal of CO_2 from both gas streams necessary)	H_2 (reformate with less than 10 ppm CO)	Water/ methanol solution	H_2 reformate	H_2, CO, natural gas	H_2, CO natural gas

Note: Reformate is pure hydrogen, or a hydrogen-rich gas, produced by 'reforming' fossil fuels.
Source: based on Laughton, 2002.

and drinking water. It uses as its electrolyte a solution of potassium hydroxide and is fuelled by pure hydrogen.

(2) **Proton Exchange Membrane Fuel Cell (PEMFC)** – this is one of a type of **solid polymer fuel cells (SPFC)** which use as their electrolyte specially treated polymers that only allow the passage of positive ions. The PEMFC uses hydrogen as its input fuel and operates at relatively low temperatures. The PEMFC technology has been refined and developed from the early spacecraft designs and has now been adopted by some leading motor manufacturers such as Daimler-Chrysler and Honda as the basis of a new generation of 'zero-emission' cars (described later in the chapter).

(3) **Direct Methanol Fuel Cell (DMFC)** – this is another type of solid polymer fuel cell, which can also run at a low temperature, but is fuelled by a water/methanol solution. This is at a less advanced stage of development compared to the PEMFC.

(4) **Phosphoric Acid Fuel Cell (PAFC)** – this uses phosphoric acid as its electrolyte with hydrogen as its fuel and operates at moderate temperatures. PAFC technology has become established as a reliable source of electricity and heat in buildings and other stationary applications. For example, in 2003 a 200 kW unit was installed in a combined heat and power generation scheme at Woking in Surrey, fuelled by hydrogen produced from natural gas.

(5) **Molten Carbonate Fuel Cell (MCFC)** – this uses molten lithium, sodium or potassium carbonate as its electrolyte and requires a high temperature for operation. It can run on hydrogen, but also directly on fuels such as carbon monoxide and natural gas. In this case, unlike other fuel cells running on hydrogen alone, it will also produce CO_2.

(6) **Solid Oxide Fuel Cell (SOFC)** – this uses a solid oxide or ceramic as the electrolyte. Like the MCFC it runs at a high temperature and can potentially use fuels such as natural gas, steam reforming it within the device. Both MCFCs and SOFCs are still at the development stage but have high efficiencies and could be promising for future large-scale, low-emission electricity generation.

More than a century and a half after Sir William Grove's invention, the fuel cell could be on the verge of more widespread deployment in the world's energy systems. The efficiencies of current systems are between 30% and 60%. They could supplant cruder, combustion-based technologies for converting the energy content of fossil fuels into useful electricity and heat. This is also a key technology underpinning an emerging 'hydrogen economy' of the future.

A fossil-fuel-based hydrogen economy

In the 1874 novel *The Mysterious Island*, by the science fiction writer Jules Verne, two characters discuss what will happen when the world's supplies of coal run out:

> 'And what will they burn instead of coal?'
>
> 'Water,' replied Harding.

> 'Water!' cried Pencroft, 'water as fuel for steamers and engines! water to heat water!'
>
> 'Yes, but water decomposed into its primitive elements,' replied Cyrus Harding, 'and decomposed doubtless, by electricity, which will then have become a powerful and manageable force. [...] Yes, my friends, I believe that water will one day be employed as fuel, that hydrogen and oxygen which constitute it, used singly or together, will furnish an inexhaustible source of heat and light, of an intensity of which coal is not capable. Some day the coalrooms of steamers and the tenders of locomotives will, instead of coal, be stored with these two condensed gases, which will burn in the furnaces with enormous calorific power.'
>
> (Verne, 1874)

Verne, understandably, left open the question of where the electricity required to 'decompose' water into hydrogen and oxygen would come from.

It has been suggested that the history of human fuel use over the twentieth century can be viewed as a process of gradual de-carbonization: a movement from coal to oil and more recently to natural gas – though all three fuels are still in use.

It looks increasingly likely that by the middle of the twenty-first century further de-carbonization will have taken place. It is possible that hydrogen, a zero-carbon fuel, could be playing a more important role, along with electricity, in the world's energy systems.

Hydrogen, however, is not like fossil fuels that can simply be extracted from the ground. Before it can be used, hydrogen needs to be extracted from the other compounds within which it is normally bonded in nature – principally water, or the hydrocarbons – and this separation requires energy. Hence hydrogen is not a primary but a secondary energy source. It is perhaps best viewed as an energy *carrier* or *vector* – a convenient way of transporting or storing the energy originating from primary sources. Its function is thus similar to that of town gas in the nineteenth century, which was a more convenient form of coal.

Some of this hydrogen could be generated by electrolysis from renewable electricity sources such as hydro, wind or solar power. But initially it seems more likely that it would be produced from fossil fuels such as natural gas or coal, with capture and sequestration of the CO_2 produced as part of the conversion process. These processes have been described in Section 7.10 and the second section of this chapter. Figure 14.10 illustrates a possible future hydrogen energy system using natural gas. The hydrogen could be used for electricity generation, distributed for cooking and heating, as well as for transport uses on land and in the air.

There are perhaps three main reasons for converting fossil fuels to hydrogen. Firstly, it is likely to be easier and cheaper to separate the CO_2 from fossil fuels during the process of hydrogen production than it is to capture CO_2 after the combustion of fossil fuels in conventional systems such as boilers. 'Pre-combustion capture' is likely to be preferable to 'post-combustion capture'.

Secondly, as discussed earlier, the hydrogen can be converted in a fuel cell into electricity and heat with relatively high efficiency (see Table 14.3). It

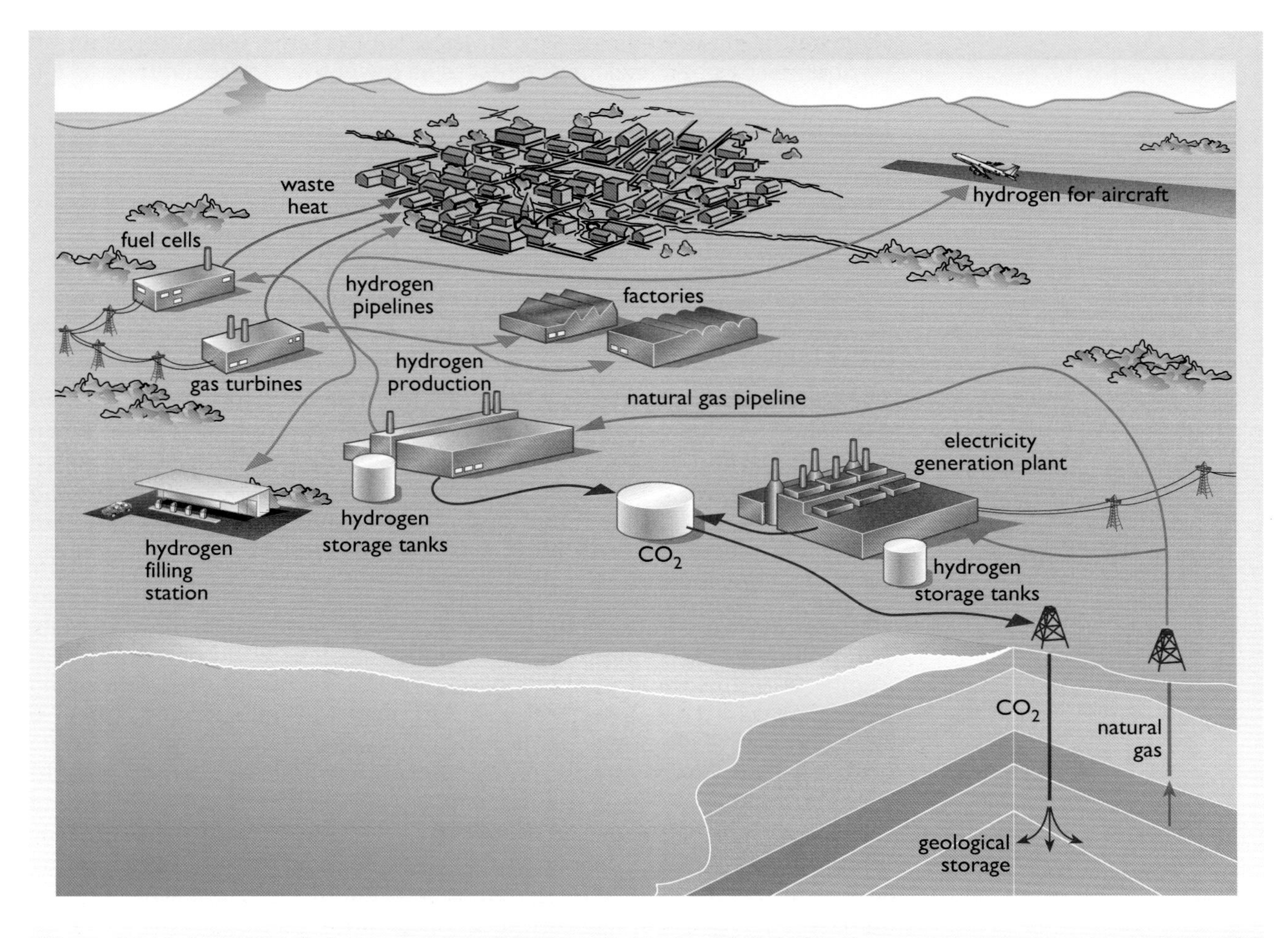

Figure 14.10 Schematic of a hydrogen energy system supplied by fossil fuels with carbon capture and storage (source: adapted from IEA, 1999)

can also, like conventional fuels, be simply burned in air to produce heat: if the combustion temperature is high enough some NO_x may be produced, but otherwise there are far fewer pollutants at the point of use than with fossil fuel combustion.

Thirdly, hydrogen is a potential transport fuel that can be made from a range of sources. This may become an important factor in dealing with the problems of 'peak oil' (Chapter 7).

Hydrogen is already produced in substantial quantities worldwide, mainly by 'steam reforming' natural gas. World production is estimated at about 500 billion cubic metres per year, with an energy content of around 6.5 EJ (Harvey, 2010a). At present, hydrogen is seldom used for energy purposes – with the exotic exception of its use as a fuel for space rockets. Its main use is in producing ammonia for fertilizer production, in oil refining and in various chemical processes. Hydrogen has for many years been transported reliably and safely over long distances. There are some 1500 km of pipelines carrying hydrogen in Europe, and some 700 km in the USA. Pipelines carrying pure hydrogen need to be made of special steel alloys to avoid the hydrogen degrading the steel, but such materials are well established and understood.

Over long distances, the energy losses involved in hydrogen distribution by pipeline can in some circumstances be lower than those incurred when electricity, the most popular energy carrier, is used for distribution (Lovins, 2005).

Hydrogen is also currently produced by electrolysis. For example, Norwegian fertilizer production is based on hydrogen produced in this way from the country's low-cost hydroelectric plants. The electrolysis process is only 70–80% efficient. Thus, at present, electrolytically produced hydrogen is normally more expensive than that derived from fossil fuels. However, this route is likely to produce high-purity hydrogen, which is required by many fuel cell designs.

The use of renewable energy to produce hydrogen in what has been called the 'solar hydrogen economy' could be feasible in the longer term, when technologies and economics are more favourable than at present; but in the medium term a hydrogen economy based on fossil fuels with CCS might be a valuable transitional stage, complementing an increased use of renewable energy and facilitating the eventual emergence of a renewables-based energy economy.

Hydrogen-powered CCGTs could also prove a good complement to variable renewable electricity supplies from sources such as wind or tidal power. Figure 14.4 illustrated methods of generating electricity using hydrogen produced from coal or natural gas. Although the 'integrated gasifier' is shown as part of the power plant, it is perhaps best thought of as a separate unit. It is likely that hydrogen production and carbon sequestration would run continuously, storing hydrogen as necessary. The CCGTs could then run intermittently as required to supply a varying electricity demand. This intermittent operation creates problems of thermal stresses from heating and cooling, particularly in the steam section of the plant. These could be reduced by using waste heat from the gasifier to keep the CCGT plants continuously 'warmed up' and ready for operation (Starr et al., 2005).

Hydrogen storage and use in transport

Properties and storage

How does hydrogen compare as a potential fuel with conventional oil and gas? Section 7.8 discussed why those fuels are regarded as so special and Table 7.6 gave some comparative properties of conventional fuels. Table 14.4 shows the properties of some of these compared to hydrogen.

The energy content of a kilogram of hydrogen is nearly three times that of a kilogram of jet fuel, petrol or even natural gas. There is also a large difference between its higher and lower heating values. This means that, if possible, the water vapour produced from its combustion should be condensed (as it might be in a condensing boiler).

However, being a gas, it has a very low energy content per unit *volume*. At normal pressures and temperatures energy storage using hydrogen involves relatively large volumes.

The principal methods of storage of hydrogen at present are (a) as a compressed gas, and (b) as a liquid. Compressed hydrogen gas can be stored in tanks made of steel or composite materials at pressures as high as 700 bar (70 MPa). In liquid form, hydrogen requires relatively less storage volume,

Table 14.4 Properties of petrol, natural gas and hydrogen

Fuel	Lower heating value /MJ kg^{-1}	CO_2 emissions LHV /g CO_2 MJ^{-1}	Higher heating value /MJ kg^{-1}	CO_2 emissions HHV /g CO_2 MJ^{-1}
Aviation jet fuel (kerosene)	43.9	72	–	–
Petrol	44.7 (9.1 kWh $litre^{-1}$)	70 (2.3 kg $litre^{-1}$)	–	–
Natural gas/CNG	47.8	57	52.8	51
Hydrogen	121.0	Nil	141.8	Nil

but has to be stored at temperatures below –253 °C, which involves highly insulated tanks and significant cooling energy use. However, as described in Chapter 7, large insulated storage tanks are already used for liquefied natural gas.

Hydrogen can also be stored by allowing it to react with certain metal alloys to form metal hydrides. The hydrogen, when required, is then released by heating the hydride. But this method at present involves substantial expense and weight.

Research continues into other, more cost-effective methods of storing hydrogen, for example by adsorbing it to finely powdered carbon particles.

Use in transport

Although hydrogen has considerable potential for use in buildings, particularly in fuel cells for CHP, it is its potential for replacing petroleum-based fuels in transport that is of particular interest.

One recent example is the Honda FCX Clarity fuel cell car, introduced in 2008 and available for lease in limited numbers in the USA, Japan and Europe (Figure 14.11). The hydrogen is stored on board as a compressed gas and, of course, requires appropriate filling stations. The car is basically an electric car, with motors driven from a 100 kW proton exchange membrane fuel cell (PEMFC) supplemented by a lithium-ion battery that stores energy from braking and deceleration. It might thus be described as a fuel-cell 'hybrid', but of course very different from the petrol–electric hybrid cars described in Section 9.4.

The combination of a fuel cell with a battery raises a basic question: which is better – a car that normally runs on a hydrogen fuel cell or a battery–electric car charged from the mains? This is discussed in Box 14.4.

Hydrogen has been introduced in a number of bus schemes. Some have used internal combustion engines, others fuel cells with an on-board reformer for fossil fuels. Figure 14.12 shows one of a fleet of ten in London which use fuel cells. These have hydrogen storage tanks on the roof and are refuelled with pressurized hydrogen at a specially built filling station.

Hydrogen may also be able to be used as a zero-carbon fuel for aircraft. Jet engines can be run on hydrogen, as Russia demonstrated in 1988 when a

Figure 14.11 The Honda FCX Clarity fuel cell car

BOX 14.4 Batteries vs hydrogen fuel cells for vehicles

As described in Section 9.4, battery electric vehicles have a long history dating back to the nineteenth century. The basic technology is well developed, including control systems and regenerative braking. The weight of the batteries, the limited range and the time taken to recharge the batteries have always been limiting factors.

New designs of rechargeable batteries, such as those using nickel–metal hydrides and lithium-ion technologies have been developed for other applications, such as laptop computers and mobile phones. Modern electric cars use very large quantities of essentially the same battery cells. However, there are possible problems about the large-scale supply of critical rare elements such as lanthanum and lithium for these batteries.

Another key factor for the future is the availability of adequate low-carbon sources of electricity. Replacing the existing stock of petroleum-fuelled vehicles with electric ones is likely to increase the total electricity demand.

Fuel cells, in contrast, have only seriously been developed since the 1950s. They are currently expensive and commercial designs rely on supplies of high purity hydrogen. Many designs require expensive platinum as a catalyst, although in very small quantities.

The range of a fuel cell car is only limited by the amount of fuel that it can carry. Given that hydrogen has a high energy density, the problem is more one of the size and weight of the pressurized tank rather than the weight of the fuel.

Future critical factors will be the development of a hydrogen filling station network and the availability of high-purity hydrogen.

modified Tupolev 155 airliner made a brief flight with one of its engines fuelled by hydrogen (see Hoffman, 2001). Storing either pressurized or liquefied hydrogen on an aircraft poses some technical problems. However, as pointed out above, hydrogen's energy density is a factor of nearly three times better than jet fuel. Given that fuel can make up 40% of the take-off weight of a long-haul flight its use offers possibilities of carrying more passengers

Figure 14.12 A Mercedes London Transport bus, powered by a hydrogen fuel cell; the storage tanks can be seen on the roof

and/or improved fuel economy. The European aircraft manufacturer Airbus Industrie is studying designs for a hydrogen-powered airliner.

Hydrogen-powered aircraft are not likely to be totally climate-friendly. Although they would not emit CO_2 they would still emit another greenhouse gas, water vapour, and do so high in the atmosphere where it can have full effect.

Hydrogen safety

Hydrogen has been considered highly unsafe by many ever since the disastrous fire on the German airship *Hindenburg* at Lakehurst, New Jersey, in 1937 (Figure 14.13). But this reputation may be unjustified.

Figure 14.13 The German hydrogen-filled airship *Hindenburg* burst into flames at Lakehurst, New Jersey in May 1937, killing 36 people. Recent research has suggested that hydrogen was not the primary cause of the disaster

In 1997 Addison Bain, former manager of the hydrogen programme at NASA's Kennedy Space Center, pointed out that the cause of the disaster may have been highly inflammable materials used to paint the airship's skin. If these had not been used then the disaster might not have happened (Hoffman, 2001).

Hydrogen is the lightest element in nature; it is highly reactive and difficult to handle. Fire and explosion are the main dangers, but on the other hand hydrogen dissipates very rapidly in confined spaces. In cars and buildings where hydrogen is already used, sensors are employed to detect concentrations of over 4% and open windows automatically.

Hydrogen flames are invisible, which increases the risk of individuals inadvertently being burned. But their higher flame speed means that hydrogen fires burn out rapidly. A vehicle fire involving petrol may take 20–30 minutes to burn out; with hydrogen, the equivalent time would be only 1–2 minutes.

The risks entailed in hydrogen fuel use, in short, are different from those associated with other fuels, but most experts consider hydrogen to be no more dangerous overall (IEA, 1999).

The pros and cons of a hydrogen economy are discussed in detail in, for example, IEA (1999), Hoffman (2001), Lovins (2005), Harvey (2010b) and Ekins (2011).

14.5 Reducing energy demand

This book has concentrated on energy supply, particularly from fossil fuels and nuclear power. It has looked at various methods to improve the efficiency of fossil fuel use, for example in various types of engines in Chapters 6 and 8 and electricity generation in Chapters 5 and 9.

If global CO_2 emissions are to be drastically cut in the next few decades, then more attention needs to be paid to what can be done to improve the efficiency of energy use at the *demand* side – that is, within our buildings, industries and society at large. This is often called 'energy conservation', but that term carries implications of 'doing without'. A better phrase would be 'demand-side energy efficiency'.

There are two distinct approaches: one *technological*; the other *social*.

The technological approach involves installing improved energy conversion (or distribution) technologies that require less energy input to achieve a given level of useful energy service. For example, Chapters 3, 5 and 9 have described the amazing technical improvements in artificial lighting over the past 200 years.

The social approach involves rearranging our lifestyles, individually and collectively, perhaps in minor ways, perhaps major, in order to reduce the amount of energy required to provide a given service. For example, it may be necessary to create (and recreate) *walkable cities* where it is not necessary to use a car to reach shops or schools.

Two particular areas are worthy of mention here: the energy used for space heating (i.e. warming the interior of rooms) in the domestic and commercial sectors of the economy, and the urgent need to reduce petroleum use in the transport sector.

Saving space heating energy

A glance at the lowest bar of Figure 3.10 shows that the UK uses just over 2 EJ y^{-1} of energy to provide space and water heating and low temperature drying. Of this, about 1.5 EJ y^{-1} is just for space heating in the domestic and commercial sectors, 25% of the total delivered energy use (DECC, 2010a). Is this heating energy use really necessary? There are two ways that it can be cut.

The first is through the use of combined heat and power generation, using the waste heat from power stations (large and small). This has been described in both Chapter 3 and Chapter 9 with particular reference to its extensive use in Denmark.

The second method is by the use of better insulation. The UK building stock has been notoriously poorly insulated in the past; indeed, the requirement for loft insulation in new houses was only introduced in building regulations in 1974. Insulation standards in UK buildings have improved considerably since the 1990s, both for new buildings and retrofits to existing buildings. Loft insulation and cavity wall insulation are currently available, heavily subsidized by energy utilities.

In Germany, the **PassivHaus standard** was developed during the 1990s and has been used in over 30 000 buildings across the world to date. This involves using very thick insulation, good-quality windows and airtight construction to reduce the space heating demand of a building to a low level. It can then be heated mainly by solar gains and heat from appliances and the occupants themselves. A '3-litre house' retrofit project in Ludwigshafen, Germany, using this concept, demonstrated a sevenfold reduction in space heating energy use (see Box 14.5).

BOX 14.5 The 3-litre house

The thermal performance of existing buildings can be radically improved if they are adequately insulated. In the late 1990s an estate of apartment blocks in Ludwigshafen in south-west Germany (Figure 14.14) was given a thorough thermal modernization including:

At least 200 mm thickness of foam insulation on the roof and in the walls (see Figure 14.15).

Triple-glazed windows with argon filling and low emissivity coatings on the glass to cut the heat loss.

Mechanical Ventilation with Heat Recovery (MHVR). This uses heat recovered from outgoing air to preheat incoming fresh air.

Monitoring showed that the net space heating energy use (i.e. that to be supplied by the heating system) fell by a factor of seven from 210 kWh per square metre of floor area per year to only 30. This is equivalent to an annual consumption of only 3 litres of heating oil per m^2 – hence the project name, the 3 Litre House (Luwoge, 2008).

The project also included a fuel cell CHP unit.

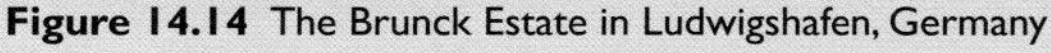

Figure 14.14 The Brunck Estate in Ludwigshafen, Germany

Figure 14.15 Laying blocks of foam insulation on the roof

The transport sector

The rise of UK transport energy demand (and that mainly involves petroleum) has been described in Chapters 3 and 7. This is largely a product of increased car ownership (Figure 8.8) and increased travel in passenger kilometres (Figure 1.24). Given the twin problems of 'peak oil' and climate change, reducing transport energy demand is an urgent task that will require both social and technological changes.

Social changes

Reducing transport energy use raises a number of questions. The first is: 'Is your journey really necessary?' The ultimate energy service of transport is mobility; however, this is not quite the same as physical transportation. Given the modern multiple modes of communication: letter, telephone, email and video conferencing, the answer may often be 'no'. Modern telecommunications also offer a form of mobility.

The next question is, perhaps: 'Does suburbia have a future?' The sprawling suburbs of many of the world's cities do require more physical travel to reach shops and facilities than more compact 'walkable' environments, i.e. more travel to achieve the same standard of mobility. While suburban living may be seen as wasteful of energy, it should be borne in mind that many suburbs date from the first half of the twentieth century. They predate the modern 'oil age' and were, even in the USA, the product of efficient public transport (often in the form of electric trams, trolleybuses and railways).

The last question is 'Can your journey be done in a less energy intensive (and CO_2 intensive) manner?' Figure 14.16 illustrates the energy intensity of different modes of transport in the UK. The various modes vary enormously in their energy requirements per passenger-kilometre travelled. Cycling and walking, of course, require no fuel input apart from food.

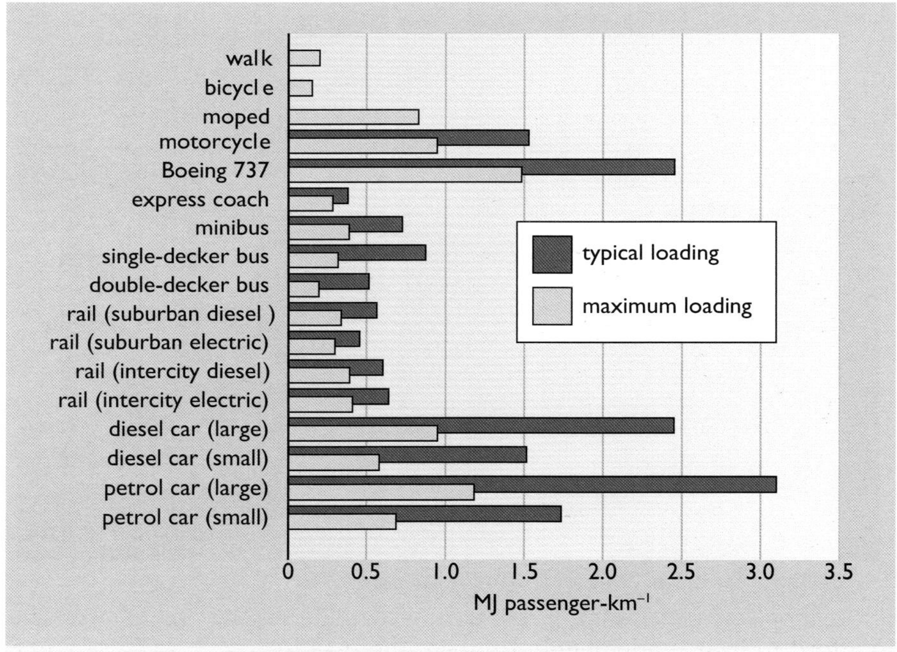

Figure 14.16 Energy efficiency of different modes of transport in the UK (source: Hughes, 1993)

One social method to reduce the energy intensity of travel is **modal shift**, i.e. moving journeys away from the energy-intensive modes and towards the more energy-frugal modes. This could be achieved without necessarily reducing the total number of journeys or overall distance travelled. For example, if a greater proportion of long-distance journeys within Europe were made by inter-city train rather than by air, the overall energy demand involved could be reduced substantially. One particular target for improvement is the urban commuter who drives to work. The energy intensity could be significantly improved by making sure that cars contain more than just one person. Even better, the journey could be made by rail or bus.

The enforcement of speed limits on motorways and trunk roads could also play a significant role in reducing energy demand. As shown in Figure 8.5, the power requirements of a car increase dramatically with speed. On UK motorways in 2009, 52 per cent of cars exceeded the 70 mph speed limit and 16 per cent of cars were recorded as travelling at 80 mph or faster (DfT, 2010).

Technological measures

Improving vehicle fuel economy is one obvious measure for reducing transport energy use. The average fuel consumption (in miles per gallon, or litres per 100 km) has improved considerably, largely as a result of CO_2 emissions legislation, particularly by the European Commission.

In order to get the maximum fuel efficiency, a car should be light (though sufficiently strong to meet crash test regulations), and to cut the air resistance it should be streamlined and have a small frontal area. These requirements have been incorporated into the Tesla electric car shown in Figure 9.24, which is both aerodynamic and uses light carbon fibre construction.

Figure 14.17 2009 prototype Volkswagen L1 with a 1 litre per 100 km fuel consumption

Where it is powered by fossil fuels, a hybrid petrol–electric or diesel–electric drive can be used. The 4-seater Toyota Prius (see Figure 9.25) uses a petrol–electric system. Volkswagen have been testing a 2-seater diesel–electric hybrid with a claimed fuel efficiency of 1 litre per 100 km or 280 miles per UK gallon (Figure 14.17).

However, there are unresolved design conflicts between a desire for fuel efficient cars, fast cars and large cars (both for carrying capacity and as symbols of social status). As pointed out in Box 8.2, there may be no going back to the 3.5 hp 'Petit Voiture' of 1900.

'Rebound effects'

While the benefits of supply-side energy efficiency measures may be easy to quantify, those on the demand side are much closer to the user and may be influenced by changes in their behaviour. Some of the potential energy savings can be absorbed by so-called **rebound effects**.

For example, a family living in a poorly insulated house may not be able to afford to heat it adequately (indeed, the installed heating system may not *physically* be able to do so). If it is insulated, then they may be able to live at a higher internal temperature, but still use exactly the same amount of energy. They may have a better energy service, comfort, but there is no actual reduction in energy demand.

Similarly the benefits of purchasing a more efficient car may simply be absorbed by driving it further, again producing an improvement in energy service with no reduction in demand.

In the past many programmes of cost-effective home insulation in the UK have concentrated on the money that they might save. This then raises the wider question of what that saved money might be spent on. What if a family chose to spend it on an energy-intensive flight to a far-off holiday destination?

These rebound effects are sometimes used as an argument against the effectiveness of demand-side energy efficiency programmes. However, they do illustrate the need to look more closely at behavioural aspects of energy use when formulating government programmes.

14.6 Harnessing renewable energy sources

Although, today, fossil and nuclear fuels provide the major part of the world's energy requirements, renewable energy and energy from wastes supply about 13%. Over half of this contribution is estimated to come from traditional biomass (straw, firewood, animal dung, etc.). In the past thirty years there has been a rapid rise in the development and deployment of many other renewable energy sources, driven by high fossil fuel prices and concerns about climate change. Energy from wastes is also often included in 'renewable energy' statistics though properly much municipal solid waste that goes for incineration consists of plastics that are really processed fossil fuel.

The general nature and scope of the various renewables is briefly summarized below, beginning with the most important renewable source, solar energy.

Solar energy

Solar energy makes an enormous contribution to our energy needs. The Sun has a surface temperature of 6000 °C, maintained by continuous nuclear fusion reactions between hydrogen atoms within its interior. It radiates huge quantities of power into the surrounding space. As shown in Figure 1.9, approximately 3.7 million EJ y^{-1} is potentially available for use at the surface. This is equivalent to over 8000 times humanity's 2009 rate of use of fossil and nuclear fuels.

It is the Sun's radiant energy that maintains the temperature of the Earth's surface at a temperature warm enough to support human life. But this energy is virtually ignored in national and international energy statistics, which are almost entirely concerned with consumption of commercial fuels.

Solar heating systems

Figure 14.18 A typical Mediterranean solar water heater in Greece

Solar energy, when it enters our buildings, warms them. This is known as **passive solar collection**. Buildings can be specifically design to maximize these solar gains, by facing them towards the Sun (south in the northern hemisphere or north in the southern hemisphere), avoiding over-shading and concentrating glazing on the sunny side.

Solar energy also illuminates them. The penetration of natural light into buildings can be encouraged by designing rooms with access to outside light and adequate window areas. This is known as **daylighting** and is an important method for avoiding the need for artificial lighting.

Solar power can also be harnessed by using solar thermal collectors. These are normally used to produce hot water for washing and may be free-standing (as in Figure 14.18) or roof-mounted. This is known as **active solar heating**.

In Europe, free-standing collectors are in widespread use in sunny countries such as Israel and Greece. Roof-mounted designs are increasingly used in countries such as Denmark and Germany. In Denmark very large, ground-mounted arrays of collectors are being constructed for use with district heating systems. China is currently the world's largest manufacturer and in 2008 had over 70% of the world's installed capacity of solar water heaters (REN21, 2010).

Solar thermal electricity

In sunny regions such as southern California or southern Spain, the Sun's rays are strong enough to make it practicable to generate high-temperature steam using arrays of concentrating mirrors. This can then power turbines for electricity generation. Such **concentrated solar power (CSP)** systems using large arrays of parabolic mirrors have been operational in California since the 1980s. As described in Section 8.6, small kilowatt-rated devices using Stirling engines have been tested. More recently, large 'power tower' systems using steerable arrays of flat mirrors called heliostats have been built in Spain (see Figures 14.19 and 14.20).

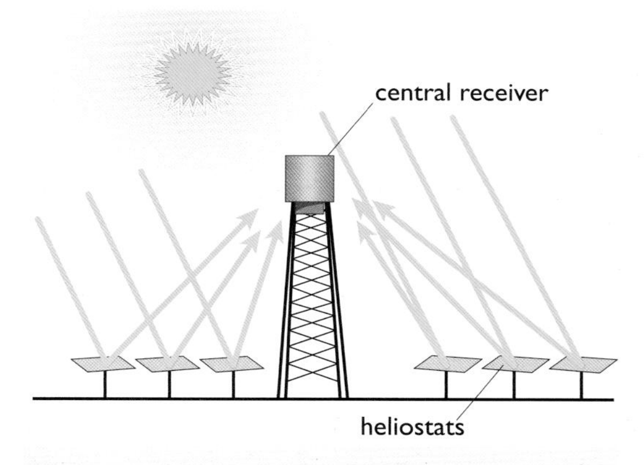

Figure 14.19 Principle of a central receiver 'power tower' solar thermal system

Figure 14.20 The Andasol PS 10 and 20 concentrating solar thermal electricity generating systems near Seville, Spain. Installed between 2009 and 2011 they have generating capacities of 10 MW and 20 MW respectively.

Although solar systems using collectors or mirrors currently only contribute under 0.1% of global primary energy, there is an enormous potential for expansion.

Photovoltaics

Solar energy can also be harnessed directly to produce electricity by using a more sophisticated technology called solar photovoltaics (PV). This has already been described briefly in Box 4.8. Photovoltaic cells are made of specially prepared layers of semiconducting materials (usually silicon) that generate electricity when photons of sunlight fall upon them. Since each cell only produces a low voltage they are normally produced in 'modules' or 'panels' containing a large number of cells in series. These can range in size from the small battery charger shown in Chapter 4 up to sizes of a metre square. Arrays of PV modules are normally mounted on the roofs or facades of domestic, commercial or industrial buildings, usually providing only some of their electricity needs (see Figure 14.21).

Figure 14.21 PV panels in the process of being installed on the roof of one of the Open University's buildings in Milton Keynes, UK in 2011

In Germany large PV arrays have also been installed on agricultural land or on disused industrial sites.

Photovoltaics produce DC electricity, which is convenient for battery charging, but where the power is to be exported to the grid it needs to be converted to AC using an electronic inverter.

Photovoltaic technology is growing very rapidly and several countries have initiated major implementation programmes, often using 'feed-in tariffs' giving guaranteed prices for exported electricity. Germany is the leading country and in 2009 it had nearly half of the world's installed capacity. In 2010 PV supplied 2% of Germany's electricity from an installed capacity of 17 GW (BMU, 2011).

Although PV's share of world electricity production is currently small, it may well make a significant contribution to world needs in coming decades. Large-scale mass production has reduced the cost of PV modules, and hence the cost of PV electricity. In countries with plenty of solar radiation PV may soon become competitive with fossil-fuelled generation. PV systems are obviously also in competition with solar thermal electricity generation. PV systems can be conveniently fitted onto existing buildings and do not require full bright sunlight. Concentrating solar thermal systems require continuous clear sunlight for operation and are much larger, with outputs of 10–100 MW, needing large areas of open land or desert.

Indirect use of solar energy

The above examples illustrate the *direct* harnessing of the Sun's radiant energy to produce heat and electricity. But the Sun's energy can also be harnessed via other forms of energy that are *indirect* manifestations of its power. Principally, these are bioenergy and hydropower, together with wind energy and wave power.

Bioenergy

From prehistoric times, human beings have harnessed the power of fire to create warmth and light and to cook food by burning wood and straw. These are created by photosynthesis, a process powered by solar energy in which water and atmospheric CO_2 are converted into carbohydrates (mixtures of carbon, hydrogen and oxygen) in a plant's leaves and stems. These, in the form of wood or other 'biomass', can be used as fuels which are sources of **bioenergy**. Biomass can also be processed into specific biofuels such as ethanol and biodiesel, which can be used for transportation purposes.

Traditional biomass

Wood and straw are still very widely used as a fuel in many parts of the 'developing' world. In some countries, other biofuels such as animal dung (ultimately also derived from the growth of plants) are also used. The actual global consumption of traditional biomass is uncertain, since little of it is 'traded' and doesn't enter into national energy statistics. IEA estimates suggest that it makes up about 6% of global primary energy use (see Figure 1.5).

The use of wood and straw raises questions of its sustainability. Firstly, the land must be managed sustainably. Excessive extraction may lead to deforestation and desertification. If the woods and forests that provide fuel are re-planted at the same rate as they are cut down, then such fuel use should in principle be 'sustainable'. The CO_2 absorbed in growing replacement trees should equal the CO_2 given off when the original ones are burned.

Secondly, the fuel must be burned cleanly and completely. The best wood stoves and furnaces can achieve near-complete combustion. However, most open fires are not so efficient and may produce large amounts of particulate wood smoke and also methane, which, as described in Chapter 13, is a much more powerful greenhouse gas than CO_2.

Methane is also produced by traditional charcoal manufacture (see Figure 5.1) which is widely practised in many countries to make a 'clean' and 'smokeless' fuel for urban dwellers.

As described in Section 13.5, in countries such as India, the twin problems of unsustainable wood use and wood smoke pollution have led to the promotion of fossil fuels for cooking in the form of liquefied petroleum gas.

Traditional biomass still has its place in developed countries. Despite its large nuclear industry, domestic wood use made up nearly 3% of France's primary energy consumption in 2009 (DGEC, 2010).

Modern biomass

A significant contribution to world supplies now comes from power plants fuelled by straw, forestry wastes or wood chips from trees grown in special plantations. This may take the form of co-firing with coal in conventional power stations such as Drax in the UK (Figure 9.33) or in purpose built dual-fuel plants such as Avedore 2 in Denmark (Figure 9.50). In Denmark straw is widely used in small-scale CHP units. In 2009 biomass-fuelled electricity generating capacity was estimated at 54 GW, mainly installed in the USA, Japan and Germany (REN21, 2010).

Municipal solid wastes (MSW) are widely incinerated for heat or electricity production. They contain a large proportion of material which is of biological origin. However, they also contain large amounts of plastics produced from fossil fuels. These can contain considerable amounts of sulfur and chlorine and there is considerable concern over the production of toxic dioxins (see Section 13.5) in their combustion. The construction of new, large waste-to-energy plants have also been opposed, partly on the grounds that they need to be fed with a steady stream of waste, thus discouraging material reuse or recycling.

Landfill gas, produced from the slow decomposition of municipal waste in landfill waste sites, is widely used for the generation of electricity. This consists of a mixture of methane and CO_2 and is burned in large internal combustion engines at the landfill sites. In 2009 its use contributed 1.5% of the UK electricity supply. Given the potency of methane as a greenhouse gas, the proper capping of landfill sites and use of the methane is an important method of reducing overall greenhouse gas emissions.

Animal wastes and wet food waste can similarly be turned into a methane/ CO_2 mixture called **biogas** by anaerobic digestion in closed containers. This is widely used for cooking in rural parts of China. Biogas can also be used for electricity generation, and even be purified so that the methane can be injected into the conventional natural gas grid. The anaerobic digestion of urban sewage is widely carried out to produce sewage gas. This is usually used to provide electricity and heat for the operation of sewage and water treatment works.

Biofuels

Biomass can also be used for transport applications in spark ignition (i.e. petrol) engines in the form of methanol, ethanol and butanol. These have been discussed briefly in Section 8.2. Methanol can be produced from wood. Ethanol and butanol can be produced by the fermentation of sugars and starch. Ethanol is the most commonly used, usually blended with petrol. In 2009 world ethanol production for transport was 76 billion litres, 88% of it in the USA and Brazil (REN21, 2010). This is equivalent to about 1% of global fossil fuel oil use.

Biodiesel can be produced simply from vegetable oils and used in conventional road diesel engines (Section 8.3). World production in 2009 was 17 billion litres. It is usually sold blended with conventional diesel fuel.

Biofuel use is potentially controversial. It can be seen as being in competition with the use of crops for food. There are questions about the overall EROEI (energy return over energy invested – see Section 13.8) and even the overall carbon savings, particularly if fossil fuels are used to process the biofuels. However, best practice should produce acceptable results (see Harvey, 2010a).

Hydroelectricity

Water power is another energy source that has been harnessed for many centuries for pumping, milling corn and driving machinery. The original source of this water flow is solar energy, warming the world's oceans and causing evaporation. In the atmosphere, this forms clouds of moisture which eventually fall back to Earth in the form of rain or snow. The water then flows down through streams and rivers, where its energy can be harnessed using water wheels or turbines to generate power.

Since the beginning of the 20th century, its main use has been in the generation of hydroelectricity, and this has grown to become one of the world's principal electricity sources. It currently provides some 2.3% of world primary energy and over 17% of world electricity supplies (see Table 2.2). Some large schemes have been described in Section 9.6 and the world's largest, the 18 GW Three Gorges Dam in China, rated at 18.2 GW, is described in Section 9.9.

When harnessed on a small scale, hydroelectric plants create few adverse environmental impacts. However, many modern installations have been built on a very large scale, involving the creation of massive dams and the flooding of extremely large areas. The flooding can result in methane emissions from rotting vegetation, meaning that hydro power is not necessarily a totally 'climate friendly' technology.

Wind energy

When solar radiation enters the Earth's atmosphere it warms different regions of the atmosphere to differing extents – most at the equator and least at the poles. Since air tends to flow from warmer to cooler regions, this causes what we call winds, and it is these air flows that are harnessed in windmills and wind turbines to produce power.

Wind power, in the form of traditional windmills used for grinding corn or pumping water, has been used for centuries. But the use of modern wind turbines for electricity generation has been growing rapidly since the 1970s, following pioneering work in Denmark. The size and power of turbines has increased from machines producing 100 kW in the 1980s up to sizes of 5 MW or more today. These can have rotor diameters and tower heights in excess of 100 m.

At present, most of these turbines have been installed on land, and in some cases there are problems of visual intrusion into the landscape (see Figure 14.22).

Within Europe, Denmark, Germany and Spain have installed large numbers of land-based turbines. In 2009 Denmark obtained 19% of its electricity from wind power. In the UK the figure was nearly 3%. The performance of wind turbines is highly dependent on the average wind speed, and there are considerable advantages in siting them out at sea. The UK, Denmark and Germany all have plans for large offshore wind farms (see Figure 14.23). Currently (2011), the UK has over 1 GW of offshore wind power capacity and plans to install 30 GW by 2020.

Figure 14.22 Ardrossan Wind Farm, Scotland

Figure 14.23 UK offshore turbines at Burbo Bank at the mouth of the River Mersey

Globally, the leading countries for wind turbine installation are the USA and China. Wind power is now one of the world's fastest-expanding sources of energy, having achieved a growth rate of roughly 27% per annum between 2004 and 2009 (REN21, 2010).

Wave power

When winds blow over the world's oceans, they cause surface waves. The power in such waves, as they gradually build up over very long distances, can be very great – as anyone watching or feeling that power being dissipated on a beach will know.

Harnessing wave power is a new technology. Various prototype devices have been developed over the past few decades, of which the 'Pelamis' (meaning sea snake) wave energy converter is one recent example (see Figure 14.24).

The device is a semi-submerged, articulated structure composed of cylindrical sections linked by hinged joints. It is deployed in the sea facing into the oncoming waves. The wave-induced bending motion of the joints is resisted by hydraulic rams, which pump high-pressure fluid through hydraulic motors via smoothing accumulators. The hydraulic motors drive electrical generators. Power from the generators is fed down a single umbilical cable to a junction on the seabed. Several devices can be connected together and linked to shore through a single seabed cable.

Three early machines were put into operation in 2008 for a short period off the coast of Portugal. The current prototype machines are 180 m long and 4 m in diameter, with four power conversion modules per machine. Each machine is rated at 750 kW and has an estimated annual average capacity factor of 25–40%, depending on conditions. They have been undergoing sea tests in Scotland.

Figure 14.24 The E-On Pelamis 2 Wave Energy Converter, pictured off Orkney, Scotland

Wave energy technology is not as developed as wind power but has enormous potential, for both the UK and other countries bordering the world's major oceans.

Non-solar renewable energy

The renewable energy sources described above are either direct or indirect forms of solar energy. However, there are two other renewable sources, tidal and geothermal energy, that do not depend on solar radiation.

Tidal energy

The energy that causes the slow but regular rise and fall of the tides around our coastlines is not the same as that which creates waves. Tides are caused principally by the gravitational pull of the Moon (and to a lesser extent the Sun) on the world's oceans.

There are essentially two types of tidal power system: **tidal barrages** and **tidal stream systems**. Tidal barrages use the change in height of water to extract the *potential energy* (they are also sometimes called tidal range systems); tidal stream systems extract the *kinetic energy* of the moving water.

Tidal barrages

The principal technology for harnessing tidal energy involves building a low dam, or barrage, across the estuary of a suitable river. The barrage has inlets that allow the rising sea levels to build up behind it. When the tide has reached maximum height, the inlets are closed and the impounded water is allowed to flow back to the sea in a controlled manner, via a turbine-generator system very similar to that used in hydroelectric schemes. There have also been proposals to build tidal lagoons involving circular barrages in shallow sea areas with large tidal ranges.

The world's largest tidal energy scheme is at La Rance in France, which has a capacity of 240 MW and has been operational since 1966 (see Figure 14.25).

There are other, smaller, tidal plants in various countries, including Canada, Russia and China. The UK has one of the world's best potential sites for a tidal energy scheme, in the Severn estuary. A large scheme was proposed in the 1970s. If built, its capacity would be around 8600 MW, much larger than any other single UK power plant, and it could provide about 4.5% of UK electricity demand. However, there are major concerns about possible effects on wildlife in the Severn estuary.

Tidal stream devices

This relatively new technology involves the use of underwater turbines (rather like submerged wind turbines) to harness the strong tidal and oceanic currents that flow in certain coastal regions. The Pentland Firth between the Scottish mainland and the Orkney Isles has a considerable potential, as does the Race of Alderney between the north-west coast of France and the island of Alderney.

A 'Seagen' 1.2 MW prototype marine current turbine (Figure 14.26) has been operating in Strangford Lough, Northern Ireland, since 2008. It is designed so that the turbines can be raised clear of the water level for maintenance. The technology is still under development, but its prospects seem good.

Figure 14.25 The 240 MW tidal barrage on the Rance estuary in France

Figure 14.26 An illustration of the Seagen Marine Current Turbine, installed in Strangford Lough in Northern Ireland in 2008

Geothermal energy

The source of geothermal energy is the Earth's internal heat. This was originally created by the gravitational contraction of the planet during its formation and is continuously augmented by the heat from the decay of radioactive substances within the Earth's core.

In some parts of the world, hot rocks are fairly near the Earth's surface and they warm underground aquifers. The resulting hot water, or in some cases steam, is used in regions where the geological conditions are favourable, such as parts of Iceland, Italy, France and the USA, for heating buildings and for electricity generation. Figure 14.27 shows the scheme at Larderello in Italy, which produces about 10% of the world's geothermal electricity. In the UK geothermal heat supplements a CHP district heating scheme at Southampton.

Figure 14.27 One of the geothermal power plants at Larderello, Italy, used to provide electricity and hot water

If geothermal heat is extracted in a particular location at a rate that does not exceed the rate at which it is being replenished from deep within the Earth, it is a renewable energy source. But in many cases this is not so: the geothermal heat is effectively being 'mined' and will 'run out' locally in perhaps a few years or decades.

The technology of ground source heat pumps (mentioned in Box 9.9) is often referred to as 'geothermal'. Although such heat pumps do physically use the ground to site their heat exchanger coils they actually draw heat from the air rather from deep below the Earth.

Sustainability of renewable energy sources

Renewable energy sources are generally sustainable in that they cannot 'run out' – although, as noted above, both biomass and geothermal energy need wise management. For all of the other renewables, almost any conceivable rate of exploitation by humans is unlikely even to approach their rate of replenishment by nature, though of course the use of all renewables is subject to various practical constraints.

Renewable energies are also relatively 'sustainable' in the additional sense that their environmental and social impacts are generally more benign than those of fossil or nuclear fuels. However, the deployment of renewable energy sources in some cases entails significant environmental and social impacts. Renewable energy sources are generally much less concentrated than fossil or nuclear fuels, so large areas of sea, land or building surfaces are often required if substantial quantities of energy are to be collected. This can lead to a significant visual impact, as in the case of wind turbines.

Also, the capital costs of many renewable sources are at present higher than those of conventional systems using fossil fuels. This balance is changing as the further development of renewable sources reduces their costs. Equally, the need to incorporate carbon capture and storage may force up the costs of systems using fossil fuels. Renewables seem attractive in many ways, but how large a contribution might they make to world energy needs in the future? This is an important question, addressed in the next section.

14.7 Towards a sustainable world energy future

> Are these the shadows of the things that Will be, or are they the shadows of things that May be, only?
>
> (Dickens, 1843)

Chapter 1 has described some of the issues surrounding what might be called a 'sustainable world energy future'. A key imperative is the need to reduce greenhouse gas emissions to limit the rise in global temperature to under 2 °C. This will require cuts in global anthropogenic greenhouse gas emissions of at least 50% from current levels by 2050 and continuing reductions beyond that date. Since CO_2 is the most important of these greenhouse gases this can equally be taken to read '50% reduction in CO_2 emissions'.

What will this mean for future energy use in the UK and globally?

Sustainable energy for the UK

It could be argued that emissions cuts should be shared equally amongst all the countries of the world. However, as shown in Table 2.3, the world's energy use is very unevenly distributed. Developed countries use far more energy per capita than developing ones. If developing countries are to be allowed to increase energy use for economic growth, then developed ones will need to cut their emissions by more than 50%. The UK government has accepted the need to cut the country's greenhouse gas emissions by 80% from 1990 levels by 2050. This has been set out in an Act of Parliament, the Climate Change Act (2008), which also set up an independent Committee on Climate Change (CCC) to advise the government and set periodic 'carbon budgets' to reach the 80% goal. This actually requires CO_2 emission cuts of *greater* than 80% because of the difficulties of achieving large cuts in the emissions of methane and N_2O emissions in agriculture.

The previous chapters have described past UK energy use in detail. Figure 14.28 shows UK CO_2 emissions from 1970 to 2010, broken down by fuel. These emissions have been declining slowly. The chart also shows the CCC's intended carbon budget trajectory to 2050.

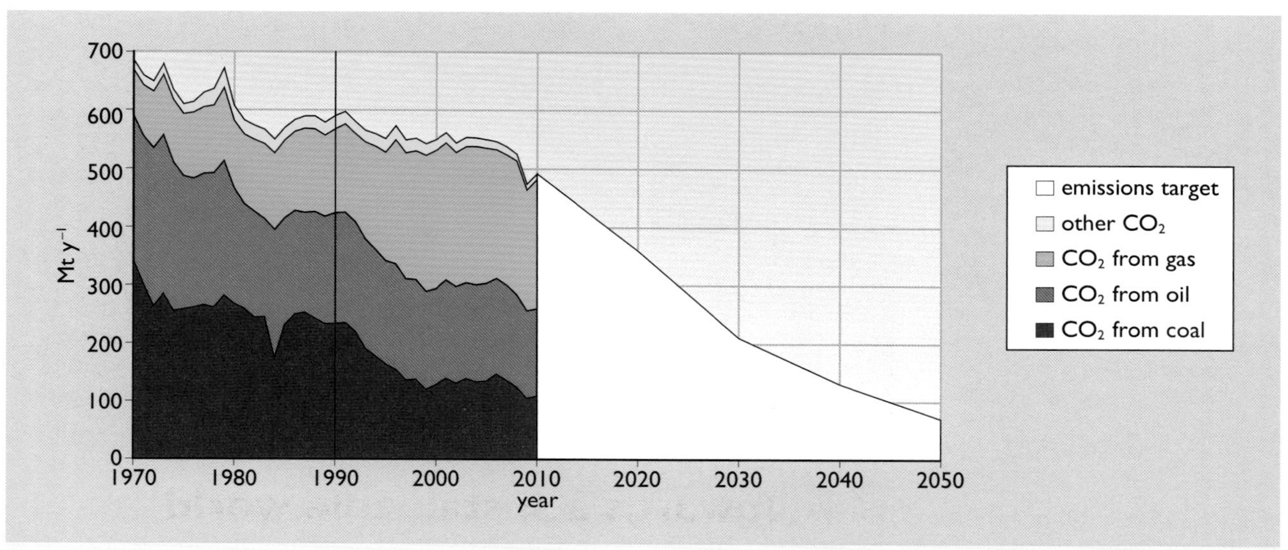

Figure 14.28 UK CO_2 emissions 1970–2010 and the CCC's 'intended carbon budget' to 2050 (sources: DECC, 2010b; DECC, 2011; CCC, 2010)

The challenge is to increase the current rate of emissions reduction. It is not the task of the Committee on Climate Change to specify exactly how this is to be done. There are obviously a wide range of possible technologies (and social measures) that can be deployed:

- energy efficiency – on both the supply side and the demand side
- fuel switching
- nuclear power
- carbon capture and storage
- renewable energy.

There is much heated debate about which are best. Given the decline in UK North Sea oil and gas production, there are important issues of energy security to be taken into account. Also, continued economic growth and any resurgence of UK manufacturing industry are likely to provide a continous pressure to increase energy demand and potential CO_2 emissions.

In order to assist with policy decisions, the UK Department of Energy and Climate Change (DECC) has carried out a '2050 Pathways Analysis' (DECC, 2010c). This sets out a range of plausible future trajectories for each energy supply and demand sector. Each pathway could reduce greenhouse gas emissions by 80%, while maintaining a secure system where energy supply meets demand.

Figure 14.29 shows the breakdown of UK energy use in 2009 and also two possible future breakdowns for 2050. 'Pathway Alpha' uses a spread mixture of options and 'Pathway Gamma' is a low nuclear approach. Obviously, if one particular technology is not used, then more effort will need to be put into all of the others. Thus, removing nuclear power as an option means that energy demand will need extra reduction and an increased contribution from renewables and CCS.

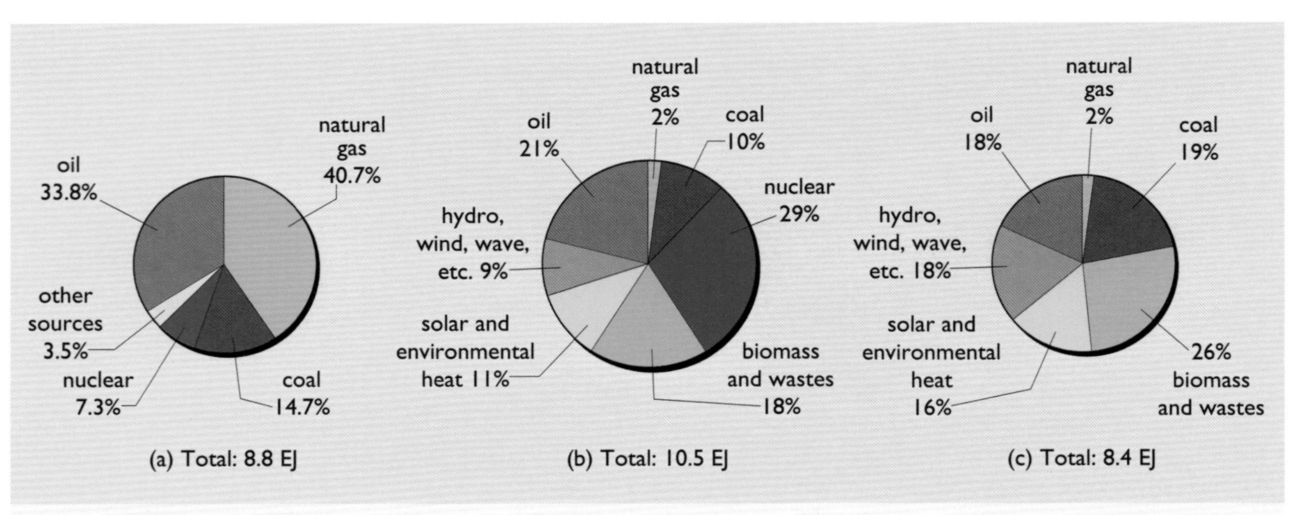

Figure 14.29 (a) UK primary energy use, 2009; (b) possible usage in 2050 via 'Pathway Alpha' – spread effort; (c) possible usage in 2050 via 'Pathway Gamma' – low nuclear (sources: DECC, 2010d; DECC, 2010c)

A key element of the pathway analysis is the decarbonization of electricity supply. Both of the pathways illustrated see a considerable continued use of oil for transport, and of coal with carbon capture and storage (CCS). What is perhaps surprising is the virtual disappearance of natural gas as a fuel. There is obviously an enormous expansion in the use of renewable energy, though this does, rather controversially, involve the import of large amounts of biomass.

These are, of course, only possible futures. Such analysis immediately opens the way to a host of possible questions: 'What if oil was very expensive indeed?', 'How are houses going to be heated without natural gas?', 'What if importing biomass was unacceptable?', etc. At the time of writing (2011) the Pathways Analysis is available as an interactive tool on the DECC website, (DECC, n.d.) allowing some of these questions to be explored.

IEA Blue Map world scenario

The International Energy Agency is usually associated with 'business as usual' projections of world energy use, for example Figure 1.10, which shows increasing fossil fuel use and implies increasing CO_2 emissions. However, in the 2010 issue of its annual *Energy Technology Perspectives* report (IEA, 2010) it described its 'Blue Map' scenario, showing how the world as a whole could achieve a 50% reduction in CO_2 emissions by 2050 from current (i.e. 2010) levels. Global CO_2 emissions would need to peak in about 2016. The projected energy use for this scenario has been shown in Figure 1.22. While there is still a projected increase in global energy use, fossil fuel use declines.

Figure 14.30 illustrates the projected CO_2 emissions for these two different scenarios and the possible effect of different options for reductions. About half of the CO_2 savings come from energy efficiency measures.

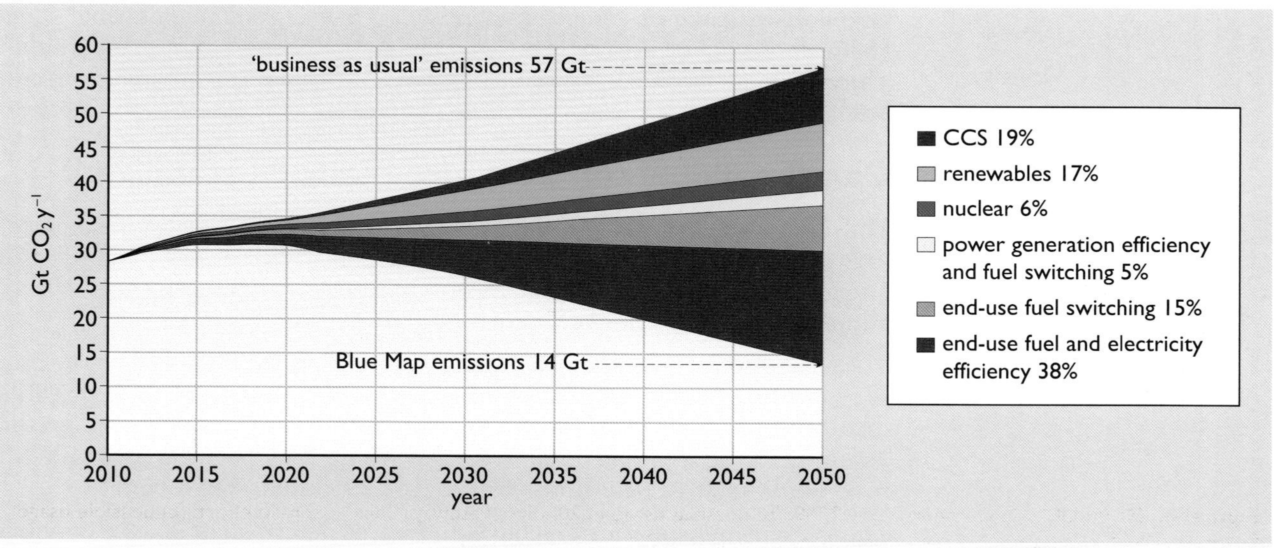

Figure 14.30 Key technologies for reducing CO_2 emissions in the IEA Blue Map scenario (source: IEA, 2010)

Figure 14.31 compares the 2009 world primary energy use breakdown (total 502 EJ as already shown as Figure 1.5) with the IEA's projected primary energy use of nearly 660 EJ in 2050.

As with the UK scenarios above, decarbonization of the electricity supply is seen as very important, particularly since this currently consumes two-thirds of the world's coal production. Coal still has a significant role, though being used with CCS. Obviously, there is still seen to be a large world demand for oil and (unlike in the UK scenarios above) natural gas.

There is an enormous expansion in the use of biomass and wastes, and in 'hydro and other' sources. 'Other' in this case includes a wide mix of renewable energy sources. Many other organizations have produced scenarios, particularly showing that mixtures of energy efficiency and renewable energy can potentially provide all of the world's future energy needs (see 'Further reading').

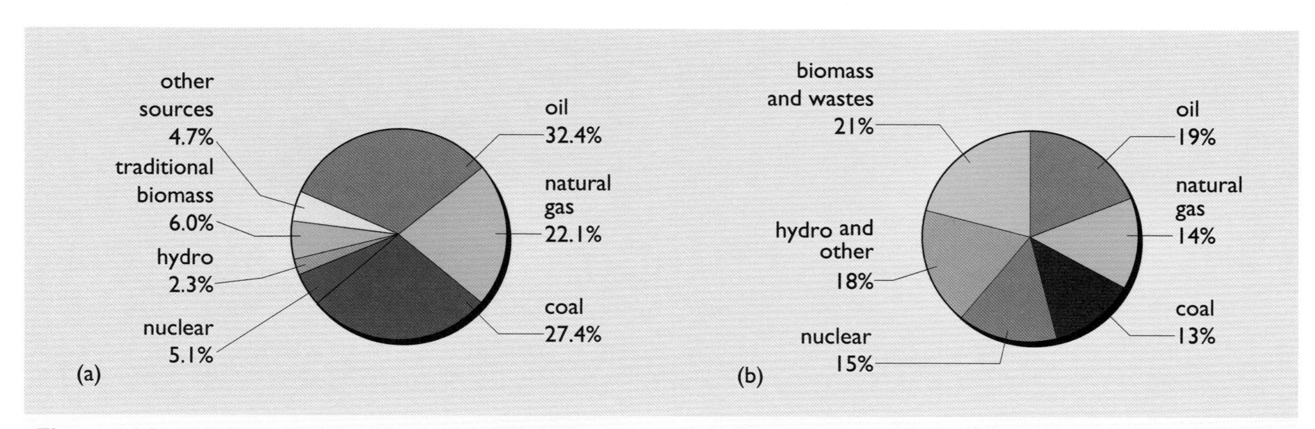

Figure 14.31 World energy use: (a) percentage contributions of various energy sources to world primary energy consumption, 2009; (b) in IEA Blue Map scenario, 2050 (sources: (a) principal sources: BP, 2010; IEA, 2009; Maddison, 2005; WWEA, 2009; (b) IEA, 2010)

One consequence of a rapid switch away from fossil fuels is that large amounts of what are currently regarded as 'reserves' will be left in the ground. This may remove some 'peakist' concerns about long term energy supplies, but may not exactly be welcomed by the current owners of these reserves.

Can the global temperature really be stabilized?

The energy scenarios above aim to stabilize the global temperature at under 2 °C warmer than it was in pre-industrial times. Will a 50% reduction in global CO_2 emissions by 2050 be enough? This takes us to the realm of atmospheric science and climate modelling.

The UK Committee on Climate Change has carried out modelling of a range of possible future greenhouse gas emission trajectories and their possible resulting global temperature effects (CCC, 2008). Figure 14.32(a) shows the assumed CO_2 emissions under a profile where they peak in 2016 and fall to 41% of the 2010 emissions (64% of the 1990 figure) by 2050.

The current atmospheric CO_2 concentration, as described in Box 1.6, is about 390 ppmv and rising. According to the climate modelling, if the emissions are reduced then this concentration should stabilize at about 450 ppmv (as shown in Figure 14.32(b)). However, there are many uncertainties in the science. Any answers can thus only be given in probabilistic terms. The central line is estimated to be the most likely outcome, but the upper and lower lines represent 10% and 90% probability limits.

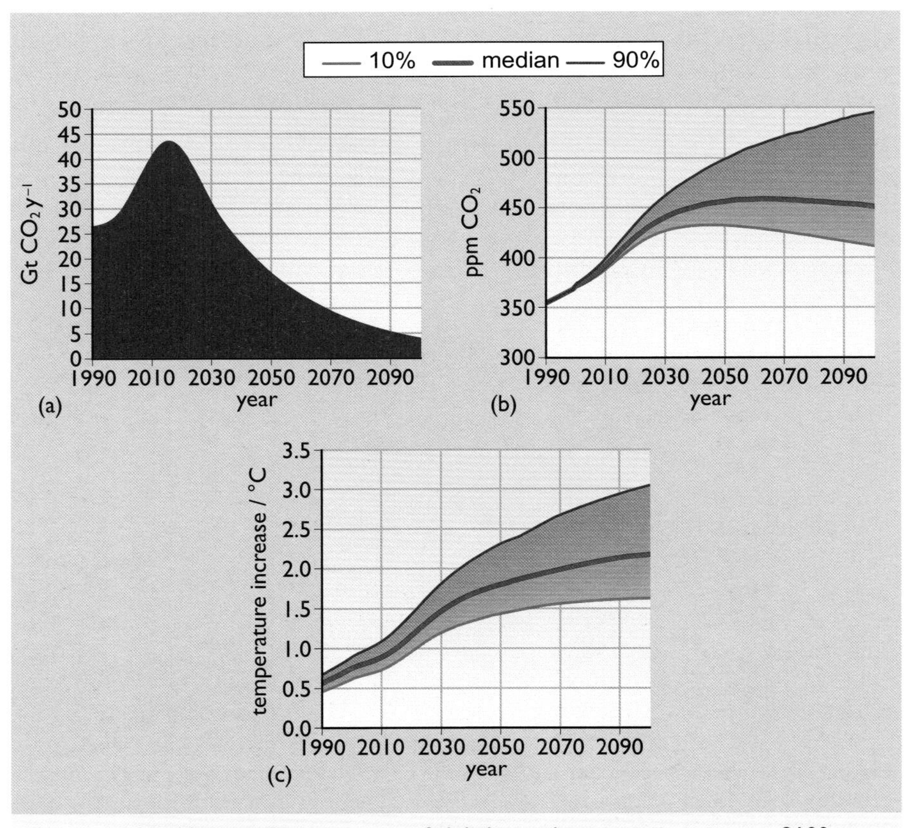

Figure 14.32 (a) a possible trajectory of global greenhouse gas emissions to 2100; (b) the estimated resulting changes in atmospheric greenhouse gas concentration; (c) estimated change in global surface temperature (source: CCC, 2008)

Figure 14.32(c) shows the estimated effect on global surface temperature. As in the previous chart, the estimates are probabilistic. The central line suggests that the most likely outcome will be a stabilization slightly above a 2 °C warming. The upper line shows a 90% confidence level that the warming will be kept below 3 °C by 2100. On the more optimistic side, the lower one shows that there is a 10% probability that the stabilization will occur with only a 1.5 °C warming. Put another way, this emission trajectory has only a one in three chance of keeping the warming below 2 °C by the year 2200. It may be that more drastic cuts will be required.

In the long term it is likely that CO_2 emissions will have to be reduced to the possible natural take-up level of the planet, which is estimated to be under 4 $GtCO_2$ y^{-1} (House et al., 2008).

Change and investment

Obviously, curbing CO_2 emissions will require changes in global energy use and these will need financial investment: in energy efficiency and new energy sources rather than continued investment in fossil fuels. In 2011 the United Nations Environment Programme published a major report on the Green Economy (UNEP, 2011). It urged a reduction in 'inefficient and wasteful' worldwide subsidies to conventional energy sources, which it estimated at some US$500–700 billion per annum, compared to support for renewable energy and biofuels, estimated at US$57 billion in 2009. It highlighted the increasing attractiveness of investment in renewable sources, not only in financial terms but also in terms of energy security, climate change mitigation and job creation. It calculated that to build a Green Economy for the world, including a sustainable energy system but extending well beyond it, would require an investment of US$1.2 trillion a year – 2% of world GDP. This does not seem an excessive price to pay to achieve such a desirable outcome.

14.8 Summary

This final chapter has attempted to bring together some of the threads running through the book. Many of the chapters have dealt with the history of particular technologies and celebrated various inventors and innovators. However, innovation is a continuous process and more will be needed to tackle the problems of climate change and fossil fuel depletion.

Section 14.2 brought together many of the methods for cleaning up the use of fossil fuels that have been described in previous chapters, involving reductions in the emissions of particulates, sulfur dioxide and nitrogen oxides. It then turned to the problems of reducing CO_2 emissions. 'Fuel switching' is one possibility, but then there are a range of 'clean coal' technologies. The Integrated Gasification Combined Cycle (IGCC) power plant is one of these, but this is only a step towards something more radical – carbon capture and storage (CCS). This requires removing the CO_2 produced by burning fossil fuels and 'sequestering' it, i.e. locking it away underground, where it cannot enhance the greenhouse effect.

Four 'capture' systems were described:

- post-combustion capture from flue gases
- oxyfuel systems

- coal-fuelled pre-combustion capture
- gas-fuelled pre-combustion capture.

All of these involve certain amounts of the 'industrial chemistry' described in Chapter 7. Two of the processes also involved the production of hydrogen, revisited in Section 14.4.

Section 14.3 described some 'geo-engineering' solutions to dealing with climate change, particularly reforestation, carbon sequestration in the oceans, and solar radiation management using sulfate aerosols and cloud seeding.

Section 14.4 moved on to the use of hydrogen as an energy carrier, starting with the theory of the fuel cell and describing its many current variants. It then described a possible natural-gas-fuelled hydrogen economy using pre-combustion capture and sequestration of its CO_2 production. Hydrogen has possibilities as a transport fuel both on land and in the air. Future fuel-cell-powered road vehicles may be in competition with battery–electric ones, but there are many considerations to be taken into account before it is possible to judge which will be best.

This book has concentrated on energy supply, but that is not to say that reducing energy demand is not equally important. Section 14.5 looked very briefly at the large potential for cutting space heating energy use in buildings. It then considered some of the possibilities, both social and technological for reducing transport energy use, possibly a matter of urgency if 'peak oil' has genuinely arrived.

Section 14.6 gave a brief tour of the many renewable energy sources.

The final section briefly considered where our future energy use might be heading given the need for drastic cuts in CO_2 emissions. There are five main approaches that can be used:

- energy efficiency – on both on the supply side and the demand side
- fuel switching
- nuclear power
- carbon capture and storage
- renewable energy.

It considered two scenarios for future UK energy use in 2050. Both of these envisage a considerable expansion of renewable energy use, but, perhaps more surprisingly, the virtual disappearance of natural gas from the UK as a fuel. It then considered the IEA's Blue Map scenario which suggests ways of cutting global CO_2 emissions from their present levels by 2050. Although this may seem a bold departure from 'business as usual', a look at calculations by the UK Committee on Climate Change suggests that even this might not be adequate to keep global surface temperatures from increasing by more than 2 °C above pre-industrial levels.

References

BMU (2011) *Renewable Energy Sources 2010*, Bundesministerium für Umwelt, Naturschutz und Reaktorsicherheit; available at http://www.bmu.de (accessed 25 May 2011).

BP (2010), *BP Statistical Review of World Energy*, London, The British Petroleum Company; available at http://www.bp.com (accessed 10 December 2010).

CCC (2008) *Building a low-carbon economy – the UK's contribution to tackling climate change*, Committee on Climate Change; available at http://www.theccc.org.uk; (accessed 23 May 2011).

CCC (2010) *The Fourth Carbon Budget: Reducing emissions through the 2020s*, Committee on Climate Change; available at http://www.theccc.org.uk; (accessed 23 May 2011).

DECC (2010a) *Energy Consumption in the United Kingdom*, statistical tables, Department of Energy and Climate Change; available at http://www.decc.gov.uk; (accessed 21 May 2011).

DECC (2010b) *UK greenhouse gas emissions: final data tables 2009*, Department of Energy and Climate Change; available at http://www.decc.gov.uk; (accessed 21 May 2011).

DECC (2010c) *2050 Pathways Analysis*, Department of Energy and Climate Change; available at http://www.decc.gov.uk; (accessed 21 May 2011).

DECC (2010d) *UK Energy in Brief 2010*, Department of Energy and Climate Change, available at http://www.decc.gov.uk (accessed 15 April 2011).

DECC (2011) *UK greenhouse gas emissions: provisional data tables 2010*, Department of Energy and Climate Change; available at http://www.decc.gov.uk (accessed 21 May 2011).

DGEC (2010) *Repères: Chiffres clés de l'énergie: Édition 2010*; available at http://www.developpement-durable.gouv.fr (accessed 24 July 2011).

DfT (2010) 'Road Statistics 2009: Traffic, Speeds and Congestion' [online], Department for Transport, http://www.dft.gov.uk (accessed 21 May 2011).

Dickens, C. (1843) *A Christmas Carol*; available at http://www.gutenberg.org/cache/epub/46/pg46.txt (accessed 22 May 2011).

Ekins, P. (ed) (2010) *Hydrogen Energy: Economic and Social Challenges*, London, Earthscan.

Harvey, L. D. D. (2010a) *Carbon-Free Energy Supply: Energy and the New Reality 2*, London, Earthscan.

Harvey, L. D. D. (2010b) *Energy Efficiency and the Demand for Energy Services: Energy and the New Reality 1*, London, Earthscan.

Hoffman, P. (2001) *Tomorrow's Energy: Hydrogen, Fuel Cells and the Prospects for a Cleaner Planet*, Cambridge, MA, MIT Press.

House, J., Huntingford, C., Knorr, W., Cornell, S. E., Cox, P. M., Harris, G. R., Jones, C. D., Lowe, J. and Prentice, C. (2008) 'What do recent advances in quantifying climate and carbon cycle uncertainties mean for climate policy?', *Environmental Research Letters*, vol. 3, no. 4, 044002.

Hughes, P. (1993) *Personal Transport and the Greenhouse Effect: a Strategy for Sustainability*, London, Earthscan.

IEA (1999) *Hydrogen – Today and Tomorrow*, IEA Greenhouse Gas R&D Programme Report, Paris, International Energy Agency, p. 34.

IEA (2009) *Key World Energy Statistics*, Paris, International Energy Agency; available at http://www.iea.org (accessed 11 December 2010).

IEA (2010) *Energy Technology Perspectives 2010, Parts 1 and 2*, Paris, International Energy Agency; available at http://www.iea.org (accessed 7 June 2011).

IPCC (2005) *Carbon Dioxide Capture and Storage,* Intergovernmental Panel on Climate Change, New York, NY, Cambridge University Press; available at http://www.ipcc.ch (accessed 15 April 2011).

IPCC (2007) *Climate Change 2007: Mitigation of Climate Change*, Intergovernmental Panel on Climate Change, New York, NY, Cambridge University Press; available at http://www.ipcc.ch (accessed 7 June 2011).

Laughton, M. (2002) 'Fuel cells', *Engineering Science and Education Journal*, vol. 11, no. 1, pp. 7–16.

Lovins, A. (2005) *Twenty Hydrogen Myths*, Report E03-05, Snowmass, CO. Rocky Mountain Institute.

Luwoge (2008) *Das 3-Liter-Haus*; available from http://www.luwoge.de (accessed 21 May 2011).

Maddison, A. (2005) 'World Development and Outlook 1820–2030: Evidence submitted to the Select Committee on Economic Affairs, House of Lords, London, for the inquiry into "Aspects of the Economics of Climate Change"'; available at http://www.ggdc.net (accessed 10 December 2010).

MIT (2007) *The Future of Coal: options for a carbon constrained world*, Cambridge, MA, Massachusetts Institute of Technology; available at http://web.mit.edu/coal/ (accessed 19 March 2011).

Mott MacDonald (2010) *UK Electricity Generation Costs Update*; available at http://www.decc.gov.uk (accessed 19 March 2011).

REN21 (2010) *Renewables 2010 Global Status Report*, Renewable Energy Policy Network for the 21st Century, Paris Secretariat; available at http://www.ren21.net (accessed 24 May 2011).

Royal Society (2009) *Geoengineering the Climate Science, Governance and Uncertainty*, London, The Royal Society.

Scottish Executive (2006) *Scottish Forestry Strategy*, SE/2006/155, Edinburgh, Forestry Commission Scotland; available at http://www.forestry.gov.uk (accessed 18 May, 2011).

Starr, F., Tzimas, E., Steen, M. and Peteves, S.D. (2005) *Flexibility in the production of hydrogen and electricity from fossil fuel power plants*,

Proceedings of the International Hydrogen Energy Congress and Exhibition IHEC2005; available at http://ie.jrc.ec.europa.eu/publications/scientific_publications/2005/P2005-096%20PUBSY%20Request%201287.pdf (accessed 19 May 2011).

UNEP (2011) *Towards a Green Economy: Pathways to Sustainable Development and Poverty Eradication*, New York, United Nations Environment Programme.

Verne, J. (1874) *Mysterious Island*; available at http://www.gutenberg.org/cache/epub/1268/pg1268.txt (accessed 19 May 2011).

WWEA (2009) *World Wind Energy Report 2009*, World Wind Energy Association; available at http://www.wwindea.org (accessed 11 December 2010).

Further reading

Boyle, G. (ed) (forthcoming), *Renewable Energy: Power for a sustainable future*, 3rd edn, Oxford, Oxford University Press/Milton Keynes, The Open University.

DECC (n.d.) *2050 Pathway Analysis*, Department of Energy and Climate Change; available at http://www.decc.gov.uk (accessed 21 May 2011).

Greenpeace/EREC (2010) *Energy (r)evolution: A Sustainable World Energy Outlook*, Amsterdam, Greenpeace International/Brussels, European Renewable Energy Council.

WWF (2011) *The Energy Report: 100% Renewable Energy by 2050*, Worldwide Fund for Nature, Gland, Switzerland/Ecofys, Utrecht, Netherlands/OMA, Rotterdam, Netherlands; available at http://www.ecofys.com (accessed 7 June 2011).

Appendix

Energy arithmetic – a quick reference

This appendix is designed as a quick reference source for 'energy arithmetic'. Section A1 offers a brief summary of the arithmetic itself. A2 gives conversion factors between the main units used for energy and power throughout this book; it includes conversions to and from some commonly used US units. Section A3 relates some older or non-SI units to their SI equivalents. (There is a brief introduction to the SI system in 'Energy Arithmetic', in Chapter 2.) Formal definitions and more detailed accounts of the basis of the SI units can be found in *Quantities, Units and Symbols* (The Royal Society, 1975).

A1 Orders of magnitude

Discussions of energy consumption and production frequently involve very *large* numbers, and accounts of processes at the atomic level often need very *small* numbers. Two solutions to the problem of manipulating such numbers are described here: the use of a shorthand form of arithmetic and the use of prefixes.

Powers of ten

Two million is two times a million, and a million is ten times ten times ten times ten times ten times ten – six of them in all. This can be written mathematically:

$$2\,000\,000 = 2 \times 10 \times 10 \times 10 \times 10 \times 10 \times 10 = 2 \times 10^6$$

The quantity 10^6 is called *ten to the power six* (or *ten to the six* for short), and the 6 is known as the **exponent**. The advantage of using this power-of-ten form is particularly obvious for very large numbers. World primary energy consumption in the year 2009, for instance, was 502 000 000 000 000 000 000 joules, and it is certainly easier to write (or say) 502×10^{18} joules than to spell out all the zeros.

The method can also be used for very small numbers, with the convention that one tenth (0.1) becomes 10^{-1}; one hundredth (0.01) becomes 10^{-2}, etc. The separation of two atoms in a typical metal, for instance, might be about 0.25 of a billionth of a metre, which is 0.000 000 000 25 m. In more compact form, it becomes 0.25×10^{-9} m.

Scientific notation is a way of writing numbers that is based on this concept, but with a more specific rule. Any number, whatever its magnitude, is written in scientific notation as a number between 1 and 10 multiplied by the appropriate power of ten. So in this form, the numbers above would be written slightly differently:

502×10^{18} becomes 5.02×10^{20}

0.25×10^{-9} becomes 2.5×10^{-10}

In practice, styles like those on the left in each of these are commonly used, and a number in this form will be accepted by computers and scientific calculators – but it will be reproduced by them in either 'long decimal' or strict scientific notation, depending on the mode selected.

Prefixes

The sequence of the first six rows in this table may be conveniently remembered (in reverse order) with the phrase 'King Midas's Golden Touch Poisoned Everything'.

The powers of ten provide the basis for the prefixes used to indicate multiples (including sub-multiples) of units. The following table, parts of which appeared in the main text as Table 2.1, shows these in decreasing order.

Table A1

Symbol	Prefix	Multiply by	... which is
E	exa-	10^{18}	one quintillion
P	peta-	10^{15}	one quadrillion
T	tera-	10^{12}	one trillion
G	giga-	10^{9}	one billion
M	mega-	10^{6}	one million
k	kilo-	10^{3}	one thousand
h	hecto-	10^{2}	one hundred
da	deca-	10	ten
d	deci-	10^{-1}	one tenth
c	centi-	10^{-2}	one hundredth
m	milli-	10^{-3}	one thousandth
μ	micro-	10^{-6}	one millionth
n	nano-	10^{-9}	one billionth
p	pico-	10^{-12}	one trillionth

The terms 1 billion = 10^9 and 1 trillion = 10^{12} are as used in this book. However, alternative long forms where 1 billion = 10^{12} and 1 trillion = 10^{18} may be found in older English books and in other languages.

In US usage the prefix M can be used to denote 1000 and (more commonly) MM to denote 1 million.

Indian statistics also use the terms 1 lakh = 100 000 and 1 crore = 10 million.

A2 Units and conversions

Chapter 2 discussed some of the units used in specifying quantities of energy or power in the 'real world'. The tables in this section summarize the relationships between some of these units.

Energy

The main energy units used in this book are the joule, the kilowatt-hour and the tonne of oil or coal equivalent. Tables A2 and A3 give the conversion factors between the most frequently used multiples of these units, on the 'household' scale (A2) and on the larger scale of national or world data (A3). The oil and coal equivalents are here expressed to two significant figures only. Since different data sources may use slightly different values and there may be confusions between lower and higher heating values, you are advised always to take note of the precise conversion factors used for the energy content of fuels adopted by any source you use. Note also that the factors for kWh and TWh assume a 100% energy conversion efficiency, and not those taking into account the electricity generation efficiency of power stations discussed in Chapter 2.

Table A2 Energy conversions at the household scale

	MJ	GJ	kWh	toe	tce
1 MJ =	1	0.001	0.2778	2.4×10^{-5}	3.6×10^{-5}
1 GJ =	1000	1	277.8	0.024	0.036
1 kWh =	3.60	0.0036	1	8.6×10^{-5}	1.3×10^{-4}
1 toe =	42 000	42	11 667	1	1.5
1 tce =	28 000	28	7778	0.67	1

Table A3 Energy conversions at the national scale

	PJ	EJ	TWh	Mtoe	Mtce
1 PJ =	1	0.001	0.2778	0.024	0.036
1 EJ =	1000	1	277.8	24	36
1 TWh =	3.60	0.0036	1	0.086	0.13
1 Mtoe =	42	0.042	11.667	1	1.5
1 Mtce =	28	0.028	7.778	0.67	1

Power

The kilowatt-hour is the standard unit of *energy* used for UK gas and electricity bills. *Power* is the rate at which energy is used, consumed, transferred or transformed. Its normal units are the watt and its multiple the kilowatt. A power of one kilowatt is equal to a rate of use of energy of one kilowatt-hour per hour. A power of one watt is equal to a rate of use of energy of one joule per second. Table A4 shows the quantities of energy per hour and per year for different constant rates in watts. Note that it can also be used to show that, for instance, 1 kWh is 3.6 MJ and 1 TWy (terawatt-year) is 31.54 EJ or 750 Mtoe.

Table A4 Power and rate of use of energy

Rate	Joules per hour	Joules per year	Kilowatt-hours per year	Tonnes of oil equivalent per year	Tonnes of coal equivalent per year
1 W	3.6 kJ	31.54 MJ	8.76	0.75×10^{-3} toe*	1.13×10^{-3} tce*
1 kW	3.6 MJ	31.54 GJ	8760	0.75 toe	1.13 tce
1 MW	3.6 GJ	31.54 TJ	8.76×10^{6}	750 toe	1130 tce
1 GW	3.6 TJ	31.54 PJ	8.76×10^{9}	0.75 Mtoe	1.13 Mtce
1 TW	3.6 PJ	31.54 EJ	8.76×10^{12}	750 Mtoe	1130 Mtce

*i.e. the energy equivalent of 0.75 kg of oil or 1.13 kg of coal

US energy units

Energy statistics in the USA are quoted in British Thermal Units (BTU). Until the 1990s UK statistics also used the BTU and a related unit, the therm.

Table A5 gives conversion factors at the 'household scale'.

Table A5 Energy conversions at the household scale

	MJ	Thousand BTU	kWh	therm	toe
1 MJ =	1	0.948	0.2778	9.48×10^{-3}	2.4×10^{-5}
1000 BTU =	1.055	1	0.293	0.01	2.5×10^{-5}
1 kWh =	3.60	3.412	1	0.034	8.6×10^{-5}
1 therm =	105.5	100	29.3	1	2.5×10^{-3}
1 toe =	42 000	40 000	11 667	400	1

The power unit of the BTU h^{-1} is commonly used for the ratings of domestic boilers and air conditioning plant; 1000 BTU h^{-1} = 0.293 kW.

Table A6 gives energy conversion factors at the 'national' level.

Table A6 Energy conversions at the national scale

	EJ	Quadrillion BTU (Quad)	TWh	Billion therms	Mtoe
1 EJ =	1	0.948	277.8	9.48	24
1 Quad =	1.055	1	293.1	10	25
1 TWh =	3.60×10^{-3}	3.41×10^{-3}	1	0.0341	0.086
1 billion therms =	0.1055	0.1	29.3	1	2.5
1 Mtoe =	0.042	0.040	11.667	0.40	1

A3 Other quantities

Table A7 gives the SI equivalents of a few other metric units and some older units that remain in common use. The final column shows the inverse relationships. For brevity, scientific notation is used for numbers greater than 10 000 or less than 0.1.

Table A7

Quantity	Unit	SI equivalent	Inverse
Mass	1 oz (ounce)	$= 2.835 \times 10^{-2}$ kg	1 kg = 35.27 oz
	1 lb (pound)	= 0.4536 kg	1 kg = 2.205 lb
	1 ton (= 2240 lb)	= 1016 kg	$1 \text{ kg} = 0.9842 \times 10^{-3}$ ton
	1 short ton (= 2000 lb)	= 907 kg	$1 \text{ kg} = 1.102 \times 10^{-3}$ short tons
	1 t (tonne)	= 1000 kg	$1 \text{ kg} = 10^{-3}$ t
	1 u (unified mass unit)	$= 1.660 \times 10^{-27}$ kg	$1 \text{ kg} = 6.024 \times 10^{26}$ u
Length	1 in (inch)	$= 2.540 \times 10^{-2}$ m	1 m = 39.37 in
	1 ft (foot)	= 0.3048 m	1 m = 3.281 ft
	1 yd (yard)	= 0.9144 m	1 m = 1.094 yd
	1 mi (mile)	$= 1.609 \times 10^{3}$ m	$1 \text{ m} = 6.214 \times 10^{-4}$ mi
Speed	1 km h^{-1} (kph)	= 0.2778 m s^{-1}	1 m s^{-1} = 3.600 kph
	1 mi h^{-1} (mph)	= 0.4470 m s^{-1}	1 m s^{-1} = 2.237 mph
Area	1 in^2	$= 6.452 \times 10^{-4} \text{ m}^2$	$1 \text{ m}^2 = 1550 \text{ in}^2$
	1 ft^2	$= 9.290 \times 10^{-2} \text{ m}^2$	$1 \text{ m}^2 = 10.76 \text{ ft}^2$
	1 yd^2	$= 0.8361 \text{ m}^2$	$1 \text{ m}^2 = 1.196 \text{ yd}^2$
	1 acre	$= 4047 \text{ m}^2$	$1 \text{ m}^2 = 2.471 \times 10^{-4}$ acre
	1 ha (hectare)	$= 10^{4} \text{ m}^2$	$1 \text{ m}^2 = 10^{-4}$ ha
	1 mi^2	$= 2.590 \times 10^{6} \text{ m}^2$	$1 \text{ m}^2 = 3.861 \times 10^{-7} \text{ mi}^2$
Volume	1 in^3	$= 1.639 \times 10^{-5} \text{ m}^3$	$1 \text{ m}^3 = 6.102 \times 10^{4} \text{ in}^3$
	1 ft^3	$= 2.832 \times 10^{-2} \text{ m}^3$	$1 \text{ m}^3 = 35.31 \text{ ft}^3$
	1 yd^3	$= 0.7646 \text{ m}^3$	$1 \text{ m}^3 = 1.308 \text{ yd}^3$
	1 litre	$= 10^{-3} \text{ m}^3$	$1 \text{ m}^3 = 1000$ litres
	1 gallon (UK)	= 4.546 litres	$1 \text{ m}^3 = 220.0$ gallons (UK)
	1 gallon (US)	= 3.785 litres	$1 \text{ m}^3 = 264.2$ gallons (US)
	1 barrel	= 159 litres	$1 \text{ m}^3 = 6.3$ barrels
	1 bushel	$= 3.637 \times 10^{-2} \text{ m}^3$	$1 \text{ m}^3 = 27.50$ bushels
Force	1 kgf (weight of 1 kg mass)	= 9.807 N	1 N = 0.102 kgf
	1 lbf (weight of 1 lb mass)	= 4.448 N	1 N = 0.2248 lbf

(Continued)

Table A7 *(Continued)*

Quantity	Unit	SI equivalent	Inverse
Pressure	1 bar (≈1 atmosphere)	= 10^5 Pa (pascals)	1 Pa = 10^{-5} bar
	1 kgf m^{-2}	= 9.807 Pa	1 Pa = 0.102 kgf m^{-2}
	1 lbf in^{-2} (or psi)	= 6 895 Pa	1 Pa = 1.450×10^{-4} psi
Energy	1 barrel of oil equivalent (boe)	= 5.7 GJ (lower heating value)	1 GJ = 0.175 boe
	1 cal (calorie)	= 4.2 J	1 J = 0.24 cal
	1 ft lb (foot pound)	= 1.356 J	1 J = 0.7375 ft lb
	1 eV (electron-volt)	= 1.602×10^{-19} J	1 J = 6.242×10^{18} eV
	1 MeV	= 1.602×10^{-13} J	1 J = 6.242×10^{12} MeV
Power	1 HP (horse power)	= 745.7 W	1 W = 1.341×10^{-3} HP

Reference

The Royal Society (1975) *Quantities, Units and Symbols*, London, The Royal Society.

Acknowledgements

Grateful acknowledgement is made to the following sources:

Chapter 1

Figures

Figure 1.6: AW2 David, B. Hudson/AP/Press Association Images: Figures 1.7, 1.16 and 1.17: British Petroleum 2010 *Statistical Review of World Energy*, British Petroleum: Figure 1.18: TV/AP/Press Association Images: Figure 1.20: Houghton, J. T. et al. (2001) *Climate Change 2001: The Scientific Basis*, Published for the Intergovernmental Panel on Climate Change by Cambridge University Press: Figure 1.21: Mohammed Saneen/AP/Press Association Images: Figure 1.23a: USCG/Science Photo Library: Figure 1.23b: Julie Dermansky/Science Photo Library.

Chapter 2

Figures

Figures 2.1 and 2.2: © Getty Images.

Chapter 3

Tables

Table 3.2: Smil, V. (1994) *Energy in World History*, Perseus Books Group;

Figures

Figure 3.2: Smil, V. (1994) *Energy in World History*, Perseus Books Group: Figures 3.3, 3.4 and 3.5: Dr Bob Everett: Figure 3.6: Mary Evans Picture Library: Figures 3.7 and 3.9b: Dr Bob Everett: Figure 3.11: Department of Trade and Industry (2001) Digest of UK Energy Statistics. Crown copyright material is reproduced under Class Licence Number C01W0000065 with the permission of the Controller, Office of Public Sector Information (OPSI): Figure 3.12: Department of Trade and Industry (2004) *Energy Consumption in the UK 2004*, Crown copyright material is reproduced under Class Licence Number C01W0000065 with the permission of the Controller, Office of Public Sector Information (OPSI): Figure 3.14: Martin Bond/Science Photo Library: Figure 3.17: Smil, V. (1994) *Energy in World History*, Perseus Books Group.

Chapter 4

Figures

Figure 4.5: Courtesy of Scottish Power: Figure 4.6: Sheila Terry/Science Photo Library: Figure 4.8: Science Photo Library: Figure 4.12: Dr Bob Everett: Figure 4.13: Sheila Terry/Science Photo Library: Figure 4.14: Sheila Terry/

Science Photo Lirary: Figure 4.15 (VHF): © P M Northwood: Figure 4.15 (microwave): © Chris Schmidt/iStockphoto.com: Figure 4.15 (infrared): Dr Arthur Tucker/Science Photo Library: Figure 4.15 (x-ray): Natallia Yaumenenka/iStockphoto.com: Figure 4.16: W F Meggers Collection/ American Institute of Physics/Science Photo Library: Figure 4.17: Dr Bob Everett: Figure 4.22: Prof. Peter Fowler/Science Photo Library: Figure 4.25: Science Source/Science Photo Library.

Chapter 5

Figures

Figure 5.1: © Rob Cousins/Alamy; Figure 5.2: © George Green/iStockphoto.com; Figure 5.3: The Yorck Project – used under The GNU Free Documentation Licence; Figures 5.7 and 5.8: Dr Bob Everett; Figure 5.10: Science Photo Library; Figure 5.11: Dr Bob Everett; Figure 5.12: Courtesy of Corus; Figure 5.13: Thurston Hopkins/Hulton Archive/Getty Images; Figure 5.14: MRP Photography; Figure 5.15: © RWE; Figure 5.16: BGR Energy Resources 2009, BGR Energy; Figure 5.19: © Gawrav Sinha/iStockphoto.com; Figure 5.21: Courtesy of Hitachi, Japan.

Chapter 6

Figures

Figure 6.6: Science Museum Pictorial; Figure 6.8: Science Photo Library; Figure 6.9: Science Photo Library; Figure 6.12: Science Photo Library; Figure 6.13: Science Photo Library; Figure 6.14: Science Photo Library; Figure 6.15: Getty Images; Figure 6.17: Dr Bob Everett; Figure 6.19: Science Photo Library; Figure 6.20a: Siemens; Figure 6.20b: Hitachi; Figure 6.22a: Dr Bob Everett.

Chapter 7

Figures

Figure 7.2: Getty Images; Figure 7.3: Getty Images; Figure 7.8: Getty Images; Figure 7.9: © Dave Jones/iStockphoto.com; Figure 7.11: BP plc; Figure 7.14: Courtesy of ExxonMobil; Figure 7.16: Courtesy of Transco; Figure 7.19: The National Grid; Figure 7.20: Adapted from *EuroGas Annual Report 2001*, courtesy of Euro Gas; Figure 7.21: Hook, M. (2009) 'Depletion and decline curve analysis in crude oil production', Licentiate thesis, Uppsala University; Figure 7.23: Getty Images; Figure 7.24: Basin Electric Power Cooperative; Figure 7.25: Getty Images; Figure 7.26: Adapted from Crawford, H.B., Eisler, W. and Strong, L. (1977) *Energy Technology Handbook*, McGraw Hill, Inc; Figure 7.27: SSPL via Getty Images; Figure 7.28: Courtesy of Sasol Ltd; Figure 7.29: Courtesy of David Crabbe; Figure 7.30: Courtesy of Suncor Energy; Figure 7.32a: Hubbert, M. K. (1956) *Nuclear Energy and the Fossil Fuels*; Figure 7.32b: EIA *Annual Energy Review 2009*, U.S. Energy Information Administration (EIA); Figure 7.33: Farrell, A. E. and A. R. Brandt. (2006). 'Risks of the oil transition'

Environmental Research Letters,1. IOP Publishing; Figure 7.34: IEA *World Energy Outlook Report 2010.*

Chapter 8

Figures

Figure 8.1: Science Photo Library; Figure 8.2: Ramage, J. (1999) *Energy Guidebook*, Oxford University Press; Figure 8.3: Rogers & Mayhew (1980) *Engineering Thermodynamics*, Longman; Figure 8.4: Science Photo Library; Figure 8.6a: Library of Congress; Figure 8.6b: Dr Bob Everett; Figure 8.11: Dr Marcus Enoch; Figure 8.12: Clive Fetter; Figure 8.14: Rolls Royce Plc; Figure 8.15: Air Force Research Laboratory Propulsion Directorate; Figure 8.16: Science Museum Pictorial; Figure 8.17: Getty Images; Figures 8.18 and 8.19: Rolls Royce Plc; Figure 8.21: Dr Bob Everett; Figure 8.22: National Renewable Energy Laboratory; Figure 8.23: Image copyright www.microchap.info; Figure 8.24: Adapted from Smil, Vaclav (1994) *Energy in World History*, Perseus Books Group.

Chapter 9

Figures

Figure 9.1: Mary Evans Picture Library; Figure 9.2: GeorgiosArt/iStock.com; Figure 9.3: HultonArchive/istockphoto.com; Figure 9.4: Courtesy of Janet Ramage; Figure 9.5: Courtesy of Bob Everett; Figure 9.6: Science Photo Library; Figure 9.7: Science Photo Library; Figure 9.9: Ashrae Centennial Collection HC; Figure 9.12d: zig4photo/istockphoto; Figure 9.13: Science Photo Library; Figure 9.16: Science Photo Library; Figure 9.17: Kriando Design/istockphoto.com; Figure 9.19: Courtesy of Bob Everett; Figure 9.20: Courtesy of Bob Everett; Figure 9.21: Courtesy of Bob Everett; Figure 9.22: Crown Copyright/Health & Safety Laboratory/Science Photo Library; Figure 9.23: Getty Images; Figure 9.24: Tesla Motors; Figure 9.25: Courtesy of Toyota (GD) plc; Figure 9.27: Illustrated London News; Figure 9.28: Courtesy of Bob Everett; Figue 9.29: Courtesy of Bob Everett; Figure 9.32: NASA; Figure 9.33: Drax Power Limited; Figure 9.34: United States Federal Government Department of Reclamation; Figure 9.36: Courtesy of Bob Everett; Figure 9.38: DECC Energy in Brief 2010, Energy Trends 2011; Figure 9.40: Courtesy of Cogenco Limited; Figure 9.42: Courtesy of Bob Everett; Figure 9.43: Courtesy of Bob Everett; Figure 9.44: Courtesy of Bob Everett; Figure 9.47: National Grid (2010) *National Grid Seven Year Statement*, National Grid; Figure 9.49: Courtesy of Bob Everett; Figure 9.50: Courtesy of Dong Energy A/S; Figure 9.51: Getty Images.

Chapter 10

Figures

Figure 10.2: Schneider, M. et al. (2009) *The World Nuclear Industry Status Report 2009.* Commissioned by The German Federal Ministry of the Environment, Nature Conservation and Reactor Safety; Figure 10.7: American Institute of Physics/Science Photo Library; Figure 10.8: Eric

Findlay; Figure 10.9: Emilio Segre Visual Archives/American Institute of Physics/ Science Photo Library; Figure 10.11: Science Photo Library; Figure 10.15: Adapted from: Hanh, O. (1950) New Atoms, Progress and some Memories. Elsevier Publishing Company Inc.

Chapter 11

Figures

Figure 11.2: NERAC/GIF (2002) *A Technology Roadmap for Generation IV Nuclear Energy Systems.* United States Department of Energy Research Advisory Committee/Generation IV International Forum; Figure 11.4: British Energy; Figure 11.5: Cameco Corporation; Figure 11.7: URENCO Limited Figure 11.8: © PA Photos/EPA; Figure 11.9a: IAEA, International Atomic Energy Agency; Figure 11.9b: The Tokyo Electric Power Company, Incorporated; Figure 11.10: Adapted from Harvey, L. D. D. (2010) Carbon Free Energy Supply. Earthscan; Figure 11.11: Sellafield Limited; Figure 11.12: United Kingdom Atomic Energy Agency; Figure 11.13: Swedish Nuclear Fuel and Waste Management Figure 11.15: PMBR South Africa www.pbmr.com; Figure 11.16: Simon Ledingham; Figure 11.19: Jeri Laber/ Physics Today Collection/American Institute OF Physics/Science Photo Library; Figure 11.21: EFDA JET; Figure 11.22: AREVA, P. Bourdon.

Chapter 12

The author would like to thank Susan Walker and Horace Herring for work on previous versions of this chapter.

Figures

Figure 12.6: Getty Images; Figure 12.7: *UK Energy in Brief* July 2001, Energy Policy of the DTI. Crown copyright material is reproduces under Class License Number C01W0000065, with the permission of he Controller of Her Majesty's Stationery Office and the Queen's printer for Scotland; Figure 12.17: Bob Everett; Figure 12.18: SSPL via Getty Images.

Chapter 13

Figures

Figure 13.1: European Environment Agency (2008) *EEA Energy and Environment Report.* European Environment Agency; Figure 13.2: McKinsey and Company (2009) Pathways to a low carbon economy. McKinsey and Company; Figure 13.11: IPCC (2007b) *Climate Change 2007: the Scientific Assessment*, Intergovernmental Panel on Climate Change, Cambridge, Cambridge University Press; Figure 13.12: IPCC, (2007a) *Climate Change 2007: Synthesis Report An Assessment of the Intergovernmental Panel on Climate Change*, Intergovernmental Panel on Climate Change, Cambridge, Cambridge University Press; Figure 13.14: The European Environment Agency (EEA). Used under a Creative Commons Licence; Figure 13.16: NASA/Goddard Space Flight Center Scientific Visualization Studio.

Chapter 14

Figures

Figure 14.2: Leonard G. Used under a Creative Commons Licence share alike Licence; Figure 14.5: Courtesy IEA Greenhouse Gas R&D programme; Figure 14.7: Science Photo Library; Figure 14.9: Laughton, M.A. (2002) 'Fuel cells', Engineering and Science Journal, vol. 11, no. 1, Feb 2002 © Institution of Electrical Engineers; Figure 14.10: Adapted from Hart, D. et al. (1999) *Hydrogen – Today and Tomorrow*, Greenhouse Gas R&D Programme; Figure 14.11: Bbqjunkie, used under a Creative Commons Licence; Figure 14.13: NY Daily News via Getty Images; Figure 14.14: Courtesy of Luwoge; Figure 14.15: Courtesy of Luwoge; Figure 14.16: Blunden, J. and Reddish, A. (1991) Energy Resources and Environment. Hodder and Stoughton; Figure 14.17: Rudolf Simon, used under a creative Commons Licence; Figure 14.18: Courtesy of Bob Everett; Figure 14.20: Koza1983, used under a Creative Commons Licence; Figure 14.22: Vincent van Zeijst, used under a Creative Commons Licence; Figure 14.23: Getty Images; Figure 14.24: Pelamis Wave Power; Figure 14.25: Martin Bond/ Science Photo Library; Figure 14.26: Peter Fraenkel, Marine Current Turbines Limited; Figure 14.27: © Atlantide Phototravel/Corbis.

Every effort has been made to contact copyright holders. If any have been inadvertently overlooked the publishers will be pleased to make the necessary arrangements at the first opportunity.

The editors would like to thank the following for their assistance with the production of this book:

Wendy Berndt
Sophia Braybrooke
Matthew Cook
Jonathan Crowe
Dewi Jackson
Marie Lacy
John Litt
Martin Keeling
Carol Morris
Suresh Nesaratnam
Sally Organ
Nathalie Richard
Yvonne Slater
Geoff Symons
David Vince
Geoff Walker
Andy Whitehead

Index

C

D

E

F

G

N

O

Q

R

S

Y

Z